Grasses

SYSTEMATICS AND EVOLUTION

SURREY W. L. JACOBS AND JOY EVERETT (EDITORS)

CSIRO
PUBLISHING

National Library of Australia Cataloguing-in-Publication entry

International Symposium on Grass Systematics and Evolution (3rd : 1998).

Grasses : systematics and evolution.

Bibliography.
Includes index.
ISBN 0 643 06438 9.
ISBN 0 643 06393 5 (set).

1. Grasses – Congresses.
I. Jacobs, S.W.L. (Surrey W. L.)
II. Everett, Joy.
III. Title.

584.9

This book is available from:
CSIRO PUBLISHING
PO Box 1139 (150 Oxford Street)
Collingwood VIC 3066
Australia

Tel: (03) 9662 7666   Int: +(613) 9662 7666
Fax: (03) 9662 7555   Int: +(613) 9662 7555
Email: sales@publish.csiro.au
http://www.publish.csiro.au

Printed in Australia by Brown Prior Anderson

Contents

PHYSIOLOGY/ECOLOGY

BREEDING SYSTEMS

BIOGEOGRAPHY

INDEX

PREFACE

The grass family is the single most important family of organisms for the survival of mankind. Of the top ten crops we use to sustain ourselves (directly or indirectly as stock feed), the first ten are all grasses. An understanding of the relationships and characteristics of grasses has played an important role so far in our history, and the importance of that role can only increase as we struggle to balance exploitation, sustainability and conservation.

There is some discussion as to when the 'first' international meeting on the systematics and evolution of the grasses was held. A meeting was held in 1959 as a special session of the Ninth International Botanical Congress in Montreal. It took 27 years before the next International Symposium occurred at the Smithsonian Institution in 1986 and, while the Symposium was advertised as 'The First …' the publication did not include that word. In the period between the Montreal and Smithsonian meetings, work on molecular evolution, physiology and numerical methods had revolutionised many areas of grass research. The Smithsonian meeting was an inspiration to the 150 participants from around the world, and the published proceedings, including many review papers, took its place as a new benchmark and major reference for further work in all of the tribes. And so it seemed essential to hold the next meeting on the grasses without such a long gap. There was a further little-publicised Conference held at the Main Botanical Garden, Moscow in 1994 but international attendance at this conference was very restricted.

It was 13 years after the Smithsonian Symposium before the Sydney commitment to hold the 2nd International Monocot Symposium coincided with a request from some of the organisers of the Smithsonian Symposium for us to host the next International Grass Symposium. It quickly became clear that the most efficient way of doing this, from both a pragmatic and scientific point of view, was to hold them jointly. This presented great opportunities to form new research associations and generally have a good time with others interested in similar research subjects. If all goes well the next grass symposium should obey the inverse exponential rule and occur approximately 6 years from Monocots II, such is the rate of research on this important economic group.

The Third International Symposium on Grass Systematics and Evolution was held in Sydney in the week beginning 28th September 1998, concurrently with the Second International Conference on Monocot Systematics. This was the first time that the two areas of Grass and Monocot research had joined together. So many of the methods and philosophies are now universally adopted that it seemed ideal that researchers had the opportunity to attend a wide range of presentations. With increasing interest in evolutionary studies, few families can be studied in isolation. To this end the first session of each day's grass program was held jointly with the general monocots session, with the latter part of the day devoted to grasses.

The grasses are a family with confusing relationships, with different data sets suggesting different arrangements. The development of techniques to obtain and analyse molecular data has meant the dawning of a new age in the study of the grasses. It seems we have been given a key to a whole new suite of rooms previously hidden behind the dark curtains of hybridisation, polyploidy and reticulate evolution. Numerical and DNA analyses are developing to a point where they are contributing to the interpretation of relationships in the cloudiest of tribes. Even though we have these new data, the role of more traditional information on breeding systems, genetic control, developmental studies and biogeography is becoming increasingly important as we aim to assemble corroborative and/or multi-sourced data matrices. The development of statistical methods for testing molecular hypotheses still has some way to go, and corroboration with independent data sets still has an important role to play.

There were 42 lectures on grass systematics and evolution, 34 of which are published here, along with two others presented in the 'general' sessions of Monocots II, and the grass presentation on composition of cell walls was combined to make a single presentation in the first volume. Of the 75 posters at Monocots II, presenters of six of the many high quality contributions presenting research on grasses were invited to elaborate and present them as papers to be included in this volume. All manuscripts submitted were refereed.

Joy Everett and Surrey Jacobs

ORGANISING COMMITTEE OF *MONOCOTS II*

Royal Botanic Gardens Sydney: Barbara Briggs, Joy Everett, Gwen Harden, Alistair Hay, Surrey Jacobs, Peter Weston, Karen Wilson

University of New England: Jeremy Bruhl

University of New South Wales: Christopher Quinn

University of Sydney: Murray Henwood

University of Technology, Sydney: David Morrison

ACKNOWLEDGEMENTS

Our thanks go to our sponsors, whose assistance made the conference feasible:

- University of New South Wales

- Royal Botanic Gardens Sydney

- Orlando Wyndham Wines

- Australian Systematic Botany Society

- Roads and Traffic Authority of New South Wales

- Swane's Garden Centre

- Ansett Australia Airlines

ROYAL BOTANIC GARDENS SYDNEY

THE UNIVERSITY OF
NEW SOUTH WALES

Australian Systematic Botany Society

JACOB'S CREEK®

The Organising Committee is grateful to all of the following for help in many different ways:

- Organisers of conference sessions, including Nigel Barker, Mary Barkworth, Spencer Barrett, Tatyana Batygina, Peter Bernhardt, Geeta Bharathan, Peter Boyce, Mark Chase, Lynn Clark, John Conran, John Dowe, John Dransfield, Soejatmi Dransfield, Mike Fay, Khidir Hilu, Steve Hopper, Toby Kellogg, Helen Kennedy, Bruce Kirchoff, John Kress, Peter Linder, Terry Macfarlane, Kathy Meney, Paul Peterson, Prakash, Alec Pridgeon, Jim Quinn, Paula Rudall, Bryan Simon, Mike Simpson, Rob Soreng, Dennis Stevenson, Ian Staff, An Van den Borre, Wal Whalley, Paul Wilkin, Fernando Zuloaga.

- Royal Botanic Gardens Sydney staff and volunteers, including Diana Adams, Georgina Bassingthwaighte, Geoff Breen, Gary Chapple, Bob Coveny, Kim Cubbin, Jane Dalby, Kathi Downs, Zonda Erskine, Dianne Godden, John Gregor, Jane Halsham, Clare Herscovitch, Dorothy Holland, Frank Howarth, Peter Jobson, Kirsten Knox, Alan Leishman, Kristina McColl, Debbie McGerty, John Matthews, Lynne Munnich, Linda Newnam, Andrew Orme, Lorraine Perrins, Karen Rinkel, Chris Simpson, Randy Sing, Jane Stafford, Leonie Stanberg, Helen Stevenson, Jenny Trustrum, Rachael Wakefield, Barbara Wiecek, Yuzammi.

- University of New South Wales staff and students, including Edward Cross, Jan De Nardi, Jeffrey Drudge, Claudia Ford, Margaret Hesslewood, Nik Lam, Hannah MacPherson, Helena Mills, Marcelle O'Brien, David Orlovich, Stefan Rose, Nikola Streiber, Joe Zuccarello.

- University of New England staff and volunteers, including Ian Telford, Wal Whalley and John Williams.

- University of Sydney staff and students, including Andrew Perkins.

- Other helpers, including John Clarkson, Carlo De Nardi, Tony Irvine, Kevin Jeans, Meredith Lewis, Amelia Martyn, John Neldner, Bruce Robertson, Bill Rudman, Susanne Stenlund, Elwyn Swane, Bruce Wannan and Alastair Wilson.

- The conference website was hosted by the Faculty of Science, University of Technology, Sydney.

REFEREES

All papers in this volume have been refereed. The editors thank the following people who generously agreed to review manuscripts:

Arthur Bailey, Nigel Barker, Mary Barkworth, Lynn Clark, John Clarkson, Travis Columbus, Henry Connor, Thomas Cope, Jerrold Davis, Soejatmi Dransfield, Bryan Hacker, Murray Henwood, Khidir Hilu, Elizabeth James, Elizabeth Kellogg, Peter Linder, Bao-rong Lu, Mark Lyle, James Quinn, Terry Macfarlane, Michelle Murphy, James Mant, Juan-Javier Ortiz-Diaz, Paul Peterson, Thomson Pizzolato, Bryan Simon, Robert Shaw, Robert Soreng, Russ Spangler, Amelia Torres, An Van den Borre, Anthony Verboom, Lesley Watson, Peter Weston, Wal Whalley, Karen Wills, Ping Yu.

Themeda quadrivalvis, a native of Malesia and spreading as a weed in other tropical areas.

Grasses: Systematics and Evolution. (2000). Eds S.W.L. Jacobs and J. Everett. (CSIRO: Melbourne)

GRASSES

A PHYLOGENY OF THE GRASS FAMILY (POACEAE), AS INFERRED FROM EIGHT CHARACTER SETS

The Grass Phylogeny Working Group[1] (in alphabetical order):

Nigel P. Barker (Department of Botany, Rhodes University, P. O. Box 94, Grahamstown, 6140, South Africa); Lynn G. Clark* (Department of Botany, Iowa State University, Ames, IA 50011 U. S. A.); Jerrold I Davis (L. H. Bailey Hortorium, Cornell University, 462 Mann Library, Ithaca, NY 14853 U. S. A.); Melvin R. Duvall (Department of Biological Sciences, Northern Illinois University, DeKalb, IL 60 115–2861 U. S. A.); Gerald F. Guala (Fairchild Tropical Garden, 11935 Old Cutler Road, Miami, FL 33156 U. S. A.); Cathy Hsiao (6005 Crossmont Court, San Jose, CA 95120 U. S. A.); Elizabeth A. Kellogg (Department of Biology, University of Missouri-St. Louis, 8001 Natural Bridge Road, St. Louis, MO 63121 U. S. A.); H. Peter Linder (Bolus Herbarium, University of Cape Town, Private Bag, Rondebosch 7700, South Africa); Roberta Mason-Gamer (Department of Biological Sciences, University of Idaho, Moscow, ID 83844 U. S. A.); Sarah Mathews (Harvard University Herbaria, 22 Divinity Avenue, Cambridge, MA 02138 U. S. A.); Robert Soreng (Department of Botany, Natural History Museum, Smithsonian Institution, Washington, D. C. 20560-0166 U. S. A.); and Russell Spangler (Harvard University Herbaria, 22 Divinity Avenue, Cambridge, MA 02138 U. S. A.).

[1] Recommended citation, abbreviated as "GPWG, 2000".
* Corresponding author

Abstract

The Grass Phylogeny Working Group (GPWG) undertook a combined analysis of eight existing data sets (one morphological, four plastid and three nuclear) for 57 grasses and four outgroup taxa. The data provide robust support for a number of major clades within the grass family. *Anomochloa* + *Streptochaeta*, Phareae, and *Puelia* + *Guaduella* are supported as the early-diverging lineages, with the remainder of the family resolved as a clade in which there is a sister-group relationship between the PACC (Panicoideae, Arundinoideae, Chloridoideae + Centothecoideae) and BOP (Bambusoideae, Oryzoideae + Pooideae) clades. A summary of the discussion regarding the GPWG proposal for a revised subfamilial classification of the grasses based on the results of this analysis is included.

Key words: Poaceae, phylogeny, combined data sets, classification, taxon sampling.

INTRODUCTION

The grass family (Poaceae) is of pre-eminent importance in the human economy, and it also includes many ecosystem dominants in a wide variety of habitats throughout the world. Widespread interest in the phylogeny and classification of the grasses reflects the unparalleled economic and ecological importance of this family. The grasses long have been recognised as a natural group, but continued advances in our understanding of the evolutionary history of the family require periodic reassessments of relationships among the major lineages of grasses and concomitant changes in classification.

The Grass Phylogeny Working Group (GPWG) was established in 1996 with the goal of combining a series of existing data sets to produce a comprehensive phylogeny for the grass family. Another more general goal of this collaboration was to help focus the further development (i.e., taxon sampling) of existing data sets. Detailed phylogenetic analyses, as well as a revised subfamilial classification for the family, will be presented in a separate manuscript (GPWG, in prep.). This paper therefore serves as a progress report, and includes a summary of discussion from the Third International Symposium on Grass Systematics and Evolution.

REVIEW OF GRASS PHYLOGENY AND CLASSIFICATION

Many comprehensive classifications of the family have been proposed through the years. Some of these, although not employing cladistic methods, were based on implicit or explicit hypotheses of phylogenetic structure (e.g., Stebbins 1956; Reeder 1957, 1962; Tateoka 1957; Prat 1960; Stebbins and Crampton 1961; Butzin 1965; Clifford 1965; Clifford *et al.* 1969; Sharma 1979; Caro 1982; Clayton and Renvoize 1986; Soderstrom and Ellis 1987; Tzvelev 1977, 1989) while others were based on phenetic analyses (Hilu and Wright 1982; Watson *et al.* 1985; Watson and Dallwitz 1992). Several additional contributions to the phylogeny of the family, using cladistic methods, and based on a variety of character sets, also have appeared in recent years (e.g., Baum 1987; Kellogg and Campbell 1987; Hamby and Zimmer 1988,

1992; Doebley *et al.* 1990; Liang and Hilu 1996; Davis and Soreng 1993; Yaneshita *et al.* 1993; Cummings *et al.* 1994; Nadot *et al.* 1994; Barker *et al.* 1995; Clark *et al.* 1995; Kellogg and Linder 1995; Duvall and Morton 1996; Mathews and Sharrock 1996; Morton *et al.* 1996; Hsiao *et al.* 1998; Kellogg 1998; Soreng and Davis 1998; Mason-Gamer *et al.* 1998). Sampling in these studies was influenced profoundly by hypotheses of relationships proposed in earlier, non-cladistic treatments, and authors of this latter set of papers therefore owe a considerable debt to those who made the first attempts to produce either a comprehensive taxonomic structure or a phylogeny of the grasses.

Phylogenetic analyses of the grasses have, of course, resolved clades, but it is possible to distinguish two aspects of the progress that has been made, both of which are required before a comprehensive systematic treatment of the family can be established. First, the monophyletic status of several major groups has been repeatedly affirmed (e.g., Bambuseae, Panicoideae). These major groups, often marked by long-recognised morphological and anatomical characteristics, and corresponding in overall membership to traditionally recognised taxa, represent most of the diversity of the family. Second, the affiliations of several anomalous genera and small groups of genera (e.g., Olyreae, Phareae, *Buergersiochloa, Streptogyna, Lygeum, Nardus*) have been established (albeit with varying degrees of confidence). The placement of these less speciose groups among the larger groupings has helped to establish the limits of the major clades that can be recognised as subfamilies.

Some of the key findings, in terms of higher-level groupings, are: 1) The Poaceae have been placed consistently with Flagellariaceae, Restionaceae, Anarthriaceae, Ecdeiocoleaceae, Centrolepidaceae, and Joinvilleaceae in a group that has been recognised as the order Poales (Dahlgren *et al.* 1985; Linder 1987; Doyle *et al.* 1992; Kellogg and Linder 1995); the Angiosperm Phylogeny Group (1998) proposed an expanded Poales, but did not question the close relationship of the core Poalean families. Within this group a sister-group relationship between Poaceae and Joinvilleaceae has been repeatedly supported (Campbell and Kellogg 1987; Clark *et al.* 1995; Soreng and Davis 1998). 2) A small set of taxa (*Anomochloa, Streptochaeta,* Phareae, *Guaduella,* and *Puelia*) has been found to represent a varying number of early-diverging lineages within the Poaceae (Clark *et al.* 1995; Duvall and Morton 1996; Soreng and Davis 1998; Clark *et al.*, in press). Although several analyses have not included these taxa, they appear as sisters to the remainder of the family when they are included. 3) Three widely recognised subfamilies, Chloridoideae, Panicoideae, and Bambusoideae *sensu stricto*, have been resolved by almost every analysis with sufficient sampling to have the potential to support these relationships (Kellogg and Campbell 1987; Barker *et al.* 1995; Clark *et al.* 1995; Duvall and Morton 1996; Mathews and Sharrock 1996; Soreng and Davis 1998). 4) A "PACC" clade, comprising a monophyletic Panicoideae, a monophyletic Chloridoideae, plus various arrangements of elements often assigned to Arundinoideae and Centothecoideae, has been resolved by most analyses (e.g., Davis and Soreng 1993; Clark *et al.* 1995; Duvall and Morton 1996; Soreng and Davis 1998).

Various other traditionally recognised groupings within Poaceae either have been refuted by phylogenetic studies, or have been substantiated only in limited ways. For example, some elements traditionally assigned to Bambusoideae *sensu lato* have been placed near Bambuseae, while other elements have been placed within other subfamilies entirely (Clark *et al.* 1995; Soreng and Davis 1998; Clark *et al.*, in press). The unity of a broadly circumscribed Arundinoideae also is refuted by most data sets (Kellogg and Campbell 1987; Barker *et al.* 1995; Clark *et al.* 1995; Barker 1997; Soreng and Davis 1998). Some genera are placed far from the bulk of the group (e.g., *Anisopogon* with Pooideae or *Gynerium* with Centothecoideae), but most arundinoid elements usually are associated, not as a single monophyletic group, but as a small number of closely related lineages. Hsiao *et al.* (1998), however, provide some evidence for monophyly of a relatively broad arundinoid clade although they did not sample widely enough from the PACC clade to establish this with confidence. The fate of the Pooideae has been somewhat different. On the one hand, the monophyly of a "core" Pooideae (including such tribes as Aveneae, Poeae, Triticeae, and Bromeae) has been substantiated by most studies (Soreng *et al.* 1990; Davis and Soreng 1993; Soreng and Davis 1998). However, this group is found to be nested within a larger monophyletic group that comprises disparate elements, including some that traditionally have been assigned to Arundinoideae and Bambusoideae (e.g., *Anisopogon*, as mentioned above, and other taxa such as *Diarrhena* and Stipeae) (Davis and Soreng 1993; Clark *et al.* 1995; Soreng and Davis 1998). Thus, the pooid group (i.e., our perception of the pooid group) has grown by the accretion of elements from the traditional Arundinoideae and Bambusoideae, and a large monophyletic group has emerged that includes some taxa not formerly associated with the Pooideae.

One area of controversy has been the "BOP" clade, which, when resolved, includes Bambusoideae, Ehrhartoideae (= Oryzoideae), and Pooideae, and is sister of the PACC clade (Clark *et al.* 1995; Mathews and Sharrock 1996; Mathews *et al.*, in press). The highly inclusive PACC + BOP clade (as resolved in these cases) includes all grasses except the few early-diverging elements (*Anomochloa*, etc.) mentioned above. An alternative to the BOP clade is supported by another set of analyses. These analyses, instead of placing Pooideae with Bambusoideae and Ehrhartoideae, place Pooideae as sister of the PACC clade, and thus within a larger grouping that excludes Bambusoideae and Ehrhartoideae (Duvall and Morton 1996; Soreng and Davis 1998). When a PACC + Pooideae clade is resolved, Bambusoideae and Ehrhartoideae are placed as additional lineages among the early-diverging elements.

SAMPLING AND COMBINATION OF DATA SETS

The comparability (or lack thereof) in taxon sampling is a major issue in combining disparate data sets. Sampling in the cladistic studies cited above has been variable, for some authors have emphasized particular groupings within the family, and even where the same subfamilies or tribes have been sampled, different genera and species often have been used as exemplars of the larger groups. Therefore, comparison of the results of two or more such analyses requires the assumption that the various exemplars of a tribe or subtribe really are closely related. Assumptions of this sort can be tested only by including the exemplars in a more comprehensive analysis. Therefore, the placement of one taxon in one region of a tree by one data set, and the placement of a putative relative in another region of a tree by another data set is not compelling evidence of conflict between the two data sets.

Another problem faced in the combination of different data sets is the tendency to generate "conglomerate taxa," these being single terminals in an analysis that include data actually collected from different taxa. Available data for small, taxonomically isolated genera such as *Lygeum* or *Anomochloa* are unlikely to be problematical, since they reflect focused sampling from a set of closely related species, or in the case of monotypic genera, the sampling of one species. These data, although acquired by persons working separately, can be combined *post facto* with little fear that the resulting string of character scores represents a heterogeneous assemblage. Problems may arise, however, when a variety of character sets have been collected from a more varied assemblage of species or genera (e.g., Triticeae represented by *Hordeum* in one study, and by *Triticum* in another). One of the greatest challenges in assembling our data set has been to determine which portions of the available data sets could be combined safely. A compromise that is acceptable at one time may prove troublesome later when additional taxa are added. For example, the tribe Bambuseae is represented in the present analysis by only two strings of data (i.e., terminals, or operational taxonomic units); one of these lines represents data drawn exclusively from *Chusquea*, while the other is a conglomerate of data from *Arundinaria*, *Bambusa*, and *Pseudosasa*. With two terminals that are almost certainly sister taxa, this use of a conglomerate taxon may be an acceptable means for increasing the amount of data in a matrix, but in future years, as the sampling of this group becomes more intensive, the latter grouping may prove to be untenable even if the monophyly of the tribe continues to be well supported (Zhang and Clark, this volume). For example, if *Arundinaria* is less closely related to *Bambusa* than it is to some third genus, and if that other genus were included in a future analysis, the conglomerate taxon might be a source of character incongruence that could have far-reaching effects on the overall analysis.

GPWG ANALYSIS

Our current taxon set includes four exemplar genera representing families Flagellariaceae, Restionaceae (two genera), and Joinvilleaceae, as outgroups, plus 57 exemplar grass genera. The sampling of grass genera includes representatives of all conventionally recognised subfamilies (i.e., Anomochlooideae, Arundinoideae, Bambusoideae, Centothecoideae, Chloridoideae, Ehrhartoideae, Panicoideae, Pharoideae, and Pooideae), plus genera of controversial placement (e.g., *Amphipogon*, *Anisopogon*, *Brachyelytrum*, *Buergersiochloa*, *Danthonia*, *Danthoniopsis*, *Lygeum*, *Micraira*, *Nardus*, *Pariana*, *Phaenosperma*, *Puelia*, *Streptogyna*, *Thysanolaena*, and *Zeugites*). Our current character set includes eight subsets. The first of these is a 'morphological' data set, a varied set of characters representing variation in macromorphology, anatomy, and biochemistry, plus structural variants such as deletions and inversions in the plastid and nuclear genomes. The seven remaining character sets are molecular, these being mapped restriction sites from throughout the plastid genome, plus nucleotide sequences of

six genes, three from the plastid genome (*ndh*F, *rbc*L, and *rpo*C2), and three from the nuclear genome (ITS, phytochrome B, and *waxy*). These eight character sets yield a total of 1915 cladistically informative characters.

Parsimony analysis of the data set, with all characters weighted equally, and with clades recognised as resolved only when support is unambiguous (using the setting *amb-* in Nona and PAUP*) (Goloboff, 1993; Swofford, 1998), yields one most-parsimonious tree (7343 steps, CI = 0.39 with uninformative characters removed, RI = 0.55; Fig. 1). There is only one polytomy in this tree, involving *Eragrostis*, *Pappophorum*, and *Uniola*; when settings are altered so that clades are accepted on the basis of either ambiguous or unambiguous support (*amb=*), three trees are recognised, differing among themselves only in the relationships among the same three genera, and the consensus is identical to the single tree that is resolved with *amb-*.

This analysis (Fig. 1) largely confirms the relationships described above. *Joinvillea* is resolved as sister of a monophyletic Poaceae. Three small lineages, accounting for five genera in the study, diverge in succession near the base of the grass family, and all other grasses fall within two large clades, identifiable as the PACC and BOP clades, which are sisters of each other (Fig. 1). Within the PACC clade, monophyletic groups identifiable as Chloridoideae, Centothecoideae, and Panicoideae are resolved (the latter two as sisters of each other), while arundinoid elements form a paraphyletic assemblage within which the Chloridoideae are nested. Within the BOP clade, a broadly-circumscribed subfamily Pooideae is sister of Bambusoideae (including Bambuseae and a variety of herbaceous olyroid elements), and this larger clade is sister of Ehrhartoideae (including *Ehrharta* and *Streptogyna*).

We have investigated support for this structure in terms of robustness to the removal of various character sets. The alternative structure described previously (PACC as sister of Pooideae, with Bambusoideae and Ehrhartoideae placed with other early-diverging elements) is resolved when either the *ndh*F, phytochrome, or ITS character set is removed. Removal of the restriction site data set results in loss of resolution of the BOP clade, but PACC + Pooideae is not resolved either; instead, relationships among the various components of these two conflicting groupings (PACC, Pooideae, Bambusoideae, and Ehrhartoideae) are unresolved. Thus, if the current data set had not included any one of these four character sets, the BOP clade would not have been resolved. Strength of support for the BOP clade, relative to that for a PACC + Pooideae clade, also was examined by analyzing the entire data set with the latter relationship constrained. The resulting single most-parsimonious tree is 7345 steps long, i.e., 2 steps or 0.03% longer than the most parsimonious tree obtained by unconstrained analysis. Further investigation, involving additional sampling of characters and taxa, will be conducted with attention to these and alternative structures.

REVISED SUBFAMILIAL CLASSIFICATION

Both the results of the GPWG phylogenetic analysis and a proposed subfamilial reclassification of the family were presented at the Third International Symposium on Grass Systematics and Evolution in Sydney. Because of the comprehensive nature of the analysis, and the proposed changes in subfamilial align-

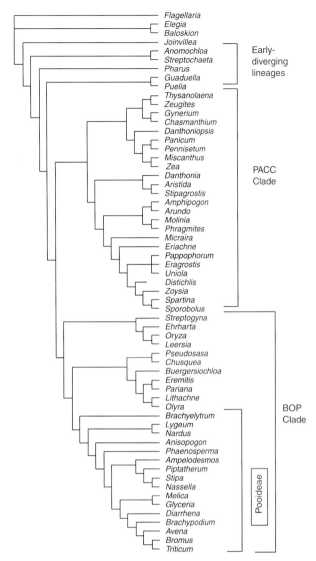

Fig 1. The single most-parsimonious tree inferred from combined analysis of eight data sets (7343 steps, CI = 0.39 with uninformative characters removed, RI = 0.55). PACC = Panicoideae, Arundinoideae, Centothecoideae and Chloridoideae; BOP = Bambusoideae, Ehrhartoideae (= Oryzoideae) and Pooideae.

tered on how to treat the PACC clade. Poor sampling in the danthonioid grasses as well as paraphyly of traditionally arundinoid elements led to a debate about the advisability of splitting the PACC clade into five subfamilies and several taxa of uncertain placement as proposed by the GPWG. Although defensible on the grounds of the robust monophyly of the PACC clade, the option of recognising this clade as a single subfamily was rejected because of the nomenclatural changes required. Another alternative suggested was to recognise two subfamilies, the Panicoideae (including the Centothecoideae and the traditional Panicoideae) and the Chloridoideae (including the various arundinoid elements, *Eriachne*, and *Micraira* as well as the traditional Chloridoideae). The third alternative involved recognising five or six subfamilies in the PACC clade, more or less along the lines proposed by the GPWG, but uncertainties as to the placement of the danthonioid grasses made it impossible to delimit subfamilies that were generally acceptable. The general sense was that either alternative two or three was workable, but that monophyletic groups should be recognised with as little disruption in nomenclature as possible.

Further discussion among members of the GPWG resulted in a decision to include a limited number of taxa from the danthonioid-arundinoid alliance for which some sequence data are already available, and to generate additional sequence data for a few species already in the data set but of uncertain placement in the current phylogenetic analyses. A revised classification will be proposed based on the results of the phylogenetic analyses including these taxa. While the placement of some genera is likely to remain uncertain pending further study, the robust resolution of major clades and the production of a subfamilial classification reflecting the current phylogenetic understanding of the family will enhance the continued study of grass evolution.

ACKNOWLEDGEMENTS

National Science Foundation Grant DEB-9806584 to LGC provided travel support for the majority of the GPWG to attend the Third International Symposium on Grass Systematics and Evolution in Sydney, Australia.

REFERENCES

Angiosperm Phylogeny Group. (1998). An ordinal classification for the families of flowering plants. *Annals of the Missouri Botanical Garden* **85**, 531-553.

Barker, N. P. (1997). The relationships of *Amphipogon, Elytrophus* and *Cyperochloa* (Poaceae) as suggested by *rbc*L sequence data. *Telopea* **7**, 205-213.

Barker, N. P., Linder, H. P. and Harley, E. F. (1995). Polyphyly in the Arundinoideae (Poaceae): evidence from *rbc*L. *Systematic Botany* **20**, 423-435.

Baum, B. R. (1987). Numerical taxonomic analyses of the Poaceae. In 'Grass systematics and evolution'. (Eds T. R. Soderstrom, K. W. Hilu, C. S. Campbell, and M. E. Barkworth.) pp. 334-342. (Smithsonian Institution Press: Washington, D. C.)

Butzin, F. (1965). Neue Untersuchungen Über die Blüte der Gramineae. Doctoral dissertation, Freie Universität Berlin, Berlin.

Campbell, C. S. and Kellogg, E. A. (1987). Sister group relationships of the Poaceae. In 'Grass systematics and evolution'. (Eds T. R. Soderstrom, K. W. Hilu, C. S. Campbell, and M. E. Barkworth.) pp. 217-224. (Smithsonian Institution Press: Washington, D. C.)

ments, a special session (led by R. Soreng and L. G. Clark) to discuss these results was held during the conference. Participants included: A. Anton, N. Barker, M. Barkworth, A. Brown, E. Canning, L. G. Clark, H. Connor, T. Cope, J. DeNardi, M. Duvall, R. Gómez-Martinez, G. Guala, B. Hacker, K. Hilu, S. Jacobs, E. A. Kellogg, P. Linder, H. Longhi-Wagner, T. Macfarlane, K. Mallett, R. Mason-Gamer, A. McCusker, T. Mejia-Saules, O. Morrone, C. Morton, J. J. Ortiz-Diaz, P. Peterson, T. Pizzolato, S. Phillips, M. Rahman, B. Simon, R. Soreng, R. Spangler, A. Van den Borre, J. Veldkamp, S. Wang, C. Weiller, R. D. B. Whalley, and F. Zuloaga.

The consensus was that the classification should reflect phylogeny, although some concern was expressed about proliferation in the number of subfamilies. Recognising each of the three early-diverging lineages and the three major lineages of the BOP clade as a subfamily was noncontroversial, and most discussion cen-

Caro, J. A. (1982). Sinopsis taxonómica de las gramineas argentinas. *Dominguezia* **4**, 1-51.

Clark, L. G., Zhang, W. and Wendel, J. F. (1995). A phylogeny of the grass family (Poaceae) based on *ndh*F sequence data. *Systematic Botany* **20**, 436-460.

Clark, L. G., Kobayashi, M., Mathews, S., Spangler, R. E. and E. A. Kellogg. (in press). The Puelioideae, a new subfamily of grasses. *Systematic Botany*.

Clayton, W. D. and Renvoize, S. A. (1986). 'Genera Graminum, Grasses of the World'. (Her Majesty's Stationery Office: London.)

Clifford, H. T. (1965). The classification of Poaceae: a statistical study. *Papers, Dept. of Biology, University of Queensland* **4**, 243-253.

Clifford, H. T., Williams, W. T. and Lance, G. N. (1969). A further numerical contribution to the classification of the Poaceae. *Australian Journal of Botany* **17**, 119-131.

Cummings, M. P., King, L. M. and Kellogg, E. A. (1994). Slipped strand mispairing in a plastid gene: *rpo*C2 in grasses (Poaceae). *Molecular and Biological Evolution* **11**, 1-8.

Dahlgren, R. M. T., Clifford, H. T. and Yeo, P. F. (1985). The families of the monocotyledons. (Springer-Verlag: New York.)

Davis, J. I. and Soreng, R. J. (1993). Phylogenetic structure in the grass family (Poaceae) as inferred from chloroplast DNA restriction site variation. *American Journal of Botany* **81**, 1444-1454.

Doebley, J., Durbin, M., Golenberg, D. M., Clegg, M. T. and Din Pow Ma. (1990). Evolutionary analysis of the large subunit of carboxylase (*rbc*L) nucleotide sequence among the grasses (Gramineae). *Evolution* **44**, 1097-1108.

Doyle, J. D., Davis, J. I., Soreng, R. J., Garvin, D. and Anderson, M. J. (1992). Chloroplast DNA inversions and the origin of the grass family (Poaceae). *Proceedings of the National Academy of Science* U. S. A. **89**, 7722-7726.

Duvall, M. R. and Morton, B. R. (1996). Molecular phylogenetics of Poaceae: An expanded analysis of *rbc*L sequence data. *Molecular Phylogenetics and Evolution* **5**, 352-358.

Goloboff, P. (1993). *Nona*, version 1.16 (computer software and manual). Distributed by the author.

Hamby, R. K. and Zimmer, E. A. (1988). Ribosomal RNA sequences for inferring phylogeny within the grass family Poaceae. *Plant Systematics and Evolution* **160**, 29-38.

Hamby, R. K. and Zimmer, E. A. (1992). Ribosomal RNA as a phylogenetic tool in plant systematics. In 'Molecular Systematics of Plants'. (Eds P. S. Soltis, D. E. Soltis and J. J. Doyle.) pp. 50-91. (Chapman and Hall: New York.)

Hilu, K. W. and Wright, K. (1982). Systematics of Gramineae: a cluster analysis study. *Taxon* **31**, 9-36.

Hsiao, C., Jacobs, S. W. L., Barker, N. P., and N. J. Chatterton. (1998). A molecular phylogeny of the subfamily Arundinoideae (Poaceae) based on sequences of rDNA. *Australian Systematic Botany* **11**, 41-52.

Kellogg, E. A. (1998). Relationships of cereal crops and other grasses. *Proceedings of the National Academy of Science* **95**, 2005-2010.

Kellogg, E. A. and Campbell, C. S. (1987). Phylogenetic analyses of the Gramineae. In 'Grass Systematics and Evolution'. (Eds T. R. Soderstrom, K. W. Hilu, C. S. Campbell and M. E. Barkworth.) pp. 310-322. (Smithsonian Institution Press: Washington, D.C.)

Kellogg, E. A. and Linder, H. P. (1995). Phylogeny of the Poales. In 'Monocotyledons: Systematics and Evolution, Vol. 2. (Eds P. J. Rudall, P. J. Cribb, D. F. Cutler and C. J. Humphries.) pp. 511-542. (Royal Botanic Gardens: Kew, U.K.)

Liang H. and Hilu, K. W. (1996). Application of the *mat*K gene sequences to grass systematics. *Canadian Journal of Botany* **74**, 125-134.

Linder, H. P. (1987). The evolutionary history of the Poales-Restionales: a hypothesis. *Kew Bulletin* **42**: 297-318.

Mason-Gamer, R. J., Weil, C. F. and Kellogg, E. A. (1998). Granule-bound starch synthase: structure, function, and phylogenetic utility. *Molecular Biology and Evolution* **15**, 1658-1673.

Mathews, S. and Sharrock, R. A. (1996). The phytochrome gene family in grasses (Poaceae): a phylogeny and evidence that grasses have a subset of the loci found in dicot angiosperms. *Molecular Biology and Evolution* **13**, 1141-1150.

Morton, B. R., Gaut, B. S. and Clegg, M. T. (1996). Evolution of alcohol-dehydrogenase genes in the palm and grass families. *Proceedings of the National Academy of Sciences, U.S.A.* **93**, 11735-11739.

Nadot, S., Bajon, R. and Lejeune, B. (1994). The chloroplast gene *rps4* as a tool for the study of Poaceae phylogeny. *Plant Systematics and Evolution* **191**, 27-38.

Prat, H. (1960). Vers une classification naturelle des Graminées. *Bulletin de la Societe Botanique de France* **107**, 32-79.

Reeder, J. (1957). The embryo in grass systematics. *American Journal of Botany* **44**, 756-768.

Reeder, J. (1962). The bambusoid embryo: a reappraisal. *American Journal of Botany* **49**, 639-641.

Sharma, M. L. (1979). Some considerations on the phylogeny and chromosomal evolution in grasses. *Cytologia* **44**, 679-685.

Soderstrom, T. R. and Ellis, R. P. (1987). The position of bamboo genera and allies in a system of grass classification. In 'Grass Systematics and Evolution'. (Eds T. R. Soderstrom, K. W. Hilu, C. S. Campbell and M. E. Barkworth.) pp. 225-238. (Smithsonian Institution Press: Washington, D.C.)

Soreng, R. J. and Davis, J. I. (1998). Phylogenetics and character evolution in the grass family (Poaceae): simultaneous analysis of morphological and chloroplast DNA restriction site character sets. *The Botanical Review* **64**, 1-85.

Soreng, R. J., Davis, J. I. and Doyle J. J. (1990). Phylogenetic analysis of chloroplast DNA restriction site variation in Poaceae subfam. Pooideae. *Plant Systematics and Evolution* **171**, 83-97.

Stebbins, G. L. (1956). Cytogenetics and the evolution of the grass family. *American Journal of Botany* **43**, 890-905.

Stebbins, G. L. and Crampton, B. (1961). A suggested revision of the grass genera of temperate North America. *Recent Advances in Botany* **1**, 133-145.

Swofford, D. L. (1998). PAUP* 4.0 (beta version; Web-distributed software). Smithsonian Institution, Washington, D. C.

Tateoka, T. (1957). Miscellaneous papers on the phylogeny of Poaceae, X: proposition of a new phylogenetic system of Poaceae. *Journal of Japanese Botany* **32**, 275-287.

Tzvelev, N. N. (1977). Zlaki S.S.S.R. [Grasses of the Soviet Union. English translation for the Smithsonian Institution. 1983. Amerind Publishing Co., New Delhi].

Tzvelev, N. N. (1989). The system of grasses (Poaceae) and their evolution. *The Botanical Review* **55**, 141-203.

Watson, L., Clifford, H. T. and Dallwitz, M. J. (1985). The classification of Poaceae: subfamilies and supertribes. *Australian Journal of Botany* **33**, 433-484.

Watson, L. and Dallwitz, M. J. (1992). 'The Grass Genera of the World'. (C.A.B. International: Wallingford, Oxon, U.K.)

Yaneshita, M., Sasakuma, T. and Ogihara, Y. (1993). Phylogenetic relationships of turfgrasses as revealed by restriction fragment analysis of chloroplast DNA. *Theoretical and Applied Genetics* **87**, 129-135.

Grasses: Systematics and Evolution. (2000). Eds S.W.L. Jacobs and J. Everett. (CSIRO: Melbourne)

A SYSTEMATIC VIEW OF THE DEVELOPMENT OF VASCULAR SYSTEMS IN CULMS AND INFLORESCENCES OF GRASSES

Thompson Demetrio Pizzolato

Delaware Agricultural Experiment Station, Department of Plant and Soil Sciences, College of Agriculture and Natural Resources, University of Delaware, Newark, DE 19717-1303, USA.

Abstract

Although developmental vascular anatomy within the leaf blade and the root has been related to grass systematics, there have been relatively few attempts to correlate vascular development in the rest of the plant with systematics. This review strives to show the promise of procambial and xylem and phloem development in uncovering systematic relationships in the axis. Studies in vascular development can correct errors made in interpreting mature vascular systems without a developmental study. Developmental vascular anatomy also reveals that vascular systems along the axis are built from procambial components initiated in relation to insertions on the axis. Comparisons of the development of these systems frequently are useful in distinguishing among the subfamilies of grasses.

Key words: Culm, development, inflorescence, phloem, procambium, vascular, systematics, systems, spikelet, xylem.

INTRODUCTION

Although developmental vascular anatomy and grass systematics have thrived separately, few attempts except in leaves (Ellis 1987; Dengler *et al.* 1997) and in roots (Esau 1957; Rechel and Walsh 1985; Clark and Fisher 1987; Raechal and Curtis 1990) have been made at a union of knowledge. The present review is such an attempt for the grass axis. This revelation of patterns in vascular development in the axis may stimulate others to add to this meager but promising union linking ontogeny of vascular systems and systematics.

INITIATION OF THE VASCULAR SYSTEM IN THE CULM

The pattern of procambial initiation in culms of the Pooideae is remarkably consistent across the investigated genera: *Deschampsia* and *Melica* (Philipson 1935), *Triticum* (Sharman and Hitch 1967; Patrick 1972a,b), *Dactylis* (Hitch and Sharman 1968a), *Poa* (Hitch and Sharman 1971), and *Hordeum* (Dannenhoffer

and Evert 1994). The median procambial trace is initiated in isolation in the insertion of its leaf primordium in the node. Acropetal and basipetal differentiation at this site away from all other procambial strands causes the procambial median trace to ascend the leaf primordium and to descend the culm (M4 in Fig. 1). After median trace initiation, the lateral procambial traces differentiate like the median trace excepting that the laterals, particularly the later laterals, tend to be initiated somewhat more distally in the leaf primordium than the median trace. Isolated differentiation of the lateral procambial traces successively interpolates them between the median trace and the earliest, major lateral trace at each leaf margin and then on both sides of earlier lateral traces in leaf and culm. Since the leaf insertions alternate distichously along the culm, a young internode contains medians opposing and laterals alternating with the corresponding traces from two adjacent distal nodes (n, n+1). Since the leaves alternate distichously, their traces do not obstruct each other in the culm until the proximal node (n+2) of the third adjacent leaf (N7 in Fig. 1). At this node (n+2), interconnections between the leaf traces of the first (n) and of the third (n+2) leaf begin the nodal plexus, which will be discussed in the following section.

Two studies of *Oryza* describe leaf trace initiation for the Oryzoideae (Kaufman 1959; Inosaka 1962). Both agree that, just as in the Pooideae, the vascular system of the culm is initiated in the insertions of the leaf primordia as isolated procambial strands which will become the major leaf traces. As in the Pooideae, bidirectional procambial differentiation from these isolated sites brings the traces distally into the leaf simultaneously with their descent in the culm. The developmental study of Kaufman (1959) indicates that the major procambial traces, in the pooid manner, descend the culm from the leaf insertion (n) and pass through the next node (n+1) before they link with other procambial strands in the next node (n+2). Kaufman (1959) suggests an oryzoid variation from the pooid pattern when he shows that some traces descending from leaf (n) merge with each other as they pass through node (n+1). Although the two studies of *Oryza* agree on the origin of the leaf trace near the leaf insertion (n), the study of Inosaka (1962) indicates that each major trace has a second isolated site of origin, in the subjacent internode opposite node (n+1). Inosaka (1962) contends that the two sites of initiation grow together. Whether the extra site of Inosaka (1962), which is lacking in the Pooideae, occurs in the Oryzoideae must be confirmed by future study. Perhaps the merger of the bundles from leaf (n) near node (n+1) noted by Kaufman (1959) is an aspect of the union of the two sites described by Inosaka (1962).

The developmental study of Hsü (1944) will be the representative pattern of leaf trace initiation for the Bambusoideae. As in the Pooideae and Oryzoideae, isolated procambial strands differentiate in the insertions of the leaf primordia of *Sinocalamus*. From these sites acropetal differentiation brings the traces higher in the primordium, and basipetal differentiation causes the procambial strands to descend the culm. As in the other subfamilies, the traces of the youngest primordia have no connections to proximal strands in the culm. Traces of older leaf primordia become continuous with older procambial strands in the culm. Hsü (1944) did not disclose where in the culm these basipetal mergers occur. Therefore, any differences in vascular initiation in bambusoid culms and in pooid and oryzoid culms remain undetected.

The pattern of procambial initiation in culms of *Digitaria* (Tiba *et al.* 1988) and *Arundinella* (Dengler *et al.* 1997) must represent the Paniceae of the Panicoideae. The median trace is initiated first in the primordial insertion; the lateral traces differentiate later and higher in the primordium. Although the situation is more certain in *Digitaria* than in *Arundinella*, a consensus is that procambia for median and lateral traces are initiated isolatedly in the insertion of the leaf they will serve, and from this site the traces differentiate bidirectionally. Although the location of the basipetal mergers is still unknown in *Digitaria*, the traces in *Arundinella* usually descend through more nodes than is typical of pooid and oryzoid grasses.

In contrast to the initiation of the procambium of the median leaf trace in the insertion of the leaf primordium of the other investigated subfamilies, in *Zea* (the representative of the Andro-

pogoneae of the Panicoideae) the procambium of the median leaf trace is first recognized two or three nodes proximal to its related leaf primordium (Sharman 1942; Esau 1943; Kumazawa 1961). Kumazawa (1961) maintains that the median traces differentiate acropetally as a continuous branch from an older leaf trace in the culm. However, Sharman (1942) and Esau (1943) present evidence that better supports the idea of a distant isolated site that bidirectionally differentiates toward the related leaf primordium and away from it to join a major lateral procambial trace in the culm usually six to nine nodes proximal to the related leaf primordium. All three studies agree that the major lateral traces are initially isolated in the leaf insertion and that the traces differentiate bidirectionally from that site. The major laterals descend distances in the culm as long as do the median traces before merging with other major lateral traces. Sharman (1942) shows that the major lateral procambial strands develop successively from both sides of the median trace toward each margin. Tiba *et al.* (1988) and Dengler *et al.* (1997) also demonstrated this pattern of lateral trace initiation for the panicoid grasses. That this pattern of lateral trace initiation differs from the interpolative one of pooid grasses is emphasized by Dannenhoffer and Evert (1994). The smallest lateral traces descend the periphery of the culm from their leaf and fuse with each other after one or two nodes (Esau 1943). These small, most recently formed traces comprise the outer system of the culm (Kumazawa 1961).

The literature upon procambial initiation in culms is nearly unanimous that leaf traces begin as isolated sites related to their primordia and that the procambial traces must differentiate basipetally to join other traces in the culm. The Andropogoneae differ from the Paniceae of the Panicoideae and from the other subfamilies because the andropogonoid median trace is begun many internodes away from its intended insertion instead of at the insertion. In order to accomplish their basipetal unions, the major traces usually traverse many nodes in the Andropogoneae, fewer in the Paniceae, but only three in the Pooideae (Fig. 1) and in the Oryzoideae. The Oryzoideae may differ from the Pooideae and Andropogoneae by having two isolated sites initiating each major trace for a leaf.

COURSE OF THE LEAF TRACES IN THE INTERNODES AND DIFFERENTIATING NODAL PLEXUS

The longitudinal course of the leaf traces in the culm and their interconnection by the development of the nodal plexus is understood best for the Pooideae (Hitch and Sharman 1971; Patrick 1972a, b; Busby and O'Brien 1979). The median and principal lateral traces descend through nodes and intervening internodes before being blocked by the corresponding trace three nodes (n+2) proximal to the node of entrance (n). At its node of entrance, a major trace is an obstructing trace because it blocks the descent of a corresponding trace from two nodes distal (Figs 1, 2). The entering major trace becomes an alternating trace as it passes through the next node (n+1). The alternate trace then becomes an obstructed trace in the next node (n+2), as the entering corresponding trace blocks further direct descent. Upon being blocked, the obstructed trace in its further basipetal differentiation forks and straddles the obstructing trace or deviates around the obstructing trace. Wherever such forking or

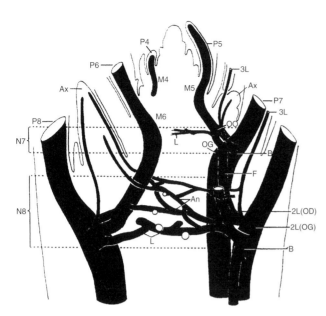

Fig. 1. Reconstruction of *Poa* (Pooideae) shoot apex, in median region, in plane of leaves. Shown are the course and extent of the median traces of successive leaves (M4, M5, *etc.*) and the branches of the two youngest nodal plexi (N7, N8) that connect these. Also shown are a third lateral (3L) of the sixth and seventh primordia (P6, P7) as well as traces from buds (Ax) in the axils of the seventh and eighth leaves. Second lateral traces (2L) of the sixth and eighth primordia illustrate how the forked part (F) of M5 merges with the bridging strand (B) between them. Other horizontal procambial strands besides B are designated by An and L. OD is a leaf trace becoming an obstructed strand; OG is a leaf trace becoming an obstructing strand. Reprinted with permission of The University of Chicago Press from Hitch and Sharman (1971).

deviation occurs, the meristematic tissue persisting in the node differentiates into horizontal procambial strands that begin linking the adjacent vertical traces with each other and their forks or deviations, as well as linking horizontally more distant vertical strands across the node (Figs 1, 2). The forking or deviating strands from the obstructed trace usually descend through the next internode into the fourth node (n+3) from the entry node (n) where the strands merge with a variety of strands in the nodal plexus developing at that node (n+3). Minor lateral traces usually descend through the periphery of one internode before they merge with the nodal plexus differentiating one node proximal (n+1) to their entrance (Fig. 1).

An understanding of the origin of the pooid nodal plexus aids in an understanding of the arrangement of the bundles in the internode. The inner ring of larger bundles are realized to be proximal continuations of the major leaf traces after their obstructing and alternate phases (Hitch and Sharman 1971; Patrick 1972a). The peripheral (outer) ring of small bundles are the basipetal continuations of the smallest lateral leaf traces, which differentiate too late to pass through another nodal plexus without ending by merger.

The great value of the study by Hitch and Sharman (1971) is its recognition that the nodal plexus is a consistent result of procambial differentiation stimulated by the obstruction of descending major leaf traces. All other studies of the nodal plexus rely upon a topographical description about a single node, which cannot reveal the underlying pattern of procambial differentiation.

Thus, the study of Hitch and Sharman (1971) makes comparison of nodal plexi among the subfamilies possible. Unfortunately, no other comparable developmental studies have been completed. By their study, Hitch and Sharman (1971) were able to unravel the pooid pattern of nodal plexus development from the topographical descriptions of Arber (1930). Once the underlying patterns of descending leaf trace obstruction are understood by developmental studies in other subfamilies, comparison of nodal plexi among subfamilies will be facilitated.

A limited comparison between the development of the nodal plexus in *Oryza* (Oryzoideae) with that in the Pooideae is possible because of the work of Kaufman (1959). The nodal plexus of *Oryza* appears to be induced by leaf trace obstructions like those in the Pooideae (Hitch and Sharman 1971). Kaufman (1959) shows that 'transversely oriented procambial strands' are initiated in the proximal levels of a node (n+2) as soon as the entering procambial traces approach leaf traces descending from leaves distal by two nodes (n). Unfortunately, by not following the descent of particular leaf traces in nodes with more developed plexi, Kaufman's (1959) study makes the assignment of roles of contribution of the leaf traces from different distal nodes less certain than by means of the study of Hitch and Sharman (1971). Nevertheless, the study by Kaufman (1959) strongly suggests that the role of obstructing and obstructed traces in nodal plexus initiation in the Oryzoideae follows the pooid pattern. However, Kaufman's (1959) as well as Inosaka's (1962) studies reveal that the development of the nodal plexus is slower for *Oryza* than is general for the Pooideae. A slowly developing nodal plexus allows more traces to descend through it to the next internode than a rapidly developing nodal plexus does (Hitch and Sharman 1971). As in internodes of the Pooideae, those of *Oryza* contain a peripheral (outer) ring of the smallest bundles, which are the basipetal continuations of the smallest leaf traces. Majumdar and Saha (1956) examined a nodal plexus of *Oryza* at a stage of maturity near that at which Kaufman (1959) ceased closely following the descending leaf traces. Unfortunately, Majumdar and Saha (1956), who did not study procambial development, assumed that procambial trace differentiation was acropetal and therefore described the nodal plexus as an ascending anastomosis that is difficult to use comparatively. Arber (1930) made the same developmental assumptions in *Leersia* (Oryzoideae). However, perhaps because she described a node with fewer traces, a mental reversal of her descriptions and drawings is easy. Such basipetal re-interpretation of Arber's (1930) study of *Leersia* reveals median and major lateral leaf traces entering the node, then being obstructed by leaf traces from two leaves distal. The nodal plexus appears at the same level that the forking strands branch from the obstructed trace, in a pattern like that described by Hitch and Sharman (1971) for the Pooideae.

The paucity of developmental data upon basipetal procambial differentiation from the node into proximal internodes and nodes in the Bambusoideae minimizes systematic comparisons with the Pooideae and Oryzoideae. Although the study of Hsü (1944) demonstrates that major traces from leaf primordia merge basipetally in the culm with traces from proximal leaves of *Sinocalamus*, it does not consider the position of the contributing leaves relative to each other nor the resulting initiation of the nodal plexus. The survey of mature nodes and internodes of many other large bamboos by Grosser and Liese (1971) shows a few differences and similarities with the Pooideae and Oryzoideae. As in these subfamilies the periphery of the internode contains the smallest bundles, and the largest bundles occur to the inside of the small bundles in the bamboos. But the bundle number is much higher in the internode, and the innermost bundles, as they descend and become peripheral, diminish in transectional area in the Bambusoideae.

Unfortunately, Grosser and Liese (1971) did not consider procambial development in the nodal plexus. They did indicate that mergers of vertical bundles do not occur in the internode but only in the nodal plexus. Most bundles entering the plexus from the immediate node (n) merge with the outer bundles comprising the nodal plexus, but a few entering bundles pass horizontally into the inner regions of the nodal plexus. This last feature may be a deviation from the pooid nodal plexus. Many bundles also enter the nodal plexus from the internode it subtends, but Grosser and Liese (1971) are unclear whether some of the large bundles from one internode pass through the nodal plexus into the next internode as do the alternating bundles of the Pooideae. However, Zee (1974) notes in *Arundinaria* that some bundles pass through the node from one internode to the next. Zee (1974) also notes that many of the traces in the node come from the attached leaf. Nevertheless, until developmental studies relate procambial development in the node and internode to the leaves stimulating the procambial leaf traces, further systematic comparisons of nodal plexi should wait.

No developmental comparisons of the nodal plexi of the Paniceae of the Panicoideae are yet possible. Tiba *et al.* (1988) did not consider the location of the basipetal leaf trace mergers that they described in a species of *Digitaria*. The topographical description of the mature nodal plexi of other species in the Paniceae by Deshpande and Sarkar (1961), who assume continuous acropetal leaf trace differentiation in the culm, results in such a complex anastomosis that the comparative value of this detailed study cannot be appreciated until a developmental study of the same material is undertaken.

The studies of Sharman (1942), Esau (1943), and Kumazawa (1961) in *Zea* permit valuable comparisons of the longitudinal courses of the leaf traces in the culms of the Andropogoneae (Panicoideae) with the longitudinal courses in the Pooideae (Hitch and Sharman 1971) and Oryzoideae (Kaufman 1959). A major difference is that the proximal paths of the major leaf traces of *Zea* from leaf insertion to merger six to nine nodes away are horizontally across the first node toward the center, then downward gradually through the intervening internodes and nodes tending toward the periphery of the opposite side of the stem. At the end of its proximal extent, the median trace merges with the first or second lateral trace of a leaf whose insertion is a few nodes proximal from the leaf supplying the median. The descent of the major laterals is like that of the medians, but it may be less abrupt. Major laterals merge like medians except that major laterals also merge with medians. Recall that in the Pooideae and probably in the Oryzoideae, the major traces descend, without moving to the center, only three nodes before

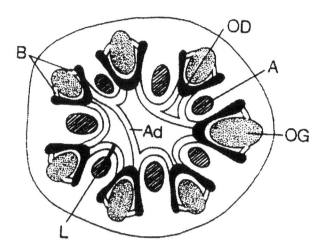

Fig. 2. The pooid pattern formed by the nodal plexus procambial strands, as seen in transverse view. A=alternating strand (cross-hatched); OD=obstructed strand (black); OG=obstructing strand (stippled). Ad, B, and L indicate horizontal procambial strands linking OD and OG. Reprinted with permission of The University of Chicago Press from Hitch and Sharman (1971).

mergers begin on the same side of the culm where the traces entered (Figs 1, 2).

Cross sections of internodes of *Zea* and other andropogonoid grasses look different from those of the Pooideae and Oryzoideae because of the larger number of bundles throughout the stem. The large bundles in the center of the stem are those major traces that are in the middle stages of descent. Many of the somewhat smaller sub-peripheral bundles are those near their point of merger (Sharman 1942; Esau 1943). These last are the bundles Kumazawa (1961) considered to be the inner system. Among the reasons for more bundles throughout a cross section of an andropogonoid internode in comparison to one of the Pooideae and Oryzoideae must be the greater number of traces differentiated by a leaf in the culm, their descent across the center of the stem, their longer descent before merger, and the delay in initiating the nodal plexus until about the tenth internode (Sharman 1942; Esau 1943).

Kumazawa (1961) with *Zea* gives the most detailed description of nodal plexus development in the Andropogoneae (Panicoideae). The horizontal procambial strands, which will become the horizontal bundles of the nodal plexus, first differentiate in the peripheral and subperipheral region of the node from persisting meristematic tissue. Recall that this is the region of merger of the major traces. The outer system of minor lateral traces and the subperipheral (inner system) of compound bundles (a bundle comprising two major traces merged near the end of their descent) become interconnected by the horizontal procambial strands differentiating among them. Although the horizontal procambial strands also differentiate among the major traces in the center of the node in their mid-descent, the major traces do not become interconnected by the horizontal strands because the major traces are now mature vascular bundles. In agreement with Kumazawa's (1961) description, Sharman (1942) shows that the nodal plexus begins in the periphery of the node; and Esau (1943) shows major traces passing vertically through the plexus in *Zea*. Mentally reversing Arber's (1930) acropetal topographical description of a differentiating nodal plexus of a *Coix* seedling shows that she observed that 1) the major traces in the beginning of their descent of the culm are unattached to the differentiating horizontal strands of the nodal plexus, that is, the descending traces of *Coix* like those of *Zea* pass through more nodal plexi than do pooid traces; and that 2) procambial attachments differentiate in the periphery of the node to join horizontally the descending vertical traces only after the latter (medians with laterals from different leaves) have merged.

This meager literature on procambial leaf trace differentiation in the culm must be augmented before comparisons among subfamilies can be ascertained. The literature as it is now suggests that leaf trace differentiation in node and internode of the Pooideae and Oryzoideae differs greatly from that of the Andropogoneae of the Panicoideae. The 'manner of production of the plexi of the Panicoideae may be different from the festucoid grasses' (Hitch and Sharman 1971) since in the andropogonoid grasses, horizontal procambial linkages form first at the node where major traces merge at the ends of their descents but, in the pooid grasses, horizontal procambial linkages form first at the node where major traces avoid each other at the beginning of the descent of one of them (the obstructing strand) and near the end of the descent of the other (the obstructed strand). Since nodal plexus differentiation is delayed in the Andropogoneae (Panicoideae) relative to the Pooideae, more of the basipetally differentiating traces of the first subfamily than of the second differentiate though more sites of nodal plexi before the horizontal procambial strands differentiate there (Sharman 1942; Esau 1943; Kumazawa 1961; Hitch and Sharman 1971).

Comparative data for the Bambusoideae is meager. Nevertheless, the following observations suggest that the stem vascular system of the Bambusoideae is more like that of the Andropogoneae than that of the Oryzoideae and Pooideae: the large number of large traces in the center of the internode; a conspicuous outer system like Kumazawa's (1961); some major traces appear to pass through the center of the plexus; and some cross abruptly to the center of the plexus from the attached leaf (Grosser and Liese 1971). However, Grosser and Liese's (1971) topographic description of the mature nodal plexus of bamboos also indicates abundant horizontal mergers between most of the traces entering from the immediate leaf and the vertical bundles in the periphery of the plexus, as in the pooid pattern. Information upon nodal plexi of other subfamilies (Fisher and French 1976) must be supplemented with developmental studies of the nodal plexus comparable to those of Hitch and Sharman (1971) for the Pooideae (Figs. 1,2).

AXILLARY BUD TRACES

The pattern underlying the procambial unions of the leaf traces developing from the axillary buds with the traces of the principal (parent) culm is more difficult to see than the pattern of mergers, in the principal culm, of traces from the principal leaves making the pooid nodal plexus. Part of the apology for the less predictable nature of bud trace mergers is that the procambium of the nodal plexus of the principal culm already has begun differentiating in the node of the leaf subtending the axillary bud (Fig. 1) as its procambial traces are differentiating into

the culm (Hitch and Sharman 1968b, 1971). Recall that when the principal leaf traces enter the culm the nodal plexus has not yet begun to differentiate.

Another source of variation is the variety of procambial leaf traces that differentiate basipetally into the principal culm from the axillary bud. The first leaf of the bud (the prophyll) has two major isolated strands in the corners of its insertion on the bud. The first (P1) is initiated a little earlier than the second (P2). The median trace (M) of the first normal bud leaf develops thirdly opposite P1 and closer to P2. The median trace of the second normal leaf (2M) is initiated later opposite M and closer to P1 (Hitch and Sharman 1968b). Besides these four major traces of the bud, isolated lateral traces for the prophyll and normal leaves are initiated soon afterwards. Moreover, the bud in the axil of the prophyll soon produces traces which will enter the nodal plexus of the principal culm (Bell 1976a).

One more source of variation in the mergers occurring between the nodal plexus of the principal culm and the traces from the axillary bud is related to the likelihood that the buds developing into branches will do so in the basal portions of the principal culm where the internodes do not elongate resulting in juxtaposed nodal plexi. Traces from the contributing axillary bud join the major traces of the subtending leaf (n) or the horizontal and vertical traces of the close nodal plexi (Hitch and Sharman 1968b, 1971; Bell 1976a). Recall that the vertical strands of a nodal plexus (Figs 1, 2) are the traces from distal leaves n (obstructing), n-1 (alternating), n-2 (obstructed) and n-3 (forking). The P1, P2, M, and 2M traces of the bud usually merge with the nodal plexus of the subtending leaf (Ax between P6 and P8 with N8 in Fig. 1) instead of joining the major traces of the subtending leaf distal to their incorporation into the nodal plexus (Hitch and Sharman 1968b, 1971; Bell 1976a). The principal bud traces also are more likely to join with the nodal plexus of the principal culm than with the nodal plexi developing within the axillary bud (Bell 1976a). When traces from the leaves of the bud reach a principal culm with elongated internodes, the bud traces behave like the lesser laterals of the culm. They may pass the node of their insertion (n) like peripheral (outer) bundles, but they will merge with the procambia of the top of the plexus at the next node (n+1). Although the traces from the axillary buds merge with the nodal plexus, the bud traces do not influence the basic pattern of its formation (Hitch and Sharman 1971).

The initiation of procambial traces in the axillary buds of pooid and oryzoid grasses is similar according to the study of Inosaka (1962). In *Oryza* the two prophyll traces and the median trace of the first normal leaf are initiated in isolation at their insertions and differentiate basipetally from the bud into the parent culm as in the pooid pattern, and merge with the traces of the culm. At this point a significant difference between pooid and oryzoid grasses may exist. In *Oryza* the major traces from the bud leaves do not tend to merge with the plexus of the node of the subtending leaf, as in pooid grasses (Hitch and Sharman 1971), but the principal bud traces of *Oryza* merge with the traces in the portion of the internode immediately distal to the nodal plexus of the node of the subtending leaf (Inosaka 1962). A result of the basipetal mergers in this region is that the traces of the axillary bud are more intimately connected to the traces of the two leaves distal to the subtending leaf than to the traces from the subtending leaf and the nodal plexus (Inosaka 1958). Perhaps this difference in bud trace mergers in the Oryzoideae results from the delay, relative to pooid grasses, in nodal plexus formation in *Oryza* (Kaufman 1959).

Studies of Grosser and Liese (1971) and Zee (1974) indicate extensive vascular connections between the axillary bud and the nodal plexus of the subtending leaf in the bamboos. However, since no developmental studies in the Bambusoideae of traces from the bud leaves like those of Hitch and Sharman (1968b, 1971) for the Pooideae yet occur, comparisons cannot be made with the Pooideae, Oryzoideae, and Andropogoneae.

A little more comparative data is available on axillary bud vasculature for the Andropogoneae (Panicoideae). Although Arber (1930) was skeptical of isolated leaf trace initiation, she observed two leaf traces in a young axillary bud of *Coix* that were still separate from the nodal plexus. Kumazawa (1961) did not describe the traces of the axillary bud of *Zea* in terms of the leaves of the bud, but as a system of outer small bundles and as a system of inner large bundles. However, other sections of the study indicate that the outer small bundles are probably related to the lesser lateral traces of the bud leaves and that the inner large bundles are probably related to the principal traces of the bud leaves, in the pooid pattern. Also in the pooid pattern, the bud traces link with the immediate nodal plexus. However, a striking difference with the pooid grasses is that the bud traces of *Zea* link in such a way with the nodal plexus that there is no direct contact of major bud traces with the major leaf traces of the subtending leaf (Kumazawa 1961). This type of interconnection is reminiscent of that noted in *Oryza* by Inosaka (1958, 1962).

NODAL ROOT INSERTIONS

A comparison of nodal root vascular systems will begin in the Oryzoideae with Kaufman's (1959) study of *Oryza* and Arber's (1930) study of *Leersia* and *Zizania*, since their studies are the amplest accounts of procambial initiation outside of the Andropogoneae. Root primordia differentiate from the culm cortex of the earlier nodes. In *Oryza*, primordia develop as meristematic masses of root cap, protoderm, and cortex surrounding a procambial central cylinder before any procambial connections occur with the vascular system differentiating simultaneously in the culm. Thus, the nodal root primordium of *Oryza* is like the axillary bud primordium of the Pooideae (Hitch and Sharman 1968b) and perhaps of other subfamilies in that both primordia differentiate their own procambia before becoming linked with the parent culm. Outside the differentiating nodal plexus of *Oryza*, horizontal procambial strands differentiate a few cells away from the root primordia after they initiate their procambial central cylinders. These peripheral, horizontal procambial strands then differentiate acropetally so as to advance to the procambial central cylinders of the root primordia, forming a ring of transversely oriented procambial strands (Kaufman 1959) that Arber (1930) describes as the root girdle. Arber (1930) and Kaufman (1959) agree that for the Oryzoideae, although procambial anastomoses will form to link the root girdle with the nodal plexus, the root girdle initially is independent of and external to the nodal plexus.

Although Bell (1976b) with *Lolium* (Pooideae) studied older nodal root insertions than those used to characterize this region for the Oryzoideae, the studies of *Lolium* show a marked similarity in the region for both subfamilies. The nodal roots of *Lolium* are clustered on the basal nodes with little internodal elongation, hence the pooid nodal plexi in this region merge with each other. External to the congested, juxtaposed nodal plexi is a perforated vascular cylinder Bell (1976b) calls the peripheral plexus. Bell (1976b) identifies the horizontal strands of the peripheral plexus as the root girdle of Arber (1930). Bell (1976b) designates the nearly vertical strands in the peripheral plexus as diffuse bundles. The description of the types of mergers and the timing of their formation strongly suggests that the diffuse bundles are the vascular linkages that Arber (1930) and Kaufman (1959) describe as differentiating between superposed root girdles, between a root girdle and the nodal plexus, and between the root girdle and the root primordium after the root primordia, root girdles, and nodal plexi themselves have differentiated. As a root primordium continues to develop, its vascular connections to the peripheral plexus are enhanced by procambial differentiation between the root primordium and the nodal plexus, bringing the root into vascular continuity with some of the leaf traces (Bell 1976b). The vascular system of the nodal root insertions of the Oryzoideae and Pooideae appear similar except for the possibility that the component of diffuse bundles (Bell 1976b) may be more elaborate in the Pooideae.

Procambial initiation of the peripheral plexus of *Zea* (Andropogoneae) (Kumazawa 1961; Hoppe *et al.* 1986) must represent the Paniceae. In the distal regions of a node, where nodal root primordia do not yet occur but adjacent to proximal regions of primordial formation, the peripheral traces of the outer and inner systems become interconnected by differentiating procambial strands. Simultaneously, the outer and inner systems are also being connected to a horizontal system of procambial root girdles differentiating in the cortex immediately surrounding the outer and inner vascular traces. Hemispherical masses of meristematic tissue differentiate to the outside of the differentiating root girdles. When the root primordia first appear, they lack differentiated regions. The procambial central cylinder of each root primordium develops later as a conical outgrowth from adjacent portions of the procambial root girdle. Recall that the initiation of procambium in the root primordium of *Oryza* differs from that of *Zea* since the procambial central cylinder of the nodal root primordium of *Oryza* is first independent of the root girdle (Kaufman 1959).

Nevertheless, the sparse literature suggests a similarity in the differentiation of the vascular system related to the nodal roots. The root primordia are initiated independently of any vascular system. The procambium of the peripheral plexus forms afterwards and externally links the nodal roots to each other and internally links them to the systems of leaf traces and nodal plexi. That systematic differences do occur among the subfamilies in nodal root insertions, at least between the Pooideae and the Panicoideae, is indicated by the findings of Aloni and Griffith (1991). The pattern of xylem differentiation in *Zea* and *Sorghum* suggests continuous procambial differentiation between nodal root primordia and the peripheral plexus. The pattern of xylem differentiation in *Triticum*, *Hordeum*, *Secale*, and *Avena* suggests that procambia of nodal roots and procambia of the peripheral plexus are initially isolated but later merge from their separate sites. The developmental studies of Kaufman (1959) and Hoppe *et al.* (1986) can serve as models to comparatively dissect the peripheral plexus (Bell 1976b).

VASCULAR CONNECTIONS BETWEEN CULM AND INFLORESCENCE

The literature allows tentative comparisons between the andropogonoid and pooid grasses only. Procambial strands differentiate in isolation at the proximal mid-regions of the inflorescence axis of *Hordeum* (Pooideae) after the appearance of the spikelet primordia along it but before the spikelet primordia contain procambia (Kirby and Rymer 1974). The strands differentiate acropetally, and basipetally to join the vascular system of the culm. The study of *Hordeum* does not indicate which procambial strands from the inflorescence join which leaf traces in the culm. Neither does a related study of *Triticum* (Pooideae). However, that study (Patrick and Wardlaw 1984) does show that traces from the inflorescence merge with the nodal plexus of the distal normal leaf (n) of the culm (the flag leaf) in the same obstructing (n), alternate (n+1) and obstructed-trace (n+2) relationship typical of pooid culms (Hitch and Sharman 1971; Patrick 1972a, b).

In the andropogonoid *Zea*, according to Kumazawa (1961), the major traces of the inflorescence axis and of its principal branches arise differently. The major traces differentiate acropetally in the inflorescence. The major traces of the inflorescence axis are acropetal branches of traces in the culm. Recall that Kumazawa (1961) differs from Sharman (1942) and Esau (1943) by maintaining that major traces in *Zea* arise as acropetally differentiating branches of other traces. Even if Kumazawa (1961) missed the basipetal mergers of traces from the inflorescence with the traces in the periphery of the culm, the study still points to a basic difference in this union of culm and inflorescence traces with the pooid grasses. Kumazawa (1961) states that the major traces from the staminate inflorescence 'descend through many foliar nodes without uniting with the leaf trace strands and move gradually toward the periphery of the stem.' Although this sparse evidence must be augmented before conclusions can be made, observations now suggest that, in pooid grasses, mergers of major traces from the inflorescences occur with major traces of the first normal leaf (the flag leaf), but in the andropogonoid grasses, mergers of inflorescence traces occur with traces from leaves inserted much more proximally. How the other subfamilies stand in this regard is unexplored.

INNER AND OUTER SYSTEMS OF INFLORESCENCE AND CULM

There is an apparent uniformity to the arrangement of vascular bundles in the internodes of the inflorescence axis and of the culm across the subfamilies. Cross sections show the inner bundles to be large and the outer bundles to be small (Kumazawa 1961; Patrick 1972a; Maze 1977). However, apparent similarity in this feature can be deceptive for two reasons. First, there are a number of subperipheral bundles that are intermediate in size and hence difficult to assign to either system by size or position.

Second, the bundles of the outer system have different origins depending on the subfamily. In the Pooideae at least (Patrick 1972a, b; Patrick and Wardlaw 1984), the outer smaller bundles may represent the brief descent of lesser leaf traces as well as the endings of the long descent of the forking extensions of major leaf traces. At least in the andropogonoid Panicoideae (Esau 1943; Maze 1977; Hoppe *et al.* 1986), the outer small bundles are the brief descent of minor leaf traces of culm or inflorescence as well as the final phase of descent of major leaf traces before merger with other bundles behaving similarly. Although Kumazawa (1961) separated the peripheral bundles in culm and inflorescence axes of *Zea* into outermost small ones (minor traces in brief descent) and subperipheral larger ones (major traces from the center after long descent), the regions are difficult to discern. The concept of inner and outer systems of traces in the culm and inflorescence may mask differences in the procambial origins of the traces among the subfamilies.

VASCULAR CONNECTIONS AMONG INFLORESCENCE BRANCHES AND THE SPIKELETS

Only two developmental studies of the vascular systems in inflorescence axes are complete. However, since these studies demonstrate for the Pooideae and the Andropogoneae of the Panicoideae quite different methods of building the vascular system, they are sufficient to suggest taxonomic differences. Maze (1977) studied procambial development for the vascular system in the inflorescence axes of *Andropogon* (Andropogoneae: Panicoideae). Compared to a cross section of an andropogonoid culm, the number of traces in the axes are fewer but consist of small peripheral traces and larger more interior traces (Fig. 3). More significantly Maze (1977) showed that all of the inner and outer traces are initially isolated and that they are initiated by the influence of spikelet primordia in the andropogonoid pattern; that is, the traces begin at some distance from their spikelet primordia. The distant stimulant to procambial trace initiation in the inflorescence axis is either the spikelet primordium itself or the differentiating glumes of the primordium (Fig. 3). Thus, the vascular system of the inflorescence axis may begin as major traces of a glume. As more and more spikelet primordia appear distally on the inflorescence axes, the procambial strands, which are initiated separately under the influence of particular primordia, progressively merge axially with each other to make sympodia (Fig. 3). The sympodium, the union of blended spikelet traces, belongs to the outer system for some of its course and to the inner for some of the rest of its course, in the andropogonoid pattern. Apparently nodal plexi do not differentiate in the inflorescence axis. An appreciation of the vascular systems of the inflorescence axes of eighteen other genera in the Andropogoneae (Chandra and Saxena 1964) must await a study like Maze's (1977) on *Andropogon* when the mature system presented topographically is developmentally unravelled.

Maze (1977) uncovered in the inflorescence axis of *Andropogon* an axial system built by basipetal mergers of isolated procambial strands initiated as traces for the spikelet primordia (Fig. 3). As the glumes and lemmas of a spikelet primordium differentiate, they too stimulate traces at a distance which will contribute to their sympodium. Maze and Scagel (1982) confirmed the influ-

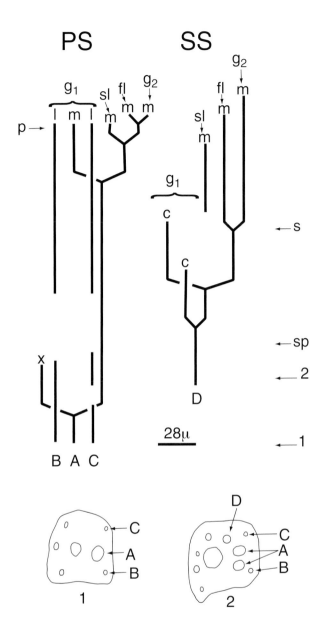

Fig. 3. Procambial strands in the axis of a spikelet pair of *Andropogon* (Andropogoneae). Cross sectional diagrams designated 1 and 2 indicate configuration at levels indicated in the diagrams. Trace designation is the same in cross sections and diagrams. PS=pedicellate spikelet; SS=sessile spikelet; sp=base of spikelet pairs; s=base of first glume of sessile spikelet; p=base of first glume of pedicellate spikelet; g1=first glume; g2=second glume; sl=sterile lemma; fl=fertile lemma; m=median trace; l=lateral trace; c=corner traces of first glume of sterile lemma; x=trace continuing up the axis. Reprinted with permission of the Botanical Society of America from Maze (1977).

ence of the spikelet primordia and their component leaves on trace initiation in the axis by noting that no other leaves (bracts) occur with the spikelet primordia on the inflorescence axis.

The contrasting pattern of procambial development initiating the vascular system of the inflorescence of pooid grasses is based upon a developmental study by Kirby and Rymer (1974) on *Hordeum*, augmented by one of Pizzolato (1997) on *Triticum*. Following the pooid pattern for the culm (Hitch and Sharman

1971), the vascular system for the inflorescence axis is initiated in the insertion of a leaf on the axis (Fig. 4). All of the major and lesser traces of the pooid inflorescence axis begin as isolated traces for the leaf primordium (the lower ridge leaf) subtending the spikelet primordium (Pizzolato 1997). The procambial traces for the glumes of the spikelet primordium merely merge in the inflorescence axis with traces already established there in response to the leaf primordium subtending each spikelet primordium. Recall that in the andropogonoid pattern, the glume traces may initiate the sympodium (Fig. 3).

Even though inflorescence axes of pooid grasses have an inner and an outer system of bundles looking like that of andropogonoid grasses (Figs 3, 4), the systems in the subfamilies are not homologous. In the inflorescence axis of *Triticum*, the outer system of small bundles is comprised of the minor lateral traces (W in Fig. 4C) of the leaf primordium (lower ridge leaf) subtending the spikelet (Pizzolato 1997). On the other hand, in the andropogonoid inflorescence axis, the outer system comprises minor lateral traces directly from the spikelet leaves as well as the proximal ends of merged spikelet traces in sympodia (Maze 1977). This last statement affirms previous statements about misleading apparent similarities in the inner and outer systems across subfamilies in their inflorescences and culms. A genuine similarity between the inflorescence axes of andropogonoid and pooid grasses is the absence of the nodal plexus. The orderly pattern of leaf trace obstruction discovered by Hitch and Sharman (1971) in the pooid culm is replaced by merging of the isolated strands so that the horizontal component of the nodal plexus does not form in the inflorescence axis (Pizzolato 1997).

Thus Maze (1977) and Pizzolato (1997) suggest a new way of comparing the vascular systems of the inflorescence axis, at least in some inflorescences of some pooid and andropogonoid grasses. However, earlier Arber (1928) had a notion that when leaves subtend spikelets, such axillant leaves influence the construction of the vascular system in the inflorescence axis. Arber's (1928) study carries the idea of this type of construction to other inflorescences in the Pooideae and in the Oryzoideae. Arber (1931) observed axillant leaves and associated bundles along the axis of the inflorescence of the pooid *Alopecurus*, and she and Pizzolato (1987) saw them on the axis just proximal to the glumes; in these cases, the bundles in the axis appeared to be related to the axillant leaves. Arber (1928) states that, in the inflorescence axis, bundles of the oryzoid *Luziola*, which has axillant leaves, and of pooid *Dactylis*, which apparently lacks them, 'behave in a way which one associates with leaf-traces rather than with branch-traces.' In more recent support of Arber (1928), Xu and Vergara (1986) found bracts along the inflorescence axis primordium of *Oryza*, and Fraser and Kokko (1993) found that the lower ridge leaf primordium of *Dactylis* is initiated but later converted into an inflorescence branch. Support for the idea of systematically different vascular systems in inflorescence axes must not rest upon what the mature system looks like or upon seeing or not seeing axillant leaves along the inflorescence, for they may occur in some andropogonoid grasses (Sundberg and Orr 1996). Support must rest on a study of procambial development in the inflorescence primordia like that of Maze (1977).

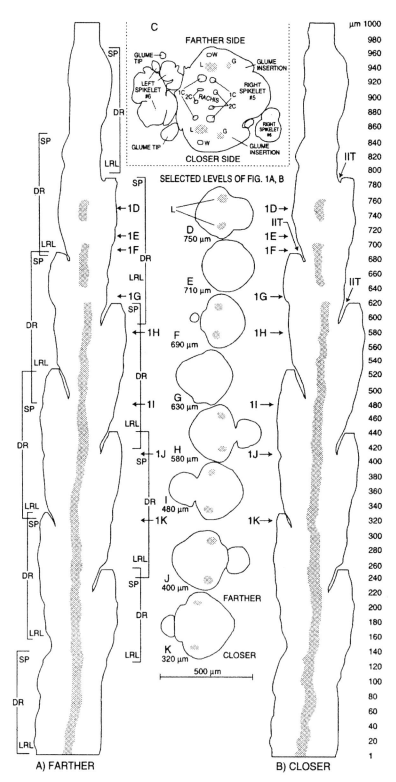

Fig. 4. Tracings of spike primordia of *Triticum* (Pooideae). A, B, Sagittal section of rachis and double ridges reconstructed from serial 2 µm-thick transections. The procambial lateral traces of the lower ridge leaves in the portions of the rachis farther and closer to the reader are stippled. C, Transection from a spike older than A and B demonstrating component appendages and procambial traces. D-K, Transections selected from those used to reconstruct A and B. DR=double ridge comprised of the spikelet primordium and the lower ridge leaf; G=glume procambial trace; IIT=insertion of the lower ridge leaf that influences the isolated initiation of the lateral procambial traces; L= lateral procambial trace (one of the first pair of traces of a lower ridge leaf); LRL=lower ridge leaf; SP=spikelet primordium; W=wing procambial trace (a peripheral bundle of a lower ridge leaf); IC=first central procambial trace (one of the first pair of the lesser traces produced by a lower ridge leaf); 2C=second central procambial trace (one of the first pair of central traces produced by a lower ridge leaf after the first central traces). Reprinted with permission of The University of Chicago Press from Pizzolato (1997).

One more aspect of the connection of spikelet to inflorescence axis must be emphasized in order to distinguish the pooid and andropogonoid inflorescences. In both types of axes, the major traces initiated by the glumes are the primary connections between the vascular system of the inflorescence axis and the vascular system of the spikelet (Maze *et al.* 1972; Maze 1977; Pizzolato 1997, 1998). As far as the glume traces are concerned, the distinction between the types rests upon the fact that, in the andropogonoid inflorescence, the glume traces are initiated proximally to the glume insertion, in the inflorescence axis (Maze 1977); but in the pooid type, the glume traces are initiated in the glume insertion in the spikelet (Maze *et al.* 1972; Pizzolato 1997, 1998). In the pooid type and perhaps in the oryzoid type, the glume traces join sympodia formed in the inflorescence axis by the leaves subtending the spikelets (G in Fig. 4C). In the andropogonoid type, the glume traces comprise the sympodia of the inflorescence axis (Fig. 3).

VASCULAR CONNECTIONS IN THE RACHILLA

Surveys of the number of vascular bundles in a rachilla segment indicate that one and three commonly occur throughout the subfamilies (Arber 1934; Belk 1939; Butzin 1965). Arber (1927) knew of the artificiality of considering the rachilla an individual organ since she realized that the vascular bundles of the rachilla are related to traces of the leaves inserted on the rachilla, particularly to the traces of the lemmas. Nevertheless, the few studies of procambial leaf trace development in rachillae suggest the pooid and chloridoid/arundinoid subfamilies have taxonomically distinct patterns. A decision as to whether the 'strong evolutionary relationship between the Chloridoideae and Arundinoideae' (Hilu and Esen 1993) extends to the vascular system of the rachilla depends upon continued study of trace development in rachillae of all subfamilies.

Procambial trace development in the rachilla of the pooid *Triticum* (Pizzolato 1998) allows comparison with similar events in the rachilla of the arundinoid *Phragmites* (Pizzolato 1994a). The three major traces of the lemma – the median trace (M), the major lateral trace of the left margin (LL), and the major lateral trace of the right margin (RR) – are the three major traces of the rachillae of *Triticum* and *Phragmites*. These traces are initiated as isolated procambial strands. A median trace (nM) tends to merge basipetally with one or both of the major lateral traces from the subtending lemma (n-1LL and/or n-1RR). Median traces never merge with median traces (Figs 5-7). Each major lateral trace tends to merge basipetally with the opposing major lateral trace of the subtending lemma (nLL with n-1RR; nRR with n-1LL).

Most of the differences in the rachilla stem from the fact that the three major lemma traces in *Triticum* are initiated at the insertion (node) of the lemma they serve, whereas such traces are initiated in *Phragmites* in the node subjacent to the lemma they serve. With acropetal differentiation the major traces of *Triticum* immediately enter the lemma they serve. Basipetal differentiation of the major traces of the rachilla of *Triticum* occurs through one internode until merger with the traces of the subtending node. On the other hand, the major lemma traces of *Phragmites* must ascend a full internode before entering the insertion of the lemma they serve. During this ascent the median trace is not

blocked by any descending traces, but the major lateral traces are blocked in mid course by mergers with the opposite lateral traces and the median trace descending from the distal lemma (Fig. 5). Compensation for such blockage and merger requires the initiation of the major lateral trace capturing bundles (LL' and RR'), which connect the lemma margins with their lemma lateral traces blocked in the rachilla in *Phragmites* (Fig. 7). Such compensation is unnecessary in *Triticum* (Fig. 6), and there are no capturing bundles. Zimmermann and Tomlinson (1972) proposed the existence of capturing bundles in the axes of other monocotyledonous stems, and now capturing bundles are revealed in the spikelets of the arundinoid *Phragmites* and the chloridoid *Muhlenbergia* (Pizzolato 1994a, 1995).

The roles of the lemma intermediate lateral traces (L and R) and of the palea traces (PL and PR) in *Triticum* and *Phragmites* are also strikingly different. In *Triticum* these four traces connect the pistil/stamen vascular systems with the median and major lateral traces of the lemma in the floret. In *Phragmites* these four traces (L, R, PL, PR) connect the pistil/stamen vascular systems with only the lemma median trace in the rachilla (Figs 6, 7). Since the ascending lemma intermediate lateral traces (L, R) merge with the descending stamen trace complex and since the palea traces function like their capturing bundles once the intermediate lateral traces enter the floret, no intermediate laterals exist in the lemma of *Phragmites* in contrast to the situation in *Triticum*.

The rachillae of *Triticum* and *Phragmites* are similar in that both have interfloret connections by means of mergers of their major lemma lateral traces. Major differences result because initiation of the three major lemma traces is at the insertion in *Triticum* but one node away in *Phragmites* (Figs 5-7). The rachillae of *Anthoxanthum* (Pooideae) and *Muhlenbergia* (Chloridoideae) are similar in that both have interfloret connections by means of mergers of their lemma median traces (Pizzolato and Robinson-Beers 1987; Pizzolato 1995). Major differences result because the median lemma trace of *Anthoxanthum* begins, as in the pooid pattern, at the lemma insertion whereas *Muhlenbergia*'s lemma median trace begins, in the arundinoid/chloridoid pattern, in the subjacent node (Fig. 8). A result of this difference is that the major lateral traces of a lemma of *Anthoxanthum* can merge at the same node directly with their own median trace. However, the major lateral traces of a lemma of *Muhlenbergia* join the median trace associated with the distal lemma with the result that the major lateral traces need to be redirected into the proximal lemma by capturing bundles (LL', RR') comparable to those of *Phragmites* (Figs 7, 8).

Vascular patterns discerned in the rachilla from studies by Arber (1927), Butzin (1965), Maze *et al.* (1971, 1972), Robinson-Beers and Pizzolato (1987a), Pizzolato and Robinson-Beers (1987) and Pizzolato (1994a, 1995, 1998) suggest that interfloret connections occur in the rachillae of pooid and arundinoid/chloridoid grasses either primarily via lemma median traces or primarily via lemma major lateral traces. Significant developmental and taxonomic differences between rachillae of pooid and arundinoid/ chloridoid grasses are reflected by the capturing bundles only in the latter group, which return distant procambial traces to their florets.

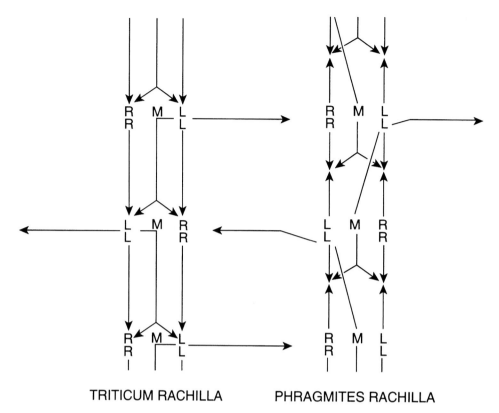

TRITICUM RACHILLA   PHRAGMITES RACHILLA

Fig. 5. Diagrams of the rachillae of *Triticum* (Pooideae) and *Phragmites* (Arundinoideae) emphasizing similarities and differences in their vascular patterns. M=lemma median trace; LL=major left lateral trace of lemma; RR=major right lateral trace of lemma. The site of the letters designating the trace is the site of isolated procambial initiation. Arrows show the directions of procambial differentiation from those sites. Only the lemma median trace is shown differentiating from the rachilla into the lemma. Lemma median traces never merge with each other. They merge with lemma lateral traces which merge with each other. Lemma traces of *Phragmites* are initiated farther from their intended lemma than are the lemma traces of *Triticum*.

VASCULAR SYSTEM OF THE FLORET: CONTRIBUTIONS FROM LEMMA AND PALEA TRACES

Although many descriptions of the floret vascular system as it relates to the lemma and palea of most subfamilies are available for comparative study (Long 1930; Arber 1934; Belk 1939; Chandra 1962a, b; Butzin 1965; Pizzolato 1987, 1988, 1989a, b, 1990a, b, 1991, 1992, 1993a, b), these do not reveal the pattern by which the vascular system is assembled, since the florets were sampled only near and after anthesis when the vascular system is almost mature. The lemma's and palea's (and stamen's and pistil's) contributions to the vascular system become so blended by maturation that errors are made by inferring developmental patterns from a study of the mature vascular system in the absence of developmental study (Pizzolato 1990a, 1993b, 1994a, 1995). New approaches to the study of the vascular system of the floret should include observation of the construction of the mature vascular system from its procambial trace components. These new studies will lead to a greater appreciation and use of the earlier studies of floret vascular systems by helping developmentally to unravel them. The developmental approach to the study of vascular systems of florets is slow; but as developmental observations are becoming available, interesting conclusions are appearing. Results from the few complete developmental studies of floret vascular systems will be presented first to be combined with a little of the more abundant information on the mature vascular system in order to make a few tentative conclusions that may be of systematic value.

Procambial traces for the median and then for the two major laterals (LL, RR) of the lemma are initiated in isolation at the lemma insertion in the rachilla of the pooid *Stipa*, *Triticum* (Fig. 6), and *Anthoxanthum* (Fig. 8). Subsequently procambial traces for the two lesser laterals (those between the median and the major laterals) arise a little higher in the lemma insertion (Maze *et al.* 1972; Pizzolato 1984; Robinson-Beers and Pizzolato 1987a; Pizzolato and Robinson-Beers 1987; Frick and Pizzolato 1987; Pizzolato 1998). Recall that this is the sequence typical of vegetative leaves of the pooid culm (Hitch and Sharman 1968a; Dannenhoffer and Evert 1994). On the opposite (posterior) side of the rachilla and a little later than the initiation of the lemma median, the lone palea trace of *Anthoxanthum* and the more typically paired palea traces of *Stipa* and *Triticum* are initiated in isolation. A portion of the distal regions of the lemma median and the major lateral traces in the rachilla differentiates acropetally into the lemma, but other more distal portions of these three traces in the rachilla enlarge and merge into a procambial mass. Although Maze *et al.* (1972) and Pizzolato (1998) did not continue their studies of procambial differentiation into the phloem and xylem, Robinson-Beers and Pizzolato (1987a, b) revealed for *Anthoxanthum* that a vertical then a horizontal network of phloem (the sieve-element plexus) and later of xylem differentiates within this procambial mass without orientation to the lemma or palea traces (Figs 8, 9). Thus, the sieve-element plexus and the associated xylem differentiate in

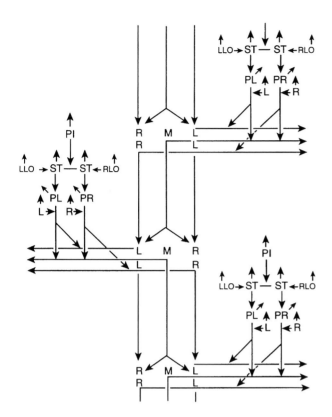

TRITICUM RACHILLA AND FLORETS

Fig. 6. Diagram of the vascular system of the florets and rachilla of *Triticum* (Pooideae). L=left intermediate lateral trace of lemma; LL=major left lateral trace of lemma; LLO=left lodicule traces; M=lemma median trace; PI=pistil trace complex; PL=left trace of palea; PR=right trace of palea; R=right intermediate lateral trace of lemma; RLO=right lodicule traces; RR=major right lateral trace of lemma; ST=stamen trace complex. Sites of abbreviations designating traces are the sites of isolated procambial initiation. Arrows indicate either acropetal differentiation from these sites into organs or basipetal mergers of these traces with other procambial traces.

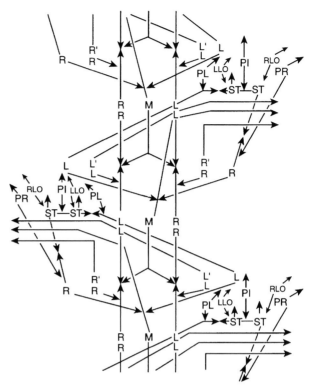

PHRAGMITES RACHILLA AND FLORETS

Fig. 7. Diagram of the vascular system of the florets and rachilla of *Phragmites* (Arundinoideae). Capturing traces (LL′, RR′) must redirect the major lateral traces for the lemmas from the rachilla into the lemmas because the major lemma traces are initiated at sites distant from the lemmas. Abbreviations are explained in Fig. 6. Sites of abbreviations are the sites of isolated procambial initiation. Arrows indicate either acropetal differentiation from these sites or their basipetal mergers with other procambial traces. Notice that the lemma intermediate lateral traces (R, L) are distracted from the lemma.

the merging distal portions of procambia originally initiated in the rachilla as the three major traces of the lemma. The sieve-element plexus and its associated xylem differentiate with orientation to the stamen traces.

After the sieve-element plexus begins differentiating in *Anthoxanthum*, the procambial trace for the palea and for the two lesser laterals of the lemma differentiate basipetally and join the procambial mass containing the sieve-element plexus (Robinson-Beers and Pizzolato 1987a, b). When all of the lemma and palea traces are proximally continuous with the procambial mass originating as the three major procambial lemma traces, phloem and xylem differentiate in the lemma and palea traces (Pizzolato and Robinson-Beers 1987; Robinson-Beers and Pizzolato 1987a, b). However, even when the vascular tissues of the lemma and palea traces differentiate next to the vascular tissues of the sieve-element plexus and its associated xylem, the vascular tissues of the lemma and palea traces probably never share sieve plates and perforation plates with the sieve-element plexus and its xylem (Pizzolato 1984; Robinson-Beers and Pizzolato 1987a, b; Frick and

Pizzolato 1987). Recall the functionally discontinuous xylem that Aloni and Griffith (1991) noted between nodal roots and culms of pooid grasses.

No developmental studies besides those on *Stipa, Anthoxanthum,* and *Triticum* are available to prove the role of the principal lemma and palea traces in initiating the proximal vascular system for other pooid florets. Without a developmental study, some papers (e.g. Pizzolato 1988) suggest that other pooid grasses manifest the *Anthoxanthum, Stipa,* and *Triticum* pattern of the isolated origin of the major lemma and palea traces and their mergers to form the vascular system of the floret. If this pattern is typical of pooid grasses, systematic contrasts in the proximal regions of the floret are particularly promising with four other subfamilies. Bambusoid grasses such as *Raddia* have small supernumerary bundles closely associated with the lemma median (Pizzolato 1990b), a feature lacking in the Pooideae and other subfamilies outside of the Bambusoideae. Developmental studies (Maze *et al.* 1971, 1972; Robinson-Beers and Pizzolato 1987a; Pizzolato 1998) upon *Stipa, Anthoxanthum* (Fig. 8) and *Triticum* (Fig. 6), without contradiction from observations of mature florets of other pooid grasses (Pizzolato 1988; 1989a, b), as well as

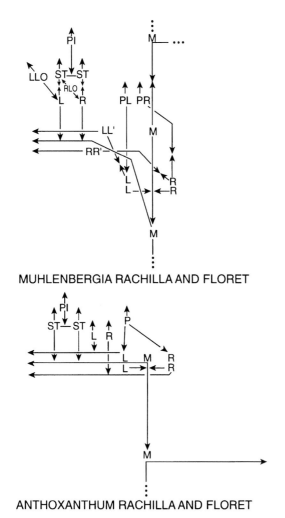

MUHLENBERGIA RACHILLA AND FLORET

ANTHOXANTHUM RACHILLA AND FLORET

Fig. 8. Diagrams of the rachillae and florets of *Muhlenbergia* (Chloridoideae) and *Anthoxanthum* (Pooideae) emphasizing similarities and differences in their vascular patterns. P=sole trace of palea of *Anthoxanthum*. All other trace abbreviations are explained in Figs 6 and 7. *Muhlenbergia* but not *Anthoxanthum* has capturing bundles (LL', RR') which redirect the major lemma lateral traces into the lemma, as also occurs in *Phragmites* but not in *Triticum*. The lemma intermediate lateral traces (R, L) are distracted from the lemma of *Muhlenbergia* but not from the lemma of *Anthoxanthum*. The major lemma traces (M, LL, RR) are initiated at the node of insertion of *Anthoxanthum* but at more distant sites in *Muhlenbergia*. Interfloret connections in the rachilla occur via the lemma median traces (M) of *Muhlenbergia* and *Anthoxanthum* unlike the condition in *Phragmites* and *Triticum* (Fig. 5).

developmental studies (Pizzolato 1994a, 1995) among arundinoid and chloridoid grasses (Figs 7, 8), all reveal the basipetal merger of the palea traces with the vascular system formed from merged lemma traces. However, in panicoid grasses, the palea traces distinctly descend the rachilla unobstructed by the lemma traces (Chandra 1962a,b; Butzin 1965; Clifford 1987; Pizzolato 1992, 1993a). Moreover, developmental studies in the Arundinoideae and Chloridoideae reveal other striking differences involving the proximal vascular components of florets.

Recall that in pooid grasses the lemma median trace and the two major lateral traces are initiated in isolation in the lemma insertion (Figs 5, 6, 8) but in arundinoid and chloridoid grasses (Figs

5, 7, 8) the lemma median trace is initiated in isolation in the node subjacent to the insertion of the rachilla (Pizzolato and Robinson-Beers 1987; Pizzolato 1994a, 1995, 1997). The two major lemma laterals of the arundinoid and chloridoid grasses are initiated in isolation also, usually near the same level as their median traces (Figs 5, 7, 8). A second significant distinction between pooid and arundinoid-chloridoid florets is that in the latter the major lemma laterals have, near the lemma insertion, a second isolated site of trace initiation (LL' and RR' in Figs 7, 8). By bidirectional differentiation the second sites unite the lemma margins with the first, distant procambia for the major lemma laterals (Pizzolato 1994a, 1995). A third important difference between pooid and arundinoid-chloridoid florets is that, although the lesser lemma laterals of both groups are initiated similarly without second sites and approach the lemma (Robinson-Beers and Pizzolato 1987a; Pizzolato 1994a, 1995), the lesser laterals (L, R) of the arundinoid-chloridoid lemma are prevented from ascending the lemma by being distracted by descending procambia (Figs 7, 8) related to the stamen and lodicule traces. Here is a striking example of using developmental studies to explain a taxonomic peculiarity of many arundinoid and chloridoid grasses: the three-nerved lemma (Decker 1964; Jacobs 1987; Hilu and Esen 1993).

Just as lemma trace procambial development differs among subfamilies so does palea trace differentiation. Recall that in the Pooideae, the isolated palea traces differentiate basipetally to join the posterior of the three distally merged procambia initiated in the rachilla as the lemma median and the major lemma lateral traces (Maze *et al.* 1972; Robinson-Beers and Pizzolato 1987a) and that in the Panicoideae, the basipetally differentiating palea traces avoid all other traces of the floret (Pizzolato 1992). The pattern of basipetal differentiation of the palea traces is different in the arundinoid-chloridoid florets since the isolated palea traces in their descent unite with either the major or the lesser lemma laterals (Pizzolato 1994a, 1995). Thus in the chloridoid *Muhlenbergia* the major lemma laterals, above the union of the second-site lemma laterals, appear to flow directly into the palea (LL and RR in Fig. 8). Thus in the arundinoid *Phragmites* (L, R in Fig. 7), and in the chloridoid *Eragrostis* and *Cynodon*, the distracted lesser lemma laterals appear to diverge into the palea (Butzin 1965; Pizzolato 1990a, 1994a, 1995).

VASCULAR SYSTEM OF THE FLORET: CONTRIBUTIONS FROM STAMEN AND LEMMA TRACES TO THE SIEVE-ELEMENT PLEXUS AND ASSOCIATED XYLEM

Although studies upon procambial and vascular differentiation leading to the formation of the vascular system of the androecium are few and restricted to the pooid, arundinoid, and chloridoid subfamilies, these studies have uncovered developmental distinctions that should stimulate comparable study in the other subfamilies. In the genera studied in the Pooideae (Maze *et al.* 1971; Aziz 1978; Pizzolato and Robinson-Beers 1987; Pizzolato 1998), in *Phragmites* of the Arundinoideae (Pizzolato 1994a), and in *Muhlenbergia* of the Chloridoideae (Pizzolato 1995), isolated procambium is initiated near the insertion of each stamen. Bidirectional procambial differentiation occurs from the initial site (Figs 6-8). The next stage of differentiation is

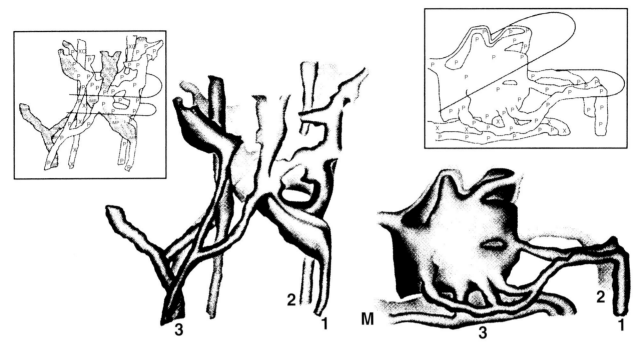

Fig. 9. Perspective drawings of an earlier (left) and later (right) stage of development of the sieve-element plexus and associated xylem discontinuity of a floret of *Anthoxanthum* (Pooideae). The inserts explain the locations of the vascular components. IMP=immature sieve elements; P=mature sieve elements; X=mature tracheary elements; XD=anomalous tracheary elements of xylem discontinuity; 1=phloem of lemma's major left lateral trace in the rachilla; 2=phloem of lemma's major right lateral trace in the rachilla; 3=xylem and phloem of lemma's median trace (M) in the rachilla. Arcs in inserts emphasize the displacement of the upper and lower portions of the sieve-element plexus during its development. Reprinted with permission of The University of Chicago Press from Robinson-Beers and Pizzolato (1987b).

also the same, with incidental differences, for the four investigated genera of the three subfamilies (Pizzolato and Robinson-Beers 1987; Robinson-Beers and Pizzolato 1987a; Maze *et al.* 1972; Pizzolato 1994a, 1995). A mass of procambium differentiates between the level of the isolated stamen traces and the median trace to the lemma. In the Pooideae the lemma median is the major contributor to the procambial mass, but the major lemma laterals and particularly the minor lemma laterals contribute little to the procambial mass as they differentiate (Maze *et al.* 1971; Pizzolato and Robinson-Beers 1987; Robinson-Beers and Pizzolato 1987a). In the Arundinoideae (Pizzolato 1994a) and the Chloridoideae (Pizzolato 1995), on the other hand, the lemma median and the lemma major laterals do not contribute to the procambial mass between the stamen insertions and the lemma median; instead the procambial lemma minor traces become the major contributors to the procambial mass as they differentiate. As a result of this distinction, the xylem and phloem to the stamens of the Pooideae are a direct continuation of the portion of the lemma median that does not enter the lemma; in fact, much more vascular tissue is directed toward the stamens from the lemma median than enters the lemma median (Figs 6, 8, 9). In the Arundinoideae and the Chloridoideae, all of the xylem and phloem to the stamens must pass through the minor lemma lateral traces, which do not enter the lemma (L, R in Figs 7, 8). Discovery of this contrast in bridging the gap between the lemma median trace and the stamen traces should stimulate a quest to see if there are other bridging patterns and to determine if they are systematically important.

The vascular tissue which differentiates from the procambial core in the rachilla between the lemma median and the stamen insertions is known as the sieve-element plexus and associated xylem (Fig. 9). Unfortunately, its vascular differentiation has been followed only in the pooid *Anthoxanthum*. This development will be outlined in order to stimulate comparative studies. Recall that the three procambial traces at the base of the floret of *Anthoxanthum* are initiated as the median trace of the lemma and its two major laterals (Pizzolato and Robinson-Beers 1987; Robinson-Beers and Pizzolato 1987a). Phloem and xylem for the lemma median and the two major laterals differentiate from these three procambial sites. The distal portions of these three initially isolated procambial strands merge, primarily through the enlargement of the procambium of the distal portion of the trace initiated as the lemma median. Recall that this is the enlargement creating the procambial mass bridging the gap between the lemma median and the stamen insertions (Figs 8, 9). Although the phloem and xylem for the lemma medians and the two major lemma laterals differentiate from the three original procambial traces, most of their vascular differentiation is independent of the lemma and is directed toward the stamens (Robinson-Beers and Pizzolato 1987a, b). Recall that even where sieve and tracheary elements related to the lemma become contiguous with elements related to the stamens, the lemma elements touching the stamen elements are probably not continuous by sieve plates and perforation plates. Phloem and xylem differentiate acropetally from the original procambia of the lemma major lateral traces into the distally merged regions of the three procambia to become the lower vascular component of the sieve-element plexus and associated

21

xylem (Pizzolato 1984; Robinson-Beers and Pizzolato 1987a, b). The result is a massive vertical, then distally horizontal sieve-element network (Fig. 9) making the lemma median trace contiguous, but not continuous by means of sieve plates, to the stamen traces.

A sieve-element plexus supplying the stamen traces in the manner of *Anthoxanthum* appears to occur in the other investigated pooid and bambusoid genera (Zee and O'Brien 1970; O'Brien *et al.* 1985; Pizzolato 1987, 1988, 1989a, 1990b). However, confirmation must await studies in them of procambial and phloem differentiation like that done for *Anthoxanthum*. Sieve-element plexi related to the stamens occur also in the arundinoid *Phragmites* and in the chloridoid *Cynodon* and *Muhlenbergia* (Pizzolato 1990a, 1993b, 1994a, 1995). Recall, however, that vascular ontogeny may be different from the pooid pattern since procambial initiation leading to the linking of the median trace and the stamens (Figs 6-8) is different indeed (Pizzolato 1994a, 1995). Unfortunately, no developmental studies for the stamen vascular system have gone beyond procambial initiation in any grass but *Anthoxanthum*. A sieve-element plexus like that of pooid, bambusoid, arundinoid, and chloridoid grasses probably does not occur in the investigated oryzoid, panicoid and andropogonoid grasses (Pizzolato 1983b, 1989b, 1991, 1993a). However, systematic conclusion must wait for studies of procambial and vascular development in these three subfamilies.

Emphasis now shifts from the phloem component differentiating between the lemma median and the stamen traces (the sieve-element plexus) of pooid, arundinoid, and chloridoid grasses to the xylem component: the xylem discontinuity. Although the xylem discontinuity is a taxonomically significant character, its ontogeny has been described only for the pooid *Anthoxanthum*. Its development will be briefly described and then its systematic distribution reviewed.

The xylem discontinuity consists of squat, anomalous tracheary elements checked in various stages of their maturation (Zee and O'Brien 1970; Pizzolato 1984). The xylem discontinuity differentiates only in the procambial mass related to the lemma median trace, not in the procambial traces related to the lemma laterals (Robinson-Beers and Pizzolato 1987b). Cells of the xylem discontinuity first appear inside of the distal horizontal portion of the sieve-element plexus at the distal end of normal tracheary elements directed toward the anterior stamen insertion. More distal and closer to the stamen insertions are two more sites of the xylem discontinuity, which are continuous with normal xylem in the stamens (XD in Fig. 9). Because of subsequent differentiation of more cells of the xylem discontinuity among them, the separate sites of the xylem discontinuity become contiguous. Thus, a core of cells of the xylem discontinuity becomes interposed between normal tracheary elements directed to the lemma and normal tracheary elements in distal parts of the stamens. The xylem discontinuity is augmented and propagated beyond the filaments because more cells of the xylem discontinuity differentiate on the flanks, no longer in response to the stamen traces but to traces related to the pistil (Robinson-Beers and Pizzolato 1987b,c).

Only the pooid *Anthoxanthum* was studied enough to prove that the xylem discontinuity develops as the complement to the distal sieve-element plexus in response to influences of the stamen traces (Pizzolato 1984; Robinson-Beers and Pizzolato 1987a,b,c). However, the distribution of the xylem discontinuity among the subfamilies is better documented. Zee and O'Brien (1970) discovered the xylem discontinuity in the florets of four pooid genera: *Triticum*, *Lolium*, *Avena*, and *Bromus*. Pizzolato (1983a, 1987, 1988, 1989a) found it in four more pooid genera: *Dactylis*, *Alopecurus*, *Phleum*, and *Phalaris*. Zee (1972) and Pizzolato (1989b) reported the absence of the xylem discontinuity from *Oryza* and *Leersia* of the Oryzoideae. In the Panicoideae, Zee and O'Brien (1971) and Pizzolato (1983b, 1991, 1993a) also noted its absence from the panicoid *Echinochloa* and *Panicum* and from the andropogonoid *Sorghum*. These authors noted that modified tracheary elements might be associated with the placental bundle, but a xylem discontinuity related to the stamen traces was not found in these genera. Neither does a xylem discontinuity occur in the only bamboo studied sufficiently: *Raddia* (Pizzolato 1990b). Judziewicz *et al.* (1999) lament the paucity of studies of spikelet vascular anatomy in the bamboos. The only studies uncovering the xylem discontinuity in florets outside of the Pooideae found it in the chloridoid *Cynodon* (Pizzolato 1990a) and *Muhlenbergia* (unpublished observations incidental to Pizzolato 1996) and in the arundinoid *Phragmites* (Pizzolato 1993b). If differentiation of the xylem discontinuity is a primitive feature lost in two lines of descent, then the distribution of this character agrees well with the 'second approximation' of Clayton and Renvoize (1986), which was constructed without using the xylem discontinuity. However, no systematic conclusions can be trusted without more study of its distribution enhanced by more knowledge of its ontogeny as well as that of the associated sieve-element plexus.

VASCULAR SYSTEM OF THE FLORET: CONTRIBUTION OF THE LODICULE TRACES

This section will stress the aspects of development and structure of vascular tissues of lodicules not covered in the reviews of Hsu (1965), Jirásek and Jozífová (1968), Guédès and Dupuy (1976), and Clifford (1987). Procambial initiation will be discussed, then the distribution of the vascular tissues differentiating from them in the lodicules.

Since *Anthoxanthum* lacks lodicules, there are only two developmental studies proving that the procambium of the lodicule originates at an isolated site near the insertion of the primordium and differentiates acropetally into it and basipetally to merge with procambia in the rest of the floret (Pizzolato 1994a, 1995). However, in the light of these two studies, the appearance of the union of the mature lodicule trace with the rest of the floret vasculature strongly suggests that a basipetal merger occurred before maturity (Butzin 1965; Pizzolato 1980, 1983a, b, 1988, 1989a, b, 1990a, b, 1991, 1993a).

The number of procambial traces initiated near the lodicule primordium may have systematic import. That number is almost always one for pooid (Craig and O'Brien 1975; Pizzolato 1983a, 1988, 1989a) chloridoid (Butzin 1965; Pizzolato 1990a, 1995), arundinoid (Pizzolato 1993b) and panicoid (Butzin

1965; Pizzolato 1983b, 1993a) genera. The number is two per lodicule in the normal position in oryzoid (Butzin 1965; Pizzolato 1990a) and bambusoid (Butzin 1965; Pizzolato 1990b) genera; the bambusoid posterior lodicule has one trace. Andropogonoid genera (Butzin 1965; Pizzolato 1991) initiate several procambial traces. *Zea* appears to be exceptional by having but one trace (Pizzolato 1980).

The precise site of basipetal mergers of the originally isolated procambial trace can be ascertained only for arundinoid *Phragmites* and chloridoid *Muhlenbergia* since they were subjects of developmental study (Pizzolato 1994a, 1995). Their lodicule traces (RLO, LLO) merge directly or indirectly with the acropetally differentiating minor lemma lateral traces (R, L) and thus become part of the reason that these laterals cannot reach the lemma (Figs 7, 8). In future developmental studies, the point of merger of the basipetally differentiating lodicule procambia with the rest of the floret's procambial components should be noted. If xylem and phloem differentiate in the lodicule trace procambia as they do in the originally isolated lemma and palea trace procambia (Robinson-Beers and Pizzolato 1987a, b), then the mature vascular tissues of the lodicule and of the rest of the floret probably are not continuous by sieve plates or perforation plates even when the vascular tissues originating from both moieties are contiguous.

The merger of the procambial trace of the lodicule with the rest of the procambial system of the floret is smooth, even if the resulting vascular tissues do not become continuous. Pizzolato (1994a) was misled by the smooth merger in the mature floret of *Phragmites* into concluding that the lodicule traces passed through the floret into the rachilla. However, developmental studies of procambial origins revealed later that the lodicule traces merely blend into the lemma minor lateral traces (Pizzolato 1994a, 1995). Conclusions from mature floret vasculature are prone to error if they lack concomitant studies of procambial development. Conclusions from earlier works, which stress with systematic inferences the major role of the lodicule traces in the formation of the floret vasculature of the bambusoid *Raddia* and of the andropogonoid *Sorghum* (Pizzolato 1990, 1991), are no longer tenable without support from developmental studies.

Attention will now be directed to a study whose systematic conclusions have been overlooked. After fixing the living lodicules of 64 species representing almost all of the subfamilies, Pissarek (1971) concluded that, with surprisingly few exceptions, the lodicule traces have tracheary elements and sieve elements in all the subfamilies but the Pooideae, which have sieve elements without tracheary elements. With one exception, all of the other studies cited in this section confirm the conclusions of Pissarek (1971). He did not include bamboos in his survey, but Butzin (1965) and Pizzolato (1990b) saw tracheary elements and sieve elements in the species of the three bambusoid genera they studied. The only conflict is that Pissarek (1971) reported the absence of tracheary elements and sieve elements from lodicules of the pooid *Phleum*, but Pizzolato (1988) reported the presence of sieve elements in *Phleum* lodicules. A peculiarity is that Pissarek (1971) reported that lodicules of two species of the pooid *Phalaris* are exceptional for having tracheary elements besides sieve elements. In a way Pizzolato (1989a) confirmed this exception by finding, besides sieve elements in their normal extent, a few tracheary ele-

ments resembling a xylem discontinuity in the insertion of some of the lodicules of a species of *Phalaris*. Pissarek's (1971) finding of tracheary elements and sieve elements in lodicules of the section of the Pooideae including *Stipa* is an exception to the pattern typical for pooid grasses that deserves deeper study. The generalization of Pissarek (1971), supported by the research of a few other students, that the lodicules of the Pooideae contain sieve elements but not tracheary elements whereas the lodicules of all other subfamilies contain both conducting cells deserves closer attention from systematists.

VASCULAR SYSTEM OF THE FLORET: CONTRIBUTION OF THE PISTIL

The vascular bundles have been appreciated as components of a unit better for the pistil than for any other organ of the grasses. The primary objective of this system-view, an approach used infrequently to study the other organs, was to let the vascular system outline the phylogeny of the pistil (Arber 1934; Belk 1939; Chandra 1963; Butzin 1965; Philipson 1985). Whether the vascular system in the pistil can be used to trace its history is incidental to this review. However, facts uncovered by workers seeking the origin of the grass pistil are valuable for themselves because they describe the vascular components of the mature pistil. This information will be used to describe the vascular system, around the time of anthesis, of a typical pistil of the grasses with most of its variations. Then patterns of development will be described with their systematic possibilities.

A series of transections acropetally cutting the ovary reveals the ovule and the encircling ovary wall containing the vascular bundles of the pistil (Chandra 1963; Maze *et al.* 1971; Pizzolato 1984, 1989a). Batygina (1971), Aziz (1972), and Lersten (1987) present three-dimensional reconstructions from these kinds of transections in *Triticum*. In the posterior (palea-side, ventral to Maze *et al.* (1971, 1972), dorsal to most authors) ovary wall occurs the large placental bundle, which ascends the placenta to merge with the chalaza of the ovule. Opposite the placental bundle, in the anterior (lemma-side, dorsal to Maze *et al.* (1971, 1972), ventral to most authors) ovary wall, many pistils have an anterior bundle that ends blindly before reaching the top of the ovary. In each flank of the ovary, in the portions of the walls closer to the placental bundle, occurs a posterolateral bundle in most grasses. Since the posterolaterals usually approach and usually enter the styles, the posterolaterals are often designated as the stylar bundles. Another pair of flanking bundles occurs less frequently than the stylar bundles and occurs closer to the anterior bundle than the stylar bundles: the anterolateral bundles. All the bundles form a unit below (proximal to) the ovule (Batygina 1971; Aziz 1972; Calderón and Soderstrom 1973; Lersten 1987).

Although discussion of the development of the bundles in the pistil will be deferred, the taxonomic distribution of the bundles is mentioned now. The anterior bundle, which lacks tracheary elements, is a promising taxonomic character because it is virtually restricted to the Pooideae where it is almost never absent (Belk 1939; Chandra 1963; Butzin 1965; Philipson 1985; Pizzolato 1984, 1987, 1988, 1989a). Almost without exception is the absence of the anterior bundle from the oryzoid, chloridoid, arundinoid, and panicoid genera (Belk 1939; Butzin 1965; Hsu

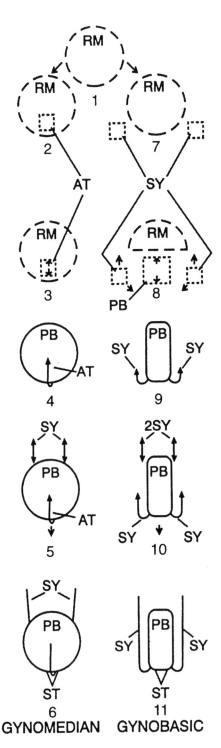

Fig. 10. Schematic sequence of procambial development in the grass pistil contrasting the gynomedian pattern of *Anthoxanthum* (Pooideae) with the gynobasic pattern of *Muhlenbergia* (Chloridoideae). The pistil is viewed from the anterior (dorsal, lemma) side of the floret. Arrows show direction of procambial differentiation. Dashed areas are residual meristem (RM) or recently initiated procambium. The blank background of the figures is ground parenchyma. In the gynomedian pattern, a procambial trace (anterior trace, AT) begins at the anterior of the placental bundle (PB) and differentiates basipetally to the placental bundle and acropetally (1-4). In the contrasting stages of the gynobasic pattern, stylar procambial traces (SY) begin at the flanks of the residual meristem and differentiate basipetally to the recently initiated placental bundle and acropetally (1,7-9). In the gynomedian pattern, the placental bundle differentiates directly from the residual meristem (2-4). In the contrasting stages of the gynobasic pattern, the placental bundle differentiates from some of the ground parenchyma which differentiated from the residual meristem (7-9). In the gynomedian pattern, stylar procambial traces begin distal to the placental bundle and differentiate basipetally to its flanks (5,6). In the contrasting stages of the gynobasic pattern, the distal sites of the stylar procambia differentiate basipetally to the acropetally differentiating proximal stylar procambia, not to the flanks of the placental bundle (10,11). Basipetal differentiation of the linked placental bundle joins it to the differentiating procambia associated with the complex (ST) of stamen traces (6,11). Reprinted with permission of the Torrey Botanical Society from Pizzolato (1996).

1965; Philipson 1985; Pizzolato 1989b, 1990a, 1993a,b, 1996) and from the herbaceous bamboos (Butzin 1965; Calderón and Soderstrom 1973; Pizzolato 1990b; Paisooksantivatana and Pohl 1992). However, one or a few anterior and probably collateral bundles occur in the ovaries of the large bamboos (Arber 1926, 1929; Butzin 1965, Calderón and Soderstrom 1973; Rudall and Dransfield 1989) and in the andropogonoid grasses (Chandra and Saxena 1964; Butzin 1965; Pizzolato 1991). The tenuous appearance of the anterior bundles in these last two groups of grasses requires further study to ascertain their distribution.

Although the ontogeny of the stylar bundles will continue to be deferred, mention is made now that, according to the references cited in the previous paragraph, the stylar bundles in all sub-families are probably collateral but for those of the Pooideae which, like their anterior bundles, lack xylem. An exception may be the stylar bundles of the Chloridoideae where xylem is discontinuous (Pizzolato 1990a) or absent (Pizzolato 1996).

The pistil bundle whose features remain to be discussed is the placental bundle. With a few exceptions, the anatomy of the placental bundle is uniform across the subfamilies. It is a large vascular bundle of parenchyma containing an outer arc of sieve elements and a core of tracheary elements between the sieve elements and the ovule. The parenchyma cells of the placental bundle, not the tracheary elements and sieve elements which distally fade out, merge with the ovule at the chalaza. This type of placental bundle is the type of the Pooideae (Chandra 1963; Butzin 1965; O'Brien *et al*. 1985; Pizzolato 1984, 1987, 1988, 1989a), Oryzoideae (Butzin 1965; Pizzolato 1989b), Chloridoideae (Butzin 1965; Pizzolato 1990a, 1995), Arundinoideae (Pizzolato 1993b), Panicoideae (Butzin 1965; Pizzolato 1993a) and the herbaceous bamboos (Butzin 1965; Calderón and Soderstrom 1973; Pizzolato 1990b). Exceptions to this type occur in some of the woody bamboos (Arber 1926, 1929; Chandra 1963; Rudall and Dransfield 1989) and in the andropogonoid grasses (Chandra and Saxena 1964; Pizzolato 1991) when the placental bundle appears in cross section not as a unit but as an arc of strands of which some pass beyond the chalaza of the ovule instead of merging with it.

In the placental bundle, the presence or absence of the xylem discontinuity is systematically noteworthy and follows the subfamilial distribution of the xylem discontinuity associated with the stamen traces. Although the placental bundle of panicoid and andropogonoid grasses lacks the xylem discontinuity, the distal tracheary elements of the placental bundle of these two subfamilies have weak walls and degenerated protoplasts that may have a xylem blocking function akin to that of the tracheary elements of the xylem discontinuity (Zee and O'Brien 1971; Frick and Pizzolato 1987; Pizzolato 1991, 1993a). When the xylem discontinuity occurs in the pistil, it occurs only in the placental bundle and in the vascular prongs joining the placental bundle to the vasculature related to the stamens (6, 10 in Fig. 10).

Ontogeny of the phloem and xylem in the placental bundle and other bundles of the pistil is described for only the pooid genera *Triticum* (Aziz 1981) and *Anthoxanthum* (Robinson-Beers and Pizzolato 1987b,c). Blending the accounts, which are in agreement, gives a base upon which to make future comparisons. Only after the procambial placental bundle becomes continuous by basi-

petally differentiating procambial prongs with the sieve-element plexus and associated xylem discontinuity recently differentiated in relation to the stamen traces, do sieve elements differentiate in isolation in the procambial placental bundle. The first site is at the juncture of the anterior bundle with the placental bundle. Two later isolated sites of sieve-element differentiation begin in the posterolateral portions of the procambial placental bundle above the first site. These three sites initiate the anterior and posterolateral traces respectively. Basipetal sieve-element differentiation from the second sites connects the posterolateral traces through the procambial pistil prongs with the sieve-element plexus related to the stamen trace system. Acropetal sieve-element differentiation from the second sites brings the sieve elements into the posterolateral (stylar) traces. Acropetal sieve-element differentiation from the first site vascularizes the anterior bundle. Basipetal sieve-element differentiation from the first site toward the sieve elements descending from the second sites connects the anterior bundle to the nascent posterolateral traces becoming continuous with the sieve-element plexus. After all of the phloem of the pistil becomes continuous, the xylem discontinuity differentiates as a companion to the sieve elements in the pistil prongs and in the placental bundle, thereby connecting the xylem discontinuity associated with the sieve-element plexus with that of the placental bundle (Robinson-Beers and Pizzolato 1987b,c).

Although differentiation of xylem and phloem in the pistil has been thus partially described for two pooid grasses only, descriptions of procambial initiation in the pistil are more nearly complete and are wider ranging. Moreover, patterns of procambial initiation in the pistil may have taxonomic value because they follow at least two basic patterns (Fig. 10). Such a conclusion rests upon developmental studies focused upon the pistil procambium in the pooid *Anthoxanthum* (Robinson-Beers and Pizzolato 1987c), in the arundinoid *Phragmites* (Pizzolato 1994b), and in the chloridoid *Muhlenbergia* (Pizzolato 1996) supplemented by broader developmental studies of procambium in the pooids *Stipa* (Maze *et al*. 1972) and *Triticum* (Barnard 1957; Aziz 1978) and in *Andropogon* (Maze 1977). Conclusions from these developmental studies are supplemented with some risk by study of the mature pistils of other pooid (Pizzolato 1987, 1988, 1989a), oryzoid (Pizzolato 1989b), bambusoid (Pizzolato 1990b), andropogonoid (Pizzolato 1991), and panicoid (Pizzolato 1993a) grasses.

In both patterns of procambial initiation that emerge so far, the procambia of the pistil are initiated in isolation from a residual meristem derived from the apical meristem (1 in Fig. 10) after all other procambia of the floret have been initiated. In the gynomedian pattern, the procambial placental and anterior bundles simultaneously differentiate directly from the residual meristem so that the placental bundle retains its broad shape (2-4 in Fig. 10). In the gynobasic pattern, the residual meristem first differentiates into ground parenchyma from which dedifferentiate a narrower procambial placental bundle and two procambial stylar (posterolateral) bundles in the same position relative to the gynomedian anterior bundle but laterally displaced (7-9 in Fig. 10). A second site of stylar procambial initiation for the gynobasic pattern (10 in Fig. 10) and a corresponding first site for the gynomedian pattern (5 in Fig. 10) are initiated distally isolated near the stylar insertions on the ovary. Simultaneously, one, two, or three procambial pistil prongs descend by differentiation from the placental bundle of both patterns to join the procambium of the future sieve-element plexus, which is related to the stamen traces (Figs 6-8; ST in 6, 11 of Fig 10). In the gynobasic pattern, procambial differentiation bridges the gap on each side of the narrow placental bundle between the stylar procambial sites (10, 11 in Fig. 10). In the gynomedian pattern, basipetal differentiation of the stylar procambia brings them into union with the flanks of the broad placental bundle (5, 6 in Fig. 10). Thus, in the pistil at anthesis after gynomedian differentiation, the stylar (posterolateral) traces attach to the placental bundle near its middle and above the base of the locule; and the anterior bundle attaches near the base of the placental bundle below the base of the locule. Thus, in the pistil at anthesis after gynobasic differentiation, there is no anterior bundle; and the stylar traces attach near the base of the placental bundle below the base of the locule. In general, the gynomedian pattern occurs in pooid grasses and the gynobasic pattern occurs among the other subfamilies.

However, there may be other basic patterns; and there certainly are lesser variations of these two patterns. Although the pooid *Triticum* appears to have the gynomedian pattern, the merger of the stylar traces with the placental bundle appears somewhat lower than just described for the gynomedian pattern (Barnard 1957; Batygina 1971; Aziz 1972, 1978, 1981; O'Brien *et al.* 1985). The pooid *Phalaris* has the gynomedian pattern (Pizzolato 1989a) but the additional, anterolateral bundles may have been initiated as a first or second set of isolated distal stylar traces. How can the anterolateral traces typical of many bamboos (Arber 1929; Chandra 1963; Rudall and Dransfield 1989) be explained in the context of the gynobasic pattern of the bamboos that lack anterolateral traces (Calderón and Soderstrom 1973; Pizzolato 1990b)? Pattern of procambial differentiation may be another variable of grass pistils that Theile *et al.* (1996) are seeking to correlate with taxonomy.

A last comment upon variations or additions to the two basic developmental patterns for pistil procambia will emphasize the potential of vascular development as an aid to grass systematics. Belk (1939) discovered for panicoid and andropogonoid grasses and Butzin (1965) and Pizzolato (1991, 1993a) confirmed her discovery that the stylar traces, which at anthesis have the gynobasic pattern (Pizzolato 1991, 1993a, 1996), appear to have a more anterior merger with the placental bundle than the merger of the stylar traces with the placental bundle in the other subfamilies. Is this a variation of the gynobasic pattern distinct enough to be recognized as a third basic pattern? Are stylar procambia for panicoid and andropogonoid grasses indeed initiated at a site more anterior than that for other grasses? Might anterolateral traces in some grasses and posterolateral traces in others supply the styles, even when both sets are not present together in the same pistil? Or, as Maze and Scagel (1982) ask 'Are the appendages of the spikelet similar to apparently comparable appendages in other grasses?' A systematic view of vascular development will help to answer fundamental questions.

ACKNOWLEDGEMENTS

I thank Mary Barkworth for asking me to make this review. Elaine Eiker typed the manuscript, Barbara Broge improved the illustrations, and Elizabeth, Jane, and Karen Pizzolato gave me encouragement. This review is dedicated to the late Bernard C. Sharman, whose scholarship and kindness encouraged my interest in the vascular systems of grasses. This review is published as Paper No. 1648 in the Journal Series of the Delaware Agricultural Experiment Station.

REFERENCES

Aloni, R., and M. Griffith (1991). Functional xylem anatomy in root-shoot junctions of sixcereal species. *Planta* **184**, 123-129.

Arber, A. (1926). Studies in the Gramineae. I. The flowers of certain Bambuseae. *Annals of Botany* **40**, 447-469.

Arber, A. (1927). Studies in the Gramineae. III. Outgrowths of the reproductive shoot, and their bearing on the significance of lodicule and epiblast. *Annals of Botany* **41**, 473-488.

Arber, A. (1928). Studies in the Gramineae. V. 1. On *Luziola* and *Dactylis*. 2. On *Lygeum* and *Nardus*. *Annals of Botany* **42**, 391-407.

Arber, A. (1929). Studies in the Gramineae. VII. On *Hordeum* and *Pariana*, with notes on 'Nepaul barley.' *Annals of Botany* **43**, 507-533.

Arber, A. (1930). Studies in the Gramineae. IX. 1. The nodal plexus. 2. Amphivasal bundles. *Annals of Botany* **44**, 593-620.

Arber, A. (1931). Studies in the Gramineae. X. 1. *Pennisetum*, *Setaria*, and *Cenchrus*. 2. *Alopecurus*. 3. *Lepturus*. *Annals of Botany* **45**, 401-420.

Arber, A. (1934). 'The Gramineae: A Study of Cereal, Bamboo, and Grass.' (The University Press: Cambridge, England.)

Aziz, P. (1972). Histogenesis of the carpel in *Triticum aestivum* L. *Botanical Gazette* **133**, 376-386.

Aziz, P. (1978). Initiation of procambial strands in the primordia of stamens and carpel of *Triticum aestivum* L. *Pakistan Journal of Scientific and Industrial Research* **21**, 12-16.

Aziz, P. (1981). Initiation of primary vascular elements in the stamens and carpel of *Triticum aestivum* L. *Botanical Journal of the Linnean Society* **82**, 69-79.

Barnard, C. (1957). Floral histogenesis in the monocotyledons. I. The Gramineae. *Australian Journal of Botany* **5**, 1-20.

Batygina, T. B. (1971). Some peculiar features of macrosporogenesis and development of the female gametophyte in wheat. *Annals de l'Universite' et de l'Association Regionale pour l'Etude la Recherche Scientifiques, Reims* **9**, 46-50.

Belk, E. (1939). Studies in the anatomy and morphology of the spikelet and flower of Gramineae. 183 pages, 510 figures. Doctoral dissertation. (Cornell University, Ithaca, USA.)

Bell, A. D. (1976a). The vascular pattern of Italian ryegrass (*Lolium multiflorum* Lam.). 3. The leaf trace system, and tiller insertion, in the adult. *Annals of Botany* **40**, 241-250.

Bell, A. D. (1976b). The vascular pattern of Italian ryegrass (*Lolium multiflorum* Lam.). 4. The peripheral plexus, and nodal root insertion. *Annals of Botany* **40**, 251-259.

Busby, C. H., and T. P. O'Brien (1979). Aspects of vascular anatomy and differentiation of vascular tissues and transfer cells in vegetative nodes of wheat. *Australian Journal of Botany* **27**, 703-711.

Butzin, F. (1965). Neue Untersuchungen über die Blüte der Gramineae. 183 pages, 35 figures, 8 plates. Doctoral dissertation. (Freie Universität, Berlin.)

Calderón, C. E., and Soderstrom, T. R. (1973). Morphological and anatomical considerations of the grass subfamily Bambusoideae based on the new genus *Maclurolyra*. *Smithsonian Contributions to Botany* **11**, 1-55.

Chandra, N. (1962a). Morphological studies in the Gramineae. I. Vascular anatomy of the spikelet in the Pooideae. *Proceedings of the National Institute of Sciences of India* **28B**, 545-562.

Chandra, N. (1962b). Morphological studies in the Gramineae. II. Vascular anatomy of the spikelet in the Paniceae. *Proceedings of the Indian Academy of Sciences Section B* **56**, 217-231.

Chandra, N. (1963). Morphological studies in the Gramineae. III. On the nature of the gynoecium in the Gramineae. *Journal of the Indian Botanical Society* **42**, 252-259.

Chandra, N., and Saxena, N. P. (1964). Morphological studies in the Gramineae. V. Vascular anatomy of the spike and spikelets in the Andropogoneae. *Proceedings of the Indian Academy of Sciences Section B* **59**,1-23.

Clark, L. G., and Fisher, J. B. (1987). Vegetative morphology of grasses: shoots and roots. In 'Grass Systematics and Evolution'. (Eds T. R. Soderstrom, K. W. Hilu, C. S. Campbell, and M. E. Barkworth.) pp. 37-45. (Smithsonian Institution Press:Washington, USA.)

Clayton, W. D., and Renvoize, S. A. (1986). 'Genera Graminum: Grasses of the World.' (Her Majesty's Stationery Office: London.)

Clifford, H. T. (1987). Spikelet and floral morphology. In 'Grass Systematics and Evolution'. (Eds T. R. Soderstrom, K. W. Hilu, C. S. Campbell, and M. E. Barkworth.) pp. 21-30. (Smithsonian Institution Press: Washington, USA.)

Craig, S., and O'Brien T. P. (1975). The lodicules of wheat: pre- and post-anthesis. *Australian Journal of Botany* **23**, 451-458.

Dannenhoffer, J. M., and Evert R. F. (1994). Development of the vascular system in the leaf of barley (*Hordeum vulgare* L.). *International Journal of Plant Sciences* **155**, 143-157.

Decker, H. F. 1964. An anatomic-systematic study of the classical tribe Festuceae (Gramineae). *American Journal of Botany* **51**, 453-463.

Dengler, N. G., Woodvine, M. A., Donnelly, P. M. and Dengler R. E. (1997). Formation of vascular pattern in developing leaves of the C₄ grass *Arundinella hirta*. *International Journal of Plant Sciences* **158**, 1-12.

Deshpande, B. D. and Sarkar, S. (1961). Nodal anatomy and the vascular system of some member of the Gramineae. *Proceedings of the National Institute of Sciences of India* **28B**, 1-12.

Ellis, R. P. (1987). A review of comparative leaf blade anatomy in the systematics of the Poaceae:The past twenty-five years. In 'Grass Systematics and Evolution'. (Eds T. R. Soderstrom, K. W. Hilu, C. S. Campbell, and M. E. Barkworth.) pp. 3-10. (Smithsonian Institution Press: Washington, USA.)

Esau, K. (1943). Ontogeny of the vascular bundle in *Zea mays*. *Hilgardia* **15**, 325-368.

Esau, K. (1957). Anatomic effects of barley yellow dwarf virus and maleic hydrazide on certain Gramineae. *Hilgardia* **27**, 15-69.

Fisher, J. B., and French, J. C. (1976). The occurrence of intercalary and uninterruped meristems in the internodes of tropical monocotyledons. *American Journal of Botany* **63**, 510-525.

Fraser, J., and Kokko, E. G. (1993). Panicle, spikelet, and floret development in orchardgrass (*Dactylis glomerata*). *Canadian Journal of Botany* **71**, 523-532.

Frick, H., and Pizzolato, T. D. (1987). Adaptive value of the xylem discontinuity in partitioning of photoassimilate to the grain. *Bulletin of the Torrey Botanical Club* **114**, 252-259.

Grosser, D., and Liese, W. (1971). On the anatomy of Asian bamboos, with special reference to their vascular bundles. *Wood Science and Technology* **5**, 290-312.

Guédès, M., and Dupuy, P. (1976). Comparative morphology of lodicules of grasses. *Botanical Journal of the Linnean Society* **73**, 317-331.

Hilu, K. W., and Esen, A. (1993). Prolamin and immunological studies in the Poaceae. III. Subfamily Chloridiodeae. *American Journal of Botany* **80**, 104-113.

Hitch, P. A., and Sharman, B. C. (1968a). Initiation of procambial strands in leaf primordia of *Dactylis glomerata* L. as an example of a temperate herbage grass. *Annals of Botany* **32**, 153-164.

Hitch, P. A., and Sharman, B. C. (1968b). Initiation of procambial strands in axillary buds of *Dactylis glomerata* L., *Secale cereale* L., and *Lolium perenne* L. *Annals of Botany* **32**, 667-676.

Hitch, P. A., and Sharman, B. C. (1971). The vascular pattern of festucoid grass axes, with particular reference to nodal plexi. *Botanical Gazette* **132**, 38-56.

Hoppe, D. C., McCully, M. E., and Wenzel, C. L. (1986). The nodal roots of *Zea*: Their development in relation to structural features of the stem. *Canadian Journal of Botany* **64**, 2524-2537.

Hsu, C.-C. (1965). The classification of *Panicum* (Gramineae) and its allies, with special reference to the characters of lodicule, style-base and lemma. *Journal of the Faculty of Science, University of Tokyo, section 3 (Botany)* **9**, 43-150.

Hsü, J. (1944). Structure and growth of the shoot apex of *Sinocalamus beecheyana* McClure. *American Journal of Botany* **31**, 404-411.

Inosaka, M. (1958). Vascular connection of the individual leaves with each other and with the tillers in rice plant. *Proceedings of the Crop Science Society of Japan* **27**, 191-192.

Inosaka, M. (1962). Studies of the development of vascular system in rice plant and the growth of each organ viewed from the vascular connection between them. *Bulletin of the Faculty of Agriculture of the University of Miyazaki* **7**, 15-89.

Jacobs, S. W. L. (1987). Systematics of chloridoid grasses. In 'Grass Systematics and Evolution'. (Eds T. R. Soderstrom, K. W. Hilu, C. S. Campbell, and M. E. Barkworth.) pp. 277-286. (Smithsonian Institution Press: Washington, USA.)

Jirásek, V., and Jozífová, M. (1968). Morphology of lodicules, their variability and importance in the taxonomy of the Poaceae family. *Boletín de la Sociedad Argentina de Botánica* **12**, 324-349.

Judziewicz, E. J., Clark, L. G., Londoño, X., and Stern, M. J. (1999). 'American Bamboos.' (Smithsonian Institution Press: Washington, USA.)

Kaufman, P. B. (1959). Development of the shoot of *Oryza sativa* L.-III. Early stages in histogenesis of the stem and ontogeny of the adventitious root. Phytomorphology **9**, 382-404.

Kirby, E. J. M., and Rymer, J. L (1974). Development of the vascular system in the ear of barley. *Annals of Botany* **38**, 565-573.

Kumazawa, M. (1961). Studies of the vascular course in maize plant. *Phytomorphology* **11**, 128-139.

Lersten, N. R. (1987). Morphology and anatomy of the wheat plant. In 'Wheat and Wheat Improvement, Second Edition'. (Ed E. G. Heyne.) pp. 33-75. (American Society of Agronomy, Crop Science Society of America, Soil Science Society of America: Madison, USA.)

Long, B. (1930). Spikelets of Johnson grass and Sudan grass. *Botanical Gazette* **89**, 154-168.

Majumdar, G. P., and Saha, B. (1956). Nodal anatomy and the vascular system of the shoot of rice plant (*Oryza sativa* L.). *Proceedings of the National Institute of Sciences of India* **22B**, 236-245.

Maze, J. (1977). The vascular system of the inflorescence axis of *Andropogon gerardii* (Gramineae) and its bearing on concepts of monocotyledon vascular tissue. *American Journal of Botany* **64**, 504-515.

Maze, J., and Scagel, R. K. (1982). Morphogenesis of the spikelets and inflorescence of *Andropogon gerardii* Vit. (Gramineae) and the relationship between form, information theory, and thermodynamics. *Canadian Journal of Botany* **60**, 806-817.

Maze, J., Dengler, N. G., and Bohm, L. R. (1971). Comparative floret development in *Stipa tortilis* and *Oryzopsis miliacea* (Gramineae). *Botanical Gazette* **132**, 273-298.

Maze, J., Bohm, L. R., and Beil, C. (1972). Studies on the relationships and evolution of supraspecific taxa utilizing developmental data. 1. *Stipa lemonii* (Gramineae).*Canadian Journal of Botany* **50**, 2327-2352.

O'Brien, T. P., Sammut, M. E., Lee, J. W., and Smart, M. G. (1985). The vascular system of the wheat spikelet. *Australian Journal of Plant Physiology* **12**, 487-511.

Paisooksantivatana, Y., and Pohl, R. W. (1992). Morphology, anatomy and cytology of the genus *Lithachne* (Poaceae: Bambusosideae). *Revista de la Biologica Tropical* **40**, 47-72.

Patrick, J. W. (1972a). Vascular system of the stem of the wheat plant. I. Mature stem. *Australian Journal of Botany* **20**, 49-63.

Patrick, J. W. (1972b). Vascular system of the stem of the wheat plant. II. Development. *Australian Journal of Botany* **20**, 65-78.

Patrick, J. W., and Wardlaw, I. F. (1984). Vascular control of photosynthate transfer from the flag leaf to the ear of wheat. *Australian Journal of Plant Physiology* **11**, 235-241.

Philipson, W. R. (1935). The development and morphology of the ligule in grasses. *The New Phytologist* **34**, 310-325.

Philipson, W. R. (1985). Is the grass gynoecium monocarpellary? *American Journal of Botany* **72**, 1954-1961.

Pissarek, H.-P. (1971). Untersuchungen über Bau und Funktion der Gramineen-Lodiculae. *Beiträge zur Biologie der Pflanzen* **47**, 313-370.

Pizzolato, T. D. (1980). On the vascular anatomy and stomates of the lodicules of *Zea mays*. *Canadian Journal of Botany* **58**, 1045-1055.

Pizzolato, T. D. (1983a). Vascular system of lodicules of *Dactylis glomerata* L. *American Journal of Botany* **70**, 17-29.

Pizzolato, T. D. (1983b). A three-dimensional reconstruction of the vascular system to the lodicules, androecium, and gynoecium of a fertile floret of *Panicum dichotomiflorum* (Gramineae). *American Journal of Botany* **70**, 1173-1187.

Pizzolato, T. D. (1984). Vascular system of the fertile floret of *Anthoxanthum odoratum* L. *Botanical Gazette* **145**, 358-371.

Pizzolato, T. D. (1987). Vascular system of the floret of *Alopecurus carolinianus* (Gramineae). *Canadian Journal of Botany* **65**, 2592-2600.

Pizzolato, T. D. (1988). Vascular system of the floret of *Phleum pratense*. *Canadian Journal of Botany* **66**, 1818-1829.

Pizzolato, T. D. (1989a). Vascular system of the fertile floret of *Phalaris arundinacea*. *Canadian Journal of Botany* **67**, 1366-1380.

Pizzolato, T. D. (1989b). Vascular system of the floret of *Leersia virginica* (Gramineae-Oryzoideae). *American Journal of Botany* **76**, 589-602.

Pizzolato, T. D. (1990a). Vascular system of the fertile floret of *Cynodon dactylon* (Gramineae-Eragrostoideae). *Botanical Gazette* **151**, 477-489.

Pizzolato, T. D. (1990b). Vascular system of the male and female florets of *Raddia brasiliensis* (Poaceae: Bambusoideae: Olyreae). *Smithsonian Contributions to Botany* **78**, 1-32.

Pizzolato, T. D. (1991). Vascular system of the fertile spikelet of *Sorghum* (Gramineae: Panicoideae). *Canadian Journal of Botany* **69**, 656-670.

Pizzolato, T. D. (1992). Distinct descent of the palea trace in *Panicum dichotomiflorum* (Gramineae: Panicoideae). *Bulletin of the Torrey Botanical Club* **119**, 125-130.

Pizzolato, T. D. (1993a). Vascular system of the fertile floret of *Panicum* (Gramineae: Panicoideae: Paniceae). *American Journal of Botany* **80**, 53-64.

Pizzolato, T. D. (1993b). Vascular system of the male and perfect florets of *Phragmites australis* (Gramineae-Arundinoideae). *International Journal of Plant Sciences* **154**, 119-133.

Pizzolato, T. D. (1994a). Procambial initiation for the vascular system of florets of *Phragmites* (Gramineae-Arundinoideae). *International Journal of Plant Sciences* **155**, 445-452.

Pizzolato, T. D. (1994b). Procambial initiation in pistils of *Phragmites* (Arundinoideae, Gramineae). *International Journal of Plant Sciences* **155**, 445-452.

Pizzolato, T. D. (1995). Patterns of procambial development in rachilla and floret of *Muhlenbergia torreyana* (Gramineae, Chloridoideae). *International Journal of Plant Sciences* **156**, 603-614.

Pizzolato, T. D. (1996). Gynobasic procambial initiation in the pistil of *Muhlenbergia torreyana* (Gramineae, Chloridoideae) in contrast to the gynomedian pattern. *Bulletin of the Torrey Botanical Club* **123**, 126-134.

Pizzolato, T. D. (1997). Procambial initiation for the vascular system in the spike of wheat. *International Journal of Plant Sciences* **158**, 121-131.

Pizzolato, T. D. (1998). Procambial initiation for the vascular system in the spikelet of wheat. *International Journal of Plant Sciences* **159**, 46-56.

Pizzolato, T. D., and Robinson-Beers, K. (1987). Initiation of the vascular system in the fertile floret of *Anthoxanthum odoratum* (Gramineae). *American Journal of Botany* **74**, 463-470.

Raechal, L. J., and Curtis, J. D. (1990). Root anatomy of the Bambusoideae (Poaceae). *American Journal of Botany* **77**, 475-482.

Rechel, E. A., and Walsh, M. A. (1985). Gross phloem anatomy in the roots of selected C_4, C_3, and C_4-C_3 intermediate *Poaceae* species. *Crop Science* **25**, 1068-1073.

Robinson-Beers, K., and Pizzolato, T. D. (1987a). Development of the vascular system in the fertile floret of *Anthoxanthum odoratum* L. (Gramineae). I. Traces to the fertile lemma and palea. *Botanical Gazette* **148**, 51-66.

Robinson-Beers, K., and Pizzolato, T. D. (1987b). Development of the vascular system in the fertile floret of *Anthoxanthum odoratum* L. (Gramineae). II. Sieve-element plexus, stamen traces, and the xylem discontinuity. *Botanical Gazette* **148**, 209-220.

Robinson-Beers, K., and Pizzolato, T. D. (1987c). Development of the vascular system in the fertile floret of *Anthoxanthum odoratum* L. (Gramineae). III. Vasculature supplying the gynoecium. *Botanical Gazette* **148**, 346-359.

Rudall, P., and Dransfield, S. (1989). Fruit structure and development in *Dinochloa* and *Ochlandra* (Gramineae-Bambusoideae). *Annals of Botany* **63**, 29-38.

Sharman, B. C. (1942). Developmental anatomy of the shoot of *Zea mays* L. *Annals of Botany* **6**, 245-283.

Sharman, B. C., and Hitch, P. A. (1967). Initiation of procambial strands in leaf primordia of bread wheat, *Triticum aestivum* L. *Annals of Botany* **31**, 229-243.

Sundberg, M.D., and Orr, A. R. (1996). Early inflorescence and floral development in *Zea mays* land race Chapalote (Poaceae). *American Journal of Botany* **83**, 1255-1265.

Theile, H. L., Clifford, H. T., and Rogers, R. W. (1996). Diversity in the grass pistil and its taxonomic significance. *Australian Systematic Botany* **9**, 903-912.

Tiba, S. D., Johnson, C. T., and Cresswell, C. F. (1988). Leaf ontogeny in *Digitaria eriantha*. *Annals of Botany* **61**, 541-549.

Xu, X.-B., and Vergara, B. S. (1986). Morphological changes in rice panicle development: A review of literature. *International Rice Research Institute Research Paper Series* **117**, 1-13.

Zee, S.-Y. (1972). Vascular tissue and transfer cell distribution in the rice spikelet. *Australian Journal of Biological Sciences* **25**, 411-414.

Zee, S.-Y. (1974). Distribution of vascular transfer cells in the culm nodes of bamboo. *Canadian Journal of Botany* **52**, 345-347.

Zee, S.-Y., and O'Brien, T. P. (1970). A special type of tracheary element associated with 'Xylem Discontinuity' in the floral axis of wheat. *Australian Journal of Biological Sciences* **23**, 783-791.

Zee, S.-Y., and O'Brien, T. P. (1971). Aleuron transfer cells and other structural features of the spikelet of millet. *Australian Journal of Biological Sciences* **24**, 391-395.

Zimmermann, M.H., and Tomlinson, P.B. (1972). The vascular system of monocotyledonous stems. *Botanical Gazette* **133**, 141-155.

Grasses: Systematics and Evolution. (2000). Eds S.W.L. Jacobs and J. Everett. (CSIRO: Melbourne)

THE GRASS INFLORESCENCE

C. Vegetti[AB] and Ana M. Anton[AC]

[A]Carrera del Investigador, CONICET.
[B]Facultad de Ciencias Agrarias, Universidad Nacional del Litoral, Kreder 2805, 3080 Esperanza, Santa Fe, Argentina.
[C]IMBIV (Universidad Nacional de Córdoba-CONICET), C. C. 495, 5000 Córdoba, Argentina.

Abstract

The structural variation of the grass inflorescence can be explained by the analysis of some processes that have operated independently or combined in different ways which principally affect two areas -the flowering unit and the enrichment zone.

Key words: Poaceae, grass, evolution, processes, inflorescence, typology.

THE GRASS INFLORESCENCE

The inflorescence has been much used in the study of the grasses as a source of diagnostic characters for identification, and for its phylogenetic value. Inflorescence typology has been applied to some grass groups, for example by Rua and Weberling (1995) and Vegetti and Weberling (1996) who concluded that polytelic inflorescences are a rule in the grasses. There the main axis lacks a terminal flower (Fig. 1) and ends in a group of flowers, the spikelet or the main florescence (MF), and lateral branches, paraclades, that repeat the structure of the main stem by producing distal flowering elements, coflorescences.

Below the coflorescences, the primary paraclades can ramify generating paraclades of different order. In the paracladial zone there are short paraclades (sP), those reduced to the coflorescences, and long paraclades (lP), those bearing secondary branches.

Sell (1976) described the flowering unit (FU) as the ensemble of the main florescence and paraclades. The FU bears only long flowering paraclades without leaves; the main axis with its leaves below the FU forms the trophotagma. The trophotagma comprises a proximal innovation zone (InZ), an intermediate inhibition zone (IZ) and a distal enrichment zone (EZ). The development of each of these is influenced by environmental conditions, and a specific zone may occasionally be absent.

When there is an enrichment zone, the paraclades themselves bear a proximal trophotagma (paraclades with trophotagma, lPt) and end in a flowering unit. Additional paraclades of consecutive order may develop in the axil of leaves other than the prophyll, enriching even more the branching system, which is proleptic or sylleptic in its development.

The wide diversification exhibited by the grasses can be explained by analyzing the different processes that, departing from a very simple panicle, have taken place throughout the evolutionary history of the family. These processes seem to have operated, independently or combined, more than once, and in different ways. Some of the processes have been previously treated (Cámara Hernández and Rua, 1991; Vegetti, 1991; Vegetti and Anton, 1995; Rua and Weberling, l. c.); others are mentioned here for the first time. In the flowering unit they have affected the main florescence and the paracladial zone. The main florescence may or may not be present; if present, the variation is expressed in the number of flowers per spikelet, rachilla prolongation, morphology of glumes and lemmas, and sexuality of the flowers.

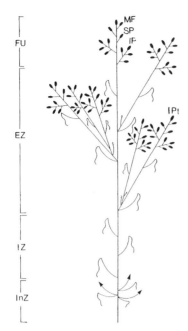

Fig. 1. Parts of the grass plant and terminology used. FU, flowering unit with main florescence (MF), short (sP) and long (lP) paraclades. On the main axis trophotagma the inhibition (InZ), innovation (IZ) and enrichment (EZ) zones are delimited, the last with long paraclades with trophotagma (lPt).

The most important variation, however, is found in the paracladial and enrichment zones; the processes that could have generated these are listed below, and can be identified in Figs 2-4 by their numbers.

1. Increase in the degree of ramification of the long paraclades.

2. Development of more branches per node, which may result from i) early branching of the primary paraclade; ii) failure of intercalary growth of the basal internode of the primary paraclade; or iii) abbreviation of some of the internodes of the inflorescence axis, leading to the formation of pseudo-verticils.

3. Partial sterility in long paraclades.

4. Truncation of the main florescence; here the axis may become sterile, or a lateral spikelet may adopt a pseudo-terminal position.

5. Homogenization; short paraclades are regularly reduced in a distal region of the inflorescence. This homogenization is generally synchronous with process 4; the distal portion of the short paraclades forms a pseudoflorescence.

6. Branching of short paraclades.

7. Truncation of the primary axis of the short paraclades; this occurs when paraclades are branched.

8. Truncation of the long paraclades (the coflorescence does not develop); this process is similar to process 4.

9. Truncation of the pseudoflorescence leaving only long paraclades.

10. Reduction of the number of long paraclades.

11. Simplification of the primary paraclades, which are reduced to coflorescences.

12. Reduction of the coflorescence internode generating sessile spikelets on the main axis and long paraclades.

13. Reduction in the length of the main axis internodes.

14. Reduction in the length of the paracladial zone as a consequence of processes 10, 11 and 13 (Fig.3). At an extreme, only few short paraclades remain.

15. Development of long paraclades with trophotagma.

16. Ramification of long paraclades with trophotagma.

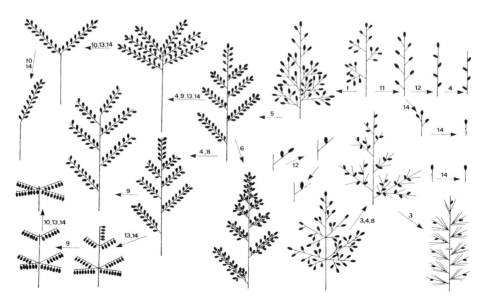

Fig. 2. Different patterns recognized in the grass inflorescences; numbers indicate the processes that have operated (see text).

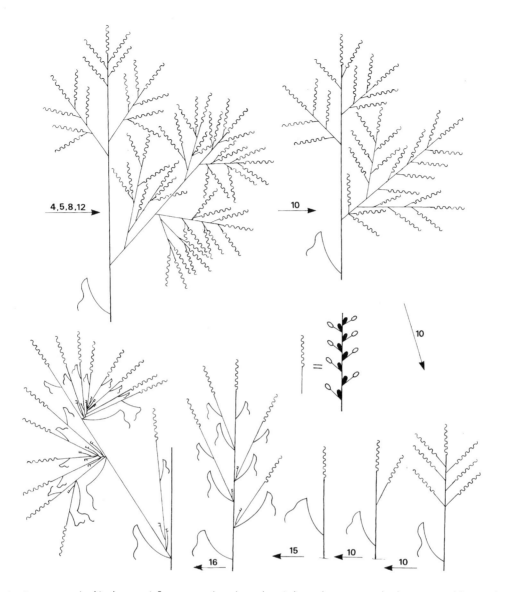

Fig. 3. Different patterns recognized in the grass inflorescences (cont.); numbers indicate the processes that have operated (see text).

CONCLUSIONS AND RECOMMENDATIONS

It is imperative to register a suite of characters — especially in zones and regions such as the flowering unit and the enrichment zone — to deal satisfactorily with general inflorescence form. Developmental studies of the vegetative and reproductive shoots are essential to the interpretation of its characteristics and structure. Sixteen processes have been identified and described as responsible for the great diversity of the grass inflorescence.

REFERENCES

Cámara Hernández, J., and Rua, G. (1991). The synflorescence of Poaceae. *Beiträge zur Biologie der Pflanzen* **66**, 297–311.

Rua, G. H., and F. Weberling. (1995). Growth form and inflorescence structure of Paspalum L. (Poaceae). *Beiträge zur Biologie der Pflanzen* **69**, 363–431.

Sell, Y. (1976). Tendances évolutives parmi les complexes inflorescentiels. *Revue Générale de Botanique* **83**, 247–267.

Vegetti, A. C. (1991). Sobre la politelia en las inflorescencias de Poaceae. *Kurtziana* **21**, 267–274.

Vegetti, A. C., and Anton, A. M. (1995). Some trends of evolution in the inflorescence of Poaceae. *Flora* **190**, 225–228.

Vegetti, A. C., and Weberling, F. (1996). The structure of the paraclades zone in Poaceae. *Taxon* **45**, 453–460.

BAMBOOS

Phyllostachys aurea, a commonly cultivated weedy bamboo with leptomorph rhizomes.

Grasses: Systematics and Evolution. (2000). Eds S.W.L. Jacobs and J. Everett. (CSIRO: Melbourne)

PHYLOGENY AND CLASSIFICATION OF THE BAMBUSOIDEAE (POACEAE)

Weiping Zhang[A] and Lynn G. Clark[B]

[A]Biogenetic Services, Inc., 2308 6th Street E., Brookings, SD 57006, U. S. A.
Department of Botany, Iowa State University, Ames, IA 50011-1020, U. S. A.

Abstract

The Bambusoideae as traditionally circumscribed was a heterogeneous group of mostly perennial, rhizomatous forest grasses with broad, pseudopetiolate leaves, arm and fusoid cells in the chlorenchyma, often bracteate synflorescences, and three lodicules. This assemblage is polyphyletic as demonstrated by several lines of evidence, both molecular and morphological, but a "core" bambusoid clade is supported as monophyletic in most analyses. We conducted a parsimony analysis of 24 core bamboo species and six other grasses using *ndh*F sequence and structural data to 1) test the monophyly of the core bamboos and 2) explore the internal structure of this clade. The two data sets were analysed separately and together. The bambusoid clade was robust in the *ndh*F and combined analyses, and was also supported by the anatomical synapomorphy of well developed, asymmetrically invaginated arm cells. Within the bamboo clade, the woody bamboos (Bambuseae) and the herbaceous olyroid bamboos (Olyreae including Parianeae and Buergersiochloeae) each formed a monophyletic clade. The olyroid clade, defined by the presence of unisexual, one-flowered spikelets, was strongly supported by the *ndh*F sequence data, and *Buergersiochloa* was resolved as basal within this clade. The woody bamboos were supported by at least three morphological synapomorphies, but resolution within this clade was relatively weak even in the combined analysis. We propose a new, narrower circumscription for the Bambusoideae, as well as a classification for the subfamily in which the two tribes Bambuseae and Olyreae are recognised.

Key words: Bambusoideae, bamboo, phylogeny, *ndh*F, Bambuseae, Olyreae, arm cells.

INTRODUCTION

The Bambusoideae as traditionally circumscribed was a heterogeneous group of mostly perennial, rhizomatous forest grasses with broad, pseudopetiolate leaves, arm and fusoid cells in the chlorenchyma, often bracteate synflorescences, and three lodicules (Clark *et al.* 1995 and references cited therein; Clark 1997a, c). In nearly all classification schemes from this century, at least five tribes [Anomochloeae, Streptochaeteae, Phareae, Bambuseae (or equivalent), and Olyreae] were included in the Bambusoideae (Table 1 of Clark *et al.* 1995). Soderstrom and Ellis (1987) defined a "core" Bambusoideae consisting of the Anomochloeae, Streptochaeteae, Buergersiochloeae, Olyreae,

and Bambuseae. Clayton and Renvoize (1986) included 14 tribes in their classification of the subfamily, while Watson and Dallwitz (1992) included 15.

Even the core Bambusoideae described above, however, is a demonstrably polyphyletic assemblage based on both molecular and morphological evidence (Clark *et al.* 1995; Zhang 1996; GPWG this volume), but a bambusoid clade consisting of the woody bamboos (Bambuseae) and the herbaceous olyroid bamboos (Olyreae, Parianeae, and Buergersiochloeae) is supported as monophyletic in most analyses in which both groups were sampled (Clark *et al.* 1995; Duvall and Morton 1996; GPWG this volume). The Anomochloeae/Streptochaeteae and the Phareae constitute the two most basal lineages in the grass family, and the other ancillary tribes formerly classified as bamboos are now distributed among the Ehrhartoideae, Pooideae, Centothecoideae, and Puelioideae (Clark *et al.* 1995; Zhang 1996; Clark 1997a; Clark *et al.*, in press).

We conducted a parsimony analysis of 24 putative core bamboo species and six additional grasses using *ndh*F sequence and structural data to 1) test the monophyly of the bambusoid clade and 2) explore the internal structure of this clade. We used the results to circumscribe the Bambusoideae more narrowly and to develop a tribal level classification for the subfamily.

MATERIALS AND METHODS

Thirty species of grasses, of which 24 belong to the putative core Bambusoideae, were included in this study (Table 1). Taxa are arranged in Table 1 according to Clayton and Renvoize (1986).

Leaf material from the same vouchers as those for the molecular analyses were used whenever possible for anatomical preparations; observations were supplemented by scanning electron micrographs for some taxa and reference to the literature. Fresh or dried leaves (from silica gel-dried material or herbarium specimens) were soaked in Pohl's solution (Pohl 1965) for a few minutes. For cross sections, the lamina was sectioned with a single-edged razor blade and mounted in lactophenol/aniline blue (Sass 1958). For epidermal scrapes, a section of the lamina was gently scraped to remove either the abaxial or adaxial surface and the mesophyll, and the remaining epidermis was mounted in lactophenol/aniline blue. For scanning electron microscopy, specimens were prepared as described in Dávila and Clark (1990).

Sequences of the *ndh*F gene were generated as described in Clark *et al.* (1995) and Zhang (1996). Insertions/deletions (indels) were excluded from the *ndh*F data matrix, leaving 2091 bp of aligned sequence. Two indels were phylogenetically informative, and these were included as binary characters 67 and 68 in the structural data matrix. The total data matrix consisted of 2159 characters (*ndh*F: 2091; structural: 68), of which 285 or 13 % were phylogenetically informative; 146 cells or 0.2 % were scored as missing.

Phylogenetic analyses were performed using PAUP * Version 4.0b1 (Swofford 1998). In all analyses, characters were weighted equally and multistate characters were unordered. The *ndh*F and structural data sets were analysed separately and together. We employed the heuristic search option with stepwise data addition and TBR branch-swapping using 1000 random-addition replicates to search for islands of shorter trees (Maddison 1991). In the analyses described below, *Streptochaeta angustifolia* was used as the outgroup, but *Streptogyna*, *Oryza*, and *Brachyelytrum* were used as outgroups in some analyses to test stability of the ingroup topology. Relative support for individual clades was evaluated using Bremer support (decay) values (Bremer 1988), and bootstrap values (Felsenstein 1985) were calculated for the combined analysis.

The structural character list and the data matrices are available upon request from Clark.

RESULTS

Tree statistics for the three separate phylogenetic analyses are shown in Table 2. In the *ndh*F analysis, *Streptochaeta*, *Puelia*, and the remainder of the taxa included form a trichotomy in the strict consensus tree. Within the latter clade, there are two sets of trichotomies. The first trichotomy consists of *Streptogyna* + [(*Oryza*, *Leersia*) *Brachyelytrum*] + the bambusoid clade; in the second, the bambusoid clade is formed by a trichotomy of the herbaceous olyroid bamboos + the temperate woody bamboos + the tropical woody bamboos. The monophyly of the bambusoid clade is supported by an estimated decay value of 5+. Other clades with similarly strong support are *Oryza* + *Leersia*, the herbaceous olyroid bamboos, and the [(*Alvimia*, *Rhipidocladum*) (*Guadua*, *Otatea*)] clade. The temperate woody bamboo clade is supported by a decay value of 4.

In the strict consensus tree of the structural analysis, the same basal trichotomy is resolved. Within the large clade, *Streptogyna* is sister to the rest of the taxa sampled. Within this latter clade, a trichotomy consisting of {[(*Olyra*, *Leersia*) *Brachyelytrum*] + Olyreae - *Buergersiochloa*} + *Buergersiochloa* + woody bamboos is recovered. Support for the monophyly of the woody bamboos is moderate (decay = 2), and there is relatively little resolution within this clade. Incongruence between the *ndh*F and structural consensus trees is seen in the placement of the [(*Oryza*, *Leersia*) *Brachyelytrum*] clade. When this clade is excluded, however, the topologies of the *ndh*F and structural consensus trees are largely congruent, differing only in the amount of resolution.

One of the six most parsimonious trees obtained in the combined analysis is shown in Fig. 1. As in the other analyses, *Streptochaeta*, *Puelia*, and the remainder of the taxa sampled form a basal trichotomy. The large clade including the bamboos is well supported by 21 character state changes, including six unique mutations, and has an estimated decay value of 5+ and a bootstrap value of 97 %. Within this clade, a dichotomy between the clade including *Streptogyna*, the rices, and *Brachyelytrum*, and the clade including the bamboos is resolved. Monophyly of the bambusoid clade is strongly supported by 16 character state changes, including four unique mutations, and an estimated decay value of 5 and a 94 % bootstrap. The association of *Streptogyna* with the [(*Oryza*, *Leersia*) *Brachyelytrum*] clade collapses in the strict consensus to form a trichotomy (Fig. 2). Within the bambusoid clade, the dichotomy between the herbaceous olyroid bamboos and the woody bamboos is well marked (Fig. 1). The olyroid clade is supported as monophyletic by 23 character state changes (including five unique mutations), an estimated decay value of 5+, and a 100 % bootstrap value. Within this clade, *Buergersiochloa* is resolved as basal and sister to the remaining olyroid bamboos, and there is strong support for some sister relationships (e.g., *Pariana* + *Eremitis* and *Sucrea* + *Raddia*). The woody bamboos are resolved as monophyletic based on nine character state changes, including one unique mutation and three structural novelties; this clade has an estimated decay and bootstrap values of 5 and 84 %. Within the woody bamboos, the temperate woody bamboo clade is extremely well supported although resolution within the clade is weak or absent, while monophyly of neither the tropical woody bamboos nor the Neotropical woody bamboos is supported (Fig. 2). The Paleotropical woody bamboo

Table 1. Taxa, Vouchers, and GenBank Numbers for Species Analysed.

Taxon	Voucher Number	GenBank Number
Bambuseae		
Alvimia gracilis Soderstr. & Londoño	AC 4389	AF 182347
Arundinaria gigantea (Walter) Chapm.	WZ 8400703	U21846
Bambusa aff. *bambos* (L.) Voss	LC 1300	U22000
Cephalostachyum pergracile Munro	WZ 8400635	U21968
Chimonobambusa marmorea (Mitford) Makino	SBG 9203	U21969
Chusquea latifolia L. G. Clark	LC & XL 417	U21989
Guadua paniculata Munro	QBG s.n.	U22001
Hickelia madagascariensis A. Camus	SD 1290	
Melocanna baccifera Kurz.	XL & LC 930	AF 182348
Nastus elatus Holttum	SD s.n.	AF 182352
Neurolepis aperta (Munro) Pilger	XL & LC 913	AF 182355
Otatea acuminata (Munro) C. Calderón & Soderstr.	LC & WZ 1348	AF 182350
Phyllostachys pubescens Mazel ex J. Houz.	LC 1289	U21970
Puelia olyriformis (Franch.) Clayton & Renvoize	CCE 288	AF 182345
Rhipidocladum pittieri (Hack.) McClure	LC & WZ 1349	AF 182349
Sasa variegata E.-G. Camus	WZ 9228	AF 182344
Shibataea kumasasa (Zoll.) Makino	LC 1290	AF 182342
Yushania exilis Yi	WZ 9230	AF 182346
Brachyelytreae		
Brachyelytrum erectum (Schreb.) P. Beauv.	LC 1330	U22004
Olyreae		
Buergersiochloa bambusoides Pilg.	SD 1382	AF 182341
Lithachne pauciflora (Sw.) P. Beauv.	LC 1297	U21978
Olyra latifolia L.	XL & LC 911	U21971
Raddia distichophylla (Schrad. ex Nees) Chase	LC 1306	U22006
Sucrea maculata Soderstr.	LC & WZ 1345	AF 182343
Oryzeae		
Leersia virginica Willd.	LC 1316	U21974
Oryza sativa L.	Sugiura (1989)	X15901
Parianeae		
Eremitis sp. nov.	LC & WZ 1343	AF 182353
Pariana radiciflora Sagot ex Doell	LC & WZ 1344	AF 182354
Streptochaeteae		
Streptochaeta angustifolia Soderstr.	LC 1304	U21982
Streptogyneae		
Streptogyna americana C. E. Hubb.	RP & GD 12310	U21965

Abbreviations are as follows: AC = André Carvalho; CCE = Cambridge Congo Expedition 1959; GD = G. Davidse; LC = L. Clark; QBG = Quail Botanical Garden; RP = R. Pohl; SBG = Sichuan Academy of Forestry Botanical Garden; SD = S. Dransfield; WZ = Weiping Zhang; XL = X. Londoño.

clade has moderate support with an estimated decay value of 3 and 86 % bootstrap support, but the only other well supported clades are the sister relationships of *Alvimia* + *Rhipidocladum* and *Guadua* + *Otatea*, and the sister relationship of these two clades.

DISCUSSION

The Bambusoid Clade. A bambusoid clade including only woody bamboos and herbaceous olyroid bamboos is supported as monophyletic in several analyses (Clark *et al.* 1995; Duvall and Morton 1996; Zhang 1996; Kelchner and Clark 1997; GWPG

this volume) although Soreng and Davis (1998) did not obtain this result. In this analysis, monophyly of the clade is robust.

The presence of arm cells previously was used as one of the characters delimiting the traditional Bambusoideae, but with little or no critical evaluation. If variation in arm cells is not recognised, then, according to the most recent phylogenetic hypotheses for the family, the presence of arm cells would be synapomorphic for the family, and symplesiomorphic in true bamboos. In those taxa with arm cells, we could distinguish three discrete states. *Streptochaeta angustifolia* exhibited weakly developed arm cells with

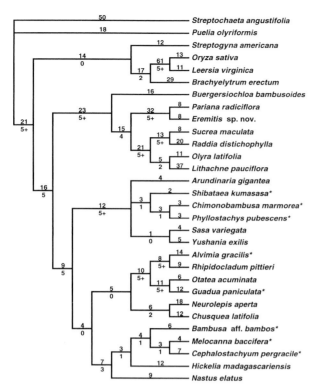

Fig. 1. One of the six most parsimonious trees inferred from analysis of the combined *ndh*F and structural data matrices. Numbers above the branches represent character support for clades. Numbers below the branches represent estimates of decay values (0 = branches that collapse in the strict consensus tree; I = branches that collapse in the strict consensus of all trees I step longer, etc.;). * = taxa with pseudospikelets.

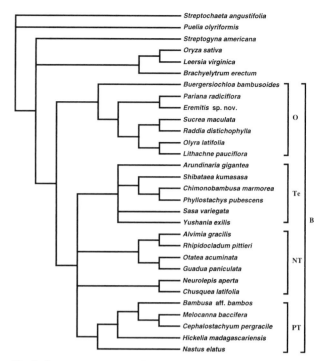

Fig. 2. Strict consensus tree of the six most parsimonious trees inferred from analysis of the combined *ndh*F and structural data matrices. B = Bambusoideae; NT = Neotropical woody bamboos; O = Olyreae; PT = Paleotropical woody bamboos; Te = Temperate woody bamboos.

irregular undulations. In *Oryza* and *Leersia*, the invaginations are distinct and occur all around the cell, forming a rosette pattern like that reported for *Phragmites australis* (Cav.) Trin. (Clifford and Watson 1977, Fig. 28b). In the majority of the bamboos, both woody and herbaceous, at least some, and usually most, of the adaxial arm cells (located between the fusoid cells and the adaxial epidermis) are distinctly invaginated on their abaxial side, giving a palisade-like appearance (Judziewicz *et al.* 1999, Fig. 15). The bambusoid clade is supported by the unambiguous gain of asymmetrically invaginated arm cells, with reversals to weakly developed arm cells in *Buergersiochloa, Guadua paniculata*, and *Nastus elatus*.

There is little or no support for the inclusion of *Streptogyna*, the Puelieae and Guaduelleae, and the rices in the bambusoid clade. In this study, *Puelia* appears in a relatively basal position within the family, and much stronger support for this placement is found in two analyses of multiple data sets (GPWG this volume; Clark *et al.*, in press). *Streptogyna* associates weakly with the rice clade here, and its position is unstable in a larger analysis (GPWG, in prep.). The presence of the outer ligule has been used to link both *Streptogyna* and the Puelieae with the woody bamboos (Clayton and Renvoize 1986; Soderstrom and Ellis 1987) but our results suggest that this character is independently derived in the three groups. The recognition of a more narrowly defined Bambusoideae, including only the herbaceous olyroid bamboos and the woody bamboos, is justified based on our results.

The Olyroid Clade. Representatives of the Olyreae *s. s.* (*Lithachne, Olyra, Raddia*, and *Sucrea*), the Parianeae (*Eremitis* and *Pariana*), and the Buergersiochloeae (*Buergersiochloa*) were included in this study. Our results are in agreement with the longstanding association of these three tribes. The herbaceous olyroid bamboos emerge as a robustly monophyletic lineage in both the *ndh*F and combined analyses, but are paraphyletic in the structural analysis, with the [(*Oryza, Leersia*) *Brachyelytrum*] clade nested within it as sister to the Parianeae + Olyreae *s. s.* In the combined analyses, the olyroid clade is supported by the presence of unisexual spikelets, which also occur independently in *Puelia*. The Parianeae and the Olyreae *s. s.* are united unambiguously by the presence of crenate and cross-shaped silica bodies. The Parianeae are characterised by the presence of flattened and indurate peduncles in the male spikelets, and the Olyreae by the presence of indurated anthecia in the female spikelets. Our results are congruent with those of Kellogg and Watson (1993), who also recovered a monophyletic olyroid clade with *Buergersiochloa* in a basal position, except that *Pharus* and *Leptaspis* nested within their olyroid clade.

Buergersiochloa has been classified either as a monogeneric tribe (Blake 1946; Soderstrom and Ellis 1987; Tzvelev 1989) or more commonly submerged into the Olyreae (Prat 1960; Fijten 1975; Clayton and Renvoize 1986; Watson and Dallwitz 1992). Blake (1946) recognised the Buergersiochloeae based on the absence of cross-veins in the leaf blades, the long-awned lemma, and the united filaments in the male spikelets. After careful reexamination, however, Fijten (1975) found that all three diagnostic characters were either not unique or uncertain in *Buergersiochloa* and she indicated that they were insufficient for the delimitation of a

Table 2. Tree statistics for the three parsimony analyses.

Tree statistics	*ndh*F data matrix	structural data matrix	combined data matrix
Number of informative characters	225	60	285
Number of trees	108	68	6
Length of most parsimonious trees	503	181	698
Consistency Index (CI)	0.539	0.376	0.486
Retention Index (RI)	0.654	0.686	0.651
Rescaled Consistency Index (RC)	0.352	0.258	0.316
Homoplasy Index (HI)	0.461	0.624	0.514

separate tribe. In spite of morphological similarities between *Buergersiochloa* and the Olyreae, Soderstrom and Ellis (1987) noted that the presence of intercostal sclerenchyma, the proximity of vascular bundles, the small fusoid cells, and the tall and narrow silica bodies, as well as the lack of the crenate and cross-shaped silica bodies so characteristic of the Olyreae (including the Parianeae), suggested a closer relationship with the woody bamboos. They considered other features, including an adaxially projecting keel with simple vasculature and symmetrical leaves, to link *Buergersiochloa* to the Olyreae, but concluded that the genus was rather isolated and should be maintained as a separate tribe. Tzvelev (1989) indicated that the structure of the gynoecium in the pistillate flowers of *Buergersiochloa*, with the two styles separate at the base and connate above as figured by Pilger (1954), was unique among grasses; Fijten (1975), however, had previously noted that this was an artifact of dissection.

The Parianeae has been treated more often as a tribe separate from the Olyreae (Prat 1960; Calderón and Soderstrom 1980; Clayton and Renvoize 1986; Tzvelev 1989) than included within the Olyreae (Soderstrom and Ellis 1987; Watson and Dallwitz 1992). *Pariana* nested within the olyroid clade of Kellogg and Watson (1993) as sister to *Maclurolyra*. Although the Parianeae exhibit leptomorph (monopodial) rhizomes, non-branching synflorescences, two to many stamens, and well developed fimbriae according to Calderón and Soderstrom (1980), more recent workers have regarded these differences as insufficient to warrant tribal recognition. It should be noted that rhizome morphology in *Pariana* is unusual, but not necessarily monopodial, and requires further study (Clark, pers. obs.). The Olyreae *s. s.* were distinguished from the Parianeae on the basis of pachymorph (sympodial) rhizomes, branching synflorescences that produce partial synflorescences, and three stamens.

If the Parianeae are recognised as a tribe, then both the Buergersiochloeae and Olyreae *s. s.* must be recognised as well. Given the strong support for the olyroid clade in our analyses, and because further study with additional sampling is needed to resolve relationships in this lineage, we adopt a conservative approach and recognize a single inclusive tribe of herbaceous bamboos, the Olyreae.

The Woody Bamboo Clade. Support for a monophyletic woody bamboo clade was relatively weak from the molecular data alone, and moderate from structural data alone, but this emerged as a robust lineage in the combined analyses (Fig. 1). In previous *ndh*F analyses of the grasses or bamboos, the temperate woody bamboos were resolved as sister to the olyroid clade, with this

clade sister to the tropical woody bamboos (Clark *et al.* 1995; Zhang 1996). Some of our *ndh*F trees presented this topology, whereas in others a monophyletic woody bamboo clade was recovered, with the temperate woody bamboos sister to the tropical woody bamboos. In the combined analysis, however, four lineages of woody bamboos formed a tetrachotomy (Fig. 2). The lack of molecular support for the woody bamboos (and for resolution within this group) may be due to a slowdown in the base substitution rate as a result of the long generation times of these plants (Gaut *et al.* 1997). The presence of woody culms, the differentiation of foliage and culm leaves (reflecting the two phases of shoot growth and development), and the gregarious flowering cycles at long intervals are all structural characters that unite the woody bamboos. As noted previously, the presence of an outer ligule is characteristic but homoplastic.

The temperate bamboo clade is supported, and often strongly so, in this and prior analyses of molecular data (Clark *et al.* 1995; Zhang 1996; Kelchner and Clark 1997), but no structural synapomorphies have been identified for it and, even with molecular data, resolution within the clade is weak or lacking. The presence of monopodial, leptomorph rhizomes that tiller or not may be a synapomorphy for this clade, but this feature also evolved in the Neotropical genus *Chusquea*, and one group of genera within the temperate bamboos possess sympodial, pachymorph rhizomes (*Fargesia, Yushania*). Pseudospikelets are found in one subclade [(*Phyllostachys, Chimonobambusa*) *Shibataea*], but also occur in two of the tropical woody bamboo clades (Fig. 1). The loss of saddle-shaped intercostal silica bodies may be worth further study as a potential synapomorphy, although this loss also appears to have occurred in the olyroid clade.

The tropical woody bamboo clade received weak support in the *ndh*F analyses, but was not supported as monophyletic in either the structural or combined analyses (Figs. 1, 2). The presence of sympodial, pachymorph rhizomes in most members of this clade is likely symplesiomorphic, as it is probable that this is the ancestral condition for the whole bamboo clade, if not the grass family. The temperate bamboo clade may be a specialised derivative from within the tropical woody bamboos, in which case the lack of monophyly of this latter group is not surprising. The Paleotropical woody bamboos, however, are moderately supported as monophyletic in this analysis by one unique mutation and the occurrence of branching in the year following shooting. Although unambiguous, this latter character has not been carefully observed and the timing of shooting may be difficult to assess among tropical bamboos relative to temperate bamboos. Watanabe *et al.* (1994), in their chloroplast restriction site study of Asiatic woody bamboos, also found support for the Paleotropical woody bamboos. The Neotropical woody bamboos emerged as two clades in our analysis, a chusqueoid clade (*Chusquea* + *Neurolepis*) and an arthrostylidioid/guaduoid clade [(*Alvimia*, *Rhipidocladum*) + (*Otatea*, *Guadua*)]. As Clark (1997b) pointed out, certain structural features suggest a different set of relationships among the tropical woody bamboos, and the conflict between molecular and structural data needs to be explored further.

In a number of recent classifications of the woody bamboos, this clade has been subdivided into several tribes (e.g., Tzvelev 1989; Zhang 1992 and references cited therein) or subtribes (Soderstrom and Ellis 1987; Dransfield and Widjaja 1995). Our analyses suggest that strict reliance on rhizome morphology and the presence or absence of pseudospikelets leads to non-monophyletic taxa, but we still do not have enough information to reconstruct relationships within the woody bamboos with much confidence. Given the strong support for the woody bamboo clade itself, we recognise a single, inclusive tribe, the Bambuseae, without recommending any subdivisions.

CLASSIFICATION

We propose a recircumscription of the Bambusoideae to include only the bambusoid clade as supported in our analyses. We exclude the following tribes: Anomochloeae, Streptochaeteae, Phareae, Puelieae, Guaduelleae, Oryzeae, Phyllorachideae, Ehrharteae, Streptogyneae, Brachyelytreae, Phaenospermatideae, Diarrheneae, and Centotheceae. The two major lineages of the bambusoid clade are here recognised as the tribes Bambuseae and Olyreae. Descriptions and synonymy for the subfamily and the two tribes are presented.

Bambusoideae Luerss., 1893: 451. Type: *Bambusa* Schreb.
Syn.: Olyroideae Pilger, 1956: 168.

Parianoideae (C. E. Hubb.) Butzin, 1965: 148.

Morphology. Perennial, rhizomatous herbaceous or woody grasses of temperate and tropical forests, tropical high montane grasslands, riverbanks, and sometimes savannas. Rhizomes strongly or weakly developed, sympodial or monopodial, and pachymorph, leptomorph or amphimorph. Culms hollow or solid. Leaves cauline, distichous; outer ligule absent (Olyreae) or present (Bambuseae); inner ligule membranous or chartaceous, fringed or unfringed; blades pseudopetiolate and usually relatively broad; sheaths often bearing lateral appendages at the summit. Synflorescences spicate, racemose or paniculate, completing development of all spikelets in one period of growth and subtending bracts and prophylls usually absent, or pseudospikelets with basal bud-bearing bracts producing two or more orders of spikelets with different phases of maturity and subtending bracts and prophylls usually present. Spikelets (or spikelets proper of the pseudospikelets) bisexual (Bambuseae) or unisexual (Olyreae), consisting of 0, 1, 2 or several glumes, 1 to many florets; lodicules usually 3 (rarely 0 or 6 to many), vascularized, often ciliate; stamens usually 2, 3, or 6 (10-40 in *Pariana*, 6-120 in *Ochlandra*); ovary glabrous or hairy, sometimes with an apical appendage; stigmas (1-) 2 or 3. Caryopsis basic, rarely bacate or nucoid, the hilum linear, extending the full length or in a few genera only

1/2 to 3/4 the length of the caryopsis, or rarely central punctate to short elliptical (*Raddiella*); endosperm hard, without lipid, containing compound starch grains; embryo small, not waisted, epiblast present, scutellar cleft present, mesocotyl internode absent, embryonic leaf margins overlapping. Base chromosome numbers: x = 7, 9, 10, 11 and 12.

Foliar anatomy. Mesophyll non-radiate, an adaxial palisade layer absent, fusoid cells usually large and well developed (sometimes absent), arm cells usually well developed with at least some of the adaxial arm cells strongly asymmetrically invaginated (usually from the abaxial side and oriented in a palisade-like fashion); kranz anatomy absent; midrib complex (bundles superposed) or simple (one bundle or an arc of bundles); adaxial bulliform cells present.

Foliar micromorphology. Stomates with dome-shaped, triangular, or parallel-sided subsidiary cells; bicellular microhairs present, panicoid-type; papillae common and abundant, often overarching the stomates.

Photosynthetic pathway. C_3.

Bambuseae Dumort., 1829: 63.
Syn.: Arthrostylidieae E.-G. Camus, 1913: 16.

Arundinarieae Ascherson & Graebner, 1902: 770.

Chusqueeae E.-G. Camus, 1913: 16.

Dendrocalameae Hackel, 1887: 92.

Melocanneae (Benth.) Keng, 1959: 31.

Oxytenanthereae Tzvelev, 1987: 22.

Shibataeeae Nakai, 1933: 83.

Yushanieae Guzmán, Anaya & Santana, 1984: 19.

Rhizomes well developed. Culms woody, usually hollow (solid in *Chusquea* and a few species of other genera); culm development occurring in two phases, first, new, unbranched shoots bearing culm leaves elongate to full height, second, culm lignification and branch development with production of foliage leaves occur; aerial vegetative branching usually complex (absent in *Glaziophyton*, *Greslania* and *Neurolepis*). Culm leaves usually well developed with expanded sheaths and reduced blades. Foliage leaves with an outer ligule; sheaths often bearing fimbriae and/or auricular appendages at the summit; blades articulated, deciduous. Flowering usually cyclical, gregarious and monocarpic. Synflorescences bracteate or not, determinate (spikelets) or indeterminate (pseudospikelets). Spikelets (or spikelets proper of the pseudospikelets) bisexual with 1 to many bisexual florets; glumes (0-) 1-4 (-7) but sometimes very reduced; lemmas multinerved, similar in texture to the glumes; paleas several-nerved with an even number of nerves, bicarinate. Caryopsis usually basic, sometimes bacate (e.g., *Alvimia*, *Melocanna*, *Olmeca*) or nucoid (e.g., *Merostachys*). Base chromosome number *x* = 10, (11), and 12.

This tribe includes 60-70 genera and at least 1,100 species worldwide (Clark 1997a, 1997c). Approximately 600 species in 40-50 genera are described from the Old World, the majority from China, and 430 described species in 21 genera are known from the New World (Clark 1997b; Judziewicz *et al.*, 1999).

New World woody bamboo diversity is estimated to include at least 70 undescribed species, however, so it is expected that the total will surpass 500 species (Clark 1997b). Members of this tribe range from 46° N to 47° S latitude and are found from sea level to 4300 m in elevation.

Olyreae Kunth ex Spenn., 1825: 172.

Syn.: Buergersiochloeae Blake, 1946: 62.

Parianeae C. E. Hubbard, 1934: 219.

Rhizomes weakly or sometimes strongly developed (*Pariana*). Culms herbaceous, vegetative branching restricted and only one phase of culm development observed. Culm leaves absent. Foliage leaves with the outer ligule absent; sheaths sometimes bearing fimbriae and/or blister-like swellings at or near the summit (*Pariana*), more often fimbriae, swellings, and auriculate appendages absent; blades not articulated, persistent or sometimes deciduous, exhibiting nocturnal folding in some genera (e.g. *Lithachne*, *Raddia*). Flowering usually annual or seasonal for extended periods, very rarely cyclical and monocarpic. Synflorescences ebracteate, apparently determinate. Spikelets unisexual and 1-flowered with no rachis extension, the plants monoecious; pistillodes or staminodes sometimes present in male or female spikelets respectively. Female spikelets with 2 glumes; lemma chartaceous to more commonly coriaceous, several-nerved, usually nonaristate except in *Buergersiochloa*; palea with few to several nerves. Male spikelets smaller than the females, glumes absent or 2 and minute; lemmas membranous, 3-nerved. Caryopsis basic. Base chromosome number x = 7, 9, 10, 11, and (12).

This overwhelmingly Neotropical tribe includes 21 genera and 107 described species distributed in the tropics and subtropics (Clark 1997c; Judziewicz *et al.*, 1999). Only the monotypic *Buergersiochloa* (New Guinea) and *Olyra latifolia* L. (tropical Africa) are represented in the Paleotropics. Members of this tribe range from 29° N to 34° S latitude and only occasionally occur above 1000 m in elevation.

ACKNOWLEDGEMENTS

This manuscript is based on dissertation research carried out by Zhang, and he thanks his family and colleagues for their support and help. Financial support for this study was derived from National Science Foundation Grant DEB-9218657 to Clark and J. F. Wendel. We thank R. Olmstead (University of Washington) for providing most of the initial primers. We are grateful to S. Dransfield and S. Renvoize (Royal Botanic Gardens, Kew) for providing leaf material of *Buergersiochloa* and *Puelia*.

REFERENCES

Blake, S. T. (1946). Two new grasses from New Guinea. *Blumea, Suppl.* **3**, 59-62.

Bremer, K. (1988). The limits of amino acid sequence data in angiosperm phylogenetic reconstruction. *Evolution* **42**, 795-803.

Butzin, F. (1965). Neue Untersuchungen über die Blüte der Gramineae. Inaugural-Dissertation, 183 pp. (Freien Universität: Berlin.)

Calderón, C. E. and Soderstrom, T. R. (1980). The genera of Bambusoideae (Poaceae) of the American continent: Keys and comments. *Smithsonian Contributions to Botany* **44**, 1-27.

Camus, E.-G. (1913). 'Les Bambusées - monographie, biologie, culture, principaux usages'. 2 volumes. (P. Lechevalier: Paris.)

Clark, L. G. (1997a). Bamboos: The centrepiece of the grass family. In 'The Bamboos'. (Ed. G. P. Chapman.) pp. 237-248. (Academic Press: London.)

Clark, L. G. (1997b). Diversity, biogeography and evolution of *Chusquea*. In 'The Bamboos'. (Ed. G. P. Chapman.) pp. 33-44. (Academic Press: London.)

Clark, L. G. (1997c). Diversity and biogeography of Ecuadorean bamboos (Poaceae: Bambusoideae) and their allies. In 'Estudios Sobre Diversidad y Ecología de Plantas, Memorias del II Congreso Ecuatoriano de Botánica'. (Eds R. Valencia and H. Balslev.) pp. 51-63. (Pontifícia Universidad Católica del Ecuador: Quito.)

Clark, L. G., Zhang, W. and Wendel, J. F. (1995). A phylogeny of the grass family (Poaceae) based on *ndh*F sequence data. *Systematic Botany* **20**, 436-460.

Clark, L. G., Kobayashi, M., Mathews, S., Spangler, R. E., and Kellogg, E. A. (In press). The Puelioideae, a new subfamily of grasses. *Systematic Botany*.

Clayton, W. D., and Renvoize, S. A. (1986). 'Genera Graminum: Grasses of the World'. (Her Majesty's Stationery Office: London.)

Clifford, H. T., and Watson, L. (1977). 'Identifying Grasses: Data, Methods and Illustrations'. (University of Queensland Press: St. Lucia.)

Dávila, P., and Clark, L. G. (1990). A scanning electron microscopy study of leaf epidermes in *Sorghastrum* (Poaceae: Andropogoneae). *American Journal of Botany* **77**, 499-511.

Dransfield, S., and Widjaja, E. A. (1995). 'Plant Resources of South-East Asia No. 7, Bamboos'. (Backhuys Publishers: Leiden.)

Dumortier, B. C. J. (1829). 'Analyse des familles des plantes'. (J. Casterman, aîné: Tournay.)

Duvall, M. R., and Morton, B. R. (1996). Molecular phylogenetics of Poaceae: An expanded analysis of *rbcL* sequence data. *Molecular Phylogenetics and Evolution* **5**, 352-358.

Felsenstein, J. (1985). Confidence limits on phylogenies: An approach using the bootstrap. *Evolution* **39**, 783-791.

Fijten, F. (1975). A taxonomic revision of *Buergersiochloa* Pilger (Gramineae). *Blumea* **22**, 415-418.

Gaut, B. S., Clark, L. G., Wendel, J. F., and Muse, S. V. (1997). Comparisons of the molecular evolutionary process at *rbcL* and *ndhF* in the grass family (Poaceae). *Molecular Biology and Evolution* **14**, 769-777.

Guzmán, M. R., Anaya C., M. C. and Santana M., F. K. (1984). El género *Otatea* (Bambusoideae) en México y Centroamérica. Bol. Inst. Bot., Univ. Guadalajara **5**, 2-20.

Hackel, E. (1887). Gramineae. In 'Die Natürlichen Pflanzenfamilien 2(2), 1-97'. (Eds A. Engler & K. Prantl.) (W. Englemann: Leipzig.)

Hubbard, C. E. (1934). Gramineae. In 'The Families of Flowering Plants, 2 (Monocotyledons), 199-229'. (Ed. J. Hutchinson.) (Macmillan: London.)

Judziewicz, E. J., Clark, L. G., Londoño, X., and Stern, M. J. 1999. 'American Bamboos'. (Smithsonian Institution Press: Washington, D. C.)

Kelchner, S. A., and Clark, L. G. (1997). Molecular evolution and phylogenetic utility of the chloroplast rpl16 intron in *Chusquea* and the Bambusoideae (Poaceae). *Molecular Phylogenetics and Evolution* **8**, 385-397.

Kellogg, E. A., and Watson, L. (1993). Phylogenetic studies of a large data set. I. Bambusoideae, Andropogonodae, and Pooideae (Gramineae). *Botanical Review* **59**, 273-343.

Keng, Y. L. (1959). Gramineae. In 'Flora illustralis plantarum primarum sinicarum'. (Ed. Y. L. Keng.) (Agency for Science Publications: Peking.)

Luerssen, C. (1893). 'Grundzüge der Botanik, ed. 5'. (Verlag von H. Hassel: Leipzig.)

Maddison, D. R. (1991). The discovery and importance of islands of most-parsimonious trees. *Systematic Zoology* **40**, 315-328.

Nakai, T. (1933). Bambusaceae in Japan Proper (II). III. *Journal of Japanese Botany* **9**, 77-95.

Pilger, R. (1954). Das System der Gramineae. *Botanische Jahrbücher* **76**, 281-384.

Pilger, R. (1956). Gramineae II. In 'Die natürlichen Pflanzenfamilien, ed. 2'. (Eds A. Engler & K. Prantl.) pp. 1-225, Bd. 14d, i-iv. (Aufl. Duncker & Humboldt: Berlin.)

Pohl, R. W. (1965). Dissecting equipment and materials for the study of minute plant structures. *Rhodora* **67**, 95-96.

Prat, H. (1960). Vers une classification naturelle des graminées. *Bulletin de la Societe Botanique de France* **107**, 32-79.

Sass, J. E. (1958). 'Botanical Microtechnique'. (Iowa State University Press: Ames.)

Soderstrom, T. R., and Ellis, R. P. (1987). The position of bamboo genera and allies in a system of grass classification. In 'Grass Systematics and Evolution'. (Eds T. R. Soderstrom, K. W. Hilu, C. S. Campbell, and M. E. Barkworth.) pp. 225-238. (Smithsonian Institution Press: Washington, D. C.)

Soreng, R., and Davis, J. I. (1998). Phylogenetics and character evolution in the grass family (Poaceae): Simultaneous analysis of morphological and chloroplast DNA restriction site character sets. *The Botanical Review* **64**, 1-85.

Spenner, F. C. L. (1825). 'Flora friburgensis'. 3 vols. (typis Friderici Wagner: Friburgi Brisoviae.)

Sugiura, M. (1989). *Oryza sativa* chloroplast DNA 134,525 bp. Nagoya, Japan: Nagoya University, Center for Gene Research.

Swofford, D. L. (1998). PAUP * Version 4.0b1 for the Mactintosh (Web-distributed). Smithsonian Institution, Washington, D. C.

Tzvelev, N. N. (1987). Systema Graminearum (Poaceae) ac earum evolutio. *Komarovskie Ctenija (Moscow & Leningrad)* **37**, 1-73.

Tzvelev, N. N. (1989). The system of grasses (Poaceae) and their evolution. *Botanical Review* **55**, 141-203.

Watanabe, M., Ito, M., and Kurita, S. (1994). Chloroplast DNA phylogeny of Asian bamboos (Bambusoideae, Poaceae) and its systematic implication. *Journal of Plant Research* **107**, 253-261.

Watson, L., and Dallwitz, M. J. (1992). 'The Grass Genera of the World'. (CAB International: Wallingford, Oxon, England.)

Zhang, W. (1992). The classification of Bambusoideae (Poaceae) in China. *Journal of the American Bamboo Society* **9**, 25-42.

Zhang, W. (1996). Phylogeny and classification of the bamboos (Poaceae: Bambusoideae) based on molecular and morphological data. Doctoral dissertation, Iowa State University, Ames.

Grasses: Systematics and Evolution. (2000). Eds S.W.L. Jacobs and J. Everett. (CSIRO: Melbourne)

WOODY BAMBOOS (GRAMINEAE-BAMBUSOIDEAE) OF MADAGASCAR

Soejatmi Dransfield

The Herbarium, Royal Botanic Gardens, Kew, Richmond, Surrey TW9 3AE, UK.

Abstract

There are about 32 species of woody bamboos native and endemic to Madagascar. Twenty species are included in the seven genera: *Cathariostachys* S. Dransf., *Decaryochloa* A. Camus, *Hickelia* A. Camus, *Hitchcockella* A. Camus, *Nastus* Juss., *Perrierbambus* A. Camus, and *Valiha* S. Dransf. Twelve species have been included in *Arundinaria* Michaux., *Cephalostachyum* Munro, and *Schizostachyum* Nees; however, the inclusion in these genera is currently being reassessed and preliminary results suggest they are misplaced. The species have inflorescences with a determinate habit. With the exception of species included in *Arundinaria* (seven species), the species have spikelets consisting of three to six glumes, one fertile floret with six stamens and three stigmas, and a reduced rachilla extension, and are arranged in paniculate or racemose inflorescences, with or without the presence of subtending bracts/sheaths and prophylls. Moreover, the species have sympodial rhizome systems, either with short or with long necks; the habit can be erect, scrambling or climbing; and the branches are borne below the supranodal ridge, at the point where the sheath scar curves downwards. The genera can be differentiated from each other mainly by the structure of the inflorescences and spikelets. They (except *Arundinaria*) form a group of bamboos that, based on the inflorescence and spikelet structure, has strong relationships with many of the New World taxa, but based on the number of stamens and stigmas, seems related to most tropical Asiatic bamboos. Except for *Arundinaria*, the genera are at present included in the subtribe *Nastineae* Soderstr. & R. P. Ellis.

Key words: Gramineae, Bambusoideae, woody bamboos, native genera, species distributions, inflorescence structure.

INTRODUCTION

Madagascar is the fourth largest island in the world, and lies about 400 km off the east coast of Africa, south of the equator. It extends 1500 km from 11° 57′ S to 25° 32′ S, and covers about 590,000 square kilometers, with an altitudinal range from sea level to 2879 m. It has a wide range of habitats, ranging from lowland and montane rain forests to spiny desert scrub and alpine grassland. The island consists of a chain of mountains that runs along the eastern-centre of the island from the north to the south. On the eastern slopes of this mountain range can be found the remains of rainforest, which at one time covered the whole eastern part of the island. The majority of the native bamboos can be found in this area.

The woody bamboos of Madagascar are poorly known, yet a better understanding of them seems crucial to the construction of bamboo phylogenies. The problems of working on the bamboos of Madagascar are that the specimens are frequently incomplete, mostly consisting of flowering branches, lacking important vegetative features, and most of the native bamboos have been described in usually Asiatic genera. The first native species to be described was *Nastus capitatus* by Kunth in 1829; in 1868 Munro described two new species *Cephalostachyum chapelieri* and *Schizostachyum parvifolium*. Since then there was no further work on Madagascar bamboos, until 1924 when A. Camus started describing new genera and new species based on herbarium specimens collected mainly by Henri Perrier de la Bâthie. Camus had never been to Madagascar, and had thus no opportunity to see the living plants; most of her descriptions on habits, size of culms etc., are based on the excellent field notes of Perrier de la Bâthie. From 1924 to 1960 Camus described five new bamboo genera: a monotypic genus *Decaryochloa* (1946), *Hickelia* with two species (1924a, 1924d, 1955), a monotypic genus *Hitchcockella* (1925c), *Perrierbambus* with two species (1924b), and another monotypic genus *Pseudocoix* (1924a, 1924d), and another 24 new species, nine of them in *Nastus* Juss. (1925a, 1937, 1947, 1951, 1957b), six in *Arundinaria* Michaux. (1924e, 1926, 1931, 1950, 1960), five in *Cephalostachyum* Munro (1925b), one in *Ochlandra* Thw. (1935) and two in *Schizostachyum* Nees (1924c, 1957a). Since then there has not been any revision of the native bamboos, partly because there were few if any new collections. In 1987 new collections were made by David Edelman, who deposited the specimens in the US National Herbarium, Smithsonian Institution, Washington D.C., USA. Between 1988 and 1996 I had the opportunity to join RBG Kew expeditions to various parts of Madagascar, and collected more specimens of some species, and so I have been able to reassess generic limits. I have recently provided revisions of *Hickelia* A. Camus, including *Pseudocoix* A. Camus (Dransfield 1994), *Decaryochloa* A. Camus (Dransfield 1997), and two new genera, *Valiha* S. Dransf. and *Cathariostachys* S. Dransf. (Dransfield 1998).

There are about 32 species of bamboos native and endemic to Madagascar; the pantropical *Bambusa vulgaris*, which could be native or not, can be found near and around villages or along rivers; introduced species include *B. multiplex*, *Dendrocalamus giganteus*, *Phyllostachys aurea*, and *Gigantochloa* sp. (See Appendix for full list.)

GENERA AND SPECIES (SEE TABLE 1)

Arundinaria Michaux. Camus described five species, mainly found growing in mountainous areas, at altitudes of 1300 to 2800 m, and so far known only from the type localities. Two of them, *A. madagascariensis* and *A. marojejyensis*, are found in mossy forest; the others, *A. ambositrensis*, *A. humbertii*, *A. ibityensis* and *A. perrieri*, are in mountain forest. These species were transferred one way or another to the genus *Sinarundinaria* Nakai (Chao and Renvoize 1989), but as *Sinarundinaria* is now regarded as a synonym of *Fargesia* Franchet, the affinity of the Madagascar species needs to be critically assessed. *A. ibityensis* is very poorly known, and its inclusion in the genus *Thamnocalamus* Munro (Chao and Renvoize, 1989) should also be reassessed.

Cathariostachys S. Dransf. The newly described genus is endemic to Madagascar, and so far comprises two species, *C. capitata* (Kunth) S. Dransf. (formerly *Nastus capitatus* Kunth) and *C. madagascariensis* (A. Camus) S. Dransf., transferred from *Cephalostachyum* (Dransfield 1998). The genus is characterised by its globose inflorescence, resembling that of

Athroostachys Bentham from Brazil (McClure 1973); it is determinate, paniculate, with a contracted main axis and very short lateral branches, forming a capitulum; each branch is subtended by a bract or sheath; prophylls are rarely present. Each of these lateral branches also bears very short branches, arranged distichously along a very short axis, with a spikelet terminating it. In *Cathariostachys* the axis is so short, that the whole branch forms a fan, sitting on the main axis. The inflorescence, therefore, consists of these neatly arranged fan-like branches, to which the generic name alludes. The genus has sympodial rhizomes with long necks up to 4 m long; the culms are erect, hollow with relatively thin walls, and the internodes are often filled with water. Branch buds are borne below the supranodal ridge at the point where the sheath scar curves downwards; primary branches are dominant, elongate and scramble over nearby trees or vegetation. Spikelets are laterally flattened, consisting of five (rarely three) glumes, a fertile floret and a vestigial rachilla extension; the floret has six stamens, and the ovary has three stigmas. The genus has a very unusual fruit/caryopsis in which the embryo is loose at the bottom of the endosperm, and seems to be attached to the endosperm at the point by the hilum; the fruit also has some sort of a stalk, or an extension of the pericarp which sticks firmly at the base of the floret (Kunth 1829; Dransfield 1998). *C. capitata* is found mainly in the east coast, growing in lowland to hill primary forest, but surviving in disturbed forest, at 5 to 700 m altitude; whereas *C. madagascariensis* can be found in lower montane forest, in Perinet and Ranomafana, at 900 to 1000 m altitude. The culms of both species grow solitarily and scattered in forest, and are utilised locally for making flutes or as water containers. The endemic species of lemurs, *Hapalemur* spp., in Ranomafana feed on *C. madagascariensis*. So far gregarious flowering habit is not known in *Cathariostachys*; usually there is only one culm in the whole forest that bears flowers, and this culm will die after flowering.

Cephalostachyum Munro. Three of the species described by Munro (1868) are found in mainland Asia, and one is from Madagascar, *C. chapelieri*, a climbing bamboo. Camus (1925b) added four new species from Madagascar, two climbing bamboos: *C. perrieri* and *C. viguieri*, and two erect ones: *C. peclardii* and *C. madagascariensis* (*C. peclardii* is now a synonym of *Cathariostachys capitata* (Kunth) S. Dransf. , and *C. madagascariensis* is *Cathariostachys madagascariensis* (A. Camus) S. Dransf.). The three climbing bamboos are currently being studied, and will eventually be excluded from *Cephalostachyum* (Dransfield 1998); they possess globose and compact inflorescences terminating leafy branches, concealed by the modified uppermost leaves; the spikelet contains one fertile floret. Camus (1925b) stated that there is also a sterile spikelet in the inflorescence. The culms are slender, and usually slightly zig-zag, the internodes are not more than one cm in diameter, and usually solid. Branches are borne below the supranodal ridge at the point where the sheath scar curves downwards. *C. chapelieri* is found in lower montane forest in Analamazoatra; *C. viguieri* is closely related to *C. chapelieri*, but is more widespread, from lower hill forest on the east coast (e.g. Masoala Peninsula) to lower montane forest in the interior (e.g. Analamazoatra and Ranomafana); *C. perrieri* is not common, found in lower hill forest in the north-east and Masoala. *C. viguieri* was found flowering gregariously and died in 1989 in Antalavia,

Table 1. Species and the distributions of native woody bamboos in Madagascar (1998) *

Species	Distributions and Habitats
Arundinaria	
ambositrensis	Ranomena (Ambositra), Ranomafana; forest, 1300–1400 m
humbertii	Andringitra; forest, 2000m
ibityensis	Mt Ibity, Antsirabe; rocky ridges, 1800-2250 m
madagascariensis	Tsaratanana; mossy forest, 2000 & 2800 m
marojejyensis	Marojejy; mossy forest, 2000 m
perrieri	Manongarivo; forest, 1000 m
Cathariostachys	
capitata	Masoala & Mananara Avaratra; forest, 50-700 m
madagascariensis	Perinet to Ranomafana; forest, 800-1200 m.
Cephalostachyum	
chapelieri	Analamazoatra; lower montane forest
perrieri	Mananara Avaratra & ?Ranomafana; forest, ± 900 m
viguieri	Masoala to Analamazoatra; forest, 50–1200 m
Decaryochloa	
diadelpha	Analamazoatra (Perinet); mountain forest, 900 m
Hickelia	
alaotrensis	Lake Alaotra; forest, 1500 m
madagascariensis	Central Highlands; mountain forest, 1000–1600 m
perrieri	Tsaratanana; mountain forest, 2400 m
Hitchcockella	
baronii	Andringitra & Manongarivo; mountain forest
Nastus	
aristatus	Manongarivo to Analamazoatra; mountain forest, 900–1200 m
elongatus	Andringitra, Ranomafana; mountain forest, 1000 m
emirnensis	Analamazoatra; mountain forest, 1000 m
humbertianus	Andohahela
lokohoensis	Lokoho; forest
madagascariensis	Central plains; forest
manongarivensis	Manongarivo; forest, 500–1600 m
perrieri	Tsaratanana; mossy forest, 1700 m
tsaratananensis	Tsaratanana; mossy forest, 2000 m
Perrierbambus	
madagascariensis	Loky, Mahajanga; lowlands, dry forest
tsarasaotrensis	Tsarasaotra
Schizostachyum	
bosseri	Fenerive; forest margins in east coast
parvifolium	from Nosy-Be, Sambava to Mananjara; white sand coastal forests
perrieri	Tsaratanana; mountain forest, 2000 m
Valiha	
diffusa	Nosy-Be, Marojejy to Ifanadiana; hill forest, open plains, 50–500m
perrieri	Andrafiamena; forest

*Mainly based on herbarium collections

Masoala Peninsula; seedlings of about 2–3 months old were found all over the forest floor (pers. obs.).

Decaryochloa A. Camus. This monotypic genus, endemic to Madagascar, is found in mountain forest at Andasibe, Moramanga. The only species, *D. diadelpha*, is remarkable in having the longest floret (4.5 – 5 cm) in the tribe *Bambuseae*. The genus is characterised by the determinate inflorescence with segmented axis, and with the presence of subtending bracts or sheaths and prophylls, bearing one to four large spikelets; the spikelet consists of 3 to 4 glumes, one floret and a vestigial rachilla extension. In describing the species, Camus (1946) noted that the arrangement of the androecium was unusual, the stamens forming two groups of three, with the filaments joined at the base. In fact this feature highlighted as diagnostic in *D. diadelpha* is found only in

one collection (*Decary 18375*), and in other collections, including my own, the filaments are free (Dransfield 1997). It seems that this species has a long flowering period, as indicated by Camus (1946) and as witnessed by me, from 1989 to 1994. It flowered gregariously over five years in Moramanga (1989 to1994) and then died, and thus seems to be monocarpic. This bamboo has sympodial rhizomes with long necks; the culms are erect with scrambling upper parts, and with a diameter of 2 – 2.5 cm, the internodes are 30 – 80 cm long, and covered by black hairs when young. The upper parts of the culms and the elongating primary branches are very long and slender, and are alike, entangled with each other and trees, so that it is difficult to measure the actual length of the culms. The branch bud is borne just below the supranodal ridge, at the point where the sheath scar curves downwards. The auricle of the culm sheath is large and pubescent adaxially. In bamboos such auricles with hairy surfaces are not common (Dransfield 1997).

Hickelia A. Camus (syn. *Pseudocoix* A. Camus). The genus is distributed mainly in Madagascar, but extends to Tanzania (Dransfield 1994). So far there are three species endemic to Madagascar; *H. madagascariensis* is found in mountain forest in Central Highlands at 1000 to 1600 m altitude; *H. alaotrensis* is known only from the type locality, and may be a small form of *H. madagascariensis*; *H. perrieri* (syn. *Pseudocoix perrieri* A. Camus) grows in mountain forest in Tsaratanana at 2400 m altitude (Dransfield 1994). They are sympodial bamboos, with relatively long necks on the rhizomes. The culms are basically erect, especially at the lower part; the upper part and the elongated dominant branches cannot support themselves and eventually scramble over nearby vegetation. The internodes are hollow with relatively thick walls, about 40 cm long. The branches are borne below the supranodal ridge at the point where the sheath scar curves downwards. The inflorescences are paniculate and determinate with segmented axes, terminating leafy or leafless branches, bearing few to several spikelets; subtending bracts and prophylls are present. The spikelets are about 20 mm long, and consist of three to five (rarely two) glumes, one fertile floret with six stamens and an ovary with three stigmas, and a vestigial rachilla extension. Anthesis and maturation of spikelets is sequential on one flowering branch; it starts from the terminal spikelet and proceeds basipetally. Superficially the lemma and palea are similar, except that the palea has two pointed tips and a vertical groove on the back. Before anthesis the palea is enclosed or wrapped in the lemma. During anthesis they pull apart so that the stamens and stigmas are exerted and exposed. After fertilization the lemma and palea become close together again; this time, however, the palea is not enclosed in the lemma except at the base, and they become adpressed to each other at the margins. From the position of the lemma and palea, it is possible to tell whether the ovary of the spikelet has or has not been fertilized (Dransfield 1994). *H. madagascariensis* has a gregarious flowering habit, and dies after flowering (Dransfield pers. obs; Ceyverel pers. comm.). In the type collection, Perrier de la Bâthie (*10787*, 1932) noted that seeds of this bamboo were sought by animals, such as rats and lemurs. The culms are used for making baskets (Dransfield pers. obs.).

Hitchcockella A. Camus. The genus is very little known; the species, *H. baronii*, was described based on one specimen collected by Baron without locality. Camus (1925c) considered it to be related to *Perrierbambus* A. Camus, and also to *Nastus* Juss. and *Chusquea* Kunth. The inflorescence consists of a single spikelet terminating a leafy branch; the spikelet is about 1 cm long, laterally flattened, and consists of two glumes, one fertile floret and a rachilla extension bearing a rudimentary floret; buds are not present at the base. The genus is currently being revised, and may contain more than one species.

Nastus Juss. The genus comprises about 19 species, distributed from Reunion and Madagascar to Java, Sumba through New Guinea and Solomon Islands, and possibly in Sumatra. Habit varies a great deal; the type species, *N. borbonicus* J. P. Gmelin from Reunion, has erect culms and sympodial rhizomes with relatively long necks. Eight species from Madagascar have climbing culms; one, *N. elongatus*, has erect culms with drooping or leaning upper parts/tips. The genus is characterised by the determinate, paniculate or racemose inflorescence with the axis not segmented or segmented only weakly, with the absence of subtending bracts/sheaths and prophylls (present in *N. hooglandii* Holtt. from New Guinea, which eventually will be excluded from the genus), and the spikelet consisting of usually five to six glumes, one fertile floret and a rachilla extension bearing a rudimentary floret. *Nastus*, however, remains poorly known and its delimitation is in need of critical reassessment. In Madagascar the species are found growing in mountain or mossy forest at 500 to 2000 m altitude, from Tsaratanana through the central region to Andohahela, near Fort Dauphin.

Perrierbambus A. Camus. The genus consists of two species and is poorly known. Soderstrom and Ellis (1987) include the genus in subtribe *Arundinariinae*. The species were known only from the type collections, until recently (1995) when flowering material of *P. madagascariensis* A. Camus was sent to Kew for identification, collected from lowland dry forest in the north-west part of the country, south of Mahajanga. The inflorescence terminates the leafy branch and is concealed in upper leaf-blades; it consists of one spikelet, which is not flattened, and the spikelet consists of two glumes and one fertile floret, with rachilla extension abortive or absent. The inflorescence in fact resembles the one-spikeleted inflorescence of *Hickelia*. This suggests that the genus is related to *Hickelia* and *Decaryochloa*, and should be removed from subtribe *Arundinariinae*, and placed together with the last two genera, at present in subtribe *Nastinae* Soderstr. and R. P. Ellis.

Schizostachyum Nees. The genus is distributed mainly in tropical Asia, extending to the Hawaiian Islands, and its occurrence in Madagascar is based on an incorrect identification. Munro (1868) described *S. parvifolium* from Madagascar and Comores; Bentham (1883) noted that its placement in the genus is doubtful, for it is more related to *Nastus* than to *Schizostachyum*. Camus was apparently not aware of Bentham's notes, and described two new species, *S. perrieri* and *S. bosseri* (A. Camus 1924c, 1957a); *S. bosseri* is in fact conspecific with *S. parvifolium*, and *S. perrieri* may be conspecific with *Arundinaria madagascariensis* A. Camus, a species found from the same locality. *Schizostachyum* has indeterminate or itecauctant inflorescences (McClure 1934), whereas in *S. parvifolium* the inflorescence is determinate (or semelauctant) with the presence of subtending bracts/sheaths and prophylls (Dransfield pers. obs.). Its placement is being investigated in the Herbarium, Royal Botanic Gardens, Kew. The bamboo can be found growing in coastal forest on white sands, from Comores to Nosy Be (north Madagascar) throughout the coastal regions of the east coast of Madagascar; it is found abundantly in the Mananara Avaratra regions. It has a scrambling habit; the culms are slender, solid and hard, and are often used locally for making rough baskets. Flowering behaviour is not known; during my collecting trip between 1988 and 1996 only one flowering specimen was found and collected from the coastal region, south of Mananara Avaratra; most of the plants that were seen and collected were sterile.

Valiha S. Dransf. 'Valiha' is a musical instrument, the tube zither, traditionally made of bamboo internodes, and the name is adopted for the generic name of the common bamboo, used in the past for making this tube zither. This bamboo, local name 'Vologaz', can be found growing wild from Nosy Be in the north throughout the eastern slopes of the main mountain range to Ifanadiana in the south, in primary forest but also on open hills growing together with *Ravenala madagascariensis*, from 50 to 700 m altitude. 'Vologaz' (or 'Volo') had been identified and long known as *Ochlandra capitata* (Kunth) E. G. Camus (E. G. Camus 1913), until I examined it in detail and found that it had no name. 'Vologaz' does not belong to *Ochlandra* or to any known genus, so I described a new genus *Valiha* to accommodate it, and named 'vologaz' as *Valiha diffusa* S. Dransf. (Dransfield 1998). *Valiha* consists of two species: *V. perrieri* (A. Camus) S. Dransf. is known only from the type collection, collected in Andrafianamena, in the north-east part of Madagascar; *V. diffusa* has sympodial rhizomes, with long necks up to 2 m long, in mixed primary forest, the culms can be found growing solitarily and scattered in between forest trees. The height and the diameter of the culms vary a great deal; in primary forest, the culms can be about 10 m tall with a diameter of 7 – 10 cm, whereas in degraded areas (open hills near villages) they are not more than 6 m tall with a diameter of 1.5 cm. It is possible that local people collect the culms almost continuously for their everyday use (light constructions, water containers), and the rhizomes are only able to produce new smaller culms (a common situation in tropical sympodial bamboos when harvesting of the culms is excessive). Branches are borne below the supranodal ridge at the point where the sheath scar curves downwards. The inflorescence is determinate with segmented axis, and with the presence of subtending bracts/sheaths and prophylls; the spikelet consists of six glumes, one fertile floret with six stamens and an ovary with three stigmas, and a vestigial rachilla extension.

KEY TO THE GENERA (EXCLUDING INTRODUCED TAXA)

1. Rhizome monopodial; inflorescence axis not segmented; spikelet with more than one fertile floret ....*Arundinaria*

   Rhizome sympodial; inflorescence axis usually segmented, not segmented or weakly segmented; spikelet with one fertile floret 2

2. Culm erect or erect then scrambling.................................. 3

   Culm climbing.. 8

3. Culm more than 5 cm in diameter (rarely 1.5 cm), usually hollow; rhizome with long necks (2 – 4 m) 4

   Culm not more than 1.8 cm in diameter, hollow or solid; rhizome with relatively long necks (about 40 cm)............... 5

4. Inflorescence capitate; prophylls usually not present; young shoot green tinged with purple, covered with black hairs; culm sheath apex ± the same width as the base, blade broadly triangular, erect, with sharp apex; dominant primary branch elongating considerably .... *Cathariostachys*

   Inflorescence racemose; prophylls present; young shoot pale green, covered with dark brown hairs; culm sheath tapering to a very narrow tip, blade lanceolate, erect then

spreading; dominant primary branch not elongating considerably ....*Valiha*

5. Spikelet 6-7 cm long, slender, cylindrical; auricles of culm sheath with long bristles, pubescent adaxially .... *Decaryochloa*

   Spikelet not more than 2 cm long; auricles of culm sheath glabrous, with or without short bristles 6

6. Inflorescence spicate; spikelet laterally compressed or flattened; culm solid and hard..... *Schizostachyum parvifolium*

   Inflorescence paniculate; spikelet round-oval; culm hollow or solid, not hard...7

7. Inflorescence bearing few to several spikelets (rarely one); spikelet 1.5 – 2 cm long; auricles of leaf-blade sheath not prominent; culm 1.8 – 2 cm in diameter, hollow ..... *Hickelia*

   Inflorescence containing one spikelet; spikelet 1 – 1.4 cm long; auricles of leaf-blade sheath present and with long bristles; culm about 1 cm in diameter, usually solid .... *Perrierbambus*

8. Inflorescence capitate; rachilla extension short with or without rudimentary floret... ....*Cephalostachyum chapelieri, C. perrieri, C. viguieri*

   Inflorescence paniculate or racemose; rachilla extension long, with rudimentary floret... 9

9. Inflorescence consisting of one spikelet .... *Hitchcockella*

   Inflorescence with more than one spikelet...................*Nastus*

DISCUSSION

The species included in *Arundinaria* possess determinate inflorescences with unsegmented or weakly segmented axes, quite different from those of the other endemic or native taxa. Moreover the rhizomes are monopodial.

The other native taxa share several similar morphological features. The rhizomes are basically sympodial; some taxa, however, have rhizomes with short necks, others with long necks. The morphology of rhizomes is widely employed in bamboo taxonomy to differentiate genera, especially in the subtropics and temperate regions of the Old World. In the Old World tropics it is less important, because in general only the sympodial type of rhizome occurs. Nevertheless, Wong (1995) recognises sympodial rhizomes with relatively long necks (5-30 cm), found in *Racemobambos setifera* Holtt. and *Soejatmia ridleyi* (Gamble) Wong, both endemic in Peninsular Malaysia. It is not known, however, whether the common occurrence of sympodial rhizomes with long necks in most Madagascar bamboos, has any significance for their position in the tribe *Bambusineae*.

It is interesting to note that the taxa with short necks in the rhizomes have mostly climbing habits, whereas those with relatively or very long necks in the rhizomes have erect culms with often scrambling upper parts. The significance of this phenomenon is, however, not known. Nine species of *Nastus*, three species of *Cephalostachyum* and *Hitchockella baronii*, have climbing habits and short-necked rhizomes. In *Decaryochloa* and *Hickelia*, and probably also in *Perrierbambus* and *Schizostachyum parvifolium*, the rhizomes have relatively long necks, and the culms are erect in the lower parts, but with scrambling upper parts and scrambling dominant elongated primary branches. The rhizomes in *Cathariostachys* and *Valiha* have very long necks, up to 4 m long, and the culms are erect with arching tips.

The lateral branches of most Madagascar genera seem to be derived from one single bud; and the buds in the majority of the taxa are borne below the supranodal ridge at the point where the sheath scars curve downwards. This characteristic feature of Madagascar native bamboos also occurs in other bamboo species, such as *Nastus productus* (Pilger) Holtt. from New Guinea (pers. obs.), and the species of *Chusquea* sect. *Longifoliae* L. G. Clark (Clark 1989). The branch structure, especially that at mid-culm nodes, is often very useful in supporting the delimitation of some bamboo genera of the New World and of the Old World (tropics and temperate zones); for example, *Bambusa magica* Holtt. from Peninsular Malaysia, was removed and transferred to a new genus *Holttumochloa* K.M. Wong, mainly based on inflorescence structure and branch morphology (Wong 1993). The branch system in the genus *Racemobambos* Holtt., from Malesia, is so characteristic that it is almost possible to recognise the genus without flowering material (Dransfield 1983). It is hoped that branch structure also will support the delimitation of bamboo genera in Madagascar, when they are investigated in more detail.

The most significant feature that the bamboo species native in Madagascar (except those included in *Arundinaria*) share in common is the structure of the inflorescence and spikelet. The spikelet contains one fertile floret and has a rachilla extension. The spikelets are arranged either in racemose or paniculate determinate inflorescences, with segmented or without (or weakly) segmented axes, and with or without subtending bracts or sheaths. It is extremely difficult to describe the structure of bamboo inflorescences accurately. McClure (1973) proposed two terms, i.e. iterauctant and semelauctant for indeterminate and determinate inflorescences respectively. The differences between them lie mainly in the structure of the basic (or ultimate) unit of the inflorescence, and the period of growth of the inflorescence. In the determinate, or semelauctant, inflorescence the basic unit is a spikelet, and the spikelets in an inflorescence mature more or less simultaneouly in one single grand period; in the indeterminate, or iterauctant, inflorescence the basic unit is a pseudospikelet, which comprises a prophyll at the base, one or more bracts each subtending a bud (or another pseudospikelet), and a proper spikelet, in this way the spikelets do not mature in one grand period. Furthermore, in indeterminate inflorescences, all axes of the branches are clearly segmented, and consequently bear a sheathing organ at each node, and each branch possesses a prophyll at the base; in determinate inflorescences, on the other hand, the branches below the spikelet are not or only weakly segmented, and sheathing organs are usually absent (McClure 1966). There are, however, some variations and anomalies in indeterminate inflorescences as McClure (1966) pointed out in two cases. One of them is that the inflorescence branch may fail to develop buds at basal nodes, and this will stop the further production of pseudospikelets. He stated further that this situation may be interpreted as a step in the direction of evolution of the bamboo inflorescence from the typical indeterminate form

towards a determinate form. Unfortunately, McClure (1966) did not mention any specific taxa possessing this type of inflorescence. On the other hand, he suggested that *Glaziophyton* Franchet (Brazil) and *Greslania* Balansa (New Caledonia), which basically have determinate inflorescences with all axes segmented and bearing a sheathing organ at each node and a prophyll, appear to be the result of a trend in this direction, and that the occasional occurrence of a dormant prophyllate bud in *Greslania* shows a relict tendency toward indeterminateness. It shows here clearly that there are two different and contradictary statements concerning the directions of evolution of bamboo inflorescences. I suggest, therefore, that it should be viewed cautiously until both types of inflorescences (i.e. indeterminate inflorescence in which the branches stop producing pseudospikelets and determinate inflorescence possessing an occasional dormant bud) are studied critically.

The inflorescence type which occurs in *Glaziophyton* and *Greslania* has been neglected and never been reviewed before, until I produced and published various genera with this type of inflorescence, such as *Nastus hooglandii* Holtt. from New Guinea (Dransfield, unpubl.), *Temburongia* S. Dransf. and K. M. Wong from Brunei (Dransfield and Wong 1996), and some of the Madagascar genera, *Decaryochloa*, *Hickelia*, *Cathariostachys* and *Valiha* (and also *Perrierbambus*). Although it is premature to suggest the position of this inflorescence type in bamboos, I prefer to treat it as a part of the determinate inflorescence, i.e. bracteate determinate inflorescence.

CONCLUSIONS

Out of about 32 species, eight of them have been revised, and seven are currently under study, mainly because more collections of these species have been made from the results of recent fieldwork. Fieldwork is essential for the understanding of the poorly known species and for more extensive knowledge of the revised species, and should be conducted before their natural habitats disappear.

Some morphological features, that commonly occur and are shared by most Madagascar native bamboos, could contribute important data for studying evolutionary trends and phylogeny of the Bambusoideae, and should be assessed more extensively.

Based on the structure of the inflorescences and the spikelets, *Cathariostachys*, *Decaryochloa*, *Hickelia*, *Perrierbambus* and *Valiha* seem to be related to each other, and should be excluded from the subtribe *Nastineae* Soderstr. and R. P. Ellis (Dransfield and Widjaja 1995). However, as I mentioned elsewhere it seems wise to defer a subtribal placement until all genera in *Nastineae* are critically revised, and subjected to a rigorous phylogenetic study. For this, more fieldwork is needed.

It seems that the native bamboos of Madagascar may be related to one clade of New World bamboos, because they possess one-flowered spikelets arranged in bracteate determinate inflorescences. However, based on the flower structure, i.e. six stamens and three stigmas, Madagascar bamboos may be closely related to most bamboos of the Old World tropics. Could they be a link between the two groups, or could they have evolved independently? The problem will be the subject of phylogenetic research for the next three years as part of collaboration between Iowa State (L. G. Clark) and Kew (S. Dransfield) on the genera possessing one-flowered spikelets.

ACKNOWLEDGEMENTS

I would like to thank Royal Botanic Gardens, Kew, for providing working facilities. Field work in 1988, 1989, 1992 was supported by Royal Botanic Gardens, Kew, in 1994 by NSF Grant through Dr L. G. Clark, Iowa, USA (National Science Foundation Grant DEB-9218657), in 1996 also supported by RBG Kew and a grant presented by the Florida Caribbean Chapter of the American Bamboo Society, in conjunction with the Kew Palm Project, led by Dr John Dransfield. I would like to thank Dr Voara Randrionosolo, Dr Albert Randrianjafy, the Director of the Department des Eaux et Forest, Dr G. Schatz (MO), Drs D. Dupuy and H. Beentje (K), and the staff of Projet Masoala in Antalaha, in helping to arrange permits and with logistic matters. Mr J. Andriantiana and Mr G. Rafamantantsoa, of Parc de Tsimbazaza, helped in the field. Drs John Dransfield (K) and Lynn Clark provided valuable comments and suggestions on this paper, and their encouragement is most appreciated.

REFERENCES

Bentham, G. & Hooker, J. D. (1883). 'Genera Plantarum'. Vol. 3 part 2. (Reeve: London.)

Camus, A. (1924a). Genre nouveaux de Bambusées malgaches. *Comptes Rendus* **179**, 478-480.

Camus, A. (1924b). *Perrierbambus*, genre nouveau de Bambusées malgaches. *Bulletin de la Société Botanique de France* **71**, 697-701.

Camus, A. (1924c). Le *Schizostachyum perrieri* A. Camus, Bambou nouveaux de Madagascar. *Bulletin de la Société Botanique de France* **71**, 780–782.

Camus, A. (1924d). *Hickelia* and *Pseudocoix*, genres nouveaux de Bambusées malgaches. *Bulletin de la Société Botanique de France* **71**, 899-906.

Camus, A. (1924e). Espèces nouvelles d'*Arundinaria* Malgaches. *Bulletin du Muséum National d'Histoire Naturelle, Paris, Série 2*, **30**, 394–396.

Camus, A. (1925a). Le genre *Nastus* Juss. *Bulletin de la Société Botanique de France* **72**, 22-27.

Camus, A. (1925b). Le genre *Cephalostachyum* à Madagascar. *Bulletin de la Société Botanique de France* **72**, 84-88.

Camus, A. (1925c). *Hitchcockella*, genre nouveau de Bambusées malgaches. *Comptes Rendus* **181**, 253–255.

Camus, A. (1926). Le genre *Arundinaria* à Madagascar. *Bulletin de la Société Botanique de France* **73**, 624–626.

Camus, A. (1931). Graminées nouvelles de Madagascar. *Bulletin de la Société Botanique de France* **78**, 8–9.

Camus, A. (1935). *Ochlandra perrieri* A. Camus, Bambou nouveau de Madagascar. *Bulletin de la Société Botanique de France* **82**, 310–311.

Camus, A. (1937). *Nastus humbertianus* A. Camus, bambou nouveaux de Madagascar. *Bulletin de la Société Botanique de France* **84**, 286.

Camus, A. (1946). *Decaryochloa* genre nouveaux de Graminées malgaches. *Bulletin de la Société Botanique de France* **93**, 242-245.

Camus, A. (1947). Graminées nouvelle de Madagascar. *Bulletin de la Société Botanique de France* **94**, 39-42.

Camus, A. (1950). *Arundinaria* et *Acroceras* de Madagascar. *Bulletin de la Société Botanique de France* **97**, 84–85.

Camus, A. (1951). "*Andropogon*" et "*Nastus*" nouveaux de Madagascar. *Notulae Systematicae. Paris.* **14**, 213-215.

Camus, A. (1955). Quelques Graminées nouvelles de Madagascar et de la Réunion. *Bulletin de la Société Botanique de France* **102**, 120–122.

Camus, A. (1957a). *Schizostachyum, Cyrtococcum* et *Sacciolepis* (Graminées) nouveaux de Madagascar. *Bulletin de la Société Botanique de France* **104**, 281–182.

Camus, A. (1957b). Contribution a l'étude des Graminées de Madagascar. *Bulletin du Muséum National d'Histoire Naturelle, Paris, Série 2*, **29**, 274-281.

Camus, A. (1960) Sur quelques Graminées magaches. *Bulletin de la Société Botanique de France* **107**, 209–211.

Camus, E. G. (1913). 'Les Bambusées — monographie, biologie, culture, principaux usages'. (P. Lechevalier: Paris.)

Chao Chi-son and Renvoize, S. A. (1988). A revision of the species described under *Arundinaria* (Gramineae) in Southeast Asia and Africa. *Kew Bulletin* **44**, 349-367.

Clark, L. G. (1989). Systematics of *Chusquea* sect. *Swallenochloa*, sect. *Verticillatae*, sect. *Serpentes*, and sect. *Longifoliae* (Poaceae-Bambusoideae). *Systematic Botany Monographs* **27**, 1 – 127.

Dransfield, S. (1983). The genus *Racemobambos* (Gramineae-Bambusoideae). *Kew Bulletin* **37**, 661-679.

Dransfield, S. (1994). The genus *Hickelia* (Gramineae-Bambusoideae). *Kew Bulletin* **49**, 429-443.

Dransfield, S. (1997). Notes on the genus *Decaryochloa* (Gramineae-Bambusoideae). *Kew Bulletin* **52**, 593-600.

Dransfield, S. (1998). *Valiha* and *Cathariostachys*, two new bamboo genera (Gramineae-Bambusoideae) from Madagascar. *Kew Bulletin* **5**, 375-397.

Dransfield, S. (unpubl.). The inflorescence structure of Malesian bamboos. Second Flora Malesiana Symposium, Yogyakarta, 7-12, September 1992.

Dransfield, S. and Widjaja, E. A. (Editors) (1995). 'Plant Resources of South-East Asia No. 7. Bamboos'. (Backhuys: Leiden.)

Dransfield, S. and Wong, K. M. (1996). *Temburongia*, a new genus of bamboo (Gramineae : Bambusoideae) from Brunei. *Sandakania* **7**, 49-58.

Kunth, C. S. (1829). Révision des Graminées. In 'Voyage aux région équinoctiales du nouveaux continent fait en 1799-1804'. (F. W. H. A. von Humboldt & A. J. A. Bonpland) (Librairie-Gide: Paris.)

McClure, F. M. (1934). The inflorescence in *Schizostachyum* Nees. *Journal of the Washington Academy of Sciences* **24**, 541 – 548.

McClure, F. M. (1966). 'The Bamboos. A fresh perspective'. (Harvard University Press: Cambridge, Massachusetts.)

McClure, F. M. (1973). Genera of bamboos native to the New World. *Smithsonian Contributions to Botany* No. **9**, 1-148.

Munro, W. (1868). A monograph of the *Bambusaceae*, including description of all species. *Transaction of the Linnean Society, London* **26**, 1-157.

Soderstrom, T. R. & Ellis, R. P. (1987). The position of bamboo genera and allies in the system of Grass classificarion. In 'Grass Systematics and Evolution'. (T.R. Soderstrom, K.W. Hilu, C. S. Campbell, and M. E. Barkworth eds).

pp. 225-238. (Smithsonian Institution Press: Washington, D.C.)

Wong, K. M. (1993). Four new genera of bamboos (Gramineae-Bambusoideae) from Malesia. *Kew Bulletin* **48**, 517-532.

Wong, K. M. (1995). 'The Morphology, anatomy, biology and classification of Peninsular Malaysian bamboos'. University of Malaya Botanical Monographs No.1. (United Selangor Press Sdn. Bhd.: Kuala Lumpur.)

Appendix

List of bamboo genera and species, native and introduced in Madagascar

Arundinaria Michaux.
  A. ambositrensis A. Camus
  A. humbertii A. Camus
  A. ibityensis A. Camus
  A. madagascariensis A. Camus
  A. marojejyensis A. Camus
  A. perrieri A. Camus
Bambusa Schreber
  B. multiplex (Lour.) Raeuschel ex J. A.. & J. H. Schultes
  B. vulgaris Schrad. ex Wendl.
Cathariostachys S. Dransf.
  C. capitata (Kunth) S. Dransf.
  C. madagascariensis (A. Camus) S. Dransf.
Cephalostachyum Munro
  C. chapelieri Munro
  C. perrieri A. Camus
  C. viguieri A. Camus
Decaryochloa A. Camus
  D. diadelpha A. Camus
Dendrocalamus Nees
  D. giganteus Munro
Hickelia A. Camus
  H. alaotrensis A. Camus

  H. madagascariensis A. Camus
  H. perrieri (A. Camus) S. Dransf.
Hitchcockella A. Camus
  H. baronii A. Camus
Nastus Juss.
  N. aristatus A. Camus
  N. elongatus A. Camus
  N. emirnensis A. Camus
  N. humbertianus A. Camus
  N. lokohoensis A. Camus
  N madagascariensis A. Camus
  N. manongarivensis A. Camus
  N. perrieri A. Camus
  N. tsaratananensis A. Camus
Perrierbambus A. Camus
  P. madagascariensis A. Camus
  P. tsarasaotrensis A. Camus
Phyllostachys Siebold & Zuccarini
  P. aurea Carr. ex A. & C. Revière
Schizostachyum Nees
  S. bosseri A. Camus
  S. parvifolium Munro
  S. perrieri A. Camus
Valiha S. Dransf.
  V. diffusa S. Dransf.
  V. perrieri (A. Camus) S. Dransf.

Grasses: Systematics and Evolution. (2000). Eds S.W.L. Jacobs and J. Everett. (CSIRO: Melbourne)

PRELIMINARY STUDIES ON TAXONOMY AND BIOSYSTEMATICS OF THE AA GENOME *ORYZA* SPECIES (POACEAE)

Bao-Rong Lu, Ma. Elizabeth B. Naredo, Amita B. Juliano, and Michael T. Jackson

Genetic Resources Center, International Rice Research Institute,
  P.O. Box 933, 1099 Manila, Philippines.

Abstract

The genus *Oryza* L. (*Oryzeae*) includes 22 wild species and two cultivated species—*O. sativa* L. (Asian rice) and *O. glaberrima* Steud. (African rice). Asian rice is an economically important crop; it is the staple food for half of the world's population. Eight taxa, including the cultivated ones, are classified in the *O. sativa* complex on the basis of their morphological similarity and their common AA genome. Wild species in this complex are the most accessible and valuable genetic resources for rice breeding. Because of considerable variation in the morphology and habitat preferences of these rice species, taxonomy in this complex has long been a problem in terms of species delimitation and nomenclature. This paper summarizes recent biosystematic studies of the AA genome *Oryza* species through interspecific hybridization and meiotic analysis of F_1 hybrids and their parental species. It concludes that most of the AA genome species in the *O. sativa* complex, such as *O. rufipogon*, *O. barthii*, *O. glumaepatula*, *O. meridionalis*, and *O. longistaminata*, are distinct species with prominent reproductive barriers between them, although differentiation of the AA genome is limited between different species in the complex. A hypothetical biosystematic relationship of the AA genome *Oryza* species is proposed.

Key words: Poaceae, *Oryza*, wild rice, taxonomy, biosystematics, meiotic pairing, differentiation.

INTRODUCTION

The genus *Oryza* L. is classified in the tribe *Oryzeae*, which contains 12 genera (Table 1), and includes 22 species widely distributed throughout the tropics. Great diversity in morphology and habitats has been observed for the various *Oryza* species. Following the taxonomic treatment of Vaughan (1989), species in this genus are classified in four complexes, with some species not yet placed in any (Table 2). Six basic genomes, AA, BB, CC, EE, FF, and GG, as well as three genomic combinations, BBCC, CCDD, and HHJJ, have been designated respectively in diploid and tetraploid *Oryza* species, although the origin of DD, HH, and JJ is still unknown. This reflects remarkable genetic variation in the genus. Because of considerable morphological variability, the frequent occurrence of intermediate types between some species, and the wide distribution of *Oryza* species, the taxonomy has long been a problem, particularly in the *O. sativa* complex, in terms of species delimitation and nomenclature.

There are two cultivated rice species that originated independently from different wild ancestral species in Asia and Africa. The Asian cultivated rice (*Oryza sativa* L.) is an economically important world cereal crop and is the staple food for about half of the world's population. This species probably originated across a broad area extending over the foothills of the Himalayas and its adjacent mountain regions in Asia, and is now grown worldwide. The African cultivated rice (*O. glaberrima* Steud.) was domesticated in West Africa and is still cultivated in some farming systems today. Rapid expansion of human populations and decreases in agricultural land require much higher rice production than that which is now achieved. Wild relatives of rice,

Table 1. Genera, number of species, distribution, and chromosome number in the tribe Oryzeae (adapted from Vaughan 1989).

Genus	No. of species	Distribution	2n
Chikusiochloa	3	China, Japan (t) [a]	24
Hygroryza	1	Asia (t + T)	24
Leersia	17	worldwide (t + T)	48, 60, 96
Luziola	11	North and South America (t + T)	24
Maltebrunia	5	tropical and southern Africa (T)	Unknown
Oryza	24	pan-tropics (T)	24, 48
Porteresia	1	South Asia (T)	48
Prosphytochloa	1	southern Africa (t)	Unknown
Potamophila	1	Australia (t + T)	24
Rhynchoryza	1	South America (t)	24
Zizania	3	Europe, Asia, N. America (t + T)	30, 34
Zizaniopsis	5	North and South America (t + T)	24

a T = tropical area, t = temperate area.

particularly those having the AA genome, are extremely valuable genetic resources that serve to broaden the genetic background of cultivated rice because the two cultivated rice species also have the same genome. They are therefore the most accessible genetic resources in the rice genepool.

Eight *Oryza* species, including the two cultigens, have the AA genome and are classified in the *O. sativa* complex (Table 2). These are *O. rufipogon*, *O. nivara*, and *O. sativa*, which are native to Asia, although the Asian cultivated rice is now grown worldwide; *O. longistaminata*, *O. barthii*, and *O. glaberrima*, which are endemic to Africa; and *O. meridionalis* and *O. glumaepatula*, which are only found in Australia and Latin America, respectively (Fig. 1). Although geographically and genetically isolated,

species in this complex from different continents have been reported to have considerable morphological similarities and intermediate types are also found between species occurring sympatrically. This has led to considerable taxonomic and nomenclature changes in the wild species in this complex.

NOMENCLATURE AND TAXONOMIC CHANGES IN AA GENOME *ORYZA* SPECIES

The two cultivated rice species have been given a large number of names and undergone considerable nomenclature revision (see Sampath 1962; Harlan and de Wet 1971; Vaughan 1989 for a review), but *Oryza sativa* L. and *O. glaberrima* Steud. have become well accepted and widely adopted names.

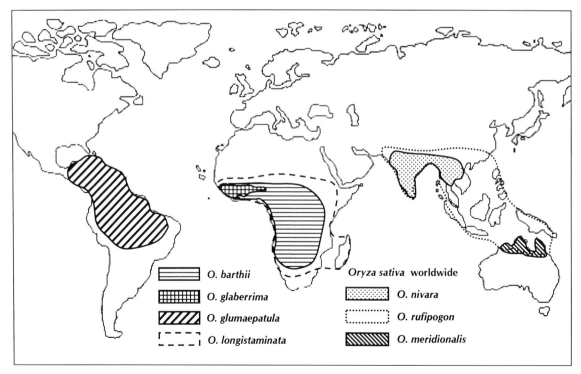

Fig. 1. Distribution of the AA genome *Oryza* species in different geographical regions.

Table 2. Chromosome number, genome content, and distribution of species in the genus *Oryza* (modified from Vaughan 1989)

Complex and species	2n	Genome	Distribution
O. sativa complex			
O. barthii A. Chev.	24	AA	Africa
O. glaberrima Steud.	24	AA	West Africa
O. glumaepatula Steud.	24	AA	South & Central America
O. longistaminata Chev. et Roehr.	24	AA	Africa
O. meridionalis Ng	24	AA	tropical Australia
O. nivara Sharma et Shastry	24	AA	tropical & subtropical Asia
O. rufipogon Griff.	24	AA	tropical & subtropical Asia, & tropical Australia
O. sativa L.	24	AA	worldwide
O. officinalis complex			
O. alta Swallen	48	CCDD	South & Central America
O. australiensis Domin.	24	EE	tropical Australia
O. eichingeri Peter	24, 48	CC [a]	South Asia & East Africa
O. grandiglumis (Doell) Prod.	48	CCDD	South & Central America
O. latifolia Desv.	48	CCDD	South & Central America
O. minuta J. S. Presl. et C. B Presl.	48	BBCC	Philippines & Papua New Guinea
O. officinalis Wall. ex Watt	24, 48	CC [a]	tropical & subtropical Asia, & tropical Australia
O. punctata Kotechy ex Steud.	24, 48	BB & BBCC	Africa
O. rhizomatis Vaughan	24	CC	Sri Lanka
O. meyeriana complex			
O. granulata Nees et Arn. ex Watt	24	GG	South & Southeast Asia
O. meyeriana (Zoll. et Mor. ex Steud.) Baill.	24	GG	Southeast Asia
O. ridleyi complex			
O. longiglumis Jansen	48	HHJJ	Irian Jaya, Indonesia, & Papua New Guinea
O. ridleyi Hook. f.	48	HHJJ	South Asia
Species not assigned to any complex			
O. brachyantha Chev. et Roehr.	24	FF	Africa
O. neocaledonica Morat	24	Unknown	New Caledonia
O. schlechteri Pilger	48	Unknown	Papua New Guinea

a A tetraploid form with 48 chromosomes has also been found in the species

The close relatives of *O. sativa* have been named differently over time. The species name *O. perennis* Moech was widely used for wild species in the *O. sativa* complex (Sampath 1962; Oka and Morishima 1967; Morishima 1969). The perennial *O. rufipogon* was referred to as Asian *O. perennis*, in line with other wild species of the complex from different continents, which were called African, American, and Oceanian *O. perennis*. This influence has been so significant that many scientists still use *O. perennis* in their recent publications (Pental and Barnes 1985; Oka 1988; Morishima *et al.* 1992; Ishii *et al.* 1996). The annual species was called *O. fatua* Koenig ex A. Chev. or *O. sativa* f. *spontanea* Roshev., and was described as *O. nivara* by Sharma and Shastry (1965) to distinguish it from *O. rufipogon*. But the name *O. nivara* has been interpreted differently. Some scientists have accepted *O. nivara* as an independent species (Chatterjee 1951; Chang 1976; Vaughan 1989, 1994) and considered it as the ancestor of the Asian cultivated rice. Others, however, treated *O. nivara* as a synonym of *O. sativa* (Duistermaat 1987) or the annual form of *O. rufipogon* (Asian *O. perennis*) (Morishima 1969; Oka 1988; Morishima *et al.* 1992). Nevertheless, *O. rufipogon* and *O. nivara* show a continuous array of intergrades

(Morishima *et al.* 1961) and intermediate perennial-annual populations have also been found in nature (Morishima 1986). Our morphological analysis also demonstrated a great degree of overlapping between these two species (Juliano *et al.* 1998).

There has also been considerable taxonomic and nomenclature confusion with the African wild rice species, although morphological differences between these species are significant. The annual *O. barthii*, which has been widely accepted as the ancestor of the African cultivated rice *O. glaberrima* (Chang 1976), was referred to as *O. breviligulata* Chev. et Roehr. (Sampath 1962; Oka and Morishima 1967; Second 1982; Ishii *et al.* 1996) or as African *O. perennis* by some authors (Pental and Barnes 1985). The perennial species *O. longistaminata* was called *O. barthii* (Sampath 1962; Clayton 1968) or *O. perennis* subsp. *barthii* (Oka and Morishima 1967; Clayton 1968). It was also common to name this perennial species as African *O. perennis* (Morishima 1969; Oka 1988; Morishima *et al.* 1992). In the Philippines we use *O. barthii* for the annual species and *O. longistaminata* for the perennial species. Our unpublished data from morphological studies show a clear separation of the two species sampled from different localities.

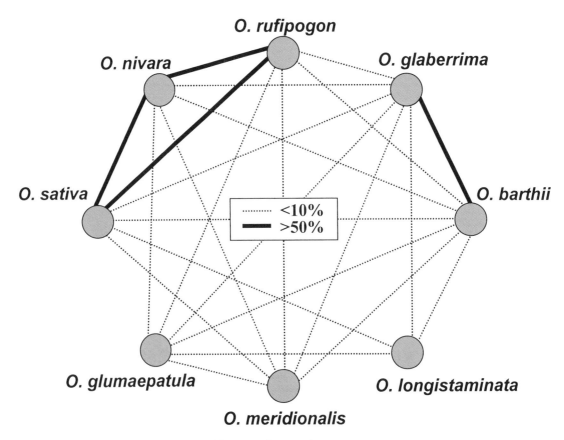

Fig. 2. Spikelet fertility (%) of F₁ hybrids between various AA genome *Oryza* species.

Oka and Morishima (1967) and Morishima (1969, 1986) suggested certain strains of wild rice from New Guinea and Australia as the Oceanian form of *O. perennis*, which Oka and Morishima (1967) indicated contained perennial and annual types. Through a comprehensive study, Ng *et al.* (1981) recognized the Australian annual strain as an independent species, *O. meridionalis*, on the basis of its unique morphological characteristics, distinct from the Asian AA genome wild *Oryza* species. In some Australian herbaria and publications, however, *O. meridionalis* was included in *O. rufipogon*. This taxonomic confusion is obviously caused by the fact that both *O. meridionalis* and *O. rufipogon* are found in Australia, and insufficient morphological characters have been used by Australian scientists to separate the two species (B. K. Simon pers. comm.).

The Latin American species *O. glumaepatula* has been clarified as the American form of *O. perennis* (Oka and Morishima 1967; Morishima 1969; Pental and Barnes 1985) or *O. cubensis* Ekman (in Vaughan 1989; Morishima *et al.*1992). Because the American *O. glumaepatula* and the Asian *O. rufipogon* have great morphological similarity and are both perennial species, Tateoka (1962) maintained no significant morphological distinction between them, and gave these two species conspecific status. In his most recent taxonomic treatment, Vaughan (1994) classified *O. glumaepatula* in *O. rufipogon*. However, through our study on morphological analysis of more than 60 samples of *O. glumaepatula* from South and Central America, and *O. rufipogon* and *O. nivara* from Asia, we have confirmed that typical *O. glumaepatula*

accessions do form a clear cluster distinct from the Asian wild rice species. We have concluded that *O. glumaepatula* warrants an independent taxonomic status (Juliano *et al.* 1998). This conclusion has gained support from our studies on interspecific hybridization and meiotic analysis (Naredo *et al.* 1998; Lu *et al.* 1998).

INTERSPECIFIC HYBRIDIZATION AND FERTILITY IN HYBRIDS

The AA genome wild *Oryza* species not only have remarkable morphological similarities but also show some introgression in nature between sympatric populations, and with cultivated species. This spontaneous intercrossing has caused the formation of many intermediate types between the cultivated and wild species, as well as between the annual and perennial forms based on our own field observations and those of others (Morishima 1969; Oka 1988; Morishima *et al.* 1992; Majumder *et al.* 1997). To estimate crossability and the degree of reproductive isolation among the different AA genome rice species, we have undertaken an extensive interspecific hybridization program, involving more than three populations of each AA genome rice species from different origins. Table 3 summarizes data on seed set for interspecific hybridization involving pollination of more than 70,000 spikelets. It is evident that crossability between the AA genome species was generally low, with a range of seed set from 4.1% to 52.0% between interspecific crosses and 11.1% to 31.5% between intraspecific crosses. Seed set of most combinations did not show significant differences between the reciprocal crosses. This indicates different degrees of pre-fertilization barriers

Table 3. Seed set (%) from intraspecific and reciprocal interspecific crosses among AA genome rice species (modified from Naredo *et al.* 1997, 1998, and our unpublished data).

Combination	O. barthii	O. glaberrima	O. glumaepatula	O. meridionalis	O. nivara	O. rufipogon
O. barthii	31.0					
O. glaberrima	27.4					
O. glumaepatula	16.1	28.4	31.5			
O. longistaminata						
O. meridionalis	16.8	52.0	11.4	11.1		
O. nivara	21.7	24.9	29.3	6.1	23.5	
O. rufipogon	22.8		19.2	7.7	21.7	13.4
O. sativa	28.2			4.1		

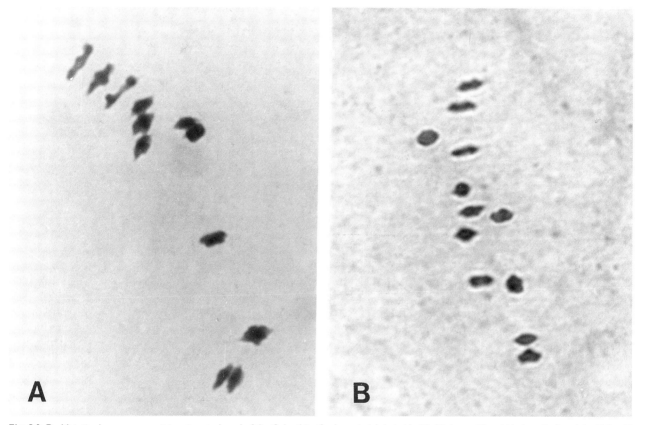

Fig. 3A-B. Meiotic chromosome pairing at metaphase I of the *O. barthii* x *O. glumaepatula* hybrid with 10 ring and 2 rod bivalents in A and the *O. barthii* x *O. nivara* hybrid with 12 ring bivalents in B.

between various AA genome rice species, particularly between those from different continents, and between certain populations of the same species. Figure 2 summarizes data for spikelet fertility of the F_1 interspecific hybrids which was generally less than 10%, except for the F_1 hybrids among *O. rufipogon*, *O. nivara*, and *O. sativa*, and those between *O. glaberrima* and *O. barthii*, for which more than 50% spikelet fertility was observed in each combination (Naredo *et al.* 1997, 1998; Naredo *et al.*, unpublished data). In contrast, spikelet fertility of the parental species included varied between 60%-80%. This indicates different degrees of reproductive isolation between different AA genome species, and comparatively strong isolation between those from different continents, supporting previous studies on hybrid fertilities of *O. perennis* from different geographical origins (Morishima 1969; Oka 1988).

GENOME RELATIONSHIP THROUGH MEIOTIC CHROMOSOME PAIRING

It is generally agreed that species in the *O. sativa* complex have essentially the same AA genome (Richharia 1960; Chu *et al.* 1969; Morishima *et al.* 1992). Some authors have used different superscripts to differentiate the AA genomes in some species, such as *O. longistaminata* (A^lA^l), *O. glumaepatula* ($A^{gp}A^{gp}$), and *O. meridionalis* (A^mA^m) (Chang 1985; Vaughan 1989). However, this differentiation cannot be supported from cytogenetic studies. To assess genomic relationships, particularly between the wild AA genome rice species, we analyzed all the available F_1 hybrids generated from the interspecific crosses. Data from chromosome configurations in meioses of different interspecific hybrids showed almost full pairing between the parental genomes (Figure 3A-B), except for some hybrids with *O. meridionalis*, in which a slightly lower value of

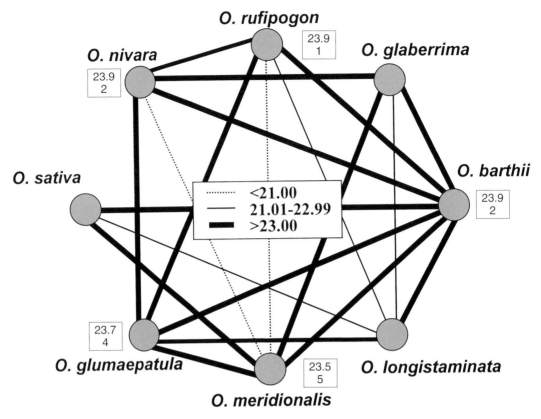

Fig. 4. Meiotic pairing of the F₁ hybrids between various AA genome *Oryza* species. Numbers in the middle box indicate chiasma frequency per pollen mother cell (PMC), and numbers in the small boxes indicate chiasma frequency per PMC of the wild parental species.

chromosome pairing was observed. Figure 4 summarizes the average chiasma frequencies per pollen mother cell (PMC) at meiotic metaphase I in the available F₁ hybrids (including the reciprocal crosses) between different AA genome rice species, and in the wild parental species (Lu *et al.* 1997, 1998; Lu *et al.*, unpublished data). The chiasma frequencies are generally higher than 20 per PMC in the F₁ hybrids (many F₁ hybrids had more than 23 chiasmata per PMC). The chromosome pairing level is almost as high as in their parental accessions, although it is slightly lower in a few combinations. This result suggests limited differentiation of the AA genome in the different species of the *O. sativa* complex, regardless of the geographical isolation of these species.

CONCLUSIONS

The wild AA genome species in the *O. sativa* complex, such as *O. rufipogon*, *O. barthii*, *O. glumaepatula*, *O. meridionalis*, and *O. longistaminata*, are distinct species with prominent degrees of reproductive isolation between them. Our interspecific hybridization studies (Naredo *et al.* 1997, 1998; Naredo *et al.*, unpublished data) strongly support the conclusion that these wild rice species warrant specific taxonomic status.

The Asian cultivated rice *O. sativa* and its putative ancestral taxa, *O. rufipogon* and *O. nivara*, have very limited reproductive isolation, although the 'typical' samples of *O. sativa*, *O. rufipogon*, and *O. nivara* are morphologically distinct. But intermediate types from introgression of the three taxa are found in nature and continuous morphological variation can be observed between

them. A similar situation is found between the African cultivated rice *O. glaberrima* and its ancestor *O. barthii*.

The differentiation of the AA genome in rice species of the *O. sativa* complex is extremely limited judging from the chromosome pairing ability of the parental genomes in interspecific hybrids. More detailed studies at the molecular level should be conducted to determine the degree of homology shared by the AA genome in different species in the *O. sativa* complex.

Based on the results from our morphological studies, interspecific hybridization, and meiotic analysis of the interspecific hybrids, in combination with reports from other molecular studies (Doi *et al.* 1996; Ishii *et al.* 1996; Martin *et al.* 1997), a tentative biosystematic relationship of the AA genome species in the *O. sativa* complex is illustrated in Figure 5. The Asian *O. sativa*, *O. nivara*, and *O. rufipogon* share close biosystematic relationships. The African *O. glaberrima* and *O. barthii* have the highest affinity, and the Latin American *O. glumaepatula* joins this African group. *O. longistaminata* and the Australian *O. meridionalis* have relatively distant relationships with the other AA genome rice species.

REFERENCES

Brar, S. D. and Khush, G. S. 1997. Alien introgression in rice. *Plant Molecular Biology* **35**, 35-47.

Chang, T. T. (1976). The origin, evolution, cultivation, dissemination, and diversification of Asian and African rices. *Euphytica* **25**, 435-411.

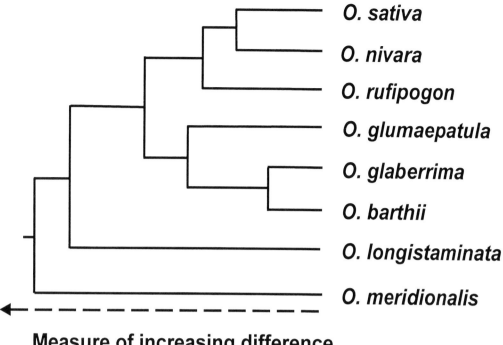

O. sativa

O. nivara

O. rufipogon

O. glumaepatula

O. glaberrima

O. barthii

O. longistaminata

O. meridionalis

Measure of increasing difference

Fig. 5. A tentative biosystematic relationship of the AA genome *Oryza* species.

Chang, T. T. (1985). Crop history and genetic conservation: Rice – a case study. *Iowa State Journal of Research* **59**, 425-455.

Chatterjee, D. (1951). Note on the origin and distribution of wild and cultivated rice. *Indian Journal of Agricultural Sciences* **18**, 185-192.

Chu, Y. E., Morishima, H., and Oka, H. I. (1969). Reproductive barriers distributed in cultivated rice species and their wild relatives. *Japanese Journal of Genetics* **44**, 225-229.

Clayton, W. D. (1968). Studies in Gramineae. XVII. West African wild rice. *Kew Bulletin* **21**, 487-488.

Doi, K. Yoshimura, A., Nakano, M., and Iwata, N. (1996). Classification of A genome species in the genus *Oryza* using nuclear DNA markers. *International Rice Research Notes* **21**, 8-10.

Duistermaat, H. (1987). A revision of *Oryza* (Gramineae) in Malaysia and Australia. *Blumea* **32**, 157-193.

Harlan, J. R., and de Wet, J. M. J. (1971). Toward a rational classification of cultivated plants. *Taxon* **20**, 509-517.

Ishii, T., Nakano, T., Maeda, H., and Kamijima, H. (1996). Phylogenetic relationships in A-genome species of rice as revealed by RAPD analysis. *Genes and Genetic Systems* **71**, 195-201.

Juliano, A. B., Naredo, M. E. B., and Jackson, M. T. (1998). Taxonomic status of *Oryza glumaepatula* Steud. I. Comparative morphological studies of New World diploids and Asian AA genome species. *Genetic Resources and Crop Evolution* **45**, 197-203.

Lu, B. R., Naredo, M. E. B., Juliano, A. B., and Jackson, M.T. (1997). Hybridization of AA genome rice species from Asia and Australia. II. Meiotic analysis of *Oryza meridionalis* and its hybrids. *Genetic Resources and Crop Evolution* **44**, 25-31.

Lu, B. R., Naredo, M. E. B., Juliano, A.B., and Jackson, M. T. (1998). Taxonomic status of *Oryza glumaepatula* Steud. III. Assessment of genomic affinity among AA genome species from the New World, Asia, and Australia. *Genetic Resources and Crop Evolution* **45**, 205-214.

Martin, C. A., Juliano, A. B., Newbury, H. J., Lu, B. R., Jackson, M.T., and Ford-Lloyd, B. V. (1997). The use of RAPD markers to facilitate the identification of *Oryza* species within a germplasm collection. *Genetic Resources and Crop Evolution* **44**, 175-183.

Majumder, N. D., Ram, T., and Sharma, A. C. (1997). Cytological and morphological variation in hybrid swarms and introgressed population of interspecific hybrids (*Oryza rufipogon* Griff. × *Oryza sativa* L.) and its impact on evolution of intermediate types. *Euphytica* **94**, 295-302.

Morishima, H. (1969). Phenetic similarity and phylogenetic relationships among strains of *Oryza perennis*, estimated by methods of numerical taxonomy. *Evolution* **23**, 429-443.

Morishima, H. (1986). Wild progenitors of cultivated rice and their population dynamics. In 'Rice Genetics.' (Ed. IRRI) pp. 3-14. (IRRI: Los Baños, Philippines.)

Morishima, H., Oka, H. I., and Chang W. T. (1961). Direction of differentiation in populations of wild rice, *Oryza perennis* and *O. sativa* f. *spontanea*. *Evolution* **15**, 326-339.

Morishima, H., Sano, Y., and Oka, H. I. (1992). Evolutionary studies in cultivated rice and its wild relatives. *Oxford Surveys in Evolutionary Biology* **8**, 135-184.

Naredo M. E. B., Juliano, A. B., Lu, B. R., and Jackson, M. T. (1997). Hybridization of AA genome rice species from Asia and Australia. I. Crosses and development of hybrids. *Genetic Resources and Crop Evolution* **44**, 17-23.

Naredo, M. E. B., Juliano, A. B., Lu, B. R., and Jackson, M. T. (1998). Taxonomic status of *Oryza glumaepatula* Steud. II. Hybridization between New World diploids and AA genome species from Asia and Australia. *Genetic Resources and Crop Evolution* **45**, 205-214.

Ng, N. Q., Chang, T. T., Williams, J. T., and Hawkes, J. G. (1981). Morphological studies of Asian rice and its related wild species and the recognition of a new Australian taxon. *Botanic Journal of Linnean Society* **16**, 303-313.

Oka, H. I. (1988). 'Origin of cultivated rice.' (Japan Scientific Societies Press: Tokyo, Japan.)

Oka, H. I., and Morishima, H. (1967). Variations in the breeding systems of a wild rice, *Oryza perennis*. *Evolution* **21**, 249-258.

Pental, D., and Barnes, S. R. (1985). Interrelationship of cultivated rice *Oryza sativa* and *O. glaberrima* with wild *O. perennis* complex. *Theoretical and Applied Genetics* **70**, 185-191.

Richharia, R. H. (1960). Origins of cultivated rices. *Indian Journal of Genetics and Plant Breeding* **20**, 1-14.

Sampath, S. (1962). The genus *Oryza*: Its taxonomy and species interrelationships. *Oryza* **1**, 1-29.

Second, G. (1982). Origin of the genetic diversity of cultivated rice (*Oryza* spp.): study of the polymorphism scored at 40 isozyme loci. *Japanese Journal of Genetics* **57**, 25-57.

Sharma, S. D. and Shastry, S. V. S. (1965). Taxonomic studies in genus *Oryza* L. III. *O. rufipogon* Griff. *sensu stricto* and *O. nivara* Sharma and Shastry *nom. nov. Indian Journal of Plant Breeding* **25**, 157-167.

Tateoka, T. (1962). Taxonomic studies of *Oryza*. II. Several species complexes. *Botanic Magazine Tokyo* **75**, 455-461.

Vaughan, D. A. (1989). The genus *Oryza* L.: current status of taxonomy. IRRI Research Paper Series 138, IRRI, Los Baños: Philippines.

Vaughan, D. A., (1994). 'The wild relatives of rice: A genetic resources handbook.' (IRRI: Los Baños: Philippines.)

POOIDS

Briza maxima, a Mediterranean species now widely naturalised around the world.

Grasses: Systematics and Evolution. (2000). Eds S.W.L. Jacobs and J. Everett. (CSIRO: Melbourne)

PHYLOGENETIC STRUCTURE IN POACEAE SUBFAMILY POOIDEAE AS INFERRED FROM MOLECULAR AND MORPHOLOGICAL CHARACTERS: MISCLASSIFICATION VERSUS RETICULATION

Robert J. Soreng[AB] *and Jerrold I. Davis*[A]

[A]L. H. Bailey Hortorium, Cornell University, Ithaca, NY, 14853.
[B]Current address, Natural History Museum, Smithsonian Institution, Washington, D.C. 20560-0166.

Abstract

Chloroplast DNA (cpDNA) restriction site (601 chars.) and morphological character (67 chars.) data were used to evaluate the phylogeny of *Pooideae*. Analysis of cpDNA variation in 79 genera and 101 species resolved the following general cladistic structure: *Brachyelytreae* ((*Nardeae Lygeeae*) (*Anisopogon* ((*Ampelodesmeae Stipeae*) ((*Brachypodieae*) (*Meliceae*) (*Diarrheneae*) ((*Bromeae Triticeae*) (*Aveneae Poeae*))))))). The addition of structural data resulted in few changes: *Brachyelytreae* ((*Nardeae Lygeeae*) ((*Anisopogon* (*Ampelodesmeae Stipeae*)) (*Diarrheneae* ((*Brachypodieae Meliceae*) ((*Bromeae Triticeae*) (*Aveneae Poeae*))))))). We suggest that conflicting placements of 14 genera of the *Poeae/Aveneae* clade, apparent between the cpDNA tree and modern classifications, sometimes signifies parallelism and convergence in structural data, and other times signifies the effects of past reticulation events between these two lineages. The distinction between tribes Poeae and *Aveneae* should be abandoned in favor of a series of small and more homogeneous subtribes within a broadly circumscribed tribe Poeae.

Key words: Chloroplast, DNA, evolution, hybridization, phylogeny, Poaceae, Pooideae, reticulation, taxonomy.

INTRODUCTION

The circumscription of subfamily *Pooideae* Benth. has changed greatly since the 19th century. Changes proceeded mainly by the recognition of new genera, and by the removal of sets of genera now considered members of other subfamilies, especially *Chloridoideae* Beilschm., *Arundinoideae* Burmeist., and *Bambusoideae* Luerss. Essentially modern treatments of the subfamily began with Avdulov (1931), Prat (1932, 1936, 1960), Pilger (1954), Stebbins (1956), Tateoka (1957), and Hubbard (1959), and culminated in three major treatments, those of Clayton and Renvoize (1986), Tzvelev (1976, 1989), and Watson and Dallwitz (1992; *Pooideae* classification based in large part on MacFarlane and Watson 1982). These classifications are based on gross morphology and on expanded knowledge of anatomy, cytology, physiology, and more recently the use of computers to analyze structural character data.

Henceforth, to avoid protracted discussions, we will confine our discussion to the three most recent of these classifications. Among these, Tzvelev's classification (1976, somewhat modified and expanded in 1989) is unique in recognizing only two subfamilies, *Bambusoideae* and *Pooideae*, the latter including tribes relegated in the systems of Clayton and Renvoize and Watson and Dallwitz to *Arundinoideae*, *Centothecoideae* Soderstr., *Chloridoideae*, *Ehrhartoideae* Link (syn. *Oryzoideae* Beilschm.), and *Panicoideae* Link. Tzvelev's classification recognizes a series of tribes that he considered to be 'festucoid' (1976, Fig. 8), and all of these tribes (except *Molinieae* Jirasek) have since been confirmed as members of a single monophyletic lineage, *Pooideae*, by cladistic studies of DNA and structural characters (DNA: cpDNA — Clark *et al.* 1995; Nadot *et al.* 1994; Catalan *et al.* 1997; Soreng and Davis 1998; nuclear ribosomal DNA (rDNA) — Hsiao *et al.* 1999. Structural characters: Kellogg and Campbell 1987, Kellogg and Watson 1993; Soreng and Davis 1998.

Simultaneous Analysis of DNA and structural characters: Soreng and Davis 1998.). In this paper we use the abbreviation 'cpDNA' to refer to analyses that reflect sequence or restriction site variation in the chloroplast genome.

These three classifications share a set of tribes in common which have come to be known as 'core' Pooideae: *Aveneae* Dumort. (*sensu* Clayton and Renvoize 1986; including *Agrostideae* Dumort.), *Bromeae* Dumort., *Triticeae* Dumort., *Meliceae* Link ex Endl., and *Poeae* R.Br. Additional minor tribes sometimes segregated from the core tribes but always placed within *Pooideae* are: *Brachypodieae* (Hack.) Hayek, *Brylkinieae* Tateoka, *Hainardieae* Greuter, *Phalarideae* Kunth, *Phleeae* Dumort., *Scolochloeae* Tzvelev, and *Seslerieae* W.D.J. Koch. All of the preceding taxa and no others were included in *Pooideae* by Watson and Dallwitz. Clayton and Renvoize (1986) included these plus: *Lygeeae* J. Presl, *Nardeae* W.D.J Koch, and *Stipeae* Dumort. Tzvelev's festucoids included all of these tribes plus *Ampelodesmeae* (Conert) Tutin (= *Poeae* in Clayton and Renvoize), *Brachyelytreae* Ohwi, *Diarrheneae* (Ohwi) C.S. Campbell, and possibly *Phaenospermatideae* Renvoize and Clayton (this was ambiguously placed between *Nardeae* (festucoid group) and *Oryzeae* Dumort. (oryzoid group). Cladistic analyses have placed all of these tribes and no others in a monophyletic lineage we call *Pooideae* (Kellogg and Campbell 1987; Nadot *et al.* 1994, Clark *et al.* 1995, Catalan *et al.* 1997, Soreng and Davis 1998, Hsiao *et al.* 1999). So far as we have been able to assess the phylogeny of the grass family, the addition of any other tribes or subtribes recognized in these three classifications to *Pooideae* would make this subfamily non-monophyletic.

The order of divergence of tribes from the most recent common ancestor of *Pooideae*, indicated by cpDNA parsimony analyses with the most complete sampling of tribes (Clark *et al.* 1995; Catalan *et al.* 1997; Soreng and Davis 1998) is (in tree notation, adding [ ] around paraphyletic lineages): *Brachyelytreae* ((*Nardeae Lygeeae*) (*Phaenospermatideae Anisopogon* (*Ampelodesmeae Stipeae*) *Diarrheneae Meliceae Brachypodieae* ((*Bromeae Triticeae*) (*Aveneae Poeae*)))). The analysis by Nadot *et al.* (1994) differs mainly in resolving *Lygeum* as the sister group of *Poeae* and *Aveneae*. As to minor tribes, genera of *Phalarideae* and *Phleeae* have been shown to align within *Aveneae* (Catalan *et al.* 1997; Soreng and Davis 1998), and of *Seslerieae* within *Poeae* (Soreng *et al.* 1990; Catalan *et al.* 1997), but other small tribes remain to be sampled. A comparable parsimony analysis of nuclear rDNA ITS sequences (Hsiao *et al.* 1999) provided a somewhat different result: *Diarrheneae* (((*Brachyelytreae Nardeae*) ([*Stipeae*] (*Anisopogon Ampelodesmos*))) ((*Meliceae* (*Brachypodieae* (*Poeae* (*Bromeae Triticeae*))))))).

Here we present an expanded analysis of *Pooideae*, based on cpDNA characters derived from mapped restriction sites, and a newly compiled set of structural characters, representing variation in gross morphology, anatomy, cytology, and physiology. Two questions are addressed. First, what is the major phylogenetic structure of *Pooideae*? To answer this question we have studied a set of species from 79 genera of *Pooideae*, representing all but two monotypic tribes (*Brylkinieae* and *Phaenospermatideae*). Second, is there evidence that that hybridization has played a role in the evolution of *Pooideae* at the tribal level? To answer the second question we compare relationships detected by phylogenetic analysis of cpDNA and structural characters with those suggested by the classifications of Clayton and Renvoize, Tzvelev, and Watson and Dallwitz. These classifications were not presented, however, as strictly phylogenetic.

An examination of alternative placements of particular taxa by characters of the plastid genome and by those of the nuclear genome (most morphological characters, presumably) may help to identify taxa that have hybrid origins. Chloroplast DNA-derived phylogenies have come to be widely accepted by the systematics community. However, enigmatic results have occurred regarding the placement of some genera in cpDNA analyses. The first sign of this in *Pooideae* was the seemingly anomalous placement of *Briza, Chascolytrum, Poidium* (syn. *Microbriza*) and *Torreyochloa* (all of which have consistently been placed in *Poeae* or its equivalent in classifications), among genera of the *Aveneae* by cpDNA data (Soreng *et al.* 1990). Since that publication, the probability of intergeneric hybridization has been explored for lineages within *Triticeae* (Kellogg *et al.* 1996; Mason-Gamer and Kellogg 1997), and here we report on additional cases which might be explained by intertribal hybridization within *Pooideae*. We argue that for some taxa placed in unanticipated phylogenetic positions by cpDNA analysis, evidence for hybrid origins are weak. However, for a few taxa, or lineages, the evidence does support hypotheses of intertribal hybrid origins.

METHODS

This study employs an exemplar species approach, with the same species sampled for both cpDNA and structural data. The main taxon sample for the cpDNA and combined data set includes 79 genera and 101 species of Pooideae (Table 1). We also obtained cpDNA data for 20 extra samples representing additional Pooideae species within genera, five more genera, and variation within three species (Table 1). The extended sample was used to confirm unusual tribal placements of genera in the main sample, and to help place the extra genera. For many of the extra taxa we did not obtain restriction site data from all restriction enzyme/taxon combinations. These extra taxa were scored for structural characters, however, and are included in some extended analyses of that portion of the data to provide greater sampling depth in that less decisive data set, and to check the placement of congeners in the cpDNA and simultaneous analyses. Complete records of the data are available on request. The cpDNA restriction site data were generated using the same restriction enzymes and methods as reported by Soreng and Davis (1998).

The data sets include 601 informative cpDNA restriction site characters and 67 structural characters.

Cladistic analysis was performed on the DNA and structural character data, separately and together. We used *Dada* (Nixon 1997) for data preparation, and *Nona* (Goloboff 1993) to find the most parsimonious trees. For each of the cladistic analyses in *Nona* we conducted at least 1000 replicate tree initiations using random taxon addition sequences and TBR swapping through 20 trees in each replicate (using the commands ho/20; mult*1000), and saved all shortest trees obtained. These trees were then swapped to completion, or to an upper limit of 10,000 trees, and strict consensus trees were produced. Results of the

Table 1. Genera mentioned in the text and lower ranked taxa sampled for cpDNA and morphology. All vouchers are at BH unless otherwise indicated. Abbreviations are: USDA = USDA Plant Introduction Station, CU = Cornell University Plantations, BHC = L.H. Bailey Hortorium Conservatory, LIT = Literature, USNHG = U.S. Natural History Greenhouse, USNZ = U.S. National Zoo. Asterisks proceeding species names indicate taxa in the extended analysis but, not included in the main cpDNA and total evidence analyses due to missing cpDNA data.

Species	Collection No.
Genera mentioned in the text: x*Achnella* Barkworth, x*Agropogon* P. Fourn., *Agrostis* L., *Aira* L., *Alopecurus* L., x*Ammocalamagrostis* P. Fourn., *Ammophila* Host, *Ampelodesmos* Link, *Amphibromus* Nees, *Anisantha* K. Koch, *Anisopogon* R. Br., *Anthoxanthum* L., *Arctagrostis* Griseb., x*Arctodupontia* Tzvelev, *Arrhenatherum* P. Beauv., *Avena* L., *Avenula* (Dumort.) Dumort., *Beckmannia* Host, *Bellardiochloa* Chiov., *Boissiera* Hochst. ex Steud., *Brachyelytrum* P. Beauv., *Brachypodium* P. Beauv., *Briza* L., x*Bromofestuca* Prodän, *Bromopsis* (Dumort.) Fourr., *Bromus* L., *Bromus*, *Calamagrostis* Adans., *Castellia* Tineo, *Catabrosa* P. Beauv., *Catapodium* Link, *Ceratochloa* P. Beauv., *Chaetopogon* Janch., *Chascolytrum* Desv., *Cinna* L., *Cutandia* Willk., *Cynosurus* L., *Dactylis* L., *Deschampsia* P. Beauv., *Desmazeria* Dumort., *Diarrhena* P. Beauv., x*Dupoa* J. Cay. & Darbysh., *Dupontia* R. Br., *Echinaria* Desf., *Elymus* L., *Festuca* L., x*Festulolium* Asch. & Graebn., x*Festulpia* Stace & Cotton, *Gastridium* P. Beauv., *Gaudinia* P. Beauv., *Glyceria* R. Br., *Hainardia* Greuter, *Helictotrichon* Besser ex Schult. & Schult.f., *Hierochloe* R. Br., *Holcus* L., *Koeleria* Pers., *Lagurus* L., *Lamarckia* Moench, *Leucopoa* Griseb., *Lolium* L., *Lygeum* Loefl. ex L., *Melica* L., *Mibora* Adans., *Microbriza* Parodi ex Nicora & Rúgolo, *Milium* L., *Molineriella* Rouy, *Muhlenbergia* Schreb., *Nardus* L., *Nassella* E. Desv., *Parapholis* C.E. Hubb., *Peridictyon* Seberg, Fred. & Baden, *Phaenosperma* Munro ex Benth., *Phalaris* L., *Phippsia* (Trin.) R. Br., *Phleum* L., *Piptatherum* P. Beauv., *Pleuropogon* R. Br., *Poa* L., *Poidium* Nees, *Polypogon* Desf., *Psilurus* Trin., *Puccinellia* Parl., x*Pucciphippsia* Tzvelev, *Rostraria* Trin., x*Schedolium* Holub, *Schedonorus* P. Beauv., *Schizachne* Hack., *Sclerochloa* P. Beauv., *Scolochloa* Link, *Scribneria* Hack., *Sesleria* Scop., *Sphenopholis* Scribn., *Sphenopus* Trin., *Stipa* L., *Torreyochloa* G.L. Church, *Triplachne* Link, *Trisetobromus* Nevski, x*Trisetokoeleria* Tzvelev, *Trisetum* Pers., *Tristachya* Nees, *Triticum* L., *Vulpia* C.C. Gmel., *Zingeria* Smirnov	
Agrostis gigantea Roth	Soreng-3429
Aira caryophyllea L.	Soreng-3810
Aira cupaniana Guss.	KEW-0027827
Alopecurus alpinus Sm.	Soreng-3514
Ammophila arenaria (L.) Link	Soreng-3389
Ampelodesmos mauritanica (Poir.) T. Durand & Schinz	Royl & Schiers s.n., **B**
Amphibromus scabrivalvis (Trin.) Swallen	K. Clay s.n.
Anisopogon avenaceus R. Br.	P. Linder 5590, **BOL**
Anthoxanthum odoratum L.	in "USDA-234745"
Arctagrostis latifolia (R. Br.) Griseb.	USDA-372661
Arrhenatherum elatius (L.) P. Beauv.	J. Davis & Soreng s.n.
Avena barbata Pott ex Link	Soreng-3625b
Avena sativa L. 'ASTRO'	CULT, CU
Avenula gervaisii Holub.	Soreng-3693
Avenula pubescens (Huds.) Dumort.	KEW-0065160
Beckmannia syzigachne (Steud.) Fernald	Soreng-3513
Bellardiochloa variegata (Lam.) Kerguelen	USDA-353455
Boissiera squarrosa (Banks & Soland.) Nevski	Collenette 4398, **E**
Brachyelytrum erectum (Schreb.) P. Beauv.	Soreng-3427a
Brachypodium distachyon (L.) P. Beauv.	USDA-422452
Brachypodium pinnatum (L.) P. Beauv.	USDA-440170
Brachypodium sylvaticum (Huds.) P. Beauv.	USDA-251102
Briza maxima L.	USDA-257681
Briza minor L.	USDA-378653
Bromus inermis Leyss.	USDA-314071 (for *Kpn* I, *Pvu* II, *Sal* I, only), others Soreng-3428
Calamagrostis canadensis (Michx.) P. Beauv.	USDA-371717
Calamagrostis purpurascens R. Br.	Soreng-3531
Castellia tuberculosa (Moris) Bor	Soreng-3632
Catabrosa aquatica (L.) P. Beauv.	Soreng-3861 (Turkey)
Catabrosa aquatica (L.) P. Beauv.	J. Davis-9 (Wyoming)
Catapodium marinum (L.) C.E. Hubb.	Soreng-3674
Catapodium rigidum (L.) C.E. Hubb. ex Dony	Soreng-3680
Chaetopogon fasciculatus (Link) Hayek	Soreng-3644
Chascolytrum erectum (Lam.) E. Desv.	USDA-282880
Cinna latifolia (Trevir ex Gopp.) Griseb.	Soreng-3383b

Table 1. Genera mentioned in the text and lower ranked taxa sampled for cpDNA and morphology. All vouchers are at BH unless otherwise indicated. Abbreviations are: USDA = USDA Plant Introduction Station, CU = Cornell University Plantations, BHC = L.H. Bailey Hortorium Conservatory, LIT = Literature, USNHG = U.S. Natural History Greenhouse, USNZ = U.S. National Zoo. Asterisks proceeding species names indicate taxa in the extended analysis but, not included in the main cpDNA and total evidence analyses due to missing cpDNA data. *(Continued)*

Species	Collection No.
Cutandia memphytica (Spreng.) K. Richt.	Boulos & Cope 17676, **E**
Cynosurus cristatus L.	KEW-005430
Cynosurus echinatus L.	Raus 14558, **B**
*Dactylis glomerata L. USDA-311033,	Soreng-3430
Dactylis hispanica Roth	Soreng-3692
Deschampsia caespitosa (L.) P. Beauv.	USDA-311043
Desmazeria sicula (Jacq.) Dumort.	KEW-0077332
Diarrhena obovata (Gleason) Brandenb.	Soreng-3426 (from Tiedye 5186, **DAO)**
Dupontia fisheri R. Br.	Darbyshire s.n., **DAO**
*Echinaria capitata (L.) Desf.	sent by H. Scholz s.n., **B**
Elymus trachycaulus (Link) Gould ex Shinners	Soreng-4291 in "M. Barkworth 88-465"
Festuca brevipila R. Tracey CULT (as F. "ovina"),	Soreng-3928
*Festuca spectabilis Jan.	USDA-383658
Gastridium ventricosum (Gouan) Schinz & Thell.	KEW-005430
Gaudinia fragilis (L.) P. Beauv.	USDA-2142496
Glyceria declinata Breb.	KEW-007674
Glyceria grandis S. Watson	J. Davis & Soreng s.n.
Glyceria striata (Lam.) Hitchc.	J. Davis & Soreng s.n.
Hainardia cylindrica (Willd.) Greuter	Royl & Schiers s.n., **B**
Helictotrichon convolutum (C. Presl) Henrard	Madrid Botanical Garden
Helictotrichon sempervirens (Vill.) Pilg.	CULT, Soreng-3927
Hierochloe alpina (Sw. ex Willd.) Roem. & Schult.	Soreng-3419-1
Holcus annuus Salzm. ex C.A. Mey.	Soreng-3609
Holcus lanatus L.	CULT BLUMELL
Koeleria macrantha (Ledeb.) Schult.	USDA-477978 (as "K. cristata")
Lagurus ovatus L.	KEW-0035662
Lamarckia aurea (L.) Moench	sent by H. Scholz s.n., **B**
Leucopoa kingii (S. Watson) W.A. Weber	Soreng-3515
Leucopoa sclerophylla (Boiss. & Hohen.) Krecz. & Bobrov	USDA-275336
Lolium perenne L. USDA-418710,	USDA-253719
*Lolium rigidum Gaudin	Soreng-3696
Lygeum sparteum L.	Soreng-3698
Melica altissima L.	USDA-325418
Melica bulbosa Geyer ex Porter & Coult.	Soreng-3382
Melica ciliata L.	USDA-383702 (as *M. cupanii* Guss. in earlier pubs.)
*Mibora minima (L.) Desv.	J.A. Devesa s.n.
Milium effusum L.	Soreng-3394
Milium vernale M. Bieb.	Soreng-3770
Molineriella laevis (Brot.) Rouy	Soreng-6313
Nardus stricta L.	Royl & Schiers s.n., **B**
Nassella viridula (Trin.) Barkworth	USDA-387938
Parapholis incurva (L.) C.E. Hubb.	KEW-004837
*Parapholis strigosa (Dumort.) C.E. Hubb.	KEW-005019
Peridictyon sanctum (Janka) Seberg, Fred. & Baden	H-6410 (Seberg et al. 1991)
Phalaris arundinacea L.	Soreng-3427
Phalaris paradoxa L.	USDA-202684

Table 1. Genera mentioned in the text and lower ranked taxa sampled for cpDNA and morphology. All vouchers are at BH unless otherwise indicated. Abbreviations are: USDA = USDA Plant Introduction Station, CU = Cornell University Plantations, BHC = L.H. Bailey Hortorium Conservatory, LIT = Literature, USNHG = U.S. Natural History Greenhouse, USNZ = U.S. National Zoo. Asterisks proceeding species names indicate taxa in the extended analysis but, not included in the main cpDNA and total evidence analyses due to missing cpDNA data. *(Continued)*

Species	Collection No.
Phippsia algida (Sol.) R. Br.	Aiken & Consaul-91-34
Phippsia algida (Sol.) R. Br.	Soreng-3520
Phleum alpinum L.	KEW-0044024
Phleum phleoides (L.) H. Karst.	KEW-0044013
Phleum pratensis L.	in "USDA-202208"
Piptatherum miliaceum (L.) Coss.	USDA-284145
Pleuropogon refractus (A. Gray) Benth.	Soreng-3427
Pleuropogon sabinii R. Br.	Aiken & Consaul-91-039
Poa eminens J. Presl	S.J. Darbyshire-85-73, **DAO**
Poa lanigera Nees	INTA
Poa palustris L.	Soreng-3354
Poidium poaemorphum (J. Presl) Matthei	USDA-353466 (as "Briza poaemorpha")
Polypogon maritimus Willd.	Soreng-3659
Polypogon monspeliensis (L.) Desf.	J. Rumely s.n.
Psilurus incurva (Gouan) Schinz & Thell.	Soreng-3600
Puccinellia distans (Jacq.) Parl. 'FULTS'	Native Plants Inc. pudi5591
Puccinellia stricta (Hook.f.) Blom	USDA-237166
Rostraria pubescens (Desf.) Tzvelev	Soreng-3793
Schedonorus phoenix (Scop.) Holub.	USDA-304844
Schizachne purpurascens (Torrey) Swallen	Soreng-3348
Sclerochloa dura (L.) P. Beauv.	Soreng-3862
Sclerochloa dura (L.) P. Beauv.	Soreng-3400
Scolochloa festucacea (Willd.) Link	B. Barker s.n.
Scribneria bolanderi (Thurb.) Hack.	J. Davis s.n.
Sesleria caerulea (L.) Ard.	sent by H. Scholz s.n., **B**
Sesleria insularis Sommier subsp. *sillingeri* (Deyl) Deyl	USDA-253719 (as "S. elongata")
Sphenopholis nitida (Biehler) Scribn.	Soreng-3398
Sphenopus divaricatus (Gouan) Rchb.	Soreng-3700
Stipa barbata Desf.	USDA-229468
Torreyochloa erecta (Hitchc.) G.L. Church	Soreng-3375
Torreyochloa pauciflora (J. Presl) G.L. Church	J. Davis-533
Triplachne nitens (Guss.) Link	Soreng-3775
Trisetum canescens Buckley	Soreng-3383a
Trisetum spicatum (L.) K. Richt.	in "Aiken-91-007-10"
Triticum aestivum L. 'SUSQUAHANNA'	Soreng s.n.
Vulpia alopecuros (Schousboe) Dumort.	USDA-238315
Vulpia ciliata Dumort.	KEW-0035068
Vulpia geniculata (L.) Link	Soreng-3619
Vulpia hispanica (Reichard) Kerguelen	Soreng-3854
Vulpia myuros (L.) C.C. Gmel.	USDA-229452

parsimony analyses were verified with the *Parsimony Ratchet* (Nixon 1998), as implemented in *Dada*. We used *Clados* (Nixon 1993) to examine and draw trees. For outgroups in analyses using cpDNA characters we used the restriction site data set described by Soreng and Davis (1998) for 34 non-*Pooideae* taxa, including three taxa from other families, and new data for *Muh-*

lenbergia and *Tristachya*. Following the results from the analyses using cpDNA characters we used *Brachyelytrum* as the outgroup for analyses of the structural character data set. Complete descriptions of the structural characters and discussion of patterns of character evolution are in a paper in preparation, hence no discussion is presented here of characters on trees.

RESULTS

Phylogenetic Structure Based on cpDNA Restriction Site Variation

This analysis detected 10,000 plus trees of 1940 steps, consistency index 0.22, retention index 0.73 (consensus tree not illustrated). Analysis of the restriction site characters alone resolved the following structure: *Brachyelytreae* ((*Lygeeae Nardeae*) (*Anisopogon* ((*Ampelodesmeae Stipeae*) (*Diarrheneae Brachypodieae Meliceae* ((*Bromeae Triticeae*) (*Aveneae Poeae*)))))). Complexes of related genera sometimes assigned in classifications to *Aveneae* (including *Agrostideae* and *Phalarideae*) were placed within *Aveneae*. Other taxa sometimes assigned to *Hainardieae*, *Milieae*, *Phleeae*, *Scolochloeae*, and *Seslerieae* were placed within *Poeae*. *Torreyochloa* and the *Briza* complex were placed within *Aveneae*. *Aira*, *Holcus*, *Molineriella*, and *Avenula* were placed among genera of *Poeae*. Genera present only in the analysis of the extended sample, *Castellia*, *Echinaria*, *Mibora*, *Psilurus*, and *Scribneria*, exhibited restriction site patterns diagnostic for *Poeae*.

Phylogenetic Structure Based on Simultaneous Analysis of Restriction Site and Morphological Characters

Simultaneous analysis of both data sets detected 2268 trees of 2655 steps, consistency index 0.20, retention index 0.69. The consensus tree is shown in Fig. 1. This analysis resolved the following tribal structure: *Brachyelytreae* ((*Lygeeae Nardeae*) ((*Anisopogon* (*Ampelodesmeae Stipeae*)) ((*Diarrheneae* ((*Brachypodieae Meliceae*) ((*Bromeae Triticeae*) (*Aveneae Poeae*)))))). The main differences between the consensus trees of the simultaneous and cpDNA analyses are in the resolution in the simultaneous analysis of a clade in which *Anisopogon* is the sister group of *Ampelodesmeae/Stipeae*, with *Diarrheneae* diverging between that clade and the now resolved *Brachypodieae/Meliceae* clade (versus a polytomy of *Diarrheneae*, *Meliceae*, *Brachypodieae*, and the clade of (*Bromeae/Triticeae* and *Poeae/Aveneae*)). The generic compositions of all tribes were identical between the two analyses.

The principal differences between the two analyses were in relationships of terminal taxa within tribes *Poeae* and *Aveneae*. However, even at that level a high degree of congruence was apparent. Groups with the same membership in *Aveneae* in both analyses are (sets arranged as in the simultaneous analysis): 1) *Helictotrichon Arrhenatherum* (*Avena Lagurus*) (*Gaudinia* ((*Trisetum canescens* (*Cinna Sphenopholis*)) (*Trisetum spicatum* (*Rostraria Koeleria*)))); 2) (*Anthoxanthum Phalaris*); 3) (*Chascolytrum Poidium*) (*Ammophila* (*Agrostis Calamagrostis Deschampsia Gastridium Triplachne* (*Chaetopogon Polypogon*)))). In the cpDNA consensus tree there is a polytomy of these clades and unaligned *Briza* species, *Hierochloe*, and (*Amphibromus Torreyochloa*). In the results of the simultaneous analysis, *Amphibromus* is the sister group of all other members of the *Aveneae* clade, and *Torreyochloa* is the sister of the following subclade within the *Aveneae* clade: ((*Hierochloe* (clade 2)) ((*Briza s.str.*) (clade 3))).

Groups with consistent membership in *Poeae* include (aligned as in the simultaneous analysis): 1) *Puccinellia* (*Sclerochloa* (*Catabrosa Phippsia*)); 2) (*Cynosurus*) (*Sphenopus* (*Cutandia* (*Catapodium* (*Desmazeria* (*Hainardia Parapholis*))))); 3) (*Festuca Vulpia*) (*Leucopoa* (*Schedonorus Lolium*)); 4) *Dupontia* ((*Bellardiochloa* (*Puccinellia stricta* (*Poa*))) (((*Arctagrostis* (*Beckmannia Alopecurus*

(*Milium*) (*Phleum*)))); 5) (*Holcus*) (*Molineriella* (*Aira*)). In *Poeae*, although placed separately in the cpDNA tree, ((*Holcus*) ((*Aira*) *Molineriella*)) and the two species of *Avenula* are united in a clade with (*Dactylis Lamarckia*) intercalated between *Avenula* and the rest of clade 5 in the simultaneous analysis tree (Fig. 1). *Dactylis*, *Lamarckia*, *Scolochloa*, and *Poeae* clades 1, 2, 3, (*Sesleria*, *Avenula pubescens* (clade 4)), and (*Avenula gervaisii* (clade 5), are in a polytomy in the cpDNA consensus tree. In the simultaneous analysis clades 1, 2 and 3 are united, and *Scolochloa* is sister of the set that consists of clade 4 plus *Sesleria*, and clade 5 plus *Avenula* and (*Dactylis Lamarckia*).

Thus, it can be seen generally that the relationships resolved within these two tribes in the simultaneous analysis were largely congruent with those resolved by the restriction sites alone. Notable exceptions are to be seen in the greater resolution in the simultaneous analysis, including the union of the separate species of *Aira*, *Avenula*, *Briza*, *Holcus*, and *Poa*, and in alternative relationships among elements within each of the clades that were resolved by both analyses.

Phylogenetic Structure Using Structural Characters Only

The main analysis of structural characters, using the same 101 taxa as in the previous analyses, detected 158 trees of 590 steps, consistency index 0.19, retention index 0.60. The consensus tree (not illustrated) was highly unresolved. The following clades were detected: 1) *Lygeeae Nardeae*; 2) *Scolochloa* ((*Boissiera Avena*) *Ampelodesmeae* ((*Triticum Peridictyon*) *Brachypodium*)); 3) *Glyceria Melica Schizachne*; 4) *Poeae* in major part (including the *Briza* complex, *Phleum* and *Beckmannia*); 5) *Aveneae* (roughly as arranged in the restriction site and simultaneous analyses, but minus *Avena*, and plus *Aira*, *Alopecurus*, *Avenula*, *Holcus*, *Hainardia*, *Milium*, *Molineriella*, *Parapholis*. Several elements always classified in *Poeae* collapsed into a polytomy in the consensus tree: *Leucopoa* spp., *Torreyochloa* spp., *Puccinellia* spp., *Festuca*, *Lolium*, and *Schedonorus*, along with *Pleuropogon* spp., *Diarrhena*, *Anisopogon*, *Stipa*, *Nassella*, *Piptatherum* and the previous 5 clades.

Because the structural characters set did not resolve a monophyletic group of *Aveneae* plus *Poeae*, an additional exploratory analysis was conducted. In this separate analysis, monophyly of a single *Poeae/Aveneae* clade was constrained, and 20 additional terminal taxa were included. This analysis allowed us to examine the affinities of various genera when they were not free to be grouped with members of other tribes. The extended and constrained analysis resulted in 1152 trees of 667 steps. In the consensus tree the forcing of *Poeae/Aveneae* into a single clade had the effects of: 1) uniting *Anisopogon* in a clade with *Stipeae* and (*Lygeeae Nardeae*); 2) supporting clade 1 as the sister group of *Diarrhena* plus all the remaining *Pooideae*; 3) uniting *Meliceae*; 4) uniting *Triticeae*; 5) resolving *Boissiera* as the sister of *Poeae/Aveneae*. One polytomy present in the consensus tree included *Ampelodesmos*, the species of *Brachypodium*, *Bromus*, and clades 3, 4, and 5. In the *Poeae/Aveneae* group, after the initial divergence of *Scolochloa* and some species of *Festuca* and allies, an extensive polytomy of all elements in each tribe was resolved in the consensus tree. However, for the taxa in the latter polytomy, only two major topologies were apparent among all

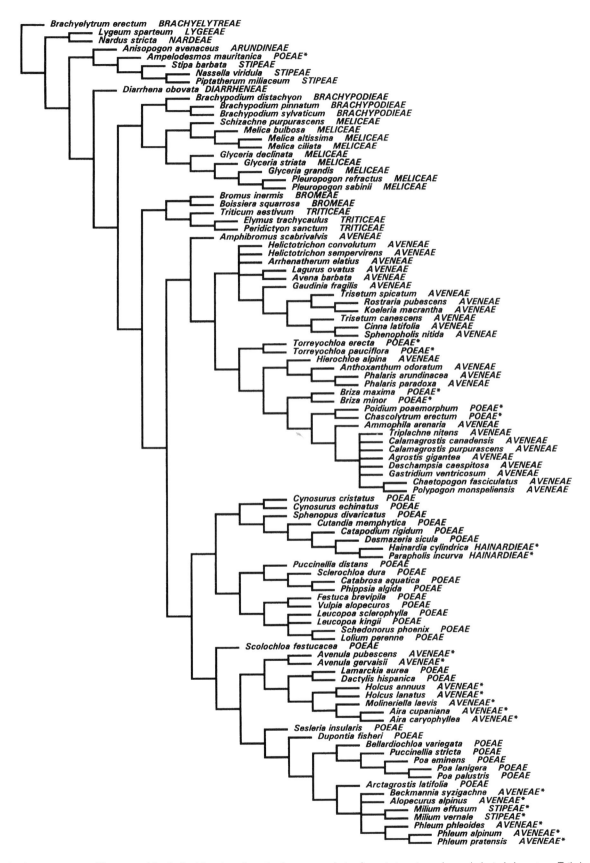

Fig. 1. Consensus tree of Poaceae subfamily Pooideae based on simultaneous analysis of restriction site and morphological characters. Tribal placements of genera follow Clayton and Renvoize (1986), and asterisks indicate those species which are either misplaced by cpDNA restriction site data, or are misclassified as to tribe.

1152 most parsimonious trees, these corresponding to *Aveneae*, and the remainder of *Poeae*. The collapse of these clades was apparently due to the unsettled affinities of elements of the *Briza* complex which were variously placed, either all in *Poeae* or divided among *Aveneae* and *Poeae* in the most parsimonious trees. *Torreyochloa* was placed within *Poeae*. *Aira, Avenula, Beckmannia, Hainardia, Holcus, Mibora, Molineriella, Parapholis, Phleum*, and *Scribneria* were placed within *Aveneae*.

Of five genera not in the main restriction site sample due to missing data, but with restriction site patterns diagnostic of the *Poeae* clade, the placements of *Mibora* and *Scribneria* were just mentioned. *Castellia* and *Psilurus* were placed in analyses of structural character data with *Festuca s.str.* and *Vulpia,* and *Echinaria* was placed with *Sesleria*.

In a third analysis with monophyly of *Poeae/Aveneae* constrained, we removed 23 taxa that were, according to most classifications, misclassified as to tribe by restriction site data. In this analysis *Poeae* and *Aveneae* were resolved as monophyletic sister clades with one exception. *Scolochloa*, of tribe *Poeae*, was placed as the sister of all other taxa in the *Aveneae* clade.

DISCUSSION

Tribal Relationships

There was little difference between the major phylogenetic structures of *Pooideae* detected by our cpDNA and simultaneous analyses. This is undoubtedly due to a strong phylogenetic signal from the cpDNA restriction site characters, in part because they are nearly ten times more numerous than the structural characters. The results of our simultaneous analysis of *Pooideae* (Fig. 1) agree with those of by Clark *et al.* (1995), and Catalan *et al.* (1997), both based on the chloroplast-encoded gene *ndh*F. All of these analyses support the position of *Brachyelytreae* as the earliest diverging line in the subfamily, followed by divergence of a clade consisting of the monotypic *Nardeae* and *Lygeeae* tribes as sister group of the remainder of *Pooideae*. These analyses also corroborate the existence of a clade consisting of *Poeae* plus *Aveneae* as sister group of another consisting of *Bromeae* plus *Triticeae* (PA/BT). However, apart from these points, tribal relationships suggested by the various analyses are not fully consistent with one another. Hsiao *et al.* (1999) presented an analysis of nuclear rDNA ITS sequences of *Poaceae* (see Introduction), but one which is not fully comparable with our analyses because a consensus tree of their parsimony analyses was not provided, and the parsimony tree presented included only a subset (11 of 37 taxa) of the representatives of *Pooideae* included in their neighbour-joining tree.

Phaenosperma, sampled in other analyses, but not in ours, was the next taxon to diverge after *Brachyelytrum* and *Nardeeae/Lygeeae* (Clark *et al.* 1995; Catalan *et al.* 1997), and this was variously followed by the divergence of either *Meliceae* or a clade including *Stipeae*. The placement of *Phaenosperma* between the *Nardeae/Lygeeae* clade and the next clade to diverge, *Stipeae,* or within a broadly circumscribed *Stipeae*, seems reasonable based on the structural features of that genus (e.g., single flowered-spikelets, three large unlobed lodicules, glabrous ovaries). A clade including *Stipeae* diverged next in our analysis and that by

Clark *et al.* (1995). In our simultaneous analysis, and the analysis by Hsiao *et al.* (1999), the clade including *Stipeae* also included *Anisopogon* and *Ampelodesmos* (neither of which were sampled in the analyses by Clark *et al.*, or Catalan *et al.*). *Diarrheneae* was the next tribe to diverge after *Stipeae* in the analyses by Catalan *et al.* (1997) and Clark *et al.* (1995; *Brachypodieae* and *Meliceae* not sampled), and in our present simultaneous analysis, but this was the first tribe to diverge within *Pooideae* in the Hsiao *et al.* (1999) analysis.

Meliceae diverged either prior to the divergence of *Stipeae* (Catalan *et al.* 1997, also detected by Nadot *et al.* 1994, and Soreng *et al.* 1990), or afterward (Hsiao *et al.* 1999), and then either before (Davis and Soreng 1993; Soreng and Davis 1998), or immediately after the divergence of *Diarrhena* (the present simultaneous analysis). In the present analysis *Meliceae* were placed in a weakly supported sister group relationship with *Brachypodieae*, whereas Catalan *et al.* (1997) and Hsiao *et al.* (1999) found *Brachypodieae* to be the sister group of the PA/BT clade. *Meliceae* and *Brachypodieae,* along with the PA/BT clade, are considered 'core' tribes of *Pooideae* in most modern classifications, and *Brachyelytreae, Diarrheneae, Lygeeae, Nardeae, Phaenospermatideae,* and *Stipeae* are regarded as remotely related to them.

We suggest the following summary of tribal relationships: *Brachyelytreae* ((*Lygeeae Nardeae*) ((*Phaenospermatideae* (*Anisopogon Ampelodesmeae Stipeae*)) (*Diarrheneae* (*Meliceae* (*Brachypodieae* ((*Bromeae Triticeae*) (*Aveneae Poeae*))))))). However, exact relationships of tribes from *Phaenospermatideae* to *Brachypodieae* must be regarded as tentative, and further investigation is warranted.

Intertribal Reticulation Versus Parallelism and Convergence

As will be readily apparent from Fig. 1 to any reader who is familiar with the Pooideae, several genera are either misplaced by the cpDNA restriction site data, or are misclassified in modern classifications (Table 2). Although all molecular analyses to date have resolved a monophyletic clade corresponding to *Aveneae* plus *Poeae*, relationships between the tribes and among the genera have not been wholly consistent. The analyses by Nadot *et al.* (1994) and Catalan *et al.* (1997) included 9 and 12 taxa of *Aveneae* and *Poeae*, respectively. Both of these detected *Poeae* as arising from within a paraphyletic *Aveneae*. Our cpDNA restriction site analyses have included 16 (Soreng *et al.* 1990), 7 (Davis and Soreng 1994), 19 (Soreng and Davis 1998), and 75 taxa of *Poeae/Aveneae* (the present data set), and have consistently resolved two well supported clades that correspond, in general, to *Poeae* and *Aveneae*. However, as stated above, there are a number of genera whose placements by cpDNA restriction site data contradict those in classifications. Given that the *Poeae* and *Aveneae* clades are among the best supported in our whole analysis (see Soreng and Davis 1998 for measures of support), and that there seems to be minimal and randomly distributed homoplasy in restriction site characters between these clades, we regard the placements of these taxa to be correct insofar as they reflect the chloroplast genome of the plants. This conclusion leaves us with two alternative hypotheses: 1) parallelism and convergence in morphological characters led agrostologists to misclassify several

Table 2. Taxa for which cpDNA relationships are incongruent with classifications: Tzvelev (1976, 1989), Clayton and Renvoize (1986) = C&R, and Watson and Dallwitz (1992) = W&D.

Taxon	Classifications Tzvelev - W&D - C&R	cpDNA	No. spp. sampled
Ampelodesmos	Ampelodesmeae - Stipeae - Poeae	Stipeae/Ampelodesmeae	1
Avenula	All Aveneae	Poeae	2 (in 2 sections)
Holcus	All Aveneae	Poeae	2
Aira	All Aveneae	Poeae	2
Molineriella	All Aveneae	Poeae	1
Deschampsia	All Aveneae	[A]Aveneae	1
Torreyochloa	All Poeae	Aveneae	2
Briza complex	All Poeae	Aveneae	6 (in 3 genera)
Phleum complex	Phleeae - Aveneae - Aveneae	Poeae	6 (in 3 genera)
Milium	Aveneae - Aveneae - Stipeae	Poeae	2
Puccinellia stricta	([B]N/A)	Poa	>80 Poa & Puccinellia
Scribneria	N/A - Aveneae - Hainardieae	Poeae	1
Mibora	All Aveneae	Poeae	1
Hainardieae	[C]Hainardieae - Poeae - Hainardieae	Poeae	3 (in 2 genera)

[A] A possible basal position in *Poeae* was detected in two cpDNA studies with low sample depth (Nadot *et al.* 1994, Catalan *et al.* 1997).
[B] In floras in which the species is treated it is always placed in *Puccinellia* not *Poa*.
[C] As *Monermeae*, but *Hainardieae* the correct name.

genera as to tribe; or 2) hybridization events allowed chloroplast genomes to be exchanged between tribes. In the latter case it is possible that chloroplasts, with their characteristic genomic characters intact, infiltrated from taxa of each tribe into species of the other tribe, while concomitant contributions of foreign nuclear genes remain hidden or were eliminated. In the following discussion we examine the evidence with respect to each hypothesis for each of the suspect genera listed in Table 2.

Before we begin to evaluate anomalous tribal placements of taxa it is appropriate to briefly review the literature supporting the existence of intergeneric hybrids in *Pooideae*. In general, taxa have been described as possible hybrids in the literature based on their intermediate or odd combinations of morphology, or unusual cytology, and many hybrids have been produced by plant geneticists. Knobloch (1968, 1972) reported more than 800 intergeneric crosses in Poaceae. Some naturally occurring intergeneric hybrids in *Pooideae* have been named (e.g., x*Achnella*, x*Agropogon*, x*Ammocalamagrostis*, x*Arctodupontia*, x*Bromofestuca*, x*Dupoa*, x*Festulpia*, x*Pucciphippsia*, x*Schedolium*, x*Trisetokoeleria*), including numerous examples in *Triticeae*. Within *Bromeae*, mostly sterile artificial crosses have been obtained between the genera (or subgenera of *Bromus*), *Anisantha*, *Bromopsis*, *Bromus*, and *Ceratochloa* (Knobloch 1968), and Stebbins (1956, 1981) postulated the origin of *Trisetobromus* as an ancient cross between *Ceratochloa* and *Bromopsis* species (discussed as subgenera of *Bromus*). Furthermore, several unnamed intertribal crosses have been reported in *Pooideae* (e.g., *Glyceria* x *Schedonorus*, *Glyceria* x *Lolium*, *Hainardia* x x*Festulolium*, *Lolium* x *Arrhenatherum*, *Lolium* x *Brachypodium*, *Lolium* x *Bromus*, *Lolium* x *Glyceria*, *Melica* x *Bromus*, *Schedonorus* x *Bromus*, *Schedonorus* x *Glyceria* (all cited in Knobloch 1968), but few if any have been confirmed. Although the majority of such wide hybrids are apparently sterile, the potential exists for the hybrids to become at least partially fertile again through polyploidization, and/or

the selective elimination of incompatible chromosomes or whole genomes.

In *Triticeae*, many intergeneric hybrids have been artificially produced, and others exist in nature. In this tribe chloroplast and nuclear DNA gene trees have been shown to be incongruent with each other, and both of these are incongruent with morphological trees (Kellogg *et al.* 1996; Mason-Gamer and Kellogg 1997). Although species considered to share whole nuclear genomes (following the results of 'genome analysis') generally resolved as close relatives in these analyses, putative relationships among the genome groups are frequently incongruent with relationships resolved by morphological data, or DNA sequence and restriction cite data from nuclear or plastid genes (but see Seberg *et al.* 1998 for a review of the limitations of genome analysis).

x*Dupoa* is an example of a well documented hybrid outside *Triticeae* (Darbyshire *et al.* 1992). It has the cpDNA of *Dupontia* and isozyme markers of its parent in *Poa*, it is somewhat intermediate between its parents in morphology, but favors its *Poa* parent, and it is sterile. *Dupontia* itself might have originated through intertribal hybridization between *Arctophila* (*Poeae*) and *Deschampsia* (Tzvelev 1976). However, *Dupontia* was placed near *Poa* in all our present analyses, as it is in classifications, and it did not cause any branches to collapse around it.

At a lower taxonomic level, the capacity for infrageneric hybridization is well known in genera such as *Poa* (Hiesey and Nobs 1982). Many wide inter-sectional hybrids have been produced artificially in *Poa*. Frequently the phenotype of the most highly polyploid parent is dominant. *Poa hartzii* subsp. *hartzii* is of particular interest in that this apomict retains each of the distinctive chloroplast types of its postulated parental species (which belong to different sections of the genus) in approximately equal numbers among individuals within each of the several populations sampled (Gillespie *et al.* 1997). That subspecies is mostly sterile-anthered and produces seed apomictically, and yet the potential exists for it to spread either type of chloroplast in its progeny.

The foregoing discussion of the well documented capacity for wide crosses in Poaceae should incline us towards caution when evaluating phylogenetic relationships using any type of data. Furthermore, we anticipate, as have Stebbins (1950) and Kellogg and Watson (1993), that additional examples of extant species and generic groups in *Poaceae* (beyond those already documented in *Triticeae*) will be discovered to be the end products of ancient hybridization events. Exposing potential new cases and proving them are different matters, however.

Kellogg and Watson (1993) suggested that the collapse of relationships in their structural character analysis of *Pooideae* was due in large part to intergeneric and intertribal hybridization. They also suggested that unusual polyploid chromosome numbers in combination with unusual morphologies be considered evidence of ancient hybridization events in taxa such as: *Ampelodesmos* (2n = 48), *Diarrhena* (2n = 38, 60), *Dupontia* (2n = 42, 44, 88, 132), *Lygeum* (2n = 40), *Nardus* (n = 13), and *Phaenosperma* (n = 12), a list to which *Deschampsia* (n = 7, 13) might be added. Thus, confirmation of unusual placements of these genera in our analyses would lend support to the general hypothesis that wide hybridization has played a significant role in the evolution of Pooideae. It should be kept in mind, however, that some of these chromosome numbers may be derivatives of base numbers that are plesiomorphic in the subfamily (Soreng and Davis 1998).

Comparisons of results of phylogenetic analyses, particularly between cpDNA-based gene phylogenies, and those based on nuclear-encoded characters, including those of morphology, are potentially useful for detecting hybrids which have retained the cpDNA type of one parent along with at least a major structural character contribution from the other parent. Because chloroplast genomes are linearly inherited and non-recombining between chloroplasts, perturbing effects to cpDNA-based phylogenies from inclusion of hybrids in analyses are likely to be minimal. Lineage sorting is most likely to occur among closely related taxa among which the chloroplast genomes have diverged little. However rare wide hybridization may have been, if chloroplasts of highly divergent lineages were present in fertile or partially fertile hybrids, new rounds of hybridization and the generation of new morphological character combinations could result in unexpected relationships in phylogenetic analyses based on that genome.

Inclusion of hybrids in simultaneous analyses of cpDNA and structural characters is likely to influence the resulting trees. Results of such inclusion would be influenced by several factors: 1) the information content in the individual data sets; 2) the relative frequency of hybrid taxa in the analysis; 3) the phylogenetic distance between parents of hybrids; 4) the patristic distance between parents of hybrids; and 5) the presence and number of non hybrid taxa intervening between parents and hybrids in the analyses (see also Funk 1985, McDade 1990). Perhaps equally important would be the consequences of loss or masking of the phenotypic characters of one of the parents. If a hybrid is intermediate, or loss or masking of one parent's nuclear genes has occurred, while that parent's chloroplast is

retained, then the hybrid should be identifiable, as in x*Dupoa* (Darbyshire *et al.* 1992). If the retained chloroplast is from a parent whose phenotype is predominantly expressed, it would be problematic to confidently identify the contributions of the other parent, if it were even possible to recognize that a hybridization event had taken place.

In the following discussion several criteria are considered in identifying and evaluating possible hybridization events: 1) high levels of support for conflicting relationships are needed to demonstrate incongruence between phylogenetic trees; 2) equivalent sampling of taxa among different analyses is important; 3) adequate depth of sampling is important; 4) chloroplast genomes are linearly inherited and non-recombinant between chloroplasts, and thus cpDNA genealogies are the least likely of relationships to be compromised by the inclusion of hybrids; 5) each case of possible hybridization should be considered separately; 6) attempts should be made to duplicate anomalous results, and vouchers checked (polymerase chain reaction technology involved in gene sequencing is susceptible to contamination by foreign DNAs, laboratory and greenhouse mix-ups happen, and identifications of original materials are sometimes mistaken).

Once the above factors are accounted for, several analytical results suggest the presence of hybrids in phylogenetic analyses. Classifications may be consistent with structural data analyses, but be incongruent with other data, especially cpDNA data, or phylogenetic analyses of different data sets may be incongruent. Combining cpDNA and structural data in analyses including hybrid taxa may cause consensus trees to collapse, or may result in hybrids grouping together within the phylogenetic framework enforced by the most decisive data set. Additional evidence of intermediacy in structural characters or of unusual assortments of characters may also be apparent. Evidence for hybridization must be weighed against the alternative hypothesis of homoplasy in one or both of the data sets.

It is important at this point to make some general observations about our cpDNA and simultaneous analyses. There is repeated molecular evidence of a monophyletic group that includes only elements conventionally assigned to *Aveneae* and *Poeae*. All previous molecular studies that have included elements from each tribe have resolved this clade as a monophyletic lineage (one exception to this general finding lies in the placement of *Ampelodesmos* among early diverging elements of the *Pooideae* by molecular data, but in *Poeae* in the classification of Clayton and Renvoize 1986). Resolution of two distinct cpDNA types, characterized as '*Poeae*' or '*Aveneae*', and generally corresponding to groups of genera belonging to each individual tribe, is consistent with all our previous studies (Soreng *et al.* 1990; Davis and Soreng 1994; Soreng and Davis 1998). In the present analyses that included cpDNA data, all species of the inclusive *Poeae/Aveneae* lineage clearly exhibited one or the other cpDNA type; no intermediates were found. Also, no change in tribal affinity of any species of *Aveneae* or *Poeae* was detected between the analyses of cpDNA and simultaneous analysis data sets. However, phylogenetic structure **within** each of these two chloroplast lineages was less well resolved by the cpDNA data. Relationships among the infratribal subgroups were generally supported by few cpDNA characters, and relationships among several clades and taxa collapsed in the cpDNA consensus tree. The addition of structural characters increased the resolution of phylogenetic structure overall in the simultaneous analysis, often uniting taxa in more traditionally accepted relationships. Corroboration of our cpDNA restriction site data for the species used in the present analyses was obtained by mapping restriction sites of the extended set of species, or second individuals within species, for 37 genera in our study, including several of the genera that appear to be misclassified, on the basis of their cpDNA types (Table 2). In the following paragraphs we present discussions of individual cases of conflict among and between different phylogenetic analyses and classifications.

Ampelodesmos. Various classifications have placed this monotypic genus in *Stipeae,* or in its own tribe near *Stipeae,* or in *Poeae,* where it was considered it to be an early diverging element (Clayton and Renvoize 1986). This genus resembles *Stipeae* in anatomy, in its number of lodicules, in the entire margins of the lodicules, and in chromosome number and size, and it has been placed near or within Stipeae in an analysis of nuclear rDNA ITS sequences (Hsiao *et al.* 1999). It is unlike *Stipeae* in number of florets, type of awn, and in hairiness of the anthers and ovaries. The latter features are all like those of *Poeae, Bromeae, Brachypodieae,* and *Triticeae.* The main analysis of structural character data in the present study placed it in a clade with the genus *Avena,* and the tribes *Bromeae, Triticeae,* and *Brachypodieae.* In analyses of structural character data, with or without a forced *Poeae/Aveneae,* the partially collapsed basal branches in consensus trees were in part due to instability in the placement of *Ampelodesmos.* The incongruent placements of *Ampelodesmos* in positions intermediate between *Stipeae* and *Poeae,* among *Brachypodieae, Bromeae,* and *Triticeae,* in the main analysis of structural character data, and in a clade with *Stipeae* in the cpDNA and simultaneous analyses are consistent with an intertribal hybrid origin hypothesis. *Ampelodesmos* may tentatively be classified in a separate tribe related to, or in a subtribe within, *Stipeae,* and not in *Poeae,* or, if it is proved to be to be a hybrid it could be placed *incertae sedis.*

Avenula. This genus has been classified in *Aveneae,* or its equivalent, by all authors, and usually has been included within *Helictotrichon.* Analyses of structural character data placed *Avenula* within *Aveneae,* and near *Helictotrichon.* Spikelets of *Avenula* are quite similar to those of *Helictotrichon,* but the two genera can be separated by a number of features, including leaf anatomy, degree of fusion of the leaf sheaths, inflorescence structure, and details of the spikelet itself. Anatomically, however, *Avenula* differs from other *Aveneae,* but closely resembles some elements of *Poeae. Avenula pubescens* combines an odd assortment of characteristics of *Avenula s.str.* and *Helictotrichon,* suggestive of an intergeneric hybrid origin. Both species of *Avenula* studied have *Poeae*-type cpDNA restriction site profiles, but were not united in the cpDNA analysis. In the cpDNA analysis *Avenula gervaisii* was placed in *Poeae* in a clade with *Holcus, Aira* and *Molineriella,* and no other taxa. The simultaneous analysis resolved a clade consisting of these taxa plus two genera of *Poeae,* and united the two species of *Avenula.* The cpDNA and simultaneous analyses united *Helictotrichon* with other closely allied genera sampled (*Avena, Arrhenatherum, Cinna, Gaudinia, Koeleria,*

Rostraria, Sphenopholis, Trisetum), within one subclade of *Aveneae.* Phylogenetic analysis of nuclear rDNA ITS sequences also supported a remote relationship between *Helictotrichon* and *Avenula* (Grebenstein *et al.* 1998). The evidence suggests that *Avenula* is derived from an intertribal hybridization event. Furthermore, the history of this event may be connected to that of the next set of genera.

Holcus, Aira and *Molineriella. Aira* and *Molineriella* are consistently classified with *Deschampsia* in *Aveneae,* usually together in the same group, subtribe *Airinae* Fr. (although see Albers and Butzin 1977). *Holcus* is consistently classified in *Aveneae,* sometimes in a monotypic subtribe, *Holcinae* Dumort. ex Tzvelev. Analysis of structural characters placed all three of these genera within *Aveneae,* with *Aira* and *Molineriella* as sister taxa, but not in a close relationship with *Holcus.* However, the cpDNA and simultaneous analyses placed these three genera in a subclade with *Avenula* (see previous paragraph) within *Poeae.* Phylogenetic analysis of nuclear rDNA ITS sequences (Grebenstein *et al.* 1998) also placed *Holcus* near *Avenula* and distant from core *Aveneae.* We are aware of no structural characters that support the placement of these three genera with *Poeae.* The evidence appears to support a hypothesis that these three genera represent the end results of the same intertribal hybridization event that produced *Avenula,* their phenotypes generally favoring their parent in the *Aveneae,* and their chloroplasts derived from their parent in *Poeae.*

Deschampsia. Classifications consistently place *Deschampsia* in *Aveneae.* The analysis of structural characters supported the placement of this genus in *Aveneae,* in relationships similar to those indicated by arrangements of genera in the three classifications. Our cpDNA and simultaneous analyses also support its placement in *Aveneae.* Two independent cpDNA gene analyses, however (*rps*4, Nadot *et al.* 1994, and *ndh*F, Catalan *et al.* 1997), placed *Deschampsia* in a paraphyletic *Aveneae* from which *Poeae* arises, as the closest relative of *Poeae,* or among the genera most closely related to *Poeae.* Our study and the *ndh*F study sampled *Deschampsia caespitosa,* and the *rps*4 study used *D. flexuosa.* Both of the latter studies have low taxon density in *Poeae/Aveneae,* and their results, though suggestive, should be considered with this in mind. We are aware of no structural characters that link *Deschampsia* to *Poeae.* Does *Deschampsia* retain two different chloroplast types? Further study of this question is needed, but on the whole there is little evidence of intertribal hybridization in the ancestry of any *Deschampsia* other than in the odd combination of chromosome numbers in this genus ($n = 7, 13$).

Torreyochloa. Classifications have consistently placed this genus in *Poeae* since it was recognized as distinct from *Glyceria.* In the consensus tree resulting from the main analysis of structural character data alone, only part of *Poeae* was resolved as monophyletic, and *Torreyochloa* was among genera excluded from that part. In the forced *Poeae/Aveneae* analysis of structural character data, *Torreyochloa* was placed among genera of *Poeae* in a clade that included *Festuca, Leucopoa* and *Lolium.* However, the analysis of cpDNA data placed *Torreyochloa* among genera of *Aveneae* as sister group of *Amphibromus.* The simultaneous analysis placed *Torreyochloa* in the same general position as did the analysis of cpDNA data alone, but *Amphibromus* was moved to the earliest-diverging position within *Aveneae*. As *Amphibromus* occurs naturally only in the southern hemisphere (recently introduced to North America), and *Torreyochloa* only in the northern hemisphere, it is highly improbable that *Torreyochloa* cpDNA arose from a hybridization event involving *Amphibromus* (or its antecedents) and some member of Poeae. We argued elsewhere (Soreng *et al.* 1990) that *Torreyochloa* may represent a case of convergent evolution in spikelet structure among wetland grasses, or that it retains a plesiomorphic spikelet structure. The latter possibility is intriguing in light of the early-diverging positions of *Amphibromus*, *Scolochloa*, and *Torreyochloa* within the *Poeae/Aveneae* clade, and their geographic occurrence in wetlands, all of which suggest that these tribes may have originated in temperate wetlands. Spikelets of *Torreyochloa* are reminiscent of those of some genera of *Aveneae* (e.g., *Amphibromus* and *Deschampsia*) in the continuation of the prominent lateral nerves of the lemmas to the margin of a somewhat truncated lemma apex. It may be argued that *Torreyochloa* has only lost the dorsal, geniculate awn (the main structural synapomorphy of the *Aveneae*). At this point we favor a hypothesis of convergence or plesiomorphy over one of a hybrid origin for *Torreyochloa*.

The *Briza* complex: *Briza, Chascolytrum, Poidium*. Classifications have consistently placed these genera in *Poeae*, sometimes all in *Briza*. The analysis of structural character data placed these genera all together within *Poeae*, or partly in *Poeae* and partly in *Aveneae*, resulting in a near total collapse of structure of *Poeae/Aveneae* in the consensus tree. The cpDNA and the simultaneous analyses placed these genera as closely related elements within *Aveneae*, but not in a monophyletic group by themselves. Analysis of nuclear rDNA ITS sequences (Hsiao *et al.* 1995, 1999) also placed *Briza* within the *Aveneae*. We suspect that *Briza s.lat.* is non-monophyletic, resulting from parallel evolution of a 'brizoid' lemma in different lineages in Eurasia and South America, presumably as an adaptation to wind dispersal, and not from a wind dispersal crossing of the Atlantic ocean and the tropics. Species of *Chascolytrum* and *Poidium* frequently exhibit a minute sub-apical awn, which is uncharacteristic of *Poeae*, and not uncommon in *Aveneae* (e.g. *Apera*, and some species of *Deschampsia* and *Koeleria*). The placement of the *Briza* complex within *Aveneae* in some analyses of structural character data leads us to suggest that these genera may simply be misclassified in *Poeae*.

The *Phleum* complex: *Beckmannia, Alopecurus,* and *Phleum*. Classifications usually place these genera in *Aveneae* or *Agrostideae*, but sometimes put them in a separate tribe, *Phleeae*, in one case with *Sesleria* and allies (Tzvelev 1989). Analysis of structural character data placed these genera in *Poeae*, whereas the analysis of the cpDNA data placed them within the *Poeae* subclade that also included *Sesleria, Milium, Poa* and other genera. Simultaneous analysis placed them in the previous larger subclade of *Poeae*, in this case in a terminal clade including only *Arctagrostis, Milium* and the *Phleum* complex. Analysis of rps4 (Nadot *et al.* 1994) placed *Phleum* and *Alopecurus* as separate members of a paraphyletic *Aveneae*, from among which *Poeae* arise (but again that sample size was small). We suggest the *Phleum* complex is misclassified in *Aveneae*, and is best treated as subtribe *Phleinae* within *Poeae*.

Milium. Classifications place this genus in *Stipeae* (based on the similarity of the spikelets to *Piptatherum*, but contradicted by the occurrence of $x = 7$ [though 4, 5, and 9 are also known], large chromosomes, endosperm with lipid, lodicule form and number, and relatively short hilum), or *Aveneae*, or infrequently in its own tribe, *Milieae*. The analyses of structural character data placed *Milium* in *Aveneae*. However, the cpDNA and the simultaneous analyses placed it among members of the *Phleum* complex (see above). The genus is unusual enough that reclassification in subtribe *Miliinae,* among one-flowered *Poeae* near *Arctagrostis* and *Phleum,* does not seem unreasonable, as there seems to be no evidence of a hybrid origin for it.

Puccinellia stricta. In consensus trees of the unforced analyses of structural character data this species was among genera in a basal polytomy including *Puccinellia distans*. Among the most parsimonious trees from the analysis of structural character data, *Puccinellia stricta* usually was placed as the sister taxon of *P. distans*. The analysis of cpDNA data placed *Puccinellia stricta* within *Poa,* and the simultaneous analysis placed it as the sister of *Poa*. We suspect that *Puccinellia stricta* is an intergeneric hybrid, for broad surveys of cpDNA restriction sites in *Poa* (Soreng 1990) and *Puccinellia* (Meng *et al.* 1994) (Table 2) have indicated that the two genera have quite distinct cpDNA types, and have detected no other misplaced taxa. In the present analysis, however, *Puccinellia stricta* did not cause trees in the simultaneous analysis to collapse around it, nor did its presence force a relationship between *Puccinellia* and *Poa,* two genera that are widely separated within *Poeae*.

Mibora and *Scribneria*. In classifications *Scribneria* is placed in *Aveneae* or *Hainardieae*, and *Mibora* is consistently placed in *Aveneae*. In the analyses of structural character data with a forced *Poeae/Aveneae* clade *Scribneria* was placed among the early-diverging members of *Aveneae*, and *Mibora* among members of subtribe *Agrostidinae* Fr., near *Milium* and *Polypogon*. Neither taxon was included in the main cpDNA or simultaneous analyses, since there were no data from these genera for several restriction enzymes. However, a separate analysis of available restriction site data placed both genera in *Poeae*, with *Scribneria* in the clade with *Avenula, Holcus, Aira,* and *Molineriella,* and with *Mibora* not associated with any particular group of taxa. The restriction site profile of *Scribneria* is dissimilar to that found in *Hainardieae* (where it was placed by Clayton and Renvoize, 1986), nor does it resemble those of the closest relatives of *Hainardieae*, as resolved in the present cpDNA and simultaneous analyses. The placement of *Scribneria* within the clade including *Avenula, Holcus, Aira,* and *Molineriella,* may indicate that it is one of the end products of the postulated hybrid origin of this group. The geographically isolated occurrence of *Scribneria* in North America, far from Europe where the other elements of the clade including *Avenula* (or *Hainardieae*) are native, remains a puzzle. We suggest that *Mibora* could be satisfactorily placed with *Milium* and *Zingeria* in subtribe *Miliinae* Dumort., within *Poeae*.

Hainardieae: *Hainardia* and *Parapholis*. Classifications place *Hainardia* and *Parapholis* either in *Poeae* or *Hainardieae*. The analyses of structural character data placed these as members of *Aveneae* allied to *Gaudinia*. However, the cpDNA and simultaneous analyses placed them as members of a subclade within *Poeae,* with *Cynosurus, Sphenopus, Catapodium, Cutandia,* and *Desmazeria.* We suggest that this cpDNA clade represents a natural alliance of taxa, most of which are annuals, and among which spicate inflorescences that disarticulate along the axis are interpreted as resulting from a trend toward panicle reduction and branch disarticulation, as is especially evident among species of *Catapodium, Desmazeria,* and *Cutandia.* The spike in *Gaudinia* represents convergence on this form in *Aveneae.* As such, the *Hainardieae* could be placed within subtribe *Parapholinae* in *Poeae,* or if *Ammochloa* is included in this lineage as we suggest, in subtribe *Ammochloinae.*

CONCLUSIONS

We suggest the following summary of tribal relationships for *Pooideae*: Brachyelytreae ((*Lygeeae Nardeae*) ((*Phaenospermatideae* (*Anisopogon Ampelodesmeae Stipeae*)) (*Diarrheneae* (*Meliceae* (*Brachypodieae* ((*Bromeae Triticeae*) (*Aveneae Poeae*))))))).

This is quite similar to tribal relationships detected in other molecular analyses. However, different tribal representation and incongruent results among the present and other studies leave us with an unresolved picture of relationships among tribes situated between 'Brachyelytreae ((*Nardeae Lygeeae*)' and '((*Bromeae Triticeae*) (*Aveneae Poeae*))'.

A number of genera of *Poeae* and *Aveneae* appear to be either misplaced by cpDNA restriction site data or misclassified in contemporary taxonomic treatments. Out of 82 genera of *Pooideae* sampled, nine that are usually classified in one tribe, either *Poeae* (*Briza, Chascolytrum, Poidium, Torreyochloa*), or its sister tribe *Aveneae* (*Aira, Avenula, Holcus, Mibora, Molineriella*), were unexpectedly grouped with members of the other tribe by cpDNA alone, and by the simultaneous analyses. Five other genera that had never been placed in *Poeae,* but sometimes had been classified in tribes other than *Aveneae* (i.e., *Scribneria* in *Hainardieae*; *Milium* in *Milieae* or *Stipeae*; *Alopecurus, Beckmannia,* and *Phleum* in *Phleeae*), were unexpectedly placed among members of *Poeae* by the cpDNA data and simultaneous analyses. In our analysis of structural characters, including these 14 genera, the genera of the *Phleum* complex were placed as members of *Poeae,* whereas most of the other genera were placed in more or less traditional tribal arrangements within *Poeae* or *Aveneae.*

The simultaneous analysis supported the tribal placements resolved by the cpDNA analysis, but also provided additional resolution of relationships within and among the tribes beyond that produced by cpDNA alone. We examined each of these genera using a set of criteria by which hybrids might be detected. We conclude that the strongest cases for intertribal hybrid origin are those of *Ampelodesmos* and *Avenula.*

Our results suggest that *Avenula* may have arisen from hybridization between *Helictotrichon* and some unidentified member of *Poeae.* In addition, either a second hybridization event took place giving rise to the immediate ancestors of *Holcus, Aira, Molineriella,* and relatives, or this set of genera may have evolved from the products of the original hybridization event which produced *Avenula.* It is further possible that *Avenula pubescens* represents a

back-cross to *Helictotrichon.* In any case, this otherwise typical group of genera of *Aveneae* are placed among genera of *Poeae* in the cpDNA and simultaneous analyses.

All other cases examined for hybridization as an explanation for the cpDNA placement in a tribe other than that (or those) in which the genera are usually classified are considered weaker. The apparently anomalous morphological character combinations in these genera may equally or more plausibly be considered cases of convergent evolution.

Two considerations compel us to agree with the suggestion of Tzvelev (1989) that the taxonomic distinction between *Poeae* and *Aveneae* be abandoned in favor of the recognition of one tribe, *Poeae,* with a series of subtribes. First, intertribal hybridization remains a plausible argument for the conflicting placements of genera resolved by the various data sets. Extensive interfertility is hardly a desirable attribute among genera, much less among tribes. Second, even if intertribal hybridization is discounted, the best alternative explanation for the present results is that parallel and convergent evolution have played such a significant role in blurring the morphological distinctions between the two tribes that a two tribe system is unjustified. There may be additional misplaced taxa still to be discovered, and a greater number of individually less inclusive groupings may provide a better summary of our understanding of this complex. We propose the following modified classification:

Tribe *Poeae* R.Br.: Accepted groups, subgroups, and subtribes.

Group 1:

- subgroup 1: *Amphibromus*[A].

- subgroup 2: subtr. *Aveninae* J. Presl.

- subgroup 3: *Torreyochloa*[A], subtrr. *Brizinae*[B] Tzvelev, *Phalaridinae* Fr., *Agrostidinae* Fr. (including *Deschampsia s.lat.*).

Group 1 x Group 2: *Avenula,* subtrr. *Airinae* Fr., *Holcinae* Dumort. ex Tzvelev.

Group 2:

- subgroup 1, subtrr. *Cynosurinae* Fr., *Ammochloinae* Tzvelev (or *Parapholiinae*?[C] Caro), *Loliinae* Dumort. (incl. *Festucinae* J. Presl), *Puccinellia* complex[A].

- subgroup 2, subtrr. [A]*Scolochloa, Dactylidinae* Stapf, *Sesleriinae* Parl., *Poinae* Dumort., *Alopecurinae* Dumort., *Miliinae* Dumort.

[A]Possible new subtribes.

[B]Provisionally including South American genera, *Chascolytrum* and *Poidium.*

[C]Former *Hainardieae* plus related *Poeae,* the name depending on the inclusion or exclusion of *Ammochloa.*

REFERENCES

Albers, F, and Butzin, F. (1977). Taxonomie und Nomenklatur der Subtriben *Aristaveninae* und *Airinae* (Gramineae – Aveneae). *Willdenowia* **8**, 81-84.

Avdulov, N. P. (1931). Kario-sistematicheskoe issledovanie semeystva zlakov. *Trudy po Prikladnoj Botanike, Gentike i Selekcii*, supplement **44**.

Catalan, P., Kellogg, E. A., and Olmstead, R. G. (1997). Phylogeny of *Poaceae* subfamily *Pooideae* based on chloroplast *ndh*F gene sequences. *Molecular Phylogenetics and Evolution* **8**, 150-166.

Clark, L. G., Zhang W.-p., and Wendel, J. F.. (1995). A phylogeny of the grass family (Poaceae) based on *ndh*F sequence data. *Systematic Botany* **20**, 436-460.

Clayton, W. D., and Renvoize, S. A.. (1986). Genera Graminum, grasses of the world. *Kew Bulletin* (additional series) XIII.

Darbyshire, S. J., Cayoutte, J., and Warwick, S. I. (1992). The intergeneric hybrid origin of *Poa labradorica* (Poaceae). *Plant Systematics and Evolution* **181**, 57-76.

Davis, J. I, and Soreng, R. J. (1993). Phylogenetic structure in the grass family (*Poaceae*) as inferred from chloroplast DNA restriction site variation. *American Journal of Botany* **80**, 1444-1454.

Funk, V. A. (1985). Phylogenetic patterns and hybridization. *Annals of the Missouri Botanical Garden*, **72**, 681-715.

Gillespie, L.J., Consaul, L. L., and Aiken, S. G. (1997). Hybridization and the origin of the arctic grass *Poa hartzii* (*Poaceae*): evidence from morphology and chloroplast DNA restriction site data. *Canadian Journal of Botany* **75**, 1978-1997.

Goloboff, P. (1993). *Nona,* version 1.16 (computer software and manual). Distributed by the author.

Grebenstein, B. Röser, M, Sauer W., and Hemleben, V. (1998). Molecular phylogenetic relationships in Aveneae (Poaceae) species and other grasses as inferred from ITS1 and ITS2 rDNA sequences. *Plant Systematics and Evolution* **213**, 233-250.

Hiesey, W. M., and Nobs, M. A. (1982). Experimental studies on the nature of species VI. Interspecific hybrid derivatives between facultatively apomictic species of bluegrasses and their responses to contrasting environments. Carnegie Institute of Washington, Publication No. 636.

Hilu, K.W., and Liang H.-p. (1997). The *mat*K gene: sequence variation and application in plant systematics. *American Journal of Botany* **84**, 830-839.

Hsiao, C., Chatterton, N. J., and Asay, K. H. (1995). Molecular phylogeny of the *Pooideae* (*Poaceae*) based on nuclear *r*DNA (ITS) sequences. *Theoretical and Applied Genetics* **90**, 389-398.

Hsiao, C., Jacobs, S. W. L., Chatterton, N. J., and Asay, K. H. (1999). A molecular phylogeny of the grass family (*Poaceae*) based on the sequences of nuclear ribosomal DNA (ITS). *Australian Systematic Botany* **11**, 667-688.

Hubbard, C. E. (1959). *Gramineae*. In 'The families of flowering plants' (2nd. ed.) (Ed. J. Hutchinson) vol. 2, pp. 710-741. (Clarendon Press: Oxford).

Kellogg, E. A., and Juliano, N. D. (1997). The structure and function of *RuBisCo* and their implications for systematic studies. *American Journal of Botany* **84**, 413-428.

Kellogg, E. A., and Appels, R. (1995). Intraspecific and interspecific variation in 5S RNA genes are decoupled in diploid wheat relatives. *Genetics* **140**, 325-343.

Kellogg, E. A., and Campbell, S. C. (1987). Phylogenetic analyses of the *Gramineae*. In 'Grass systematics and evolution'. (Eds T. R. Soderstrom, K. W. Hilu, C. S. Campbell, and M. E. Barkworth) pp. 310-322. (Smithsonian Institution Press: Washington, D.C.)

Kellogg, E. A., and Watson, L. (1993). Phylogenetic studies of a large data set. I. *Bambusoideae, Andropogonodae*, and *Pooideae* (*Gramineae*). *The Botanical Review* **59**, 273-243.

Kellogg, E. A., Appels, R., and Mason-Gamer, R. J. (1996). When gene trees tell different stories, incongruent gene trees for diploid genera of the *Triticeae* (*Gramineae*). *Systematic Botany* **21**, 321-347.

Knobloch, I. W. (1968). A check list of crosses in the *Gramineae*. Privately published.

Knobloch, I. W. (1972). Intergeneric hybridization in flowering plants. *Taxon* **21**, 97-103.

MacFarlane, T. D., and Watson, L. (1982). The classification of *Poaceae* subfamily *Pooideae*. *Taxon* **31**, 178-203.

Mason-Gamer, R.J., and Kellogg, E.A. (1997). Testing for phylogenetic conflict among molecular data sets in the tribe *Triticeae* (*Gramineae*). *Systematic Biology* **45**, 524-545.

McDade, L. (1990). Hybrids and phylogenetic systematics. I. Patterns of character expression in hybrids and their implications for cladistic analysis. Evolution **44**, 1685-1700.

Meng K.-c., Soreng, R. J., and Davis, J. I. (1994). Phylogenetic relationships among *Puccinellia* and allied genera of *Poaceae* as inferred from chloroplast DNA restriction site variation. *American Journal of Botany* **81**, 119-126.

Nadot, S., Bajon, R., and Lejeune, B. (1994). The chloroplast gene *rps*4 as a tool for the study of *Poaceae* phylogeny. *Plant Systematics and Evolution* **191**, 27-38.

Nixon, K. C. (1993). *CLADOS*, version 1.4.98 (computer software and manual). Distibuted by the author.

Nixon, K.C. (1998). The Parsimony Ratchet, a new method for rapid parsimony analysis and broad sampling of tree islands in large data sets. [Abstract of presentation at Hennig XVII, Sao Paolo, September 1998, available on the website of the Willi Hennig Society]

Nixon, K. C. (1997). DADA, version 1.1.4 (computer software and manual). Distibuted by the author.

Pilger, R. (1954). Das system der *Gramineae*. *Botanische Jahrbücher für Systematik* **74**, 199-265.

Prat, H. (1960). Vers une classification naturelle des Gramineé's. Bulletin de la Société Botanique de France **107**, 32-79.

Prat, H. (1932). L'épiderme des Graminées. Etude anatomique et systématique. *Annales des Sciences Naturelles* (ser. 10) **14**, 118-325.

Prat, H. (1936). La systématique des Graminées. *Annales des Sciences Naturelles, Botanique* (series 10) **18**, 165-258.

Seberg, O., and Petersen, G. (1998). A critical review of concepts and methods used in classical genome analysis. . *The Botanical Review* **64**, 372-417.

Soreng, R. J. (1990). Chloroplast-DNA Phylogenetics and biogeography in a reticulating group: Study in *Poa* (*Poaceae*). *American Journal of Botany* **77**, 1383-1400.

Soreng, R. J., and Davis, J. I. (1998). Phylogenetics and characters evolution in the grass family (*Poaceae*): Simultaneous analysis of morphological and chloroplast DNA restriction site character sets. *The Botanical Review* **64**, 1-85.

Soreng, R. J., Davis, J. I, and Doyle, J. J. (1990). A phylogenetic analysis of chloroplast DNA restriction site variation in *Poaceae* subfam. *Pooideae*. *Plant Systematics and Evolution* **172**, 83-97.

Stebbins, G. L. (1956). Cytogenetics and the evolution of the grass family. *American Journal of Botany* **43**, 890-905.

Stebbins, G. L. (1981). Chromosomes and evolution in the genus *Bromus* (*Gramineae*). *Botanische Jahrbücher für Systematik*. **102**, 359-379.

Tateoka, T. (1957). Miscellaneous papers on the phylogeny of *Poaceae*, X: Proposition of a new phylogenetic system of *Poaceae*. *Journal of Japanese Botany* **32**, 275-287.

Tzvelev, N. N. (1976). 'Zlaki S.S.S.R.' [1983. 'Grasses of the Soviet Union'. (English translation for the Smithsonian Institution, Amerind Publishing Co.: New Delhi.)]

Tzvelev, N. N. (1989). The system of Grasses (*Poaceae*) and their evolution. *The Botanical Review* **55**, 141-204.

Watson, L. and Dallwitz, M. J. (1992). 'The Grass Genera of the World'. (CAB International: Wallingford, UK.)

Grasses: Systematics and Evolution. (2000). Eds S.W.L. Jacobs and J. Everett. (CSIRO: Melbourne)

RELATIONSHIPS WITHIN THE STIPOID GRASSES (GRAMINEAE)

S.W.L. Jacobs[A], Joy Everett[A], Mary E. Barkworth[B] and Cathy Hsiao[C]

[A]National Herbarium, Royal Botanic Gardens, Sydney, NSW 2000, Australia.
[B]Intermountain Herbarium, Dept of Biology, Utah State University, Logan, Utah 84322-5305, USA.
[C]USDA Agricultural Research Service, Forage and Range Research, Utah State University, Logan Utah 84322-6300, USA.

Abstract

The relationships within the Stipoid grasses were examined using two data sets: (i) sequence data of the internal transcribed spacer (ITS) region of nuclear rDNA; and (ii) data from 30 anatomical and morphological characters. The ITS analysis used 67 species from nine different genera (*Austrostipa, Anemanthele, Achnatherum, Stipa s. str., Nassella, Oryzopsis, Piptochaetium, Hesperostipa,* and *Piptatherum*) as well as one species each of *Nardus, Anisopogon, Brachyelytrum, Ampelodesmos* and *Diarrhena. Nardus* and/or *Brachyelytrum* were used as the outgroup. The data were analysed using the neighbour-joining option of MEGA. The anatomical/morphological analysis was based on 30 characters using 38 species groups. Most (23) were monophyletic groups of *Austrostipa*. Other groups/genera were *Achnatherum, Ptilagrostis,* 'Boreobtuseae', *Piptatherum* sect. *Virescentia, Stipa, Hesperostipa, Piptochaetium, Stipa* sect. *Podopogon, Nassella, Stephanostipa, Oryzopsis, Piptatherum* sect. *Piptatherum, Nardus, Nicoraella* (as 'Obtusae'), and *Piptatherum miliaceum.* The data were analysed using PAUP with *Joinvillea* as the outgroup.

These and other analyses indicate that *Anisopogon, Ampelodesmos, Diarrhena, Brachyelytrum* and *Nardus* are all associated with the Stipoid grasses but their relationships are unstable between analyses. *Nardus* and *Brachyelytrum* are frequently inserted at the base of the clade. Our morphological and ITS analyses more or less agree on: (i) the wider concept of *Nassella* (e.g., the inclusion of *Stephanostipa*); (ii) the distinction of *Austrostipa* and *Achnatherum* and the sister group relationship of these two genera; (iii) the distinctness of *Hesperostipa*; (iv) the unsatisfactory nature of the current circumscription of *Oryzopsis* and *Piptatherum*; (v) the distant relationship of *Piptatherum miliaceum* to other species currently assigned to either *Piptatherum* or *Oryzopsis*; and (vi) the subgeneric classification in *Austrostipa*. The ITS analyses further indicated that: (i) *Anemanthele* may not be worth recognising as a distinct genus; (ii) *Achnatherum* may not be monophyletic; and (iii) *Stipa s. str.* may not be monophyletic. The morphological analysis further supported: (i) the inclusion of *Podopogon* in *Piptochaetium*; and a sister group relationship for *Nassella* and *Piptochaetium*.

Key words: Stipeae, ITS, morphology, anatomy, *Achnatherum, Nassella, Stipa, Austrostipa, Piptatherum, Anemanthele, Oryzopsis, Hesperostipa.*

INTRODUCTION

There has been considerable resistance to recognising generic offshoots of *Stipa*. Some of the generic names now being resurrected were first proposed more than 150 years ago. It is a familiar story, with resistance to the breaking up of genera such as *Panicum, Poa* and *Festuca* still being common. The reasons for such resistance are many. Often we simply do not wish to change. Sometimes the suggested changes mean that a local intuitively satisfying classification is replaced by one that may make the overall picture clearer at the expense of clarity at the local scale. All of these problems need to be resolved before a new classification is widely accepted, and progress invariably happens piecemeal because of the different scales involved.

Barkworth (1990) details the history of use of generic names in the Stipeae. We will briefly summarise that here but concentrate on changes since that time.

Bentham (1882) reduced what we now recognize as the Stipeae to two genera, *Stipa* and *Oryzopsis*, but taxonomists in other parts of the world often recognised more genera, particularly for species with short florets. For instance, Hitchcock (1935) accepted both *Piptochaetium* and *Nassella*, albeit treating each as including only species with short, relatively stout, florets. Parodi (1947) documented the need to expand *Piptochaetium* to include species with elongated spikelets at the time treated as members of *Stipa* sect. *Podopogon*. This treatment was quickly accepted in South America, where the genus is most abundant. But, because of the pervasive influence of Hitchcock (1935, 1951), and the failure to read non-North American literature, the wisdom of Parodi's recommendation was not accepted in North America until after Thomasson (1978) showed that it was also supported by lemma epidermal morphology.

Nassella was restricted to species with short florets (Duran and Rosengurtt 1956; Caro 1966; Muñoz-Schick 1990). Thomasson (1976) demonstrated that it too should be expanded to include several species with long florets that were then included in *Stipa*, a change that is now reflected in several treatments (Vickery and Jacobs 1980; Wheeler *et al.* 1982; Barkworth and Everett 1987; Barkworth 1990; Torres 1997a).

Oryzopsis has either been interpreted as a genus that occurs in both Eurasia and North America (Bentham 1882; Hitchcock 1935, 1951; Johnson 1945), or as a strictly North American genus, with the Eurasian taxa then being placed in *Piptatherum* (Freitag 1975; Tzvelev 1976), or as a monotypic genus (Kuo *et al.* 1983; Barkworth and Everett 1987; Everett 1990). If it is treated as a monotypic genus, the problem becomes one of determining where the excluded species belong. Barkworth (1993) placed most in *Piptatherum*, some in *Achnatherum*, and left the status of three North American species (*O. pungens*, *O. canadensis*, and *O. exigua*) unresolved.

After comparing *O. asperifolia* with Eurasian species of *Oryzopsis sensu lato*, Freitag (1975) concluded that the Eurasian species constituted a separate genus, *Piptatherum*, with three sections. He also drew attention to the isolation of *P. miliaceum* from other members of the genus.

Tzvelev (1976) recognized three stipoid genera in the Soviet flora, *Stipa sensu stricto*, *Achnatherum*, and *Ptilagrostis*. A year later, after reviewing arguments presented by Keng (1965), Tzvelev (1977) expanded *Achnatherum* to include several of the species that he had previously included in *Stipa*. This change was supported by Barkworth and Everett (1987) and expanded by Barkworth (1993) to several North American species. Although there are many Mexican and South American species that appear to belong in the genus, as interpreted by these authors, the nomenclatural changes have not been made because of problems that exist at both the generic and species level. Recognition of

Achnatherum is not supported by all taxonomists. For instance, Freitag (1985) and Curto (1998) have advocated adopting a broad concept of *Stipa*, one that is essentially identical to that adopted by Bentham (1882).

Freitag (1985) classified Asian species of *Stipa* by treating *Achnatherum* and *Ptilagrostis* as sections. While a sectional treatment works well in a confined geographical area, the total variation led Barkworth (1993) to follow Tzvelev (1976), and to describe a further genus *Hesperostipa*. Since Barkworth (1993), the genus *Jarava* has been resurrected (Jacobs and Everett 1997), and new genera *Nicoraella* (Torres 1997b) and *Austrostipa* (Jacobs and Everett 1996) described, with probably more to come (Vazquez, pers. comm.).

Given the recent rapid changes, and the potential for several more, it is appropriate to review the current situation and to use two different data sets to demonstrate where problems exist.

MATERIALS AND METHODS

Plant Material
DNA was extracted from 67 species of *Stipa s. lat.* (*Achnatherum, Anemanthele, Austrostipa, Nassella, Oryzopsis, Piptatherum, Piptochaetium, Stipa s. str.,* and *Hesperostipa*), as well as one species each of *Nardus, Anisopogon, Brachyelytrum, Ampelodesmos, Diarrhena, Lithacne* and *Oryza*. Vouchers are mostly lodged at NSW and UTC; details of voucher and source material are available from NSW (SWLJ) and UTC (MEB).

Thirty morphological/anatomical characters were scored for 39 taxa. The 39 taxa included 23 groups within *Austrostipa* (58 species); *Achnatherum* (77 species); 'Boreobtuseae' (3 species); *Nassella* (10 species); *Oryzopsis* (1 species); *Piptatherum* sections *Piptatherum* (9 species), *Miliacea* (2 species), *Virescentia* (2 species); *Piptochaetium* (17 species); *Ptilagrostis* (2 species); *Stipa s. str.* (17 species); *Hesperostipa* (3 species); *Podopogon* (2 species); *Stephanostipa* (40 species); *Nicoraella* (2 species) *Nardus* (1 species); and *Joinvillea* (2 species as outgroup, Campbell and Kellogg 1987). Vouchers are mainly at NSW and K, with some from US, UTC, FLAS, CANB. Details of vouchers are available from Everett (1990).

DNA Extraction, Amplification and Sequencing
Extraction methods of total DNA varied according to the condition of the material. For fresh leaves, a modified CTAB method was used (Williams *et al.* 1993). Ethanol-preserved leaves were extracted by the phenol-chloroform method (Hsiao *et al.* 1994). Herbarium specimens were extracted using a silica matrix method (Poinar 1994; C. Hsiao, unpublished data), developed to eliminate any inhibition from the dried herbarium samples, and used by Hsaio *et al.* (1998, 1999).

DNA amplification and sequencing were as described by Hsiao *et al.* (1994, 1998, 1999).

Morphological/Anatomical Characters
The characters used were selected from a group of 50 characters used in preliminary analyses (Everett 1990). The final 30 characters chosen were those with higher consistency indices that contributed most to the trees in preliminary analyses.

The characters used (and the states recorded) were: rootstock (rhizomatous/caespitose); shoots (extravaginal/intravaginal); culms (simple/branched); glume girders (absent/present); glume length (proportional to lemma); floret shape (turbinate/cylindrical); awn (centric/excentric); lemma compression (cylindrical/dorsally compressed/laterally compressed); lemma indumentum (presence/absence; overall/only on nerves); lemma rolling (convolute/involute); callus (obtuse/acute); callus disarticulation scar (round/oval/linear); awn (persistent/caducous; straight/geniculate); column (fine, short/robust, long); palea length (proportional to lemma); palea tip (rounded/keeled); palea indumentum (absent/present); lodicule number; (lemma) fundamental cells sidewall (shape: straight/sinuate/strongly wavy; thickness: thin/thickened; silicification: present/absent); (lemma fundamental cell) endwall shape (straight/wavy); (lemma) short cells (with/without hooks); (lemma) silica bodies (present/absent); silica body shape (round or oval/square or rectangular).

Specimens were examined and the data recorded directly except for *Joinvillea* where recorded data were supplemented by information from Newell (1969), and Tomlinson and Smith (1970) when specimens at NSW lacked the relevant structures.

DNA Sequence Analysis

Sequences of the entire ITS region were aligned with the LINEUP program of the University of Wisconsin Genetics Computer Group (known as GCG) and the CLUSTAL multiple sequence alignment programs (Higgins *et al.* 1992). In the complete data set, alignment is often random in the region with long contiguous gaps due to deletion or insertion events. To detect and remove random pairing a stepwise alignment strategy was used (Hsaio 1998,1999). This strategy requires aligning closely related groups first, then aligning the full sequence data set with the GCG and CLUSTAL multiple sequence alignment programs. Any new taxa or taxa of questionable affinities were tested repeatedly against other taxa. If such taxa proved unstable in these analyses, every attempt was made to find potentially pairing or closely related taxa for sampling to provide more robust results. The final alignment was optimised manually. This process eliminated most of the random pairing of the characters and is critical for sequence alignment of the variable DNA region. Regions with high deletion or insertion events were also assessed by stepwise exclusion tests and compared to the complete data set; the overall phylogenetic conclusions were not affected by their omission. Therefore, phylogenetic analysis of the complete data set was effected and no data were discarded. Gaps were treated as new states.

The sequence alignments for various data sets are too large to be included in this paper but they are available upon request to Cathy Hsiao.

Various genera were used as outgroups — *Ampelodesmos, Nardus, Lygeum, Brachyelytrum, Anisopogon, Diarrhena, Lithacne* and *Oryza* — both singly and in various combinations. Results from Hsiao *et al.* (1998) indicated all these genera were potential outgroups. In practice all were effective and produced similar results, although both *Anisopogon* and *Diarrhena* were unstable and sometimes appeared among the other core genera of Stipeae. *Brachyelytrum* and *Nardus* were the genera we used most frequently as outgroups and the ones used in the trees presented.

MEGA, version 1.01 (Kumar *et al.* 1993), was used to compute sequence statistics and bootstrap values. The neighbour-joining distance method (NJ) of MEGA was used for phylogenetic reconstruction using Kimura 2-parameter distance and pairwise gap deletion options. We also used PAUP (Swofford 1993) in our previous studies (Hsiao *et al.* 1998,1999) but we found the neighbour-joining option of MEGA produced similar but slightly more consistent results.

Morphological/Anatomical Data Analysis

The morphological/anatomical data were analysed using PAUP (Swofford 1985) and the options detailed in Everett (1990) (Global branchswapping, HTU character state optimisation, trees rooted with respect to one or more outgroups) and presented as a consensus tree.

The *Austrostipa* species were analysed separately to define monophyletic groups for inclusion in this analysis. Details of this process and composition of the groups are in Everett (1990).

These monophyletic groups of *Austrostipa* were then included in an analysis of the whole group as outlined above.

RESULTS AND DISCUSSION

There was very little or no sequence variation within a species. Small intraspecific variation (two to four bases) was observed in two species, *Jarava ichu* and *Achnatherum hymenoides,* both of which are common and widely distributed species.

Results from Hsiao *et al.* (1999) indicated that the subfamily Pooideae and, within it, the tribe Stipeae were monophyletic. The genera *Brachyelytrum, Nardus, Lygeum* and *Diarrhena* were basal to those genera normally considered to be core representatives of the tribe (Barkworth and Everett 1987). *Anisopogon* and *Ampelodesmos* were included within these core genera. The larger data set in this study (Fig. 1) more or less confirmed these placements but demonstrated that the positions of *Diarrhena* (also sometimes included among the core genera), *Anisopogon* and *Ampelodesmos* were unstable in the sense that their position in the classification changed with minor alterations in the species used to make up the data set. Fig. 1 is an example of a classification including these genera. The results of the morphological data analysis (Fig. 2) placed *Nardus* among the core genera. While our results indicate that all six genera are clearly associated with the Stipeae, it is not really clear just how close that association should be.

Lygeum and *Brachyelytrum* were never placed among the core genera in our analyses. *Nardus* was only placed within the core genera in the morphological data analysis. The confusion is not resolved here, the evidence not being strong enough to suggest removal of these three genera from their unigeneric tribes (Clayton and Renvoize 1986; Watson and Dallwitz (1994). There is, however, little evidence to support the inclusion of *Nardus* and *Lygeum* within the subfamily Arundinoideae as suggested by Watson and Dallwitz, though they do also place the Stipeae in that subfamily. In the updatable web version of Watson and Dallwitz (http://www.biodiversity.bio.uno.edu/delta/), *Nardus, Lygeum* and *Ampelodesmos* are currently included in their subfamily Stipoideae, and *Brachyelytrum* and *Diarrhena* in their Bambusoideae.

Fig. 1. Phylogram (showing relative branch lengths) produced from ITS data using CLUSTAL. Outgroups are *Brachyelytrum erectum* and *Nardus stricta*.

Anisopogon, on the other hand, does seem to be consistently associated with the Stipeae and our data support Watson and Dallwitz (1994) in their placement of that genus, and not its inclusion in the Arundineae, as suggested by Clayton and Renvoize (1986).

Our DNA data suggest positions for *Ampelodesmos* and *Diarrhena* somewhere near the Stipeae, but the multi-floreted spikelet structure in both of these genera means that their posi-

tions will remain in dispute. The lack of decisive placement for any of these genera indicates that they are not closely related to any of the other taxa sampled. It may be possible to sample other taxa to help improve the resolution but, if there are no extant close relatives, their positions may remain uncertain.

The two data sets used different taxonomic units, the ITS data (Fig. 3) using species as the basic unit, while the morphologi-

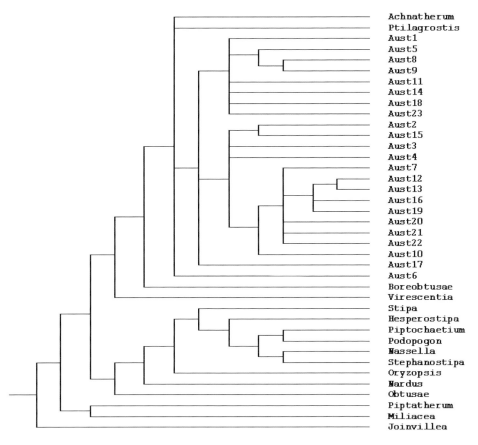

Fig. 2. Strict Consensus Tree using morphological data produced by CONTREE (Swofford 1985) from 5 trees (50 output from PAUP). Each terminal taxon was determined to be monophyletic by a series of preliminary analyses. The outgroup is *Joinvillea*.

cal/anatomical data set (Fig. 2) used predetermined monophyletic groups. Despite this, there is a surprising degree of concordance between the results obtained. Both analyses support:

(i) the wider concept of *Nassella*, i.e. the inclusion of *Stipa* subgen. *Stephanostipa* (Barkworth 1990);

(ii) the separation and distinction of *Austrostipa* from *Stipa s.str.* (Jacobs and Everett 1996);

(iii) the subgeneric classification of *Austrostipa* (Jacobs and Everett 1996);

(iv) the recognition of *Achnatherum* (Barkworth 1993);

(v) the sister group relationship of *Achnatherum* and *Austrostipa* (Barkworth and Everett 1987);

(vi) the distinctness of *Hesperostipa* from *Stipa s. str.* (Barkworth and Everett 1987; Barkworth 1993);

(vii) the restriction of *Oryzopsis* to the single species *O. asperifolia* (Kuo *et al.* 1983; Barkworth and Everett 1987); and

(viii) the unsatisfactory nature of the current circumscription of *Oryzopsis* and *Piptatherum* despite the improvements made by Freitag (1975).

The analysis of the ITS sequences further indicated that:

(i) *Anemanthele* may not be worth maintaining as a genus distinct from *Austrostipa*, unless that genus is divided even further;

(ii) *Achnatherum* may not be monophyletic;

(iii) *Jarava* probably is a monophyletic group (Jacobs and Everett 1997); and

(iv) *Stipa s. str.* may not be monophyletic.

The analysis of the morphological/anatomical data further supported:

(i) the inclusion of *Podopogon* within *Piptochaetium* (Parodi 1944; Thomasson 1977; Barkworth and Everett 1987);

(ii) a sister group relationship for *Nassella* and *Piptochaetium*; and

(iii) the distinctiveness of *Nicoraella* (Torres 1997b), the 'Obtusae Group' of Parodi (1946) and Barkworth and Everett (1987).

Achnatherum. While both analyses support the expanded concept of *Achnatherum*, the ITS analyses suggest that neither the genus as originally circumscribed, nor that based on the expanded circumscription is monophyletic. The analysis presented in Fig. 3 demonstrates that most of *Achnatherum* is monophyletic with outliers of *A. splendens* (with *Oryzopsis*),

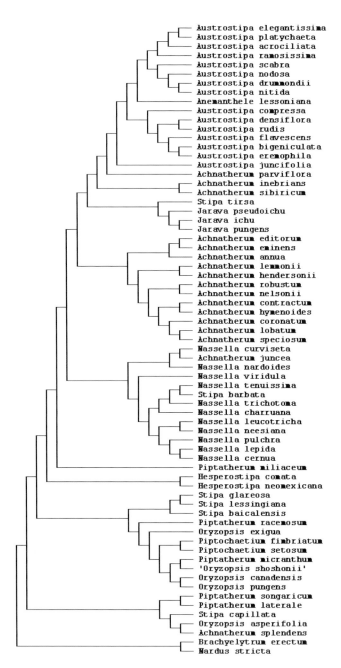

Fig. 3. Cladogram produced from ITS data using CLUSTAL with *Brachyelytrum erectum* and *Nardus stricta* as outgroups.

A. juncea (with *Nassella*) and two separate clades, one of a single species (*A. parviflora*) and one of two (*A. inebrians* and *A. sibiricum*), if *Jarava* is recognised as distinct. While the first two outliers consistently appear in the same place, the next two small clades are unstable, varying in content with different data sets. It is not immediately obvious from these analyses what the best options are, apart from the recognition of *Jarava*, as none of the outliers seem to represent coherent groups. Further work is required to sort out the optimum treatment.

Anemanthele is distinguished by its single stamen and characteristic hilum, though there is some doubt about the hilum character being as distinct as originally stated by Veldkamp (1985) (Ever-

ett 1990). Barkworth and Everett (1987) included *Anemanthele* in their concept of *Achnatherum*; Jacobs and Everett (1997) retained it as distinct. The restricted ITS analysis shows it included in *Austrostipa*, basal to all except *Austrostipa juncifolia*. The position of *Anemanthele* varied slightly with the outgroups used but was always retained within the *Austrostipa* clade. The problem is not resolved and more samples, especially of *Austrostipa* subgen. *Lobatae*, need to be included in further analyses.

Jarava was first recognised by Ruiz & Pavon (1794) but, until recently (Jacobs and Everett 1997), was treated as a section of the genus *Stipa* (Caro and Sanchez 1973), along with many of the genera now recognised as distinct. The diaspores are wind-dispersed and have a range of morphological characters that distinguish them from the rest of *Achnatherum* (Jacobs and Everett 1997). The ITS analysis supports the recognition of *Jarava* as a distinct clade.

Oryzopsis has had a variety of treatments. Freitag (1975) in his treatment of *Piptatherum* recognised several species in *Oryzopsis* and referred to a table where he compares *Piptatherum* with the type species of *Oryzopsis, O. asperifolia*, and to another group of North American species of *Oryzopsis*. Interestingly, this latter column is missing from the table. Kuo *et al.* (1983), Barkworth and Everett (1987) and Everett (1990) have all suggested that *Oryzopsis* should be monotypic and restricted to *O. asperifolia*. Barkworth (1993) moved several species to *Achnatherum*. Our results here further support those conclusions. The problem lies with where the other species should go. The 'easy' solution has been to include them within an expanded *Piptatherum* but neither of our analyses support that option. Many species of *Oryzopsis s.lat.* associate more closely with some *Piptatherum* spp. than they do with other *Oryzopsis* spp. and the options will be discussed further under *Piptatherum*.

Piptatherum was revised by Freitag (1975). While his generic circumscription is not supported by either analysis, his subgeneric classification is recognisable, if somewhat modified. Both analyses recognised four groups in the *Oryzopsis/Piptatherum* complex (excluding *Oryzopsis s. str.*). *P. miliaceum* was recognised as distinct in both analyses, and *Piptatherum* sect. *Piptatherum* recognised as a monophyletic group in both analyses. The 'Boreobtuseae' (*O. canadensis, O. pungens* and *O. exigua*) was a monophyletic group in the morphological analysis. In the ITS analysis, the branch including these three taxa also included *S. 'shoshoneana', P. micranthum* and *P. racemosum* (Fig. 3) and two species of *Piptochaetium*. The two Asian species of *Piptatherum* sampled, *P. songaricum* and *P. laterale*, appeared together elsewhere, as a sister group to *O. asperifolia*.

Piptochaetium in the ITS analysis is recovered as a monophyletic unit but is embedded among members of the *Piptatherum/Oryzopsis* complex. Only two species of *Piptochaetium* were sampled and this may explain its unexpected position in the classification. In the PAUP analysis, 17 spp. of *Piptochaetium* were sampled and the genus was recognised as a monophyletic group and a sister group to *Nassella*, a more intuitively satisfying result.

Stipa s.str. is represented by a small monophyletic clade of three species in the ITS analysis (*S. glareosa, S. lessingiana* and *S. baicalensis*) while three other species (*S. tirsa, S. barbata* and *S. capillata*) are scattered (Fig. 3), and their relative positions remained reasonably stable, altering only slightly with different data sets. This lack of monophyly is a surprising result as it has always been assumed that *Stipa s.str.* represents a closely related group and the number of species sampled was based on that assumption. Seventeen species were included in the group for the PAUP analysis (Fig. 1) and little variation recorded. The results from the ITS analysis are not adequate to suggest anything more than the need to include more species from this group in any further analyses. There is obviously a need to examine even this new narrowly-defined genus in more detail.

Phylogeny

There is little support from either analysis for the suggestion of two distinct lineages of American Stipeae (Barkworth and Everett 1987). Both analyses suggest that, of the core genera, members of the *Piptatherum/Oryzopsis* complex are basal, with taxa from the Americas and Europe/Asia then distributed through the trees. Both analyses indicate that *Austrostipa* is probably the most recently derived genus or group.

ACKNOWLEDGEMENTS

We would like to thank Les Watson and Maria Torres for valuable comments on the manuscript.

REFERENCES

Baldwin, B.G. (1992). Phylogenetic utility of the internal transcribed spacers of nuclear ribosomal DNA in plants: an example from the Compositae. *Molecular Phylogenetics and Evolution* **1**, 3-16.

Barkworth, M.E. (1982). Embryological characters and the taxonomy of the Stipeae (Gramineae). *Taxon* **31**, 233-243.

Barkworth, M.E. (1990). *Nassella* (Gramineae, Stipeae): revised interpretation and nomenclatural changes. *Taxon* **39**, 597-614.

Barkworth, M.E. (1993). North American Stipeae (Gramineae): taxonomic changes and other comments. *Phytologia* **74**, 1-25.

Barkworth, M.E., and Everett, J. (1987). Evolution in the Stipeae: identification and relationships of its monophyletic taxa. 'Grass Systematics and Evolution'. (Eds T.R. Soderstrom, K.W. Hilu, C.S. Campbell and M.E. Barkworth) pp. 251-264. (Smithsonian Institution Press: Washington.)

Bentham, G. (1882). Notes on Gramineae. *Journal of the Linnean Society, Botany* **19**, 14-134.

Campbell, C.S., and Kellogg, E.A. (1987) Sister group relationships of the Poaceae. In 'Grass Systematics and Evolution'. (Eds T.R. Soderstrom, K.W. Hilu, C.S. Campbell and M.E. Barkworth) pp. 217-224. (Smithsonian Institution Press: Washington.)

Caro, J.A. (1966). Las especies de *Stipa* (Gramineae) de la region Central Argentina. *Kurtziana* **3**, 7-119.

Caro, J.A., and Sanchez, E. (1973). Las especies de *Stipa* (Gramineae) del subgenero *Jarava*. *Kurtziana* **7**, 61-116.

Clayton, W.D., and Renvoize, S.A. (1986). 'Genera Graminum, Grasses of the World.' (Her Majesty's Stationery Office: London.)

Curto, M. (1998) A new *Stipa* (Poaceae: Stipeae) from Idaho and Nevada, *Madrono*, **45**, 57-63.

Duran, L., and Rosengurtt, B. (1956). La flechilla de los generos *Stipa* y *Piptochaetium* del Uruguay. Asociacion de Estudiantes de Agronomia *Agros Revistas* No. **141**, 9-31.

Everett, J. (1990). *Systematic relationships of the Australian Stipeae (Poaceae)*. M.Sc. thesis, School of Biological Sciences, University of Sydney.

Freitag, H. (1975). The genus *Piptatherum* (Gramineae) in Southwest Asia. *Notes from the Royal Botanic Garden Edinburgh* **33**, 341-408.

Freitag, H. (1985) The genus *Stipa* (Gramineae) in southwest and south Asia. *Notes from the Royal Botanic Garden Edinburgh* **42**, 355-489.

Higgins, D.G., Bleasby, A.J., and Fuchs, R. (1992). CLUSTAL: improved software for multiple sequence alignment. *Comparative and Applied Biosciences* **8**, 189-191.

Hitchcock, A.S. (1925a). The North American species of *Stipa*. *Contributions from the United States National Herbarium* **24**, 215-262.

Hitchcock, A.S. (1925b). Synopsis of the South American species of *Stipa*. *Contributions from the United States National Herbarium* **24**, 263-289.

Hitchcock, A.S., and Chase, A. (1951). 'Manual of the Grasses of the United States'. 2nd ed. United States Department of Agriculture Miscellaneous publication no. 200.

Hsiao, C., Chatterton, N.J., Asay, K.H., and Jensen, K.B. (1994). Phylogenetic relationships of 10 grass species: an assessment of phylogenetic utility of the internal transcribed spacer region in nuclear ribosomal DNA in monocots. *Genome* **37**, 112-120.

Hsiao, C., Chatterton, N.J., Asay, K.H., and Jensen, K.B. (1995a). Molecular phylogeny of the Pooideae (Poaceae) based on nuclear rDNA (ITS) sequences. *Theoretical and Applied Genetics* **90**, 389-398.

Hsiao, C., Chatterton, N.J., Asay, K.H., and Jensen, K.B. (1995b). Phylogenetic relationships of the monogenomic species of the wheat tribe, Triticeae (Poaceae), inferred from nuclear rDNA (internal transcribed spacer) sequences. *Genome* **38**, 211-223.

Hsiao, C., Jacobs, S.W.L., Barker, N.P., and Chatterton, N.J. (1998). A molecular phylogeny of the subfamily Arundinoideae (Poaceae) based on sequences of rDNA. *Australian Systematic Botany* **11**, 41-52.

Hsiao, C., Jacobs, S.W.L., Chatterton, N.J., and Asay, K.H. (1999). A molecular phylogeny of the grass family (Poaceae) based on the sequences of nuclear ribosomal DNA (ITS). *Australian Systematic Botany* **11**, 667-688.

Jacobs, S.W.L., and Everett, J. (1996). *Austrostipa*, a new genus, and new names for Australasian species formerly included in *Stipa* (Gramineae). *Telopea* **6**, 579-595.

Jacobs, S.W.L., and Everett, J. (1993). *Nassella*. In 'Flora of New South Wales' vol. 4. (Ed. G. Harden.) pp. 638-639. (University of New South Wales Press: Sydney.)

Jacobs, S.W.L., and Everett, J. (1997). *Jarava plumosa* (Gramineae), a new combination for the species formerly known as *Stipa papposa*. *Telopea* **7**, 301-302.

Jacobs, S.W.L., Everett, J., and Barkworth, M.E. (1997). Clarification of morphological terms used in the Stipeae (Gramineae) and a reassessment of *Nassella* in Australia. *Taxon* **44**, 33-41.

Johnson, B.L. (1945). Cytotaxonomic studies in *Oryzopsis*. *Botanical Gazette* **107**, 1-32.

Keng, Y-L. (1965). *Flora Illustrata Plantarum Sinicarum*. (Scientific Publishing Co.: Beijing.) [cited in Tsvelev 1977]

Kumar, S., Tamura, K., and Nei, M. (1993). MEGA, Molecular Evolutionary Genetics Analysis, version 1.0. Institute of Molecular Evolutionary Genetics. (The Pennsylvania State University: University Park, Pennsylvania.)

Kuo, P-C., Wang, S., and Lu, S. (1983). [Evolution of spikelet morphology and generic relationships of the tribe Stipeae in China.] *Acta Botanica Boreale-Occidentale Sinica* **3**, 18-27.

Muñoz-Schick, M. (1990). Revision del genero *Nassella* (Trin.) E. Desv. (Gramineae) en Chile. *Gayana Botanica* **47**, 9-35.

Newell, T.K. (1969). A study of the genus *Joinvillea* (Flagellariaceae). *Journal of the Arnold Arboretum* **50**, 527-555.

Parodi, L.R. (1944). Revision de las Gramineas Australes Americanas del genero *Piptochaetium*. *Revista del Museo de la Plata* **6**, 213-310.

Parodi, L.R. (1946). The Andean species of the genus *Stipa* allied to *Stipa obtusa*. *Blumea* Supplement **3**, 63-70.

Parodi, L.R. (1947). Especies de Gramineas del genero *Nassella* de la Argentina y Chile. *Darwiniana* **7**, 369-395.

Poinar, H. (1994). Glass milk, a method for extracting DNA from fossil material. *Ancient DNA News Letter* **2**, 12-13.

Ruiz, H. & Pavon, J. (1794). *Prodromus Flora Peruviana et Chiliensis*.

Swofford, D.L. (1985). 'PAUP: Phylogenetic Analysis Using Parsimony. Version 2.4.' (Computer program distributed by the Illinois Natural History Survey: Champaign, Illinois.)

Swofford, D.L. (1993). 'PAUP: Phylogenetic Analysis Using parsimony. Version 3.1.1.' (Computer program distributed by the Illinois Natural History Survey: Champaign, IL.)

Thomasson, J.R. (1976). Tertiary grasses and other angiosperms from Kansas, Nebraska and Colorado. 411 pp. Doctoral dissertation Iowa State University, Ames, Iowa.

Thomasson, J.R. (1978). Epidermal patterns of the lemma in some fossil and living grasses and their phylogenetic significance. *Science* **199**, 975-977.

Thomasson, J.R. (1987). Fossil grasses: 1820-1986 and beyond. In 'Grass Systematics and Evolution'. (Eds T.R. Soderstrom, K.W. Hilu, C.S. Campbell, and M.E. Barkworth.). pp. 159-167. (Smithsonian Institution Press: Washington, D.C.)

Tomlinson, P.B., and Smith, A.C. (1970). Joinvilleaceae, a new family of monocotyledons. *Taxon* **19**, 887-889.

Torres, M.A. (1997a). *Nassella* (Gramineae) del noroeste de la Argentina. Monografia 13. Ministerio de la Producción y el Empleo Provincia de Buenos Aires, Comisión de Investigaciones Científicas. pp. 5-45.

Torres, M.A. (1997b). *Nicoraella* (Gramineae) un nuevo genero para America del Sur. Monografia 13. Ministerio de la Producción y el Empleo Provincia de Buenos Aires Comisión de Investigaciones Científicas. pp 69-77.

Tzvelev, N.N. (1976). 'Grasses of the Soviet Union'. 2 vols. (Translation 1983, Smithsonian Institution Libraries and the National Science Foundation: Washington DC.)

Tzvelev, N.N. (1977). [On the origin and evolution of Feathergrasses (*Stipa* L.)] in *Problemi Ekologii, Geobotaniki, Botanicheskoi Geografii i Floristiki*. pp. 139-150. (Leningrad: Nauka) [Translated by K. Gonzalez for M. Barkworth. Copy on file in Intermountain Herbarium, Utah State University.]

Veldkamp, J.F. (1985) *Anemanthele* Veldk. (Gramineae: Stipeae), a new genus from New Zealand. *Acta Botanica Neerlandica*, **34**, 105–109.

Vickery, J.W., and Jacobs, S.W.L. (1980). *Nassella* and *Oryzopsis* (Poaceae) in New South Wales. *Telopea* **2**, 17-23.

Watson, L., and Dallwitz, M.J. (1994). 'The Grass Genera of the World'. 2nd ed. (C.A.B. International: Wallingford, Oxon, England.)

Wheeler, D.J.B., Jacobs, S.W.L., and Norton, B.E. (1982). Grasses of New South Wales. 306 pages. (University of New England Press: Armidale.)

Williams, J.G.K., Hanafey, M.K., Rafalski, J.A., and Tingey, S.V. (1993). Genetic analysis using random amplified polymorphic DNA markers. *Methods Enzymology* **218**, 704-740.

Grasses: Systematics and Evolution. (2000). Eds S.W.L. Jacobs and J. Everett. (CSIRO: Melbourne)

PRELIMINARY VIEWS ON THE TRIBE MELICEAE (GRAMINEAE: POOIDEAE)

Teresa Mejia-Saulés [AC] *and Frank A. Bisby* [B]

[A] Biodiversity & Ecology Research Division, School of Biological Sciences, University of Southampton, Southampton S016 7PX, UK.

[B] Centre for Plant Diversity & Systematics, School of Plant Sciences, The University of Reading, Whiteknights, Reading RG6 6AS, UK.

[C] Present address: Instituto de Ecologia A.C. Apartado Postal 63, 91000 Xalapa, Veracruz, Mexico.

Abstract

The tribe Meliceae is a group of 162 grass species of temperate regions, widely distributed around the world. It has a history of unstable classification with the inclusion or exclusion of several genera by different authors. *Melica* is one of the largest genera, with 84 species, and there is some debate as to whether certain species should be separated into a segregate genus *Bromelica*. This paper reports preliminary views on the definition of the tribe Meliceae and the status of the genus *Melica* itself. A morphological data set (including vegetative and floral characters) and three anatomical data sets (lemma epidermis, abaxial epidermis and transverse section of the leaf) have been completed for 75 species. These data sets were used in both phenetic and cladistic analyses. The preliminary results from both these analyses give a pattern in which existing genera appear to be coherent, and are suggestive of an inner group recognizable as the tribe Meliceae. However, *Melica* appears to be split, usually into two groups. A small DNA sequence comparison being made for the *trn*L chloroplast gene will be published in the future.

Key words: Meliceae, Gramineae, systematic, anatomy, cladistic, phenetic, *trn*L, *Melica*.

INTRODUCTION

A large proportion of the grasses adapted to cool climates of both hemispheres belong to the Pooideae subfamily, with approx. 160 genera and 3000 species (Macfarlane 1986). This subfamily is divided into 10 tribes (Clayton and Renvoize 1986): Nardeae, Lygeeae, Stipeae, Poeae, Hainardieae, Meliceae, Brylkinieae, Aveneae, Bromeae and Triticeae.

The tribe Meliceae, the focus of this study, is a group of 162 grass species of temperate regions that are widely distributed around the world. The tribe has a history of unstable classification with the inclusion or exclusion of several genera by different authors (Nevski 1934; Pilger 1954; Macfarlane and Watson 1980, 1982; Hilu and Wright 1982; Gould 1968; Clayton and Renvoize 1986; Tzvelev 1989; Watson and Dallwitz 1992; Kellogg and Watson 1993) (Table 1). A core element of the tribe

Meliceae has been recognized consistently by a number of authors. The largest genera are *Melica* (74-84 spp. with or without segregates), *Glyceria* (± 40 spp.) and *Briza* (20 spp.). The smaller genera are *Pleuropogon* (5 spp.), *Triniochloa* (5 spp.), *Catabrosa* (2 spp.) and the monotypic genera *Anthochloa*, *Brylkinia*, *Lycochloa*, *Neostapfia*, *Schizachne* and *Streblochaete*.

The species of *Melica* are distributed in temperate regions throughout the world, except in Australia. They are mostly found in shady woodlands on dry stony slopes. The genus has been divided into sections or subgenera on the basis of morphological features (Nyman 1878; Hitchcock 1935; Papp 1928, 1932 and 1937; Hempel 1970, 1971, 1973; Clayton and Renvoize 1986). A few authors continue to recognize the subgenus *Bromelica* as a separate genus (Farwell 1919; Nicora 1973; Torres 1980; Muñoz 1983-84).

Table 1. Genera included in the tribe Meliceae by several authors (on the basis of morphological features).

Genus	Nevski (1934)	Pilger (1954)	Gould (1968)	Macfarlane & Watson (1982)	Hilu & Wright (1982)	Macfarlane (1986)	Clayton & Renvoize (1986)	Tzvelev (1989)	Watson & Dallwitz (1992)	Kellogg & Watson (1993)
Glyceria	✔	Tribe: Festuceae Subtribe: Glyceriinae	✔	✔	✔	✔	✔	✔	✔	✔
Melica (Bromelica)	✔	Tribe: Festuceae Subtribe: Melicinae	✔	✔	✔	✔	✔	✔	✔	✔
Schizachne	✔	Tribe: Festuceae Subtribe: Melicinae	✔	✔	–	✔	✔	✔	✔	✔
Pleuropogon	✔	Tribe: Festuceae Subtribe: Glyceriinae	✔	✔	–	✔	✔	✔	✔	Pooid Group
Triniochloa	–	Tribe: Stipeae	–	✔	–	✔	✔	✔	✔	✔
Lycochloa	–	Tribe: Festuceae Subtribe: Melicinae	–	✔	–	✔	✔	✔	✔	Pooid Group
Streblochaete	–	Tribe: Festuceae Subtribe: Brominae	–	✔	–	✔	✔	✔	✔	Pooid Group
Anthochloa	–	Tribe: Festuceae Subtribe: Melicinae	–	–	–	Poeae	✔	Poeae	Poeae	Pooid Group
Catabrosa	–	Tribe: Festuceae Subtribe: Glyceriinae	✔	✔	–	✔	Poeae	Poeae		Pooid Group
Brylkinia	–	Tribe: Festuceae Subtribe: Festucinae	–	✔	–	✔	Brylkinieae	Brylkinia		Pooid Group
Briza	–	Tribe: Festuceae Subtribe: Festucinae	Poeae	–	✔	Poeae	Poeae	Poeae	Poeae	Pooid Group
Neostapfia	–	Tribe: Festuceae Subtribe: Melicinae	Orcuttieae	–	–	–	Orcuttieae	Cynodonteae	Orcuttieae	–

Minimal Hypothesis → (*Glyceria* through *Anthochloa*)

Maximal Hypothesis → (*Glyceria* through *Neostapfia*)

✔ = included in the Meliceae tribe; – = note included in this author's study; Brylkinieae, Cynodonteae, Orcuttieae, Poeae, Stipeae = excluded from the Meliceae tribe

The first objective of this study is to examine in more detail the patterns of diversity in and around the Meliceae tribe. This is to generate evolutionary hypotheses for this worldwide group, and to clarify the classification. At the extremes are two hypotheses:

1) The 'Maximal hypothesis' in which 12 genera are included in the tribe Meliceae (*Anthochloa, Briza, Brylkinia, Catabrosa, Glyceria, Lycochloa, Melica, Neostapfia, Pleuropogon, Schizachne, Streblochaete* and *Triniochloa*) as in the treatments of Gould (1968), Macfarlane and Watson (1980), Hilu and Wright (1982) and Watson and Dallwitz (1992).

2) The 'Minimal hypothesis' in which just eight genera are included (*Anthochloa, Glyceria, Lycochloa, Melica, Pleuropogon, Schizachne, Streblochaete* and *Triniochloa*) with the other four genera considered to be in other tribes as in the treatments of Clayton and Renvoize (1986).

The second objective is to find if *Melica* itself is considered monophyletic and homogeneous or whether it should be subdivided either as a subgenus or split into two genera: *Melica* and *Bromelica*.

METHODS

1) Species chosen for this study. The genera considered as candidates for the tribe Meliceae for this study were *Anthochloa, Briza, Brylkinia, Catabrosa, Glyceria, Lycochloa, Melica, Neostapfia, Pleuropogon, Schizachne, Streblochaete* and *Triniochloa*. *Ampelodesmos, Brachypodium, Diarrhena* and *Stipa* were considered as initial outgroups for the cladistic analyses.

Seventy-five species in total were chosen for this study: 63 species were considered as candidates in the Meliceae and 12 species as initial outgroups. *Melica* is represented in this study by 40 species. This ensures that our sample covers both the ranges of forms known from earlier studies and the full geographical range of the tribe and *Melica* itself.

2) Characters. The characters chosen included 19 new morphological characters and many that were based on those used by earlier authors such as Hilu and Wright (1982), Macfarlane and Watson (1980, 1982) and Kellogg and Watson (1993).

3) Morphological survey. Live material and herbarium specimens were examined under a dissecting microscope. Ten or more specimens were examined for each species.

4) Anatomical survey. The abaxial epidermis and transverse section of the leaf were surveyed following Metcalfe's (1960) technique. The lemma epidermis was examined under the Scanning Electron Microscope (SEM) following standard techniques.

5) Morphological and anatomical survey. The morphological and anatomical survey generated data for the 75 species scored for a list of 125 characters which includes 62 morphological characters covering vegetative and floral parts of the plant, 39 characters covering both abaxial epidermis and transverse section of the leaf-blade, and 24 characters from the lemma epidermis. The data were used in both phenetic and cladistic analyses.

6) Phenetic analyses. Both morphological and anatomical data sets were used as the basis of the analyses. Preliminary phenetic analyses were conducted using Multi-Variate Statistical Package (MVSP, Kovach 1995). Gower's coefficient and Unweighted Pair Group Average Method (UPGMA) were performed with the morphological and anatomical data sets.

7) Cladistic analyses. Cladistic analyses were applied to a select number of taxa in order to get an overview of the phylogeny of the tribe. A total of 32 taxa were chosen representing the 12 genera within the study of the Meliceae and related genera, and four genera considered as outgroups. Candidates as outgroups were *Ampelodesmos, Brachypodium, Diarrhena* and *Stipa*. After some careful discussion the monotypic genus *Neostapfia* was excluded from the cladistic analysis on the basis of our own phenetic results. It was placed so far outside the main cluster in all the phenetic analyses that we concluded it must be belong to some completely different tribe far from Meliceae.

The same data sets used in the phenetic analyses were used as the basis of the cladistic analyses. Preliminary cladistic analysis was performed using PAUP v.3.1.1 (Swofford 1993). Uninformative and constant characters were excluded. All characters were weighted equally and unordered. A maximum parsimony analysis consisting of a hundred replicates of random addition taxon entries was performed. Searches for strict consensus trees were conducted. Clade support was tested using both Bootstrap and AutoDecay ver.3.0.3 (Eriksson and Wikström 1997).

PRELIMINARY RESULTS

Phenetic Analyses

These are phenetic preliminary results that have not yet been completed.

The UPGMA cluster analysis separated into two main clusters: **cluster I** contains *Anthochloa, Catabrosa, Glyceria* and *Pleuropogon,* and **cluster II** includes *Lycochloa, Schizachne, Streblochaete, Triniochloa* and the largest genus *Melica* (Fig. 1).

Cluster I has two recognizable subclusters: **subcluster Ia**, includes *Anthochloa* and *Catabrosa*, and **subcluster Ib** which includes *Glyceria* and *Pleuropogon*.

Cluster II. This cluster represents the larger of the two clusters and includes *Lycochloa, Schizachne, Streblochaete, Triniochloa* and *Melica*. It is split into five subclusters: in **subcluster IIa** *Lycochloa* and *Melica macra* appear in one cluster that is linked to *Melica-Schizachne-Streblochaete-Triniochloa*. **Subcluster IIb** is remarkable in that 15 species of *Melica* are away from the other 25 species. Eleven of these 15 species are the North American species placed in *Bromelica* by some authors. **Subcluster IIc** includes *Melica onoei, Schizachne, Streblochaete* and *Triniochloa*. **Subcluster IId** includes 18 widely distributed *Melica* species. **Subcluster IIe** includes five South American species of *Melica*.

Overall, the phenogram shows several completely isolated taxa: *Briza, Brylkinia* and *Neostapfia*. *Neostapfia* shows the lowest similarity with the others in the cluster analysis and it is thus placed well outside the main cluster.

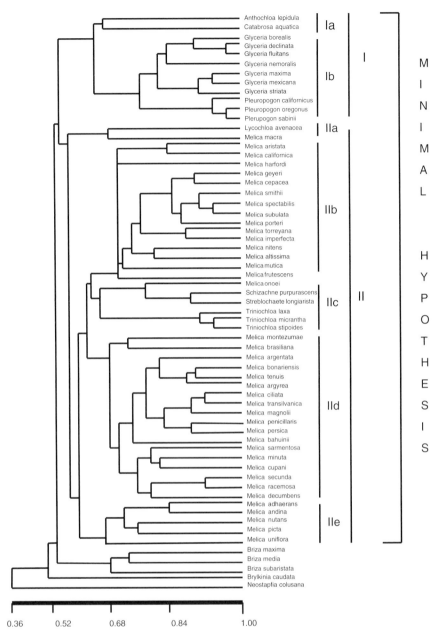

Fig. I. Phenogram represents the clustering of the Meliceae tribe and related genera.

Cladistic Analyses

These are cladistic preliminary results that have not yet been completed.

Heuristic searches yielded three most parsimonious trees with 566 length, Consistency Index (CI)= 0.288 and Retention Index (RI)= 0.496. The low level of Consistency Index indicates a high level of homoplasy. The strict consensus tree (Fig. 2) has some consistent structure and some small groups are well supported such as the main clade marked as **A** (70% Bootstrap, five steps Decay), *Stipa* (93%, seven steps Decay), *Pleuropogon* (89%, five steps Decay) and *Brachypodium* (81%, six steps Decay) and a few *Melica* species such as *M. argyrea, M. brasiliana* and *M. ciliata* (73%, three steps Decay) (Fig. 2).

The cladogram in Figure 2 shows some support (70% Bootstrap, five steps Decay) for a monophyletic ingroup clade marked as

clade **A**. Outside this clade are placed the outgroups *Ampelodesmos* and *Stipa*. **Clade A** includes two distinctive clades marked as **clade B** and **clade C** that present three steps Decay. **Clade B** includes the outgroups *Brachypodium* and *Diarrhena*, and the monotypic genus *Brylkinia*. **Clade C** includes two recognisable subclusters marked as subcluster **CI** and **CII** in figure 2 and four isolate taxa: *Triniochloa* (represented by one species in this study), and the monotypic genera *Lycochloa, Schizachne* and *Streblochaete*. **Subclade CI** (three steps Decay) includes *Anthochloa, Catabrosa, Briza, Glyceria* and *Pleuropogon*. **Subclade CII** (three steps Decay) includes *Melica*, which splits into two groups: one group represented by the upper spikelet illustration in figure 2 which includes the four North American species considered by some authors as *Bromelica,* and the second group represented by the lower spikelet illustration in figure 2, which includes seven widely distributed *Melica* species.

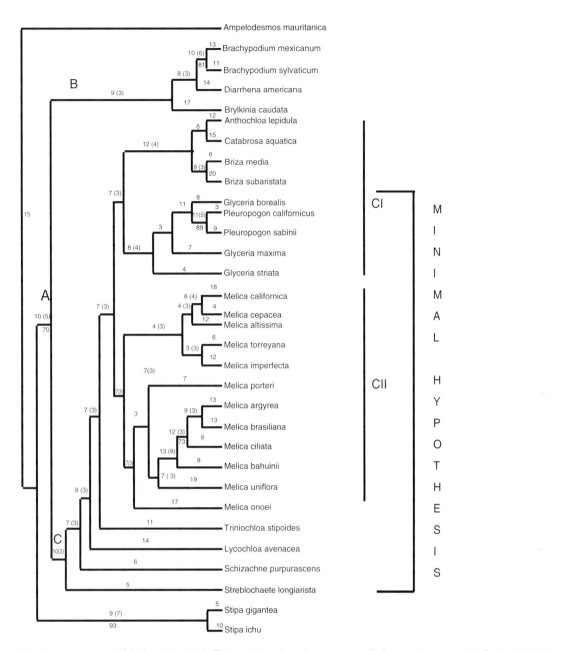

Fig. 2. One of the most parsimonious trees of the heuristic search. This tree is topologically congruent with the strict consensus tree. Figures above the branches indicate branch length. Figures below branches indicate bootstrap values (Tree length = 566, Consistency Index = 0.288, Retention Index = 0.496).

DISCUSSION

Results for both phenetic and cladistic analyses suggest that both the minimal and maximal hypotheses as to which genera are to be classified in the Meliceae should be rejected. These results show nine genera as possible candidates for inclusion in the tribe: *Anthochloa, Catabrosa, Glyceria, Pleuropogon, Melica, Lycochloa, Schizachne, Streblochaete* and *Triniochloa*. Cladistic results showed high homoplasy in the shortest tree found in the analysis performed, and recognise a monophyletic clade marked as **clade C** (Fig. 2) that includes the nine taxa proposed as members of Meliceae. These taxa are grouped by sharing only two features, a connate sheath, and lemma with a membranous apex.

It is ambiguous whether *Briza* should be included or excluded from the tribe. It was placed either inside or outside the core Meliceae group in both phenetic and cladistic analyses. *Briza* presents plesiomorphic characters such as an open sheath, and lemma with a coriaceous apex.

The monotypic genus *Brylkinia* is placed always outside the main cluster in both the cladistic and phenetic analyses performed. Cladistic analysis detected two apomorphic characters for *Brylkinia*, sheath connate and rachilla disarticulation below glumes. Plesiomorphic characters were found such as lemma with a coriaceous apex and silica bodies with a smooth outline present in the intercostal zone of the leaf-blade. Results from both phenetic and cladistic analyses suggest the exclusion of *Brylkinia* from the Meliceae and place it in a different tribe as some authors have already proposed (Clayton and Renvoize 1986; Tzvelev 1989).

87

The monotypic genus *Neostapfia* was placed far outside the main cluster in all the phenetic analyses performed. This genus presents unique morphological and anatomical features: the culm, sheath and leaves are covered by glandular macro-hairs, the ligule and glumes are absent, there is the presence of Crozier type and mushroom-button macro-hairs covering the plant and the arrangement of the chlorenchyma is radiate. These phenetic results suggest that *Neostapfia* belongs to a very different tribe to Meliceae, as some authors have already proposed (Gould 1968; Clayton and Renvoize 1986; Tzvelev 1989; Watson and Dallwitz 1992).

Melica was fragmented in both of the phenetic and cladistic analyses performed. These analyses show two recognisable groups: one group includes the North American *Melica* species considered by some authors as *Bromelica* mixed with some Eurasian species. This group shares features such as rachilla with disarticulation above the glumes, the lemma shorter than the glumes, spikelets large with several florets and the reduced florets similar in shape to the hermaphroditic florets. The second group includes widely distributed *Melica* species and they are grouped by sharing features such as rachilla disarticulation below glumes, the spikelets being reduced in both size and the number of florets (1-3 florets), and the reduced florets having a different shape to the hermaphroditic florets. There thus appear to be several *Melica* subgroupings, and the *Bromelica* group may not be monophyletic. More detailed analyses will be performed to delimit this genus.

A preliminary conclusion can be suggested on the basis of the phenetic and cladistic results. Meliceae appears to be monophyletic considering that *Anthochloa, Catabrosa, Glyceria, Lycochloa, Melica, Pleuropogon, Schizachne, Streblochaete* and *Triniochloa* are included with the main group (strongly supported in the consensus tree). *Briza* has an ambiguous position and its inclusion in the tribe is not yet clear.

Although the phenetic and cladistic preliminary results split *Melica* into two groups, the delimitation of this genus is not yet clear.

Future work includes a smaller DNA sequence comparison being made for the *trn*L chloroplast with the hope of obtaining better support for the delimitation of the tribe Meliceae, as well as the delimitation of *Melica*.

Acknowledgements

This study was supported by the British Council and the Mexican Research Council (CONACyT) studentship 85342.

References

Clayton, W.D., and Renvoize, S.A. (1986). 'Genera Graminum'. (Her Majesty's Stationary Office: London)

Eriksson, T., and Wikström, N. (1995). AutoDecay, version 3.0.3. Computer Software Programme distributed by the authors. Botaniska Institution, University of Stockholm, Sweden.

Farwell, O.A. (1919). Bromelica (Thurber): A new genus of grasses. *Rhodora* **21**, 76-78.

Gould, F.W. (1968). 'Grass Systematics'. (McGraw-Hill, Inc.: New York).

Hempel, W. (1970). Taxonomische und chorologische Untersuchungen an Arten von Melica L. subgen. *Melica. Feddes Repertorium* **81**, 131-145.

Hempel, W. (1971). Die systematische Stellung von *Melica uniflora* Retz. und *Melica rectiflora* Boiss. et Heldr. (*Melica* L. subgen. *Bulbimelica*, subg. nov.). *Feddes Reportorium* **81**, 657-686.

Hempel, W. (1973). Die systematische Stellum von *Melica altissima* L. und des *Melica minuta* L.- *Melica ramosa* Vill. Komplexes (*Melica* L. subgen. *Altimelica* Hempel). *Feddes Repertorium* **84**, 533-568.

Hilu, K.W., and Wright, K. (1982). Systematics of Gramineae: A cluster analysis study. *Taxon* **31**, 9-36.

Hitchcock, A.S. (1935). 'Manual of the grasses of the United States'. 2nd edition. (United States Department of Agriculture Miscellaneous Publications, 200: Washington DC).

Kellogg, E.A., and Watson, L. (1993). Phylogenetic studies of a large data set. I Bambusoideae, Andropogonodae and Pooideae (Gramineae). *Botanical Review* **59**, 273-342.

Kovach, W.L. 1995. MVSP- A Multivariate Statistical Package for IBM-PC, ver. 2.2. Kovach Computing Services, Pentraeth, Wales, U.K.

Macfarlane, T.D. (1986). Poaceae Subfamily Pooideae. In 'Grass Systematics and Evolution'. (Eds T.R. Soderstrom, K.W. Hilu, C.S. Campbell and M.E. Barkworth) pp. 265- 276. (Smithsonian Institution Press: Washington, D.C.)

Macfarlane, T.D., and Watson, L. (1980). The circumscription of Poaceae subfamily Pooideae, with notes on some controversial genera. *Taxon* **29**, 645-666.

Macfarlane, T.D., and Watson L. (1982). The classification of Poaceae subfamily Pooideae. *Taxon* **31**, 178-203.

Metcalfe, C.R. (1960). 'Anatomy of the Monocotyledons I. Gramineae'. (Claredon Press: Oxford).

Muñoz, S.M. (1983-84). Revision de las especies del género Melica L. (Gramineae) en Chile. *Boletin del Museo Nacional de Historia Natural* **40**, 41-89.

Nevski, A. (1934). Blepharolepis Papp. Acta *Universitatis Asiae Medial Botanica* **8**, 10-13.

Nicora, E.G. (1973). Gramineas nuevas para la Flora Argentina. *Darwiniana* **18**, 265-272.

Nyman, C.F. (1878). 'Conspectus Floræ Europææ ' (Part I). (Typis Officinæ Bohlinianæ: Örebro.)

Papp, C. (1928). Monographie der Sudamerikanischen Arten der Gattung Melica L. Feddes Repertorium **25**, 97-160.

Papp, C. (1932). Monographie der Europäischen Arten der Gattung Melica L. Botanische Jahrbücher Systematische **65**, 275-348.

Papp, C. (1937). Monographie der Asiatischen Arten der Gattung Melica L. *Academia Româna, Memoriile Sectiunei Stiintifice Seria* III. **12**, 187-267.

Pilger, R. (1954). Das system der Gramineae. *Botanische Jahrbücher Systematische* **76**, 281-384.

Swofford, D.L. (1993). Phylogenetic Analysis Using Parsimony, version 3.1.1. Computer Software Programme distributed by Illinois Natural History Survey, Champaign, Illinois.

Torres, M.A. (1980). Revision de las especies Argentinas del género Melica L. (Gramineae). *Lilloa* **29**, 1-115.

Tzvelev, N.N. (1989). The System of Grasses (Poaceae) and their Evolution. *The Botanical Review* **55**, 141-204.

Watson, L., and Dallwitz, M.J. (1992). 'The Grass Genera of the World'. (C.A.B. International: Wallingford.)

Grasses: Systematics and Evolution. (2000). Eds S.W.L. Jacobs and J. Everett. (CSIRO: Melbourne)

THE SPECIES OF *BROMUS* (POACEAE: BROMEAE) IN SOUTH AMERICA

Ana M. Planchuelo[A] *and Paul M. Peterson*[B]

[A] CONICET, Facultad de Ciencias Agropecuarias, Universidad Nacional de Córdoba,
   Casilla de Correo 509, 5000 Córdoba, Argentina.
[B] Department of Botany, National Museum of Natural History,
   Smithonian Institution, Washington, DC 20560-0166, U.S.A.

Abstract

A key to the native and introduced species of *Bromus* in South America, distribution maps, and habitat information are presented.

Key words: Bromus, Poaceae, Gramineae, grasses, South America.

INTRODUCTION

The genus *Bromus* L. comprises approximately 150 species distributed in temperate and cool regions of both hemispheres. In South America, species of *Bromus* range from sea level to alpine tundra in the Andes. We recognize 38 species in five sections as follows: Section *Bromopsis,* containing all perennials, found primarily in the Cordilleras; Section *Bromus* with all annuals from Argentina and Chile; Section *Ceratochloa,* containing annuals, biennials or perennials composed of several polyploid complexes with many infraspecific taxa; Section *Genea* with all annuals from Argentina and Chile; and Section *Neobromus,* containing only annuals. This presentation is a preliminary account of the genus in South America.

MATERIALS AND METHODS

The taxonomic work is based chiefly on studies of type specimens when available, and herbarium specimens from the following herbaria (ACOR, BA, BAA, BAB, BAF, BM, CORD, CTES, G, K, LIL, LP, MO, NY, P, SI, UC, UMO, US). Information on the habitat and ecological aspects of these species is based on personal observation of the plants in the field. The literature for these species was also used to establish the current distribution of each taxon. A complete list of specimens used in this study will appear in the final taxonomic treatment for the genus in South America.

KEY TO THE SECTIONS OF *BROMUS*

1. Spikelets usually laterally compressed; lemmas compressed and keeled; if the spikelet is not compressed near base then the lower glume is 5-nerved and the lemmas are compressed and keeled on the upper quarter with an awn less than 3 mm long..Sect. *Ceratochloa*

1′. Spikelets usually not laterally compressed; lemmas not keeled; if the spikelet is compressed then the lower glume is 3-nerved 2

2. Lemma apex bidentate..3

3. Awn of the lemma geniculate and/or twisted ....Sect. *Neobromus*

3′. Awn of the lemma straight and stiff, not geniculate and/or twisted..Sect. *Genea*

2′. Lemma apex entire, acute or subacute, but not bidentate 4

4. Plants generally perennial with or without rhizomes, the bases fibrous; lower glume 1(3)-nerved; upper glume 3(5)-nerved.. Sect. *Bromopsis*

4′. Plants annual; lower glume 5(3)-nerved; upper glume 7(5)-nerved.. Sect. *Bromus*

KEY TO THE SPECIES OF SECTION *BROMOPSIS*

1. Panicle tightly contracted, 4 cm long or less; pedicels less than 1 cm long; lemmas densely pubescent on the entire surface..2

2. Spikelet with 4 florets; lower glume 10-11 mm long; central nerve of the lemma prolonged to form an awn.................. ....*B. villosissimus*

2′. Spikelet with (4) 5 or 6 (7) florets; lower glume 5-6 mm long; central nerve not prolonged to form an awn, rounded dorsally .... *B. pellitus*

1′. Panicle loose or semi-contracted, more than 5 cm long; pedicels more than 1.5 cm long; lemmas glabrous, scabrous, or pubescent on the dorsal surface only and/or pilose on the margins 3

3. Glumes unequal in length, the lower glume approximately 10 mm long, half as long as the upper glume ......*B. segetum*

3′. Glumes subequal in length, the lower glume almost as long as the upper glume..4

4. Panicles open, nodding; pedicels flexuous 5

5. Plants 90-120 cm tall; spikelets with 4-6 florets ....*B. araucanus*

5′. Plants 10-90 cm tall; spikelets with 6-8 florets 6

6. Panicles 4-8 cm long with few spikelets; lower glume 5-7 mm long 7

7. Plants with only basal leaves; pedicels scabrous; lemma 5-nerved...*B. modestus*

7′. Plants with basal and cauline leaves; pedicels with lanulose hairs; lemma (5-) 7-nerved..................................*B. lanatus*

6′. Panicles more than 8 cm long, with many spikelets; lower glumes (8-) 9-10 mm long 8

8. Upper glume 7-8 mm long; lemma 8-10 mm long ....*B. pitensis*

8′. Upper glume more than 8 mm long; lemma 12 mm or more long 9

9. Lemma with an awn (5-) 6-8 mm long .... *B. flexuosus*

9′. Lemma awnless or mucronate .... *B. pflanzii*

4′. Panicle not open, erect; pedicels not flexuous................. 10

10. Plants rhizomatous..11

11. Lower glume 1 (-3)-nerved, upper glume 3-nerved; lemma muticous or short-awned, the awn less than 3 mm long..... ....*B. inermis*

11′. Lower glume 3-nerved, upper glume 5 (7)-nerved; lemma awned, the awn 3-7 mm long..........................*B. auleticus*

10′. Plants not rhizomatous 12

12. Lower glume 3-5 mm long; upper glume 6-9 mm long...... ....*B. brachyanthera*

12′. Lower glume 7-14 mm long; upper glume 9-15 mm long 13

13. Lower glume 9-13 mm long; upper glume 10-15 mm; lemma lanceolate with wide membranous or hyaline margins that are densely pubescent ....*B. setifolius*

13′. Lower glume 7-9 mm long; upper glume 9-11 mm long; lemma linear lanceolate with narrow non-hyaline margins.. ....*B. erectus*

KEY TO THE SPECIES OF SECTION *BROMUS*

1. Panicles 3-10 (-12) cm long, contracted, erect; pedicels 3-5 mm long...2

2. Sheaths glabrous; lemma awns borne on the upper 1/3, well below the bifid apex...*B. scoparius*

2′. Sheaths pubescent; lemma awns borne in the sinus of the bifid apex or just below...3

3. Glumes and lemmas pubescent....................*B. hordeaceous*

3′. Glumes and lemmas glabrous ....*B. racemosus*

1′. Panicles more than 12 cm long, nodding; pedicels 5 mm or more long 4

4. Glumes and lemmas tomentose; spikelets more than 25 mm long... *B. lanuginosus*

4′. Glumes and lemmas glabrous, scabrous, or with a few scattered hairs but not tomentose; spikelets less than 25 mm long..5

5. Margins of the lemmas involute below.............................6

6. Lemma awn 3-5 (-6) mm long in the upper florets, shorter or absent in the lower florets; palea about as long as the lemma ....*B. secalinus*

6′. Lemma awn 6-8 mm long in all florets; palea shorter than the lemma.. *B. squarrosus*

5′. Margins of the lemmas not involute below 7

7. Palea about as long as the lemma; anthers more than 3.5 mm long...*B. arvensis*

7′. Palea shorter than the lemma; anthers 1-1.5 mm long 8

8. Spikelets inflated, turgid; florets imbricate, rachilla not evident at maturity; lemmatal awns curved at anthesis ....*B. japonicus*

8′. Spikelets not inflated or turgid but somewhat compressed; florets slightly imbricate, rachilla evident at maturity; lemmatal awns straight .... *B. commutatus*

KEY TO THE SPECIES OF SECTION *CERATOCHLOA*

1. Lemma mucronate or with an awn less than 1.5 mm long.. 2

2. Plants perennial with short rhizomes .... *B. mango*

2′. Plants annual, biennial or short lived perennials without rhizomes..3

3. Lemmas densely pubescent, the hairs longer and more numerous near the apex .... *B. tunicatus*

3´. Lemmas glabrous or scabrous, occasionally with a few short hairs.. *B. catharticus*

1´. Lemma awned, the awn more than 1.5 mm long 4

4. Glumes and lemmas pubescent .... *B. lithobius*

4´. Glumes and lemmas glabrous, scabrous, or with a few short hairs near the base.. 5

5. Lower glume 3-nerved 6

6. Spikelets 25-30 mm long; lower glume 10 mm long, narrowly triangular .... *B. striatus*

6´. Spikelets 10-20 mm long; lower glume 7-9 mm long, lanceolate.. *B. coloratus*

5´. Lower glume 5-nerved 7

7. Lemma awn 1.5-3 (-4) mm long; anthers 1-1.5 mm long, occasionally chasmogamous florets with anthers 3-4 mm long .... *B. catharticus*

7´. Lemma awn (3-) 4-10 mm long; anthers (3-) 4-7 mm long 8

8. Lemma 15-18 mm long, the awn (3) 4-5 mm long; anthers 4-7 mm long................................... *B. bonariensis*

8´. Lemma 10-12 mm long, the awn 5-10 mm long; anthers (3-) 4-4.5 mm long.. *B. cebadilla*

KEY TO THE SPECIES IN SECTION *GENEA*

1. Panicles very contracted, erect.. 2

2. Panicles 4-8 cm long, dense and compact; pedicels less than 5 mm long; culms pubescent .... *B. rubens*

2´. Panicles (5-) 6-10 cm long, not compact; pedicels 1-1.5 (-2) cm long; culms glabrous .... *B. madritensis*

1´. Panicles loose or semi-contracted with extended branches.. 3

3. Peduncles flexuous; lower glume 4-6 mm long; upper glume 8-10 mm long... *B. tectorum*

3´. Peduncles divergent; lower glume more than 8 mm long; upper glume more than 10 mm long 4

4. Lemma awn 15-25 mm long; lower glume 7-9 mm long; ....*B. sterilis*

4´. Lemma awn 30-50 mm long; lower glume 15-20 mm long .... *B. rigidus*

KEY TO THE SPECIES OF SECTION *NEOBROMUS*

1. Glumes almost as long as the lemmas; lower glume 8-12 mm long; upper glume12-15 mm long......... *B. berteroanus*

1´. Glumes much shorter than the lemmas; lower glume 4-4.6 mm long; upper glume 5.2-6 mm long............*B. gunckelii*

DISTRIBUTION, HABITAT AND INFORMATION ON EACH SPECIES

Figures 1 to 10 show the geographic distribution and examples of spikelets of the species drawn at the scale indicated.

Section *Bromopsis* Dumort.

= Section *Pnigma* Dumort. - according to Soderstrom and Beaman (1968)

= Subgenus *Festucoides* (Cosson & Durieu) Hack. -according to Matthei (1986)

= Section *Festucaria* Gren. & Godr. - according to Parodi (1947)

= Subgenus *Zerna* (Panz) Shear - according to Shear (1900)

***Bromus araucanus* Phil.** - Southern Chile (Region IX-XII) and Argentine Patagonia where it grows along the Andes between 600-800 m. *Bromus araucanus* var. *obtusiflorus* (Hack.) J. A. Cámara is not recognized.

***Bromus auleticus* Trin. ex Nees** - Northeastern and central plain of Argentina, southern Brazil, and Uruguay. Grows in prairies, alluvial plains, and slopes. Used as natural pasture for grazing. Some ecotypes have been selected as forage in semiarid regions.

***Bromus brachyanthera* Döll** - Grows in southern Brazil, Uruguay, Argentina, and Bolivia. Found in wet places and slopes, roadsides, fields, and waste places. Two varieties are recognized as follows: *B. brachyanthera* var. *brachyanthera* with glabrous spikelets and *B. brachyanthera* var. *uruguayensis* (Arechav.) J. A. Cámara with pubescent spikelets.

***Bromus erectus* Huds.** - Introduced from Europe. Sporadically found in Chile, Argentina, and Uruguay. Grows on disturbed ground. Very closely related and hardly distinguishable from *B. riparius* (Wagnon 1952) which is not found in South America.

***Bromus flexuosus* Planchuelo** - Endemic to northwestern Argentina. Grows at more than 2000 m. Planchuelo (1983) differentiated into two altitudinal ecotypes.

***Bromus inermis* Leyss.** - Native in the Old World, introduced and widely distributed as hay and forage in North America and Canada. In South America it is introduced as forage and occasionally cultivated in Uruguay, Bolivia, and Argentina.

***Bromus lanatus* Kunth** - Native in the Andes, occurring from northern Chile and Argentina to the paramo of Colombia and Venezuela. Not very common. Grows on mountain slopes from 3000 m to the permanent snow. Closely related to *B. modestus*, *B. pitensis*, and *B. flexuosus*.

***Bromus modestus* Renvoize** - Native in the Andes of Perú and Bolivia. Grows on mountain slopes between 3600-4700 m. Renvoize (1994) published this new name for *B. frigidus* Ball. Species very closely related to *B. lanatus*. Some authors consider this a synonym of *B. pitensis*.

***Bromus pellitus* Hack.** - Endemic in the extreme south of Chile (XII Region) and in Tierra del Fuego. Not common, found on rocky or gravelled slopes and meadows.

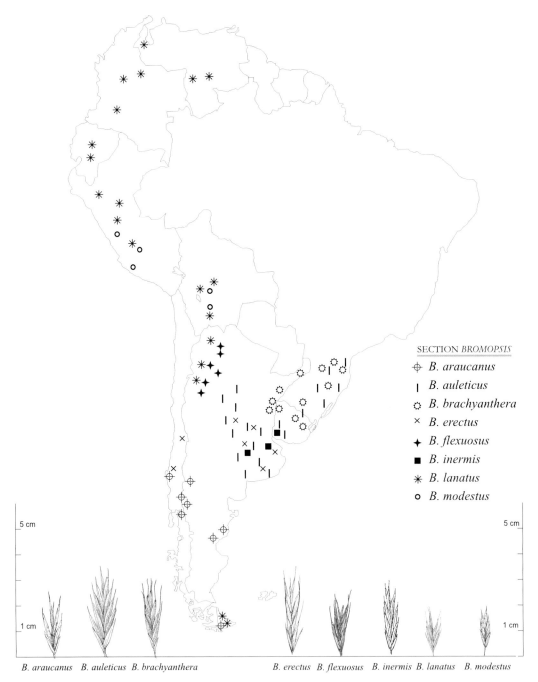

Fig. 1. Geographic distribution of: *Bromus araucanus*; *B. auleticus*; *B. brachyanthera*; *B. erectus*; *B. flexuosus*; *B. inermis*; *B. lanatus* and *B. modestus*; and examples of a spikelet of each species.

Bromus pflanzii Pilg. - Native species in Bolivia. It grows above 3500 m on waste ground of Cochabamba and La Paz. Some authors consider this a synonym of *B. lanatus*, however, the characters of the spikelets are very distinctive.

Bromus pitensis Kunth - Native in the Andes, occurring from northern Bolivia to the paramo of Colombia and Venezuela, sharing the same habitat as *B. lanatus*. Some authors consider this a synonym of *B. lanatus*.

Bromus segetum Kunth - Endemic in the central Andes, with isolated occurrences in northern Bolivia, Ecuador, and Colombia. Very similar to *B. pitensis*.

Bromus setifolius J. Presl - Widely distributed in Chile and Argentine Patagonia from the Province of Neuquén southward to Santa Cruz. Three varieties are recognized as follows: *B. setifolius* var. *setifolius* has linear blades less than 1 mm wide; *B. setifolius* var. *pictus* (Hook. f.) Skottsb. has blades 2-4 mm wide, spikelets less than 24 mm long, and lemmas less than 15 mm long; *B. setifolius* var. *brevifolius* Nees has blades 2-4 mm wide, spikelets 25 mm or more long, and lemmas 16 mm long or more long.

Bromus villosissimus Hitchc. - Endemic in the central Andes. Grows on wet slopes and in the puna between 4600-4900 m. Individuals are often very small, a typical feature in high elevation species.

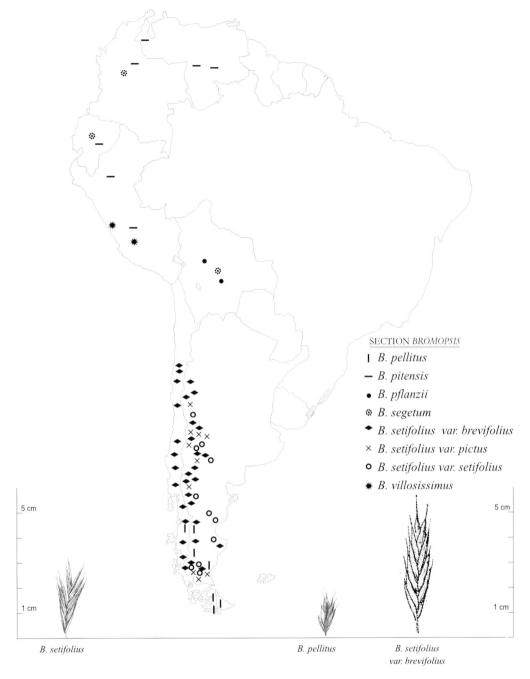

SECTION *BROMOPSIS*

	B. pellitus
—	*B. pitensis*
•	*B. pflanzii*
⊛	*B. segetum*
◆	*B. setifolius var. brevifolius*
×	*B. setifolius var. pictus*
○	*B. setifolius var. setifolius*
✳	*B. villosissimus*

B. setifolius

B. pellitus

B. setifolius var. brevifolius

Fig. 2. Geographic distribution of: *Bromus pellitus*; *B. pitensis*; *B. pflanzii*; *B. segetum*; *B. setifolius* and *B. villosissimus*; and examples of spikelet of some species.

Section *Bromus*

= Subgenus *Bromus* - according to Matthei (1986)

= Section *Bromium* Dumort. - according to Hitchock (1935)

= Section *Zeobromus* Griseb. - according to Parodi (1947)

***Bromus arvensis* L.** - Introduced in Argentina where it sporadically occurs in Buenos Aires Province. Closely related to *B. commutatus*, *B. japonicus,* and *B. secalinus* and can be separated from these by its larger anthers.

***Bromus commutatus* Schrad.** - Introduced in South America from central Europe, found sporadically in Argentina, Brazil, and Uruguay. Grows in pasture fields and disturbed ground. Closely related to *B. racemosus* and *B. japonicus* but more vigorous.

***Bromus hordeaceus* L.** - A species of European origin, introduced and adventive in North and South America. Found in central and southern Argentina, southern Brazil, Chile, and Uruguay. Several infraspecific taxa are recognized for North America and different ecotypes have been observed in Argentina. Most of the traditional literature considered this a synonym of *B. mollis* L.

***Bromus japonicus* Thunb.** - Originally described in Japan but distributed throughout temperate habitats around the

93

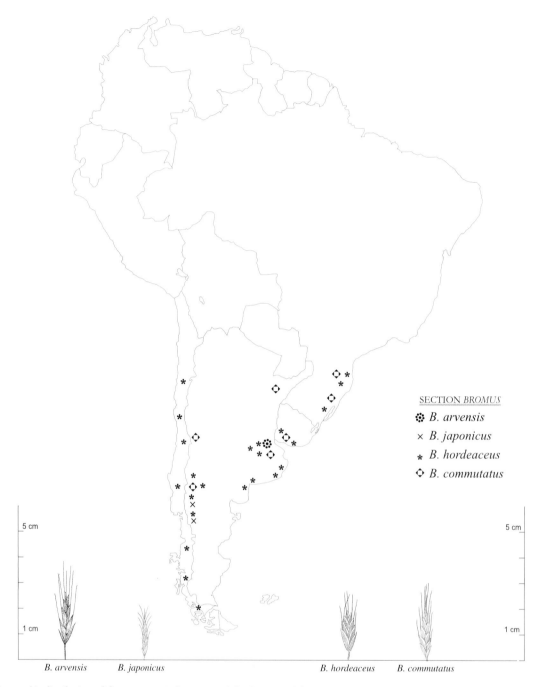

Fig. 3. Geographic distribution of: *Bromus arvensis*; *B. japonicus*; *B. hordeaceus* and *B. commutatus*; and examples of a spikelet of each species.

world. Occasionally collected in Argentina, Bolivia, and Chile where it grows in disturbed ground. Related to *B. arvensis* and *B. commutatus* but with turgid spikelets and divergent-twisted awns.

***Bromus lanuginosus* Poir.** - A species of European origin, rare in South America. The only collection record was in 1929 from Punta Ballenas, Uruguay. Grows in sandy soils.

***Bromus racemosus* L.** - A European species, introduced and now found throughout the northern USA and southern Canada. In South America it is found occasionally in the southern Andes

of Argentina and Chile. Considered by some authors as a depauperate form of *B. commutatus.*

***Bromus scoparius* L.** - Introduced from Europe, found in central Chile where two varieties are recognized as follows: *B. scoparius* var. *scoparius* has glabrous spikelets and *B. scoparius* var. *villiglumis* Maire & Weiller has pubescent spikelets.

***Bromus secalinus* L.** - Introduced from Europe and very common in USA. In South America it is adventive, found in southern Chile and Argentina. Closely related to *B. commutatus, B. racemosus,* and *B. squarrosus.* It can be differentiated from these

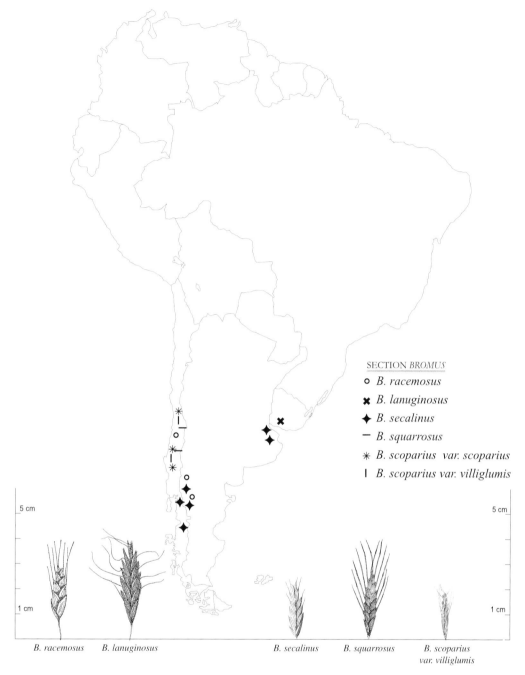

SECTION *BROMUS*

○ *B. racemosus*

✖ *B. lanuginosus*

✦ *B. secalinus*

— *B. squarrosus*

✳ *B. scoparius* var. *scoparius*

❙ *B. scoparius* var. *villiglumis*

B. racemosus   B. lanuginosus   B. secalinus   B. squarrosus   B. scoparius
                                                                       var. villiglumis

Fig. 4. Geographic distribution of: *Bromus racemosus*; *B. lanuginosus*; *B. secalinus*; *B. squarrosus* and *B. scoparius*; and examples of a spikelet of each species.

three species by having lower margins of the lemma involute and the length of the lemma awn.

***Bromus squarrosus* L.** - A European species collected only once in the Metropolitan Region of central Chile. It is closely related to *B. secalinus* and *B. arvensis* but has longer awns.

Section *Ceratochloa* (P. Beauv.) Griseb.
= Subgenus *Ceratochloa* (P. Beauv.) Hack. — according to Matthei (1986)

***Bromus bonariensis* Parodi & J. A. Cámara** - Endemic in the southern region of Buenos Aires Province and Lihuel Calel in La Pampa, Argentina. It is found in mountain slopes and hills in humid places. It is closely related to *B. catharticus* and also resembles *B. stamineus* which also has long awns.

***Bromus catharticus* Vahl** - Native in South America, widely introduced into temperate regions worldwide. Several authors recognize more than one species. It is a well known natural pasture grass and several commercial varieties are cultivated as winter forage. We recognize two varieties as follows: *B. catharticus* var. *catharticus* with lemma awn 0.5-4 (-5) mm long with young blades predominantly convolute and *B. catharticus* var. *rupestris* (Spegazzini) Planchuelo & P. M. Peterson with a

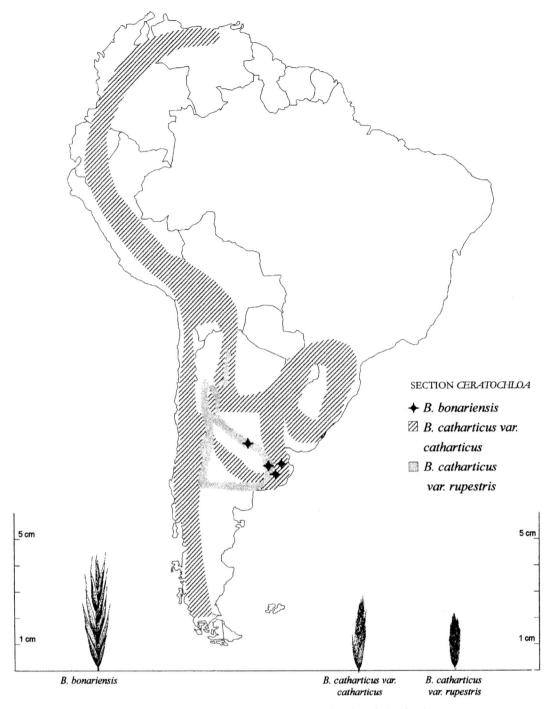

SECTION *CERATOCHLOA*

✦ *B. bonariensis*

▨ *B. catharticus var. catharticus*

▨ *B. catharticus var. rupestris*

B. bonariensis

B. catharticus var. catharticus

B. catharticus var. rupestris

Fig. 5. Geographic distribution of: *Bromus catharticus* and *Bromus bonariensis*; and examples of a spikelet of each species.

mucronate lemma or with a short awn 0.3-0.5 (1) mm long with young blades predominantly folded (Peterson and Planchuelo 1998).

***Bromus cebadilla* Steud.** - Native and widely distributed in Chile (Region II to Region XII) along the Andes and coastal plains. In Argentina this species is found in western Patagonia and on the island of Tierra del Fuego. It occurs in humid places, waste ground and roadsides up to 1000 m. Several authors have considered it a synonym of *B. stamineus* E. Desv., however,

Gutiérrez and Pensiero (1998) place *B. stamineus* in synonymy on the basis of priority.

***Bromus coloratus* Steud.** - Native in southwestern Argentina and central and southern Chile, on both sides of the Patagonian Andes. Occurs in humid places and clear areas of the austral forest. It is considered a good natural pasture.

***Bromus lithobius* Trin.** - Distributed in the southwest of Argentina and through Chile (Region IV to XI) where it occurs from

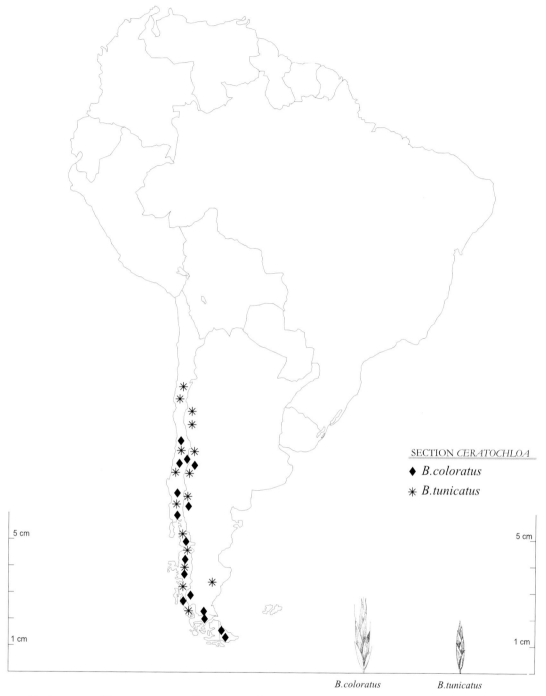

SECTION *CERATOCHLOA*
♦ *B.coloratus*
✳ *B.tunicatus*

B.coloratus   *B.tunicatus*

Fig. 6. Geographic distribution of: *Bromus coloratus* and *B. tunicatus*; and examples of spikelets of both species.

sea level up to 1800 m. This species seems closely related to *B. catharticus*, but can be differentiated by its pubescent spikelets.

***Bromus mango* E. Desv.** - Described from the Chiloé Island, Chile. It was a popular grain crop for the native Araucanos prior to the Spanish conquest. At present it is only found sporadically in southwestern Argentina and Chile (Región XI).

***Bromus striatus* Hitch.** - This is a rare species, endemic to the coastal area of central and southern Perú, where it can be found growing in sandy soil and waste places.

***Bromus tunicatus* Phil.** - Widely distributed in Chile (Region IV southward to IX) where it occurs along the Andes. In Argentina it is found in western Patagonia between 1200-3600 m.

Section *Genea* Dumort.
= Subgenus *Stenobromus* (Griseb.) Hack. - according to Matthei (1986)

***Bromus madritensis* L.** - Of European origin, introduced into North and South America where it is found sporadically in sandy soils and disturbed places. This species occurs in Chile (Regions

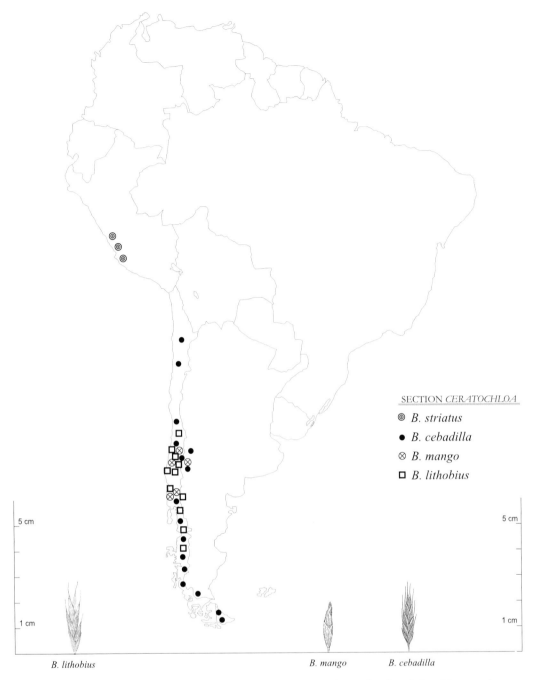

Fig. 7. Geographic distribution of: *Bromus cebadilla*; *B. lithobius*; *B. mango* and *B. striatus*; and examples of a spikelet of three species.

IV, V, and IX) and in the coastal areas of Uruguay and Argentina. This species is closely related to *B. rubens*.

***Bromus rigidus* Roth** - An Old World species, introduced into western North America. In South America it is introduced in Chile (from Region III southward to Region X) and Argentina. It occurs in disturbed areas, sand dunes, roadsides, and along railroads. Its stiff awns sometimes cause mechanical damage to sheep.

***Bromus rubens* L.** - Of European origin, introduced into western North America. In South America it is present sporadically in Argentina where it occurs in waste places, rocky slopes, and fields.

***Bromus sterilis* L.** - This weedy European species was introduced into Chile (Region V southward to Region X) and Patagonian Argentina.

***Bromus tectorum* L.** - A weedy European species, introduced and widely distributed in North America. In South America it is introduced and found in the Patagonian Andes of Argentina and Chile. It grows in disturbed places along fields, rivers, and road banks.

Section Neobromus (Shear) Hitchc.
***Bromus berteroanus* Colla** - This native species of Chile is widely spread throughout much of western North and South America. In South America it occurs in Patagonian Argentina and

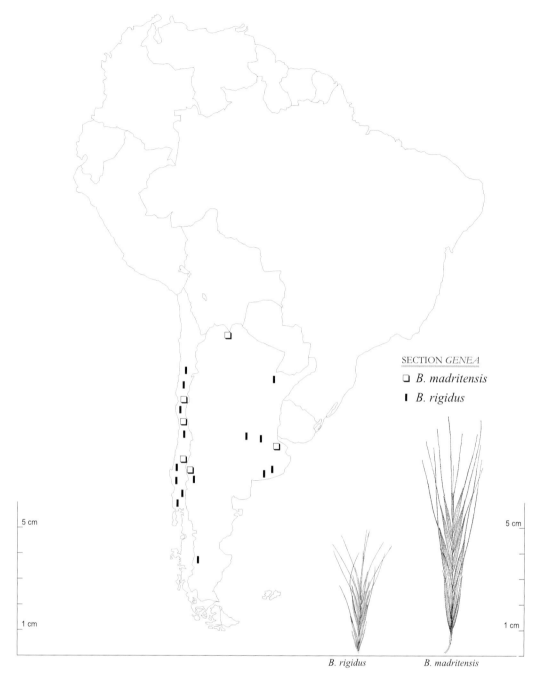

SECTION *GENEA*

□ *B. madritensis*

▌ *B. rigidus*

B. rigidus   *B. madritensis*

Fig. 8. Geographic distribution of: *Bromus madritensis* and *B. rigidus*; and examples of a spikelet of each species.

Chile, and is also found in Brazil and Uruguay. It grows in open places, mountain slopes, plains, and roadways. Several authors have treated this species as a synonym of *B. trinii* E. Desv.

***Bromus gunckelii* Matthei** - An endemic species from the Parinacota Province of Chile (Region I) where its ranges between 3200-3500 m. It is very closely related to *B. berteroanus* but differs by having smaller glumes.

Species not considered

Bromus bolivianus Renvoize; *Bromus burkartii* Muñoz; *Bromus lanceolatus* Roth; *Bromus lepidus* Holmb.

ACKNOWLEDGEMENTS

The authors are indebted to the curators of the herbaria from which the specimens used in this paper were borrowed. Special thanks are given to CONICOR, Universidad Nacional de Córdoba, Argentina and the Office of Fellowships and Grants of the Smithsonian Institution for the economic support for this research. We also extend our appreciation to Alejandro Barbeito for preparing the illustrations, and Mark Lyle for reviewing the manuscript.

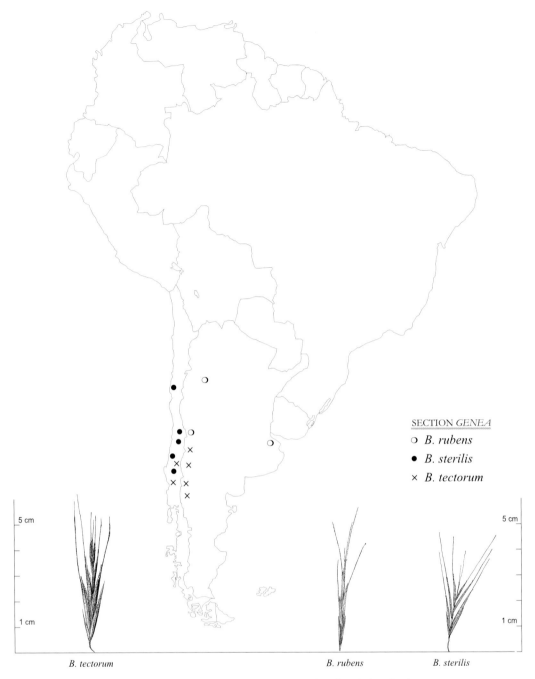

SECTION *GENEA*

○ *B. rubens*

● *B. sterilis*

× *B. tectorum*

5 cm

1 cm

5 cm

1 cm

B. tectorum

B. rubens

B. sterilis

Fig. 9. Geographic distribution of: *Bromus rubens*; *B. sterilis* and *B. tectorum*; and examples of a spikelet of each species.

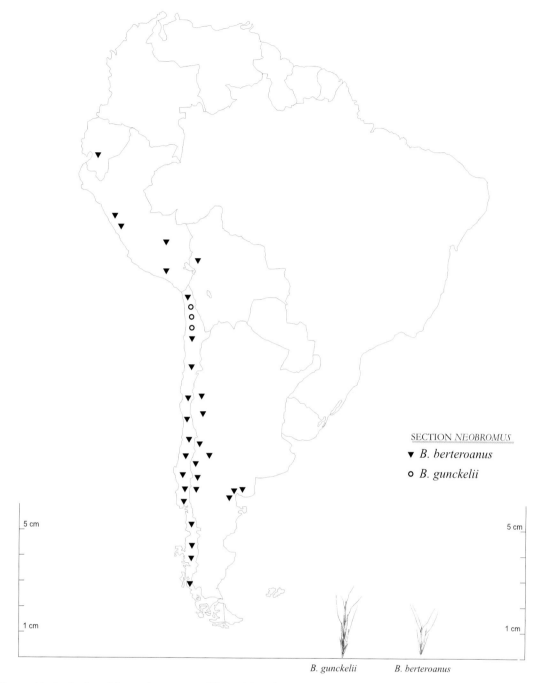

SECTION *NEOBROMUS*

▼ *B. berteroanus*

○ *B. gunckelii*

B. gunckelii   B. berteroanus

Fig. 10. Geographic distribution of: *Bromus berteroanus* and *B. gunckelii*; and examples of a spikelet of each species.

REFERENCES

Gutiérrez, H. F., and Pensiero, J. F. (1998). Sinopsis de las especies argentinas del género *Bromus* (Poaceae). *Darwiniana* **35**, 75-114.

Hitchcock, A. S. (1935). Manual of the grasses of the United States. *United States Department of Agriculture Miscellaneous Publication* **200**, 1-1040.

Hitchcock, A. S., and Chase, A. (1951). Manual of the grasses of the United States. 2nd. ed. *United States Department of Agriculture Miscellaneous Publication* **200**, 1-1051.

Matthei, O. (1986). El genero *Bromus* L. (Poaceae) en Chile. *Gayana* **43**, 47-110.

Parodi, L. R. (1947). Las gramíneas del género *Bromus* adventicias en la Argentina. *Revista Argentina de Agronomía* **14**, 1-19.

Peterson, P. M., and Planchuelo, A. M. (1998). *Bromus catharticus* in South America (Poaceae: Bromeae). *Novon* **8**, 53-60.

Planchuelo, A. M. (1983). Una nueva especie de *Bromus* (Poaceae) en la Flora Argentina. *Kurtziana* **16**, 123-131.

Renvoize, S. A. (1994) Notes on *Sporobolus* & *Bromus* (Gramineae) from the Andes. *Kew Bulletin* **49**, 543-546.

Shear, C. L. (1900). A revision of the North American species of *Bromus* occurring north of Mexico. *United States Department of Agriculture Bulletin* 23, pp 66.

Soderstrom. T. R., and Beaman, J. H. (1968). The genus *Bromus* (Gramineae) in Mexico and Central America. *Publications of the Museum, Michigan State University Biological Series* **3**, 465-520.

Wagnon, H. K. (1952). A revision of the genus *Bromus*, Section *Bromopsis*, of North America. *Brittonia* **7**, 415-480.

GRASSES

Grasses: Systematics and Evolution. (2000). Eds S.W.L. Jacobs and J. Everett. (CSIRO: Melbourne)

PHYLOGENETIC ANALYSIS OF THE TRITICEAE USING THE STARCH SYNTHASE GENE, AND A PRELIMINARY ANALYSIS OF SOME NORTH AMERICAN *ELYMUS* SPECIES

Roberta J. Mason-Gamer[AC] *and Elizabeth A. Kellogg*[B]

[A]University of Idaho, Department of Biological Sciences Moscow, Idaho, USA 83844-3051.
[B]University of Missouri, Department of Biology 8001 Natural Bridge Road, St. Louis, Missouri, USA 63121.
[C]Corresponding author; e-mail: robie@uidaho.edu

Abstract

The relationships among the intersterile diploid genera of the Triticeae have long been contested, and remain unresolved by published molecular data sets, which exhibit strong phylogenetic conflict. We have generated an additional data set for the tribe using partial (1.3 kb) sequences of the gene for granule-bound starch synthase (GBSSI or *waxy*; EC 2.4.1.11). The gene is presumed to exist in a single copy per genome, and therefore complements the clonal, maternally-inherited chloroplast genome and the highly repetitive nuclear markers that have already been examined. While analyses of exon characters presented here do not strongly support intergeneric relationships, they do highlight at least three examples of possible gene exchange among genera: 1) the presence of a divergent intron in *Aegilops searsii*, 2) the separation of two *Dasypyrum villosum* accessions, and 3) the inclusion of two *Pseudoroegneria* (genome designation **StSt**) species in the *Critesion* (**HH**) clade. Further investigation of the last of these observations includes screening of additional sequences from *Pseudoroegneria* and *Critesion*, and sequences from **StStHH** allotetraploids from the genus *Elymus*. These illustrate 1) The presence of **St** and **H** starch synthase genes in most of the *Elymus* accessions screened, 2) the hybrid nature of one of the diploid *Pseudoroegneria* individuals, and 3) diverse origins of both the **St** and the **H** genomes in *Elymus*.

Key words: Triticeae, Poaceae, *Elymus*, *Pseudoroegneria*, *Critesion*, wheat tribe, starch synthase, *waxy*, allopolyploidy.

INTRODUCTION

The evolutionary relationships among the species and genera of the wheat tribe Triticeae have long been of interest. Much of the available data on the history of this economically important tribe have come from extensive cytogenetic studies, which have been especially important to our understanding of hybridization and polyploidy. Furthermore, analyses of chromosome pairing in diploid hybrids have been used to delimit genomic genera within the Triticeae, such that hybrids between species within genera exhibit extensive chromosomal pairing at meiosis, while chromosomes in intergeneric hybrids show little or no pairing. Cytogenetic data have thus provided a means of segregating diploid and polyploid species into genera based on their genomic constitution (Dewey 1982; Löve 1984; Barkworth and Dewey 1985),

although the approach has been criticized on both methodological and theoretical grounds (Baum *et al.* 1987; Kellogg 1989; Seberg 1989; Seberg and Petersen 1998). Whatever their value for delimiting genera, chromosome pairing data do not provide the information necessary for reconstructing hierarchical relationships among genera (Baum *et al.* 1987; Kellogg 1989; Seberg 1989; Petersen and Seberg 1996; Seberg and Petersen 1998).

In the last few years, the relationships among diploid genera have been repeatedly addressed using molecular data. In four separate studies, five data sets have been produced, all of which sample all or most of the diploid genera in the tribe. These include two chloroplast DNA (cpDNA) data sets, one based on restriction site variation (Mason-Gamer and Kellogg 1996a) and the other on nucleotide sequences of the gene encoding the a-subunit of

RNA polymerase (*rpo*A; Petersen and Seberg 1997). Published nuclear molecular data sets include sequences from the internal transcribed spacer (ITS) region of the nuclear rDNA repeat (Hsiao *et al.* 1995), and sequences of intergenic spacer regions from two, independently-evolving 5S rDNA arrays (Kellogg and Appels 1995). Rather than resolving the relationships among the diploid genera, however, comparisons among these data sets highlight the unexpectedly complex history of the tribe (Kellogg *et al.* 1996; Mason-Gamer and Kellogg 1996b; Petersen and Seberg 1997).

Comparisons among molecular data sets reveal high levels of phylogenetic conflict. The only two data sets that are consistent with one another are those representing the clonal chloroplast genome. The most extensive differences are seen in comparisons between cpDNA and nuclear data (Kellogg *et al.* 1996; Mason-Gamer and Kellogg 1996b; Petersen and Seberg 1997). There are some examples of well-supported, statistically significant incongruence among nuclear data sets (Mason-Gamer and Kellogg 1996b), usually involving the misplacement of individual taxa. Comparisons among the data sets suggest that 1) the cpDNA and nuclear data sets are irreconcilably incongruent, and 2) the nuclear data sets can be combined (following the removal of a few problem taxa) to give a single estimate of a nuclear DNA phylogeny.

To further address the degree of conflict among Triticeae gene trees, we have generated an additional nuclear DNA data set using partial sequences of the presumed single-copy starch synthase gene. The gene provides a useful complement to the clonal, maternal chloroplast genome, and to the highly repetitive nuclear rDNA spacers that have already been analyzed for the same taxa. Our goals are to determine 1) whether these data provide additional, well-supported resolution at the intergeneric level, and 2) whether the relationships are consistent with those from other molecular markers.

Along with the diploid genera, we have analyzed several North American species of the allopolyploid genus *Elymus*. Worldwide, *Elymus* species exhibit a variety of genomic combinations (e.g., Dewey 1982; Löve 1984), but the North American species examined so far are derived from *Critesion* (genome designation **H**) and *Pseudoroegneria* (**St**). We are interested in 1) whether distinct copies of the starch synthase gene, representing both the **St** and the **H** genomes, can be recovered from within *Elymus* individuals and 2) whether starch synthase data can provide information about the origins of different *Elymus* species.

MATERIALS AND METHODS

Taxa are shown in Table 1. Seed sources for all accessions other than MEB (provided by M. E. Barkworth) and RJMG (collected by R. J. Mason-Gamer) are provided in Mason-Gamer and Kellogg (1996a). Previously published sequences (Mason-Gamer *et al.* 1998) are indicated on the table. DNA was isolated using the method of Doyle and Doyle (1987), and methods of PCR amplification, cloning, and sequencing are described in detail elsewhere (Mason-Gamer *et al.* 1998). In brief: a 1.3 kb fragment of the starch synthase gene was amplified using F-for (TGCGAGCTCGACAACATCATGCG) and

M-bac (GGCGAGCGGCGCGATCCCTCGCC) primers. Fragments were amplified from undiluted total DNA following denaturation at 94°C for one minute; five cycles of 94° for 45 seconds, 65° for 2 minutes, and 72° for 1 minute; 30 cycles of 94° for 30 seconds, 65° for 40 seconds, and 72° for 40 seconds; and a final 20-minute elongation at 72°.

PCR fragments were cloned into T-tailed vectors following the protocol provided with the Promega pGEM-T Easy cloning kit. For presumed polyploid taxa, three replicate PCR reactions were run and combined before cloning to help counter the effects of PCR drift and increase chances of obtaining sequences representing both genomes, when present. Plasmid isolation procedures followed ABI User Bulletin 18 from October 1991. ABI recommendations were followed for sequencing reactions and for ABI 377 sequencing gels. For the analysis of diploid genera, the 1.3 kb fragment was sequenced on both strands using the PCR primers F-for and M-bac, along with H2-for (GAGGCCAAGGCGCT-GAACAAGG), J-bac (ACGTCGGGGCCCTTCTGCTC), L1-for (GCAAGACCGGGTTCCACATGG), and L2-bac (CGCT-GAGGCGGCCCATGTGG).

For the further analysis of **HH**- and **StSt**-genome diploids (*Pseudoroegneria* and *Critesion*) and the **StStHH** polyploids (*Elymus*), we used a rapid-screening approach, sequencing with just the M-bac primer, to identify and distinguish clones. Between four and eight cloned PCR products were screened for each individual. Sequences were about 600 basepairs in length. All sequences were checked and edited in Sequencher (Gene Codes Corporation) and aligned using CLUSTAL V (Higgins *et al.* 1992).

Phylogenetic analysis of sequences from diploids were carried out with PAUP* 4.0d64 (D. Swofford 1998). Analyses were done on complete sequences and on exons only, and with all sites equally-weighted or reweighted once based on the maximum value of the rescaled consistency index. The trees and bootstrap values shown (Fig. 1) are based on the reweighted exon characters. In the analysis of the shorter sequences derived from the rapid-screening approach, introns were included, and characters were equally-weighted in the search and bootstrap analyses.

Conflict between the starch synthase data set and selected hypotheses of monophyly were tested using a Wilcoxon signed ranks (WSR) test (Siegel 1956; Templeton 1983; Larson 1994) as described in Mason-Gamer and Kellogg (1996b).

RESULTS AND DISCUSSION

Alignment of starch synthase introns among all genera was often problematic. While introns are often alignable across most taxa, there are generally some sequences that are too highly divergent to be aligned with confidence. This leads to uncertain homology assumptions, and concerns over whether tree topologies reflect the history of the gene or problems with the alignment. Therefore, we have taken the more conservative approach of using only the exon characters, which are easily aligned, for the analysis of all diploid genera. With this approach, we retain only 139 parsimony-informative characters (514 characters are constant, and 118 are variable but parsimony-uninformative). An equally-weighted parsimony analysis (result not shown) resulted in 210 trees with consistency index

Table 1. List of taxa used in phylogenetic analyses.

Taxon	Accession
Aegilops L.	
Aegilops bicornis Forsskål	Morrison s.n.[a]
Aegilops caudata L.	G758[a]
Aegilops comosa Sibth. and Smith	G602[a]
Aegilops longissima Schweinf. & Muschl.	Morrison s.n.[a]
Aegilops mutica Boiss.	Morrison s.n.
Aegilops searsii Feldman & Kislev	Morrison s.n.[a]
Aegilops speltoides Tausch	Morrison s.n.[a]
Aegilops tauschii Cosson	Morrison s.n.[a]
Aegilops umbellulata Zhuk.	Morrison s.n.[a]
Aegilops uniaristata Vis.	G1297[a]
Agropyron Gaertner	
Agropyron cristatum (L.) Gaertner	C-3-6-10[a]
Agropyron cristatum	PI281862
Agropyron cristatum	PI315357
Agropyron cristatum	RJMG133b
Agropyron cristatum ssp. *puberulum* (Boiss. ex Steudel) Tzvelev	PI229574
Agropyron mongolicum Keng	D2774
Australopyrum Á.Löve.	
Australopyrum retrofractum (Vick.) Á.Löve	Crane 86146
Australopyrum velutinum (Nees) B.Simon	D2873-2878
Bromus L.	
Bromus tectorum L.	Kellogg s.n.
Critesion Rafin.	
Critesion brevisubulatum (Trin.) Á.Löve	PI401387 (C-39-70-75)
Critesion bulbosum (L.) Á.Löve	PI440417
Critesion californicum (Covas & Stebbins) Á.Löve	MA-138-1-40[a]
Critesion jubatum (L.) Nevski	RJMG 134a
Critesion violaceum (Boiss. & Hohen) Á.Löve	PI401390 (C-39-41-45)
Dasypyrum (Cosson & Durieu) T.Durand	
Dasypyrum villosum (L.) Candargy	PI251478[a]
Dasypyrum villosum	PI470279
Elymus L.	
Elymus canadensis L.	MEB97-86
Elymus ciliaris (Trin.) Tzvelev	D2811 (R-38-56-60)
Elymus ciliaris	MA-140-21-30
Elymus glaucus Buckley	MEB97-83
Elymus glaucus	RJMG109
Elymus glaucus	RJMG121
Elymus glaucus	RJMG130
Elymus glaucus	RJMG141
Elymus hystrix L.	MEB s.n.
Elymus lanceolatus (Scribn. & J.G.Smith) Gould	D2847 (R-13-6-10)
Elymus lanceolatus	D2845 (R 14-1-5)
Elymus trachycaulis (Link) Gould ex Shinners	RJMG135a
Eremopyrum (Ledeb.) Jaub. & Spach.	
Eremopyrum bonaepartis (Spreng.) Nevski	H5554

Table 1. List of taxa used in phylogenetic analyses. *(Continued)*

Taxon	Accession
Eremopyrum bonaepartis	H5569[a]
Eremopyrum distans (C.Koch) Nevski	H5552
Eremopyrum orientale (L.) Jaub. & Spach	H5555
HENRARDIA C.E.HUBB.	
Henrardia persica (Boiss.) C.E.Hubb.	H5556[a]
HETERANTHELIUM HOCHST. EX JAUB. & SPACH.	
Heterantheliun piliferum (Banks & Sol.) Hochst.	PI402352[a]
HORDEUM L.	
Hordeum vulgare L.	b
LEYMUS HOCHST.	
Leymus racemosus ssp. **sabulosus** (M.Bieb) Tzvelev	D3489
PERIDICTYON SEBERG, FREDERIKSEN & BADEN	
Peridictyon sanctum (Janka) Seberg, Frederiksen, & Baden	KJ248[a]
PSATHYROSTACHYS NEVSKI	
Psathyrostachys fragilis (Boiss.) Nevski	C-46-6-15[a]
Psathyrostachys juncea (Fischer) Nevski	PI206684[a]
Psathyrostachys juncea	PI314521
Psathyrostachys juncea	PI499672
PSEUDOROEGNERIA (NEVSKI) Á.LÖVE	
Pseudoroegneria strigosa ssp. **aegilopoides** (Drobov) Á.Löve	PI531755
Pseudoroegneria strigosa ssp. **aegilopoides**	DT3101 (MA-109-31-50)
Pseudoroegneria spicata (Pursh) Á.Löve	PI232117[a]
Pseudoroegneria spicata	PI236681
Pseudoroegneria spicata	D2836
Pseudoroegneria spicata	D2844
Pseudoroegneria spicata	RJMG110
Pseudoroegneria spicata	RJMG112
Pseudoroegneria spicata	RJMG142
Pseudoroegneria spicata ssp. *inermis* (Scribn. & Smith) Á.Löve	D2839
Pseudoroegneria stipifolia (Czern. ex Nevski) Á.Löve	PI325181 (PW-23-41-50)
SECALE L.	
Secale montanum Guss.	PI440654[a]
Secale montanum ssp. *anatolicum* (Boiss.) Tzvelev	PI206992
Secale cereale L.	Kellogg s.n.
TAENIATHERUM NEVSKI	
Taeniatherum caput-medusae (L.) Nevski	PI208075 (MB-106-41-79)
THINOPYRUM Á.LÖVE	
Thinopyrum bessarabicum (Savul. & Rayss) Á.Löve	PI531711[a]
Thinopyrum elongatum (Host) D.R.Dewey	PI531719[a]
Thinopyrum elongatum	D3611
Thinopyrum scirpeum (C.Presl) D.R.Dewey	C-15-21-25
TRITICUM L.	
Triticum baeoticum Boiss.	Morrison s.n.[a]
Triticum monococcum L.	PI221413[a]
Triticum urartu Tumanian	Morrison s.n.[a]

Note: Accessions included in the tribal analysis (Fig. 1) are underlined; those included in the quick-screen analysis (Fig. 3) are in boldface.
[a]Mason-Gamer *et al.* 1998
[b]Rohde *et al.* 1988

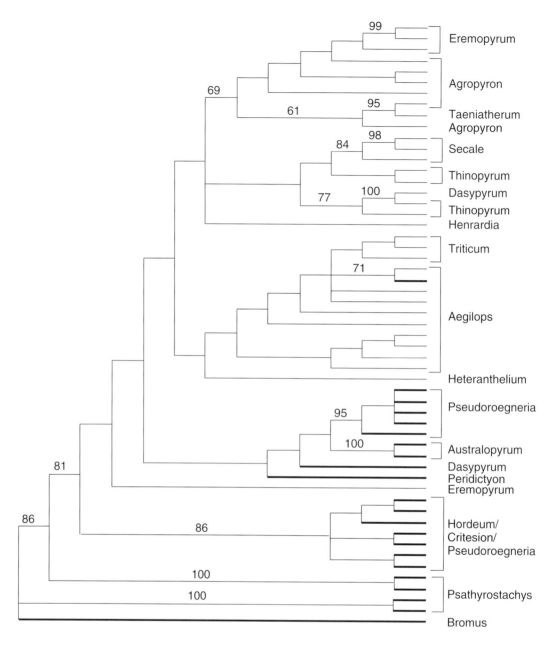

Fig. 1. Cladogram based on analysis of starch synthase exons. The tree topology and bootstrap values were obtained after characters were reweighted once based on maximum rescaled consistency index. Bold vs. light terminal branches show the distribution of the two classes of intron 10.

(CI), excluding uninformative characters, of 0.43, retention index (RI) of 0.67, and rescaled consistency index (RCI) of 0.37. The reweighted analysis resulted in 10 trees, which were used to generate a strict consensus tree (Fig. 1).

The starch synthase gene tree supports the basal position of *Psathyrostachys*, followed by the clade containing *Hordeum* and *Critesion*. (The placement of some *Pseudoroegneria* species in the *Hordeum/Critesion* clade is discussed below.) The basal positions of *Pseudoroegneria* and *Hordeum/Critesion* are in agreement with chloroplast DNA data from restriction sites (Mason-Gamer and Kellogg 1996a) and *rpo*A sequence data (Petersen and Seberg 1997). The two genera are also basal in the ITS tree (Hsiao *et al.* 1995) but with their positions reversed. The 5S short-spacer

data, as reanalyzed in Kellogg *et al.* (1996) are consistent with these results, with *Psathyrostachys, Critesion, Secale*, and *Henrardia* unresolved at the base of the tree. (The 5S long-spacer trees did not include *Psathyrostachys* or the outgroup *Bromus*; Kellogg and Appels 1995; Kellogg *et al.* 1996). The basal position of *Psathyrostachys* and *Critesion/Hordeum*, having gained further support from starch synthase analyses, is the only feature of the intergeneric relationships that is a point of strong consensus among these molecular data sets.

An example of apparent introgression of a portion of the starch synthase gene is illustrated by the phylogenetic distribution of the distinctly polymorphic intron 10. Unlike the other introns, which are more or less alignable throughout the tribe, there are

```
Aegilops       GTACATC-GTCG-TCGACCCCGCAACCCGACCCGCCATTGCTGAAACTTCGATCAAGCAGACCTAA----GGAAT--------GATCGAATGCATTGCAG
Agropyron      GTACGCCACGCCATCGACCCCGCAACCCGACC-GCCATTGCTAAAACTTCGATCAATCAGACCTAA----GGATT--------GATCGAATGCATTGCAG
Dasypyrum      GTGCGTC-----ATCGACCCCGCAACCCGACC-GCCATTGCTGAAACTTCGATCAAGCAGACCTAA----GGAAT--------GATCGAATGCATTGCAG
Secale         GTACGTC-GTCGATCGACCCGGCAACCCGACC-GCCATTGCTGAAACTTCGGTCAAGCAGACCTAATTAAGGAAT--------GATCGGATGCGTTGCAG
Thinopyrum     GTACGTC-----ATCGACCCCGCAACCCGATC-GCCATTGCTGAAACTTCGATCAAGCAGACCTAA----GGAAT--------GATCGAATGCATTGCAG
```

```
Aegilops        GTACGTCAACCGACATTGCTGACCCGTTGAGGAAAGCCTCCTGATAG------CTCGCCGTGGGGATGGATGGGTGACTGA-CTGATCGAATGCATTGCAG
Australopyrum   GTACGTCGACCGACTTTGCTGATCCATTCAGAAAAGTCTCGTGATGG------CTCGCTGTGGGGACGAATGGATGATTGA--------AATGCATTGCAG
Critesion       GTACGTCGACCGACATTGCTGATCCGTTCGAACAGGTCTCCTGATAG------CTCGCCATGGGGATGGATGGATGATTGA-GTGATCGAATGCATTGCAG
Psathyrostachys GTACGTGCACCGACATTGCTGATCCGTTCAGAAAAGTCTCCTGATAG------CTCGCCGTCGGGATGGATGGATGATTGA-ATGATCGAATGCATTGCAG
Pseudoroegneria GTACGTGGACCGACATTGCTGATCCGTTCAGAAATGTCTCCTGATAGTGATAGCTCGCCGTGGGTATGTATGGATGATTGAAGTGATCGAAT-CATTGCAG
```

Fig. 2. A selection of five examples of intron 10 sequences from each of the two distinct groups into which these sequences fall. The ends of the introns (black bars) are alignable throughout the tribe, but alignment between the two intron groups is otherwise impossible.

two very different categories of intron 10 sequences (Fig. 2). Within each category, the introns are easy to align, but alignment between the categories is impossible. The split defined by the intron 10 categories is not entirely congruent with the exon tree (Fig. 1). The distribution of the two intron 10 categories on the tree illustrates at least one apparent case of introgression of the intron into one of the species of *Aegilops* (*A. searsii*).

Several genera are polyphyletic on the starch synthase tree (Fig. 1). Two of these reflect the results of at least some previous studies: *Eremopyrum* was similarly split by the cpDNA restriction site data (Mason-Gamer and Kellogg 1996a), and the monophyly of the included *Thinopyrum* species has been a point of dispute for several years (Jauhar 1988, 1990; Wang 1989; Wang and Hsiao 1989). The paraphyly of *Agropyron* on the tree is not in agreement with other data to date, and may be an artifact of the low level of support.

Two examples of polyphyletic genera are intriguing in light of earlier data. The separation of the two *Dasypyrum villosum* accessions is in direct conflict with the cpDNA tree (Mason-Gamer and Kellogg 1996a), in which the exact same individuals were placed together. The starch synthase exon data significantly reject the monophyly of *Dasypyrum* in a WSR test (*P* < 0.01). Furthermore, the cpDNA tree contained a well-supported clade containing *Dasypyrum*, *Thinopyrum*, and *Pseudoroegneria*, and in which the three genera were unresolved. Such a clade was not found in any other existing nuclear gene trees; in those, *Dasypyrum* was not grouped with either *Thinopyrum* or *Pseudoroegneria*. The placement of these three genera together in the cpDNA tree clearly shows a past association among them, and the position of the two *Dasypyrum* sequences on the starch synthase tree, with one near *Thinopyrum* and the other near (although not sister to) *Pseudoroegneria* provides independent evidence of past gene exchange among these genera.

The inclusion of the presumably **StSt** *Pseudoroegneria stipifolia* and *P. strigosa* within the **H**-genome *Critesion* clade is also an unexpected result. Chloroplast DNA data (Mason-Gamer and Kellogg 1996a) for the exact same individuals place them in a clade with the other *Pseudoroegneria* accessions, nowhere near *Critesion*. The *waxy* exon data significantly reject the monophyly of *Pseudoroegneria* in a WSR test (*P* < 0.01). This result is particularly interesting because, although **StSt** and **HH** diploids are intersterile, there are many **StStHH** allopolyploid species, which form part of the genus *Elymus*. The placement of *Pseudoroegneria* individuals in the **H**-genome clade, though surprising, is consist-

ent with the hypothesis that allotetraploid taxa may serve as bridges to gene exchange among diploids that are otherwise unable to exchange genes (Kellogg *et al.* 1996). However, it is impossible to rule out the alternative possibility of occasional gene exchange directly among diploids.

One of the *Pseudoroegneria* accessions that appears in the **H** clade (*P. strigosa*; Fig. 1) was, in our subsequent rapid-screening procedure, shown to have a **St**-like copy of starch synthase as well. With respect to this locus, therefore, this individual appears to be a hybrid. Again, the mechanism of gene exchange is of interest but remains unknown.

Further analyses of *Critesion*, *Pseudoroegneria*, and **StStHH** allotetraploid *Elymus* were done using the rapid-screening procedure. The analysis resulted in more than 30,000 trees with a CI of 0.69, RI of 0.93, and RCI of 0.73 (Fig. 3). All trees include a monophyletic group apparently corresponding to the **H** genome. The **St** sequences form two distinct clades. In some cases, two divergent **St** genomes are found in the same plant, including both a diploid (*P. spicata* 6) and two tetraploids (*E. lanceolatus* 1 and 2). The **St** starch synthase genes from two **StStYY** *E. ciliaris* individuals, representing Japan and China, show no differentiation from the other members of the **St** clade, all of which represent North America. The **Y** genome sequences of *E. ciliaris* are grouped with the **St** sequences on the strict consensus tree (not shown), but the bootstrap support for any hypothesis of relationships among the two **St** clades and the **Y** clade is less than 50% (Fig. 3).

In many, but not all, of the *Elymus* individuals examined, we found copies of the starch synthase gene representing both **St** and **H** genomes. Assuming that 1) the two copies are present in equal frequencies in our combination of three PCR reactions, 2) both copies clone equally efficiently, and 3) clones of the two copies transform and grow equally well, then we expect about an 88% chance of obtaining both copies in a minimal screening of four clones, and a greater than 99% chance in a screening of eight clones. However, because this sampling procedure is based on numerous assumptions, it would be difficult to conclude with certainty that one or the other of the copies has been lost from an individual in cases where they are not both found.

The potential difficulty of using genome-specific markers to draw conclusions about the presence or absence of particular genomes is illustrated by *Elymus hystrix*. The lack of three out of four **St** genome-specific markers in *E. hystrix* led Svitashev *et al.*

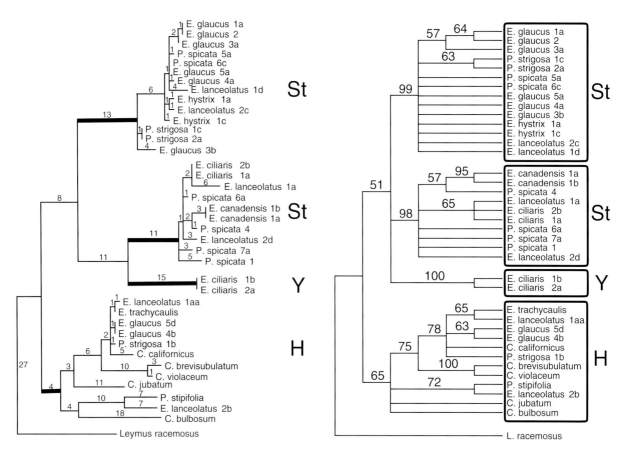

Fig. 3. Cladograms based on rapid-screening of *Elymus*, *Pseudoroegneria*, and *Critesion* species. The tree is rooted using *Leymus racemosus* ssp. *sabulosus* as an outgroup. Genus abbreviations are for *Elymus*, *Critesion*, and *Pseudoroegneria*. Species names with the same number represent the same plant (e.g., sequences labeled *E. lanceolatus* 2b, 2c, and 2d were amplified and cloned from a single individual). Designations of the genomes thought to be represented (**St**, **H**, **Y**) are listed along the right of each tree. Left: one randomly selected tree with branch lengths included. Right: 50% majority-rule bootstrap consensus tree.

(1998) to the reasonable conclusion that the **St** genome had been lost or highly modified. In this study, the **St** starch synthase gene is present in *E. hystrix* and is nearly identical to those of two other North American **StStHH** *Elymus* species (Fig. 3). This would have led us to conclude that *E. hystrix* does contain the **St** genome, with no major modifications. Difficulties with either conclusion exist at two levels. First, the presence/absence of individual markers probably should not be assumed to extend to the entire genome. There may be cases where only parts of a genome are lost or modified. Second, the lack or modification of a genome (or part of a genome) in one individual may not accurately predict its presence or absence in other individuals of the same taxonomic species. In the worst-case scenario, such conclusions can only be drawn on an individual-by-individual basis.

Even with the short sequences used in our rapid-screen analysis of **St** and **H** genomes, there may be enough variation among sequences, especially those in the **H** clade, to shed light on the origins of some of the polyploid species (Fig. 3). For example, the **H** starch synthase gene in *E. lanceolatus* 2 appears to have arisen from a different ancestor than those in the other *Elymus* accessions. This is consistent with the allozyme data of Jaaska (1992), which suggested that genetic diversity within **StStHH** *Elymus* reflects multiple independent origins of the **StStHH** genome combination.

In summary, the starch synthase data support the basal position of *Pseudoroegneria* and *Critesion*, in agreement with most molecular data sets. The data do not strongly resolve other intergeneric relationships, but do suggest a history of gene exchange involving *Aegilops*, *Dasypyrum*, and *Pseudoroegneria*. Starch synthase sequence data can provide information on the genomic constitution of *Elymus* species, and suggest multiple origins of the North American **StStHH** group.

The goal of arriving at a single hypothesis of relationships among diploid Triticeae genera is probably an unrealistic one, since different genetic markers appear to have different histories. Most phylogeny reconstruction methods estimate trees with divergent, bifurcating branching patterns. The evolutionary history of the Triticeae, on the other hand, has been shaped by reticulation at several levels. Discordance among gene trees reveals a widespread pattern of reticulation among the intersterile diploid genera. This complex pattern is compounded by the existence of numerous allopolyploid genomic combinations, which are reticulate by definition. Hybridization among polyploids (Dewey 1982) adds yet another level of complexity. Perhaps future studies focusing on allopolyploid formation and hybridization will lead to a deeper understanding of the evolutionary history of this interesting group.

ACKNOWLEDGEMENTS

The work was funded by a University of Idaho Stillinger Foundation grant to RJMG and NSF award DEB-9419748 to EAK. Thanks to Mary Barkworth for generously sending seeds, a draft key to North American *Elymus*, and encouraging words.

REFERENCES

Barkworth, M. E., and Dewey, D. R. (1985). Genomically-based genera in the perennial Triticeae of North America: identification and membership. *American Journal of Botany* **72**, 767–776.

Baum, B. R., Estes, J. R., and Gupta, P. K. (1987) Assessment of the genomic system of classification in the Triticeae. *American Journal of Botany* **74**, 1388–1395.

Dewey, D. R. (1982). Genomic and phylogenetic relationships among North American perennial Triticeae. In 'Grasses and Grasslands'. (Eds J. R. Estes, R. J. Tyrl, and J. N. Brunken.) pp. 51–88. (University of Oklahoma Press: Norman.)

Doyle, J. J., and Doyle, J. L. (1987). A rapid DNA isolation procedure for small quantities of fresh leaf tissue. *Phytochemical Bulletin* **19**, 11–15.

Higgins, D. G., Bleasby, A. J., and Fuchs, R. (1992). CLUSTAL V: improved software for multiple sequence alignment. *Computer Applications in the Biosciences* **8**, 189–191.

Hsiao, C., Chatterton, N. J., Asay, K. H., and Jensen, K. B. (1995). Phylogenetic relationships of the monogenomic species of the wheat tribe, Triticeae (Poaceae), inferred from nuclear rDNA (internal transcribed spacer) sequences. *Genome* **38**, 211–223.

Jaaska, V. (1992). Isoenzyme variation in the grass genus *Elymus*. *Hereditas* **117**, 11–22.

Jauhar, P. P. (1988). A reassessment of genome relationships between *Thinopyrum bessarabicum* and *T. elongatum* of the Triticeae. *Genome* **30**, 903–914.

Jauhar, P. P. (1990). Dilemma of genome relationship in the diploid species, *Thinopyrum bessarabicum* and *Thinopyrum elongatum* (Triticeae: Poaceae). *Genome* **33**: 944–946.

Kellogg, E. A. (1989). Comments on genomic genera in the Triticeae (Poaceae). *American Journal of Botany* **76**, 796–805.

Kellogg, E. A., and Appels R. (1995). Intraspecific and interspecific variation in 5S RNA genes are decoupled in diploid wheat relatives. *Genetics* **140**, 325–343.

Kellogg, E. A., Appels, R., and Mason-Gamer, R. J. (1996). When gene trees tell different stories: the diploid genera of the Triticeae (Gramineae). *Systematic Botany* **21**, 321–347.

Larson, A. (1994). The comparison of morphological and molecular data in phylogenetic systematics. In 'Molecular Ecology and Evolution: Approaches and Applications'. (Eds B. Shierwater, B. Streit, G. P. Wagner, and R. DeSalle) pp. 371–390. (Birkhäuser Verlag: Basel, Switzerland.)

Löve, Á. (1984). Conspectus of the Triticeae. *Feddes Repertorium* **95**, 425–521.

Mason-Gamer, R. J., and Kellogg, E. A. (1996a). Chloroplast DNA analysis of the monogenomic Triticeae: phylogenetic implications and genome-specific markers. In 'Methods of Genome Analysis in Plants'. (Ed. P. P. Jauhar.) pp. 301–325. (CRC Press: Boca Raton, Florida.)

Mason-Gamer, R. J., and Kellogg, E. A. (1996b). Testing for phylogenetic conflict among molecular data sets in the Triticeae. *Systematic Biology* **45**, 524–545.

Mason-Gamer, R. J., Weil, C. F., and Kellogg, E. A. (1998). Granule-bound starch synthase: structure, function, and phylogenetic utility. *Molecular Biology and Evolution* **15**, 1658–1673.

Petersen, G., and Seberg, O. (1996). Chromosomes, genomes, and the concept of homology. In 'Proceedings of the Second International Triticeae Symposium'. (Eds R. R.-C. Wang, K. B. Jensen, and C. Jaussi.) pp. 13–18. (Utah State University: Logan.)

Petersen, G., and Seberg, O. (1997). Phylogenetic analysis of the Triticeae (Poaceae) based on *rpoA* sequence data. *Molecular Phylogenetics and Evolution* **7**, 217–230.

Rohde, W., Becker, D., and Salamini, F. (1988). Structural analysis of the waxy locus from *Hordeum vulgare*. *Nucleic Acids Research* **16**, 7185–7186.

Seberg, O. (1989). Genome analysis, phylogeny, and classification. *Plant Systematics and Evolution* **166**, 159–171.

Seberg, O., and Petersen, G. (1998). A critical review of the concepts and methods used in classical genome analysis. *The Botanical Review* **64**, 372–417.

Siegel, S. (1956). 'Nonparametric Statistics for the Behavioral Sciences'. (McGraw-Hill: New York.)

Svitashev, S., Bryngelsson, T., Xiaomei, L., and Wang, R. R.-C. (1998). Genome-specific repetitive DNA and RAPD markers for genome identification in *Elymus* and *Hordelymus*. *Genome* **41**, 120–128.

Swofford, D. L. (1998). PAUP*: Phylogenetic analysis using parsimony, test version 4.0d64. (Sinauer Associates: Sunderland, Massachusetts.)

Templeton, A. R. (1983). Phylogenetic inference from restriction endonuclease cleavage site maps with particular reference to the evolution of humans and the apes. *Evolution* **37**, 221–244.

Wang, R. R.-C. (1989). An assessment of genome analysis based on chromosome pairing in hybrids of perennial Triticeae. *Genome* **32**, 179–189.

Wang, R. R.-C., and Hsiao, C. (1989). Genome relationships between *Thinopyrum bessarabicum* and *T. elongatum*: revisited. *Genome* **32**, 802–809.

Grasses: Systematics and Evolution. (2000). Eds S.W.L. Jacobs and J. Everett. (CSIRO: Melbourne)

CHANGING PERCEPTIONS OF THE TRITICEAE

Mary E. Barkworth

Intermountain Herbarium, Department of Biology, Utah State University,
Logan, Utah, U.S.A. 84322-5305.

Abstract

Taxonomists, both past and present, differ considerably in their generic treatment of the Triticeae. Some of the differences are attributable to differences in knowledge, but others may reflect differences in goals and cultural background. Linnaeus' treatment in *Species Plantarum* was designed primarily as an aid to identification. Bentham, who published a treatment of the tribe in 1882, also emphasized the importance of having easily recognized entities as the basic units of a classification. In contrast, Nevski's goal was to develop a classification that reflected the tribe's phylogeny. Partly because of this difference in goals, Nevski's generic treatment differed substantially from that of Bentham. It seems likely that Nevski's youth and the cultural milieu in which he worked also contributed to Nevski's willingness to suggest a radically different treatment. His treatment was quickly adopted in the Soviet Union and China, but only slowly and partially in non-communist countries. In *Zlaki CCCP*, Tzvelev followed Nevski's treatment for most annual genera, but interpreted *Elymus*, *Leymus*, and *Elytrigia* more broadly. Melderis initially recognized *Roegneria* as distinct from *Elymus*, but later included both it and *Elytrigia* in *Elymus*. In 1984, Löve and Dewey independently proposed that generic treatment in the Triticeae should be based on a single character, genomic constitution, arguing that such an approach would more accurately reflect biological and/or phylogenetic relationships within the tribe. Many oppose the resulting treatments, some because they disagree with the approach, others because some of the genera the two recognize are hard to identify in the field. A comparison of the treatments in recent floras indicates, however, that the areas of contention are relatively few. Unfortunately, the number of species affected is considerable. Tables comparing the generic treatment of the tribe in several floras, both past and present, are presented.

Keywords: Triticeae, classification, taxonomic history.

INTRODUCTION

The last few decades have seen the publication of several, evidently distinct, generic treatments of the Triticeae. This has led to debate about taxonomic philosophies and characters, and considerable frustration among those who are obliged to use the most pervasive consequence of a taxonomic treatment, its nomenclature. The purpose of this paper is to review current generic treatments of the tribe, with particular attention to the bases of their differences. An integral part of such a review is consideration of the characters, analytical procedures, and philosophies that lie behind the treatments.

My approach is historical because new taxonomic treatments are never completely new. They are modifications, major or minor, of their predecessors. The nomenclatural conservatism mandated by the Code contributes to this continuity but, more importantly, taxonomists are humans who live and work at a particular time and in a particular culture. We try to be objective in our work, but the questions that we ask, the means that we use to address them, the taxa that we focus on, and the classifications that we prefer are affected by our education, method preferences, taxonomic philosophy, location, and position.

IN THE BEGINNING

Linnaeus' (1753) *Species Plantarum* provides a convenient starting point for a historical review of generic concepts. In it, Linnaeus recognized five genera that are now included in the Triticeae: *Triticum, Hordeum, Secale, Elymus,* and *Aegilops.* One other genus, *Bromus,* contained a single species that is now included in the tribe as *Agropyron cristatum* (L.) Gaertn. Of the five genera, only *Secale* and *Hordeum* are still interpreted as they were by Linnaeus (and his predecessors). The limits of the other three genera have changed substantially, but only one of the species that Linnaeus included in the five genera, *Ae. incurva* L. [= *Parapholis incurva* (L.) C.E. Hubb.], is now placed in a different tribe.

Linnaeus did not recognize suprageneric taxa in *Species Plantarum.* As his friend, John Smith, explained (Smith 1807, cited in Stearn 1960): 'The sole aim [of *Species Plantarum*] is to help anyone learn the name and history of an unknown plant in the most easy and certain manner'. Its structure is similar to that of an identification key, with the leads employing characters that Linnaeus considered simple to use. The lowest level 'keyed out' were groups of genera that happened to be alike in the chosen characters. *Triticum, Hordeum, Secale,* and *Elymus,* together with several other genera, fell out in group 3, characterized by having perfect flowers with three stamens, but *Aegilops* was placed in group 23, a group of genera characterized as having male, female, and bisexual flowers, sometimes on the same plant, sometimes on different plants.

The tribe's current circumscription was established around 1934, the one question remaining in contention being whether or not to include *Brachypodium* (cf. Nevski 1934a; Hubbard 1934; Tzvelev 1976; Clayton and Renvoize 1986). Recent data support its exclusion (Kellogg 1992; Hsiao *et al.* 1995; Catalan *et al.* 1997).

Taxonomists soon added to Linnaeus' five genera (Table 1). Gaertner (1770) was first, describing *Agropyron* for *Bromus cristatus* L. Four other genera were added prior to 1815, but only *Agropyron* found general acceptance in the early nineteenth century (Kunth 1815; Berchtold and Presl 1820; Dumortier 1824). Dumortier, but not Kunth, also recognized *Hystrix*[1], but *Spelta, Zeocriton,* and *Elytrigia* were rarely accepted by nineteenth century taxonomists.

IMPERIAL VIEW

By 1882, 25 genera had been described in the Triticeae as we now understand it (Table 2), but Bentham (1882), who worked at Kew, the center for plant taxonomy of the British Empire, recognized only five of these. His treatment was adopted throughout the English-speaking world, including, with minor modifications, the United States. What led Bentham to such a conservative generic approach? Like Linnaeus, Bentham saw the goal of botanical classification as helping people learn the different kinds of plants: 'so long as the number of orders [= families] can be kept within, or not much beyond a couple of hundred, it may reason-

ably be expected that a botanist of ordinary capacity should obtain a sufficient general idea of their nature and characters to call them at any time individually to his mind for the purposes of comparison' (Bentham 1857, cited in Stevens 1997a). He took the same approach to the circumscription of genera, i.e., preferring a few large genera to numerous small genera.

De Jussieu expressed a similar idea in 1789, but for him the magic number was 100 (Stevens 1997a). The numbers are undoubtedly somewhat arbitrary, but it seems possible that Bentham 'grew up' knowing De Jussieu's genera, became aware of their flaws, but concluded that, by circumscribing taxa somewhat more narrowly, a satisfactory classification could be devised.

Stevens (1997b) identifies two other principles that guided Bentham's decisions: a desire to maintain the meaning of established names and to recognize 'natural' groups. What was a natural group? A group that shared important morphological characteristics. This approach differs from Linnaeus' divisive approach in that the groups were defined by their similarities in several characters rather than resulting from their differences in individual characters.

Bentham's tribes and genera of grasses are undoubtedly easy to learn and his emphasis on gross morphological features made field identification simple. He noted that 'Differences in the size of the embryo, in the form of the so-called scutellum on the caryopsis, or in the longitudinal groove or cavity frequently observable on the caryopsis, have sometimes been brought forward as absolute generic, if not tribal, characters, and they may often be really important; but we know, as yet, too little about them to test their value fairly' (Bentham 1882, p. 29). Terrell and Peterson's (1993) observations suggest that his caution was justified.

Hackel's (1889) treatment of the Triticeae, which was included in *Die Natürlichen Pflanzenfamilien,* differs from Bentham's only in the recognition of two small genera: *Heteranthelium* with one species and *Dasypyrum* with two species. These two treatments were adopted, with minor modifications, in floras throughout the world. The most frequent modifications were recognition of *Aegilops* and *Sitanion,* genera that both Bentham and Hackel included in *Triticum* and *Elymus,* respectively. The treatment of these two genera is still contentious (cf. Greuter and Rechinger 1967; Clayton and Renvoize 1986; Dewey 1984; Löve 1984).

FIRST REVOLUTIONARY PROPOSAL

In 1931 Avdulov, who worked at the Komarov Botanical Institute in what is now St. Petersberg, published his landmark study of the Gramineae. In this work, Avdulov showed that data on chromosome numbers and morphology, when combined with data from a wide range of other characters, both morphological and anatomical, supported a very different infrafamilial classification from that advocated by Bentham and Hackel. Prat (1932, 1936) independently reached somewhat similar conclusions and now we all think in terms of subfamilies and tribes that owe more to Avdulov and Prat than Bentham and Hackel.

What does this have to do with the Triticeae? Working at the same institute as Avdulov was another young taxonomist, Nevski, who began publishing a series of revisionary studies of

---

1. Here and elsewhere, I have used the valid equivalent of an invalid name to keep the focus on taxonomic, rather than nomenclatural, changes.

Table I. Publication dates for genera of the Triticeae. The identity symbol, ≡, is used to indicate the valid equivalent of an invalid name; the equals symbol, =, indicates that the listed genus is frequently treated as a synonym of the name that follows the equal sign.

Linnaeus (**1753**): *Aegilops, Elymus, Hordeum, Secale, Triticum.* Gaertner (**1770**): *Agropyron.* Wolf (**1776**): *Spelta* [= *Triticum*], *Zeocriton* [= *Hordeum*]. Humboldt (**1790**): *Asperella* [≡ *Hystrix*]. Moench (**1794**): *Hystrix.* Desvaux (**1810**): *Elytrigia.* Rafinesque (**1819**): *Critesion* [= *Hordeum*], *Sitanion.* Link (**1834**): *Crithodium* [= *Triticum*]. Nees (**1838**): *Polyantherix* [= *Sitanion*]. Seringe (**1842**): *Nivieria* [= *Triticum*]. Hochstetter (**1848**): *Leymus.* Koch (**1848**): *Roegneria.* E. Meyer (**1848**): *Critho.* Jaubert & Spach (**1851**): *Crithopsis, Eremopyrum, Heteranthelium.* Steudel (**1854**): *Anthosachne.* Turczaninov (**1862**): *Stenostachys.* Alefeld (**1866**): *Deina.* Schur (**1866**): *Haynaldia* [≡ *Dasypyrum*]. Durand (**1888**): *Dasypyrum.* Husnot (**1899**): *Goulardia* [= *Elymus*]. Eig (**1929**): *Amblyopyrum* [= *Aegilops*]. Nevski (**1932**): *Clinelymus* [≡ *Elymus*], *Terrella* [= *Elymus*]. Nevski (**1934**): *Aneurolepidium* [= *Leymus*], *Malacurus* [= *Leymus*], *Psathyrostachys, Taeniatherum.* Drobov (**1941**): *Campeiostachys* [= *Elymus*], *Semeiostachys* [= *Elymus*]. Zotov (**1943**): *Cockaynea* [= *Stenostachys*]. C.E. Hubbard (**1947**): *Henrardia.* Melderis (**1978**): *Festucopsis.* Á. Löve (**1980**): *Lophopyrum* [= *Thinopyrum*], *Pascopyrum, Pseudoroegneria, Thinopyrum.* Á. Löve (**1982**): *Aegilemma, Aegilonearum, Chennapyrum, Comopyrum, Cylindropyrum, Gastropyrum, Kiharapyrum, Orrhopygium, Patropyrum, Sitopsis* [all generally included in *Aegilops*]. Á. Löve (**1984**): *Australopyrum.* Á. Löve (**1986**): *Trichopyrum.* Seberg et al. (**1991**): *Peridictyon.* Seberg & Linde-Laursen (**1996**): *Eremium* [= *Leymus*].

the tribe in 1932 (Nevski 1932, 1934a, 1934b). His treatment differed radically from those of Bentham (1882) and Hackel (1889), both in the number of genera recognized and in their circumscription (Table 3). It bears considerable resemblance to current treatments, once one allows for his mistaken interpretation of *Leymus arenarius* (L.) Hochst. as the type of *Elymus*. Nevski explained that it was necessary to split up Bentham's and Hackel's genera in order to portray the evolutionary history of the tribe. For instance, he considered that reduction from few to solitary spikelets at each rachis node had occurred independently within each of the three lineages he identified in the tribe. This required recognition of at least three genera within Bentham's *Agropyron*. He also distinguished five genera among the perennial species that Bentham and Hackel had placed in *Elymus*. Clearly, Nevski was not concerned with the size of genera, nor with maintaining the meaning of existing names.

In preparing his revision, Nevski considered morphological, phytogeographic, and cytological data plus 'the findings of those with an applied interest in the Triticeae'. The morphological characters that he considered included some that Bentham had not used, but they were accessible with a hand lens. He was aware that the tribe had a chromosome base number of seven and that it included many polyploids, but he considered it impossible to draw any conclusion from such data at the time.

There are several parallels between Bentham's situation and that of Nevski. Like Bentham, Nevski worked at a premier botanical institution with a first class library and a major floristic project, the production of a flora of the U.S.S.R., in connection with which there was intensive botanical exploration. There were also substantial differences between the two. Bentham was 82 when he published *Notes on Gramineae*, and worked in a country that had had a stable government for a long period of time. Nevski was 26 when he published his first paper on the Triticeae and worked in a country that was trying a new form of government and at an institution where synthesis of a wide range of data had been used to support a radically different treatment of the Gramineae. His institute's botanical exploration did not cover such a large part of the globe as that emanating from Kew, but it did include central and northeastern Asia, a center of diversity for the perennial Triticeae (West *et al.* 1988).

Nevski's treatment was incorporated in the Flora of the USSR (Nevski 1934b), but it was several years before any of his generic concepts were adopted in the West. The recommendation that several of Bentham's genera be excluded from the tribe was

accepted more readily, but this, as Nevski pointed out, had already been recommended by Harz (1880), Hayek (1925), Holmberg (1926), and Avdulov (1931). Hubbard's (1934) and Pilger's (1947) adoption of the narrower tribal limits was based on Hayek's (1925) paper.

Pilger (1947) and Melderis (1953) were among the earliest Western taxonomists to adopt some of Nevski's genera in developing their own treatment of the tribe. They accepted most of his small annual genera (Table 3), but not his treatment of the perennial genera *Elymus, Roegneria, Leymus, Agropyron,* and *Elytrigia.*

Melderis (1953) was the first taxonomist to argue that *Elymus* should include the solitary-spikeleted species that Nevski had placed in *Roegneria*, a change that has been adopted in many subsequent treatments of the tribe, but not by many Chinese taxonomists (Keng 1965; Kuo 1987; Baum *et al.* 1991). He initially agreed with Nevski in recognizing *Terrella* and *Elytrigia*, but differed in the treatment of *Leymus*, expanding it to include *Aneurolepidium* and *Malacurus*, and in adopting the traditional interpretation of *Hordeum*. He later (1978, 1980) modified his treatment of *Elymus*, expanding the genus to include *Elytrigia* and *Terrella* as well as *Roegneria*.

Pilger (1954) treated *Agropyron* in almost the same sense as Bentham (1882) except that he accepted Nevski's removal of the annual species to *Eremopyrum*. He expanded *Elymus* to include *Terrella* and *Leymus* to include *Aneurolepidium* but, unlike Melderis, excluded *Malacurus* from *Leymus*. In other respects, his treatment was identical to that of Melderis.

ADVENT OF CYTOLOGICAL DATA

Melderis (1953) was the first to base his treatment of the Triticeae in part, but only in part, on cytological data. He cited papers by Stebbins (Stebbins *et al.* 1946a, 1946b; Stebbins and Walters 1949) showing that *Agropyron* and *Elymus*, as interpreted by Bentham, contained cytogenetically disparate elements, but emphasized that he constructed his treatment 'On the basis of [my] own morphological and genetical studies of members of the tribe Triticeae, as well as on the cytogenetic evidence found in the literature' (p. 854). In his later paper (Melderis 1978) he noted that 'the [Triticeae] presents complex phylogenetic and taxonomic problems. Much of the difficulty in the delimitation of some genera, especially *Elymus* L. *sensu lato* and *Agropyron* Gaertner *sensu lato*, arises from the absence of clear-cut generic

Table 2. Genera of the Triticeae as recognized by various taxonomists prior to 1930. Treatments marked by an asterisk were limited in scope and may not, therefore, reflect all the genera that the taxonomist concerned recognized. To facilitate comparison, illegitimate names have been replaced by legitimate names.

	Linnaeus 1753	Kunth 1815	Berchtold & Presl* 1820	Dumortier 1824	Spenner 1825*	Bentham 1882	Hackel 1889
Elymus	+	+	+	+		+	+
Sitanion							
Hordeum	+	+	+	+	+	+	+
Critesion							
Hystrix				+		+	+
Agropyron		+	+	+		+	+
Heteranthelium							+
Haynaldia							+
Secale	+	+	+	+	+	+	+
Triticum	+	+		+	+	+	+
Aegilops	+	+		+			

characters and from the presence of numerous intergeneric hybrids'(p. 369).

The reference to 'clear-cut generic characters' reminds one of Linnaeus' and Bentham's concern for an easily learned systematic treatment. Other parts of his paper also make it clear that Melderis considered it important that genera be readily distinguishable. Nevertheless, he cited data from leaf anatomy, chromosome morphology, and genomic analyses in support of his decisions, characters that ordinary botanists can understand but not ones that are useful for field identification.

Tzvelev's (1976) treatment of the Triticeae in *Zlaki CCCP* [Grasses of the U.S.S.R.] was similar to that of Nevski, differing primarily in the expansion of *Leymus* to include both *Aneurolepidium* and *Malacurus* and correction of Nevski's nomenclatural errors. The major distinction between Tzvelev's treatment in *Zlaki CCCP* and that of Melderis (1980) in *Flora Europaea* lies in their treatment of *Elytrigia*, which Tzvelev recognized but Melderis included in *Elymus*. Chinese taxonomists have generally accepted both *Elytrigia* and *Roegneria* (Keng 1965; Kuo 1987; Baum *et al.* 1991, 1995). Keng, but not Kuo, also recognized *Aneurolepidium* as distinct from *Leymus*, as has Baum (1979). Thus, by the early 1980s, treatments of the Eurasian Triticeae were substantially alike, differing primarily in how perennial species with one spikelet per node were treated. Treatments of the Australasian and American genera continued to be based on that of Bentham (1882) until the late 1980s, differing only in the recognition of *Aegilops* and *Sitanion*.

In 1987 Tzvelev published a synopsis of the tribe that was global in scope. It shows that he and Melderis differed more than was evident in the two floristic works cited above, for he accepted many of the non-Soviet segregates that had by then been published by Löve (see below), including *Terrella*, a genus that Nevski and he interpreted as consisting of the eastern North American species of *Elymus* with glumes that are narrow and strongly thickened at the base. [The name is apparently a misspelling of *Terrellia* Lunnell, which is invalid, having been published as a synonym of *Elymus*]. Tzvelev did not accept those of Löve's segregates that are represented in the Russian flora,

namely *Pseudoroegneria*, *Thinopyrum*, *Lophopyrum*, and the various segregates of *Aegilops*.

Between 1945 and 1955, Stebbins and his students conducted numerous studies of crossing relationships in the tribe (see references in Stebbins 1956) that led Stebbins to state that 'a systematic treatment of the tribe Hordeae [sic] which will express the true interrelationships of its species must begin by uniting all of the genera in one' (Stebbins 1956, p. 240). The simplicity of this suggestion, although not adopted in any floristic treatment, commended it to many North American taxonomists.

Between 1955 and 1984, Dewey and others increased the information on crossing relationships within the tribe enormously, as a result of which Dewey became convinced that there were discrete entities within the perennial members of the tribe that merited generic recognition (see Dewey 1984 for references). In comparing the data from such studies with existing generic treatments, he noticed that the greatest congruence was with Tsvelev's (1976) treatment. In 1979, at a symposium sponsored by the American Society of Plant Taxonomists, he proposed that North American taxonomists should adopt Tzvelev's system because 'I am convinced that in the long run a system that reflects biological relationship will prove more useful, even though less convenient, than a system based on morphology' (Dewey 1982, p. 83). There was little comment at the symposium itself, but he received a number of letters in response to his suggestion. These expressed a range of attitudes, from strong support for the recognition of additional perennial genera to equally strong support for Stebbins' 'one genus' approach.

When the symposium proceedings were published (Estes *et al.* 1982), they included a paper by Estes and Tyrl (1982) that had not been presented at the conference. In it, the authors advocated the recognition of only two genera for the native perennial North American Triticeae: *Hordeum* and *Elymus*. This, the authors claimed, 'produces a model that relates complex patterns of variation and inferred phylogeny. Ease of identification, data incorporation, interpretation, and retrieval are balanced against evolutionary considerations'. Therefore, a system based on two genera is the most useful and appropriate model' (p. 161).

113

Table 3. Representative generic treatments published after 1930. Nomenclatural errors have been corrected. Asterisked treatments are limited in either regional or taxonomic scope. Names in adjacent, similarly-shaded cells within a column were considered congeneric by the author concerned.

	Nevski 1934	Melderis 1947	Pilger 1954	Dewey 1984*	Á. Löve 1984, 1986	Tzvelev 1987	Kellogg 1989
Elymus	+	+	+	+	+	+	+
Roegneria	+						
Terrella	+	+				+	
Hystrix	+	+	+			+	
Sitanion	+	+	+			+	
Anthosachne	+	+					
Australopyrum			+		+	+	
Psathyrostachys	+	+	+	+	+	+	+
Leymus	+	+	+	+	+	+	+
Aneurolepidium	+						
Malacurus	+		+				
Pascopyrum				+	+	+	+
Stenostachys		+			+	+	
Agropyron	+	+	+	+	+	+	+
Pseudoroegneria				+	+		+
Elytrigia	+	+		+	+	+	+
Lophopyrum					+	+	+
Thinopyrum				+	+	+	+
Trichopyrum					+		
Hordeum	+	+	+	+	+	+	+
Critesion	+			+	+		
Taeniatherum	+	+	+	+	+	+	
Hordelymus	+	+	+		+	+	
Crithopsis		+	+		+		
Heteranthelium	+	+	+		+	+	+
Dasypyrum	+	+	+		+	+	+
Secale	+	+	+	+	+	+	+
Eremopyrum	+	+	+	+	+	+	+
Henrardia	+			+	+		
Amblyopyrum		+	+		+	+	
Aegilops	+	+	+		+a	+	+
Crithodium					+		+
Triticum	+	+	+		+	+	+
Festucopsis					+	+	+
Peridictyon							

a. Löve recognized *Gigachilon, Cylindropyrum, Sitopsis, Patropyrum, Gastropyrum, Aegilonearum, Orrhopygium, Chennapyrum, Comopyrum, Aegilemma, Aegilopodes,* and *Kiharapyrum* as segregates of *Aegilops* but they have rarely, if ever, been accepted.

Arnow (1987) adopted this approach, but it has not, to my knowledge, been used in any other North American flora.

SECOND REVOLUTIONARY PROPOSAL

In 1984, Löve and Dewey independently proposed that generic determination in the Triticeae be based solely on genomic constitution. This was a more radical proposal than Nevski's revision of the tribe, and than Dewey's (1982) earlier proposal, for it eliminated from consideration all but one character, genomic constitution, which, at that time, meant observations of chromo-some pairing during meiosis. Löve (1984, p. 426) acknowledged, however, that the genomic constitution of many species had not been experimentally determined and that he had, therefore, inferred the generic placement of such species (and their genomic constitution) 'on the basis of the classical morphological-geographical concept only'.

Löve applied the principle of genomic classification more rigidly than Dewey, recognizing 39 genera in the tribe, including 26 in what most taxonomists call *Aegilops.* Dewey (1984) was aware that *Elymus,* as he and Löve interpreted it, included species with at least three different genomic constitutions, **StH**, **StY**, and **StPY** [genomic designations follow Wang *et al.* 1996], but stated that 'After the genomic relationships are fully understood in *Elymus*, several new genera may have to be described' (p. 247). Interestingly, he used conventional nomenclature in referring to the annual species of *Aegilops* and *Triticum*[1] 'to make the discussion less confusing to the reader' (p. 216), a statement that acknowledges the advantage of retaining the meaning of established names.

REACTION TO THE CONCEPT OF GENOMIC GENERA

The idea that the taxonomic treatment of a group, any group, should be based solely on consideration of a single character is absurd, whether that character be inflorescence structure, growth habit, genomic constitution, or gene sequence. To Dewey, the fact that the species within the various genomic groups of Triticeae had other characteristics in common was an obvious corollary of their genomic similarity. When he studied cytology the general understanding was that, if chromosomes paired at meiosis, it was because they were generally similar throughout their length. By 1984, Dewey knew that this interpretation was no longer accepted. He argued, however, that, at least within the perennial Triticeae, genomic constitution was a more reliable predictor of crossing behavior, morphology, and physiology than any other characteristic.

Although it is absurd to base taxonomy on a single, predetermined character, it is not absurd to reconsider the taxonomy of a group based on a single character to see how well its pattern of variation correlates with that of the variation in other characters. If one finds greater corroboration from other characters for groups formed based on the 'new' character than for groups in current classifications, it is appropriate to consider changing the classification. For an example of how powerful such an approach can be, one need look no further than Avdulov's (1931) paper. I worked with Dewey in providing the combinations needed to implement his system in North America (Barkworth and Dewey 1985) because I was (and am) convinced that the validity of the genera he recognized was, in the main, better corroborated by the variation in other characters, including morphological characters, than were the genera recognized in other treatments.

Almost all taxonomists object to complete reliance on genome analysis for determining taxonomic groups, but Seberg and Peterson (1998) state that 'it has no role to play in taxonomy and is of limited value in understanding evolution. Genome analysis suffers from limited knowledge of the mechanisms behind meiosis, an imprecise handling of data, is pervaded by numerous *ad hoc* assumptions, lacks a clear connection to the concept of homology, and cannot be given a meaningful biological interpretation'. I agree with many, but not all, of their criticisms of genome analysis. It is true, as they point out, that the mechanisms that govern meiosis, and hence genomic analysis, are not fully understood, but nor are those that determine glume morphology, growth habit, or the number of spikelets at a node. If lack of understand-

ing makes a character inappropriate for use in taxonomy, there would be no taxonomic treatments. The criteria for being a useful taxonomic character are, however, different: if a character varies within the group being studied, it merits consideration in a taxonomic study; if its pattern of variation is found to be consistent with that of other characters, it has taxonomic value. In other words, it can be used in the circumscription of taxa. In this sense, even though we do not understand the mechanism behind it, genomic constitution is of taxonomic value.

It is also true that genomic constitution provides limited information about evolution, but limited information is better than no information. It is, for instance, highly improbable (but not impossible, see Mason-Gamer and Kellogg, this volume) that any diploid **H** genome species is derived from an **StH** polyploid, but highly probable that an **H** genome species, either diploid or polyploid, extinct or extant, figures in the ancestry of all **StH** polyploids. If two species of Triticeae are genomically alike, they can tentatively be considered part of the same taxon. Genomic information does not, however, provide information on the relationships between the various diploid genomic groups, nor, in the case of polyploids, whether polyploids with the same genomic constitution were derived from the same polyploid ancestor, let alone the same diploid ancestors.

Some of Seberg and Petersen's (1998) other criticisms seem to be based in part on the assumption that cladistic analysis is the only valid means of determining phylogeny. I consider it a valuable tool, but one based on a simplistic model of evolution. The phylogenetic history of the Triticeae is one in which the data suggest that reticulation has played a major role. The cladistic model does not permit this. Its use in discovering probable gene histories rather than species histories is more reasonable, but we know little about how the genes being examined influence the success of a species. We can only look for correlation in patterns of modification.

The different uses of 'homology' in cytogenetics and evolutionary biology are little more significant than differences between British and American English. Despite the problems that Seberg and Petersen cite in data analysis, there is little disagreement about genome designations. The one area of controversy (Jauhar 1990a, 1990b; Wang and Hsiao 1989), is a reflection of the fact that genome homology (using homology in the cytogenetic sense) is a continuum.

Löve's (1984) and Dewey's (1984) papers had an immediate impact. The effect was particularly marked in North America because until 1984, almost all treatments of North American Triticeae followed Hitchcock (1935, 1951) which was, in turn, based on Bentham (1882). Dewey's treatment has won widespread acceptance among plant geneticists, partly because of the respect he had earned from such scientists but largely because it provided a more useful framework within which to conduct research than any existing alternative. It is also used by most plant systematists working with the tribe, at least as a taxonomic treatment that merits evaluation. It has not been widely accepted by floristic botanists, possibly because understanding generic relationships is less important to such individuals than ease in assigning a generic name to a species.

---

1. Dewey stated that he used conventional names for *Triticum*, *Hordeum*, and *Secale*, but he treated *Hordeum* and *Critesion* as separate genera and even Löve (1984) did not change the circumscription of *Secale*.

ADVENT OF MOLECULAR DATA

In recent years, 'genomic analysis' has come to include studies in which molecular data are used to identify the genomes present in a taxon (Barkworth *et al.* 1996; Dubcovsky *et al.* 1997; Jensen and Wang 1997). The taxonomic value of such data should be evaluated in the same manner as that of any character, i.e., consistency within a group and congruence with other characters. When used in this manner, they may be used to support a particular generic treatment of a species but they, like other characters (including classical genomic analysis), should never be the only basis for determining the generic affiliation of a species. The ability to identify genomic groups with molecular characters indicates that 'genomic constitution' does have some fundamental reality, but so far it has not brought us closer to understanding what a genome is. It also means that we have rather different kinds of data being interpreted as providing the same information. There are no data, so far, to say that this is not valid, but we should be alert to the possibility that this may not always be the case.

Molecular techniques are also being used to elucidate relationships within the Triticeae to an extent that was previously impossible. The results confirm that the Triticeae has a complex phylogenetic history and offer greater insight into why recognition of some genera is particularly controversial. Trees for the diploid species based on data from nuclear and chloroplast genes are not completely concordant, and even some of the diploid species appear to have a hybrid origin (Kellogg *et al.* 1996; Petersen and Seberg 1997; Mason-Gamer and Kellogg this volume). Molecular studies of the heterogenomic polyploids support recognition of the **StH** and **StY** taxa as distinct clades (Svitashev *et al.* 1996), as suggested by their genomic constitution and morphology (Lu 1996). They also suggest that Asian and North American **StH** taxa may belong to different lineages and that *Hystrix patula* Moench, the type species of *Hystrix*, has a different version of the **St** genome from that in the other **StH** species examined (Svitashev *et al.* 1998). Other data (Dubcovsky *et al.* 1997; Jensen and Wang 1997) indicate that some species of *Hystrix* are more closely related to *Leymus* than to *Hystrix patula* and that *Eremium* should also be included in *Leymus* (Dubcovsky *et al.* 1997).

So where do we stand now? All would like to see unanimous agreement on an easy to use generic treatment of the tribe. Most, but not all, also want a taxonomy that will reflect, as closely as possible, the phylogenetic history of the tribe. Obtaining comparable data on all taxonomically useful characters for all its taxa in the tribe would undoubtedly be useful, but it will take considerable time and will not lead to universal agreement on a generic treatment. Some taxonomists will continue to prefer large, easily distinguished genera (Estes and Tyrl 1982; Clayton and Renvoize 1986; Assadi and Runemark 1995); others will continue to prefer smaller genera with a more uniform genealogy (Dewey 1982; Kellogg 1989; Barkworth 1997), even if they are harder to recognize

This basic difference in philosophy has led to remarkably little disagreement among recent floras (Table 4). Almost all the disagreement concerns the treatment of the perennial species with solitary spikelets and rhizomes or long anthers, i.e., most of the

species that Bentham (1882) and Hackel (1889) placed in *Agropyron*. These are sometimes included in *Elymus* (Melderis 1980), sometimes in several smaller genera (Tsvelev 1987; Barkworth 1997). Kuo (1987) recognizes *Roegneria*, as do Baum *et al.* (1991, 1995). Baum *et al.* (1991, 1995) argue that *Elymus*, *Roegneria*, and *Kengyilia* are morphologically readily distinguishable and merit generic recognition, but their circumscription of *Roegneria* includes many species that other taxonomists would place in *Elymus*, *Australopyrum*, and *Stenostachys* (Lu 1996; Connor 1994; Barkworth *et al.* 1996), even if *Roegneria* is accepted as a valid genus.

This apparently moderate level of disagreement may be an illusion. Many species currently bear names that reflect their classification under one of the older schemes. As such species are re-examined, some will probably be reassigned. This is to be expected in a large tribe with a global distribution. The complex history that is being revealed by studies such as Kellogg *et al.* (1996) and Svitashev *et al.* (1998) make it very unlikely that there will ever be complete agreement as to the best taxonomic treatment of the tribe. It is imperative, however, that, in proposing a change in classification, we consider a wide range of characters and taxa. By so doing there is greater assurance that, when the ordinary botanist has become familiar with several members of the tribe, he or she will have learned something of lasting value.

POSTSCRIPT

A reviewer suggested that I include my own opinion concerning the taxonomic treatment of the Triticeae in this paper. I have taken gross advantage of the suggestion by adding a column to Table 4 and the following paragraphs of explanation.

A good taxonomic treatment is one that reflects phylogenetic history, because such a treatment is more likely to have a high predictive value. I would expect to find that most taxa in such a treatment of vascular plants could be characterized morphologically, because it is the morphology and physiology of vascular plants on which selection pressures act. This does not mean that most taxa will be *obviously* distinct. Bentham's (1882) genera were obviously distinct, but were not consistent with the phylogeny of the group and had low predictive value.

Unless our understanding of the evolution of the Triticeae undergoes a major change, I do not see there ever being complete agreement on the generic treatment of the tribe because of the complexity of the phylogenetic relationships within the tribe. Even some of the diploids show evidence of a polyphyletic origin (Mason-Gamer and Kellogg this volume). Autoploidy and alloploidy merely increase the opportunities for complexity. I do not see this lack of complete agreement as a major problem. Humans can learn to appreciate different points of view, in taxonomy as in other aspects of life, and databases can be set up to interpret different generic treatments. What should be of far greater concern than the generic treatment of the tribe is clarification of the many species problems that exist among its members.

As can be seen, my personal preference would be to include *Aegilops* in *Triticum*. The problems that arise from not doing so were reflected at a workshop on the taxonomy of *Triticum* and *Aegilops* that was held in conjunction with the International

Table 4. Treatment of the Triticeae in various contemporary regional floras. The floras have been selected to provide an overview of treatments in current use. Three recently recognized genera are also shown even though they have not yet been treated in any regional flora. '*': the genus was described after publication of the flora; '+': the genus is recognized as such in the flora concerned; 'NI': the genus is not included in the flora concerned; an abbreviated generic name indicates that the genus is treated as a synonym of the abbreviated genus;. Genera in **boldface** are interpreted similarly in all the selected floras in which they are represented.

Taxon	USSR (Tsvelev 1976)	Europe (Melderis 1980)	Turkey (Davis 1985)	China (Kuo 1987)	South America (Nicora and Rúgolo de Agrasar 1987)	New Zealand (Connor 1994)	North America (Barkworth et al., in prep.)	Barkworth (See post-script)
Aegilops	+	+	+	+	NI	NI	+	[Trit]
Agropyron	+	+	+	+	+	NI	+	+
Amblyopyrum	+	NI	NI	NI	NI	NI	NI	[Trit]
Australopyrum	NI	NI	NI	NI	NI	+	NI	+
Crithopsis	NI	+	+	NI	NI	NI	NI	No opinion
Dasypyrum	+	+	+	NI	NI	NI	+	+
Elymus	+	+	+	+	+	+	+	+
Elytrigia	+	[Elym]	[Elym]	+	[Agrop]	NI	[Elym]	[Elym]
Eremium	NI	NI	NI	NI	[Elym]*	NI	NI	?[Leym]
Eremopyrum	+	+	+	+	NI	NI	+	+
Festucopsis	NI	+	NI	NI	NI	NI	NI	+
Henrardia	+	NI	+	NI	NI	NI	NI	+
Heteranthelium	+	NI	+	NI	NI	NI	NI	+
Hordelymus	+	+	+	NI	NI	NI	NI	+
Hordeum	+	+	+	+	+	NI	+	+
Hystrix	+	NI	NI	+	NI	NI	[Elym, but see text]	[Elym, but see text]
Kengyilia	[Elym]*	[Elym]	[Elym]	[Roeg]*	NI	NI	NI	[Elym]
Leymus	+	+	+	+	[Elym]	NI	+	+
Peridictyon	NI	[Fest]*	NI	NI	NI	NI	NI	+
Psathyrostachys	+	+	+	+	NI	NI	+	+
Pseudoroegneria	[Elytr]	[Elym]	[Elym]	[Elytr]	NI	NI	+	+
Roegneria	[Elym]	[Elym]	[Elym]	+	NI	NI	NI	[Elym]
Secale	+	+	+	+	+	NI	+	+
Stenostachys	NI	NI	NI	NI	NI	+	NI	+
Taeniatherum	+	+	+	NI	NI	NI	+	+
Thinopyrum	[Elytr]	[Elym]	[Elym]	[Elytr]	[Agrop]	NI	+	+
Triticum	+	+	+	+	+	NI	+	+

Triticum Mapping Initiative in Saskatoon. All those present were aware that many species have names in both genera (sometimes more than one name), several of the younger individuals present who were not trained as taxonomists commented that it had come as a shock to them. They also pointed out that, if one is in a short-term post-doctoral position, not becoming aware of such problems for a few months can be a significant problem. There is now a Web site that provides such information for these two genera.[1]

I have no opinion on *Crithopsis* because I have had no reason to look at it. The inclusion of *Amblyopyrum* in a broadly interpreted *Triticum* seems reasonable, but I am not familiar with the species involved. I accept *Australopyrum* and *Stenostachys* because they

are accepted by Connor (Connor *et al.* 1993; Connor 1994), whose papers provide evidence of careful consideration of multiple aspects of the taxa involved. I have very limited knowledge of the two genera.

My inclusion of *Elytrigia* in *Elymus* is one instance where recognition that the genomic constitution of the type species, *Elymus* (or *Elytrigia*) *repens* L. was **StStH** swayed my mind. I have never liked treating *E. repens* as generically distinct from *E. lanceolatus* because of their morphological similarity, but I followed Tzvelev in doing so. I would not, however, transfer all the species that have been included in *Elytrigia* to *Elymus*, just *E. repens*. Some of the others I include in *Thinopyrum*, others in *Pseudoroegneria*; and others I do not know enough about to comment on. I consider *Thinopyrum* a morphologically distinct genus, although Assadi and Runemark (1995) disagree. I have some trouble with

1. http://www.ksu.edu/wgrc/Germplasm/Taxonomy/taxintro.html

recognizing *Pseudoroegneria* as a genus, but I know it only from North America where there is a morphologically similar **StH** species, *Elymus wawawaiensis* Carlson & Barkworth (Carlson and Barkworth 1997). I retain it as a genus because those who are also familiar with the Asian species of *Pseudoroegneria* and *Elymus* assure me that the genus is both morphologically and ecologically distinct from other members of the tribe. Another argument in support of its generic status is that it appears to be the source of one ancestor of most alloploid Triticeae.

I retain *Roegneria* and *Kengyilia* in *Elymus* because I have not found the arguments for recognizing them sufficiently persuasive (Baum *et al.* 1991, 1995). Some of the species that Baum *et al.* (1991) place in *Roegneria* cannot be excluded from *Elymus* using the characteristics they suggest (Barkworth *et al.* 1996). I am told that, with practice, one can learn to recognize *Roegneria* in the field, but that it is very similar to *Elymus*. It seems best to treat both it and *Kengyilia* as infrageneric, rather than generic taxa, as Salomon and Lu (pers. comm.) recommend.

I currently include *Sitanion* in *Elymus*, but would have little difficulty in treating it as a separate genus. It is morphologically distinct and hybrids with *Elymus* are highly sterile. *Hystrix* is a greater problem. Baden *et al.* (1997) recognize it in the traditional sense but Church's (1967) arguments for including the type species in *Elymus* are, in my opinion, persuasive. The only other species with which I am familiar, *Elymus californicus* (Bol.) Gould has morphological, cytological, and molecular characteristics that suggest it belongs in *Leymus* (Barkworth 1997), but more closely resembles North American species of *Elymus* in its ecology and other morphological characteristics. I keep it in *Elymus* because doing so does not require a new combination, one that I would not wish to make without knowing a lot more about the Asian species that have been included in *Hystrix*. I find, however, the arguments of Dubcovsky *et al.* (1997) for including the monotypic *Eremium* (Seberg and Linde-Laursen 1996) in *Leymus* to be compelling. My treatment of the remaining genera is not controversial.

Acknowledgements

The opinions expressed in this paper are my own, but I have benefitted from discussions with numerous individuals over the years. The late D.R. Dewey was always willing to share his knowledge of the tribe. We disagreed in some matters, but I have profound respect for his work and his knowledge of the species with which he worked. John McNeill introduced me to the differing philosophical approaches to taxonomy. Peter Stevens sent me several papers that helped me understand Bentham's thinking and encouraged me to consider the cultural setting in which taxonomists work. Toby Kellogg's comments and questions have frequently caused me to ask new questions or to look at old questions in new ways. Björn Salomon and Bao Rong Lu have freely shared their knowledge of Asian Triticeae, and their ideas concerning the taxonomy of the tribe. Ole Seberg and Gitte Petersen kindly made the manuscript of their paper available to me ahead of publication. Claus Baden, Roland von Bothmer, Signe Frederiksen, and Julian Campbell have shared their insights concerning their favorite portions of the tribe. I thank all these individuals. They have made working with the Triticeae an enjoyable experience despite the refusal of the tribe's members to behave as humans might wish. I also thank Kathleen Capels for her invaluable assistance in tracking down obscure references and Karen Gonzales who translated Nevski's papers for me.

References

Avdulov, N.P. (1931). Karyosystematische Untersuchung de Familie Graminae. *Bulletin of Applied Botany, of Genetics, and Plant Breeding* [*Trudy po Prikladnoi Botanike, Genetike i Selektsii*] Supplement [*Prilozenie*] **44**, 1-428.

Arnow, L. (1987). Gramineae. In 'A Utah Flora.' (Eds. S.L. Welsh, N.D. Atwood, L.C. Higgins, and S. Goodrich.) pp. 684-788. Great Basin Naturalist Memoirs 9. (Brigham Young University: Provo, Utah.)

Assadi, M., and Runemark, H. (1995). Hybridization, genomic constitution and generic delimitation in *Elymus* s.l. (Poaceae: Triticeae). *Plant Systematics and Evolution* **194**, 189-205.

Baden, C., Frederiksen, S., and Seberg, O. (1997). A taxonomic revision of the genus *Hystrix* (Triticeae, Poaceae). *Nordic Journal of Botany* **17**, 449-467.

Barkworth, M.E. (1997). Taxonomic and nomenclatural comments on the Triticeae in North America. *Phytologia* **83**, 302-311 [imprint date 1997; publication date 1998].

Barkworth, M.E., and Dewey, D.R. (1985). Genomically based genera in the perennial Triticeae of North America: Identification and membership. *American Journal of Botany* **72**, 767-776.

Barkworth, M.E., Burkhamer, R.L., and Talbert, L.E. (1996). *Elymus calderi*: A new species in the Triticeae (Poaceae). *Systematic Botany* **21**, 349-354.

Baum, B.R. (1979). The genus *Elymus* in Canada-Bowden's generic concept and key reappraised and retypification of *E. canadensis*. *Canadian Journal of Botany* **57**, 946-951.

Baum, B.R., Yen, C., and Yang, J.-L. (1991). *Roegneria*: Its generic limits and justification for its recognition. *Canadian Journal of Botany* **69**, 282-294.

Baum B.R., Yang, J.-L., and Yen, C. (1995). Taxonomic separation of *Kengyilia* (Poaceae: Triticeae) in relation to nearest related *Roegneria, Elymus,* and *Agropyron*, based on some morphological characters. *Plant Systematics and Evolution* **194**, 123-132.

Bentham, G. (1882). Notes on Gramineae. *Journal of the Linnean Society, Botany* **18**, 14-134.

Berchtold, F., and Presl, J.S. (1820). 'O Prirozenosti Rostlin.' (Krala Wiljma Enersa: Praha.)

Carlson, J.R. and Barkworth, M.E. (1997). *Elymus wawawaiensis*: a species hitherto confused with *Pseudoroegneria spicata* (Triticeae, Poaceae). *Phytologia* **83**, 312-330.

Catalan, P., Kellogg, E.A., and Olmstead, R.G. (1997). Phylogeny of Poaceae subfamily Pooideae based on chloroplast ndhF gene sequences. *Molecular Phylogenetics and Systematics* **8**,150-166.

Church, G.L. (1967). Taxonomic and genetic relationships of eastern North American species of *Elymus* with setaceous glumes. *Rhodora* **69**, 121-162.

Clayton, W.D., and Renvoize, S.A. (1986). 'Genera Graminum: Grasses of the World.' Kew Bulletin Additional Series 13. (Her Majesty's Stationery Office: London.)

Connor, H.E. (1994). Indigenous New Zealand Triticeae: Gramineae. *New Zealand Journal of Botany* **32**, 125-154.

Connor, H.E., Molloy, B.P.J., and Dawson, M.I. (1993). *Australopyrum* (Triticeae: Gramineae) in New Zealand. *New Zealand Journal of Botany* **31**, 1-10.

Davis, P.B., Ed. (1985). 'Flora of Turkey, vol. 9.' (University Press: Edinburgh.)

Dewey, D.R. (1982). Genomic and phylogenetic relationships among North American perennial Triticeae. In 'Grasses and Grasslands:

Systematics and Ecology.' (Eds. J.R. Estes, R.J. Tyrl, and J.N. Brunken.) pp. 51-88. (University of Oklahoma Press: Norman.)

Dewey, D.R. (1984). The genomic system of classification as a guide to intergeneric hybridization with the perennial Triticeae. In 'Gene Manipulation in Plant Improvement.' (Ed. J.P. Gustafson.) pp. 209-279. (Plenum Publishing Corporation: New York.)

Dubcovsky, J., Schlatter, A.R., and Echaide, M. (1997). Genome analysis of South American *Elymus* (Triticeae) and *Leymus* (Triticeae) species based on variation in repeated nucleotide sequences. *Genome* **40**, 505-520.

Dumortier, B.C.J. (1824). 'Observations sur les Graminées de la flore Belgique.' (J. Casterman aîné: Tournay.)

Estes, J.R., and Tyrl, R.J. (1982). The generic concept and generic circumscription in the Triticeae: An end paper. In 'Grasses and Grasslands: Systematics and Ecology.' (Eds J.R. Estes, R.J. Tyrl, and J.N. Brunken.) pp. 145-164. (University of Oklahoma Press: Norman.)

Estes, J.R., Tyrl, R.J., and Brunken, J.N., Eds. (1982). 'Grasses and Grasslands: Systematics and Ecology.' (University of Oklahoma Press: Norman.)

Gaertner, J. (1770). Observaciones et descriptiones botanicae. *Novi Commentarii Academiae Scientarum Imperalis Petropolitanae* **14**, 531-547.

Greuter, W., and Rechinger, K.H. (1967). Chloris Kythereia. *Boissiera* **13**, 22-196.

Hackel, E. (1889). Gramineae. In 'Die Natürlichen Pflanzenfamilien.' (Eds. A. Engler and K. Prantl.) pp. 1-97. (Verlag von Wilhelm Engelmann: Leipzig.)

Harz, C.O. (1880). Beiträge zur Systematik der Gramineen. *Linnaea* **43**, 1-30.

Hayek, A. (1925). Zur Systematik der Gramineen. *Oesterreichische Botanische Zetschrift* **74**, 11-12.

Holmberg, 0. (1926). Ueber der Begrenzung und Einteilung der Gramineen-Tribus Festuceae und Hordeeeae. *Botaniska Notiser* **1926**, 69-80.

Hitchcock, A.S. (1935). 'Manual of the Grasses of the United States.' U.S. Department of Agriculture Miscellaneous Publication No. 200. (U.S. Government Printing Office: Washington, D.C.)

Hitchcock, A.S. (1951). 'Manual of the Grasses of the United States.' Edition 2, revised by A. Chase. U.S. Department of Agriculture Miscellaneous Publication No. 200. (U.S. Government Printing Office: Washington, D.C.) [imprint date 1950; publication date 1951.]

Hsiao, C., Chatterton, N.J., Asay, K.H. and Jensen, K.B. (1995). Molecular phylogeny of the Pooideae (Poaceae) based on nuclear rDNA (ITS) sequences. *Theoretical and Applied Genetics* **90**, 389-398.

Hubbard, C.E. (1934). Gramineae. In 'The Families of Flowering Plants II. Monocotyledons.' (Ed. J. Hutchinson.) pp. 199-229. (MacMillan and Co., Ltd.: London.)

Jauhar, P. (1990a). Dilemma of genome relationship in the diploid species *Thinopyrum bessarabicum* and *Thinopyrum elongatum* (Triticeae: Poaceae). *Genome* **33**, 944-946.

Jauhar, P. (1990b). Multidisciplinary approach to genome analysis in the diploid species, *Thinopyrum bessarabicum* and *Th. elongatum* (*Lophopyrum elongatum*), of the Triticeae. *Theoretical and Applied Genetics* **80**, 523-536.

Jensen, K.B., and Wang, R.R.-C. (1997). Cytogenetic and molecular evidence for transfering *Elymus coreanus* and *E. californicus* into the genus *Leymus*. *International Journal of Plant Sciences* **158**, 188-193.

Kellogg, E.A. (1989). Comments on genomic genera in the Triticeae (Poaceae). *American Journal of Botany* **76**, 796-805.

Kellogg, E.A. (1992). Tools for studying the chloroplast genome in the Triticeae (Gramineae): An *Eco*RI map, a diagnostic deletion, and support for *Bromus* as an outgroup. *American Journal of Botany* **79**, 186-197.

Kellogg, E.A., Appels, R., and Mason-Gamer, R.J. (1996). When genes tell different stories: The diploid genera of Triticeae (Gramineae). *Systematic Botany* **21**, 321-347.

Keng, Y.-L. (1965). 'Flora Illustrata Plantarum Primarum Sinicarum.' (Scientific Publishing Co.: Beijing.)

Kunth, C.S. (1815). Gramineae. In 'Nova Genera et Species Plantarum.' (Eds. [F.W.H.]A. Humboldt, A.J.A. Bonpland, and C.S. Kunth.) pp. 84-201. (Librairiae graeco-latino-germanicae: Paris.)

Kuo, P.-C., Ed. (1987). 'Flora Reipublicae Popularis Sinicae. Tomus 9(3).' (Science Press: Beijing.)

Linnaeus, C. (1753). 'Species Plantarum.' (Laurentii Salvii: Stockholm.)

Löve, Á. (1984). Conspectus of the Triticeae. *Feddes Repertorium* **95**, 425-521.

Löve, Á. (1986). Some taxonomical adjustments in Eurasiatic wheatgrasses. *Veröffentlichungen des Geobotanischen Institutes der Eidgenössische Technische Hochschule, Stiftung Rübel, in Zürich* **87**, 43-52.

Lu, B-R. (1996). The genus *Elymus* in Asia. In 'Proceedings of the Second International Triticeae Symposium.' (Eds R.R-C. Wang, K.B. Jensen, and C. Jaussi.) pp. 219-233. (Forage and Range Laboratory, U.S. Department of Agriculture-Agricultural Research Service: Logan, Utah.)

Melderis, A. (1953). Generic problems within the tribe Hordeeae. In 'Proceedings of the Seventh International Botanical Congress, Stockholm, July 12-20, 1950.' (Eds H. Osvald and E. Aberg.) pp. 853-854. (Almqvist & Wiksell: Stockholm, and Chronica Botanica:Waltham, Massachusetts.)

Melderis, A. (1978). Taxonomic notes on the tribe Triticeae (Gramineae), with special reference to the genera *Elymus* L. *sensu lato*, and *Agropyron* Gaertner *sensu lato*. *Botanical Journal of the Linnean Society* **76**, 369-384.

Melderis, A. (1980). Tribe Triticeae. In 'Flora Europaea, vol. 5.' (Eds. T.G. Tutin, V.H. Heywood, N.A. Burges, D.M. Moore, D.H. Valentine, S.M. Walters, and D.A. Webb.) pp. 190-206. (Cambridge University Press: Cambridge.)

Nevski, S.A. (1932). Agrostological Studies. III. *Clinelymus* (Griseb.) Nevski, a new genus of Gramineae. *Bulletin de jardin botanique de l'academié des sciences de l'URSS [Izvestiya Botanicheskogo Sada Akademii Nauk SSSR]* **30**, 637-652. [In Russian].

Nevski, S.A. (1934a). Agrostological Studies IV. The systematics of the tribe Hordeeae. *Trudy Botanicheskogo Instituta Akademii Nauk SSSR, Ser. 1, Flora i Sistematika Vysshikh Rastenii* **1**, 9-32. [In Russian].

Nevski, S.A. (1934b). Tribe XIV. Hordeae Benth. In 'Flora of the U.S.S.R., vol. 2.' (Trans. N. Landau) (Eds R.Y. Rozhevits [R.J. Roshevitz] and B.K. Shishkin.) pp. 469-579. (Israel Program for Scientific Translations, Ltd.: Jerusalem.)

Nicora, E.G. and Rúgolo de Agrasar, Z.E. (1987). 'Los Generos de gramineas de American Austral'. (Editorial Hemisferio Sur S.A.: Buenos Aires.)

Petersen, G., and Seberg, O. (1997). Phylogenetic analysis of the Triticeae (Poaceae) based on *rpoA* sequence data. *Molecular Phylogenetics and Evolution* **7**, 217-230.

Pilger, R. (1947). Addimenta agrostologica I. Triticeae (Hordeeae). *Botanisches Jahrbücher für Systematik, Pflanzengeschichte und Pflanzengeographie* **74**, 1-27.

Pilger, R. (1954). Das system der Gramineae unter Ausschluss der Bambusoideae. [Ed. and completed by E. Potztal.] *Botanisches Jahrbücher für Systematik, Pflanzengeschichte und Pflanzengeographie* **76**, 281-384. [English translation by W.C. Worsdell.]

Prat, H. (1932). L'epiderme des Graminées. Étude anatomique et systématique. *Annales des Sciences Naturelles, Botanique* **14**, 117-324.

Prat, H. (1936). La systématique des Graminées. *Annales des Sciences Naturelles, Botanique* **18**, 165-258.

Seberg, O., and Linde-Laursen, I. 1996. *Eremium*, a new genus of the Triticeae (Poaceae) from North America. Systematic Botany **21**, 3-15.

Seberg, O., and Petersen, G. (1998). A critical review of concepts and methods used in classical genome analysis. *Botanical Review* **64**, 373-417.

Spenner, F.C.L. (1825). 'Flora Friburgensis et regionium proxime adjacentum.' (Friburg Brisgoviae [Freiburg im Breisgau, Germany]: Friderici Wagner.)

Stearn, W.T. (1960). Notes on Linnaeus' 'Genera Plantarum'. In 'Genera Plantarum, ed. 5' [facsimile edition] (C. Linnaeus) pp.i-xxiv. (H.R. Engelmann [J. Cramer]: Weinheim, Germany, and Wheldon & Wesley, Ltd.: New York.)

Stebbins, G.L., Jr. (1956). Taxonomy and evolution of genera, with special reference to the family Gramineae. *Evolution* **10**, 235-245.

Stebbins, G.L., Jr., Valencia, J.I., and Valencia, R.M. (1946a). Artificial and natural hybrids in the Gramineae, tribe Hordeae. I. *Elymus*, *Sitanion*, and *Agropyron*. *American Journal of Botany* **33**, 338-351.

Stebbins, G.L., Jr., Valencia, J.I., and Valencia, R.M. (1946b). Artificial and natural hybrids in the Gramineae, tribe Hordeae. II. *Agropyron, Elymus*, and *Hordeum*. *American Journal of Botany* **33**, 579-686.

Stebbins, G.L., Jr., and Walters. M.S. (1949). Artificial and natural hybrids in the Gramineae, tribe Hordeae. III. Hybrids involving *Elymus condensatus* and *E. triticoides*. *American Journal of Botany* **36**, 291-301.

Stevens, P.F. (1997a). Mind, memory, and history: How classifications are shaped by and through time, and some consequences. *Zoologica Scripta* **26**, 293-301.

Stevens P.F. (1997b). How to interpret botanical classifications-suggestions from history. *BioScience* **47**, 243-250.

Svitashev, S., Salomon, B., Bryngelsson, T., and Bothmer, R. von (1996). A study of 28 *Elymus* species using repetitive DNA sequences. *Genome* **39**, 1093-1101.

Svitashev, S., Bryngelsson, T., Li, X.-M., and Wang, R.R.-C. (1998). Genome specific repetitive DNA and RAPD markers for genome identification in *Elymus* and *Hordelymus*. *Genome* **41**, 120-128.

Terrell, E.E. and Peterson, P.M. (1993). Caryopsis morphology and classification in the Triticeae (Pooideae: Poaceae). *Smithsonian Contributions to Botany* **83**, 1-25.

Tzvelev, N.N. (1976). 'Zlaki CCCP.' (Nauka: Leningrad [St. Petersburg].) [In Russian]; (1983) 'Grasses of the Soviet Union' (Amerind Publishing Co.: New Delhi, India.) [English translation, published for the Smithsonian Institution Libraries and the National Science Foundation, Washington, D.C.]

Tzvelev, N.N. (1987). 'Systema graminearum (Poaceae) ac earum evolutio.' Komarov Readings 371-75. (Komarov Institute, Academy of Sciences, URSS: Leningrad [St. Petersburg].)

Wang, R.R.-C., Bothmer, R. von, Dvorak, J., Fedak, G., Linde-Laursen, I., and Muramatsu, M. (1996). Genome symbols in the Triticeae (Poaceae). In 'Proceedings of the Second International Triticeae Symposium.' (Eds R.R.-C. Wang, K.B. Jensen, and C. Jaussi.) pp. 29-34. (Forage and Range Laboratory, U.S. Department of Agriculture-Agricultural Research Service: Logan, Utah.

Wang, R.R.-C. and Hsiao. C. (1989). Genome relationships between *Thinopyrum bessarabicum* and *T. elongatum*: revisited. *Genome* **32**, 802-809.

West, J.G., McIntyre, C.L., and Appels, R. (1988). Evolution and systematic relationships in the Triticeae (Poaceae). *Plant Systematics and Evolution* **160**, 1-28.

PANICOIDS

Thuarea involuta, a common pantropical strand species.

Grasses: Systematics and Evolution. (2000). Eds S.W.L. Jacobs and J. Everett. (CSIRO: Melbourne)

A CLADISTIC ANALYSIS OF THE PANICEAE: A PRELIMINARY APPROACH

Fernando O. Zuloaga, Osvaldo Morrone and Liliana M. Giussani

Instituto de Botánica Darwinion, Labardén 200, C.C. 22, San Isidro (1642), Argentina.
email: fzuloaga@darwin.edu.ar

Abstract

The present treatment investigates cladistic relationships among genera of the tribe Paniceae (Poaceae) using morphological data. Sixty-seven characters were scored across 110 taxa within the tribe which represent current and previous classification systems; tribe Isachneae was chosen as the outgroup for the analysis, following Brown (1977) and Kellogg and Campbell (1987). The analysis resulted in 20,345 equally parsimonious trees of 372 steps each, with a total of 58 clades resolved in the strict consensus; these clades are discussed, considering the characters supporting each clade. This analysis was compared with early classification systems of the tribe, and discrepancies were observed between the results and the classification proposed by several authors. Also, we found that *Panicum* is a polyphyletic genus, and should be divided into several taxa, and other genera, such as *Arthropogon* and *Streptostachys*, are not monophyletic. The findings highlight the need for a broad scale phylogenetic analysis, including molecular evidence, of the Paniceae.

Key words: Poaceae, Paniceae, cladistics, morphology, phylogeny, taxonomy.

INTRODUCTION

The earliest classifications of grasses were based primarily on inflorescence and spikelet morphology. Brown (1814) recognised two tribes within the Poaceae, the Paniceae and the Poeae, and characterised the Paniceae as having two flowers per spikelet, with the lower flower always imperfect. Later, Bentham (1881) indicated that in the Panicanae, which is equivalent to the Paniceae of Brown, the spikelet disarticulates entirely, including the glumes, from the pedicel, while in most of the Poaceae the spikelet falls without the glumes, which remain attached to the pedicel. Bentham recognised six tribes within the Panicanae, including the Paniceae, in which he described two primary groups, the 'Paniceae proper' containing eleven genera such as *Paspalum* L., *Panicum* L., *Ichnanthus* P. Beauv., *Oplismenus* P. Beauv., and *Setaria* P. Beauv., and the 'Cenchrus' group including *Cenchrus* L., *Pennisetum* Rich. and other tropical or subtropical genera, with spikelets having an involucre of bristles below the spikelets. Other

tribes recognised by Bentham were the Maydeae Matthieu (comprising mostly what is now considered to belong within the Andropogoneae Dumort. and *Pariana* Aubl., an herbaceous bambusoid grass), the Oryzeae Dumort., the Tristegineae Nees (with genera now considered in the Arundinelleae Stapf, such as *Arundinella* Raddi, and *Arthropogon* Nees, *Reynaudia* Kunth, *Rhynchelytrum* Nees and others), the Zoysieae Benth. (which includes some panicoid genera, such as *Anthephora* Schreb. and several chloridoid genera), and the Andropogoneae. *Panicum* was subdivided by Bentham into eleven sections, many of which are now recognised as genera, e.g. *Digitaria* Haller, *Brachiaria* (Trin.) Griseb., and *Echinochloa* P. Beauv. Later, Hackel (1887) published a similar classification to the one of Bentham.

During the first half of this century, new anatomical and cytological characters were used in the classifications of grasses. As a result, Pilger (1940) recognised six subtribes within the Paniceae: Panicinae Stapf, including *Isachne* R. Br. and related genera,

Melinidinae (Hitchc.) Pilg., Anthephorinae Benth., Boivinellinae (A. Camus) Pilg., Lecomtellinae Pilg., and Trachyinae Pilg.; he also considered the Arthropogoneae as a separate tribe. Pilger (1954) raised some of these sections to the tribal level, recognising the Paniceae, Melinideae (with *Melinis* P. Beauv., *Rhynchelytrum*, and *Tricholaena* Schult.) the tribe Isachneae (with *Isachne, Coelachne* R. Br., *Heteranthoecia* Stapf, and *Limnopoa* C.E. Hubb.) the Anthephoreae Potztal (with a single genus, *Anthephora*) Boivinelleae A. Camus (with the genera *Cyphochlaena* Hack., *Boivinella* A. Camus, and *Perulifera* A. Camus) the tribe Lecomtelleae Potztal (with *Lecomtella* A. Camus) Trachyeae Pilg. (with *Trachys* Pers.) and tribe Arthropogoneae, the latter including the genera *Snowdenia* C.E. Hubb., *Reynaudia, Achlaena* Griseb., and *Arthropogon*. Hsu (1965) used anatomical characters, including transverse section of the leaves, and epidermis of the upper lemma, as well as features of the lodicules, nervation, texture and form, to characterise different genera of the Paniceae and to distinguish several subgenera and sections of *Panicum*. Butzin (1970) distinguished several groups within the Paniceae by the presence or absence of cross-veins in leaves; this author created subtribe Microcalaminae Butzin to include all genera sharing this character. This scheme segregated several genera clearly related to *Panicum*, e.g. *Lasiacis* (Griseb.) Hitchc., *Acroceras* Stapf, *Oplismenus, Sacciolepis* Nash, and others; Butzin established new subtribes, with a few genera, such as Xerochloinae Butzin, Uranthoeciinae Butzin, Thuareinae Ohwi, Otachyriinae Butzin, which are clearly related to larger genera. Butzin's classification also placed genera that are often considered to be subgenera by other authors into different subtribes, for example: *Leptoloma* Chase was placed in subtribe Panicinae, while *Digitaria* was placed in subtribe Paspalinae.

In the last thirty years a significant correlation has been found between morphological structures of the grasses and anatomical, cytological and physiological characters, especially the association between leaf anatomical features and photosynthetic pathways. Brown (1977) divided the Paniceae, taking into consideration the leaf anatomy and photosynthetic pathway, into four subtribes, based on a study of a total of 86 genera and 610 species. Brown included a few heterogeneous genera with Kranz and non-Kranz species, such as *Panicum, Alloteropsis* C. Presl, and *Neurachne* R. Br., and suggested that Kranz species have evolved from non-Kranz taxa. Clayton and Renvoize (1986) used exomorphological characters, as well as anatomical features, to distinguish seven subtribes characterised by these authors mainly by spikelet characters, such as presence of bristles, and texture of the upper lemma, and also by anatomical characters. Subtribe Neurachninae (S.T. Blake) Clayton and Renvoize includes a small group of Australian genera, with indurated glumes as long as the spikelet, the upper lemma indurated or membranous. Subtribe Digitariinae Butzin was characterised by inflorescences usually composed of digitate, racemose branches with secund spikelets, the proximal glume usually small or absent, the distal lemma with flat, thin margins enclosing most or all of the distal palea; subtribe Arthropogoninae Butzin, with two American genera, was distinguished by having laterally compressed spikelets (a character discussed later in this contribution) and hyaline upper lemma and palea. Subtribe Cenchrinae (Dumort.) Dumort., including Anthephorinae Benth., has bristles or scales that are deciduous with the spikelets.

Subtribe Setariinae (including the Panicinae, Paspalinae Griseb., and Brachiarininae Butzin) includes all genera with an indurate upper anthecium, *Panicum* being the centre of the subtribe; this subtribe is heterogeneous regarding its anatomical features and physiological pathways, but shares a similar spikelet structure. Subtribe Melinidinae (Hitchc.) Pilg. was characterised by its paniculate inflorescences, laterally compressed spikelets, reduced lower glume and indurated upper lemma and palea; it includes a few genera, such as *Melinis, Rhynchelytrum*, and *Tricholaena*. Finally, subtribe Spinificinae Ohwi, with only three genera, *Xerochloa* R. Br., *Zygochloa* S.T. Blake, and *Spinifex* L., has several characteristic features in the inflorescences; also, many species are dioecious.

Additionally, several cladistic studies were conducted at the subfamily and tribal level, utilising morphological characters (Kellogg and Campbell 1987) and Kellogg and Watson (1993) for the Bambusoideae Ascher. & Graebn., Andropogoneae, and Pooideae. More recently several cladistic treatments were published using molecular data (Davis and Soreng 1993; Duvall *et al.* 1994; Soreng and Davis 1998; Hsiao *et al.* 1999, and others), which proved to be very useful for phylogenetic reconstruction. However, up to the present there has been no attempt to conduct a cladistic analysis of the Paniceae, using either morphological or molecular data. Hsiao *et al.* (1999) recently stressed the monophyly of the Panicoideae, indicating that the Paniceae and Andropogoneae are the two major monophyletic tribes of this subfamily.

Tribe Paniceae is treated here in an inclusive sense; several segregates, such as the tribe Arthropogoneae, were studied to analyse the relationships between these groups. The Paniceae contains more than 100 genera all over the world, mainly distributed in Africa, Australia, and America. The greatest diversity of genera and species is concentrated in the tropics and relatively few genera include species that extend beyond warm-temperate regions.

This study presents preliminary results of a cladistic analysis, based on morphological characters, in order to test previous classifications (in particular the ones proposed by Brown 1977, and Clayton and Renvoize 1986), to evaluate the generic status of several taxa, including *Panicum*, and to check if the proposed subtribes are based on true apomorphies.

METHOD

Data Matrix Characteristics

Sixty-seven exomorphological and anatomical characters were used in the cladistic analysis; all multistate characters were treated as unordered. Characters were directly observed from herbarium material, from a total of nearly 1,000 specimens, or in a few cases were compiled mainly from the literature. The data matrix used in this analysis is shown in Table 1.

Parsimony analyses under implied weights were carried out using Pee-Wee ver. 2.8 (Goloboff 1993), with the default weighting function (concavity = 3); settings *amb-* (clades resolved only if they have unambiguous support) and *poly=* (polytomies allowed).

In all analyses the fittest cladograms were obtained using the instruction *MULT*2500*. This instruction carries out 2500

Table 1. Taxon/character state matrix used for cladistic analysis of the Paniceae. Polymorphisms are signified as follows: A = [0 1]; B = [0 2]; C = [1 2]; D = [0 -]; E = [1 -]; F = [2 -]; G = [0 1 2]; H = [0 1 2 3]. '-' indicate character inapplicable for a taxon; '?' indicate character unobserved for a taxon.

	1	2	3	4	5	6
	1234567890	1234567890	1234567890	1234567890	1234567890	1234567
OUTGROUP	000000000A	000100000A	000000AA00	AC0000I0000	010B000001	0A000000100000
ACHLAENA	0000000000	0010000011	00100001011	02000121000	010000000010	00010--012110
ACRITOCHAETE	0000000000110	0000000110	000000010010	00111020010	1000001000	00000100000
ACROCERAS	0000000000010	0000001100	000A00002000	0010200111001	11000000A00	100000
ALEXFLOYDIA	0000000000010	000000010000	0010000200000	00001100000	0000000??--012010
ALLOTEROPSIS	0000000000010	000000000000	001000020000	110200IA0010	00100000000IAB0A0
AMPHICARPUM	000000001000	10000000011002	-00--0020000	02001000000	000000000100000
ANCISTRACHNE	0000000000010	00000001000	100100002000	0000200100000	00000000100000
ANTHAENANTIOPSIS	0000000000010	00000001100	100100002000	0110200111000	0000000--012010
ANTHENANTIA	0000000000010	00000001100	2-01--0020000	1100010000001	000000--012010
ANTHEPHORA	110001000-0010	00000000002	-00--1010001	10000100000001	10A000--012A10
ARTHRAGROSTIS	0000000000010	0000100000000010	0002000010	100000000000010	102010
ARTHROPOGON-1	0000000000010	0000001001000	010110CAAA121	000001000000010	00010--012010
ARTHROPOGON-2	0000000000010	0000000000000	010110200003	1000010000001	0000000100001
ARTHROPOGON LANCEOLATUS	0000000000010	00000010000000A	0110C0000210	000010000001	000000--012110
AXONOPUS	00000000000110	0000000001002	-00--00B00001	002001A000000	00000--012010
BAPTORHACHIS	0?100000000110	0000000011002	-00--0010110110	??010000000110000	?--012010
BRACHIARIA	0000000000110	0001000100000	I0002000000001	000000000000001	1102010
CALYPTOCHLOA	00000010000100	0000010001000000	2000031020011	10100010000000	100000
CENCHRUS	110010000-000100	0000000001000A	I000011000010	00000001	10A000--012010
CENTROCHLOA	0000000000110	0000000001002	-00--0010001002	0011000001000	000--012010
CHAETIUM	0000000001001	000000101100000	1011021000110	BB0120000001	00000IDEAA2010
CHAETOPOA	110001000-0010	0000000000000000	1100000000110	??01000000010011	00--012010
CHAMAERAPHIS	110010000-0010	00000010000000	1000020000110	0001000000010	00000--012010
CHLOROCALYMMA	1110010000-010	000000000000000	0010001000010	000000001100000	--012010
CLEISTOCHLOA	0000000100001000	0000000100CD00	0020001002001	01000001000000001	00000
CLIFFORDIOCHLOA	0000000000110	00000001100001	00002000030020010	00000001000000	100001
CYRTOCOCCUM	0000000000010	00000011100000A	I0011000101020010	010110000001000	100000
DALLWATSONIA	0000000000010	000000000000A	00002000030000010	00000000000000	100000
DIGITARIA	00000000001110	00000001100A00A0000C	000010020010000000100A000	--012010
DISSOCHONDRUS	1100100000010	000000011000001	0000200000101110	0000000000	--012110
ECHINOCHLOA	0000000001100	0000011000001	00002A0000I0000I0010111	0000000--012010
ECHINOLAENA	00100000001100	000100110010001A	00020000001001	000000000000001	00100000
ENTOLASIA	0000000000010	00000001100100000C	0000100200102000001	000000001	00000
ERIOCHLOA	0000000000110	0001001000100A0000200000102111	000000000A00001102010
FASCICULOCHLOA	0000000000010	000000000000020000C	00000200100000000I000000	101010
GERRITEA	0000000001000	0000010000000I0002000030000010	000001000000?00100000
HOLCOLEMMA	1100100000010	0000001100002000I200000I0211	0000000000000--012010
HOMOLEPIS	0000000000010	0000000011000200001021	000000000000000100100001
HOMOPHOLIS	0000000000010	0000000001000200001000001	000000000000000100000
HYDROTHAUMA	0000000001000	00000011000101000020000310000I00000I00000000100000
HYGROCHLOA	11000010000110000001000000000I2000031000010000001001000--012010
HYLEBATES	0000000000010	000000000000000020000300001000000100000200100000
HYMENACHNE	0000000000010	0000001100000000C00031010011000000000000100000
ICHNANTHUS	0000000000010	000001000001000I00020000000010000000000000100000
IXOPHORUS	1100104000011000000011000020000200000102110000000020000--012010

Table 1. Taxon/character state matrix used for cladistic analysis of the Paniceae. Polymorphisms are signified as follows: A = [0 1]; B = [0 2]; C = [1 2]; D = [0 -]; E = [1 -]; F = [2 -]; G = [0 1 2]; H = [0 1 2 3].'-' indicate character inapplicable for a taxon;'?' indicate character unobserved for a taxon. *(Continued)*

	1	2	3	4	5	6
	1234567890	1234567890	1234567890	1234567890	1234567890	1234567
LASIACIS	0000000000	0100000000	020000A000	0200000000	0011100100	0000000000100000
LEPTOCORYPHIUM	0000000000	0100000000	0002-00--0	0200003100	0010000000	01000000--012010
MEGALOPROTACHNE	0000000000	0100000001	1000001100	01000011020	0100000001	0000--012010
MELINIS	0000000000	0100000000	01010000010	211113100001	0000000010	00010011020110
MESOSETUM	0000000010	1100000000A	1000000A0A	00C0100110	0001000000	01000012--012010
MICROCALAMUS	0000000000	0100000000	0010000100	01200000102	001010011	100000000100000
NEURACHNE	0000000000	01000000110	0000011000	20000310001	010100001	0000000001AG0A0
ODONTELYTRUM	1100010000-000	100000000000	CD01DD0110	0001100001	0000000111	01000--012010
OPLISMENOPSIS	0000000000	0100000000	0100000101	102A0000100	0010000000	0000000002001 00000
OPLISMENUS	0000000000	1100000000	1100000111A	0200001000	0010000000	000000010010 0000
ORYZIDIUM	0000000000	0100010000	000000010010	200001100	00100000010	00000010102010
OTACHYRIUM	0000000010	10000000000	0000210010C	0000000000	1001000000	0000000010 00000
OTTOCHLOA	0000000000	0100000001	1000000100	1C0000002001	0000000000	00000000000100000
PANICUM*	0000000000	0100000000	00A00A0000	2000000A20A	1BB0000000	0000000100000
PANICUM sect. LAXA	0000000000	11000000001	100000A000	0200003002	0010000000	00000000000100001
PANICUM sect. LOREA	0000000000	0100000000	0000000001	000020000000	0010000000	00000000000100000
PANICUM sect. MEGISTA	0000000000	0100000000	0000000001	000020000000	0010000000	00000000000100000
PANICUM sect. PHANOPYRUM	0000000000	11000010011	000001000020	000000000010	0000000000	00000000100000
PANICUM sect. RUDGEANA	0000000000	1000001000001	0001000020	00000000101	0000000000	0000010102010
PANICUM sect. STOLONIFERA	0000000000	1100001001	1000001000020	00000000010	0000000000	00000000100000
PANICUM subg. AGROSTOIDES	00000000000A	100000000AA	00A00A0000C	0000AAAG0A1	AAA00000000	000--A12010
PANICUM subg. DICHANTHELIUM	0000000000	010000000000	000010000	200000002001	1000000000	0000000000100000
PANICUM subg. MEGATHYRSUS	0000000000	0100000000	00000010000	200000012011	0000000000	00000011102010
PANICUM subg. PANICUM	0000000000	0100001000	0000A00002	0000000AA010	0000000000	00AA102010
PARANEURACHNE	0000000000	01000000110	00000110002	00001102001	000000010	00000--012010
PARACTAENUM	1100100000111	00000001	0000010000	20000010211	00000000000	000--012010
PARATHERIA	1100100100001	1000000000	0000010013	00001102001	00000010000	00--012010
PASPALIDIUM	1100000000011	00000010000	00A0000200	00000102111	000000000A	000--012010
PASPALUM group DECUMBENTES	0000000000	11000000011	00000100002	00000002001	0000000000	000--012010
PASPALUM subg. ANACHYRIS	0000000000	110000000110	0020A0--DE	F0D001002001	0000000000	0000--012010
PASPALUM subg. CERESIA	0010000000	01100000A001	0002000--0	0200001 00B00	1A0000001	000000--012010
PASPALUM subg. PASPALUM	A0A00000001	1A00000001A	002000--00	20000000B001	A000000000	000--012A10
PENNISETUM	1100100000-001	100000000000	0A000AH0000	11000011000	0011000000	100A000--012010
PLAGIANTHA	0000000001	01000000000	0000010000	200001000201	00000000000	0000000100000
PSEUDECHINOLAENA	0000000000	1100000001	1000001100	020000000001	1000000000	0000000100000
PSEUDORAPHIS	1100000000011	000000100000	10010200011	000010000001	020000	--012010
REIMAROCHLOA	0000000000	11000000010	002-10-----	0-0021000010	0000001010	000--012010

Table 1. Taxon/character state matrix used for cladistic analysis of the Paniceae. Polymorphisms are signified as follows: A = [0 1]; B = [0 2]; C = [1 2]; D = [0 -]; E = [1 -]; F = [2 -]; G = [0 1 2]; H = [0 1 2 3]. '-' indicate character inapplicable for a taxon; '?' indicate character unobserved for a taxon. *(Continued)*

	1 1234567890	2 1234567890	3 1234567890	4 1234567890	5 1234567890	6 1234567
REYNAUDIA	0000000000	0010000000	0010000011	0211112 10	---0--000--	1110012--012010
RHYNCHELYTRUM	0000000000	010000000A	0010100A00	1021111310	0001000000	10000100110 2010
SACCIOLEPIS	0000000000	0010000000	1100010A00	0020000300	0010000000	100000000100000
SCUTACHNE	0000000000	0010000000	1100000100	0020000000	2001000000	000000001102010
SETARIA	1100100000	0010000000	000000000A	000AC00000	0120110000	0000000000--012010
SETARIOPSIS	1100100000	0010000000	0000000101	0000020000	0010211000	00000000--012010
SNOWDENIA	0000000000	0010000000	1100000100	1310003100	0010000000	01000000--012010
SPHENERIA	0000000001	0110000000	110002-00-	-002000010	0200100000	0000000--012010
SPINIFEX	1101002 00-	0010100000	0100000001	0002000011	0000100000	001000000--012010
STEINCHISMA	0000000000	0010000000	0000000020	0002000010	00201000000	000A000000101010
STENOTAPHRUM	1100000000	0110000100	01000000A0	0002000011	0000100001	0000001000000--012010
STEREOCHLAENA	0000000000	1100000001	0002D00DDAA	11-0011020	0100000011	01000--012010
STREPTOLOPHUS	1100100000	-0001000000	00000001001	1000011020	0100000001	001000--012010
STREPTOSTACHYS-I	0000000000	0010000000	0000010000	2000000200	1020000000	0002--012010
STREPTOSTACHYS ASPERIFOLIA	0000000001	1000000011	00000A1000	2000000200	1200000000	00200100001
TATIANYX	0000000001	0010000000	0000001000	0200000002	0110000000	0002--012010
THRASYA	2-1000000001	1000000001	1000001000	0C00000020	0111000000	0000A--012010
THRASYOPSIS	2-1000000001	1000000001	1000001100	0200000102	0010000000	0000?--012010
THYRIDOLEPIS	0000000000	0010000000	0000000A10	0020001102	0010000000	0100000000100000
TRACHYS	1110010000-	0001000000	0000000100	0110000110	0001000000	01001000--012010
TRICHOLAENA	0000000000	0010000010	0000010010	0001000010	0001000000	00000001102010
TRISCENIA	0000000000	0010000011	0000000000	0010000100	0010000001	000000100000
URANTHOECIUM	1100000000	-0010001000	0000001000	1200001100	2010001000	1000000--012010
UROCHLOA	0000000000	110000000AA	00000A0000	20000001BB	11100A0000	00000011102010
WHITEOCHLOA	0000000000	0010000100	1100000100	00200001A0	2A100000000	0000--012010
XEROCHLOA	1101000000	-0010000100	00BDA1DDD1	2000011000	0100000010	01000--012010
YAKIRRA	0000000000	0010000100	0001000100	0002000001	0100000000	000010102010
ZYGOCHLOA	1101002000	0010100000	0000000110	0020000110	0001000000	0000--012010

replications, randomising the order of the taxa, creating a Wagner tree and submitting it to branch-swapping by means of tree-bisection reconnection (TBR). It stores in memory the 20 most parsimonious trees (*hold/20 mult*2500*). The shortest trees retained from each of the 20 subsearches were then TBR-swapped to completion (*max**). The swapping was aborted after 20,345 most-parsimonious trees were obtained. These trees have a length of 372, with a fit = 365.8, CI= 0.19, and RI= 0.62. To evaluate the relative support of clades, we calculated branch support with instruction *bsupport* in Pee-Wee.

Taxa

All taxa included in this study were initially assumed to be monophyletic, but this assumption was then tested in later analyses. Those genera with a clear polymorphism were later subdivided into two or more units; as a result a total of 110 taxa of the Paniceae were selected for this study, based on the revisions made by Clayton and Renvoize (1986) and Watson and Dallwitz (1992).

A few genera were not included in this preliminary revision, due to the lack of adequate material; these genera were: *Poecilostachys* Hack., *Cyphochlaena* Hack., *Lecomtella* A. Camus, *Thyridachne* C.E. Hubb., *Louisiella* C.E. Hubb. and Léonard, *Eccoptocarpha* Launert, *Yvesia* A. Camus, *Thuarea* Pers., *Plagiosetum* Benth., *Tarigidia* Stent and *Ophiochloa* Davidse, Filg. and Zuloaga.

As previously mentioned, polymorphism in terminal taxa may indicate that those elements are not monophyletic, 'a possibility that may not be overlooked when using traditionally defined genera. The best way to avoid unwarranted assumptions of monophyly is to split up polymorphic taxa into monomorphic subunits' (Nixon and Davis 1991).

As a result of this preliminary analysis, the arrangement of terminal taxa in *Panicum, Paspalum, Streptostachys* Desv., and *Arthropogon*, corresponds to monophyletic taxonomic subunits (subgenera, sections or even species), defined as such by several autapomorphies.

Outgroup Selection

Kellogg and Campbell (1987), in their cladistic analysis of the Poaceae, indicated that the Panicoideae is a monophyletic group, with *Allochaete* C.E. Hubb., *Micraira* F. Muell., *Isachne, Heteranthoecia* Stapf, *Coelachne* R. Br., and the Centotheceae Ridley as sister groups. Brown (1977) also considered Isachneae as a sister group of the Paniceae. In this study, the Isachneae, with the genera *Isachne, Heteranthoecia, Coelachne*, and *Sphaerocaryum* Hook. f., was used as the outgroup of the Paniceae. This tribe is a monophyletic group defined by the presence of two hardened anthecia, usually fertile, the lemma with thread-like microhairs on its surface, and also possesses rectangular silica bodies; (Hsu 1965; Brown 1977). In all these genera the primary disarticulation point is above the glumes and below the lower lemma; the presence of simple starch grains links the Isachneae with the rest of the Paniceae and Panicoideae. The Paniceae is defined as a monophyletic group by the disarticulation point at the base of the lower glume.

Characters

The majority of characters and character states for this analysis were taken from direct examination of herbarium specimens deposited at BAA, BRI, CEN, G, MO, SI, SP, US (Holmgren *et al.* 1990) or from observation of populations in the field. Later, this information was checked against the literature; the main reference for this final analysis was the treatment of Watson and Dallwitz (1992).

Seventy-seven characters were investigated and 67 of these were used in this analysis. Potentially useful characters, such as basic chromosome number, were discarded because data were available for only a few genera. Arbitrarily, we used only characters for which 90% of the sampled taxa could be scored. Characters used in this analysis are described below. Many of these characters are readily coded as binary. However, patterns of morphological variation found in terminal taxa require the use of some multi-state characters. Therefore, those characters that were excessively polymorphic (arbitrarily designated as over 75% of the taxa) were excluded from the analysis.

Our observations were compared with the information available on several revisions, such as: *Acroceras* (Zuloaga *et al.* 1987); *Alloteropsis* (Gibbs Russell 1983); *Arthropogon* (Tateoka 1963; Filgueiras 1982, 1996); *Alexfloydia* (Simon 1992); *Axonopus* (Black 1963); *Anthaenantiopsis* (Morrone *et al.* 1993); *Brachiaria* (Morrone and Zuloaga 1992); *Chaetium* (Morrone *et al.* 1998); *Chlorocalymma* (Clayton 1970); *Cliffordiochloa* (Simon 1992); *Dallwatsonia* (Simon 1992); *Fasciculochloa* (Simon and Weiller 1995); *Gerritea* (Zuloaga *et al.* 1993); *Homolepis* (Zuloaga and Soderstrom 1985); *Homopholis* (Wills 1996); *Ichnanthus* (Stieber 1982, 1987); *Lasiacis* (Davidse 1978); *Mesosetum* (Filgueiras 1990); *Oplismenus* (Scholz 1981); *Otachyrium* (Sendulsky and Soderstrom 1984); *Panicum* (Morrone and Zuloaga 1991b; Zuloaga, 1987a, b, 1989; Zuloaga and Sendulsky 1988; Zuloaga and Morrone 1996; Zuloaga *et al.* 1989, 1992, 1993a, b), *Paspalum* (Chase 1929, inéd.; Morrone *et al.* 1995, 1996); *Pennisetum* (Türpe 1983); *Steinchisma* (Zuloaga *et al.* 1998); *Sacciolepis* (Judziewicz 1990); *Setaria* (Rominger 1962; Pensiero 1999); *Stenotaphrum* (Sauer 1972); *Streptolophus* (Hughes 1923); *Strep-*

tostachys (Morrone and Zuloaga 1991a); *Tatianyx* (Zuloaga and Soderstrom 1985); *Thrasya* (Burman 1980); *Thrasyopsis* (Burman 1980); *Urochloa* (Morrone and Zuloaga 1992, 1993; Webster 1988); *Yakirra* (Lazarides and Webster 1986).

Inflorescence

1. Rachis: terminating in a spikelet (0) terminating in a bristle (1) terminating as a foliaceous axis (2)

2. Main axis of inflorescences: terminating in a spikelet (0) terminating in a bristle (1)

3. Foliaceous rachis: absent (0) present (1)

Reduction of the terminal spikelet in the inflorescence of Paniceae, is an important character (Webster 1988); the axis can end in a naked point, as a result of the abortion of the terminal spikelet; this sterile projection can also be modified into a bristle or a foliaceous axis.

4. Bracts of the inflorescences: absent (0) present (1)

5. Involucral bristles (cauline): absent (0) present (1)

6. Involucral bracts: absent (0) present (1)

The presence or absence of bristles is a significant character within the tribe (Webster 1992). Clayton and Renvoize (1986) mentioned that bristles or scales are derived from branches of the inflorescences (Sohns 1955; Butzin 1977), forming an involucre surrounding the spikelets. This reduction and condensation of branches can form a rosette of unbranched bristles, in *Pennisetum* and related genera, can produce a cup by coalescence, in *Cenchrus* and other genera, or be reduced to a single seta, as in the genus *Paratheria*.

The relationship of *Anthephora*, and related genera, with other members of the Paniceae is not clear, due to the controversial interpretation of the bract that subtends the spikelet. Stapf (1919), Chippindall (1955), Reeder (1960) and Clayton and Renvoize (1982, 1986) considered this structure as a modified bract that retains a sterile spikelet in the apex of a bristle; this sterile spikelet is also transformed into a scale. This criterion is followed in the present treatment.

7. Sexuality (in the upper floret): perfect (0) staminate and pistillate flowers on one plant (1) staminate and pistillate flowers on different plants (2) perfect, staminate and pistillate flowers on one plant (3) pistillate (with perfect flowers in the lower floret) (4)

8. Cleistogenes in leaf axils: absent (0) present (1)

9. Rhizanthogenes: absent (0) present (1)

10. Apex of the pedicel: truncate (0) oblique (1)

Several authors (Zuloaga and Soderstrom 1985; Webster 1988, 1992; Morrone *et al.* 1998) considered the insertion of the spikelet in the pedicel a valuable taxonomic character for different genera of the tribe. This insertion was considered here truncate whenever it is perpendicular to the main axis of the pedicel, and oblique when it is diagonal.

Spikelet

11. Spikelet: truncate (0) oblique (1)

This character was defined taking into account the angle present between the main axis of the spikelet and the main axis of the pedicel; therefore the spikelet was describe as oblique when this angle was more than 45° and truncate when it was less than 45°.

12. Unilateral spikelet (presence): absent (0) present (1)

13. Disarticulation at the base of the spikelet: absent (0) present (1)

14. Disarticulation at the base of the primary branches : absent (0) present (1)

15. Disarticulation at the base of the inflorescence: absent (0) present (1)

16. Disarticulation between the lower glume and lower lemma: absent (0) present (1)

17. Disarticulation at the base of the pedicel: absent (0) present (1)

18. Disarticulation at the nodes of the main axis: absent (0) present (1)

19. Disarticulation at the base of the upper anthecium: absent (0) present (1)

The diaspore disarticulation point is one of the most significant characters for the differentiation of genera in the Paniceae (Webster 1988, 1992). Although a disarticulation point was observed in most of the genera at the base of the spikelet, a secondary disarticulation point is frequently present at the base of the upper anthecium, which therefore falls naked at maturity. Also disarticulation points were found at the base of the first order branches, at the base of the main axis of the inflorescence, or at several nodes of the main axis.

20. Stipe at the base of the spikelet: absent (0) present (1)

The stipe is here defined as a prolongation of the rachilla between the lower and upper glume, whenever this prolongation is more than 0.5 mm long.

21. Hairy callus: absent (0) present (1)

The presence of a hairy callus at the base of the spikelet was used as a taxonomic character for several genera (Filgueiras 1982; Clayton and Renvoize 1986); its value was also questioned by Webster (1992). We here score the presence of a hairy callus whenever a dense ring of hairs is present at the base of the spikelet.

22. Spikelet orientation (abaxial): absent (0) present (1)

23. Spikelet orientation (adaxial): absent (0) present (1)

Spikelet orientation has been used frequently as a valuable diagnostic character for the Paniceae (Pilger 1940; Clayton and Renvoize 1982, 1986; Zuloaga and Soderstrom 1985); however, several authors found it difficult to determine the spikelet orientation in some groups (Blake 1958; Hsu 1965; Brown 1977 and Webster 1987, 1988). To solve this problem we considered that an adaxial spikelet is homologous to a terminal spikelet but positioned in a secondary branch, with the lower glume or upper lemma orientated towards the primary branch. The abaxial spikelet represents, on the other hand, an axial spikelet on a tertiary branch, with the lower glume and upper lemma opposite to the primary branch.

24. Spikelet (compression): dorsiventrally compressed (0) laterally compressed (1) terete (2)

This is a valuable character within the Paniceae (Webster 1988). It was studied in partially included inflorescences within the leaf sheath, since mature spikelets can present alterations on their shape in herbarium material.

25 Upper flower with basal scars or appendages: absent (0) present (1)

26. Lower glume: present (0) vestigial (1) absent (0)

27. Upper glume: not saccate (0) saccate (1)

28. Upper glume: present (0) absent (0)

29. Lower palea: absent (0) present, not expanded (1) present, expanded (2)

30. Lower and upper glume: of different size (0) the same size (1)

31. Lower glume: muticous (0) awned (1)

32. Upper glume: muticous (0) awned (1)

33. Upper glume: as long as the lower lemma (0) 1/2 or less the length of the lower lemma (1)

34. Upper glume: 2 or 4-nerved (0) 3-nerved (1) 5-11-nerved (2) enerved (3)

35. Lower lemma: muticous (0) awned (1)

36. Apex of the upper glume: entire (0) bifid (1) trilobate (2)

37. Apex of the lower lemma: entire (0) bifid (1)

Characters 24 to 37 make reference to different structures of the spikelet bracts. The homologies of these characters were established following the criteria sustained by Troll (1967:1248-1290). Any prolongation above 0.5 mm was considered an awn.

Upper Anthecium

38. Upper anthecium: dorsiventrally compressed (0) laterally compressed (1) terete (2)

39. Upper anthecium: crustaceous (0) cartilaginous (1) hyaline (2) membranous (3)

40. Upper palea: tightly clasped by the lemma (0) gaping (1)

41. Upper lemma texture: smooth (0) transversely rugose (1)

42. Upper palea: without simple papillae (0) with simple papillae at the apex (1) with simple papillae all over its surface (2)

43. Upper palea: without compound papillae (0) with compound papillae at the apex (1) with compound papillae all over its surface (0)

44. Upper palea (surface texture): smooth (0) transversely rugose (1)

45. Upper palea: absent (0) present (1)

46. Upper palea (presence of microhairs): without bicellular microhairs (0) with bicellular microhairs at the apex (1) with bicellular microhairs all over its surface (2)

47. Upper palea (presence of macrohairs): without macrohairs (0) with macrohairs at the apex (1) with macrohairs all over its surface (2)

48. Upper lemma: not gibbous (0) gibbous (1)

49. Upper lemma: muticous (0) awned (1)

50. Upper lemma: not differentiated at the apex (0) differentiated at the apex (1)

51. Upper palea (apex): straight (0) recurved (1)

52. Upper palea (apex): not differentiated at the apex (0) differentiated at the apex (1)

53. Upper lemma (margins): tucked in onto the palea (0) lying flat and exposed on the palea (1)

Characters 38 to 53 involved different structures of the upper lemma and palea, which were considered diagnostic for genera of the Paniceae in different publications (Hsu 1965; Zuloaga and Soderstrom 1985; Webster 1988). They were studied in mature upper anthecia, especially when these anthecia enclosed a caryopsis.

Floret
54. Lodicules: present (0) absent (1)

55. Stamens (number): three (0) two (1) absent (2)

56. Style base: free (0) fused (1)

57. Ovary: glabrous (0) pilose (1)

Fruit
58. Caryopsis (compression): dorsiventral (0) lateral (1)

59. Caryopsis (hilum): punctiform (0) linear, up to 1/2 the length of the caryopsis (1) linear, as long as the length of the caryopsis (2)

The hilum type has been characterised as a diagnostic character for the Paniceae (Clayton and Renvoize 1986; Zuloaga and Soderstrom 1985; Sendulsky *et al.* 1987; Morrone and Zuloaga 1991a). We distinguished the hilum as punctiform when it is less than 1/4 the length of the caryopsis, and linear when the hilum is as long as the caryopsis or 1/2 as long as the caryopsis.

Leaf Anatomy
60. Centripetal chloroplasts on the parenchymatous sheath: absent (0) present (1)

61. Centrifugal chloroplasts on the parenchymatous sheath: absent (0) present (1)

62. Parenchymatous sheath: absent (0) present (1)

63. Mestome Kranz sheath: absent (0) present (1)

64. Maximum cells-distant count: more than 6 (0) 3-5 (1) 2-3 (2)

65. Distinctive Kranz cells: absent (0) present (1)

66. Mesophyll (arrangement): lax (0) compact (1)

67. Fusoid cells: absent (0) present (1)

Characters 60, 61, 64 and 65 are related with the photosynthetic pathway and therefore with the Kranz syndrome. Brown (1977), Ellis (1977), Clayton and Renvoize (1986), Hattersley (1987) and Hattersley and Watson (1992) emphasised its value in the generic delimitation of the Paniceae. In order to establish homologies for the present work, we followed the criteria sustained by Brown (1977) and Dengler *et al.* (1985); these authors considered the Kranz sheath of the MS anatomical type homologous to the mestome sheath of non-Kranz species, while the Kranz sheath of the PS anatomical type is homologous to the parenchymatous sheath of the non-Kranz species.

RESULTS AND DISCUSSION

Fifty-eight clades were resolved in the strict consensus of 20,345 equally parsimonious trees (length 372 steps, fit 365.8, Fig. 1). This consensus tree does not agree, in general terms, with the previous intuitive classification proposed by Clayton and Renvoize (1986), although a few of their groups are well supported. Clayton and Renvoize grouped, as previously stated, the Paniceae in seven subtribes. The circumscription of these subtribes was compared with our results and they are indicated on the right of the consensus tree (with its acronyms). In a similar way we compared the informal arrangement made by Brown (1977), who considered the following groups: Subtribe 1, Kranz, MS (NADP-me) genera; Subtribe 2, non-Kranz genera (including C_3/C_4 intermediates); Subtribe 3, Kranz PS (PCK) genera; and Subtribe 4, Kranz PS (NAD-me) genera.

Some of the subtribes of Clayton and Renvoize are not monophyletic groups. First, the clade *Alloteropsis* to *Cleistochloa* is defined by character #40 (upper palea gaping). Within this clade we can see two members of the Neurachninae, *Thyridolepis,* and *Neurachne*, while the other one, *Paraneurachne*, is placed in a different clade. This suggests that this subtribe is not monophyletic.

Lasiacis, Cyrtoccocum, Acroceras, and *Microcalamus*, previously included within the Setariinae, form, within the clade *Alloteropsis* to *Cleistochloa*, a monophyletic clade defined by characters #50 (upper lemma differentiated at the apex) and #53 (margin of the upper lemma tucked in onto the palea). Several authors have considered these genera as the most primitive of the tribe, due to the occurrence of several presumed primitive characters, such as bambusoid habit, cross-venation on the leaves, and pseudopetiolate blades, characters that were not used in the present study.

Entolasia, Calyptochloa and *Cleistochloa* form a monophyletic clade supported by character #26 (vestigial lower glume), with *Acritochaete* as the sister group. All these genera occur in tropical Africa and Australia and three of them were considered closely related within the subtribe Setariinae (Clayton and Renvoize 1986); *Acritochaete*, on the other hand, was placed by these authors in the subtribe Digitariinae.

The clade of *Oplismenopsis* and *Oplismenus* is supported by character #32 (awned lower glume).

Fasciculochloa, an Australian genus recently described (Simon and Weiller 1995), and *Steinchisma*, treated by several authors at the generic level (Zuloaga *et al.* 1998) or as a subgenus of *Panicum*, are sister taxa in the analysis, while their relationship to the

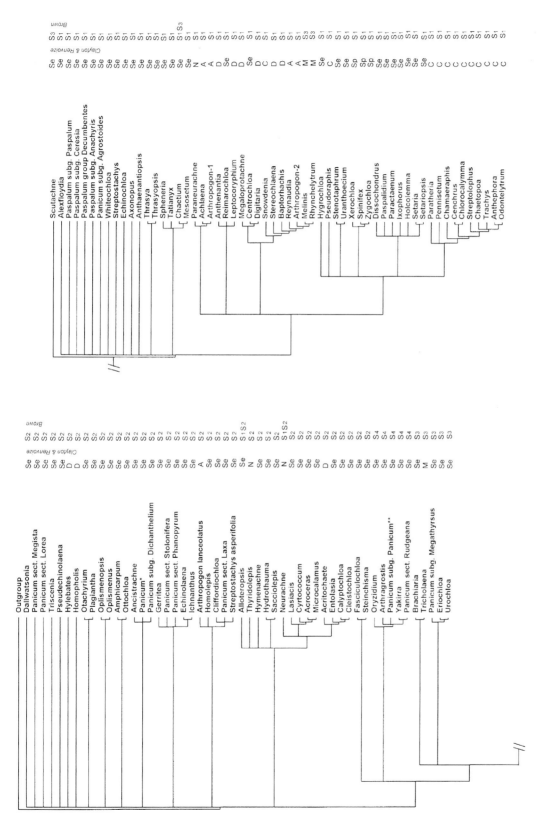

Fig. 1. Strict consensus of 20,345 equally parsimonious cladograms (372 steps, fit= 365.8, K= 3). Subtribal classifications of Clayton and Renvoize (1986) and Brown (1977) are compared. Subtribal abbreviations are signified as follow: A = Arthropogoninae; C = Cenchrinae; D = Digitariinae; M = Melinidinae; N = Neurachninae; Se = Setariinae; Sp = Spinificinae; S1 = Subtribe 1, Kranz, MS (NADP-me) genera; S2 = Subtribe 2, non-Kranz genera (including C₃/C₄ intermediates); S3 = Subtribe 3, Kranz PS (PCK) genera; S4 = Subtribe 4, Kranz PS (NAD-me) genera. *Panicum** = includes species of sections *Parviglumia, Parvifolia, Verrucosa, Verruculosa, Monticola,* and *Clavelligera; Panicum*** = includes species of subg. *Panicum,* excluding sect. *Rudgeana.*

rest of the Paniceae is not clear. This clade is supported by character #29 (expanded lower palea) and #43 (upper palea with compound papillae all over its surface). Also, both genera comprise C_3/C_4 species with an intermediate anatomical structure; the number of stamens in the upper floret is 2 in *Fasciculochloa* and 2, or occasionally 3, in *Steinchisma*.

Brachiaria and *Tricholaena* are also sister taxa and constitute a monophyletic clade, supported by character #19 (disarticulation at the base of the upper anthecium). *Tricholaena* has been classified within the Melinidinae, although the genus has spikelets dorsiventrally compressed instead of laterally compressed. This character was mentioned as being diagnostic for the subtribe. Brown (1977) also indicated that the affinity of the Melinidinae is with *Brachiaria* due to a similar PCK anatomical type. The present study supports the exclusion of *Tricholaena* from the Melinidinae and also indicates that the PCK anatomical type may have arisen more than once in the evolution of the tribe. This agrees with Ellis (1988), who also indicated a relationship between *Tricholaena* and *Brachiaria* based on anatomical characteristics.

The subgenus *Megathyrsus* of *Panicum*, *Eriochloa* and *Urochloa* form a robust clade sustained by characters #41 (transversely rugose upper lemma) and #44 (transversely rugose upper palea). Webster (1987) recently transferred *Panicum maximum* to *Urochloa*; our analysis indicates that these taxa are a monophyletic group and agrees with the realignment made by Webster.

Thrasya and *Thrasyopsis* are a single clade supported by character #1 (rachis terminating as a foliaceous axis); Webster and Valdés Reyna (1988), among others, previously supported this relation.

The clade *Anthaenantia* to *Leptocoryphium* is monophyletic, having in common character #26 (lower glume absent).

The clade *Hygrochloa* to *Odontelytrum* is a robust group sustained by the presence of a sterile prolongation of the apex of the rachis (#1) and of the main axis (character #2). Within this clade, some of the terminal taxa are well resolved, such as *Xerochloa*, *Spinifex* and *Zygochloa*, all of which appear within a monophyletic group defined by characters #4 (bracts of the inflorescence present), and #15 (disarticulation at the base of the inflorescence); all these taxa are closely related and were previously included in subtribe Spinificinae. Webster (1987, 1988) stressed the relation of *Stenotaphrum*, a pantropical genus, with *Uranthoecium*, an Australian genus; in this analysis both genera have character #17 in common (disarticulation at the base of the pedicel).

The clearly defined clade *Dissochondrus* to *Setariopsis* is, within the clade *Hygrochloa* to *Odontelytrum*, supported by characters #40 (upper palea tightly clasped by the lemma), #41 (upper lemma transversely rugose), #44 (upper palea transversely rugose), and #53 (upper lemma margins tucked in onto the palea). Clayton and Renvoize (1986) suggested that setiform ramifications in *Setaria* and other genera represents a parallelism within the tribe, a hypothesis that is not supported by this analysis.

Paratheria to *Odontelytrum* form, also within the clade *Hygrochloa* to *Odontelytrum* , a clade defined by the presence of primary articulations at the base of the primary branches #22 (spikelet orientation). Within this group, the clade *Cenchrus* to *Chlorocalymma* is supported by character #54 (lodicules absent). The

clade *Chaetopoa* to *Odontelytrum* is, on the other hand, defined by characters #5 (involucral bristles absent) and #6 (involucral bracts present), *Streptolophus* being the sister group.

All genera of the clade *Spheneria* to *Mesosetum* are defined by the presence of an oblique apex of the pedicel (character #10).

Several authors have indicated that *Panicum* is not a monophyletic entity, a hypothesis confirmed by this preliminary study. We therefore decided to segregate *Panicum*, at a world-wide level, into subgenera or sections. As a result section *Laxa* is part of a monophyletic clade, which also includes *Arthropogon lanceolatus*, *Homolepis*, *Cliffordiochloa* and *Streptostachys asperifolia*. The presence of fusoid cells, character #67, supports this clade. It should be mentioned that *Laxa*, a mostly American section of *Panicum*, is clearly related to *Cliffordiochloa*, a recently described Australian monotypic genus.

Another clade is formed by *Ancistrachne* together with several American, African, and Asian sections of *Panicum* (among them *Parviglumia*, *Parvifolia*, *Verrucosa*, *Verruculosa*, *Monticola*, and *Clavelligera*, represented by *Panicum** in Fig. 1), and the American subgenus *Dichanthelium*. This clade is supported by character #42 (upper palea with simple papillae all over the surface), but the relationships within this clade are still not clear, although a close relationship of *Dichanthelium* with the sections of *Panicum* previously mentioned is probable.

Sections *Phanopyrum* and *Stolonifera* of *Panicum*, together with *Gerritea*, *Echinolaena* and *Ichnanthus*, form a group that has in common the character of the articulation at the base of the upper anthecium (#19).

Subgenus *Panicum*, treated at a worldwide level, and excluding sect. *Rudgeana*, which is considered separately, forms a clade with sect. *Rudgeana*, *Yakirra*, *Arthragrostis*, and *Oryzidium*, and is united by character #60, presence of centripetal chloroplasts on the outer parenchymatous sheath of the vascular bundles. *Oryzidium* is the basal group of *Arthragrostis-Panicum* sect. *Rudgeana*; the latter is a robust group supported by characters #43 (upper palea with compound papillae at the apex), and #19 (articulation at the base of the upper anthecium).

Subgenus *Agrostoides*, which includes all MS species of *Panicum*, was not related to other groups or genera; a similar situation is present in sections *Megista* and *Lorea*. It is interesting that the number of MS species is small in *Panicum*; on the contrary, most of the genera of Paniceae, such as *Paspalum*, *Axonopus*, *Setaria*, *Digitaria*, and others) are MS. Sections of subgenus *Agrostoides* can be regarded as different genera, although this possibility should be confirmed by further studies.

Paspalum is one of the largest genera of the Paniceae, and was divided in three subgenera (*Paspalum*, *Ceresia*, and *Anachyris*), and one informal group, *Decumbentes*, due to its wide morphological variations. However, the affinities of these subgenera were not solved with this analysis.

Two genera were divided into different sets for this study: *Streptostachys* and *Arthropogon*. *Streptostachys asperifolia*, the type species of the genus, and *Arthropogon lanceolatus* fall, as previously stated, within a monophyletic clade together with *Cliffordioch-*

loa, *Panicum* sect. *Laxa* and *Homolepis*, the remaining three species of *Streptostachys* are grouped within another clade.

Clayton and Renvoize (1986) included the following genera in subtribe Arthropogoninae: *Reynaudia* and *Arthropogon* (including in the latter the monotypic Cuban genus *Achlaena*). The species of *Arthropogon*, excluding *A. lanceolatus*, were separated into different portions due to the great morphological variation found in their components. Those species showed an interesting arrangement in the cladogram: *Arthropogon piptostachyus* (= *Achlaena*), *A. scaber*, *A. bolivianus*, and *A. rupestris*, represented by *Arthropogon* number 1 in Fig. 1, form a solid clade, supported by characters #22 (spikelet orientation absent), #31 (awned lower glume), #32 (awned upper glume), #39 (hyaline upper anthecium), and #65 (distinctive Kranz cells present). On the other hand, *Arthropogon villosus*, *A. xerachne* and *A. filifolius*, represented by *Arthropogon* number 2, have clear affinities with *Reynaudia*, the sister group of these species. Webster and Valdés Reyna (1988) stressed the relationships of *Reynaudia* with *Melinis* and *Rhynchelytrum*.

Melinis and *Rhynchelytrum* are a monophyletic clade defined by characters # 26 (vestigial lower glume), #39 (membranous upper anthecium), #62 (parenchymatous sheath present), and #63 (mestome Kranz sheath absent). Clayton and Renvoize (1986) mentioned that these genera are derived from a PS *Panicum* ancestor; nevertheless, only *Tricholaena* is related in this analysis to *Brachiaria*. *Arthropogon p.p.* is the sister group of this clade. This result could suggest that the evolution of the PCK character occurred more than once in the Poaceae. It should be mentioned that several authors considered *Rhynchelytrum* a synonym of *Melinis*.

CONCLUSIONS

- This study supports the idea that the Paniceae is a monophyletic group.

- Discrepancies were observed between these results and the classification proposed by Clayton and Renvoize (1986).

- It is obvious that *Panicum* is a polyphyletic genus, and should be divided into several taxa.

- It is still necessary to further clarify the relationship of several Australian and American taxa, particularly *Fasciculochloa*, *Steinchisma*, *Cliffordiochloa*, and *Panicum* sect. *Laxa*.

- Other non-monophyletic genera, such as *Arthropogon* and *Streptostachys*, should be segregated into monophyletic units.

This preliminary study, based on morphological characters, elucidated several partial resolutions of groups in the Paniceae, in spite of the polymorphism and numerous homoplasies present in the data used in the analysis. Additional micromorphological characters of flowers, inflorescences and other structures will be necessary for a complete understanding of the evolutionary pattern of the Paniceae. This should be supplemented with molecular data which will surely provide important additional insights and allow a more precise definition of monophyletic groups in the tribe.

ACKNOWLEDGEMENTS

The authors would like to express their gratitude to an anonymous referee, Martín Ramirez and Lone Aagesen for comments on the manuscript, Alejandra Garbini for her help with the illustration and Consejo Nacional de Investigaciones Científicas y Técnicas, CONICET, Argentina, for financial support through grant N° 4440.

REFERENCES

Bentham, G. (1881). Notes on Gramineae. *Journal of the Linnean Society, Botany* **19**, 14-134.

Black, G. A. (1963). Grasses of the genus *Axonopus*: a taxonomic treatment. *Advancing Frontiers of Plant Sciences* **5**, 1-186.

Blake, S. T. (1958). New criteria for distinguishing genera allied to Panicum (Gramineae). *Proceedings of the Royal Society of Queensland* **70**, 15-19.

Brown, R. (1814). Genera remarks, geographical and systematical, on the botany of Terra Australis. In `A voyage to Terra Australis, Undertaken for the purpose of completing the discovery of that vast country, and prosecuted in the years 1801, 1802, and 1903´. (Ed Flinders) Vol. 2 pp 533-613. (W. Bulmer & Company: London.)

Brown, W. V. (1977). The Kranz syndrome and its subtypes in grass systematics. *Memoirs of the Torrey Botanical Club* **23**, 1-97.

Burman, A. G. (1980). Notes on the genera *Thrasya* H.B.K. and *Thrasyopsis* Parodi (Paniceae: Gramineae). *Brittonia* **32**, 217-221.

Butzin, F. (1970). Die systematische Gliederung der Paniceae. *Willdenowia* **6**, 179-192.

Butzin, D. (1977). Evolution der Infloreszenzen in der Berstenhirsen-Verwandtschaft. *Willdenowia* **8**, 67-79.

Chase, A. (1929). The Nort American species of *Paspalum*. *Contributions from the United States National Herbarium* **28**, 1-310.

Chase, A. (Inéd.). Revision of the South American species of genus *Paspalum*.

Chippindall, L. K. A. (1955). `The grasses and pastures of South Africa´ (CNA: Johannesburg).

Clayton, W. D. (1970). Studies in the Gramineae: A curious genus from Tanzania. *Kew Bulletin* **24**, 461-463.

Clayton, W. D., and Renvoize, S. A. (1982). Gramineae, Part 3. In `Flora of Tropical East Africa´ pp 1-898. Ed R. M. Polhill) (A. A. Balkema: Rotterdam.)

Clayton, W. D., and Renvoize, S. A. (1986). Genera graminum. Grasses of the world. *Kew Bulletin, Additional Series* **13**, 1--389.

Davidse, G. (1978). A systematic study of the genus *Lasiacis* (Gramineae: Paniceae). *Annals of the Missouri Botanical Garden* **65**, 1133-1258.

Davis, J. I., and Soreng, R. J. (1883). Phylogenetic structure in the grass family (Poaceae) as inferred from chloroplast DNA restriction site variation. *American Journal of Botany* **80**, 1444-1454.

Dengler, N. G., Dengler, R. E., and & Hattersley, P. W. (1985). Differing ontogenetic origins of PCR ('Kranz') sheaths in leaf blades of C_4 grasses (Poaceae). *American Journal of Botany* **72**, 284-302.

Duvall, M. R., Peterson, P. M., and Christensen, A. H. (1994). Alliances of *Muhlenbergia* (Poaceae) within New World Eragrostideae are identified by phylogenetic analysis of mapped restriction sites from plastid DNAs. *American Journal of Botany* **81**, 622-629.

Ellis, R. P. (1977). Distribution of the Kranz Syndrome in the Southern African Eragrostoideae and Panicoideae according to bundle sheath anatomy and cytology. *Agroplantae* **9**, 73-110.

Ellis, R. P. (1988). Leaf anatomy and systematics of *Panicum* (Poaceae: Panicoideae) in southern Africa. *Monographs in Systematic Botany from the Missouri Botanical Garden* **25**, 129-156.

Filgueiras, T. S. (1982). Taxonomia e distribuiçao de Arthropogon Nees (Gramineae). *Bradea* **3**, 303-322.

Filgueiras, T. S. (1990). Revisão de *Mesosetum* Steudel (Gramineae: Paniceae). *Acta Amazonica* **14**, 47-114.

Filgueiras, T. S. (1996). *Arthropogon rupestris* (Poaceae: Arthopogoneae) a new species from the Brazilian cerrado vegetation and a revised key for the genus. *Nordic Journal of Botany* **16**, 69-72.

Gibbs-Russell, G. E. (1983). The taxonomic position of C₃ and C₄ *Alloteropsis semiala* (Poaceae) in South Africa. *Bothalia* **14**, 205-213.

Goloboff, P. A. (1993). Pee-Wee versión 2.8 (32 bit). Programa y documentación distribuida por el autor. (Tucumán: Argentina)

Hackel, E. (1887). Gramineae. In `Die natürlichen Planzenfamilien´. (Eds A. Engler and K. Prantl) pp. 1-97. (Engelmann: Leipzig.)

Hattersley, P. W. (1987). Variations on photosynthetic pathway. In `Grass Systematics and Evolution´. (Eds T. R. Soderstrom, K. W. Hilu, C. S. Campbell, and M. E. Barkworth) pp. 49-64. (Smithsonian Institution Press: Washington, DC.)

Hattersley, P. W., and Watson, L. (1992). Diversification of photosynthesis. In `Grass evolution and domestication´. (Ed G. P. Chapman) pp 38-116. (Cambridge University Press: London.)

Holmgren, P., K., Holmgren, N. H., and Barnett, L. C. (1990). Index Herbariorum. Part I: the Herbaria of the World. *Regnum Vegetabile* **120**, 1-693.

Hsiao, C., Jacobs, S. W., Chatterton, N. J., and Asay, K. H. (1999). A molecular phylogeny of the grass family (Poaceae) based on the sequences of nuclear ribosonal DNA (ITS). *Australian Systematic Botany* **11**, 667-688.

Hsu, C. C. (1965). The classification of *Panicum* (Gramineae) and its allies, with special reference to the characters of lodicule, style-base and lemma. *Journal of the Faculty of Science, University of Tokyo, Section 3, Botany* **9**, 43-150.

Hughes, D. K. (1923). *Streptolophus*, *Orthachne* and *Streptachne*. *Kew Bulletin* **5**, 177-181.

Judziewicz, E. J. (1990). A new South American species of *Sacciolepis* (Poaceae: Panicoideae: Paniceae), with a summary of the genus in the New World. *Systematic Botany* **15**, 415-420.

Kellogg, E. A., and Campbell, C. S. (1987). Phylogenetic analyses of the Gramineae. In `Grass Systematics and Evolution´. (Eds T. R. Soderstrom, K. W. Hilu, C. S. Campbell, and M. E. Barkworth) pp. 310-322. (Smithsonian Institution Press: Washington, DC.)

Kellog, E. A., and Watson, L. (1993). Phylogenetic studies of a large data set. I. Bambusoideae, Andropogonadae, and Pooideae (Gramineae). *The Botanical Review* **59**, 273-343.

Lazarides, M., and Webster, R. D. (1984). *Yakirra* (Paniceae: Poaceae), a new genus for Australia. *Brunonia* **7**, 289-296.

Morrone, O., and Zuloaga, F. O. (1991a). Revisión del género *Streptostachys* (Poaceae-Panicoideae), su posición sistemática dentro de la tribu Paniceae. *Annals of the Missouri Botanical Garden* **78**, 359--376.

Morrone, O., and Zuloaga, F. O. (1991b). Estudios morfológicos en el subgénero *Dichanthelium* de *Panicum* (Poaceae: Panicoideae: Paniceae), con especial referencia a *Panicum sabulorum*. *Annals of the Missouri Botanical Garden* **78**, 915-927.

Morrone, O., and Zuloaga, F. O. (1992). Revisión de las especies sudamericanas nativas e introducidas de los géneros *Brachiaria y Urochloa* (Poaceae: Panicoideae: Paniceae). *Darwiniana* **31**, 43-109.

Morrone, O., and Zuloaga, F. O. (1993). Sinopsis del género *Urochloa* (Poaceae: Panicoideae: Paniceae) para México y América Central. *Darwiniana* **32**, 59-75.

Morrone, O., Filgueiras, T. S. , Zuloaga, F. O., and Dubcovsky, J. (1993). Revision of *Anthaenantiopsis* (Poaceae: Panicoideae: Paniceae). *Systematic Botany* **18**, 434-453.

Morrone, O., Zuloaga, F. O., and Carbonó, E. (1995). Revisión del grupo Racemosa del género *Paspalum* (Poaceae: Panicoideae: Paniceae). *Annals of the Missouri Botanical Garden* **82**, 82--116.

Morrone, O., Vega, A., and Zuloaga, F. O (1996). Revisión de las especies del género *Paspalum* (Poaceae: Panicoideae: Paniceae), grupo Dissecta. *Candollea* **51**, 103-138.

Morrone, O. Zuloaga, F. O., Arriaga, M. O., Pozner, R., and Aliscioni, S. S. (1998). Revisión sistemática y análisis cladístico del género *Chaetium* (Poaceae: Panicoideae: Paniceae). *Annals of the Missouri Botanical Garden* **85**, 404-424.

Nixon, K. C., and Davis, J. I. (1991). Polimorphic taxa, missing values and cladistic analysis. *Cladistics* **7**, 233-241.

Pensiero, J. F. (1999). Las especies sudamericanas del género *Setaria* (Poaceae, Paniceae). *Darwiniana* **37**, 37-151.

Pilger, R. (1940). Gramineae III: Unterfamilie Panicoideae. In `Die natürlichen Pflanzenfamilien´ second edition. (Eds A. Engler and K. Prantl) pp 1-208. (Engelmann: Leipzig.)

Pilger, R. (1954). Das system der Gramineae. *Botanische Jahrbücher* **76**, 281-384.

Reeder, J. R. (1960). The systematics position of the grass genus *Anthephora*. *Transaction of the American Microscopical Society* **79**, 211-218.

Rominger, J. M. (1962). Taxonomy of *Setaria* (Gramineae) in North America. *Illinois Biological Monographs* **29**, 1-132.

Sauer, J. D. (1972). Revision of *Stenotaphrum* (Gramineae: Paniceae) with attention to its historical geography. *Brittonia* **24**, 202-222.

Scholz, U. (1981). Monographie der Gattung *Oplismenus* (Gramineae). *Phanerogamarum Monographiae* **13**, 1-213.

Sendulsky, T., and Soderstrom, T. R. (1984). Revision of the South American genus *Otachyrium* (Poaceae: Panicoideae). *Smithsonian Contributions to Botany* **57**, 1-24.

Sendulsky, T., Filgueiras, T. S., and Burman, A. G. (1987). Fruits, embryos and seedlings. . In `Grass Systematics and Evolution´. (Eds T. R. Soderstrom, K. W. Hilu, C. S. Campbell, and M. E. Barkworth) pp. 31-36. (Smithsonian Institution Press: Washington, DC.)

Simon, B. K. (1992). Studies in Australian Grasses 6. *Alexfloydia*, *Cliffordiochloa* and *Dallwatsonia*, three new panicoid grass genera from Eastern Australia. *Austrobaileya* **3**, 669-681.

Simon, B. K., and Weiller, C. M. (1995). *Fasciculochloa*, a new grass genus (Poaceae: Paniceae) from south-eastern Queensland. *Austrobaileya* **4**, 369-379.

Soreng, R. J., and Davis, J. I. (1998). Phylogenetics and character evolution in the grass family (Poaceae): simultaneous analysis of morphological and chloroplast DNA restriction site character sets. *The Botanical Review* **64**, 1-84.

Sohns E. R. (1955). *Cenchrus* and *Pennisetum*: fascicle morphology. *Journal of the Washington Academy of Sciences* **45**, 135-143.

Stapf, O. (1919). Gramineae. In `Flora of Tropical Africa´. (Ed D. Prain) Vol 9 pp 422-480 (L. Reeve & Co.: London.)

Stieber, M. T. (1982). Revision of *Ichnanthus* sect. Ichnanthus (Gramineae, Panicoideae). *Systematic Botany* **7**, 85-115.

Stieber, M. T. (1987). Revision of *Ichnanthus* sect. Foveolatus (Gramineae: Panicoideae). *Systematic Botany* **12**, 187-216.

Tateoka, T. (1963). Notes on some grasses XV. Affinities and species relationship of *Arthropogon* and relatives, with reference to their leaf-structure. *Botanical Magazine (Tokyo)* **76**, 286-291.

Troll, W. (1967). `Vergleichende Morphologie der höheren Pflanzen´ 2 Teil. IV Aabschnitt. (Gebrüder Borntrager: Berlin).

Türpe, A. M. (1983). Las especies sudamericanas del género *Pennisetum* L.C. Richard (Gramineae). *Lilloa* **36**, 105-129.

Watson, L., and Dallwitz, M. J. (1992). `The Grass Genera of the World´. (C.A.B. International: Wallingford, Oxon, England.)

Webster, R. D. (1987). `The Australian Paniceae (Poaceae)´. (J. Cramer: Stuttgart.)

Webster, R. D. (1988). Genera of the North American Paniceae (Poaceae: Panicoideae). *Systematic Botany* **13**, 576-609.

Webster, R.D. (1992). Character significance and generic similarities in the Paniceae (Poaceae: Panicoideae) *Sida* **15**, 185-213.

Webster, R. D., and Valdés Reyna, J. (1988). Genera of Mesoamerican Paniceae (Poaceae; Panicoideae). *Sida* **13**, 187-221.

Wills, K. E. (1996). `Systematic studies in *Homopholis* (Poaceae: Panicoideae: Paniceae)'. MSc Thesis, University of New England.

Zuloaga, F. O. (1987a). A revision of *Panicum* Subg. *Panicum* Sect. *Rudgeana* (Poaceae: Paniceae). *Annals of the Missouri Botanical Garden* **74**, 463-478.

Zuloaga, F. O. (1987b). Systematics of the New World species of *Panicum* (Poaceae: Paniceae). In `Grass Systematics and Evolution´. (Eds T. R. Soderstrom, K. W. Hilu, C. S. Campbell, and M. E. Barkworth) pp. 31-36. (Smithsonian Institution Press: Washington, DC.)

Zuloaga, F. O. (1989). El género *Panicum* en la República Argentina. III. *Darwiniana* **29**, 289-370.

Zuloaga, F. O., and Soderstrom, T. R. (1985). Classification of the outlying species of New World Panicum (Poaceae: Paniceae). *Smithsonian Contributions to Botany* **59**, 1-63.

Zuloaga, F. O., Morone, O., and Sáenz, A. A. (1987). Estudio exomorfológico e histofoliar de las especies americanas del género *Acroceras* (Poaceae: Paniceae). *Darwiniana* **28**, 191-217.

Zuloaga, F. O., and Sendulsky, T. (1988). A revision of *Panicum* Subg. *Phanopyrum* Sect. *Stolonifera* (Poaceae: Paniceae). *Annals of the Missouri Botanical Garden* **75**, 420-455.

Zuloaga, F. O., Morrone, O., Vega, A. S., and Giussani, L. M. (1998). Revisión y análisis cladístico de *Steinchisma* (Poaceae: Panicoideae: Paniceae). *Annals of the Missouri Botanical Garden* **85**, 631-656.

Zuloaga, F. O., Morrone, O., and Dubcovsky, J. (1989). Exomorphological, anatomical, and cytological studies in *Panicum validum* (Poaceae: Paniceae). Its systematic position within the genus. *Systematic Botany* **14**, 220-230.

Zuloaga, F. O., Ellis, R. P., and Morrone, O. (1992). A revision of *Panicum* subgenus *Phanopyrum* section *Laxa* (Poaceae: Panicoideae: Paniceae). *Annals of the Missouri Botanical Garden* **79**, 770-818.

Zuloaga, F. O., Ellis, R. P., and Morrone, O. (1993a). A revision of *Panicum* subgenus *Dichanthelium* section *Dichanthelium* (Poaceae: Panicoideae: Paniceae). *Annals of the Missouri Botanical Garden* **80**, 119-190.

Zuloaga, F. O., Dubcovsky, J., and Morrone, O. (1993b). Infrageneric phenetic relations in the genus *Panicum* (Poaceae: Panicoideae: Paniceae): a numerical analysis. *Canadian Journal of Botany* **71**, 1312-1327.

Zuloaga, F.O., Morrone, O., and Killeen, T. (1993). *Gerritea*, a new genus of Paniceae (Poaceae: Panicoideae) from South America. *Novon* **3**, 213-219.

Zuloaga, F. O., and Morrone, O. (1996). Revisión de las especies americanas de *Panicum* subgénero *Panicum* sección *Panicum* (Poaceae: Panicoideae: Paniceae). *Annals of the Missouri Botanical Garden* **83**, 200--280.

Grasses: Systematics and Evolution. (2000). Eds S.W.L. Jacobs and J. Everett. (CSIRO: Melbourne)

PHYLOGENY OF THE SUBFAMILY PANICOIDEAE WITH EMPHASIS ON THE TRIBE PANICEAE: EVIDENCE FROM THE *TRN*L-F CPDNA REGION.

Rosalba Gómez-Martínez[AB] and Alastair Culham[A]*

[A]Department of Botany, The University of Reading, Whiteknights, PO Box 221, Reading, RG6 6AS, UK.
[B]Current address: Universidad Francisco de Miranda, Apdo. 7506, Coro,4101 Falcón, Venezuela.
[*]Corresponding author.

Abstract

A cladistic analysis of the subfamily Panicoideae with special emphasis in the tribe Paniceae is presented using the chloroplast DNA sequence region *trn*L-*trn*F. Support for the monophyly of the Panicoideae and its sister relationship to the Centothecoideae was found. The tribe Andropogoneae is clearly monophyletic but its relationship to the rest of the Panicoideae is ambiguous. Accordingly, the tribe Paniceae can be interpreted either as monophyletic or paraphyletic. Within the Paniceae, two major clades were identified: one formed by American native taxa with $x = 10$, and the other formed by Pantropical taxa with $x = 9$. These results suggest that the diversification of the subfamily Panicoideae and of the tribe Paniceae might have been associated with past events of vicariance coupled with karyotype rearrangements.

Key words: Poaceae, Panicoideae, Paniceae, Andropogoneae, Setariinae, Digitariinae, *trn*L, *trn*F, phylogeny, cpDNA.

INTRODUCTION

The subfamily Panicoideae is considered a natural group defined by the occurrence of two unique derived characters among grasses: the presence of 2-flowered spikelets with a reduced (male or barren) proximal floret, and an embryo of the type P-PP (Kellogg and Campbell 1987; Clayton and Renvoize 1986). Early cladistic studies of morphology found support for the monophyly of the subfamily (Kellogg and Campbell 1987). This result has been corroborated by more recent molecular studies (Barker 1997; Barker *et al.* 1995; Clark *et al.* 1995; Nadot *et al.* 1994; Doebley *et al.* 1990; Duvall and Morton 1990), although the Panicoideae resolved as paraphyletic in the chloroplast *mat*K gene phylogeny of Liang and Hilu (1996). In all cases, the Panicoideae taxon sampling has been small (six to eight species at the most, only two in some cases).

The molecular phylogenetic information available for the family has been summarised by Kellogg and Linder (1995) and Kellogg (1998a, 1998b). They conclude that the evidence gathered till now points to a monophyletic Panicoideae sister to the Centothecoideae. The monophyly of the Andropogoneae is also very well supported, but no conclusions can be reached as to the status of the rest of the tribes.

Although some groups are assigned different ranks in each of the two systems of classification currently used for the grass family (Watson and Dallwitz 1992; Clayton and Renvoize 1986), they agree on the delimitation of the Panicoideae and the major clusters of genera within it (with the exception of the tribes Eriochloeae and Steyermarkochloeae). Clayton and Renvoize (1986) opted for a detailed scheme of classification (tribes and subtribes), and suggested more informal or intuitive sets of relationships for the members of the subfamily.

Most characters used to delimit groups and define genera within the Panicoideae are reproductive (inflorescence type and presence of bracts or bristles, spikelet compression, grouping, dimorphism, disarticulation and orientation, degree of induration of the floret and type of lemma margins, among others). The leaf anatomy and photosynthetic type are regarded as very important to define major groups in this subfamily.

The proposed sets of relationships among the Panicoideae, and the Paniceae in particular, have been explored here using sequence data from the *trn*L intron, the *trn*L 3' exon and the *trn*L-*trn*F spacer regions of the cpDNA (subsequently referred as the *trn*L-F region). The *trn*L gene encodes for the transfer RNA for the amino-acid leucine, and the coding regions (exons) are interrupted by a class I intron. This gene is located in the long single copy region of the chloroplast DNA contiguously to the *trn*T and *trn*F genes. This region of the chloroplast genome has been used to address systematic issues at different levels in several groups of flowering plants (Compton *et al.* 1998; Bakker *et al.* 1998; Gielly and Taberlet 1996; Kita *et al.* 1995; Böhle *et al.* 1994). Both point and length mutations occur in this region, providing phylogenetic resolution at the intergeneric and inter-specific levels in several groups studied so far (Compton *et al.* 1998; Gómez-Martínez and Culham 1997; Freeman and Ybarra 1996; Gielly and Taberlet 1996; Hauk *et al.* 1996; Kita *et al.* 1995; Böhle *et al.* 1994; van Ham *et al.* 1994; Ferris *et al.* 1993).

MATERIALS AND METHODS

Fifty-one grass taxa were included in this study, nine outgroups, and 42 panicoid grasses (nine species of Andropogoneae, one species of Arundinelleae, and 32 species of Paniceae) (Table 1). Within the Paniceae tribe, the subtribes Setariinae, Digitariinae, Cenchrinae and Melinidinae were represented in the sampling. Standard procedures were used to extract, amplify and sequence the DNA region (Gielly and Taberlet 1996; Taberlet *et al.* 1991); cladistic parsimony analyses were performed using PAUP* version 4.064 (with kind permission of Swofford, pers. comm.). Informative insertion/deletion events were coded as multistate characters and added to the data matrix. Jackknife and bootstrap support values were estimated as implemented in PAUP* version 4.064.

RESULTS

The detailed description of the analyses performed and of the variation found in the *trn*L-F region will be presented elsewhere (Gómez-Martínez and Culham, in preparation). The percentage of informative positions was 19%, slightly less than that observed for the *ndh*F region of grasses (Clark *et al.* 1995).

The cladistic search using the maximum parsimony criterion yielded 744 most parsimonious trees (using the option of collapsing branches when minimum length equal zero), of 611 steps long. The consistency index was 0.61 and the retention index was 0.75, excluding autoapomorphic and non-variable sites.

The majority rule consensus tree is presented in Fig. 1, indicating the branches that collapse in the strict consensus tree, as well as jackknife and bootstrap support values. The Panicoideae is resolved as a monophyletic subfamily, sister to the Centothecoideae, represented here only by *Megastachya madagascariensis*, with relatively moderate support. Within the Panicoideae, three major clades are well supported:

Clade 1: Formed by all the Andropogoneae species plus *Arundinella* sp.

Clade 2: Formed by the genera *Ichnanthus, Axonopus, Paspalum, Mesosetum,* and the species of *Panicum* subgenus *Phanopyrum* section *Laxa* and *Panicum obtusum* from subgenus *Agrostoides* section *Obtusa.* The species of all these taxa are native to the American continent (with the exception of one Pantropical species of *Ichnanthus* and very few Old World species of *Paspalum*) and will be subsequently referred to as the 'American Paniceae' clade.

Clade 3: Formed by *Melinis, Brachiaria, Eriochloa, Setaria, Pennisetum, Cenchrus,* and *Panicum* subgenus *Panicum, P. bulbosum* from subgenus *Agrostoides* section *Bulbosa* and *P. maximum* from subgenus *Megathyrsus.* All these taxa are basically Pantropical in distribution being more speciose in the Old World and will be subsequently referred to as the 'Pantropical Paniceae' clade.

The representatives of *Acroceras, Echinochloa, Sacciolepis, Digitaria* and *Panicum sabulorum* (subgenus *Dichanthelium*) and *P. trichanthum* are not resolved and all of them collapse in a basal polytomy within the Panicoideae subfamily.

The relationships of each major clade in relation to the others are ambiguous. The Andropogoneae plus *Arundinella* clade appears as sister to the American Paniceae taxa in most of the trees found, but collapses in a basal polytomy in about 1/4 of the trees recovered, resulting in an unresolved backbone structure for the subfamily.

Within each clade a phylogenetic structure is also revealed, pointing to some well supported sister relationships. Within the Pantropical Paniceae clade in particular, three major sets of taxa are defined. The first group is a strongly supported clade formed by the setae-bearing species (*Cenchrus, Pennisetum* and *Setaria*) and *Panicum bulbosum.* The second clade includes all the C_4-PEP-CK taxa sampled (*Melinis, Brachiaria, Eriochloa,* and *Panicum maximum*) and is only weakly supported, although it has moderate support when *Melinis* is excluded. The third group of species within Pantropical Paniceae is formed by the species of *Panicum* subgenus *Panicum,* the only group of panicoid species sampled with the C_4-NAD-ME photosynthetic type.

DISCUSSION

The monophyly of the subfamily Panicoideae and its sister relationship to the Centothecoideae are supported by the results of this research. The monophyletic nature of the tribe Andropogoneae and its close association to *Arundinella* is also supported. However, the ambiguity in the placement of this clade in relation to the rest of the taxa sampled does not provide evidence to support or contradict the monophyly of the Paniceae. The variation found in the *trn*L-F region does not provide enough resolution to address this particular issue due to the deletion of 104 bp from the intron of the Andropogoneae clade. The characters that support the placement of the Andropogoneae plus *Arundinella* as the sister clade to the American Paniceae taxa in most of the trees found are highly homoplasious. When a successive weighting search strategy is applied the resulting tree topology places the Andropogoneae plus *Arundinella* clade as the sister group to the

Table 1. List of taxa included in this study. Details about specimens and vouchers will be supplied by the authors on request.

Subfamily Panicoideae

Tribe Andropogoneae
1. *Andropogon gerardii* Vitman
2. *Coix lacryma-jobi* L.
3. *Chrysopogon gryllus* Trin.
4. *Hyparrhenia hirta* (L.) Stapf
5. *Miscanthus* sp.
6. *Rottboellia aurita* Steud.
7. *Sorghum bicolor* (L.) Moench.
8. *Tripsacum dactyloides* (L.) L.
9. *Zea mays* L.

Tribe Arundinelleae
10. *Arundinella nepalensis* Trin.

Tribe Paniceae

Subtribe Cenchriineae
11. *Cenchrus ciliaris* L.
12. *Pennisetum alopecuroides* (L.) Spreng.
13. *Pennisetum polystachyon* (L.) Schult

Subtribe Digitariineae
14. *Digitaria ciliaris* (Retz.) Koeler
15. *Leptocoryphium lanatum* (Kunth) Nees

Subtribe Melinidinae
16. *Melinis repens* (Willd.) Zizka

Subtribe Setariinae
17. *Acroceras zizaniodes* (Kunth) Dandy
18. *Axonopus anceps* (Mez) Hitchc.
19. *Axonopus canescens* (Nees in Trin.) Pilger
20. *Axonopus compressus* (Sw.) P. Beauv.
21. *Axonopus furcatus* (Flüggé) Hitchc.
22. *Brachiaria mutica* (Forssk.) Stapf
23. *Brachiaria* sp.
24. *Echinochloa crus-galli* (L.) P. Beauv.
25. *Eriochloa punctata* (L.) Hamilton
26. *Ichnanthus pallens* (Sw.) Munro
27. *Mesosetum cayennense* Steud.
28. *Mesosetum loliiforme* (Steud.) Chase

Panicum subg. *Agrostoides*
29. *Panicum bulbosum* Kunth
30. *Panicum obtusum* Kunth

Panicum subg *Dichanthelium*
31. *Panicum sabulorum* Lam.

Panicum subg. *Megathyrsus*
32. *Panicum maximum* Jacq.

Panicum subg. *Panicum*
33. *Panicum tricholaenoides* Steud.
34. *Panicum virgatum* L.

Panicum subg. *Phanopyrum*
35. *Panicum laxum* Sw.
36. *Panicum polygonatum* Schrad. ex Schultes
37. *Panicum trichanthum* Nees
38. *Paspalum blodgettii* Chapm.
39. *Paspalum dilatatum* Poir.
40. *Sacciolepis indica* (L.) Chase
41. *Setaria palmifolia* (Koen.) Stapf.
42. *Setaria poiretiana* (Schult.) Kunth

OUTGROUPS:

Subfamily Arundinoideae
43. *Cortaderia selloana* (Schult. & Schult. f.) Asch. & Grabn.
44. *Phragmites australis* (Cav.) Trin. ex Steud.

Table 1. List of taxa included in this study. Details about specimens and vouchers will be supplied by the authors on request. *(Continued)*

OUTGROUPS: *continued*

Subfamily Bambusoideae
45. *Bambusa* sp.

Subfamily Centothecoideae
46. *Megastachya madagascariensis* (Lam.) Chase

Subfamily Chloridoideae
47. *Spartina pectinata* Link.

Subfamily Oryzoideae
48. *Oryza sativa* L.

Subfamily Pooideae
49. *Hordeum vulgare* L.
50. *Puccinellia distans* (Jacq.) Parl.
51. *Triticum aestivum* L.

whole Paniceae tribe (Gómez-Martínez and Culham in preparation). No bootstrap/jackknife support is found for this placement either. The use of a different molecular marker should provide more informative characters to test the monophyly/paraphyly of the Paniceae tribe.

Within the Paniceae tribe, several conclusions can be reached. The subtribes Setariinae and Digitariinae are paraphyletic. Clayton and Renvoize (1986) based the distinction of these subtribes on the relative degree of induration of the fertile floret and the presence of flat *vs.* inrolled lemma margins. The results obtained indicate that these characters are homoplasious. In contrast, the presence of bristles in the spikelets, a character regarded by Clayton and Renvoize as having originated independently in *Setaria* and the Cenchrinae can be used as a synapomorphy to support the 'setae bearing clade' (*Cenchrus*, *Pennisetum*, and *Setaria*).

The large genus *Panicum* is clearly polyphyletic. *Panicum maximum* is resolved as sister to *Brachiaria*, a relationship that has been previously suggested by Brown (1977) and Webster (1987), taking into account the presence of transversely rugose florets and the C_4-PEP-CK metabolic type. With the exception of *Panicum* subgenus *Panicum*, the subgenera proposed by Zuloaga (1987) included in this study (subgenus *Agrostoides* and subgenus *Phanopyrum*) are paraphyletic (Table 1, Fig. 1). Only one species of subgenus *Dichanthelium* was studied, and no conclusion can be reached as to its monophyly.

Axonopus and *Mesosetum* are resolved as monophyletic genera with a strong bootstrap support (Fig. 1). In the case of *Axonopus* additional sequence data from the *trn*L-F marker and *rpl*-16 intron cpDNA regions strongly support this finding (Gómez-Martínez, unpublished data). The big genera *Paspalum* and *Brachiaria* represented in this study only by a couple of species each are resolved as monophyletic but with only moderate bootstrap/jackknife support (Fig. 1). Further sampling of these two taxa is required in order to confirm their monophyly. *Pennisetum* is probably paraphyletic in relation to *Cenchrus* (tribe Cenchrinae), a foreseen result considering the high number of morphological gradations between these two genera.

No obvious morphological synapomorphies have been identified so far to support the two major Paniceae clades identified by this

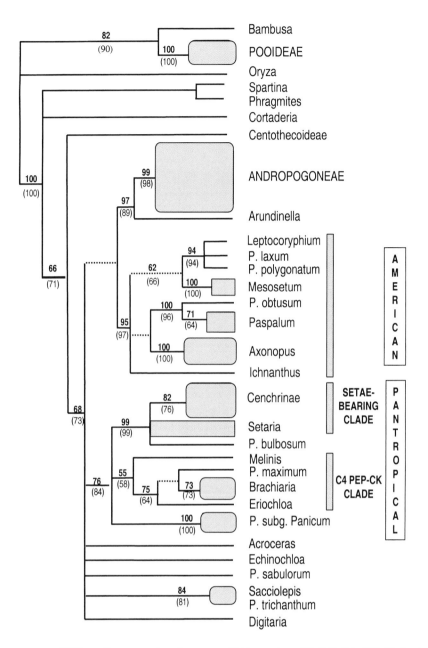

Fig. 1. Majority rule consensus tree of 744 equally most parsimonious trees of 611 steps long, C.I.=0.61, R.I.=0.75. All the characters used had equal weights. Dashed lines indicate branches that collapse in the strict consensus tree. Bold numbers above branches represent jackknife support values. Numbers between brackets represent bootstrap support values. Terminal monophyletic clades have been grouped together under shaded boxes. The complete list of taxa is detailed in Table 1. The letter 'P.' stands for *Panicum*.

study. However, they have a good correlation with general distribution pattern and basic chromosome numbers. The taxa included in the American Paniceae clade all have a basic chromosome number of x = 10, with the exception of *Mesosetum* for which one count of 2n = 16 is available. The Pantropical Paniceae taxa have a basic chromosome number of x = 9, although *Brachiaria* and *Cenchrus* also have x = 7. The unresolved Paniceae taxa also have x = 9, while *Digitaria* has been reported as x = 9, 15 or 17. This pattern suggests that the evolution within the tribe may have been related to changes in chromosome number. Phylogenetic studies performed within the Andropogoneae suggest that there is a tendency to reduction in chromosome

number within this tribe (Spangler *et al.* 1999). This could be a more generalised evolutionary trend within the whole subfamily Panicoideae. The split into New World *vs.* Pantropical taxa needs to be explored further using formal biogeographic analyses to assess the possible association to vicariance events.

The evolution of the photosynthetic metabolic pathway does not show a clear pattern. Only one clear reversal from the C_4 to the C_3 condition can be identified in the phylogeny (*Panicum* section *Laxa*). At this point of knowledge it appears to be equally parsimonious to assume a C_3 or a C_4 ancestral condition for the subfamily. The inclusion of new taxa such as *Isachne* and

Neurachne, among others, will be crucial to be able to reach a firmer conclusion in this direction.

ACKNOWLEDGEMENTS

The authors want to thank Toby Kellogg, Nigel Barker and Steve Renvoize for supplying DNA and botanical samples. Lynn Clark, Cynthia Morton and Derek Clayton provided helpful suggestions and discussions. We also thank Jim Mant for critically reviewing the manuscript. This research received financial support from Fundación Gran Mariscal de Ayacucho and Universidad Francisco de Miranda in Venezuela and also from The Linnean Society of London.

REFERENCES

Bakker, F. T., Hellbrügge, D., Culham, A., and Gibby, M. (1998). Phylogenetic relationships within *Pelargonium* sect. *Peristera* (Geraniaceae) inferred from nrDNA and cpDNA sequence comparisons. *Plant Systematics and Evolution* **211**, 273-287.

Barker, N. (1997). The relationships of *Amphipogon*, *Elytrophorus* and *Cyperochloa* (Poaceae) as suggested by *rbc*L sequence data. *Telopea* **7**, 205-213.

Barker, N., Linder, H. P., and Harley, E. H. (1995). Polyphyly of Arundinoideae (Poaceae): evidence from *rbc*L sequence data. *Systematic Botany* **20**, 423-435.

Böhle, U.-R. Hilger, H., Cerff, R., and Martin, W. F. (1994). Non-coding chloroplast DNA for plant molecular systematics at the infrageneric level. In: Molecular ecology and evolution: approaches and applications. (Eds B. Schierwater, B. Streit, G. P. Wagner and R. DeSalle). pp. 391-403. (Birkhäuser Verlag: Basel.)

Brown, W. V. (1977). The Kranz syndrome and its subtypes in grass systematics. *Memoirs of the Torrey Botanical Club* **23**, iv + 97.

Clark, L. G. , Zhang, W., and Wendel, J. F. (1995). A phylogeny of the grass family (Poaceae) based on *ndh*F sequence data. *Systematic Botany* **20**, 436-460.

Clayton, W. D., and Renvoize, S. A. (1986). *Genera graminum*: grasses of the world. (Royal Botanic Gardens Kew: London.)

Compton, J. A., A. Culham, and S L. Jury. (1998). Reclassification of *Actaea* to include *Cimicifuga* and *Souliea* (Ranunculaceae): phylogeny inferred from morphology, nrDNA ITS, and cpDNA *trn*L-F sequence variation. *Taxon* **47**, 593-634.

Doebley, J., Durbin, M., Golenberg, E. M., Clegg, M. T., and Ma, D. P. (1990). Evolutionary analysis of the large subunit of carboxylase (*rbc*L) nucleotide sequence data among the grasses (Poaceae). *Evolution* **44**, 1097-1108.

Duvall, M. R., and Morton, B. R. (1996). Molecular phylogenetics of Poaceae: and expanded analysis of *rbc*L sequence data. *Molecular Phylogenetics and Evolution* **5**, 352-358.

Ferris, C., Oliver, R. P., Davy, A. J., and Hewitt, G. M. (1993). Native oak chloroplasts reveal an ancient divide across Europe. *Molecular Ecology* **2**, 337-344.

Freeman, C. E., and Ybarra, G. R. (1996). The utility of chloroplast *trn*L intron DNA sequences in inferring phylogeny within *Keckiella* (Scrophulariaceae*). American Journal of Botany* **83**, 206-207.

Gielly, L., and Taberlet, P. (1996). Chloroplast DNA sequencing to resolve plant phylogenies between closely related taxa. In: Molecular genetics approaches in conservation (Eds. T. B. Smith and R. K. Wayne). pp. 143-153. (Oxford University Press: New York.)

Gómez-Martínez, R., and Culham, A. (1997). Systematics of the grass genus *Axonopus* (Panicoideae, Setariinae): evidence from *trn*L-*trn*F data. *American Journal of Botany* **84**, 198.

Hauk, W. D., Parks, C. R., and Chase, M. W. (1996). A comparison between *trn*L-F intergeneric spacer and *rbc*L DNA sequence data: an example from Ophioglossaceae. *American Journal of Botany* **83**, 126-127.

Kellogg, E. A. (1998a). Who's related to whom? Recent results from molecular systematics studies. *Current Opinion in Plant Biology* **1**, 149-158.

Kellogg, E. A. (1998b). Relationships of cereal crops and other grasses. *Proceedings of the National Academy of Sciences USA* **95**, 2005-2010.

Kellogg, E. A., and Campbell, C. S. (1987). Phylogenetic analyses of the Gramineae. In: Grass systematics and evolution. (Eds T. Soderstrom, K. Hilu, C. Campbell and M. Barkworth). pp. 310-322. (Smithsonian Institution Press, Washington, D. C.)

Kellogg, E. A., and Linder, H. P. (1995). Phylogeny of Poales. In: Monocotyledons: systematics and evolution. (Eds P. J. Rudall, P. J. Cribb, D. F. Cutler and C. J. Humphries). pp. 511-542. (Royal Botanic Gardens Kew: London.)

Kita, Y., Ueda, K., and Kadota, Y. (1995). Molecular phylogeny and evolution of the Asian *Aconitum* subgenus *Aconitum* (Ranunculaceae). *Journal of Plant Research* **108**, 429-442.

Liang, H., and Hilu, K. W. (1996). Application of the *mat*K gene sequences to grass systematics. *Canadian Journal of Botany* **74**, 125-134.

Nadot, S. R., Bajon, R., and Lejeune, B. (1994). The chloroplast gene *rps*4 as a tool for the study of Poaceae phylogeny. *Plant Systematics and Evolution* **191**, 27-38.

Spangler, R., Zaitchick, B., Russo, E., and Kellogg, E. (1999) Andropogoneae evolution and generic limits in *Sorghum* (Poaceae) using *ndh*F sequences. *Sysytematic Botany* **24**, 267–281.

Taberlet, P., Gielly, L., Patou, G., and Bouvet, J. 1991. Universal primers for amplification of three non-coding regions of chloroplast DNA. *Plant Molecular Biology* **117**, 1105-1109.

Watson, L., and Dallwitz, M. J. (1992). The grass genera of the World. (CAB International: Wallingford.)

van Ham, R. C. H. J., Hart, H. H., Mes, T. H. M., and Sandbrink, J. M. (1994). Molecular evolution of non-coding regions of the chloroplast genome in the Crassulaceae and related species. *Current Genetics* **25**, 558-566.

Webster, R. D. 1987. The Australian Paniceae (Poaceae). (J. Cramer: Stuttgart.)

Zuloaga, F. (1987). Systematics of the New World species of *Panicum* (Poaceae:Paniceae). In: Grass systematics and evolution. (Eds T. Soderstrom, K. Hilu, C. Campbell and M. Barkworth). pp. 287-306. (Smithsonian Institution Press: Washington, D. C.)

Grasses: Systematics and Evolution. (2000). Eds S.W.L. Jacobs and J. Everett. (CSIRO: Melbourne)

GRASSES

AUSTRAL SOUTH AMERICAN SPECIES OF *ERIOCHLOA*

Mirta O. Arriaga

Museo Argentino de Ciencias Naturales; 'Bernardino Rivadavia', A. Gallardo 470. CP1405, Buenos Aires, Argentina.

Abstract

Eriochloa, a genus of the tribe Paniceae from temperate regions of both hemispheres, comprises 16 species in South America. Ten of these are recorded from the austral areas of the continent, including Argentina, Bolivia, Brazil, Chile, Paraguay and Uruguay. A description of the genus is included, discussing and illustrating those particularly important characters such as pedicel apex, callus, lower glume, anthecium, or those representative in the delimitation of taxa, such as growth habit, inflorescence type, presence or absence of lower palea, length of spikelets, upper glume and lower lemma features, presence or absence of an awn in the upper floret. Additional characters were analyzed in order to complete a total picture of the genus: culm and rachis anatomy, salt glands, and micromorphological characteristics of anthecia, as well as habitat and geographical distribution. Finally, a description of those characters considered most important in the delimitation of taxa, is given for all species of the genus. On the basis of these characters North-, Meso-, South American and Australian species are compared.

Key words: *Eriochloa*, South America, morphology, biogeography.

INTRODUCTION

Eriochloa Kunth, a genus of the grass tribe Paniceae, which includes 25 to 30 species growing in tropical, subtropical and warm temperate regions of both hemispheres (Clayton and Renvoize 1986; Nicora and Rugolo 1987), is very well documented for North America by Shaw and Smeins (1981) and Shaw and Webster (1987). For South American species the information available is fragmentary, and limited to new taxa or species from regional floras. Renvoize's (1995) key to South American species did not include some taxa described from Argentina and/or Brazil, and provided little detail. *Eriochloa* comprises 16 species in South America, including 10 species recorded for the South American austral areas; Argentina, Bolivia, Brazil, Chile, Paraguay and Uruguay. These are:

E. boliviensis Renvoize, *E. distachya* Kunth, *E. gracilis* (E. Fourn.) Hitch., *E. grandiflora* (Trin.) Benth., *E. montevidensis*

Griseb., *E. nana* Arriaga, *E. polystachya* Kunth, *E. pseudoacrotricha* (Thell.) S.T. Blake, *E. punctata* (L.) Ham. and *E. tridentata* (Trin.) Kuhlm (Renvoize 1995; Arriaga 1994, 1995). This paper documents the austral South American species, their intraspecific relationships in this area and relationships with species from other areas.

The genus has some economic value as forage species in native pastures.

Generic Description

All species hermaphrodite, annual to perennial, caespitose or stoloniferous, erect or decumbent. Flowering culms terminated by a solitary paniculate inflorescence. Leaves not obviously differentiated, without auricles. Ligule usually a fringe of hairs with a minute membranous rim at the base. Nodes hairy or glabrous, internodes hollow. Inflorescences with primary branches situated on one side or alternating on both sides of the main axis. Rachis terminating in a spikelet. Pedicels disk-like and concave at the apex. Disarticulation at the spikelet base. Callus differentiated, not prolonged into a stipe, not flared to form a discoid receptacle, glabrous. Spikelets dorsally compressed, ovate-acute. Lower glume very reduced. Upper glume present, glabrous or hairy, muticous or awned. Lower lemma muticous, mucronate or awned. Palea of lower floret present or absent. Upper lemma cartilaginous, with or without an awn at the tip, minutely transversely rugose, palea of similar texture. Lodicules present, not fused, glabrous. Caryopsis not longitudinally grooved. Endosperm solid, containing only simple starch grains. Base chromosome number x=9.

Homogeneous leaf anatomy within the genus: first, second and third order vascular bundles (VB) alternating in transverse sections, no more than four cells apart. First and second order VB surrounded by two bundle sheaths: the inner one, a continuous sclerenchymous sheath; the outer, a Kranz sheath (PS type), formed by very large parenchyma specialized cells with slightly thickened walls and the outer tangential wall inflated. Specialized, large, centrifugal chloroplasts in Kranz sheath. Starch storage only in the outer sheath. First and second VB, frequently interrupted abaxially and occasionally adaxially also by sclerenchyma girders; marginal sclerenchyma also present. Chlorenchyma radiating conspicuously or inconspicuously, from VB. Big groups of fan-shaped bulliform cells present in the adaxial side of the leaf blade, over third order VB or between first or second order bundles.

Macrohairs present, with sunken bases, associated with bulliform cells on adaxial epidermis, or with colourless cells that form cushions around the hair bases. Inflated outer tangential cell wall on adaxial epidermis, giving a papillose aspect. Long cells with sinuous walls sometimes alternating with short cells, transversely long, or silico-suberous pairs on intercostal zones. Short suberous cells and siliceous cells of dumbbell shape, short or elongated, alternating on costal zones. Triangular to dome-shaped stomata in two rows along each side of the costal zones. Two-celled panicoid type microhairs present.

On the basis of leaf anatomy characters and $\delta^{13}C$ ratios (Shaw and Smeins 1981; Arriaga 1990) it may be inferred that the genus uses the C_4 photosynthetic pathway, with PEPck decarboxylating enzymes (Gutierrez *et al.* 1976).

METHODS

Transverse sections were made from both herbarium and fresh material. Sections were obtained either freehand or the material was embedded in wax and sectioned on a rotary microtome. The sections were stained with Alcian Blue/Safranin.

Fluorescence microscopy was used for sections of herbarium material. Using Acridin Orange and Methylene Blue as fluorochromes in simple fluorochrome techniques and Acridin Orange/calcofluor in a combined technique, it was possible to deduce the nature of the wall of the salt gland (Dizeo de Strittmatter 1986).

I also used fluorescence microscopy for identifying specialized chloroplasts from the Kranz sheath in culms, rachises, leaf, by differential autofluorescence (Elkin and Park 1975).

The nature of ions excreted by salt glands and microbodies included in the outer tangential walls of the warty papillae was analysed and measured by X-ray microanalysis. Scanning Electron Microscopy was also used with specimens coated with carbon and gold-paladium.

RESULTS

Generic Characters:
Pedicel apex: The zone of disarticulation of the spikelet is located between the spikelet base and the pedicel, and is favoured by the form of the apex of the pedicel. It enlarges here, by the overgrowth of parenchymatic tissue, presenting a flat disk-like form, concave at the apex (Fig. 1: A).

Callus: The callus, a cup-like structure at the spikelet base is the most significant diagnostic characteristic of the genus (Fig. 1: G-H). Following Thompson *et al.* (1990), I consider this unusual structure not derived from the first glume as previously interpreted. Anatomically the callus has a non-vascularized structure, formed by large parenchymatic cells, surrounding a column of tissue, that is continuous with the pedicel. The central column, with an amphicribal stele (3 bundles partially fused) in the middle, and the cup are fused only at the base of the cup (Fig. 1: B-C). In almost all species, the body of the callus bends inward on itself towards the apex of the cup, forming a false rim.

It appears that this unusual structure is formed by a proliferation of the ground tissue at the spikelet base (Fig. 1: B,C,E).

According to Davidse (1986), the cup functions as an ant-attractor, as lipids produced in elaiosomes located in the bead-like swelling are dropped into the cup. This might be the explanation of its presence, I did not research this aspect in this present study.

Lower glume: Present inside the cup-like base (Fig. 1: C), very reduced, resembling a scale; when bigger it may be observed coming out above the callus top (Fig. 1: D,F). It has the same epidermal characteristics as the upper glume.

Fertile anthecium: Upper lemma cartilaginous, minutely transversely rugose (Fig. 3: G), the margins not inrolled, covering the edges of the palea of similar texture (Fig. 1: H; Fig. 3: A). Germination lid conspicuous at the base of the palea (Fig. 1: G).

Diagnostic Specific Characters
Growth habit: annual in *E. boliviensis, E. gracilis, E. nana,* and perennial in *E. distachya, E. grandiflora, E. montevidensis, E. polystachya, E. pseudoacrotricha, E. punctata, E. tridentata.*

Inflorescence type: The inflorescence presents a wide diversification within the *Poaceae* family, with taxonomic value at the subfamilial, tribal and even genus level, but it is not usual to find this characteristic taxonomically useful at species level as in *Eriochloa* (Fig. 2).

There are two types of inflorescences within the genus: bilateral inflorescences, with primary paracladia (branches) alternating on both sides of the main axis (*E. gracilis, E. montevidensis, E. nana,*

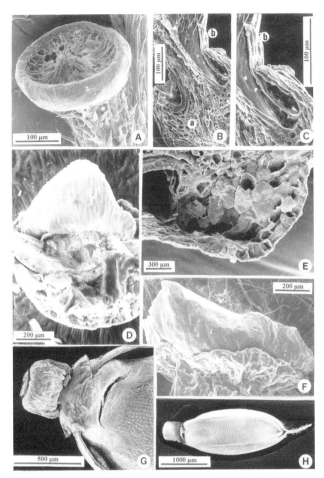

Fig. 1. Photomicrographs obtained by the use of SEM. A, *Eriochloa montevidensis* (BA 11266). Concave pedicel apex view. B, C, *E. punctata* (L. Parodi 5532, BAA). Longitudinal view of base of spikelet: a) callus; b) lower glume. D, *E. polystachya* (J. Valls 1123, CTES). Upper view of lower glume. E, *E. montevidensis* (BA 11266). Inner structure of callus. F,G,H, *E. punctata* (BAB 10902). F, Lateral view of lower glume. G, Germination lid at the base of the anthecia. H, Fertile anthecia.

E. polystachya, E. pseudoacrotricha, E. punctata and *E. tridentata*); or unilateral inflorescences, with primary paracladia situated only on one side of the principal axis (*E. boliviensis, E. distachya, E. grandiflora*). In all cases only one paracladium per node. A pattern of reduction in the number of paracladia, in the length of the paracladia and in the number of spikelets on them, may be inferred from extreme and intermediate cases. Following the general criterion in Poaceae of Vegetti and Anton (1995) the reductive process seems to have started in a 'panicle' and followed two ways: 1) a reduction in extension of the inflorescence, the number and length of primary branches, and in the number of spikelets, maintaining always the bilateral position of paracladia on the axis and spikelets on the racemes (Fig. 2); 2) a reduction of similar characteristics to (1) but accompanied by a total reduction (truncation) of paracladia on the axis and spikelets on paracladia to one side of the inflorescence, resulting in an unilateral position of elements (Fig. 2). Sometimes the inflorescence is reduced to a single paracladium (*E. distachya*) (Fig. 2).

Presence or absence of lower palea: Lower palea present (in *E. boliviensis, E. distachya, E. grandiflora, E. polystachya*) or absent

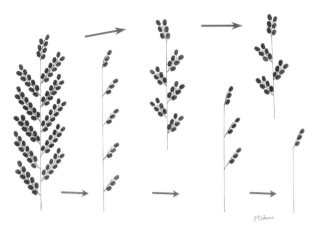

Fig. 2. Trends of reductive characteristics in *Eriochloa* inflorescences.

(*E. gracilis, E. montevidensis, E. nana, E. pseudoacrotricha, E. punctata, E. tridentata*); when present, fully developed (*E. boliviensis, E. grandiflora*) or vestigial (*E. distachya, E. polystachya*), in all cases hyaline (Fig. 3: A).

Length of spikelets: from 3 mm to 7(-8) mm, depending on the species: *E. boliviensis* 4-5 mm, *E. distachya* 4.5-5.5 mm, *E. gracilis* 4 mm, *E. grandiflora* 6-7 mm, *E. montevidensis* 3-3.7 (-4) mm, *E. nana* 4-5.5 mm, *E. polystachya* 3-3.7 mm, *E. punctata* 4-4.5 (-5) mm, *E. pseudoacrotricha* 6-7 (-8) mm, *E. tridentata* 4-5 mm.

Upper glume apex: It may be acute, acuminate, awned, muticous, subulate or with three little teeth. *E. montevidensis, E. polystachya, E. grandiflora* have acute, muticous apex (Fig. 3: F); *E. boliviensis, E. distachya* acute, awned (Fig. 3: A); *E. gracilis* subulate (Fig. 3: E); *E. tridentata* three terminal teeth, the central longest (Fig. 3: C); *E. nana* acute, awned (Fig. 3: B); *E. pseudoacrotricha* long-awned (Fig. 3: D); *E. punctata* acute to long-acute.

Lower lemma apex: It may be acute, acuminate, aristulate or muticous. Within the studied species *E. pseudoacrotricha* has acuminate, aristulate lower lemma apex; *E. nana* (Fig. 3: B) *E. tridentata, E. punctata* and *E. gracilis* have an acuminate, muticous one, while in *E. polystachya, E. grandiflora* (Fig. 3: F), *E. distachya, E. boliviensis* and *E. montevidensis* it is acute and muticous.

Upper glume/lower lemma relation: An important character to observe is the relation between the length of the upper glume and the lower lemma. In *E. montevidensis, E. polystachya, E. grandiflora* the upper glume and lower lemma are almost equal in length. In *E. pseudoacrotricha* the lower lemma is 50 to 60% as long as the upper glume. In *E. gracilis, E. punctata, E. tridentata, E. distachya, E. boliviensis, E. nana* the lower lemma is 80 to 90% as long as the upper glume (Fig. 4).

Anthecia/total spikelet length proportion: In *E. gracilis* and *E. pseudoacrotricha* the anthecia are almost 50% the length of the spikelet. In *E. montevidensis, E. polystachya* and *E. grandiflora* the anthecia are 85% or more the length of the spikelet. In *E. punctata, E. tridentata, E. distachya, E. boliviensis*, and *E. nana* the anthecia are almost 60 to 75% the length of the spikelet (Fig. 4).

Presence or absence of an awn in the upper floret: The upper floret may be muticous, mucronate, or awned. The upper lemma of

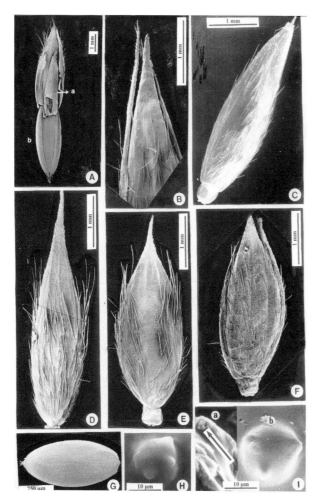

Fig. 3. Photomicrographs obtained by the use of SEM. A. *E. distachya* (Steinbach 1940, BA). Lower palea (a), fertile anthecia (b), B, *E. nana* (Petetin et al 1920, BAB). Upper glume and lower lemma acuminate apex. C, *E. tridentata* (Krapovickas et al 29419, CTES). Tri-dentate upper glume apex. D, *E. pseudoacrotricha* (Lahitte y Castro 47614, BAB). Long aristate glume apex. E, *E. gracilis* (A. Hunziker 10990, CORD). Subulate upper lemma apex. F, *E. grandiflora* (Hassler 2247,G). Upper glume and lower lemma acute apex. G, *E. gracilis* (Hitchcock 1574, LIL). Anthecia view. H, *E. grandiflora* (LIL 45978). Four verrucose papillae of anthecia. I, *E. punctata* (BA 31/880). a) Arrow points to the siliceous microbody inside the tangential cell wall of the papillae; b) three verrucose papilla of anthecia.

E. gracilis is mucronate or muticous with a rim of short stiff hairs (Fig. 3: G); *E. polystachya* is shortly mucronate; *E. distachya*, *E. boliviensis* and *E. grandiflora* muticous with a rim of short stiff hairs (Fig. 3: A); *E. montevidensis*, *E. nana*, *E. punctata*, *E. pseudoacrotricha* and *E. tridentata* have an awn at the top of the upper lemma (Fig. 1: H).

Anatomical Characters

Anatomy of the culm and rachis: Both culm and rachis show Kranz structure. At the top of the culm, below the inflorescence, where it is well exposed to the light, the maximum development of Kranz structure is found. There the Kranz parenchymatous sheath of the vascular bundles and chlorenchyma represents a continuous structure, surrounding the peripheric vascular bundles, all around the culm. No sclerenchyma is present in the subepidermal position. The continuous structure also occurs in the axis and ramifications (Arriaga 1990). The functionality of the

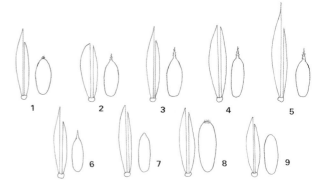

Fig. 4. Relations between upper glume, lower lemma and anthecium length. 1- *E. gracilis*; 2- *E. montevidensis*; 3- *E. punctata*; 4- *E. tridentata*; 5- *E. pseudoacrotricha*; 6- *E. nana*; 7- *E. polystachya*; 8- *E. grandiflora*; 9- *E. distachya*, *E. boliviensis*.

C_4 photosynthetic pathway in these culms was detected on the basis of the differential autofluorescence of the chloroplasts of the parenchymatous vascular bundle sheath (Kranz sheath) and those of the chlorenchyma; the starch accumulation in the Kranz sheath; and the $\delta^{13}C$ ratios of the stems of *E. punctata*, *E. pseudoacrotricha* and *E. montevidensis* (Arriaga 1990).

Salt glands: They occur on the culm, immediately below the inflorescence, on the rachis and ramifications of it and represent a new morphological type, different from the graminoid salt glands, resembling microhairs, previously described (Liphschitz and Waisel 1982; Oross and Thomson 1982; Amarasinghe and Watson 1988). Salt glands in *Eriochloa* are similar to the macrohairs usually present in this genus, but are longer (more than 700 µm), and consist of bicellular structures. These bicellular structures have a rounded basal cell, sunken into the chlorenchyma, and an elongated apical cell. The walls of both cells are heavily cutinized. There is a specialized tissue around the base of the basal cell, the endodermal tissue, with heavy cutinization on its walls. At the top of the apical cell, there is a collecting chamber between the cellulose layer of the wall and the cuticle, with only one large pore at the distal part, which opens when hydrostatic pressure increases within the collecting chamber.

These salt glands were first described in *E. montevidensis*, *E. pseudoacrotricha*, *E. punctata* (Arriaga 1992) and then confirmed in *E. distachya*, *E. tridentata*, *E. boliviensis*, *E. grandiflora*, *E. nana*.

X-ray microanalysis detected the elements present in the glands as Na, Mg, P, S, Cl and K, with K and Cl the dominant elements. An increase of the elements from the endodermal tissue to the cap cell, was also detected.

The siting of salt glands coincides with the zone of maximal development of Kranz structure (zone of maximal efficiency in photosynthesis), and would correspond to a need for high amounts of energy to transport symplastically and excrete salts.

Micromorphological characteristics of anthecia: The external epidermis of anthecia is characterised by narrow long cells with sinuous walls. Each one has a papilla, located excentrically. These papillae have three prominences in the external tangential wall (Fig. 3: Ib), corresponding to inorganic microbodies placed into the wall (Fig. 3: Ia), in almost all the species examined, with the

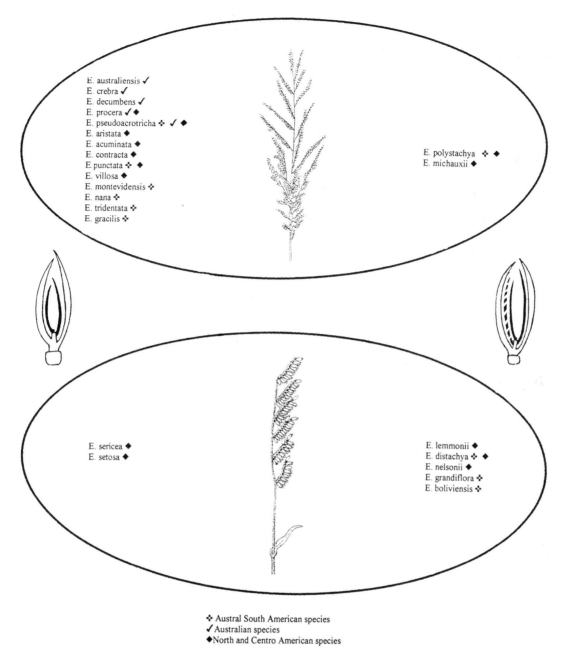

E. australiensis ✓
E. crebra ✓
E. decumbens ✓
E. procera ✓ ◆
E. pseudoacrotricha ✢ ✓ ◆
E. aristata ◆
E. acuminata ◆
E. contracta ◆
E. punctata ✢ ◆
E. villosa ◆
E. montevidensis ✢
E. nana ✢
E. tridentata ✢
E. gracilis ✢

E. polystachya ✢ ◆
E. michauxii ◆

E. sericea ◆
E. setosa ◆

E. lemmonii ◆
E. distachya ✢ ◆
E. nelsonii ◆
E. grandiflora ✢
E. boliviensis ✢

✢ Austral South American species
✓ Australian species
◆ North and Centro American species

Fig. 5. Comparison of Australian, North, Meso and South American species of the genus, taking into account inflorescence type and presence or absence of lower palea.

exception of *E. grandiflora* and *E. polystachya*, that show four or five prominences (Fig. 3: H) with the same origin. The siliceous nature of these microbodies was confirmed by energy dispersive X-ray microanalysis (Arriaga 1987). Taking into account the morphology of these structures (warts on the wall of a papillae), they are named warty papillae, and give a particular texture to upper lemma and palea (Fig. 1: H; Fig. 3: G).

Distribution and Biogeography:

E. boliviensis, swampy places in grasslands of Bolivia (Beni) and probably Brazil (Mato Grosso).

E. distachya, savannas and wet grasslands of Costa Rica, Guatemala, Panama, Venezuela, Colombia, Brazil, Peru, Bolivia, Paraguay.

E. gracilis, a native as a weed in ditches, fields and disturbed areas from southwestern USA and most of Mexico, and widespread in the southeastern US, introduced (probably accidentally) in Argentina (Salinas Grandes, Cordoba Province) where it was recorded only once near the railways.

E. grandiflora, wet grasslands or hill slopes in Bolivia, Brazil, Paraguay and Argentina (Misiones Province).

E. montevidensis, wet to swampy grasslands, sometimes on saline soils from Uruguay, Argentina and Paraguay.

E. nana, native in wet grasslands in Argentina (Corrientes and Santiago del Estero Provinces).

145

E. polystachya, coastal wet soils and grasslands in Cuba, Jamaica, Puerto Rico, Trinidad, Mexico, Costa Rica, Honduras, Guyana, Surinam, Venezuela, Colombia, Ecuador, Peru and Brazil, introduced in the south of US.

E. pseudoacrotricha, native of Australia, introduced as forage in Argentina, and considered naturalized (Corrientes, Chaco, La Rioja, Salta and San Luis Provinces).

E. punctata, weed in fields and wet grasslands and coastal areas or saline soils in Central America, Uruguay, Paraguay, Chile and widespread in temperate regions in Argentina.

E. tridentata, weed in fields and wet grasslands in Brazil (Mato Grosso) and Argentina (Corrientes, Chaco, Entre Rios, Formosa and Santa Fe Provinces).

Key to Austral South American Species:

1   Unilateral inflorescence type. Lower palea present.

2   Annual growth habit. Inflorescence 3-6 paracladia.
....*Eriochloa boliviensis*

2'- Perennial growth habit.

3-   Inflorescence 3-10-(14) paracladia. Spikelets 5-7 mm long, apex acute, muticous.....................................*E. grandiflora*

3'- Inflorescence 1-2-(3) paracladia. Spikelets 3.2-4.4 mm long, apex acute, aristate. ....*E. distachya*

1'- Bilateral inflorescence type.

4   Upper floret muticous or apiculate.

5   Upper glume acute. Lower palea present. Spikelet 3.2-3.7 mm long. ....*E. polystachya*

5'- Upper glume subulate. Lower palea absent. Spikelets 4 mm long. ....*E. gracilis*

4'- Upper floret aristate. Lower palea always absent.

6   Upper glume apex three dentate........................*E. tridentata*

6'   Upper glume acuminate or acute.

7   Upper glume long aristate. Spikelets 6-7-(8) mm long. ....
....*E. pseudoacrotricha*

7'   Upper glume not aristate.

8   Upper glume acute. Upper glume and lower lemma subequal in length. Anthecia 85-90% total length of spikelet.
....*E. montevidensis*

8'   Upper glume acuminate. Upper glume longer than lower lemma.

9   Plants less then 40 cm tall, monocaule, inflorescence conspicuously emergent..*E. nana*

9'   Plants more than 40 cm tall, pluricaule, inflorescence not conspicuously emergent. ....*E. punctata*

Relationships to other American and Australian species

A total of 16 species are recorded for North and Central America (Shaw and Webster 1987; Crins 1991; Davidse and Pohl 1992).

They are: *E. acuminata* (Presl) Kunth, *E. aristata* Vasey, *E. contracta* Hitchc., *E. distachya* Kunth, *E. fatmensis* (Hochst. and Steud.) W.D. Clayton, *E. lemmonii* Vasey and Scribner, *E. michauxii* (Poir.) Hitchc., *E. nelsonii* Scribner and Smith, *E. pacifica* Mez, *E. polystachya* Kunth, *E. pseudoacrotricha* (Thell.) S.T. Blake, *E. punctata* (L.) Ham., *E. sericea* (Scheele) Vasey, *E. setosa* (A. Rich.) Hitchc., *E. stevensii* Davidse, *E. villosa* (Thunb.) Kunth. There are also 16 species described for South America: *E. boliviensis*, *E. distachya*, *E. eggersii* Hitchc., *E. gracilis*, *E. grandiflora*, *E. montevidensis*, *E. multiflora* Renvoize, *E. nana*, *E. pacifica*, *E. peruviana* Mez, *E. polystachya*, *E. procera* (Retz.) Hubb., *E. pseudoacrotricha*, *E. punctata*, *E. tridentata* and *E. weberbaueri* Mez (Davidse and Pohl 1992; Renvoize 1995; Arriaga 1994, 1995). The five Australian species of Eriochloa are *E. pseudoacrotricha*, *E. decumbens* Bailey, *E. australiensis* Stapf., *E. crebra* S.T. Blake and *E. procera* (Webster 1987).

From all important infrageneric diagnostic characters, there are some that are very conspicuous: growth habit, number of inflorescences per plant, bilateral or unilateral inflorescence type, number of paracladia in each panicle, length of spikelet, shape of upper glume apex, presence or absence of lower palea, presence or absence of awn in upper lemma, habitat and geographical distribution. For comparative purposes I have recorded them for all species of the genus, compiled from personal observations and/or cited bibliography.

E. acuminata: annual, pluricaule, bilateral inflorescence type, 5-20 paracladia, spikelet length 3.8-5-(5.5) mm, upper glume apex acuminate or acute, lower palea absent, upper lemma mucronate; disturbed areas in south-east of USA and Mexico.

E. aristata: annual, pluricaule, bilateral inflorescence type, 16-30 paracladia, spikelet length 4-8 mm, upper glume apex acuminate, lower palea absent, upper lemma mucronate or aristate; south-east and central USA, Mexico, La Antigua, El Salvador, Guatemala, Colombia.

E. australiensis: annual (biennial), bilateral inflorescence type, 12-25 paracladia, spikelet length 5.7-15 mm, upper glume apex acuminate, lower palea absent, upper lemma aristate; arid zones of Australia.

E. contracta: annual, pluricaule, bilateral inflorescence type, 10-20-(28) paracladia, spikelet length 3.5-4.5 mm, upper glume apex acuminate, lower palea absent, upper lemma aristate; disturbed areas in south and central USA.

E. crebra: perennial, pluricaule, bilateral inflorescence type, spikelet length 3.9-5.5 mm, upper glume apex acute to acuminate, mucronate, lower palea absent, upper lemma aristate; wet and semiarid zones in Australia.

E. decumbens: perennial, pluricaule, bilateral inflorescence type, spikelet length 2.7-3.3 mm, upper glume apex acuminate, aristate, lower palea absent, upper lemma aristate; wet areas and coastal lands in Australia.

E. distachya: perennial, pluricaule, unilateral inflorescence type, 1-3 paracladia, spikelet length 3.2-4.4 mm, upper glume apex acuminate to acute, lower palea present, upper lemma muticous;

savannas and wet grasslands Costa Rica, Guatemala, Panama, Bolivia, Peru, Brazil, Colombia, Paraguay, Venezuela.

E. eggersii: annual, pluricaule, bilateral inflorescence type, 14-15 paracladia, spikelet length 4-7 mm, upper glume apex acuminate, lower palea absent, upper lemma muticous; Ecuador.

E. fatmensis: annual, pluricaule, bilateral inflorescence type, 3-10 paracladia, spikelet length 2.7-3.3 mm, upper glume apex acuminate, aristate, lower palea absent, upper lemma aristate; wet lands and grasslands in Africa, Arabia, India, introduced USA.

E. lemmonii: annual, pluricaule, unilateral inflorescence type, 3-8 paracladia, spikelet length 3.8-4.5 mm, upper glume apex acute, lower palea present, upper lemma muticous or apiculate; rocky soils south USA, Mexico.

E. michauxii: perennial, pluricaule, bilateral inflorescence type, 10-25 paracladia, spikelet length 4-5.5 mm, upper glume apex acute, muticous, lower palea present, upper lemma mucronate; wet lands and prairie south-east of USA.

E. multiflora: annual, pluricaule, bilateral inflorescence type, 40-80 paracladia, spikelet length 3-5 mm, upper glume apex acuminate, lower palea absent, upper lemma muticous or mucronate; wet lands in Ecuador.

E. nelsonii: annual, pluricaule, bilateral inflorescence type, 3-7-(10) paracladia, spikelet length 5-8 mm, upper glume apex acute to obtuse, lower palea vestigial or absent, upper lemma mucronate; disturbed lands, hill slopes in Mexico, Guatemala, Honduras, Nicaragua.

E. pacifica: annual, pluricaule, bilateral inflorescence type, 3-16 paracladia, spikelet length 5-7 mm, upper glume apex acuminate, lower palea absent, upper lemma aristate; wet lands, sandy maritime coasts in Ecuador, Peru, Galapagos Is., adventive in Nicaragua.

E. peruviana: annual, pluricaule, 3-17 paracladia, spikelet length 3 mm; Peru.

E. procera: perennial, pluricaule, bilateral inflorescence type, spikelet length 3-3.8 mm, upper glume apex acute, muticous, lower palea absent, upper lemma aristate; wet and semiarid, coastal lands in Australia.

E. sericea: perennial, pluricaule, bilateral inflorescence type, 4-8 paracladia, spikelet length 4-5 mm, upper glume apex acute muticous, lower palea absent, upper lemma mucronate; clay soils, prairie, south USA, Mexico.

E. setosa: perennial, pluricaule, bilateral inflorescence type, 1-6 paracladia, spikelet length 3-4 mm, upper glume apex acuminate, muticous or aristate, lower palea absent, upper lemma aristate; hills, savannas, sandy and disturbed areas in Cuba.

E. stevensii: annual, pluricaule, bilateral inflorescence type, 17-57 paracladia, spikelet length 3.4-5.5 mm, upper glume apex acuminate, lower palea absent, upper lemma apiculate; river sides and ditches, clay soils in Nicaragua, savannas in Ecuador, Peru and Venezuela.

E. villosa: annual, pluricaule, bilateral inflorescence type, spikelet length 2-8 mm, upper glume apex acute muticous, lower palea absent, upper lemma apiculate; disturbed lands in the east of Russia, North and South Korea, Japan and China, introduced in central south USA.

E. weberbaueri: perennial, pluricaule, 8-12 paracladia, spikelet length 2.5-3.5 mm, upper lemma awnless or mucronate; Ecuador, Peru.

Following criteria devised by Stebbins (1982) and Vegetti and Anton (1995) for determining the primitive or advanced states of the previous characters given we can see that perennial species, with a bilateral inflorescence type, reduced number and length of paracladia in the inflorescences, reduced number of spikelets per paracladia, aristate glume, presence of lower palea, aristate upper lemma, are considered primitive.

From these characters we find that all Australian species are included in the same group, with almost all primitive characters (except by the absence of lower palea). In this group there are also the majority of South, Central and North American species. However, within the Central and North American species it is shown that very few (*E. polystachya* and *E. michauxii*) maintain almost all characters in the primitive stage, and only two (*E. sericea* and *E. setosa*) evolved almost all characters to an advanced stage. Of the South American species, only one (*E. polystachya*) shows almost all characters primitive, none show all characters in advanced stage, the majority showing a mixture.

In Fig. 3 two conspicuous morphological characters, taxonomically important at species level, are compared according to their occurrence and geographical distribution.

Studying the micromorphology of the upper floret in these species, I found all of them possess papillae of excentrical position on epidermal cells, each one showing three warty siliceous prominences, with the exception of *E. polystachya* and *E. grandiflora* that show four to five warts on the papillae. Shaw and Smeins (1981) have also observed more than three prominences on papillae in North American *E. punctata*.

DISCUSSION

Panicoids are thought to have had a Gondwanan origin (Clayton 1981) and to have spread early throughout Gondwanaland where *Eriochloa* must have originated (Simon and Jacobs 1990). Taking into account the actual distribution of the species with most primitive characters (South America, Africa and Australia), and their complete absence from Europe, it can be inferred that the genus might have its origin during the mid- or early Upper Cretaceous, before the fragmentation of Gondwana. From there it spread to North Africa and Indo-China, radiating into Central and North America out from South America, later.

The existence of salt glands in species which at present occupy non-saline habitats indicates that *Eriochloa* probably originated as a halophyte and that, some time in the past, its ancestors occupied coastal or saline habitats (Liphschitz *et al.* 1974), migrating then to non-saline habitats. Such migration probably occurred not too long ago, as those plants still retain semisunken salt glands. Few species maintain their original coastal habitats but because of drastic climatic changes, others diversified adapting to savannas, wet lands, swampy soils and arid or semiarid conditions.

147

CONCLUSIONS

It is well established that the swollen callus base present on *Eriochloa* spikelets, might not be considered as a reduced and modified lower glume, on account of the presence of a very reduced lower glume inside the basal cup of the spikelet, sometimes appearing above the callus top.

The best characters to delimit species are: their inflorescence type, upper glume and lower lemma characteristics, presence or absence of a lower palea, spikelet and anthecia length relationships and presence or absence of an awn in the upper lemma. Anatomical characters from leaves and culms are not useful in the delimitation of species because of their homogeneity within the genus.

Although *Eriochloa* is not considered to be an obligate halophytic genus, plants of this genus sometimes live in saline environments or saline patches, sometimes cohabiting with euhalophytic genera. On the existence of salt glands it can be inferred that *Eriochloa* is a facultative halophytic genus, and an important candidate for economic utilization on saline environments.

I have proposed a Gondwanan origin for *Eriochloa*, probably from a halophytic ancestor.

It is clear that further work, on primitive and advanced characters of the genus, is needed. Future work includes a cladistic and biogeographical analysis of all the species to show the evolutionary trends followed by these taxa.

ACKNOWLEDGEMENTS

I am grateful to Dr Surrey Jacobs (Royal Botanic Gardens Sydney, Australia) and collaborators for all facilities and patience. My appreciation to Mrs Maria Dolores Montero (CONICET-MACN, Argentina) for technical assistance. Supported by funds of a PEI 0139/97 CONICET.

REFERENCES

Amarasinghe, V. and Watson, L. (1988). Comparative ultrastructure of microhairs in grasses. *The Botanical Journal of the Linnean Society* **98**, 303-319.

Arriaga, M.O. (1987). Interpretacion del ornamento del antecio de *Eriochloa* (Poaceae). *Boletin de la Sociedad Argentina de Botanica* **25**, 31-141.

Arriaga, M.O. (1990). Desarrollo de la estructura Kranz en tallo de especies de *Eriochloa* (Paniceae-Poaceae). *Boletin de la Sociedad Argentina de Botanica* **26**, 177-185.

Arriaga, M.O. (1992). Salt glands in flowering culms of *Eriochloa* species (Poaceae). Bothalia **22**, 111 - 117.

Arriaga, M.O. (1994). *Eriochloa*. In 'Flora del Paraguay-23. Gramineae V'. (Eds R. Spichiger and L. Ramella) pp. 153-159. (Conservatoire et Jardin botaniques, Ville de Geneve: Switzerland and Missouri Botanical Garden, St. Louis, USA.)

Arriaga, M.O. (1995). *Eriochloa*. In 'Flora Fanerogamica Argentina, fasc. 12'. (Ed. A. Hunziker) pp.41-45. (PROFLORA, CONICET: Argentina).

Clayton, W.D. (1981) Evolution and distribution of grasses. *Annals of the Missouri Botanical Garden* **68**, 5-14.

Clayton, W.D., and Renvoize, S.A. (1986). 'Genera Graminum.' (Royal Botanic Gardens: Kew.)

Crins, W.J. (1991). The genera of Paniceae (Gramineae: Panicoideae) in the Southeastern United States. *Journal of the Arnold Arboretum*, Suppl. **1**, 171-312.

Davidse, G. (1987). Fruit dispersal in the Poaceae. In 'International Grass Symposium' (Eds T.R. Soderstrom, K.W. Hilu, C.S. Campbell and M.E. Barkworth) pp. 143-155. (Smithsonian Institution Press: Washington.)

Davidse, G. and Pohl, R.H. (1992). New species of *Festuca*, *Sporobolus* and *Eriochloa* from Mesoamerica and South America. *Novon* **2**, 322-328.

Dizeo de Strittmatter, C. (1986). Uso de tecnicas de fluorescencias en materiales vegetales. *Parodiana* **4**, 213-220.

Elkin, L. and Park, R.B. (1975). Chloroplast fluorescence of C_4 plants. I. Detection with infrared color film. *Planta* (Berlin). **127**, 243-250.

Gutierrez, M.; Edwards, G.E. and Brown, W.V. (1976). PEP Carboxikinase containing species in the *Brachiaria* group of the subfamily *Panicoideae*. *Biochemic Systematics and Ecology* **4**, 47-49.

Liphschitz, N. and Waisel, Y. (1982). Adaptation of plants to saline environments: excretion and glandular structure. In 'Tasks for vegetation science' **2**. (Eds D.N. Sen and K.S. Raypurohit. (Junk: The Hague, Holland.)

Liphschitz, N., Shomer-Ilan, A., Eshel, A. and Waisel, Y. (1974). Salt glands on leaves of Rhodes Grass (*Chloris gayana* Kunth). *Annals of Botany* **38**, 459-462.

Nicora, E.G. and Rugolo, Z. (1987). 'Los generos de gramineas de America austral', pp. 611. (Edit. Hemisferio Sur: Argentina.)

Oross, J.W. and Thomson, W.W. (1982). The ultrastructure of the salt glands of *Cynodon* and *Distichlis* (Poaceae). *American Journal of Botany* **69**, 939-949.

Renvoize, S.A. (1995). Two new species of *Eriochloa* (Gramineae) from South America. *Kew Bulletin* **50**, 343-347.

Shaw, R.B. and Smeins, F.E. (1981). Some anatomical and morphological characteristics of the North American species of *Eriochloa* (Poaceae-Paniceae). *Botanical Gazette* **142**, 534-544.

Shaw, R.B. and Webster, R.D. (1987). The genus *Eriochloa* (Poaceae, Paniceae) in North and Central America. *Sida* **12**, 165-207.

Simon, B.K. and Jacobs, S.W.L. (1990). Gondwanan grasses in the Australian Flora. *Austrobaileya* **3**, 239-260.

Stebbins, G.L. (1982). Major trends of evolution in the Poaceae and their possible significance. In Estes, J.R., Tyrl, R.J. and Brunken, J.N. 'Grasses and grasslands. Systematics and Ecology'. (University Oklahoma Press: Norman.)

Thompson, R.A., Tyrl, R.J. and Estes, J.R. (1990). Comparative anatomy of the spikelet callus of *Eriochloa*, *Brachiaria* and *Urochloa* (Poaceae, Paniceae, Setariineae). *American Journal of Botany* **77**, 1463-1468.

Vegetti, A. and Anton, A.M.(1995). Some evolution trends in the inflorescence of *Poaceae Flora*. **190**, 225-228.

Webster, R.D. (1987). 'The Australian Paniceae (Poaceae)', (J. Cramer: Berlin, Stuttgart.)

MOLECULAR AND MORPHOLOGICAL EVOLUTION IN THE ANDROPOGONEAE

Elizabeth A. Kellogg

Department of Biology, University of Missouri-St. Louis, 8001 Natural Bridge Road,
St. Louis, Missouri 63121, USA.

Abstract

Many of the characters suggested to support the monophyly of the Andropogoneae are in fact characters shared with members of the Paniceae, including the spikelet structure, mode of male flower formation, photosynthetic pathway, inflorescence branching pattern and paired spikelets. Molecular data support the monophyly of a group including *Arundinella* plus Andropogoneae, and show that the lineages within Andropogoneae were established rapidly. Characters that correlate with the origin of the traditional Andropogoneae are presence of a disarticulating rachis and differentiation of the spikelets of a pair. Within the tribe, there is variation in presence of distinctive cells in the mesophyll, formation of branch complexes on the upper part of the culm (the anthotagma), timing of inflorescence branching, sex expression of spikelets, induration of glumes, and formation of awns. The different states of these characters do not correlate with the molecular phylogeny. Arundinelleae and the subtribes of Andropogoneae are polyphyletic, and should be abandoned.

Key words: Andropogoneae, maize, evolution, classification, grass.

INTRODUCTION

Evolution occurs by tinkering with structures and processes that already exist, such that descendants generally have simple modifications of, rather than dramatic departures from, characteristics of their immediate ancestors. Sometimes structures arise that permit rapid diversification of lineages, a pattern that is sometimes attributed to adaptation. In this paper I will show that features of the grasses of the tribe Andropogoneae can be viewed as modifications of those of their close relatives in the tribe Paniceae. I will then show that the tribe constitutes a sudden burst of morphological evolution, and will describe some of the characters that vary among the genera. Finally, I will speculate on the adaptive significance of the andropogonoid innovations and comment on the implications for classification.

I. KEY CHARACTERS OF PANICOIDEAE, RETAINED IN ANDROPOGONEAE

Andropogoneae and Paniceae are members of the monophyletic subfamily Panicoideae, itself part of the large and well-supported panicoid-arundinoid-chloridoid-centothecoid (PACC) clade of the Poaceae (Davis and Soreng 1993; Kellogg 1998; see also GPWG this volume). Robert Brown (1810, 1814) was the first to recognize a group corresponding closely to the current Panicoideae, including species with paired florets, the lower of which is reduced. Tateoka (1962) also noted that members of the subfamily all have simple starch grains in the endosperm. Kellogg and Campbell (1987) postulated that the spikelet and starch grain characters were uniquely derived and indicated monophyly of the subfamily.

Panicoideae are shown to be monophyletic by data from the chloroplast genes for the large subunit of ribulose 1,5-bisphosphate carboxylase/oxygenase (*rbc*L; Doebley *et al.* 1990; Barker *et al.* 1995; Duvall and Morton 1996), NADH-dehydrogenase, subunit F (*ndh*F; Clark *et al.* 1995), the ß" subunit of RNA polymerase II (*rpo*C2; Cummings *et al.* 1994), and ribosomal protein 4 (*rps4*; Nadot *et al.* 1994), and also by chloroplast restriction site variation (Davis and Soreng 1993). In addition, sequences of the nuclear genes for ribosomal RNA (rRNA; Hamby and Zimmer 1988), phytochrome B (*phyB*; Mathews and Sharrock 1996; Mathews *et al.* in press), and granule bound starch synthase I (GBSSI or *waxy*; Mason-Gamer *et al.* 1998) also indicate monophyly of the subfamily. Members of the Panicoideae also share a unique arrangement of the nuclear genome (Kellogg 1998). There is thus little doubt that the subfamily is monophyletic.

The oldest fossils that can be confidently identified as Panicoideae date to about 15 mya (Nambudiri *et al.* 1978; Whistler and Burbank 1992). This is consistent with molecular clock estimates that place the divergence of *Pennisetum* and maize at about 28 mya (Gaut and Doebley 1997), and of other members of the Panicoideae at 20-23 mya (Kellogg and Russo unpublished).

Spikelet Development

All panicoid grasses have two florets per spikelet, with the lower one male or sterile. Spikelet and early floret development are uniform throughout the subfamily (LeRoux and Kellogg 1999). Glumes initiate as ridges on the spikelet primordium, the outer one forming first (Fig. 1a). Next the lemmas initiate, followed by the three stamen primordia and the gynoecial primordium. The palea and lodicules form later. The gynoecium forms a wall or ridge around the developing nucellus. At this stage the anthers are about the same length as the gynoecium (Fig. 1b), and in awned species the lemma awn generally extends early in development (Fig. 1c). The gynoecial ridge then extends upward to form styles and stigmatic areas, and the anthers continue to elongate (Fig. 1d). Flowers that will become pistillate initiate both stamens and gynoecium, but the stamens cease developing after anther thecae form (Fig. 1e). Flowers that will become male also initiate both stamens and a gynoecium, but the gynoecium ceases developing when the gynoecial ridge clearly surrounds the nucellus (Fig. 1f). This correlates with death of the subepidermal cells in the gynoecium, as seen by loss of cytoplasm and nuclei (Fig. 2; LeRoux and Kellogg 1999).

Both the development and histology of staminate flower formation are unique to the Panicoideae. The distantly related oryzoid grass, *Zizania aquatica*, has staminate flowers, but these appear to be derived independently from those in the Panicoideae. In *Z. aquatica*, the gynoecium develops stylar arms before development ceases, thus progressing well beyond the stage of the gynoecium in male flowers in Panicoideae (Zaitchik, LeRoux, and Kellogg unpublished data). Although cells in gynoecia of *Z. aquatica* do eventually die, their death is not accompanied by loss of nuclei or breakdown of DNA. At the same time, a darkly staining substance is deposited in many cells of the nucellus. These histological differences suggest that either the mechanism of gynoecial abortion is entirely different in oryzoids and panicoids, or that a common mechanism is deployed at different developmental times and in different ways.

In maize, gynoecial abortion and cell death are controlled directly or indirectly by the product of the gene *Tasselseed2* (*Ts2*). Mutations in *Ts2* lead to the formation of seeds in the normally male flowers of the tassel (Dellaporta and Calderon-Urrea 1993, 1994; DeLong *et al.* 1993). In normal staminate flowers of maize, Calderon-Urrea and Dellaporta (1999) have

observed that expression of *Ts2* correlates with a pattern of cell death similar to that seen in other Andropogoneae. In addition, Li *et al.* (1997) have shown that the ortholog of *Ts2*, *gynomonoecious sex form* (*gsf*), controls the formation of staminate flowers in *Tripsacum dactyloides*, again causing cell death in the gynoecium. Controlled cell death in the subepidermal layers of the gynoecium is apparently the histological signature for *Ts2*. We therefore suspect that *Ts2* controls formation of male flowers throughout the Panicoideae, and may be the genetic basis of the subfamilial synapomorphy.

Photosynthetic Pathway

Many panicoid genera use the C_4 photosynthetic pathway, with some using NADP-malic enzyme (NADP-ME) to decarboxylate the four carbon compound, some NAD-malic enzyme (NAD-ME) and some phosphoenolpyruvate carboxykinase (PCK; Hattersley and Watson 1992). These biochemical subtypes correlate with leaf anatomical characteristics. The Andropogoneae are all C_4 NADP-ME and, like other NADP-ME panicoids, have a single bundle sheath. This sheath is derived from procambial tissue and is developmentally like the mestome sheath (the inner bundle sheath) of other grasses (Dengler *et al.* 1985). The single bundle sheath is the site of photosynthetic carbon reduction, and therefore expresses RuBisCO (Langdale *et al.* 1988; Ueno 1992; Sinha and Kellogg 1996); the bundle sheath plastids are agranal, reflecting a reduction in the amount of light harvesting chlorophyll a,b binding proteins (Sheen and Bogorad 1986; Sinha and Kellogg 1996).

The histological and physiological similarities among all C_4 NADP-ME panicoids, including Andropogoneae, suggest that all the NADP-ME species may be descended from a common ancestor. This has not yet been tested rigorously, although preliminary data on chloroplast gene sequences suggest that all C_4 NADP-ME taxa could form a single clade (Gómez-Martínez this volume). The physiology of the Andropogoneae, therefore, may be ancestral and shared with their panicoid sister taxa. This also means that the characteristic andropogonoid leaf anatomy - closely spaced veins with a single bundle sheath - is shared with its panicoid ancestors.

Inflorescences with Multiple Orthostichies

The female inflorescence of maize is considered to be unusual in the grasses and has presented a problem for those studying the evolution of the maize inflorescence because the spikelets of the maize ear are borne in multiple orthostichies, rather than two (distichous; e.g., Doebley 1995). Developmental studies of other Andropogoneae and Paniceae, however, have shown that many Andropogoneae and Paniceae have non-distichous inflorescences and this represents the ancestral condition for the Andropogoneae (LeRoux and Kellogg unpublished data). Although the phyllotaxy in early development is not a perfect spiral, there are clearly multiple orthostichies. Sampling of this character outside Andropogoneae is limited to *Setaria* P. Beauv., *Panicum* L., and *Paspalum* L. (LeRoux and Kellogg unpublished; Rua and Weberling 1995). In Andropogoneae the ear and central axis of the tassel of maize are known to exhibit spiral phyllotaxis, and we have also observed multiple orthostichies in branching patterns of *Chrysopogon* Trin., *Capillipedium* Stapf, *Bothriochloa* Kuntze,

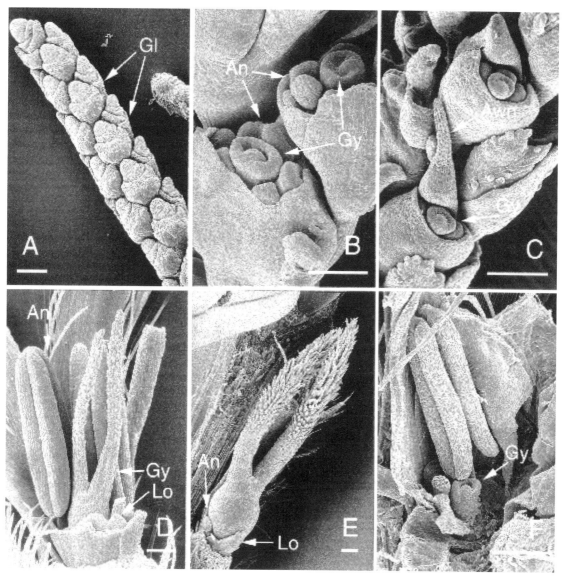

Fig. 1. Floral development of members of the Andropogoneae. Gl = glume; An = anther; Gy = gynoecium; Lo= lodicule. Scale bars=100μm. A. *Coelorachis selloana* showing initiation of glumes and lemmas, well before formation of any floral primordia. B. A sessile and pedicellate spikelet of *Bothriochloa bladhii*, both with anthers and a gynoecium forming a ridge around the nucellus. The gynoecium in the pedicellate spikelet will abort. C. Early development of *Heteropogon contortus* showing formation of awns on lemmas of pistillate florets. D. Hermaphrodite floret of *Bothriochloa bladhii*. E. Pistillate floret of *Heteropogon contortus*. F. Staminate floret of *Hyparrhenia hirta*. Note that gynoecium has ceased to develop at gynoecial ridge stage.

Dichanthium Willemeet, *Andropogon* L., and *Sorghum* Moench. (Note that branching pattern is not to be confused with spikelet arrangement.) Spiral phyllotaxis is also found in rice (Itoh *et al*. 1998), which suggests that this may be the ancestral condition in most, if not all, grasses.

Paired Spikelets

Andropogoneae are usually defined as having spikelets paired, but this characteristic appears in many other panicoid genera as well (Watson and Dallwitz 1992). In early development, a single primordium enlarges and then divides to give rise to a pair of spikelets, one sessile, or nearly so, and the other pedicellate (Fig. 3).

In comparing taxa with unpaired spikelets to those with paired spikelets, it is meaningless to ask whether the unpaired spikelet is homologous to the sessile or the pedicellate spikelet of the pair. The primordium divides and its products differentiate, so neither member of the pair is developmentally the same as the ancestral condition. Spikelet pairing is thus potentially synapomorphic for a large group of Paniceae including Andropogoneae.

II. ANDROPOGONEAE ARE A SINGLE RAPID RADIATION

The Andropogoneae are monophyletic, based on data from morphology (Kellogg and Watson 1993), *ndh*F (Spangler *et al*. 1999), GBSSI (Mason-Gamer *et al*. 1998), and phytochrome B (Mathews *et al*. in press). All data show that *Arundinella* Raddi is sister to Andropogoneae, and the support for the monophyly of the *Arundinella*/Andropogoneae clade is even stronger than for the monophyly of the Andropogoneae itself.

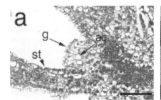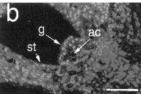

Fig. 2. Section through the gynoecium of a staminate floret of *Heteropogon contortus* at the stage of gynoecial abortion. g = aborting gynoecium; st = stamen; ac = aborting cell . A. Toluidine blue stained section, photographed with transmitted light. Note loss of cytoplasm in subepidermal cells. B. DAPI-stained section, photographed with reflected UV light. Note loss of nuclei in gynoecial cells.

The three gene trees differ somewhat in their sample of taxa, yet they all lead to similar conclusions. The GBSSI tree is shown in Fig. 4. All three molecules exhibit enough variability to distinguish genera and congeneric species, but branches are very short along the 'backbone' of the tree. These short branches do not reflect homoplasy or ambiguity, but rather reflect a small number of highly consistent mutations. When data from multiple genes are combined in a single analysis, the general pattern does not change (not shown). This pattern appears in three different genes, reflecting the history of both the chloroplast and the nuclear genomes, so we infer that the tribe diversified over a very short period of time. Similar patterns in other groups (e.g. columbines, Hodges and Arnold 1994) have led authors to suggest that the radiation must reflect a 'key innovation' that permitted diversification. This possibility will be discussed at the end of this paper.

Despite uncertainty in the order of branching, certain major clades appear in all gene trees. The genera *Chrysopogon*, *Cymbopogon* Sprengel and *Zea* L. are monophyletic. *Schizachyrium scoparium* (Michx.) Nees and *Hyparrhenia hirta* (L.) Stapf are sisters. *Cymbopogon* and *Hyparrhenia* Andersson ex Fournier are not closely related, despite similar inflorescence morphology. *Bothriochloa*, *Capillipedium*, and *Dichanthium* form a clade, which is not surprising given their complex relationships and interfertility (DeWet and Harlan 1970). *Ischaemum* L., *Coelorachis* Brongn., and *Coix* L. are not consistently placed with any other clade in any of the gene trees.

There are many differences among the gene trees, but these are generally poorly supported. For example, the GBSSI phylogeny in Fig. 4 shows that a few base pairs link *Andropogon gerardii* Vitman with *Zea*, but *ndh*F places *A. gerardii* with *Schizachyrium* Nees (Spangler this volume; Spangler *et al.* 1999). *Sorghum bicolor* (L.) Moench is placed with *Chrysopogon* by GBSSI (Fig. 4), but the two genera are only distantly related in *ndh*F (Spangler this volume; Spangler *et al.* 1999) or *phyB* (Mathews and Kellogg, unpublished data) phylogenies. Even among GBSSI phylogenies the tree topology varies depending on which taxa are included as outgroups. The poorly supported portions of the tree in Fig. 4 show relationships different from those found by Mason-Gamer *et al.* (1998), who used a larger set of outgroups. These differences are almost certainly caused by differing resolution of the very small number of characters linking major clades.

The short branches along the backbone of the tree suggest that the early evolution of the Andropogoneae proceeded rapidly and

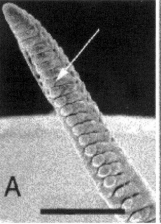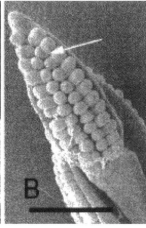

Fig. 3. Formation of spikelet pairs in a member of the Andropogoneae (*Tripsacum dactyloides*, A) and a member of the Paniceae (*Paspalum* sp., B). Spikelet pair primordia begin as a single primordium and then divide (arrows). Scale bar = I mm.

left little evidence of the order of events. This means that there is little evidence for or against any supergeneric groups. In their work using *ndh*F, Spangler *et al.* (1999) constrained the analysis to create a monophyletic Maydeae (i.e., to put *Coix* with *Zea*, *Tripsacum* L., and *Chionachne* R. Br.). The constrained trees were not significantly different from unconstrained trees. This means that we cannot reject the hypothesis that the traditional Maydeae form a clade. Similarly, trees constrained to place the awned taxa together were not significantly different from unconstrained trees.

As discussed above, many of the putative morphological synapomorphies of the Andropogoneae - C_4 NADP-ME photosynthesis, a single bundle sheath, spiral phyllotaxy, and spikelet pairing - are actually shared with other Panicoideae, and thus represent symplesiomorphies. There are no obvious characters correlating with the rapid burst of cladogenesis seen in the Andropogoneae. In many, but not all, members of the tribe the sessile and pedicellate spikelets differ in sex expression, glume morphology, and presence or absence of an awn, and this capacity to differentiate is not seen in the Paniceae. The glumes of many taxa are indurate, but this too is far from universal (see below). Many taxa have a disarticulating rachis, which gives rise to the spikelet pair as the dispersal unit. This may be responsible for the world-wide distribution of the group, but does not by itself explain the rapid radiation.

III. Morphological Variation in Andropogoneae

Leaf Anatomy; Distinctive Cells

In its mesophyll, the genus *Arundinella* produces cells that form longitudinal files parallel to and between the veins. These cells appear circular in cross section and have abundant chloroplasts. Their unusual appearance led some workers to call them 'distinctive cells' (Tateoka 1956a, 1956b, 1958, 1963). Immunolocalization has shown that 'distinctive cells' are physiologically equivalent to bundle sheath cells, express RuBisCO, and are the site of carbon reduction (Hattersley *et al.* 1977; Sinha and Kellogg 1996). They connect eventually with vascular tissue (Dengler and Dengler 1990).

'Distinctive cells' function like bundle sheath cells and keep the distance between mesophyll cells and bundle sheath cells low, as necessitated by the C_4 pathway (Dengler *et al.* 1990). They were thought by some to represent a step on the way to developing C_4 photosynthesis. Phylogenetic data, however, show that they represent losses rather than gains, as hypothesized earlier by Brown (1977). They appear not only in *Arundinella*, but also in *Garnotia* Brongn. (Tateoka 1958), *Loudetia* Hochst. & Steud. (Brown 1977), *Trichopteryx* Nees (Brown 1977), *Arthraxon* P. Beauv. (Ueno 1995), and *Microstegium vimineum* (Trin.) A. Camus (Ueno 1995). *ndh*F data (Spangler this volume; Spangler *et al.* 1999) place the latter species near the base of the Andropogoneae but quite separate from *Arundinella*, indicating that distinctive cells have appeared in parallel multiple times. In addition, production of distinctive cells is reportedly variable within *Microstegium* Nees (Ueno 1995). We do not know whether their production is environmentally influenced, or can be affected easily by selection.

Branch Complexes on the Upper Culm

The term inflorescence is ambiguous in Andropogoneae. The stem can form branches from the lowest nodes, in which case they are considered tillers; the next several nodes generally produce only a single axillary bud and this rarely develops, and then the upper nodes on the plant can produce branches that will terminate in inflorescences. Hagemann (1990) calls this upper part of the plant the anthotagma, the region in which branches terminate in inflorescences. In the taxonomic literature, this entire upper region is called the inflorescence, and then the terminal spikelet-bearing portion the raceme. Here I reserve the term inflorescence for the structure subtended by a single leaf, or terminal on the main axis.

In the anthotagma, genera of Andropogoneae vary as to the number of branches produced at a node. Some genera, such as *Chrysopogon* or *Sorghum*, produce only a single axillary bud or branch (Kellogg, unpublished observations), whereas others, such as *Hyparrhenia*, *Schizachyrium*, or *Coelorachis*, produce large branch complexes. These complexes constitute a series of branches and their prophylls, and have been described by many authors (Jacques-Félix 1961; Clayton 1969; Campbell 1983; Vegetti and Tivano 1991, among others). Only the leaf subtending the first branch of the complex develops; other leaves do not form. A similar branch complex may also form on the first or second branch, thus reiterating the branching pattern and leading to the complex broom-like 'inflorescences' (i.e. anthotagma) of some Andropogoneae. This character changes repeatedly in evolutionary time, as seen when it is mapped on a molecular phylogeny (Fig. 4).

Inflorescence Branching

Timing of branching varies considerably in terminal, spikelet-pair-bearing branches. This will be described in detail by LeRoux and Kellogg (in preparation), and will only be summarized here. In many Andropogoneae, inflorescence branches initiate early in development and all differentiate and elongate simultaneously. In a few taxa, however, the central axis elongates and spikelets develop extensively before branching occurs at the base of the main axis. These branches are not in the same orthostichies as the spikelet pairs, so the entire inflorescence is polystichous. Despite their late appearance developmentally, when the branches do elongate they rapidly catch up to the main axis. In still other taxa, branches do not form at all.

Inflorescences can thus be described as branched or unbranched and, if branched, branching early or branching late. When these characters are mapped onto a phylogeny it is clear that they change repeatedly in evolutionary time (Fig. 4). I speculate that the genetic control of inflorescence branching is easily modified by selection, such that small changes in gene action result in shifts between structures classed as panicles, digitate racemes, or spike-like racemes. Multiple genes in maize, such as *barren inflorescence*, the *ramosa* genes, *tasselseed 4*, and *tasselseed 6*, all affect inflorescence branching (Neuffer *et al.* 1997; Irish 1998) and might be candidates for genes controlling morphological change in evolutionary time.

Sex Expression Based on Floret and Spikelet Position

In the great majority of Andropogoneae, only the distal floret of the spikelet is fertile; the lemma of the proximal floret forms, but other organs never initiate. This appears to be the derived condition in Andropogoneae. Among all the early-diverging lineages, spikelets are bifloral. In bifloral spikelets, the lower floret is staminate, except in the ear (pistillate inflorescence) of maize, in which the lower floret develops as pistillate, but ultimately aborts.

In most cases, the distal floret of the sessile spikelet is hermaphrodite, whereas the distal floret of the pedicellate spikelet is staminate. There are some notable exceptions, however. For example, the pedicellate spikelet is hermaphrodite and the sessile spikelet is staminate in *Agenium* Nees and *Homozeugos* Stapf. In some taxa such as *Coelorachis*, both spikelets produce hermaphrodite flowers.

It is common for the pedicellate spikelet to become reduced or to abort entirely. This often varies among members of the same species, so that some plants will have fully developed, staminate pedicellate spikelets, whereas others will have only sterile pedicels.

Glume Induration

Andropogoneae are often described as having indurate glumes, contrasting with the hyaline glumes of Paniceae. The most strikingly indurated glumes appear in pistillate spikelets of species of *Zea*, in which induration is created by an unusually thick layer of lignin. Dorweiler *et al.* (1993) and Dorweiler and Doebley (1997) have shown that this is due to the action of a single allele of the gene *Teosinte glume architecture* (*Tga*). *Tga* affects only the glumes of the pistillate spikelets. The staminate spikelets have ordinary membranous glumes. We have attempted to determine the distribution of lignin in glumes of other Andropogoneae and find that the majority of species have glumes that look like those of tassels of *Zea* (Bradford and Kellogg unpublished observations). Glumes were sectioned free-hand, and then viewed with UV light. Under these conditions, lignin fluoresces. In the glumes of most Andropogoneae examined, a single lignified layer appeared in the center of the glume (Bradford and Kellogg, unpublished observations). There is variation in number of non-lignified cell layers, with rather more in *Phacelurus* Griseb. than other taxa examined, but no species had glumes approaching the

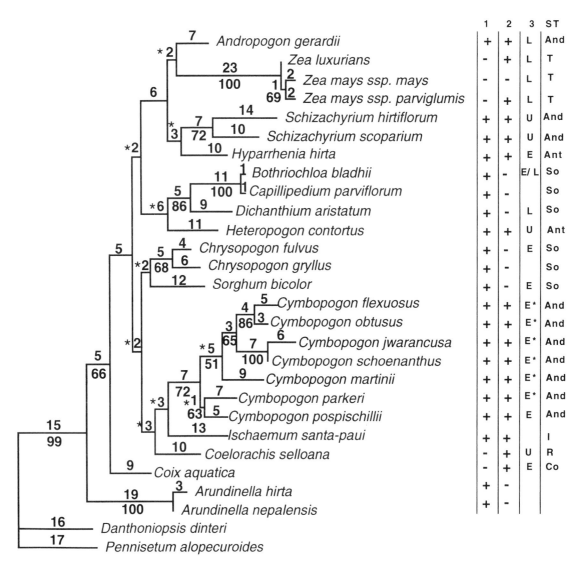

Fig. 4. One of 153 trees produced from phylogenetic analysis of granule bound starch synthase I (GBSSI or *waxy*) based on sequences from Mason-Gamer *et al.* (1998). Length = 344, CI=0.488, RI=0.599 with uninformative characters excluded. Numbers above branches indicate branch lengths, below branches are bootstrap values. Branches with asterisks do not appear in the strict consensus of all trees. Columns at the right of the phylogeny show the distribution of 1, awns (+ = present); 2, proliferation of axillary branches (+ = present); 3, timing of inflorescence branch elongation (L = late, E = early, U = unbranched, E* = inferred early based on adult morphology); ST = subtribe according to Clayton and Renvoize (1986).

thickness of the lignified layer in pistillate spikelets of teosinte. Comparative studies with Paniceae will be necessary to determine if the small line of lignin observed in glumes of Andropogoneae is unique or shared with other Panicoideae.

Awns

Awns form early in development on the lemmas of many Andropogoneae. In most taxa with awns, the body of the lemma barely develops and instead is a hyaline membrane at the base of the awn. This contrasts with the situation in many other awned grasses in which the body of the lemma develops into a relatively large, often photosynthetic structure. Awns, when present, generally form only on lemmas of pistillate or hermaphrodite florets, suggesting that sex expression and awn production may be genetically linked in

the tribe. This also means that the genetic signals that differentiate staminate and hermaphrodite spikelets must occur long before formation or abortion of floral organ primordia.

Andropogoneae were divided informally by Clayton (1972, 1973) into the 'awned' and 'awnless' taxa. Although a morphological cladistic analysis suggested that these groups might represent separate clades (Kellogg and Watson 1993), this has not been supported by molecular data (Fig. 4; see also Mason-Gamer *et al.* 1998; Spangler this volume; Spangler *et al.*1999).

IV. CLASSIFICATION

Andropogoneae are strongly supported as monophyletic only if *Arundinella* is included, a result found in three different molecular phylogenetic studies (Spangler *et al.* 1999; Mathews and Kellogg unpublished; Mason-Gamer *et al.* 1998). One recommendation of this paper therefore, is that *Arundinella* be henceforth considered a member of Andropogoneae. This will not necessitate much emendation to descriptions of the tribe, as *Arundinella* also has paired spikelets and firm glumes although, unlike most Andropogoneae, it has a tough rachis.

Arundinelleae, on the other hand, are unquestionably polyphyletic. *Danthoniopsis* Stapf, conventionally placed in Arundinelleae, is only distantly related to *Arundinella* and Andropogoneae and, in some molecular data sets, is sister to all remaining Panicoideae (see GPWG this volume). Leaf cross sections of *Danthoniopsis dinteri* (Pilg.) C. E. Hubb. reveal a conventional parenchymatous bundle sheath and mestome sheath, as is typical of an NAD-ME-type C_4 member of Panicoideae (N. Dengler unpublished observations). *Rattraya* J. B. Phipps, which Clayton and Renvoize (1986) synonymize with *Danthoniopsis*, is more closely related to members of the Paniceae (Spangler *et al.* 1999; Spangler, this volume). *Tristachya* Nees is placed by *rbc*L data as sister to *Hyparrhenia* (Barker *et al.* 1995), and thus is included in Andropogoneae.

Although we still lack phylogenetic data on eight members of Arundinelleae, the fact that *Danthoniopsis*, *Rattraya*, and *Tristachya* are unrelated to each other and to *Arundinella* suggests that the tribe should simply be abandoned. We therefore recommend that *Arundinella* and *Tristachya* be placed in the Andropogoneae, *Danthoniopsis* and *Rattraya* in the Paniceae, and the remaining genera *incertae sedis*.

None of the subtribes of Andropogoneae, most of which date to the early nineteenth century, is monophyletic. This was first noted in morphological cladistic studies (Kellogg and Watson 1993), and is further supported by the molecular phylogenies available to date. Until there is a comprehensive and robust phylogeny for Andropogoneae, subtribal classification should perhaps be avoided. Specific comments on individual subtribes follow.

Saccharinae Griseb.

The *Saccharum* L. group itself, including *Miscanthus* Andersson and *Saccharum*, is probably monophyletic (Sobral *et al.* 1994; Mathews and Kellogg unpublished). No molecular data are available on other members of the subtribe, except that the genus *Microstegium* may be polyphyletic. *ndh*F sequences place *M. vimineum* in a clade with *Chrysopogon*, whereas *M. nudum* (Trin.) A. Camus is closer to *Miscanthus* and *Sorghum* (Spangler *et al.* 1999; Spangler this volume). This needs to be tested by broader sampling. *M. nudum* has been segregated as the genus *Leptatherum* Nees.

Germainiinae Clayton

The Germainiinae is polyphyletic based on morphological cladistic studies (Kellogg and Watson 1993), which place *Germainia* Balansa & Poitr. in a clade with *Euclasta* Franch. and *Spathia* Ewart, *Trachypogon* Nees with *Homozeugos* and *Agenium*, and *Apocopis* Nees with *Apluda* L. The monotypic Dimeriinae Hackel includes only *Dimeria* R. Br.; morphological data place it

sister to *Cleistachne* Bentham, with both genera included in Saccharinae. No molecular data are available. Similarly, the monotypic Coicinae Reichenb. includes only *Coix*, and is placed with *Zea* and *Tripsacum* in morphological phylogenies (Kellogg and Watson 1993), with the 'core Andropogoneae' by *ndh*F (Spangler *et al.* 1999), and sister to all other Andropogoneae by GBSSI (Mason-Gamer *et al.* 1998; also Fig. 4).

Sorghinae Bluff

The Sorghinae are definitely polyphyletic based on sequences of *ndh*F (Spangler *et al.* 1999; Spangler this volume), GBSSI (Mason-Gamer *et al.* 1998), phytochrome B (Mathews and Kellogg unpublished), and morphology. All analyses place *Sorghum* separate from *Sorghastrum* Nash, *Dichanthium*, *Bothriochloa* and *Capillipedium*. For additional discussion of this subtribe, see Spangler (this volume).

Ischaeminae Presl

The Ischaeminae is polyphyletic. Morphological data place *Ischaemum* and its sister *Digastrium* (Hack.) A. Camus sister to many genera of Rottboelliinae (Kellogg and Watson 1993) whereas *Apluda* is sister to *Apocopis*. The *ndh*F phylogeny (Spangler *et al.* 1999; Spangler, this volume) also places *Ischaemum* as an early diverging branch, whereas *Apluda* is part of the 'core' Andropogoneae.

Andropogoninae Presl, Anthistiriinae Presl, Rottboelliinae Presl

The Andropogoninae, Anthistiriinae, and Rottboelliinae are all polyphyletic, as demonstrated by morphology (Kellogg and Watson 1993), *ndh*F data (Spangler *et al.* 1999; Spangler this volume), and GBSSI (Fig. 4; Mason-Gamer *et al.* 1998).

Tripsacinae Dumort

The Tripsacinae, including only *Zea* and *Tripsacum*, is monophyletic in all molecular studies to date, including *ndh*F (Spangler *et al.* 1999; Spangler this volume), *phy*B (Mathews and Kellogg unpublished data), and morphology (Kellogg and Watson 1993). The inflorescence structure of the two genera is unique, in that pistillate spikelets are unpaired in all wild species. This occurs because of suppression of the pedicellate spikelet (Sundberg and Orr 1986, 1990; Kellogg and Huang unpublished observations).

Chionachninae Clayton

Chionachninae, including *Chionachne*, *Polytoca* R. Br., *Sclerachne* R. Br., and *Trilobachne* Henrad is apparently polyphyletic based on studies of morphology (Kellogg and Watson 1993), but molecular data are available only on *Chionachne*.

Clayton (1973, 1981) and Clayton and Renvoize (1986) observe the considerable differences among the monoecious members of Andropogoneae, and argue convincingly against placing them together in the Maydeae as had been done by many authors, including Watson and Dallwitz (1992). Clayton notes in particular the peculiar inflorescence structure of *Coix* and of *Chionachne* which is distinct from that of *Zea* or *Tripsacum*. Nonetheless, phylogenetic analyses cannot rule out the possibility that *Zea*, *Tripsacum*, *Coix* and *Chionachne* form a clade. Although *Coix* does not appear with the other three in the shortest trees, when the four genera are forced together, the trees are not significantly longer by any of several statistical tests (Spangler *et al.* 1999).

To summarize, the following are recommendations for classification:

1. Andropogoneae be emended to include *Arundinella* and *Tristachya*.

2. Arundinelleae be abandoned, the genera *Danthoniopsis* and *Rattraya* placed in Paniceae, and the genera *Loudetia*, *Loudetiopsis* Conert, *Trichopteryx* Nees, *Gilgiochloa* Pilg., *Jansenella* Bor, *Chandrasekharania* Nair, Ramachandran & Sreekumar, *Zonotriche* (C. E. Hubb.) Phipps, and *Garnotia* be considered panicoid *incertae sedis*.

3. Subtribes of Andropogoneae be abandoned until a robust phylogeny of the tribe can be achieved. If, as we suspect, the major lineages of the tribe were established rapidly, a fully resolved and well-supported phylogeny may not be discernable. In this case, designation of subtribes may be arbitrary.

V. Speculations on Radiation and Adaptation

'...the exuberant display of form in Andropogoneae cries out for some explanation of its function' (Clayton 1987).

The apparently rapid radiation of Andropogoneae leads one to wonder if there might be some adaptive explanation, as indicated by the preceding quote from Clayton. Such adaptations are not as obvious in wind-pollinated species as in insect-pollinated ones. However, it is possible that Andropogoneae have developed an unusually flexible system for deploying florets of different sexes and this allows a rapid response to selection for different allocation to male and female functions. McKone *et al.* (1998) have shown that *Andropogon gerardii* and *Sorghastrum nutans* (L.) Nash have a significantly male-biased sex allocation, as is predicted by theory if pollen is not limiting and plants are largely self-incompatible.

This speculation requires that most Andropogoneae be self incompatible, whereas those that have lost male flowers (e.g. *Coelorachis*) should be self-compatible. It also predicts that if there are changes in compatibility systems, then appropriate sex expression changes should occur rapidly. Furthermore, it predicts that the pattern of evolution in the sister group - Paniceae - should be more conventionally tree-like rather than shrub-like. These predictions can be tested by additional phylogenetic study combined with studies of breeding systems.

Yet another possibility is that there is selection on the architecture of the inflorescence itself. Testing would require biomechanical studies to determine if different inflorescence forms had different effects on pollen dispersal or capture.

A third possibility is that the radiation occurred because selection was relaxed for a time, and the different morphological structures are in fact all equivalent. This hypothesis would be supported by negative results from any of the studies described in the previous paragraphs, but would be difficult to distinguish from a fourth and final hypothesis, that the radiation of Andropogoneae was driven by some character that had only indirect effects on inflorescence structure.

Acknowledgements

I thank H. P. Linder and R. E. Spangler for helpful comments on the manuscript. The work described herein was supported in part by NSF grant #DEB-9419748.

References

Barker, N. P., Linder, H. P., and Harley, E. (1995). Phylogeny of Poaceae based on *rbcL* sequences. *Systematic Botany* **20**, 423-435.

Brown, R. (1810). 'Prodromus Florae Novae Hollandiae.' (J. Johnson & Co.: London.)

Brown, R. (1814). 'General remarks, geographical and systematical, on the Botany of Terra Australis.' (G. and W. Nicol: London.)

Brown, W. V. (1977). The Kranz syndrome and its subtypes in grass systematics. *Memoirs of the Torrey Botanical Club* **23**, 1-97.

Calderon-Urrea, A., and Dellaporta, S. L. (1999). Cell death and cell protection genes determine the fate of pistils in maize. *Development* **126**, 435-441.

Campbell, C. S. (1983). Systematics of the *Andropogon virginicus* complex (Gramineae). *Journal of the Arnold Arboretum* **64**, 171-254.

Clark, L. G., Zhang, W., and Wendel, J. F. (1995). A phylogeny of the grass family (Poaceae) based on *ndhF* sequence data. *Systematic Botany* **20**, 436-460.

Clayton, W. D. (1969). A revision of the genus *Hyparrhenia*. *Kew Bulletin Additional Series* **11**, 1-196.

Clayton, W. D. (1972). The awned genera of Andropogoneae. Studies in the Gramineae: XXXI. *Kew Bulletin* **27**, 457-474.

Clayton, W. D. (1973). The awnless genera of Andropogoneae. Studies in the Gramineae: XXXIII. *Kew Bulletin* **28**, 49-58.

Clayton, W. D. (1981). Notes on the tribe Andropogoneae (Gramineae). *Kew Bulletin* **35**, 813-818.

Clayton, W. D. (1987). Andropogoneae. In 'Grass systematics and evolution.' (Eds T. R. Soderstrom, K. W. Hilu, C. S. Campbell and M. E. Barkworth.) pp. 307-309. (Smithsonian Institution Press: Washington, DC.)

Clayton, W. D., and Renvoize, S. A. (1986). 'Genera graminum.' (Her Majesty's Stationery Office: London.)

Cummings, M. P., King, L. M., and Kellogg, E. A. (1994). Slipped-strand mispairing in a plastid gene: *rpoC2* in grasses (Poaceae). *Molecular Biology and Evolution* **11**, 1-8.

Davis, J. I., and Soreng, R. J. (1993). Phylogenetic structure in the grass family (Poaceae), as determined from chloroplast DNA restriction site variation. *American Journal of Botany* **80**, 1444-1454.

Dellaporta, S. L., and Calderon-Urrea, A. (1993). Sex determination in flowering plants. *Plant Cell* **5**, 1241-1251.

Dellaporta, S. L., and Calderon-Urrea, A. (1994). The sex determination process in maize. *Science* **266**, 1501-1505.

DeLong, A., Calderon-Urrea, A., and Dellaporta, S. L. (1993). Sex determination gene *TASSELSEED2* of maize encodes a short-chain alcohol dehydrogenase required for stage-specific floral organ abortion. *Cell* **74**, 757-768.

Dengler, R. E., and Dengler, N. G. (1990). Leaf vascular architecture in the atypical C_4 NADP-malic enzyme grass *Arundinella hirta*. *Canadian Journal of Botany* **68**, 1208-1221.

Dengler, N. G., Dengler, R. E., and Grenville, D. J. (1990). Comparison of photosynthetic carbon reduction (Kranz) cells having different ontogenetic origins in the C_4 NADP-malic enzyme grass *Arundinella hirta*. *Canadian Journal of Botany* **68**, 1222-1232.

Dengler, N. G., Dengler, R. E., and Hattersley, P. W. (1985). Differing ontogenetic origins of PCR ("Kranz") sheaths in leaf blades of C_4 grasses (Poaceae). *American Journal of Botany* **72**, 284-302.

DeWet, J. M. J., and Harlan, J. R. (1970). Apomixis, polyploidy, and speciation in *Dichanthium. Evolution* **24**, 270-277.

Doebley, J. (1995). Genetics, development, and the morphological evolution of maize. In 'Experimental and molecular approaches to plant biosystematics.' (Eds P. C. Hoch and A. G. Stephenson) pp. 57-70. (Missouri Botanical Garden: St. Louis.)

Doebley, J., Durbin, M., Golenberg, E. M., Clegg, M. T., and Ma, D. P. (1990). Evolutionary analysis of the large subunit of carboxylase (*rbcL*) nucleotide sequence among the grasses (Gramineae). *Evolution* **44**, 1097-1108.

Dorweiler, J. E., and Doebley, J. (1997). Developmental analysis of *Teosinte glume architecture 1*: a key locus in the evolution of maize (Poaceae). *American Journal of Botany* **84**, 1313-1322.

Dorweiler, J., Stec, A., Kermicle, J., and Doebley, J. (1993). *Teosinte glume architecture 1*: a genetic locus controlling a key step in maize evolution. *Science* **262**, 233-235.

Duvall, M. R., and Morton, B. R. (1996). Molecular phylogenetics of Poaceae: an expanded analysis of *rbcL* sequence data. *Molecular Phylogenetics and Evolution* **5**, 352-358.

Irish, E. E. (1997). Class II tassel seed mutations provide evidence for multiple types of inflorescence meristems in maize (Poaceae). *American Journal of Botany* **84**, 1502-1515.

Gaut, B. S., and Doebley, J. F. (1997). DNA sequence evidence for the segmental allotetraploid origin of maize. *Proceedings of the National Academy of Sciences, USA.* **94**, 6809-6814.

Hagemann, W. (1990). Comparative morphology of acrogenous branch systems and phylogenetic considerations. II. Angiosperms. *Acta Biotheoretica* **38**, 207-242.

Hamby, R. K., and Zimmer, E. A. (1988). Ribosomal RNA sequences for inferring phylogeny within the grass family (Poaceae). *Plant Systematics and Evolution* **160**, 29-37.

Hattersley, P. W., and Watson, L. (1992). Diversification of photosynthesis. In 'Grass evolution and domestication.' (G. P. Chapman, ed.) pp. 38-116. (Cambridge University Press: Cambridge.)

Hattersley, P. W., Watson, L., and Osmond, C. B. (1977). In situ immunofluorescent labelling of ribulose-1,5 bisphosphate carboxylase in leaves of C_3 and C_4 plants. *Australian Journal of Plant Physiology* **4**, 523-539.

Hodges, S. A., and Arnold, M. L. (1994). Columbines: A geographically widespread species flock. *Proceedings of the National Academy of Sciences, USA* **91**, 5129-5132.

Itoh, J.-I., Hasegawa, A., Kitano, H., and Nagato, Y. (1998). A recessive heterochronic mutation, *plastochron1*, shortens the plastochron and elongates the vegetative phase in rice. *Plant Cell* **10**, 1511-1521.

Jacques-Félix, H. (1961). Observations sur la variabilité morphologique de *Coix lacryma-jobi. Journal d'Agriculture Tropicale et de Botanique Appliquée* **8**, 44-56.

Kellogg, E. A. (1998). Relationships of cereal crops and other grasses. *Proceedings of the National Academy of Sciences, USA* **95**, 2005-2010.

Kellogg, E. A., and Campbell, C. S. (1987). Phylogenetic analyses of the Gramineae. In 'Grass systematics and evolution.' (Eds T. R. Soderstrom, K. W. Hilu, C. S. Campbell and M. E. Barkworth.) pp. 310-322. (Smithsonian Institution Press: Washington, DC.)

Kellogg, E. A., and Watson, L. (1993). Phylogenetic studies of a large data set. I. Bambusoideae, Andropogonodae, and Pooideae (Gramineae). *Botanical Review* **59**, 273-343.

Langdale, J. A., Rothermel, B. A., and Nelson, T. (1988). Cellular pattern of photosynthetic gene expression in developing maize leaves. *Genes and Development* **2**, 106-115.

LeRoux, L. G., and Kellogg, E. A. (1999). Floral development and the formation of unisexual spikelets in the Andropogoneae (Poaceae). *American Journal of Botany* **86**, 354-366.

Li, D., Blakey, A., Dewald, C., and Dellaporta, S. L. (1997). Evidence for a common sex determination mechanism for pistil abortion in maize and in its wild relative *Tripsacum. Proceedings of the National Academy of Sciences, USA* **94**, 4217-4222.

Mason-Gamer, R. J., Weil, C. F., and Kellogg, E. A. (1998). Granulebound starch synthase: structure, function, and phylogenetic utility. *Molecular Biology and Evolution* **15**, 1658-1673.

Mathews, S., and Sharrock, R. A. (1996). The phytochrome gene family in grasses (Poaceae): a phylogeny and evidence that grasses have a subset of the loci found in dicot angiosperms. *Molecular Biology and Evolution* **13**, 1141-1150.

Mathews, S., Tsai, R. C., and Kellogg, E. A. In press. Phylogenetic structure in the grass family (Poaceae): evidence from the nuclear gene phytochrome B. *American Journal of Botany*.

McKone, M. J., Lund, C. P., and O'Brien, J. M. (1998). Reproductive biology of two dominant prairie grasses (*Andropogon gerardii* and *Sorghastrum nutans*, Poaceae): male-biased sex allocation in wind-pollinated plants? *American Journal of Botany* **85**, 776-783.

Nadot, S., Bajon, R., and Lejeune, B. (1994). The chloroplast gene *rps4* as a tool for the study of Poaceae phylogeny. *Plant Systematics and Evolution* **191**, 27-38.

Nambudiri, E. M. V., Tidwell, W. D., Smith, B. N., and Hebbert, N. P. (1978). A C_4 plant from the Pliocene. *Nature* **276**, 816-817.

Neuffer, M. G., Coe, E. H., and Wessler, S. R. (1997). 'The mutants of maize.' (Cold Spring Harbor Laboratory Press: Cold Spring Harbor).

Rua, G. H., and Weberling, F. (1995). Growth form and inflorescence structure of *Paspalum* L. (Poaceae, Paniceae): a comparative morphological approach. *Beiträge zur Biologie der Pflanzen* **69**, 363-431.

Sheen, J.-Y., and Bogorad, L. (1986). Differential expression of six light-harvesting chlorophyll a/b binding protein genes in maize leaf cell types. *Proceedings of the National Academy of Science, USA.* **83**, 7811-7815.

Sinha, N. R., and Kellogg, E. A. (1996). Parallelism and diversity in multiple origins of C_4 photosynthesis in grasses. *American Journal of Botany* **83**, 1458-1470.

Sobral, B. W. S., Braga, D. P. V., LaHood, E. S., and Keim, P. (1994). Phylogenetic analysis of chloroplast restriction enzyme site mutations in the Saccharinae Griseb. subtribe of the Andropogoneae Dumort. tribe. *Theoretical and Applied Genetics* **87**, 843-853.

Spangler, R. E., Zaitchik, B., Russo, E., and Kellogg, E. (1999). Andropogoneae evolution and generic limits in *Sorghum* (Poaceae) using *ndhF* sequences. *Systematic Botany* **24**, 267-281.

Sundberg, M. D., and Orr, A. R. (1986). Early inflorescence and floral development in *Zea diploperennis*, diploperennial teosinte. *American Journal of Botany* **73**, 1699-1712.

Sundberg, M. D., and Orr, A. R. (1990). Inflorescence development in two annual teosintes: *Zea mays* subsp. *mexicana* and *Z. mays* subsp. *parviglumis. American Journal of Botany* **77**, 141-152.

Tateoka, T. (1956a). Notes on some grasses. I. *Botanical Magazine (Tokyo)* **69**, 311-315.

Tateoka, T. (1956b). Reexamination of anatomical characteristics of the leaves in Eragrostoideae and Panicoideae (Poaceae). *Journal of Japanese Botany* **31**, 210-218.

Tateoka, T. (1958). Notes on some grasses. VIII. On leaf structure of *Arundinella* and *Garnotia. Botanical Gazette* **120**, 101-109.

Tateoka, T. (1962). Starch grains of endosperm in grass systematics. *Botanical Magazine*, Tokyo **75**, 377-383.

Tateoka, T. (1963). Notes on some grasses. XV. Affinities and species relationships of Arthropogon and relatives, with reference to their leaf structure. *Botanical Magazine* **76**, 286-291.

Ueno, O. (1992). Immunogold localization of photosynthetic enzymes in leaves of *Aristida latifolia*, a unique C_4 grass with a double chlorenchymatous bundle sheath. *Physiologia Plantarum* **85**, 189-196.

Ueno, O. (1995). Occurrence of distinctive cells in leaves of C_4 species in *Arthraxon* and *Microstegium* (Andropogoneae-Poaceae) and the structural and immunocytochemical characterization of these cells. *International Journal of Plant Sciences* **156**, 270-289.

Vegetti, A. C., and Tivano, J. C. (1991). Synflorescence in *Schizachyrium microstachyum* (Poaceae). *Beiträge zur Biologie der Pflanzen* **66**, 165-178.

Watson, L., and Dallwitz, M. J. (1992). 'The grass genera of the world'. (CAB International: Wallingford.)

Whistler, D. P., and Burbank, D. W. (1992). Miocene biostratigraphy and biochronology of the Dove Spring Formation, Mojave Desert, California, and characterization of the Clarendonian mammal age (late Miocene) in California. *Geological Society of America Bulletin* **104**, 644-658.

Grasses: Systematics and Evolution. (2000). Eds S.W.L. Jacobs and J. Everett. (CSIRO: Melbourne)

THE RELATION OF SPACE AND GEOGRAPHY TO CLADOGENIC EVENTS IN *AGENIUM* AND *HOMOZEUGOS* (POACEAE: ANDROPOGONEAE) IN SOUTH AMERICA AND AFRICA

Gerald F. Guala

Fairchild Tropical Garden, 11935 Old Cutler Rd, Miami, FL 33156-4299 U.S.A.

Abstract

Homozeugos and *Agenium* are andropogonoid savanna grasses from central Africa and central South America respectively. They form a clade with the widely distributed genus *Trachypogon*. Cladograms from revisions of both genera are used here to determine the correlation of spatial environmental parameters with speciation events in the group. A Geographic Information System (GIS) was compiled using 45 layers of spatial environmental data. These data were integrated with the distributions of the species to predict the species' possible distributions using coincidence of layers in the GIS. The spatial data were then integrated with the cladogram to determine the degree of correlation of different environmental parameters with cladogenesis. The environmental parameters that are correlated with speciation events will potentially be useful in studies of evolution, historical biogeography and conservation management. Mean monthly precipitation (for several months), hours of sunshine (for several months), temperature, and dry season vegetation were correlated to varying degrees with cladogenesis in *Homozeugos*.

Key words: Cladistic biogeography, *Homozeugos*, GIS, spatial correlation, Poaceae, Africa, *Agenium*.

INTRODUCTION

This study examines some of the factors that are correlated with cladogenesis, the creation of biodiversity, in two genera of tropical savanna grasses. It provides the groundwork for much more detailed examinations of speciation and biogeography in the future. For many years, investigators have attempted to determine factors that produce biodiversity as well as those that determine distribution. These investigations have generally examined the relationship between geographic areas or habitats and the number of species occurring within them. Those with higher numbers of species were deemed more conducive to the production, or maintenance, of biodiversity, depending on the philosophical leanings of the investigator. These studies have often suffered from inexplicit and outdated taxonomy as well as a horizontal structure that examines only a horizontal slice (usually at the species level) through the phylogeny of the organisms in question. In this study I have taken a different route by generating explicit, well supported, phylogenetic hypotheses, and then examining the correlation of specific spatially varying habitat characteristics with all of the possible clades within those phylogenetic hypotheses. This method yields a group of spatial characteristics that are correlated with the factors that play a role in the actual production of biodiversity, rather than just a group of areas where biodiversity resides.

Environmental history clearly plays a role in the diversification of clades. One useful method of applying environmental data to cladograms, 'historical ecology' (Brooks and Mclennan 1991, 1993) requires a reasonable historical model of the geographic area or process being examined. However, it is not necessary to know the exact sequence of a large set of combined environmental events which have occurred in a specific geographic place, to be able to relate this history to phylogenetic history. This is true

Table 1. Spatial parameter layers included in the GIS.

Layer Name	Resolution	Description
CloudJan - CloudDec (12 files)	1 Deg.	Percentage of actual bright sunshine hours as a percentage of potential sunshine hours per month.
TempJan -TempDec (12 files)	1 Deg.	Mean monthly temperature expressed as the number of 10ths of a degree Celsius.
RainJan -RainDec (12 files)	1 Deg.	Mean monthly precipitation expressed in millimeters.
Albedo	1 Deg.	Surface albedo in October expressed as percentage of possible X 100.
Elevation	5 Min.	Elevation in meters.
Rain	30 Min.	Mean annual precipitation expressed in millimeters.
Temperat	30 Min.	Mean annual surface temperature expressed in 10ths of a degree Celsius.
Sand	1 Deg.	Percentage of sand in the top 1 meter of soil.
Silt	1 Deg.	Percentage of silt in the top 1 meter of soil.
Clay	1 Deg.	Percentage of clay in the top 1 meter of soil.
GVIJan	10 Min.	Calibrated generalized vegetation index for January 1990.
GVIJul	10 Min.	Calibrated generalized vegetation index for July 1990.

because in the absence of a good fossil record, there is no reason to assume that a given taxon evolved and diversified in a given geographic place just because it happens to reside there now. A more realistic model is evolution within a given 'niche' that may or may not have been spatially stationary in history. An example is the model of 'habitat plates' advanced by Vrba (1993) as part of her 'habitat theory' (Vrba 1992). This spatial movement of habitats is the reason that I am making only historical argument here that can be overlayed onto paleogeography to test the hypotheses that result from it, not a historical biogeographic one. The investigations of two large areas of different continents with very different post-cretaceous climatic histories (Coetzee 1993; Romero 1993) allow the study of the interaction of enormously different scales of mobility and change in environmental parameters (e.g. migration of precipitation isoclines vs. relatively horizontally stationary elevational change). Originally, I wanted to compare the spatial variables correlated with speciation in two related genera of grasses growing in similar habitats on South America and Africa. An extensive survey of the distributions of grass genera showed that *Agenium* Nees and *Homozeugos* Stapf are closely related, grow in very similar savanna habitats, and have enough species to have a sufficient number of internal nodes in their cladograms but not so many species as to be unmanageable. They are also wind pollinated and widely distributed within the savanna habitats on their respective continents, thus insuring a large sample of the spatial environment on each continent without the confounding effects of pollinator ranges and behaviors. They form a well supported clade, along with *Trachypogon* Nees (Kellogg and Watson 1993; Guala 1998) which is widely distributed in Africa, Central America, and South America.

Cladistic analysis has brought about a revolution in systematics by allowing the generation of explicit and testable phylogenetic hypotheses that are required for this type of study. The phylogenetic hypotheses in this investigation were derived from cladistic analyses using morphology and anatomy within the genera, with the addition of DNA (ITS I & II) sequences to independently test Kellogg and Watson's (1993) hypothesis of generic relationships, which had been derived from morphology and anatomy (see Guala 1998). The Phylogenetic Species Concept (Donoghue 1985; Mishler 1985; Mishler and Brandon 1987) provided the explicit and robust delimitation of species that was required to get comparable monophyletic units for comparison.

Geographic Information Systems (GIS's) are computerized matrix or vector databases of spatial data and associated nonspatial data. They have very advanced display and analysis capabilities and have brought about a revolution in the spatial analysis of data by combining the analytical power of relational database structure with highly accurate georeferenced spatial displays. A very large number of existing data sets of worldwide coverage have been incorporated into Geographic Information Systems and have been made readily available on common desktop computer systems (e.g. NOAA-EPA Global Ecosystems Database Project 1992; National Climatic Data Center 1992, 1994).

The ranges of species may have been studied for as long as species have been recognized, but it wasn't until the 1950's that important work on the direct linkage of these ranges to specific spatial parameters began to emerge. Dahl (1951) examined the limitation of the range of alpine plants in Scandinavia, and Pigott (1954, 1974, 1975) critically examined elements of the flora of Britain. The first work on potential spatial distributions to be done in a modern, integrated way with a rudimentary GIS using multiple spatial parameters was that of Nix (1975) and his students and coworkers (Nix and Kalma 1972; Nix *et al.* 1977) in Australia. This work evolved into the present state of the art in prediction of species ranges using elaborate GIS-based models, with programs such as BIOCLIM (Nix 1986; Busby 1991; McMahon 1996) and the GARP model currently under development at the Environmental Resources Information Network (ERIN) in Australia.

Two types of data are analyzed here. The most extensive data are a series of 45 layers of environmental parameters including temperature, precipitation and several physical parameters of the soil and habitat (Table 1). Numerous studies have pointed to the importance of climatic factors in limiting the distribution of species (Cary *et al.* 1995; Dahl 1951; Nix 1975; 1986; Nix and Kalma 1972; Pigott 1954, 1974, 1975). Soil properties also have been shown to be useful in delimiting the ranges of grass species at the mega-scale (Carey *et al.* 1995; Guala 1992, 1995) as well as the micro-scale (Bradshaw 1959; Jowett 1959; McNeilly 1968; Shackleton and Shackleton 1994). Both scales are addressed in the present study. The GIS includes soil variables at the mega-scale and the analyses of soil samples taken with each collection made for the study address micro-scale variation. I have been able to build a globally comprehensive integrated system of environmental variables to predict where the species of *Agenium* and *Homozeugos* can live and to identify a few tantalizing spatial environmental variables that are correlated with cladogenesis. This is in addition to the very comprehensive habitat descriptions that the system automatically produces.

MATERIALS AND METHODS

Phylogenetic Analyses

The integrity of any investigation of cladogenesis relies on a sound cladistic analysis. Full revisions of both genera with complete cladistic analyses are given in Guala (1998) and separate publications are in preparation that describe the phylogenetic analyses in depth.

GIS Analyses

Data Layers

A geographic information system (GIS) was compiled from the data set, NOAA-EPA Global Ecosystems Database Project (1992), and converted to ArcView 3.0a (ESRI 1996, 1997) format. Although several other data sets were consulted and examined over the course of the study, only data from the NOAA-EPA CD-ROM were included in the final version of the GIS used for the analyses and maps presented here. The data sets were copied from the CD-ROM, converted from integer/binary format to real/binary format using the CONVERT module of IDRISI vers. 4.0 (Eastman 1992) and then converted to ArcView 3.0a format. Conversion of files from IDRISI to Arcview grid format was done manually on the layer of surface Albedo and the 13 surface temperature layers. All other layers were converted using an ArcView 3.0a Avenue script written and provided by Justin Moat of the Madagascar Biodiversity Project, Royal Botanic Gardens, Kew. Forty-five data layers (or themes) were used in the GIS and each has inherent characteristics that may influence its accuracy and applicability. Surface temperature, precipitation and cloudiness data (each with 12 monthly means) are from the data set of Leemans and Cramer (1992) which was originally derived from Leemans and Cramer (1991). The annual summaries of precipitation and surface temperature in the data set of Legates and Willmott (1989, 1992) were also used in the GIS because none were included in the Leemans and Cramer set. Preliminary studies showed albedo to be only moderately definitive for species distributions so only one layer was selected for inclusion in the GIS. I used the October layer because it falls in the least variable period of the year for which data were available throughout most of the study area. The data are from (Matthews 1992) and are derived from a combination of cultivation intensity and vegetation as well as published sources and extensive satellite imagery using the techniques described in Matthews (1983a, 1983b). Global Vegetation Index (GVI) layers for January and July of 1990 were included because they present a relatively high resolution assessment of the vegetation density and greenness over the study area. January and July were chosen because they are always part of the wet and dry seasons respectively. Values from 1990 were chosen because they

most closely match the mean in the study area for the years available (1985-1990). The "monthly experimental calibrated GVI" data layers were used from the data set of EDC-NESDIS (1992). They were developed using methods described in Kineman and Ohrenschall (1992). The FAO/UNESCO soil units map (UNEP/GRID 1992) which is derived from the FAO Soil Map of the World in Digital Form (UNEP/GRID 1986), was also included in the GIS but was too big to manipulate with the other layers (230 MB in ArcView Format). The soil types that each species occurs on were not definitive for any species and the layer was excluded.

Each specimen examined was mapped to a single point which ranged from an informed guess, in the case of weakly documented specimens, to exact localities determined with a Trimble Ensign global positioning system (GPS) for all of my own collections. These were saved as degrees latitude and degrees longitude, to six decimel places. Distributions of each species were saved from the spreadsheet as a Dbase file with three fields (latitude, longitude and a value field which contained the collection number). There were no identical collection numbers in the study set. These Dbase files were then imported into ArcView as tables and added to the View as an event theme. Thus data could be retrieved and compared for each collection based on the collection number. Each file was also subsequently converted (as a copy) to both a shape file and a grid to facilitate some manipulations and output within the GIS. A copy of each table was also made with the collection number in the value field replaced by "1", forming a single attribute data file for making species comparisons.

Predicted Potential Distribution Methods

To produce a predicted potential distribution of each species, the value of each of the 45 layers in the GIS was summarized for the points in each species distribution table. Maximum, minimum, range, mean and standard deviation were transcribed to a spreadsheet for later use. Each layer was then reclassified so that all cells with values falling between the previously determined maximum and minimum for the species in question were given a value of "1" and all other cells were given a value of "0".

All of the layers were then added together. Thus, a cell with a value of 1 in all of the reclassified layers would have a value of 45 in the sum of the maps. Each cell was then multiplied by 100 and divided by the total possible sum to give a percentage of the layers intersecting within each cell.

Characters Correlated with Speciation Methods

A method for determining and comparing the different levels of evolutionary significance, or correlation with speciation events, in the distributions of spatial parameters among species was developed in 1993 (Guala 1996) and further refined here. The existing GIS and a database of physical and chemical analyses of soil samples that were taken along with every specimen collected for the present study, were used for this analysis. All soil samples from South America were analyzed at the Centro de Pesquisa Agropecuária dos Cerrados (CPAC-EMBRAPA) in Brasilia, Brazil. All soil samples from Africa were analyzed at the Soil Laboratory, Box X505, Causeway, Harare, Zimbabwe. Color value, hue, and chroma were determined by the author on wet and dry samples in full sun using a Munsell Soil Color Chart with the Tropical Soils Supplement. Bulk density was determined using a 100ml 3:5 aspect aluminum tube sampler (Guala 1998) and an Ohaus portable balance.

The primary method for comparing the ranges of spatial parameters to the cladogram is as follows.

1.) The GIS was queried for the values at each point of occurrence for each species (giving a range). This was done for all of the spatial environmental layers in the GIS.

2.) The component of the cladogram topology for all species (Nelson and Platnick 1981; Michevich and Platnick 1989) was calculated. This is the maximum possible component value.

4.) The ranges of GIS layer values were then mapped onto the cladogram as any character distribution would be. Ranges are additive down the tree. Note that values at the nodes are not character states of the hypothetical ancestor, only a sum of the clade.

5.) Nodes that showed a discontinuity between sistergroups were labelled as informative and counted (the component value of the environmental parameter contained in the GIS layer).

6.) The component value of the parameter was then divided by the total possible component value to produce an index value.

The determination of which nodes were informative was influenced by the delimitation of character states. Although every method of dividing characters into states is laden with pitfalls (Stevens 1991; Stevens and Gift 1997) in this case a gap based approach relying on discontinuities of one data unit or more seemed to be the most defensible. Because the data were, for the most part, originally continuous data extrapolated from the geographic positions of the total number of known populations there should be no missing data that can be estimated or assumed. Furthermore, a large number of variables in a large number of different units were considered, making impossible any kind of standardization of units needed to justify a specific gap size larger than the resolution of the data set. What are really being examined here are correlations of environmental parameter distributions (at the mapped spatial resolution currently available) with cladogenic events that are "diagnosable" at the current level of resolution.

RESULTS

Predicted Potential Distribution Maps

The predicted potential distribution maps for each species are given in Figure 1.

Characters Correlated with Speciation

There were no completely discontinuous spatial parameter distributions between the species of *Agenium* in the GIS. Among species of *Homozeugos*, there were 18 spatial parameters in the GIS that showed discontinuities. The component and index value of each of these parameters is given in Table 2.

A second independent set of spatial data was the soil analysis for samples from all of the specimens collected. The complete soil data were included in the habitat descriptions in prior taxonomic

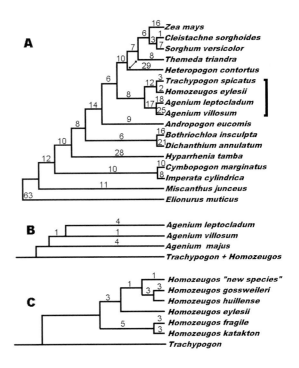

Fig. 1. Cladograms used in the present study (from Guala 1998). Branch lengths are indicated. A. Cladogram of several Andropogoneae based on ITS data. Two cladograms differing only in the placement of *Themeda triandra* (indicated by an arrow) were found. 428 steps, CI=0.708, HI=.292, RI=0.573, RC=0.406. Bootstrap support for the *Agenium+Trachypogon+Homozeugos* clade = 73%. B. Cladogram of *Agenium* based on anatomy and morphology. 12 characters, 12 steps, no homoplasy. C. Cladogram of *Homozeugos* based on anatomy and morphology. 17 characters, 20 steps, CI=0.9, RI=0.83, RC=0.75.

treatments (Guala 1998). In *Agenium* there was a significant sample for both *A. leptocladum* (Hackel) Clayton and *A. villosum* (Nees) Pilger and there were no discontinuities between them. In *Homozeugos*, only a single collection of *H. katakton* Clayton was made, in contrast to the 27 collections of *H. eylesii*. There were discontinuous differences between the two species in the silt, fine sand, medium sand and total sand fractions of the soil as well as the pH as measured with $CaCl_2$ buffer. The pH $CaCl_2$ difference (2.9-3.2 in *H. katakton* vs. 3.7-5.1 in *H. eylesii*) may directly reflect the physical characteristics in this case because of the paucity of clay (2%) and silt (0%) in the *H. katakton* samples. The fraction data in the GIS showed no discontinuities. There are data for only two of the six species in the genus, so it is not realistic to apply these data to the cladogram. However, these preliminary results do emphasize the importance of both physical and chemical analyses of the soil in future work.

DISCUSSION

Predicted Potential Distribution Maps

The priority of the present study was to be able to easily dissect and interpret the terms responsible for distribution. This required a simplified system so that individual parameters and their roles would be easily recognized. The model used here to predict distribution is very rudimentary but its simplicity makes it very flexible and widely applicable. Parameters can easily be

Table 2. Index values for spatial parameters associated with speciation events.

GIS Layer	Node Score	Index Value
CloudApr	2	0.4
CloudAug	3	0.6
CloudFeb	1	0.2
CloudJan	1	0.2
CloudJul	2	0.4
CloudMay	2	0.4
CloudNov	2	0.4
CloudOct	4	0.8
CloudSep	3	0.6
Rain	2	0.4
RainApr	3	0.6
RainJan	3	0.6
RainNov	3	0.6
RainOct	3	0.6
RainSep	2	0.4
TempNov	2	0.4
Temperat	2	0.4
GVIJul	2	0.4

included or excluded in order to assess their significance in the distribution and any spatial parameter can be employed as a raw or weighted parameter. All parameters are equally weighted in the analyses shown here. The percentage output allows standardization with differing numbers of layers. This is a useful index because some layers contribute little or nothing to the delimitation of the predicted distribution. The percentage output is not a probability, only a description of the data. For the percentage to be comparable to a probability, one would have to assume that the layers constitute a random sample of all possible layers in niche space and that all layers are of equal importance within that space. This is clearly not the case.

Characters Correlated with Speciation

The goal of this analysis is to determine spatial environmental parameters that are correlated with cladogenesis and hence, possibly associated with such events. It is not entirely clear why there were no discontinuous spatial variables among the species of *Agenium*. No two species of *Agenium* were ever found growing together in the field (Guala 1998), although *Agenium leptocladum* and *A. villosum* were sometimes found within a few kilometers of each other, and *A. villosum* is known from the same general region as *A. majus*. Thus all but the most fine scale spatial variables will overlap in values. In *Homozeugos*, even though *H. fragile* Stapf and *H. gossweileri* Stapf were mapped to the same point and *H. eylesii* has been collected near *H. katakton*, there are twice as many species and hence five internal nodes, rather than two (Fig. 2), in the cladogram giving a much greater chance for discontinuity to occur.

The chart of component values shows that relative cloudiness and rainfall are clearly correlated with cladogenesis in *Homozeugos* (Table b). It is not clear whether it is the amount of sun or rain per month that is important because of the obvious correla-

tion between cloudiness and rain. Rainfall limits plant growth in much of southern Africa, so it is quite reasonable that it could also be a driving force in cladogenesis. Another important point is that November is the only month in which temperature is significant, but it is neither the warmest nor the coldest month of the year. It is, however, the beginning of the growing season and may have a complex interaction with precipitation. The appearance of July GVI among the significant parameters is interesting because it is a measure of greenness in the driest month of the year, when all of the grasses are brown. This may actually signal a primary habitat shift between areas with differing levels of woody cover. Some species of shrubs and trees, such as *Acacia albida* Delile, *Cassia abbreviata* Oliver, and *Colophospermum mopane* (Kirk ex Benth) Kirk ex J. Léonard are green in the study region during the dry season.

The effect of the the mapping of *Homozeugos fragile* and *H. gossweileri* at the same point should also be noted. Although there are five nodes in the cladogram of *Homozeugos*, the maximum possible component value in this data set is four because, in this specific data set both *H. fragile* and *H. gossweileri* will have the same value for any layer in the GIS. Thus there can be no discontinuity in their common node at the base of the cladogram.

Sampling of soils was incomplete, but there are weak indications that the physical properties of the soil may play a role in cladogenesis in *Homozeugos*. In contrast to the weakness of the soil data, the GIS environmental layers provide a clear indication that precipitation (or the lack of it) and the associated layers describing the percentage of sunshine hours per month are parameters associated with cladogenic events. There is an indication that temperature, at least in the beginning of the growing season, may be linked to cladogenic events as well. The July GVI is also interesting because it may signal a habitat shift during a speciation event.

Distributions with respect to environmental parameters have other uses as well. For example, *Homozeugos huillense* (Rendle) Stapf was found to be a metaspecies in the cladistic analysis (Guala 1998). The spatial data examined in the present study provides support, albeit weak, from an independent data set for the recognition of the taxon as a species. *Homozeugos huillense* has a unique range of values in the GIS layers: CloudAug, CloudDec, CloudJan, CloudJul, CloudSep, Rain, RainJan and GVIJul. This means that it occupies a definably unique niche that is likely to require unique, although possibly very subtle, physiological adaptations to it. These adaptations can be taken as autapomorphies for the species.

In summary, the clear phylogenetic hypotheses and distributional data included in a modern systematic revision are the raw data that can be integrated with a GIS to precisely predict ranges, determine environmental parameters associated with cladogenesis (and ultimately biodiversity) and to supplement phylogenetic hypotheses.

ACKNOWLEDGEMENTS

I wish to acknowledge the help of everyone who provided assistance, advice or encouragement during my work, I am grateful to all. Prominent among these are Drs. Walter Judd, Dana Griffin,

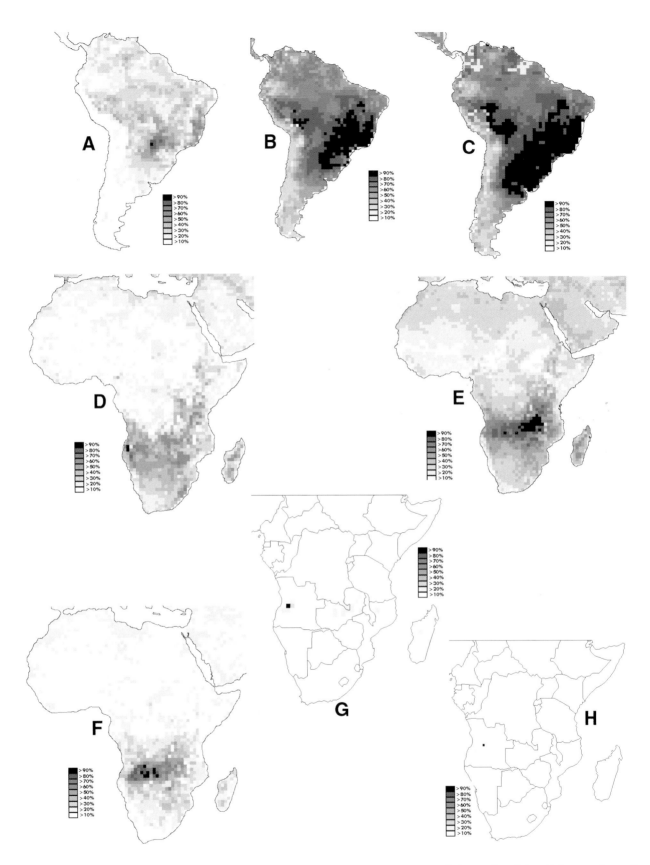

Fig. 2. Predicted potential distributions of the species of *Agenium* and *Homozeugos*. A. *Agenium majus*. B. *Agenium leptocladum*. C. *Agenium villosum*. D. *Homozeugos huillense*. E. *Homozeugos eylesii*. F. *Homozeugos katakton*. G. *Homozeugos "n. sp."* H. *Homozeugos gossweilerii* and *H. fragile*. The legend indicates the percentage of layers intersecting. Darker areas indicate the intersection of more layers and thus, a higher probability of suitable habitat.

Jon Reiskind, Hugh Popenoe, Elizabeth Kellogg, Tarciso Filgueiras, Lynn Clark, Mark Whitten, Norris Williams, Robert Scholes and Greg Erdos. I thank the staff of all of the herbaria that loaned specimens for my study, especially those of of R, RB, IBGE, PRE, SRGH, MPR, J, and US who provided both specimens and work space. I am especially indebted to Ms. Lyn Fish, Dr. Roberta Mendonça, Mr. Sergio Seraiba, Mr. Godfrey Mwila, Ms. Marjorie Knowles, and Dr. Kevin Balkwill. Many people generously provided lodging or other assistance during my two years of fieldwork including Dimitri Sucre, Jorge Gomes, Peter Johnstone and family, and Pierce Taylor and family, Peter Crawshaw, Tim & Pene Truluck and Charles and Teddy Brightman. Special thanks go to Thia Hunter. I also wish to thank the National Geographic Society, the National Science Foundation (Grant #9423452), The United States Department of Agriculture, The Garden Clubs of America, The World Wildlife Fund, The American Bamboo Society, and The American Society of Plant Taxonomists for their generous monetary support of this project.

REFERENCES

Bradshaw, A.D. (1959). Population differentiation in *Agrostis tenuis* Sibth. I. Morphological differentiation. *New Phytologist* **58**, 208-227

Brooks, D.R., and McLennan, D.A. (1993). Historical ecology: examining phylogenetic components of community evolution. pp. 141- 175. In: R. Ricklefs and D. Schluter, eds, 'Species Diversity in Ecological Communities' (University of Chicago Press: Chicago.)

Brooks, D.R. and. McLennan, D.A. (1991). 'Phylogeny, Ecology and Behaviour: a research program in comparative biology' (University of Chicago Press: Chicago.)

Busby, J.R. (1991). BIOCLIM - A Bioclimatic Analysis and Prediction System. pp. 64-68. *In*: C.R. Margules and M.P. Austin, eds 'Nature Conservation: Cost Effective Biological Surveys and Data Analysis' (CSIRO: Canberra.)

Cary, P.D., Watkinson, A.R., and Gerard, F.F. (1995). The determinants of the distribution and abundance of the winter annual grass *Vulpia ciliata ssp. ambigua*. *Journal of Ecology* **83**, 177-187.

Coetzee, J.A. (1993). African Flora since the Terminal Jurassic. pp. 37-61. In P. Goldblatt ed., 'Biological Relationships Between Africa and South America' (Yale University Press: New Haven.)

Dahl, E. (1951). On the relation between summer temperatures and the distribution of alpine plants in the lowlands of Fennoscandia. *Oikos* **3**, 22-52.

Donoghue, M.J. (1985). A critique of the biological species concept and recommendations for a phylogenetic alternative. *Bryologist* **88**, 172-181.

Eastman, J.R. (1992). 'IDRISI version 4.0, Graduate School of Geography' (Clark University: Worcester .)

EDC-NESDIS (1992). Monthly Global Vegetation Index from Gallo Bi-Weekly Experimental Calibrated GVI (April 1985 - December 1990). Digital raster data on a 10-minute Geographic (lat/long) 1080X2160 grid. *In*: 'Global Ecosystems Database Version 1.0: Disc A.' (NOAA National Geophysical Data Center: Boulder.)

ESRI. (1996). 'ArcView 3.0 with Spatial Analyst'. (Environmental Systems Research Institute: Redlands .)

ESRI. (1997). 'ArcView 3.0a patch for ArcView 3.0 '(Environmental Systems Research Institute: Redlands .)

Guala, G.F. (1992). 'All about *Apoclada* (Poaceae: Bambusoideae: Bambusodae): a monograph of the genus.' Msc. Thesis (University of Florida: Gainesville .)

Guala, G.F. (1995). A cladistic analysis and revision of the genus *Apoclada* (Poaceae: Bambusoideae: Bambusodae). *Systematic Botany* **20**, 207-223.

Guala, G.F. (1996). Notes on the role of cladistic phylogenetic theory in the management of biodiversity. *Kirkia* **16**, 1-9.

Guala, G.F. (1998). Revisions of *Agenium* and *Homozeugos*: integrating cladistic analysis and geographic information sysytems. Doctoral Dissertation. (University of Florida: Gainesville .) University microfilms # 9837398.

Jowett, D. (1959). Adaptation of a lead-tolerant population of *Agrostis tenuis* to low soil fertility. *Nature* **184**, 43.

Kellogg, E.A., and Watson, L. (1993). Phylogenetic studies of a large data set. I Bambusoideae, Andropogonodae, and Pooideae (Gramineae). *Botanical Review* **59**, 273-343.

Kineman, J., and Ohrenschall, M. (1992). 'Global Ecosystems Database Version 1.0: Disc A, Documentation Manual. Key to Geophysical Records, Documentation No. 27.' (USDOC/NOAA National Geophysical Data Center: Boulder.)

Leemans, R., and Cramer, W.P. (1991). 'The IIASA Database for Mean Monthly Values of Temperature, Precipitation and Cloudiness on a Global Terrestrial Grid. Digital Raster Data on a 30 minute Geographic (lat/long) 360X720 grid'. (IIASA: Laxenburg, Austria.)

Leemans, R., and Cramer, W.P. (1992). The IIASA Database for Mean Monthly Values of Temperature, Precipitation and Cloudiness on a Global Terrestrial Grid. Digital Raster Data on a 30 minute Geographic (lat/long) 360X720 grid. *In*: 'Global Ecosystems Database Version 1.0: Disc A'. (NOAA National Geophysical Data Center: Boulder CO.

Legates, D.R., and Willmott, C.J. (1989). 'Monthly Average Surface Air Temperature and Precipitation' (National Center for Atmospheric Research: Boulder CO.)

Legates, D.R., and Willmott, C.J. (1992). Monthly Average Surface Air Temperature and Precipitation. Digital Raster Data on a 30 minute Geographic (lat/long) 360X720 grid. *In*: 'Global Ecosystems Database Version 1.0: Disc A'. (NOAA National Geophysical Data Center: Boulder CO.

Matthews, E. (1983a). Vegetation, land-use and seasonal albedo data sets: Documentation of archived data tape. NASA Technical Memorandum 86107 (NASA: Washington DC.)

Matthews, E. (1983b). Global vegetation and land use: new high resolution data bases for climate studies. *Journal of Climatology and Applied Meteorology*. **22**, 474-487.

Matthews, E. (1992). Global vegetation, land-use, and seasonal albedo. Digital Raster Data on a 1-degree Geographic (lat/long) 180X360 grid. *In*: 'Global Ecosystems Database Version 1.0: Disc A'. (NOAA National Geophysical Data Center: Boulder CO.)

McMahon, J.P., Hutchinson, M.F., Nix, H.A., and Ord, K.D. (1996). 'ANUCLIM Version 1 User's Guide' (ANU,CRES: Canberra .)

McNeilly, T. (1968). Evolution in closely adjacent plant populations. III. *Agrostis tenuis* on a smal copper mine. *Heredity* **23**, 99-108.

Mickevich, M.F., and Platnick, N.I. (1989). On the information content of classifications. *Cladistics* **5**, 33-47.

Mishler, B.D. (1985). The morphological, developmental and phylogenetic basis of species concepts in bryophytes. *Bryologist* **88**, 207-214.

Mishler, B.D., and Brandon, R.N. (1987). Individuality, pluralism and the phylogenetic species concept. *Biology and Philosophy* **2**, 397-414.

National Climatic Data Center. (1992). 'International Station Meteorological Climate Summary. Vers. 2' (Naval Oceanography Command Detachment, National Climatic Data Center: Asheville.)

National Climatic Data Center. (1994). 'Global Daily Summary of Temperature and Precipitation Vers. 1' (National Climatic Data Center: Asheville.)

Nelson, G., and Platnick, N. (1981). 'Systematics and Biogeography: Cladistics and Vicariance' (Columbia University Press: New York.)

Nix, H.A. (1975). The Australian climate and its effect on grain yield and quality. pp. 183-226. In: A. Lazenby and E.M. Matheson eds, 'Australian Field Crops 1: Wheat and Other Temperate Cereals' (Angus and Robertson: Sydney.)

Nix, H.A. (1986). A biogeogaphic analysis of Australian Elapid snakes, In: Longmore, R. ed., Atlas of Australian Elapid Snakes. *Australian Flora and Fauna Series* **8**, 4-15.

Nix, H.A., and Kalma, J.D. (1972). Climate as a dominant control in the biogeography of Northern Australia and New Guinea. pp. 22-70 In: D. Walker ed., Bridge and Barrier - The Natural and Cultural Heritage of the Torres Strait. Australian National University Department of Biogeography, Geomorphology Publication BG/3.

Nix, H.A., McMahon, J.P., and Mackenzie, D. (1977). Potential areas of production and the future of Pigeon Pea and other grain legumes in Australia. pp. 110-119. In: E.S. Wallis and P.C. Whiteman Eds, 'The Potential for Pigeon Pea in Australia, Proceedings of Pigeon Pea (*Cajanus cajan* (L.) Millsp.) Field Day' (Department of Agriculture; University of Queensland: Australia.)

NOAA-EPA Global Ecosystems Database Project. (1992). 'Global Ecosystems Database Version 1.0. User's Guide, Documentation, Reprints, and Digital Data on CD-ROM' (USDOC/NOAA National Geophysical Data Center: Boulder CO.)

Pigott, C.D. (1954). On the interpretation of the discontinuous distributions shown by certain British species of open habitats. *Journal of Ecology* **42**, 95-116.

Pigott, C.D. (1974). The response of plants to climate and climate change. pp. 11-17. *In*: F. Perring ed., 'The Flora of a Changing Britain' (Classey: Hampton.)

Pigott, C.D. (1975). Experimental studies on the influence of climate on the geographical distribution of plants. *Weather* **30**, 82-90.

Romero, E.J. (1993). South American paleofloras. pp. 62-85 In: P. Goldblatt ed., 'Biological Relationships Between Africa and South America' (Yale University Press: New Haven.)

Shackleton, S.E., and Shackleton, C.M. (1994). Habitat factors influencing the distribution of *Cymbopogon validus* in Mkambati Game Reserve, Transkei. *African Journal of Range and Forage Science* **11**, 1-6.

Stevens, P.F. (1991). Character states, continuous variation and phylogenetic analysis: a review. *Systematic Botany* **16**, 553-583.

Stevens, P.F. and N. Gift. (1997). Vagaries in the delimitation of character states in quantitative variation - an experimental study. *Systematic Biology* **46**, 112-125.

Swofford, D.L. (1991). 'PAUP: Phylogenetic analysis using parsimony. vers. 3.11' (Illinois State Natural History Survey: Champaign-Urbana).

UNEP/GRID. (1986). 'FAO Soil Map of the World in Digital Form. Digital Raster Data on a 2-minute Geographic (lat/long)10800X5400 grid' (UNEP/GRID: Carouge, Switzerland.)

UNEP/GRID. (1992). Global gridded FAO/UNESCO Soil Units. Digital Raster Data on a 2-minute Geographic (lat/long)10800X5400 grid. *In*: 'Global Ecosystems Database Version 1.0: Disc A' (NOAA National Geophysical Data Center: Boulder CO.)

Vrba, E.S. (1992). Mammals as the key to evolutionary theory. *Journal of Mammology.* **73**, 1-28.

Vrba, E.S. (1993). Mammal evolution in the African neogene and a new look at the great American interchange. In: P. Goldblatt ed., 'Biological Relationships Between Africa and South America' (Yale University Press: New Haven.)

Grasses: Systematics and Evolution. (2000). Eds S.W.L. Jacobs and J. Everett. (CSIRO: Melbourne)

ANDROPOGONEAE SYSTEMATICS AND GENERIC LIMITS IN *SORGHUM*

Russell E. Spangler

Harvard University Herbaria, 22 Divinity Avenue, Cambridge, MA 02138.

Abstract

Generic limits of *Sorghum* were analyzed by looking at an *ndh*F phylogeny for the grass tribe Andropogoneae. Low variation across the tribe made it difficult to clearly identify traditional subtribes, suggesting that a rapid radiation led to tremendous morphological variability in a short time period. Overlapping morphological character states also obscure boundaries between higher taxa. *Sorghum* is not monophyletic in the analysis, and generic boundaries could either be expanded to accommodate *Cleistachne*, *Miscanthus*, and *Microstegium nudum*, or the genus could be split into several lineages. Subgenus *Stiposorghum* presents difficulty in species identification due to overlap of key characters of the spikelet.

Key words: Andropogoneae, generic limits, *ndh*F, phylogeny, *Sorghum*.

INTRODUCTION

Detailed systematic and evolutionary studies in a genus require a sound knowledge of the generic limits of the taxa in question. This knowledge can be particularly difficult when clinal variation rather than disjunct variation exists between genera. Generic concepts in the grass tribe Andropogoneae have continually challenged systematists, in large part because of a gradient of variation between groups across the tribe (Clayton 1987). The genus *Sorghum* in tribe Andropogoneae has an obscure generic limit when examined with morphologically similar genera in the rest of subtribe Sorghinae (*sensu* Clayton and Renvoize 1986). Unique, distinguishing characters for *Sorghum* have not yet been found, but the paniculate inflorescence, sessile and pedicellate spikelets differing in sex expression, solid rachis internodes, and a coriaceous, rounded lower glume on the sessile spikelet are a combination of features found in *Sorghum* species. Some of these characters, however, are present in all Sorghinae, and others are present in some Sorghinae, but not in the same combination as in *Sorghum*. One step in a mono-

graphic revision of *Sorghum* currently underway was to define generic limits by examining the phylogeny of tribe Andropogoneae to determine whether the current set of species in the genus is monophyletic, or whether there are alternative groupings to represent *Sorghum* taxa. Sequences for the chloroplast gene *ndh*F were generated for phylogeny estimation (Fig. 1) in the Andropogoneae (Spangler *et al.* 1999), and the implications of that study on morphological work associated with the monographic work on *Sorghum* are presented here.

SORGHINAE AND ANDROPOGONEAE SYSTEMATICS

Determining the monophyly of *Sorghum* requires evaluation of the placement of *Sorghum* species in subtribe Sorghinae. The morphological characters used to define Sorghinae (Clayton and Renvoize 1986) are, in abbreviated form: usually terminal digitate or paniculate racemes; slender raceme rhachis and internodes; dissimilar spikelet pairs; sessile spikelet lower glume firm, usually convex on back; upper lemma awned; and the pedicelled spikelet staminate or barren, or much reduced. Aside from the

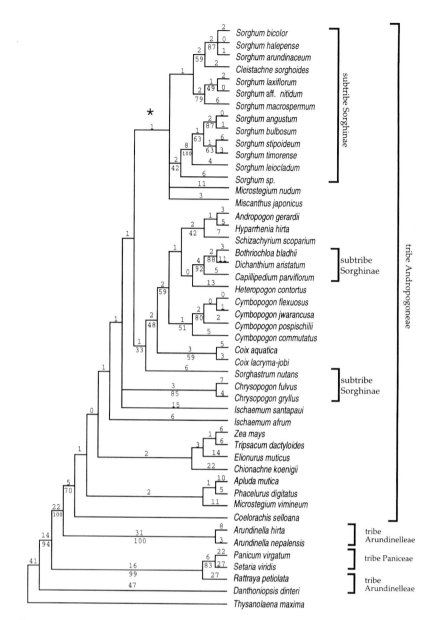

Fig. 1. One of 5661 equally most parsimonious trees based on *ndh*F sequences (data from Spangler *et al.* 1999). Branch lengths are indicated above each branch and bootstrap percentages are listed below nodes that are resolved in the strict consensus of all equally most parsimonious trees. The asterisk marks an expanded *Sorghum* clade.

spikelet pairs being different and the sessile spikelet lower glume hardness, these characters are also used to define subtribe Saccharinae. In fact, many of these features are also found in other subtribes of Andropogoneae in varying combinations. Doubts about the monophyly of Sorghinae led me to look at the entire tribe Andropogoneae to determine whether Sorghinae was indeed a monophyletic group within which *Sorghum* formed its own monophyletic genus.

Morphological discontinuities useful in taxon delimitation are essentially absent in Andropogoneae. Hackel (1889) recognized a small number of genera in Andropogoneae and included most species in the single genus *Andropogon*, reflecting the difficulty in distinguishing discrete groups within the tribe. Subsequent major classifications (Stapf 1917-1919; Pilger 1954) have narrowed Hackel's original generic concepts, and produced the current 11

subtribes of 85 genera and 960 species (Clayton and Renvoize 1986). The information used to produce recent classifications came from phenetic studies, primarily using inflorescence and spikelet characters (Clayton 1972, 1973), and from cytotaxonomic studies (Celarier 1956, 1957, 1958, 1959). Recent molecular studies used nuclear genes (Mason-Gamer *et al.* 1998 for *waxy*; J. Spies and E. Kellogg, unpublished for ITS; S. Mathews and E. Kellogg, unpublished for phytochrome B) and the chloroplast gene *ndh*F (Spangler *et al.* 1999) to reconstruct phylogeny on a sample of Andropogoneae. All the sequencing studies found short internal branch lengths within the tribe (as in Fig. 1), suggesting a continuum of taxa.

Subtribal recognition in Andropogoneae is tenuous and can be misleading due to overlapping character states for features defining these taxa. Sorghinae, represented in Fig. 1 by *Sorghum*,

Sorghastrum, Cleistachne, Bothriochloa, Dichanthium, Capillipe- dium, and *Chrysopogon,* clearly does not form a monophyletic group in the *ndh*F tree. Statistical tests constraining these taxa together indicated it is an artificial grouping (Spangler *et al.* 1999). Other groupings of taxa within Andropogoneae based on morphological characters, such as the presence or absence of awns (Clayton 1972, 1973) or monoecy in *Zea, Tripsacum, Chionacne,* and *Coix* (Watson and Dallwitz 1992), are also not supported by the *ndh*F topology (Fig. 1). Indeed, the nearly star- like phylogeny of the Andropogoneae may reflect a rapid evolu- tionary radiation leading to the tribal diversity (Kellogg, this vol- ume). We cannot confidently define subtribes until we can find characters that accurately identify internal lineages.

THE GENUS *SORGHUM*

The *ndh*F data give us insight into the generic limit problem in *Sorghum,* helping us identify the monophyletic group of taxa for generic recognition, and to exclude other species that are not part of that group. Inclusion of *Cleistachne, Miscanthus,* and *Microste- gium nudum* make *Sorghum* a paraphyletic group in the molecu- lar tree (asterisk). The *ndh*F placement of *Cleistachne* in *Sorghum* agrees with previous molecular studies (Duvall and Doebley 1990; Sun *et al.* 1994), and preliminary morphological observa- tions using leaf epidermal anatomy of *Cleistachne* show it to have similarities to *Sorghum halepense.* Conversely, leaf epidermal pat- terns in *Miscanthus japonicus* and *Microstegium nudum* did not show close resemblance to other *Sorghum* taxa, but quantifica- tion of epidermal characters is incomplete. *Microstegium nudum* and *Miscanthus* also differ from *Sorghum* species morphologically by having a digitate inflorescence, and the pedicellate and sessile spikelets are similar to each other. The grouping of *Microstegium nudum* and *Miscanthus japonicus* with *Sorghum* + *Cleistachne* may be an artifact of *ndh*F, so further investigation looking at addi- tional genes is necessary to resolve this. *Sorghum trichocladum* is the only new world species of *Sorghum* , found in isolated loca- tions of central and southern Mexico, Honduras, and Guate- mala. It was not sampled for the *ndh*F study. A recent study examined this species in detail and determined that it was dis- tinct enough to be considered its own genus, *Mesosorghum* (Dávila, pers. comm.). The remaining taxa of subtribe Sorghinae sampled with *ndh*F are not part of the clade containing *Sorghum* species, but are placed into a separate clade containing *Andropo- gon* and other members of what have been previously considered subtribes Andropogoninae and Anthistiriinae. The molecular data suggest most of the taxa placed in subtribe Sorghineae are not, in fact, close relatives of *Sorghum. Cleistachne* is an excep- tion, and should be considered part of *Sorghum* as currently cir- cumscribed. One scenario may be to tentatively accept the clade marked with an asterisk (Fig. 1) as being the genus *Sorghum,* modified pending the outcome of further investigation on *Mis- canthus* and *Microstegium nudum.*

Alternatively, there is *ndh*F support for recognition of three well supported lineages in *Sorghum* (bootstrap percentages over 70%). These could actually be considered distinct genera, especially given that the individual lineages within *Sorghum* have as many molecular characters supporting them as most generic level line- ages in the tribe. The lineage composed of *Sorghum bicolor, S.*

halepense, and *S. arundinaceum,* forms a clade of African and Mediterranean taxa, corresponding to the previously defined sub- genus Eu-Sorghum (Garber 1950). *Sorghum laxiflorum* and *S. macrospermum* are each monotypic members of their own respec- tive subgenera (Garber 1950), and overlap in range with the Aus- tralian and South-East Asian *Sorghum nitidum* (represented here by *Sorghum* aff. *nitidum*), forming an 'Austral-Asian clade'. The well supported clade containing *Sorghum angustum, S. bulbosum, S. stipoideum, S. timorense,* and *S. leiocladum* is an Australian clade, referred to below as such. Weak support for the African col- lected taxon, *S. sp.* , and the Australian clade as sister taxa may hint at the biogeographic origin of the Australian clade. There is moderate *ndh*F support for *Cleistachne* as sister to the Eu-Sor- ghum clade. Morphological differences in the degree of pedicel- late spikelet development, chromosome number, and epidermal patterns exist between the three major *Sorghum* clades. However, the few morphological characters examined so far are often homo- plasious. Cladistic analysis of these morphological characters will be necessary to compare with the branch support from the molec- ular data to determine whether these lineages are truly distinct.

Species determinations within the Eu-Sorghum lineage and within the Austral-Asian lineage (*S. laxiflorum* + *S. nitidum* + *S. macrospermum*) are generally not difficult. The individual of *S.* aff. *nitidum* sampled here may represent a possible hybrid given that its chloroplast (*ndh*F) and nuclear (*waxy*; R. Spangler, unpublished data) DNA sequences differ in their placement of this taxon. Investigation of this problem is ongoing. A tremen- dous amount of work has been done on Eu-Sorghum species limits (Snowden 1936, 1955). *Sorghum bicolor* is highly variable, but justification for the recognition of a single species that includes all the morphological variation was done by Harlan and DeWet (1972). *Sorghum halepense* is distinguished by its strongly rhizomatous habit and a chromosome number of n=20. The Austral-Asian lineage also has characters with non-overlap of character states sufficient for accurate species recognition. For example, the sessile spikelet of *S. macrospermum* is over twice the length of the other two species, and the pedicellate spikelet is well developed in *S. nitidum* compared to the reduced condition in *S. macrospermum* and *S. laxiflorum.*

The Australian clade, however, presents many problems in spe- cies recognition due to overlap of character states in features con- sidered important for species delimitation. In particular, taxa in subgenus *Stiposorghum* (Garber 1950), including many newly described species (Lazarides *et al.* 1991), are difficult to distin- guish. Key characters used to differentiate these species include size of the keel on the lower glume of the sessile spikelet, caryop- sis shape, and presence or absence of pruinosity, and show nearly continuous variation between species in the group. *Stiposorghum* species are all found in the northern part of Australia, often with multiple species overlapping extensively in their ranges. Collec- tion records and brief field examinations indicated that, in some cases, two or more defined species occur growing together in mixed populations. This suggests some of the named specimens may simply be variant individuals of extensive populations of *Sorghum* that occur across northern Australia. Field botanists have commented to me about the difficulty in identifying many *Stiposorghum* collections to species. Clearly, additional work on the crossing behavior and morphological taxonomy is necessary to decipher the actual species status of taxa in *Stiposorghum*.

CONCLUSIONS

Morphological characters have proved difficult to use in defining groups in Andropogoneae. Molecular evidence is accumulating to show the differences between lineages are few in number, possibly a reflection of a rapid evolutionary radiation in this tribe. Data from molecules have indicated *Sorghum* is not a monophyletic genus as traditionally perceived, and revising generic limits for the group is necessary. Expanding *Sorghum* to include *Miscanthus* and *Microstegium nudum* is not considered a strong alternative due to conflicting morphological characters and lack of additional molecular studies supporting that conclusion. The concept of three distinct lineages emerging from *Sorghum* is the preferred hypothesis, supported by both molecular and morphological characters, though only preliminarily at this point. Future cladistic work on additional molecules as well as morphology should help to resolve this generic-limits problem and give insights into morphological characters that may help distinguish species within the separate lineages.

ACKNOWLEDGEMENTS

Thanks to the organizing committee for the opportunity to present this work, Ben Zaitchik and Elizabeth Russo for tremendous efforts in the data collection, and Elizabeth Kellogg for being a fountain of ideas and help in all phases of this work. Funding for this project was supported by National Science Foundation grants to Elizabeth Kellogg (DEB 9419748) for research and Lynn Clark (DEB-9806584) for travel.

REFERENCES

Celarier, R. P. (1956). Cytotaxonomy of the Andropogoneae I. Subtribes Dimeriinae and Saccharinae. *Cytologia* **21**, 272-291.

Celarier, R. P. (1957). Cytotaxonomy of the Andropogoneae II. Subtribes Ischaeminae, Rottboelliinae, and the Maydeae. *Cytologia* **22**, 160-183.

Celarier, R. P. (1958). Cytotaxonomy of the Andropogoneae III. Subtribe Sorgheae, genus *Sorghum*. *Cytologia* **23**, 395-418.

Celarier, R. P. (1959). Cytotaxonomy of the Andropogoneae IV. Subtribe Sorgheae. *Cytologia* **24**, 285-303.

Clayton, W. D. (1972). The awned genera of Andropogoneae. Studies in the Gramineae: XXXI. *Kew Bulletin* **27**, 457-474.

Clayton, W. D. (1973). The awnless genera of Andropogoneae. Studies in the Gramineae: XXXIII. *Kew Bulletin* **28**, 49-58.

Clayton, W. D. (1987). Andropogoneae. In 'Grass Systematics and Evolution'. (Eds T. R. Soderstrom, K. W. Hilu, C. S. Campbell, and M. E. Barkworth.) pp. 307-309. (Smithsonian Institution Press: Washington, D. C.)

Clayton, W. D., and Renvoize, S. A. (1986). 'Genera Graminum'. (Her Majesty's Stationery Office: London.)

Duvall, M. R., and Doebley, J. F. (1990). Restriction site variation in the chloroplast genome of *Sorghum* (Poaceae). *Systematic Botany* **15**, 472-480.

Garber, E. D. (1950). Cytotaxonomic studies in the genus *Sorghum*. *University of California Publications in Botany* **23**, 283-361.

Hackel, E. (1889). 'Andropogoneae'. Monographiae Phanerogamarum. (Sumptibus G. Masson: Paris.)

Harlan, J. R., and de Wet, J. M. J. (1972). A simplified classification of cultivated sorghum. *Crop Science* **12**, 172-176.

Lazarides, M., Hacker J. B., and Andrew, M. H. (1991). Taxonomy, cytology and ecology of indigenous Australian sorghums (*Sorghum* Moench: Andropogoneae: Poaceae). *Australian Systematic Botany* **4**, 591-635.

Mason-Gamer, R. J., Weil, C. F., and Kellogg, E. A. (1998). Granule-bound starch synthase: structure, function, and phylogenetic utility. *Molecular Biology and Evolution* **15**, 1658-1673.

Pilger, R. (1954). Das System der Gramineae. *Botanische Jahrbücher* **76**, 281-384.

Snowden, J. D. (1936). 'The Cultivated Races of Sorghum'. (Adlard & Son, Ltd.: London.)

Snowden, J. D. (1955). The wild fodder Sorghums of the section Eu-Sorghum. *Journal of the Linnean Society of London* **55**, 191-260.

Spangler, R. E., Zaitchik, B., Russo, E., and Kellogg, E. (1999). Andropogoneae evolution and generic limits in *Sorghum* (Poaceae) using *ndh*F sequences. *Systematic Botany* **24**, in press.

Stapf, O. (1917-1919). 'Gramineae'. Flora of Tropical Africa. Volume 9 parts1-3. (Reeve and Co.: London.)

Sun, Y., Skinner, D. Z., Liang, G. H. and Hulbert, S. H., (1994). Phylogenetic analysis of *Sorghum* and related taxa using internal transcribed spacers of nuclear ribosomal DNA. *Theoretical and Applied Genetics* **89**, 26-32.

Watson, L., and M. J. Dallwitz. (1992). 'The Grass Genera of the World'. (CAB International: Wallingford, UK.)

CHLORIDOIDS

Chloris gayana, a southern African species now widely cultivated as a pasture grass.

Grasses: Systematics and Evolution. (2000). Eds S.W.L. Jacobs and J. Everett. (CSIRO: Melbourne)

PHYLOGENETIC RELATIONSHIPS IN SUBFAMILY CHLORIDOIDEAE (POACEAE) BASED ON *MAT*K SEQUENCES: A PRELIMINARY ASSESSMENT

K. W. Hilu[A] and L. A. Alice[B]

[A] corresponding author; Department of Biology, Virginia Polytechnic Institute and State University, Blacksburg VA 24061-0406, USA. email: hilukw@vt.edu
[B] Department of Biology, Western Kentucky University, Bowling Green KY 42101, USA.

Abstract

Sequence data from the chloroplast gene *mat*K were used to assess phylogenetic affinities and address taxonomic questions within the Chloridoideae. The monophyly of the subfamily is strongly supported. A major clade comprising two *Eragrostis* species and *Pappophorum* in one subclade and the Uniolinae (Eragrostideae) in another is sister to all other Chloridoideae. The two major tribes Chlorideae and Eragrostideae do not appear monophyletic, although several clades are well supported such as Boutelouinae, *Chloris* and related genera, Sporobolinae plus *Eragrostis advena*, and Triodiinae. The two large genera *Eragrostis* and *Chloris* do not seem monophyletic and a detailed analysis of their species could provide insight into the systematics of the subfamily. This study demonstrates the utility of *mat*K sequences in addressing existing taxonomic problems and identifying lineages in the Chloridoideae.

Key words: Chloridoideae, Poaceae, grasses, phylogeny, systematics, *mat*K, sequences.

INTRODUCTION

The Chloridoideae are part of the well-defined PACC group (Panicoideae, Arundinoideae, Centothecoideae, Chloridoideae; Davis and Soreng 1993), and are considered closely related to the Arundinoideae (Jacobs 1987). The subfamily has had a history of unsettled taxonomic problems at the tribal and generic levels. Jacobs (1987) noted that 'taxonomists agreed to disagree' on its classification. The number of tribes varies from two to eight and boundaries between the two major tribes, Chlorideae (Cynodonteae) and Eragrostideae, have been considered arbitrary (Hilu and Wright 1982; Campbell 1985; Hilu and Esen 1993). Campbell (1985), and Watson and Dallwitz (1992) merged the two tribes, emphasizing inconsistencies in the morphological and anatomical characters that distinguish them. Systematic problems in the Chloridoideae extend beyond these two major tribes, as suspected homoplasy and lack of consistency in characteristic features have resulted in disagreements over the taxonomic rank of the majority of its smaller tribes (e.g. Orcuttieae, Sporoboleae, Spartineae, Unioleae, and Zoysieae). Phylogenetic relationships among the chloridoid tribes are not well understood (reviewed in Hilu and Wright 1982), and proposed trends have not been subjected to rigorous testing. The monophyly of the Chloridoideae has also been questioned repeatedly (Campbell 1985; Kellogg and Campbell 1987; Jacobs 1987).

Traits often used in circumscribing the Chloridoideae are the three-nerved lemmas, chloridoid embryo (P+PF), inflated bicellular microhairs, and either the PCK or NAD-ME C_4 carboxylation pathway and associated ultrastructure or the XyMS+ leaf anatomy (Prat 1936; Reeder 1957; Tateoka *et al.* 1959; Carolin *et al.* 1973; Hattersley and Watson 1976; Brown 1977). Inflorescence type, floret number, lemma and glume features, and leaf epidermis characters are some of the main taxonomic traits used in tribal classification of the Chloridoideae. Morphological characters used in Chloridoideae systematics are extremely variable and often lack consistency at the tribal or even the generic level.

Clayton and Renvoize (1986), emphasizing the latter point, indicated that it is likely that physiological differences might often outweigh the morphological distinctions. These factors have contributed to obfuscating the taxonomic treatments of various tribes, raising questions about homoplasy in these characters and making decisions on character polarity and phylogenetic trends difficult to support.

This study explores the utility of sequence variation of the plastid gene *mat*K in the systematics of the Chloridoideae. The *mat*K coding region was sequenced for representative species from all chloridoid tribes. In grasses, the *mat*K gene is approximately 1540 base pairs (bp) and is nested within the *trn*K intron (Hilu and Liang 1997; K. Hilu *et al.* in press). This latter feature makes Polymerase Chain Reaction (PCR) amplification using primers for the *trn*K exons practical. *mat*K also shows greater variability than other chloroplast genes used in molecular systematics (Olmstead and Palmer 1994; Johnson and Soltis 1995; Hilu and Liang 1997) and, thus, has the potential for providing more characters and increased phylogenetic resolution.

MATERIALS AND METHODS

Taxa and Outgroups

A total of 30 species from 26 genera representing all tribes and subtribes of the Chloridoideae was sampled (Table 1). Our classification above the genus level generally follows Clayton and Renvoize (1986) whereas circumscription of *Chloris* and related genera largely follows Anderson (1974). *Phragmites* (Arundinoideae), *Zeugites* (Centothecoideae), and *Andropogon* (Panicoideae), members of the PACC group, were selected for rooting. Our choice of outgroups was based on results from preliminary analyses with representatives of non-PACC subfamilies or Joinvilleaceae as outgroups. These analyses showed no major differences in tree topologies between those and the PACC outgroups, but better resolution was obtained with PACC subfamilies as outgroups.

DNA Isolation, PCR Amplification, and Sequencing

Leaf tissue was harvested from either greenhouse-grown or field-collected plants. Total cellular DNA was isolated following Hilu *et al.* (1997). Because the *mat*K gene is part of the *trn*K intron we used two primers (MG1 and MG15), located in the *trn*K 5' and 3' exons respectively for PCR amplification. For sequencing, *trn*K region PCR products were electrophoresed in 0.8% agarose gels and DNA fragments of appropriate size excised and purified using a QIAquick gel extraction kit following the manufacturer's instructions (QIAGEN, Inc., Valencia, California). For each accession, the entire *mat*K coding region was sequenced with three to six primers (Table 2). Sequencing reactions were carried out using an ABI Prism™ Dye Terminator Cycle Sequencing Ready Reaction Kit with Amplitaq DNA polymerase FS (Perkin Elmer, Norwalk, Connecticut). Samples were electrophoresed in an ABI 373A automated DNA sequencer with a stretch gel (Applied Biosystems, Inc., Foster City, California). Chromatograms were manually edited using Sequence Navigator 1.0 software (Applied Biosystems, Inc., Foster City, California).

Sequence Alignment and Phylogenetic Analysis

Alignment of all sequences was unambiguous and, thus, done visually. We tested for phylogenetic signal by using the Random Trees option in PAUP*4.0b1 (Swofford 1998) and comparing the g_1 value for the distribution of tree lengths of 100,000 random trees using the critical value (at $\alpha = 0.05$) for 500 variable characters and 25 taxa. Beyond 15 taxa g_1 critical values change very little, allowing them to be used in a conservative test with more taxa (Hillis and Huelsenbeck 1992). Phylogenies were generated using Fitch parsimony as implemented in PAUP employing heuristic searches consisting of 1000 replicates of random stepwise addition of taxa with Mulpars on and tree-bisection-reconnection (TBR) branch swapping. No gaps were required to align the *mat*K sequences examined, and all characters were weighted equally. Sets of equally parsimonious trees were summarized by strict consensus. Decay indices (Bremer 1988; Donoghue *et al.* 1992) and bootstrap values (Felsenstein 1985), based on 100 replicates, were calculated as measures of support for individual clades. Decay analyses were performed with Auto-Decay (Eriksson and Wikstrom 1996) and the reverse constraint option in PAUP.

RESULTS

*mat*K Length, Sequence Divergence, and Nucleotide Site Variation

The *mat*K coding region is 1542 bp in length in the outgroups and Chloridoideae taxa examined. Three species had missing data: *Plectrachne* (11 bp), *Trichloris* (224 bp), and *Monodia* (276 bp). Ambiguous base calls in *Lepturus* were scored using the IUPAC (International Union of Pure and Applied Chemistry) symbols. Pairwise divergence of sequences ranges from 2.9-6.9% between the outgroups and Chloridoideae and 0.0% (*Chloris gayana* and *Eustachys mutica*) - 6.1% (*Bouteloua curtipendula* and *Eleusine indica*) within the Chloridoideae. Of the 1542 aligned characters, 407 (26.4%) are variable and 189 (12.3%) are parsimony informative. The third position of codons comprise 41.0% of the variable sites and account for 45.0% of the parsimony informative sites.

Phylogeny of Subfamily Chloridoideae

Phylogenetic signal in the Chloridoideae *mat*K data set is significant ($P < 0.01$) based on the g_1 value (-0.820). Cladistic analysis yielded three equally parsimonious trees, 643 steps in length (strict consensus in Fig. 1). The Consistency Index (CI) and Retention Index (RI), excluding uninformative characters, were 0.562 and 0.703, respectively. Monophyly of the Chloridoideae is strongly supported with a 98% bootstrap value and a decay index of 10. Within the Chloridoideae, the best supported clade is the most basal one that includes two well supported subclades. One of these subclades contains two genera of subtribe Uniolinae (100% bootstrap value and decay index of 5) and the second subclade contains *Eragrostis echinochloidea*, *E. minor* and *Pappophorum* (100% bootstrap value and decay index of 9). The sister group relationship of this clade to the remaining chloridoids is reasonably well supported with a bootstrap value of 83% and a decay index of 3. The next lineage in the Chloridoideae includes the apparently paraphyletic subtribe Sporobolinae. This clade of *Sporobolus*, *Muhlenbergia*, and *Eragrostis advena* has a bootstrap

Table 1. List of accessions sampled.
Origin is cited by country except for the United States where abbreviations of state names are included. Accessions sampled from the U.S. Department of Agriculture (USDA), Agricultural Research Service-National Plant Germplasm System are listed by their PI numbers. For others, collector name, number, and herbarium code (Holmgren *et al.* 1990) are noted. V = specimen vouchered at VPI. Other species were identified from leaves, spikelets, and seeds. *USDA collection PI 275326 sent as *Eleusine compressa* (Forssk.) Asch. & Schweinf. ex C. Chr. was identified as *Dinebra retroflexa* and PI 365011 sent as *Schmidtia pappophoroides* Steud. ex J.A. Schmidt was identified as *Eragrostis echinochloidea*.

Taxon	Origin	Voucher	GenBank
CHLORIDOIDEAE			
Chlorideae			
Boutelouinae			
Bouteloua curtipendula (Michx.) Torr.	TX-USA	PI 216213	AF144578
Pleuraphis jamesii Torr.	NM-USA	PI 476995[V]	AF144579
Chloridinae			
Chloris gayana Kunth	Uganda	PI 205251[V]	AF144582
Chloris mossambicensis K. Schum.	South Africa	PI 365060[V]	AF144583
Cynodon dactylon (L.) Pers.	South Africa	PI 224149	AF144584
Enteropogon acicularis (Lindl.) Lazarides	Australia	PI 238258	AF144585
Eustachys distichophylla (Lag.) Nees	Brazil	PI 404298	AF144586
Eustachys mutica (L.) Cufod.	South Africa	PI 410113[V]	AF144587
Trichloris crinita (Lag.) Parodi	Argentina	PI 265569[V]	AF144588
Pommereullinae			
Astrebla lappacea (Lindl.) Domin	Australia	PI 284733	AF144589
Zoysiinae			
Perotis rara R. Br.	Australia	PI 238348[V]	AF144590
Tragus berteronianus Schult.	South Africa	N. Barker 1128	AF144591
Eragrostideae			
Eleusininae			
Coelachyrum yemenicum (Schweinf.) S. M. Phillips	South Africa	PI 364502	AF144581
Eleusine indica (L.) Gaertn.	China	PI 408801	AF144580
Eragrostis advena (Stapf) Phillips	Australia	S. Jacobs 7960, NSW	AF144592
*Dinebra retroflexa (Vahl) Panz.	India	PI 275326[V]	AF144594
*Eragrostis echinochloidea Stapf	South Africa	PI 365011[V]	AF144605
Eragrostis minor Host	Afghanistan	PI 223263	AF144593
Trichoneura grandiglumis (Nees) Ekman	South Africa	PI 365064[V]	AF144595
Tridens brasiliensis Nees ex Steud.	Brazil	PI 310319[V]	AF144596
Monanthochloinae			
Aeluropus littoralis (Gouan) Parl.	ex Soviet Union	PI 392332	AF144597
Sporobolinae			
Muhlenbergia montana (Nutt.) Hitchc.	NM-USA	PI 477979[V]	AF144600
Sporobolus indicus (L.) R. Br.	Brazil	PI 310309[V]	AF144601
Triodiinae			
Monodia stipoides S. Jacobs	Australia	S. Jacobs 8032, NSW	AF144602
Plectrachne pungens (R. Br.) C. E. Hubb.	Australia	S. Jacobs 8031, NSW	AF144603
Uniolinae			
Fingerhuthia sesleriformis Nees	South Africa	PI 299968[V]	AF144606
Uniola paniculata L.	VA-USA	Knepper s. n[V].	AF144607
Leptureae			
Lepturus repens (G. Forst.) R. Br.	Australia	Latz 10843, MO	AF144598
Orcuttieae			
Orcuttia californica Vasey	CA-USA	O. Mistretta s. n., RSA	AF144599
Pappophoreae			
Pappophorum bicolor E. Fourn.	Mexico	PI 216526[V]	AF144604
OUTGROUPS			
Arundinoideae			
Phragmites australis (Cav.) Trin. ex Steud.	VA-USA	G. Fleming s. n.	AF144575
Centothecoideae			
Zeugites pittieri Hack.	Costa Rica	L. Clark 1171, ISC	AF144576
Panicoideae			
Andropogon gerardii Vitman	SD-USA	PI 315661	AF144577

175

Table 2. List of primers used in *mat*K sequencing.
Location is based on the 1542 character data matrix. In primer *mat*K 7B forward, R = A/G.

Primer name	Sequence (5' --> 3')	Location
*trn*K S5-1 forward	ACCCTGTTCTGACCATATTG	93 bp upstream of *mat*K
*mat*K 1210 reverse	GTAGTTGAGAAAGAATCGC	543-561
*mat*K W forward	TACCCTATCCTATCCAT	464-480
*mat*K 9 reverse	TACGAGCTAAAGTTCTAGC	1321-1339
*mat*K 7B forward	GATTTATCRGATTGGGAT	1195-1212
*trn*K MG15 reverse	AACTAGTCGGATGGAGTAGAT	*trn*K 3' exon

value of 100% and a decay index of 11. Subtribe Boutelouinae of the Chlorideae, represented by *Bouteloua* and *Pleuraphis*, also appears monophyletic although the current analysis does not include *Melanocenchris*, the only Old World genus in the subtribe. Many of the remaining nodes in the *mat*K phylogeny are not well supported and represent taxa of the Eragrostideae and the small tribes Leptureae and Orcuttieae. Of the three genera in which we sampled two or more species (*Chloris*, *Eragrostis*, and *Eustachys*), none appears monophyletic.

DISCUSSION

The monophyletic origin of the Chloridoideae has been questioned. In a cladistic analysis using primarily morphological and anatomical characters in the Poaceae, Kellogg and Campbell (1987) indicated that the Chloridoideae may be monophyletic but the evidence is less conclusive than in other subfamilies. Although they defined the Chloridoideae by the inflated distal microhair cell, they felt that this character was one of the weakest synapomorphies in their analysis. Jacobs (1987) further indicated that it would be wise not to discard the possibility of a polyphyletic origin of the Chloridoideae. He indicated that it is easier to explain the observed variation in physiological, anatomical and morphological characters on the basis of polyphyly. On the other hand, Clayton and Renvoize (1986) believe that the subfamily can be considered as a monophyletic unit whose adoption of efficient C_4 photosynthesis has led to successful proliferation in the tropics. The monophyly of the Chloridoideae is strongly supported in our *mat*K-based phylogeny (Fig. 1), a result that is consistent with data obtained from prolamin and immunological investigations (Hilu and Esen 1993).

Although the major tribes Chlorideae and Eragrostideae appear polyphyletic, various clades corresponding to subtribes or other tribes are evident. The lack of support for the Chlorideae and Eragrostideae as distinct lineages was also reflected in Van den Borre and Watson's (1997) morphological-anatomical study of the subfamily. The emergence of the Pappophoreae and Uniolinae in a strongly supported clade is an intriguing finding. Phylogenetic affinities between these two groups have not been previously suspected. Clayton and Renvoize (1986) considered the Pappophoreae as an isolated tribe in the subfamily but placed the Uniolinae as a basal group in their schematic representation of the relationships among Chloridoideae taxa. Van den Borre and Watson (1997) demonstrated the near-basal position of the Pappophoreae in their phylogeny, but the Uni-

olinae was internal. The Pappophoreae are delimited on the basis of the many-awned, many-nerved lemmas and the presence of elongated, bulbous tip microhairs in all of its genera except *Pappophorum* (Renvoize 1985). The Uniolinae are characterized by the many-nerved lemmas and leaf epidermis characters (Campbell 1985). Disagreement still exists on whether *Uniola* along with a few related genera should be considered as a subtribe in the Eragrostideae or given tribal status (reviewed in Hilu and Wright 1982 and Campbell 1985). Outstanding features common to the Pappophoreae and Uniolinae are the many-nerved lemma, tough rachilla, paniculate inflorescence, and 1-several florets per spikelet. This pappophoroid-unioloid affinity and its possible basal position in the Chloridoideae needs further study with other genes. If this phylogenetic position is substantiated by other independent data, it might provide some valuable insight into the evolution of the Chloridoideae and its relationship to the Arundinoideae as the Uniolinae are suspected to be a link between the two subfamilies (Clayton and Renvoize 1986).

The association of *Sporobolus*, *Muhlenbergia* and *Eragrostis advena* in a well supported clade is another important feature of the *mat*K phylogeny. *Sporobolus*, *Muhlenbergia* and related genera were segregated in a separate tribe, Sporoboleae, based on the paniculate inflorescence, 1-flowered spikelet and lack of culm pulvinus (Brown *et al.* 1959). Variation in these characters has led some to lump the tribe with the Eragrostideae (Stebbins and Crampton 1961; Hilu and Wright 1982; Gould and Shaw 1983; Clayton and Renvoize 1986; Peterson 1998). The association of *Sporobolus* and *Eragrostis* is not surprising. *Sporobolus* is distinguished from *Eragrostis* by the 1-flowered spikelet, 1-nerved lemma, and free pericarp, although intermediates do exist. *Eragrostis advena* fits well in the intermediate group with the suppressed lateral veins of its lemma and the free pericarp. This *mat*K phylogeny, thus, points to an alliance with the sporoboloid group and provides evidence for the polyphyly of *Eragrostis*.

The *Eleusine-Coelachyrum* clade received strong support. *Eleusine* has the several florets per spikelet characteristic of the Eragrostideae and most species have the digitate inflorescence of the Chlorideae. The taxonomic affinity between *Eleusine* and *Coelachyrum* is evident in the several florets per spikelet and the free pericarp. The inflorescence is variable in *Coelachyrum*, but the *C. yemenicum* inflorescence is in the form of racemes along a central axis, a type that approaches some of the *Eleusine* species. Clayton and Renvoize (1986) stated that *Coelachyrum* provides a link between

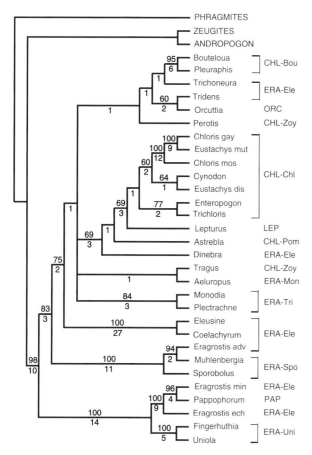

Fig. 1. *mat*K-based phylogeny of subfamily Chloridoideae. Bootstrap values are above branches and decay values are below branches. For each accession, tribes are denoted by three uppercase letters (CHL = Chlorideae; ERA = Eragrostideae; LEP = Leptureae; ORC = Orcuttieae; PAP = Pappophoreae). Subtribal classification codes are the following: Chlorideae (Bou = Boutelouinae; Chl = Chloridinae; Pom = Pommereullinae; Zoy = Zoysiinae) and Eragrostideae (Ele = Eleusininae; Mon = Monanthochloinae; Spo = Sporobolinae; Tri = Triodiinae; Uni = Uniolinae). Species names in genera where more than one species was sampled are abbreviated using the first three letters. Outgroup taxa are in uppercase.

Tridens, *Eleusine*, and *Eragrostis*. However, *Tridens* occurred in a separate clade sister to *Orcuttia*, and the *Eragrostis* species sampled were in distant clades (Fig. 1).

Among the remaining Chloridoideae taxa, *Monodia* and *Plectrachne* of subtribe Triodiinae formed a well supported clade. This subtribe includes four genera endemic to Australia and is defined by morphological and anatomical characters associated with its xerophytic habitat (Clayton and Renvoize 1986; Jacobs 1987). Clayton and Renvoize (1986) suspected some 'superficial resemblance' to certain members of the Arundineae. The internal position of the Triodiinae in the phylogeny does not support that notion. Conclusive assessment of its affinity awaits the inclusion of additional members of the Arundineae.

The Boutelouinae, represented by *Bouteloua* and *Pleuraphis*, formed a strongly supported alliance in an overall weakly supported clade that included *Trichoneura*, *Tridens*, *Orcuttia*, and *Perotis* (Fig. 1). A Boutelouinae clade was also evident in the

study of Van den Borre and Watson (1997). The Boutelouinae are New World in distribution except for *Melanocenchris* and defined by the usually deciduous raceme-type inflorescence with unilateral spikelets and the presence of trilobed lemmas. *Trichoneura* shares some inflorescence characteristics with the Boutelouinae, having appressed racemes on a central axis. *Orcuttia*, on the other hand, is one of three genera that has been treated as a separate tribe, Orcuttieae. The tribe is defined by many-nerved lemmas and absence of ligules (Clayton and Renvoize 1986), and is confined in distribution to vernal pools in California and Baja California. Although only one genus is represented here, the similarities between Orcuttieae and Pappophoreae based on the many-nerved lemmas appear superficial because the two tribes occurred in distant clades (Fig. 1).

Perotis, along with other genera such as *Zoysia* and *Tragus*, has been placed in the tribe Zoysieae based on inflorescence and spikelet characters. However, Clayton and Renvoize (1986) asserted that parallelism in some of these characters undermines the tribal status of the Zoysieae, and consequently the group has been included in the Chlorideae as a subtribe. Based on *mat*K data, the Zoysiinae (*Tragus* and *Perotis*) may be polyphyletic but bootstrap and decay support is low. The position of *Perotis* sister to the Boutelouinae gains support from other studies. In Hilu and Wright's (1982) morphological and anatomical study of the Poaceae, *Perotis* appeared close to *Bouteloua*. In the immunoblot experiments of the Chloridoideae (Hilu and Esen 1993), antiserum to prolamins of *Bouteloua gracilis* (Kunth) Lag. ex Griffiths reacted as strong with antigens of *Perotis* as it did with the homologous antigen (*Bouteloua* antigen), indicating very strong affinities between the two species.

The presence of the Chloridinae genera *Chloris*, *Cynodon*, *Enteropogon*, *Eustachys*, and *Trichloris* in a clade is taxonomically sound, although support is low. A similar assemblage of genera was also apparent in Van den Borre and Watson's (1997) study. *Chloris* and *Eustachys* are barely separable morphologically (Clayton and Renvoize 1986) and in some treatments are considered congeneric. The *mat*K data do not support monophyly of the two genera but do point to a close alliance as *Chloris gayana*, *C. mossambicensis* and *Eustachys mutica* formed a clade supported with a 100% bootstrap value and a decay index of 12. *Cynodon* is very closely related to *Chloris* and intergeneric hybrids between the two have been reported (Anderson 1974; Clayton and Renvoize 1986). The distinction between *Chloris* and *Trichloris* becomes very blurred when the Australian species of the latter are considered; in fact, Anderson (1974) united the two genera under *Chloris*. *Enteropogon* differs from *Chloris* and *Trichloris* primarily by the very rounded lemma and the narrow and dorsally compressed grain. Therefore, the clade includes a tightly knit group of genera that often are not well differentiated from each other.

Astrebla is an Australian genus that has been placed along with *Lintonia* and *Pommereulla* in the Chlorideae subtribe Pommereullinae (Clayton and Renvoize 1986). Material was not available for the other two genera to test the monophyly of the group, however the position of *Astrebla* in this *mat*K phylogeny suggests affinity to the Chlorideae.

Lepturus with its single, bilateral raceme, 1-flowered, sunken spikelet, and seashore habitat belongs to the monotypic Leptureae, a tribe with contentious affinities (Clayton and Renvoize 1986). In our *mat*K phylogeny, *Lepturus* was sister to the Chloridinae (69% bootstrap value and decay index of 3).

Dinebra, Eleusine, Coelachyrum, Tridens, Trichoneura, and *Eragrostis* are assigned to the Eleusininae (Eragrostideae) by Clayton and Renvoize (1986). Support from *mat*K data is lacking for the unity of this group. The apparent polyphyly of this subtribe is substantiated by the sister position of *Dinebra* to the Chloridinae plus *Lepturus* and *Astrebla* and by the association of the three *Eragrostis* species with the Sporobolinae and Pappophoreae.

Aeluropus and *Tragus* (Zoysiinae) formed a poorly supported clade (Fig. 1). *Aeluropus* belongs to the Monanthochloinae, a group of seven halophytic genera characterized by rolled, distichous, and often pungent leaves. Material for other species has been requested for a detailed analysis with *mat*K.

It is evident from this study that the major tribes Chlorideae and Eragrostideae are not monophyletic. However, some groups of closely knit taxa emerged such as Boutelouinae, *Eleusine-Coelachyrum* and Triodiinae. Emphasis on the number of florets per spikelet as a trait separating the Chlorideae from the Eragrostideae is not consistent with the *mat*K phylogeny. The two genera *Eragrostis* and *Chloris* do not appear monophyletic, and a more comprehensive analysis of the two could provide insight into the evolution of some lineages in the subfamily. The *mat*K gene appears useful in clarifying the systematics of this subfamily. A detailed study of the Chloridoideae is in progress.

ACKNOWLEDGEMENTS

The authors thank the U.S. Department of Agriculture, Agricultural Research Service-National Plant Germplasm System, Nigel Barker, Lynn Clark, Gary Fleming, Surrey Jacobs, and David Knepper for providing plant material. The authors also thank Patty Singer at the University of Maine DNA sequencing facility for her service, and Thomas Borsch, Irene Boyle, and Tracey Bodo Slotta for comments on the manuscript. This work was supported by NSF grant number DEB 9634231 to KWH.

REFERENCES

Anderson, D. E. (1974). Taxonomy of the genus *Chloris* (Gramineae). *Brigham Young University Science Bulletin, Biological Series* **19**, 1-133.

Bremer, K. (1988). The limits of amino acid sequence data in angiosperm phylogenetic reconstruction. *Evolution* **42**, 795-803.

Brown, W. V. (1977). The Kranz Syndrome and its subtypes in grass systematics. *Memoirs of the Torrey Botanical Club* **23**, iv + 97 pages.

Brown, W. V., Harris, W. E., and Graham, J. D. (1959). Grass morphology and systematics I. The internode. *Southwestern Naturalist* **4**, 115-125.

Campbell, C. S. (1985). The subfamilies and tribes of grasses in the southeastern United States. *Journal of the Arnold Arboretum* **66**, 123-199.

Carolin, R. C., Jacobs, S. W. L., and Vesk, M. (1973). The structure of the cells of the mesophyll and parenchymatous bundle sheath of the Gramineae. *Botanical Journal of the Linnean Society* **66**, 259-275.

Clayton, W. D., and Renvoize, S. A. (1986). 'Genera graminum.' (HMSO publications: London.)

Davis, J. I., and Soreng, R. J. (1993). Phylogenetic structure in the grass family (Poaceae) as inferred from chloroplast DNA restriction site variation. *American Journal of Botany* **80**, 1444-1454.

Donoghue, M. J., Olmstead, R. G., Smith, J. F., and Palmer, J. D. (1992). Phylogenetic relationships of Dipsacales based on *rbcL* sequences. *Annals of the Missouri Botanical Garden* **79**, 333-345.

Eriksson, T., and Wikstrom, N. (1996). 'AutoDecay 3.0.' (Stockholm University: Stockholm.)

Felsenstein, J. (1985). Confidence limits on phylogenies: an approach using the bootstrap. *Evolution* **39**, 783-791.

Gould, F. W., and Shaw, R. B. (1983). 'Grass Systematics.' 2nd ed. (Texas A & M University Press: College Station.)

Hattersley, P. W., and Watson, L. (1976). C$_4$ grasses: an anatomical criterion for distinguishing between NADP-malic enzyme species and PCK or NAD-malic enzyme species. *Australian Journal of Botany* **24**, 297-308.

Hillis, D. M., and Huelsenbeck, J. P. (1992). Signal, noise and reliability in molecular phylogenetic analyses. *Journal of Heredity* **83**, 189-195.

Hilu, K. W. (1981). Taxonomic status of the disputable *Eleusine compressa* (Gramineae). *Kew Bulletin* **36**, 559-563.

Hilu, K.W., ALice, L.A., and Liang, H. (in press). Phylogeny of Poaceae inferred fom *mat*K sequences. *Annals of the Missouri Botanical Garden*.

Hilu, K. W., and Esen, A. (1993). Prolamin and immunological studies in the Poaceae: III. Subfamily Chloridoideae. *American Journal of Botany* **80**, 104-113.

Hilu, K. W., de Wet, J. M. J., and Seigler, D. (1978). Flavonoids and the systematics of *Eleusine*. *Biochemical Systematics and Ecology* **6**, 247-249.

Hilu, K.W., and Liang, H. (1997). The *mat*K gene: sequence variation and application in plant systematics. *American Journal of Botany* **84**, 830-839.

Hilu, K. W., M'Ribu, K., Liang, H., and. Mandelbaum, C. (1997). Fonio Millets: ethnobotany, genetic diversity and evolution. *South African Journal of Botany* **63**, 185-190.

Hilu, K. W., and Wright, K. (1982). Systematics of Gramineae: A cluster analysis study. *Taxon* **31**, 9-36.

Holmgren, P. K., Holmgren, N. H., and Barnett, L. C. (1990). 'Regnum Vegetabile 120, Index Herbariorum. Part 1, The herbaria of the world, 8th ed.' (New York Botanical Garden: New York.)

Jacobs, S. W. L. (1987). Systematics of the chloridoid grasses. In 'Grass Systematics and Evolution'. (Eds T. R. Soderstrom, K. W. Hilu, C. S. Campbell, and M. E. Barkworth.) pp. 277-286. Smithsonian: Washington.

Johnson, L. A., and Soltis, D. E. (1995). Phylogenetic inference in Saxifragaceae sensu stricto and *Gilia* (Polemoniaceae) using *mat*K sequences. *Annals of the Missouri Botanical Garden* **82**, 149-175.

Kellogg, E. A., and Campbell, C. S. (1987). Phylogenetic analysis of the Gramineae. In 'Grass Systematics and Evolution'. (Eds T. R. Soderstrom, K. W. Hilu, C. S. Campbell, and M. E. Barkworth.) pp. 310-322. Smithsonian: Washington.

Olmstead, R. G., and Palmer, J. D. (1994). Chloroplast DNA systematics: a review of methods and data analysis. *American Journal of Botany* **81**, 1205-1224.

Peterson, P. M. (1998). Systematics of the Muhlenbergiinae (Chloridoideae: Eragrostideae). *American Journal of Botany* **85**, Suppl., 150.

Peterson, P. M., Webster, R. D., and Valdés-Reyna, J. (1995). Subtribal classification of the New World Eragrostideae (Poaceae: Chloridoideae). *Sida* **16**, 529-544.

Prat, H. (1936). La systématique des Graminées. Annales des Science Naturelles, Botanique, Ser. 10, **18**, 165-258.

Reeder, J. R. (1957). The embryo in grass systematics. *American Journal of Botany* **44**, 756-769.

Renvoize, S. A. (1985). A survey of leaf-blade anatomy in grasses. VII. Pommereulleae, Orcuttieae and Pappophoreae. *Kew Bulletin* **40**, 737-44.

Stebbins, G. L., and Crampton, B. (1961). A suggested revision of the grass genera of temperate North America. *Recent Advances in Botany* **1**, 133-145.

Swofford, D. L. (1998). 'PAUP*—Phylogenetic Analysis Using Parsimony. 4.0b1.' (Sinauer: Sunderland.)

Tateoka, T., Inoue, S., and Kawano, S. (1959). Notes on some grasses. IX. Systematic significance of bicellular microhairs of leaf epidermis. *Botanical Gazette* **21**, 80-91.

Van den Borre, A., and Watson, L. (1997). On the classification of the Chloridoideae. *Australian Systematic Botany* **10**, 491-531.

Watson, L., and Dallwitz, M. J. (1992). 'The grass genera of the world.' (C. A. B. International: Wallingford.)

GRASSES

Grasses: Systematics and Evolution. (2000). Eds S.W.L. Jacobs and J. Everett. (CSIRO: Melbourne)

ON THE CLASSIFICATION OF THE CHLORIDOIDEAE: RESULTS FROM MORPHOLOGICAL AND LEAF ANATOMICAL DATA ANALYSES

An Van den Borre[A] and Leslie Watson[B]

[A] Bioinformatics Group, Research School of Biological Sciences, Australian National University, GPO Box 475, Canberra ACT, 2601, Australia.
[B] 78 Vancouver St, Albany, WA, 6330, Australia.

Abstract

Throughout the history of the Chloridoideae a great variety of tribes and subtribes has been recognised. A predominant theme, however, has been the continued recognition of two main tribes centered on the large genera *Eragrostis* (Eragrosteae) and *Chloris* (Chlorideae, Cynodonteae). The taxonomic merit of this main division has been seriously questioned in recent times.

Recent investigations (Van den Borre & Watson 1997) have analysed detailed descriptions of the Grass Genera of the World database (Watson and Dallwitz 1992 and onwards) that were revised to incorporate characters resulting from a previous investigation of the largest Chloridoid genus, *Eragrostis* (Van den Borre and Watson 1994).

For this investigation a series of phenetic and cladistic analyses was conducted on 166 generic descriptions of Chloridoideae using 120 morphological and leaf anatomical characters (see Fig. 1 for results of ISS-flexible analysis on all genera of the Chloridoideae and outgroup genera *Rytidosperma* and *Schismus*, from Van den Borre and Watson 1997; results from other analyses and a complete list of characters and their states can be found in the same publication). The results were not supportive of the traditional main division, but instead indicated five quite different high level groups. By contrast the results afforded strong support for the small tribes Pappophoreae, Orcuttieae, and Triodieae. Based on these findings a new classification model for the subfamily was proposed that presented eight high level groups in the Chloridoideae (Table 1).

The results also highlighted the special problems posed by polymorphic taxa, such as *Eragrostis*, in planning and interpreting taxonomic and phylogenetic studies. *Eragrostis* posed a major problem in the interpretation of the results since its two subgenera emerged in two different high level groups.

Despite all uncertainties, the new groupings set out in the classification model permit better, more coherent descriptions than do their equivalents in earlier classifications. They also provide sampling guidance (Fig. 2) for continuing classificatory and phylogenetic studies, ranging from more of the same kind (using better phenetic and cladistic methods) to molecular sequencing, as well as for comparative ecophysiological and experimental work.

Key words: Chloridoideae, morphology, anatomy, classification.

Table 1. Informal classification of the Chloridoideae

TRIODIEAE

Monodia S. W. L. Jacobs, *Plectrachne* Henrard, *Symplectrodia* Lazarides, *Triodia* R. Br.

ORCUTTIEAE

Neostapfia Davy, *Orcuttia* Vasey, *Tuctoria* Reeder

PAPPOPHOREAE

Cottea Kunth, *Enneapogon* Desv. ex P. Beauv., *Kaokochloa* De Winter, *Pappophorum* Schreb., *Schmidtia* Steud. ex J.A.Schmidt

ZOYSIA GROUP

Boutelouinae

Bouteloua Lag., *Buchloë* Engelm., *Buchlomimus* Reeder, C. Reeder & Rzed., *Cathestecum* J. Presl, *Cyclostachya* Reeder & C. Reeder, *Griffithsochloa* G. J. Pierce, *Opizia* Presl, *Pentarrhaphis* Kunth, *Pringleochloa* Scribn., *Soderstromia* C.V. Morton

Zoysiinae

Aegopogon Humb. and Bonpl. ex Willd., *Decaryella* A. Camus, *Dignathia* Stapf, *Farrago* Clayton, *Hilaria* Kunth, *Leptothrium* Kunth, *Lopholepis* Decne., *Lycurus* Kunth, *Melanocenchris* Nees, *Monelytrum* Hack. ex Schinz, *Mosdenia* Stent, *Perotis* W. Aiton, *Pseudozoysia* Chiov., *Schaffnerella* Nash*, *Tetrachaete* Chiov., *Tragus* Haller f., *Zoysia* Willd.*

CHLORIS GROUP

Acrachne Wight and Arn. ex Chiov., *Afrotrichloris* Chiov., *Apochiton* C. E. Hubb., *Astrebla* F. Muell., *Austrochloris* Lazarides, *Brachyachne* (Benth.) Stapf, *Brachychloa* S.M. Phillips, *Chloris* Sw., *Chrysochloa* Swallen, *Cynodon* Rich., *Cypholepis* Chiov., *Dactyloctenium* Willd., *Daknopholis* Clayton, *Diplachne* P. Beauv., *Eleusine* Gaertn., *Enteropogon* Nees ex P. Beauv., *Eustachys* Desfv., *Gouinia* E. Fourn. ex Benth., *Harpochloa* Kunth, *Ischnurus* Balf. f., *Leptochloa* P. Beauv., *Lepturopetium* Morat, *Lepturus* R. Br., *Lintonia* Stapf, *Neobouteloua* Gould, *Neostapfiella* A. Camus, *Ochthochloa* Edgew., *Oxychloris* Lazarides, *Pereilema* J. Presl, *Pommereulla* L. f., *Pterochloris* (A. Camus) A. Camus, *Rendlia* Chiov., *Saugetia* Hitchc. Chase, *Schoenefeldia* Kunth, *Tetrapogon* Desf.

ERAGROSTIS GROUP I

Eragrostis subg. Caesiae group:

Acamptoclados Nash, *Ectrosia* R. Br., *Ectrosiopsis* Ohwi ex Jansen, *Eragrostis* Wolf subg. Caesiae, *Harpachne* Hochst. ex A. Rich, *Heterachne* Benth., *Planichloa* B. K. Simon, *Steirachne* Ekman, *Viguierella* A. Camus

Uniola group:

Entoplocamia Stapf, *Desmostachya* (Hook. f.) Stapf, *Leptochloöpsis* H.O. Yates, *Myriostachya* (Benth.) Hook. f., *Sclerodactylon* Stapf*, *Tetrachne* Nees, *Uniola* L.

Monanthochloë group:

Distichlis Raf.., *Jouvea* E. Fourn., *Monanthochloë* Engelm, *Psilolemma* S.M. Phillips, *Reederochloa* Soderstr. & H. F. Decker

Spartina group:

Calamovilfa (A. Gray) Hack. ex Scribn. & Southw., *Crypsis* Aiton, *Fingerhuthia* Nees, *Spartina* Schreb., *Sporobolus* R. Br.*, *Urochondra* C. E. Hubb.

ERAGROSTIS GROUP II

Allolepis Soderstr. & H. F. Decker, *Blepharidachne* Hack., *Cladoraphis* Franch., *Coelachyrum* Hochst. & Nees, *Dasyochloa* Willd. ex Rydb., *Diandrochloa* De Winter, *Eragrostis* subg. Eragrostis, *Erioneuron* Nash, *Hubbardochloa* Auquier, *Munroa* Torr., *Neeragrostis* Bush, *Neyraudia* Hook. f., *Odyssea* Stapf, *Piptophyllum* C. E. Hubb., *Psammagrostis* C. A. Gardner & C. E. Hubb., *Redfieldia* Vasey, *Scleropogon* Phil., *Stiburus* Stapf, *Swallenia* Soderstr. & H. F. Decker, *Thellungia* Stapf, *Tridens* Roem. & Schult., *Triplasis* P. Beauv., *Triraphis* R. Br., *Vaseyochloa* Hitchc., *Vietnamochloa* Veldkamp. & Nowack*.

MUHLENBERGIA GROUP

Aeluropus Trin., *Bealia* Scribn., *Bewsia* Goossens, *Blepharoneuron* Nash, *Chaboissaea* F. Fourn., *Chaetostichium* C. E. Hubb., *Coelachyropsis* Bor, *Craspedorhachis* Benth., *Ctenium* Panz., *Dinebra* Jacq., *Drake-Brockmania* Stapf, *Eragrostiella* Bor, *Gymnopogon* P. Beauv., *Halopyrum* Stapf, *Indopoa* Bor, *Kampochloa* Clayton, *Kengia* Packer, *Leptocarydion* Hochst. ex Stapf, *Lepturella* Stapf, *Lophacme* Stapf, *Microchloa* R. Br, *Muhlenbergia* Schreb., *Neesiochloa* Pilg., *Orinus* Hitchc., *Oropetium* Trin., *Pogonarthria* Stapf, *Pogoneura* Napper, *Pogonochloa* C. E. Hubb., *Polevansia* de Winter, *Richardsiella* Elffers & Kenn. O'Byrne, *Schedonnardus* Steud., *Schenckochloa* J. J. Ortiíz, *Silentvalleya* V. J. Nair, Sreek., Vajr. and Bhargavan, *Sohnsia* Airy Shaw, *Trichoneura* Andersson, *Tripogon* Roem. and Schult., *Willkommia* Hack.

---

*Indicates genera very tentatively allocated.

REFERENCES

Clayton, W. D., and Renvoize, S. A. (1986). 'Genera Graminum.' Kew Bulletin Additional Series XIII (HMSO: London.)

Van den Borre, A., and Watson, L. (1994). The infrageneric classification of *Eragrostis* (Poaceae). *Taxon* **43**, 383-422.

Van den Borre, A., and Watson, L. (1997). On the classification of the Chloridoideae (Poaceae). *Australian Systematic Botany* **10**, 491-531.

Watson, L., and Dallwitz, M.J. (1993). 'The Grass Genera of the World: Interactive Identification and Information Retrieval.' (URL http://biodiversity.bio.uno.edu/delta/grass/)

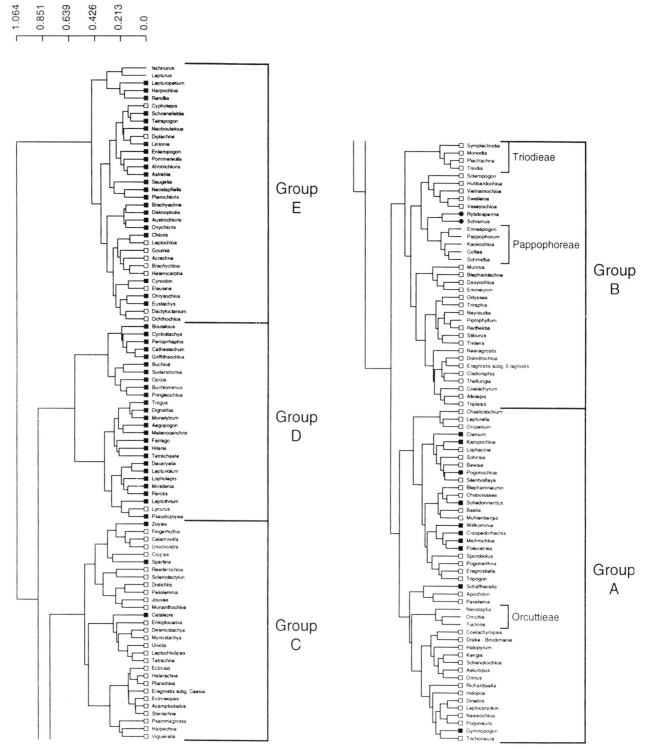

Fig. I. Analysis 3: ISS-flexible (s = 0.5) on all genera of the Chloridoideae and outgroup genera *Rytidosperma* and *Schismus* (●).
■: the genera traditionally (Clayton & Renvoize 1986) placed with the Cynodonteae
□: those placed with the Eragrostideae.

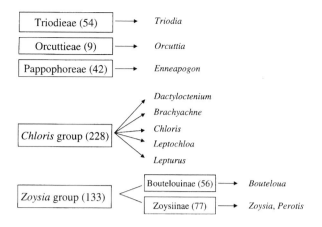

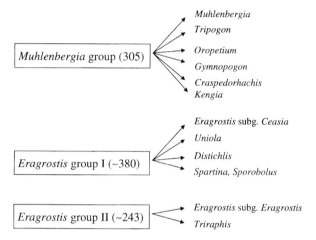

Fig. 2. Suggested representative genera for the Chloridoideae allocated to the informal groups specified in Table 1. The estimated number of species for each of these groups in parentheses.

GRASSES

Grasses: Systematics and Evolution. (2000). Eds S.W.L. Jacobs and J. Everett. (CSIRO: Melbourne)

PHYLOGENETIC RELATIONSHIPS OF THE GENUS *SPOROBOLUS* (POACEAE: ERAGROSTIDEAE) BASED ON NUCLEAR RIBOSOMAL DNA ITS SEQUENCES

Juan-Javier Ortiz-Diaz[A,B] *and Alastair Culham*[A]

[A] Centre for Systematics and Plant Diversity , Department of Botany, School of Plant Science, University of Reading, P. O. Box 221, Reading RG6 6AS. United Kingdom.

[B] Departamento de Botánica, Licenciatura en Biología, FMVZ, Universidad Autónoma de Yucatán, A. P. 4-116 Itzimná, Mérida, Yucatán, 97000 Mexico.

Abstract

The genus *Sporobolus* (L.) R. Br. (Poaceae:Eragrostideae) comprises approximately 160 species distributed throughout the tropics and subtropics. Morphological characters used to recognise *Sporobolus* are shared by other members of the subtribe Sporobolinae, thus the generic boundaries are unclear. In addition, Baaijens and Veldkamp's classification of the genus is incomplete, because they considered largely Malesian species. In this paper nuclear ribosomal DNA ITS sequences were used to assess the global relationships of the genus. The ITS sequence data rendered *Sporobolus* paraphyletic with *Calamovilfa gigantea, Crypsis alopecuroides* and two *Eragrostis* (*E. advena* and *E. megalosperma*) nested within *Sporobolus*. These genera have a follicoid caryopsis and a ciliate ligule as common features. The outgroups *Muhlenbergia montana, Blepharoneuron tricholepis* and *Eragrostis curvula* possess a true caryopsis and a membranous ligule, except the last taxon which also has a ciliate ligule. Sections *Sporobolus* and *Triachyrum* of Baaijens and Veldkamp's classification plus novel monophyletic clades are supported by high jackknife values.

Key words: *Sporobolus*, ITS, DNA.

INTRODUCTION

Sporobolus is a subcosmopolitan genus and one of the largest genera of the tribe Eragrostideae. It is found mostly in tropical and subtropical areas of the world: widely throughout Africa (73 spp.), North America (45 spp.) and Asia (34 spp.); less commonly in Australia (21 spp.) and South America (18 spp.) and sparsely in the Atlantic Islands (5 spp.), Japan (3 spp.), Indian Ocean Islands (3 spp.) and southern Europe (1 sp.). In natural habitats *Sporobolus* occurs under dry, saline, alkaline sandy and clayey conditions. Their elevational range is from sea level up to 3500 m in open grasslands, woodlands, savannahs, tropical forests and on roadsides. They show various degrees of tolerance to shading and can be found growing under the canopy of tropical forest (Ortiz pers. obser.). *S. heterolepis* is a codominant species in the temperate grasslands of the United States (Gould and Shaw 1983). They are very abundant elements in the African savan-nahs and a distinctive component of the halophilous and gypsophilous floras of the world (*S. airoides, S. coromandelianus, S. ioclados, S. rigens, S. spiciformis, S. virginicus, S. nealleyi, S. tremulus*). *S. indicus*. and *S. pyramidalis* colonise secondary habitats, replacing primary species after disturbance (Clayton 1964; Ortiz pers. obser.).

Sporobolus is currently placed in subfamily Chloridoideae, tribe Eragrostideae, subtribe Sporobolinae (Clayton and Renvoize 1986, 1992). Its main diagnostic characters are:

- the ligule is a line of hairs (ciliate).

- the inflorescence is an open or contracted panicle.

- the spikelets are one-flowered.

- the lemma is one-nerved.

Table 1. Morphological characters of *Sporobolus* and related genera.

Taxa/Char.	Ligule	Spikelet/ no. of flowers	Lemma/ no. of nerves	Type of caryopsis
Blepharoneuron	Membranous	1	3	True
Calamovilfa	Hairy	1(-2)	1	Follicoid
Crypsis	Hairy	1	1	Follicoid
Eragrostis	Hairy or membranous	2-many	3	True or follicoid
Muhlenbergia	Membranous	1(-2)	3	True
Sporobolus	Hairy	1(-2)	1(-3)	Follicoid
Urochondra	Hairy	1	1	Follicoid

• the fruit is a follicoid caryopsis (the pericarp peels off when it is moistened

• and the seed can either be expelled or remain attached to the spikelet).

None of the above characters is exclusive to *Sporobolus* since other genera of the tribe Eragrostideae share one or more of the characters (Table 1). Clayton and Renvoize (1986), Duvall *et al.* (1994), Watson and Dallwitz (1996) have pointed out that the boundaries of *Sporobolus* are unclear. Two-flowered spikelets are occasionally present in some species of *Sporobolus* as well as in *Calamovilfa* and *Muhlenbergia*. *S. subtilis*, the only representative of the subgenus *Chaetorhachia*, has a well developed rhachilla, which resembles that of the Cynodonteae genera.

Generally, the lemma in *Sporobolus* is one-nerved but three-nerved lemmas occur in several species. One-nerved lemmas are also present in *Calamovilfa*, *Crypsis* and *Urochondra* and three-nerved lemmas occur in many eragrostoid grasses. In addition, the follicoid caryopsis characteristic of *Sporobolus* is shared by these three genera.

Sporobolus is thought to be a derivative of *Eragrostis*. Clayton and Renvoize (1986) have pointed out the close relationships between *Eragrostis* and *Sporobolus* suggesting that the boundaries between the genera are blurred. Their data distinguished *Eragrostis* by its multi-flowered spikelets and three-nerved lemmas, and fruit with an adherent pericarp, but there are several species such as *E. advena*, *E. megalosperma* and *E. stapfianus* with intermediate characteristics (many-flowered or rarely one-flowered spikelets, one-to three-nerved lemmas, and a free pericarp) which have arbitrarily been assigned to *Eragrostis*. In addition, *E. advena* sensu Clayton and Renvoize (1986) has been recognised as *Thellungia* by Watson and Dallwitz (1996) on the basis of possession of several-flowered spikelets and three-nerved lemmas. Jacobs in 1987 suggested that these two genera should be included in *Eragrostis s.l.* and then the whole group be reclassified.

Taxonomically, the genus *Sporobolus* presents many problems. This genus is, after *Festuca*, one of the most problematic genera within the Poaceae (Renvoize pers. comm.). Although regional works have been conducted in this century (Parodi 1928; Bor 1960; Clayton 1964; Baaijens and Veldkamp 1991) there is no global treatment for the genus. Baaijens and Veldkamp (1991) made the first attempt to disentangle the complex relationships within *Sporobolus*. They proposed five sections: *Sporobolus*, *Triachyrum*, *Virginicae*, *Agrosticula* and *Fimbriatae*. Their sectional classification based on morphology and anatomy clarified the infrageneric relationships, but they considered largely Malesian species. The aim of this paper is to examine the generic limits of the genus *Sporobolus* within a global perspective based on ITS data. More specifically, a) to test the monophyly of the genus, and b) to assess the infrageneric classification of Baaijens and Veldkamp (1991).

MATERIALS AND METHODS

Forty-two taxa of *Sporobolus* were used in this study to represent the morphological variation within the genus, having at least one taxon for each of the sections of Baaijens and Veldkamp (1991). Three species of *Eragrostis* (subtribe Eragrostidinae), and one each of *Blepharoneuron*, *Muhlenbergia* (subtribe Muhlenbergiinae), *Calamovilfa* and *Crypsis* (subtribe Sporobolinae) were selected as outgroups to encompass a wide range of non-*Sporobolus* morphology.

Standard procedures for DNA extraction, amplification and automated sequencing of the nuclear ribosomal DNA ITS region were employed (Doyle and Doyle 1987; White *et al.* 1990; Sun *et al.* 1994). An alignment was constructed and employed as a data matrix. Two data sets were used for the cladistic analyses: data set A with insertion-deletion (indels) events treated as missing data and data set B with indels coded as binary characters. Data matrices were analysed using Phylogenetic Analysis Using Parsimony (PAUP)* version 4d64 (Swofford pers. comm.). Characters (nucleotide sites) were treated as unordered, weighted equally and optimised via accelerated transformation. For a particular taxon, a site having a multiple nucleotide was interpreted as a polymorphism. The heuristic option was used to search for all most parsimonious trees. Starting trees were obtained via random stepwise addition, with one tree held at each step. Tree bisection-reconnection was employed as the branch swapping algorithm. The steepest descent option was chosen. Zero-length branches were collapsed. One hundred replicates were performed on both data sets. Clade support was tested with jackknife values (1000 replicates, 33.33% deletion and emulating jackknife resampling) in PAUP* 4d64. Also computed using PAUP were the strict consensus trees, consistency index (CI) and retention index (RI).

RESULTS

As the strict consensus trees of both analyses were topologically congruent only results from data set B are shown. The heuristic analysis yielded 1236 most parsimonious trees of 1410 steps with a CI = 0.46 and RI = 0.70. The strict consensus tree (Fig. 1)

185

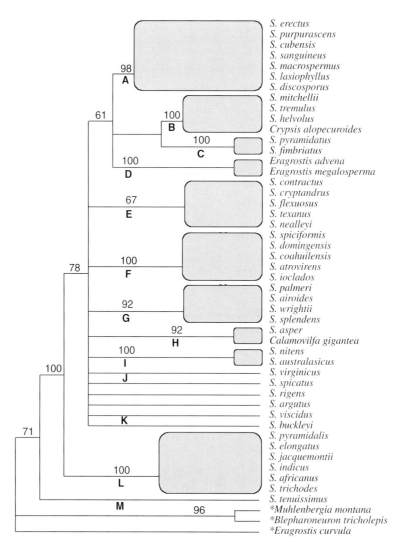

Fig. 1. Strict consensus of 1236 trees obtained from heuristic search of ITS region sequences of representatives of *Sporobolus* and related genera. Asterisks precede species designated as the outgroup. Bold characters represent the clades recognised here, including the Baaijens and Veldkamp's (1991) sections. Numbers above branches indicate jackknife support values. Tree length = 1410, CI = 0.46, RI = 0.70.

displays a polytomy, but several internal clades are resolved that have high jackknife support values, including sections *Sporobolus* (100%), *Triachyrum* (98%) and the *S. airoides* complex (92%).

Discussion

This study presents the first molecular phylogenetic hypothesis for relationships among species of *Sporobolus* and related genera. As a result of the above analysis the genus is resolved as paraphyletic (71% jackknife support) having two species of *Eragrostis*, *Calamovilfa* and *Crypsis* nested within *Sporobolus*.

These results also show a group consisting of all *Sporobolus* species studied, *Calamovilfa gigantea*, *Crypsis alopecuroides*, *Eragrostis advena*, and *E. megalosperma*. All members of this group share a follicoid caryopsis and a ciliate ligule. *Muhlenbergia* and *Blepharoneuron*, members of the subtribe Muhlenbergiinae, possess a true caryopsis and a membranous ligule (Peterson *et al.* 1995, 1997). *Eragrostis curvula* (subtribe Eragrostidinae) also possesses a true caryopsis, but its ligule is ciliate. This phylogeny suggests

that the evolution of a free pericarp was a single event. Further sampling within the tribe Eragrostidae would be needed to confirm this hypothesis.

Although the ITS data provide support for the sections *Sporobolus* and *Triachyrum* (Baaijens and Veldkamp 1991) the morphological characters used to recognise them, such as longevity, leaf-blades conspicuously keeled, panicle branches with glands and seeds more less ellipsoid in tranverse section are inconsistent. Therefore, the limits of the sections *Sporobolus* and *Triachyrum* have been expanded and refined to suit *S. trichodes*, *S. macrospermus*, *S. erectus* and *S. lasiophyllus*. In addition, six novel monophyletic clades were obtained from the ITS analyses, not previously treated by Baaijens and Veldkamp (1991), which may require formal recognition.

Clade A is the monophyletic section *Triachyrum* which has 98% jackknife support (Fig. 1). This section comprises *S. cubensis*, *S. discosporus*, *S. erectus*, *S. lasiophyllus*, *S. macrospermus*, *S. purpurascens* and *S. sanguineus*. These taxa are characterised by inflorescences with branches uniformly whorled, second glume as long as the spikelet, and fruits strongly laterally compressed.

Clade B comprises four taxa (*Crypsis alopecuroides, Sporobolus helvolus, S. mitchellii* and *S. tremulus*) with 100% jackknife support. Clayton and Renvoize (1986) pointed out that *Crypsis* is related to *Sporobolus*, especially to *S. spicatus*, and that *Crypsis* differs from *Sporobolus* mainly in the laterally compressed spikelets. Morphological characters that define *Crypsis* such as annual plants, spikelets laterally compressed, one-nerved lemmas and follicoid caryopsis are present elsewhere in *Sporobolus*. *Crypsis* is a small genus, with eight annual species distributed throughout the Mediterranean basin. The ITS tree shows that *Crypsis alopecuroides* is closely related to *S. helvolus, S. tremulus* and *S. mitchellii.*

Clade C corresponds with section *Fimbriatae* recognised by Baaijens and Veldkamp (1991). In our analysis *S. fimbriatus*, from Africa, is associated with *S. pyramidatus*, from America with 100% jackknife support. Morphological characters of *S. pyramidatus* simply do not fit in with those used to circumscribe the section, thus relationships between these two species remain obscure. It is noteworthy that *S. argutus*, not resolved in the tree, has been considered as *S. pyramidatus* by Hitchcock (1935, 1937) and Gould (1976). Close examination of *S. argutus* specimens suggests its close relationship to *S. pyramidatus.*

Clade D comprises two *Eragrostis* species. Clayton and Renvoize (1986) have pointed out that *E. advena* and *E. megalosperma* are intermediate species between *Eragrostis* and *Sporobolus*, by their possessing one-to several-flowered spikelets, one-to three-nerved lemmas, and follicoid caryopses. There are ten more *Eragrostis* species with a follicoid caryopsis (Clayton pers. comm.). The ITS tree suggests that these species are derived from within *Sporobolus* (100% jackknife support) and may show reversals or *de novo* development of those characters which give the impression of intermediacy. In fact, close examination of herbarium specimens of different *Sporobolus* species have shown variability in the number of nerves from one to three. I have also observed two-flowered spikelets in a sample of *S. buckleyi.*

Clade E consists of five species, *S. contractus, S. cryptandrus, S. flexuosus, S. nealleyi* and *S. texanus*. The first three have been considered traditionally as a part of the *S. cryptandrus* complex (Hitchcock 1935, 1937; Cronquist 1977). Morphological characters, such as spikelets laterally compressed, glumes with a scabrous nerve and a translucent yellow or orange endosperm makes this group consistent. All five species are confined to North America except for *S. cryptandrus* which is an amphitropical taxon occurring also in Argentina. These species were not treated by Baaijens and Veldkamp (1991).

Clade F comprises five species (*S. atrovirens, S. coahuilensis, S. domingensis, S. ioclados* and *S. spiciformis*) which has 100% jackknife support, but it seems that there are no morphological characters to define it. Ecological data show that all five species are confined to saline and alkaline soils. *S. ioclados*, from Africa is sister to four Middle American-Caribbean species. Its presence within this clade would suggest a possible African origin. Species of this group were not recognised by Baaijens and Veldkamp (1991).

Clade G is a novel clade comprising four species, *S. airoides, S. palmeri, S. splendens* and *S. wrightii*, and has 92% jackknife support. These species are characterised by being the tallest and most densely tufted of the genus. Moreover, they possess dorsally compressed spikelets and plump fruits. They inhabit alkaline-saline soils of xerophytic grasslands and are distributed in the Pacific coastal areas of Mexico and North America. *S. airoides* and *S. wrightii*, have been considered traditionally as a part of the *S. airoides* complex (Hitchcock 1935, 1937; Gould 1976; Cronquist *et al.* 1977; Gould and Moran 1981; Wipff and Jones 1994). Although the ITS phylogeny does not group *S. buckleyi* with any of the members of this clade, this species does have morphological characters that would justify its being part of this clade. None of the species of this clade was treated by Baaijens and Veldkamp (1991).

Clade H groups *Calamovilfa gigantea* and *Sporobolus asper* (92% jackknife support). Reeder and Ellington (1960) proposed that *Calamovilfa* is closely allied to *Sporobolus*. The genus is characterised by possession of one-flowered spikelets, one-nerved lemmas, follicoid caryopsis, and ciliate ligules. These characteristics are also found in *Sporobolus* and *Crypsis*. Fig. 1 shows that these three genera and the two follicoid caryopsis *Eragrostis* taxa seem to share a common ancestor and *Calamovilfa* is the only exclusively New World genus included in the subtribe Sporobolinae.

Clade I is a novel clade with 100% jackknife support, comprising only two species, *S. australasicus* and *S. nitens*. Superimposed morphological data suggests that they share the character of spherical fruit in common.

Clade J is a single taxon, *S. virginicus* which is a halophytic perennial grass with a worldwide distribution in latitudes ranging from temperate to tropical. Morphological data suggests that *S. spicatus* might be related to *S. virginicus*, but the lack of resolution in the tree does not resolve the relationship. Consequently, the monophyly of this section could not be evaluated. Further sampling among *S. virginicus* allied species (not available for this study) would test the monophyly of section *Virginicae*. Nevertheless, it can be seen that *S. virginicus* is not a member of the other sections shown.

Clade K consists solely of *S. viscidus*. This species is endemic to Central Mexico, has unique morphology, stalked-crateriform glands on sheaths and blades, a transluscent reddish-orange endosperm and a bulgy embryo underneath. Like *S. virginicus*, it is not included in any of the other clades.

Clade L comprises the section *Sporobolus* of Baaijens and Veldkamp (1991). It has 100% jackknife support. The species sampled here have been traditionally considered as a part of the *S. indicus* complex (Clayton 1964; Jovet and Guédes 1968; Baaijens and Veldkamp 1991; Manderet 1992). Morphological characters, such as glumes short, obtuse and fruits oblong-quadrangular define this group. Representatives of this section are pantropical in distribution suggesting an ancient origin.

S. trichodes is sister to this group. This taxon shares the common features of spikelet and fruit morphology with the other members of this group. However, *S. trichodes* has an open diffuse panicle (other members of this section have a densely to moderately contracted, occasionally rather diffuse panicle). According to Vegetti and Antón (1995) the grass inflorescence has undergone transformational stages from open diffuse panicles to contracted ones, and perhaps the presence of an open inflorescence in *S. trichodes* may represent a plesiomorphic character for the group. Morphology and ITS data sequences support its inclusion in Baaijens and Veldkamp's (1991) section *Sporobolus*.

Clade M is the monotypic section *Agrosticula* represented by the annual species *S. tenuissimus*. According to the ITS phylogeny this section occupies a basal position within the genus. There are shared morphological characters between sections *Agrosticula* and *Sporobolus*, but this section is clearly distinct. The scattered occurrence of annual species throughout the phylogeny indicates that longevity has little phylogenetic importance in this genus.

CONCLUSION

The data presented here cast serious doubt on the current generic circumscription of *Sporobolus*, *Calamovilfa* and *Crypsis*. It is clear that *Calamovilfa* and *Crypsis* form monophyletic clades within *Sporobolus* clades, so rendering *Sporobolus* paraphyletic. However, it is considered unwise to make any formal changes to the generic boundaries of these taxa based solely upon the nuclear ribosomal DNA ITS data set as it could it risk unnecessary nomenclatural instability. Systematists have already pointed out the very close affinities of *Sporobolus*, *Calamovilfa* and *Crypsis* (Clayton and Renvoize 1986, 1992; Watson and Dallwitz 1996; Peterson *et al.* 1997), but further data from morphology may provide a basis on which nomenclatural changes could be justified. The Baaijens and Veldkamp (1991) classification is not comprehensive because of the limited number of taxa covered and the small number of morphological characters used to define sections. A new system at the global level is needed for the genus.

ACKNOWLEDGEMENTS

We would like to thank the curators of the herbaria ANSM, CIIDIR, K and RNG for allowing the dissection of samples for DNA extraction. We thank the directors of the Royal Botanic Gardens, Kew and Regional Plant Introduction Station, Washington State University, USA for allowing access to living material from their garden and seed collections. S. L. Jacobs and J. P. Wipff have also kindly provided *Sporobolus* and *Eragrostis* seeds from Australia and USA respectively. We are grateful to many people for their assistance with field work. Particularly, J. Tun, E. Ucán, L. M. Ortega, S. Avendaño, Y. Herrera, S. González and E. Estrada. This research was possible thanks to the Consejo Nacional de Ciencia y Tecnologia (CONACYT) Grant no. 88944.

REFERENCES

Baaijens, G. H., and Veldkamp J. F. (1991). *Sporobolus* (Gramineae) in Malesia. *Blumea* **35**, 393-458.

Bor, N. L. (1960). 'Grasses of India, Burma and Ceylon'. (Pergamon: Oxford.)

Clayton, W. D. (1964). Studies in Gramineae: VI. *Kew Bulletin*. **17**, 287-296.

Clayton, W. D., and Renvoize S. A. (1986). Genera Graminum. *Kew Bulletin Additional Series* **13**, 1-398.

Clayton, W. D., and Renvoize S. A. (1992). A system of classification for the grasses. In 'Grass evolution and domestication'. (Ed. G. P. Chapman) pp. 338-353. (University Press: Cambridge.)

Cronquist, A., Holmgren, A., Holmgren, N. H., Reveal, J. L., and Holmgren P. (1997). The monocotyledons. Intermountain flora, vascular plants of the intermountain west, U. S. A. Vol. 6. (The New York Botanical Garden, Columbia University Press: New York.)

Doyle, J. J., and Doyle J. L. (1987). A rapid DNA isolation procedure for small quantities of fresh leaf tissue. *Phytochemistry Bulletin of the Botanical Society of America* **19**, 11-15.

Duvall, M. R., Peterson, P. M., and Christensen A. H. (1994). Alliances of *Muhlenbergia*(Poaceae) within New World Eragrostideae are identified by phylogenetic analysis of mapped restriction sites from plastid DNA. *American Journal of Botany* **81**, 622-629.

Gould, F. W. (1976). 'Grasses of Texas'. (Texas A and M University Press: College Station.)

Gould, F. W., and Moran R. (1981). 'The grasses of Baja California, Mexico'. (San Diego Society of Natural History. Memories 12: San Diego.)

Gould, F. W., and Shaw R. B. (1983). 'Grass systematics' 2nd. ed. (Texas A and M University Press: College Station.)

Hitchcock, A. S. (1935). *Sporobolus* (Poales) Poaceae (pars). *North American Flora* **17**, 481-482.

Hitchcock, A. S. (1937). *Sporobolus* (Poales) Poaceae (pars). *North American Flora* **17**, 483-496.

Jacobs, S. W. L. (1987). Systematics of the Chloridoid grasses. In 'Grass Systematics and Evolution'. (Eds T. R. Soderstrom, K. W. Hilu, C. S. Campbell and M. E. Barkworth.) pp. 277-286. (Smithsonian Institution Press: Washington, DC.)

Jovet, P., and Guédes M. (1968). Le *Sporobolus indicus* (L.) R. Br. var. *fertilis* (Steud.) Jov. and Guéd. naturalisé en France, avec une revue du *Sporobolus indicus* dans le monde. *Bulletin de Centre d'Études et de Recherches Scientifiques, Biarritz,* France **7**, 47-75.

Manderet, G. (1992). 'Étude de la variation phenotipique dans le groupe *Sporobolus indicus*' (L.) R. Br. Ph. D. Thesis. (Museum Nationale d'Histoire Naturalle: Paris).

Parodi, L. (1928). Revisión de las especies argentinas del género *Sporobolus*. *Revista de la Facultad de Agronomia y Veterinaria* **2**, 115-168.

Peterson, P. M., Webster, R. D., and Valdes-Reyna J. (1995). Subtribal classification of the New World Eragrostideae (Poaceae:Chloridoideae). *Sida* **16**, 529-544.

Peterson, P. M., Webster, R. D., and Valdes-Reyna J. (1997). Genera of New World Eragrostideae (Poaceae: Chloridoideae). *Smithsonian Contributions to Botany*. **87**, 1-57.

Reeder, J. R., and Ellington M. A. (1960). *Calamovilfa*, a misplaced genus of Gramineae. *Brittonia* **12**, 71-77.

Sun, Y., Skinner, D. Z., Lang, G. H., and Hulbert S. H. (1994). Phylogenetic analysis of *Sorghum* and related taxa using internal transcribed spacers of nuclear ribosomal DNA. *Theoretical and Applied Genetics* **89**, 26-32.

Vegetti, A., and Antón A. M. (1995). Some evolution trends in the inflorescence of Poaceae. *Flora* **190**, 225-228.

Watson, L., and M. Dallwitz. (1996). 'The grass genera of the world'. (CAB International: Wallingford.)

White, T. J., Burns T., Lee S., and Taylor J. (1990). Amplification and direct sequencing of fungal ribosomal RNA genes for phylogenetics. In 'PCR protocols: a guide to methods and applications' (Eds M. A. Innis, D. H. Gelfand, J. J. Sninski and T. J. White) pp. 315-322. (Academic Press: San Diego.)

Wipff, J. K., and Jones S. D. (1994). *Sporobolus potosiensis* (Poaceae:Eragrostideae): A new rhizomatous species from San Luis Potosi, Mexico and a new combination in *Sporobolus airoides*. *Sida* **16**, 163-169.

Grasses: Systematics and Evolution. (2000). Eds S.W.L. Jacobs and J. Everett. (CSIRO: Melbourne)

GRASSES

PHYLOGENETICS OF *BOUTELOUA* AND RELATIVES (GRAMINEAE: CHLORIDOIDEAE): CLADISTIC PARSIMONY ANALYSIS OF INTERNAL TRANSCRIBED SPACER (NRDNA) AND *TRN*L-F (CPDNA) SEQUENCES

J. Travis Columbus, Michael S. Kinney, Maria Elena Siqueiros Delgado and J. Mark Porter

Rancho Santa Ana Botanic Garden, 1500 North College Avenue, Claremont, California 91711-3157, U.S.A.

Abstract

Cladistic parsimony analyses of DNA sequences were carried out for *Bouteloua* and genera thought to be related: *Aegopogon, Buchloë, Buchlomimus, Cathestecum, Cyclostachya, Griffithsochloa, Opizia, Pentarrhaphis, Pleuraphis, Pringleochloa,* and *Soderstromia* (Gramineae: Chloridoideae). The outgroup consisted of *Chloris, Cynodon, Microchloa,* and *Tragus*. In all, 41 species were represented, including 25 species of *Bouteloua*. The nucleotide sequences comprised the internal transcribed spacers and intervening 5.8S exon of nuclear ribosomal DNA and *trn*L intron, *trn*L 3' exon, and *trn*L-*trn*F intergenic spacer of chloroplast DNA. Sequences from the two genomes were analyzed separately and in combination. The phylogenies were found to be highly, though not perfectly, concordant. Neither *Bouteloua* nor either of its subgenera (*Bouteloua* and *Chondrosium*) were found to be monophyletic. Sexual dimorphism of the spikelet and inflorescence appears to have evolved repeatedly and different pathways are implied. The remarkably similar *Chondrosium*-like staminate inflorescences of *Buchloë*, *Opizia,* and *Buchlomimus-Pringleochloa* likely do not represent the retention of an ancestral state, but instead an architecture that has evolved independently in these lineages through selection for male function.

Key words: *Bouteloua,* Boutelouinae, Chloridoideae, cladistics, DNA sequences, Gramineae, internal transcribed spacers (ITS), phylogeny, sexual dimorphism, *trn*L-*trn*F.

INTRODUCTION

The American grass genus *Bouteloua* (Chloridoideae) displays considerable variation in inflorescence form. When Lagasca described *Bouteloua* (nom. cons., originally *Botelua*) in 1805 he listed only five species, but among these were *B. racemosa* Lag. (= *B. curtipendula*) and *B. simplex* Lag., which represent extremes in inflorescence form even among all 42 species currently recognized (Gould 1980; Reeder and Reeder 1981; Beetle 1986; Columbus 1996). *Bouteloua curtipendula* (the type species) possesses numerous, short, deciduous branches per inflorescence, each branch bearing few, appressed spikelets, while *B. simplex* has a single (rarely more), long, persistent branch with numerous, spreading (pectinate) spikelets that disarticulate at the base of the fertile (proximal) floret. The latter species is the type of subgenus *Chondrosium*, which has been treated by some authors, including

Clayton and Renvoize (1986), as a separate genus. Griffiths (1912) and Gould (1980) followed Lagasca's concept of *Bouteloua* in their revisions of the genus, employing *Chondrosium* as one of two sections and subgenera, respectively.

Genera thought to be related to *Bouteloua* were grouped by Clayton and Renvoize (1986) into subtribe Boutelouinae of tribe Cynodonteae (Table 1). While most of these genera have long been linked to *Bouteloua*, including segregates *Buchlomimus*, *Cyclostachya,* and *Neobouteloua,* some genera, in particular *Aegopogon, Hilaria, Pleuraphis, Schaffnerella,* and *Soderstromia,* have been associated less frequently with *Bouteloua* and in fact have been placed by many authors (e.g., Bentham 1881; Hackel 1887–1888, 1897; Beal 1896; Nash 1912; Hitchcock 1913, 1951; Bews 1929; Roshevits 1937) in a tribe separate from the other genera, Zoysieae Benth.

Table 1. Genera comprising subtribe Boutelouinae (Clayton and Renvoize 1986, for the most part), numbers of species, and natural geographic distributions.

Genus	No. spp.	Distribution
Aegopogon Humb. & Bonpl. ex Willd.	4	Americas, Papua New Guinea
Bouteloua Lag., nom. cons.		
subg. *Bouteloua*	25	Americas, West Indies
subg. *Chondrosium* (Desv.) Gould[A]	17	Americas
Buchloë Engelm., nom. cons.	1	North America
Buchlomimus Reeder, C. Reeder, & Rzed.	1	Mexico
Cathestecum J. Presl	4	North America
Cyclostachya Reeder & C. Reeder	1	Mexico
Griffithsochloa G. J. Pierce	1	Mexico
Hilaria Kunth	7	North America
Melanocenchris Nees	2	NE Africa, SW Asia
Neobouteloua Gould	2	Chile, Argentina
Opizia J. Presl	2	Mexico
Pentarrhaphis Kunth	3	Mexico, Central America, Colombia, Venezuela
Pleuraphis Torr.[B]	3	U.S.A., Mexico
Pringleochloa Scribn.	1	Mexico
Schaffnerella Nash	1	Mexico
Soderstromia C. V. Morton	1	Mexico, Central America

[A]Treated as a genus by Clayton and Renvoize (1986).
[B]Treated as a synonym of *Hilaria* by Clayton and Renvoize (1986).

An outstanding feature common to all genera except *Bouteloua, Melanocenchris, Neobouteloua, Pentarrhaphis,* and *Schaffnerella* is sexual dimorphism of the spikelet and often inflorescence. In *Aegopogon, Cathestecum, Griffithsochloa, Hilaria,* and *Pleuraphis,* borne on each inflorescence branch are a central (terminal) spikelet and two lateral spikelets (one often not developed in *A. bryophilus* Döll), the central differing nearly always in sex and to some degree morphologically from the laterals. The central spikelet contains a hermaphrodite or carpellate floret, with or without additional staminate, neuter, and/or rarely hermaphrodite florets, while the lateral spikelets normally possess staminate florets and, in *Aegopogon, Cathestecum,* and *Griffithsochloa,* differ primarily from the central in being smaller and developing fewer florets (in *Cathestecum* the first glumes also differ). Spikelets of *Hilaria* and *Pleuraphis,* on the other hand, are quite dimorphic. With hermaphrodite central spikelets, *Aegopogon* and *Pleuraphis* are andromonoecious, whereas *Hilaria* has a carpellate central spikelet and is monoecious. Sexuality in *Cathestecum* and *Griffithsochloa* is more complicated, as described by Pierce (1978, 1979). Both andromonoecious and monoecious plants are found in a majority of species, but *C. varium* Swallen also exhibits trimonoecism (staminate, carpellate, and hermaphrodite flowers all present) and *C. brevifolium* is predominantly dioecious, though sometimes monoecious and very rarely andromonoecious plants are encountered.

The staminate and carpellate spikelets of *C. brevifolium,* therefore, are usually distributed on separate inflorescences (and on separate plants). Of further interest, the staminate and carpellate lemmas are short- and long-awned, respectively, while awns of monoecious plants are typically intermediate in length (Pierce 1979). Dimorphic inflorescences are also characteristic of *Buchloë, Buchlomimus, Cyclostachya, Opizia, Pringleochloa,* and *Soderstromia.* As in *Cathestecum brevifolium,* the carpellate lemmas of all of these genera except *Buchloë* are long-awned compared to their short- or unawned staminate counterparts. Awn length is but one of a number of traits that may vary according to gender; others include inflorescence height, number of branches per inflorescence, branch length, number of spikelets per branch, spikelet length, number of florets per spikelet, bract texture (firmness) and vestiture, and lodicule presence/absence. Pierce (1979), in his detailed study of sexual dimorphism in *Cathestecum brevifolium,* even detected vegetative differences. None of the taxa exhibits all of the traits. Dimorphism is most conspicuous in *Buchloë* and *Pringleochloa,* which have burr-like carpellate inflorescences branches usually contained within or only slightly exceeding the foliage and *Chondrosium*-like staminate branches exserted well above the foliage. *Buchlomimus* and *Cyclostachya* appear to be obligately dioecious, whereas both monoecism and dioecism are known to be expressed in the remaining taxa. Hermaphrodite flowers are rare. Also, except for the short-lived *Opizia bracteata,* all are clonal (stoloniferous) perennials. Interestingly, Reeder and Reeder (1966) reported 'dioecy (or gynodioecy)' in some populations of *Bouteloua chondrosioides,* but the plants 'are essentially indistinguishable except that in some, flowers with only one type of sex organ are developed.' Later Reeder (1969) indicated that the populations are likely gynodioecious, having hermaphrodite and carpellate plants, a conclusion confirmed by H. E. Conner (personal communication), who raised seedlings from three Mexican localities. Fifty-seven plants were hermaphrodite and 118 were carpellate, which conforms to a 3:5 ratio ($\chi^2 = 1.814$) indicating recessive gene control of male sterility similar to that in *Cortaderia selloana* (Schult. & Schult. f.) Asch. & Graebn. (Arundinoideae) (Conner and Charlesworth 1989).

Recently, Columbus *et al.* (1998) carried out a cladistic parsimony analysis of internal transcribed spacer and 5.8S (hereafter

Table 2. Taxa, collections/vouchers, and origin of collections used in cladistic parsimony analysis of ITS and *trn*L-F sequences Asterisks denote outgroup taxa. In bold are members of *Bouteloua* subg. *Chondrosium*. Unless indicated otherwise, collection numbers are those of Columbus and vouchers are deposited in RSA.

Taxon	Number	Origin
Aegopogon cenchroides Humb. & Bonpl. ex Willd.	2383	Mexico: Mexico
Bouteloua americana (L.) Scribn.	Worthington 22775	Grenada: St. George
B. annua Swallen	2434	Mexico: Baja California Sur
B. aristidoides (Kunth) Griseb. var. *aristidoides*	2444	U.S.A.: Arizona
B. barbata Lag. var. ***barbata***	2229	U.S.A.: Arizona
B. chasei Swallen	2861	Mexico: Nuevo León
B. chihuahuana (M. C. Johnst.) J. T. Columbus	2824	Mexico: Chihuahua
B. chondrosioides (Kunth) Benth. ex S. Watson	2422	Mexico: Oaxaca
B. curtipendula (Michx.) Torr. var. *caespitosa* Gould & Kapadia	2500	U.S.A.: Arizona
B. elata Reeder & C. Reeder	2358	Mexico: Jalisco
B. eludens Griffiths	2272	U.S.A.: Arizona
B. eriopoda (Torr.) Torr.	2461	U.S.A.: Arizona
B. gracilis (Kunth) Lag. ex Griffiths	2460	U.S.A.: Arizona
B. hirsuta Lag.	2453	U.S.A.: Arizona
B. johnstonii Swallen	2851	Mexico: Coahuila
B. juncea (Desv. ex P. Beauv.) Hitchc.	Axelrod 8856 (UPRRP)	Puerto Rico: Guánica
B. karwinskii (E. Fourn.) Griffiths	2208	Mexico: Zacatecas
B. media (E. Fourn.) Gould & Kapadia	2420	Mexico: Oaxaca
B. parryi (E. Fourn.) Griffiths var. ***parryi***	2299	Mexico: Sonora
B. pectinata Feath.	2899	U.S.A.: Texas
B. ramosa Scribn. ex Vasey	2287	Mexico: Coahuila
B. rigidiseta (Steud.) Hitchc.	2231	U.S.A.: Texas
B. scorpioides Lag.	2344	Mexico: Mexico
B. triaena (Trin.) Scribn.	2357	Mexico: Jalisco
B. trifida Thurb. ex S. Watson	2465	U.S.A.: Arizona
B. williamsii Swallen	2353	Mexico: Jalisco
Buchloë dactyloides (Nutt.) Engelm.	2198	Mexico: Nuevo León
Buchlomimus nervatus (Swallen) Reeder, C. Reeder, & Rzed.	2336	Mexico: Mexico
Cathestecum brevifolium Swallen	2520	U.S.A.: Arizona
*Chloris virgata Sw.	2455	U.S.A.: Arizona
Cyclostachya stolonifera (Scribn.) Reeder & C. Reeder	3044	Mexico: Zacatecas
*Cynodon dactylon (L.) Pers.	2691	U.S.A.: California
Griffithsochloa multifida (Griffiths) G. J. Pierce	2417	Mexico: Oaxaca
*Microchloa kunthii Desv.	2345	Mexico: Mexico
Opizia bracteata McVaugh	2373	Mexico: Michoacán
Opizia stolonifera J. Presl	2375	Mexico: Michoacán
Pentarrhaphis scabra Kunth	2424	Mexico: Chiapas
Pleuraphis rigida Thurb.	2443	U.S.A.: Arizona
Pringleochloa stolonifera (E. Fourn.) Scribn.	2642	Mexico: Puebla
Soderstromia mexicana (Scribn.) C. V. Morton	2398	Mexico: Oaxaca
*Tragus racemosus (L.) All.	2228	U.S.A.: Arizona

referred to as ITS) sequences of nuclear ribosomal DNA representing 51 species and all of the genera above except *Melanocenchris*, *Neobouteloua*, and *Schaffnerella*. Neither *Bouteloua* nor its two subgenera were found to be monophyletic. These findings also suggested homoplasy in morphological, anatomical, and breeding system traits, including sexual dimorphism of the spike-let and inflorescence. Seeking a phylogenetic estimate from the nonrecombining chloroplast genome, Columbus *et al.* (in prep.) sequenced and analyzed the *trn*L intron, *trn*L 3' exon, and *trn*L-*trn*F intergenic spacer (hereafter simply *trn*L-F). The *trn*L-F phylogeny was found to be largely, though not perfectly, concordant with the ITS phylogeny and the same conclusions were reached.

In this study we explore the conflict between the phylogenies, analyze the two data sets in combination, and focus on the evolution of sexual dimorphism.

MATERIALS AND METHODS

The species and specimens common to the ITS and *trn*L-F studies (Columbus *et al.* 1998; in prep.) were selected for this investigation. In all, 37 ingroup species in 12 genera plus four outgroup genera/species were included (Table 2). Nucleotide sequences, alignments, and identification of insertions/deletions (indels) are provided in Columbus *et al.* (1998; in prep.). Indels shared by two or more taxa were coded and appended to the matrices. PAUP* version 4.0 beta 1 (Swofford 1998) was used for the analyses. All characters (nucleotide sites and indels) were weighted equally. Heuristic searches were employed to locate the most parsimonious (minimum-length) trees. To assess support for clades, bootstrapping was performed using 10,000 replicates and the 'fast' stepwise-addition option, and decay indices (Bremer 1988; Donoghue *et al.* 1992) were calculated as described in Columbus *et al.* (1998). Analyses were carried out for the ITS and *trn*L-F data sets separately and in combination.

RESULTS AND DISCUSSION

The searches yielded four, 20,803, and six most parsimonious trees from ITS, *trn*L-F, and the combined data, respectively. Shown in Figs 1–3 are the strict consensus trees, including bootstrap percentages and decay indices. Figure 4 is one of the most parsimonious trees from the combined analysis, drawn as a phylogram. Although there are differences between the ITS (Fig. 1) and *trn*L-F (Fig. 2) strict consensus trees, much agreement exists, and nearly all of the discordance involves poorly supported nodes. Clades having the same composition of taxa in all minimum-length trees from all three analyses are identified by bullets in Figs 1–4. There are no well-supported clades resulting from the combined analysis that are not also present and well supported in the ITS and/or *trn*L-F phylogenies. In only one case, involving *Opizia* and *Bouteloua chondrosioides*, do relationships well supported in both the ITS and *trn*L-F phylogenies conflict. *Opizia bracteata* is sister to *O. stolonifera* in the ITS phylogeny but is sister to *B. chondrosioides* in the *trn*L-F phylogeny. Monophyly of *Opizia* is strongly supported when the data are combined (Fig. 3).

It is apparent from these molecular data that neither *Bouteloua* nor either of its subgenera are monophyletic. Of particular note, *B. chondrosioides* forms a clade with *Opizia*; *B. eludens* forms a clade with *Buchloë, Cathestecum, Griffithsochloa, Pentarrhaphis,* and *Soderstromia*; and *B. annua* and *B. aristidoides* (subg. *Bouteloua*) form a clade with *B. eriopoda, B. hirsuta,* and *B. pectinata* (subg. *Chondrosium*). See Columbus *et al.* (1998) and Columbus (1999) for discussion of these and other clades.

It is also apparent that spikelet unisexuality has arisen repeatedly (Figs 1–3). Clearly, this assemblage of species is predisposed to this condition. As described above, unisexual spikelets are found either in monomorphic inflorescences having a triad of dimorphic spikelets per branch or in dimorphic inflorescences having various numbers (taxon dependent) of spikelets per branch. The spikelet triad arrangement evidently evolved in parallel on at least

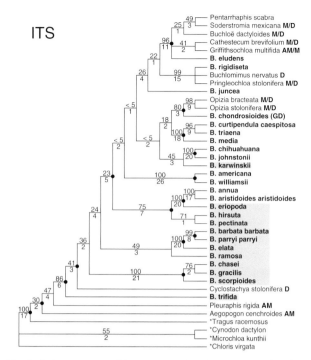

Fig. 1. Strict consensus of four most parsimonious trees resulting from a PAUP* analysis of ITS sequences of representatives of *Bouteloua* and related and outgroup genera. Asterisks denote outgroup taxa. In bold are members of *Bouteloua*. Shaded are members of *Bouteloua* subg. *Chondrosium*. Numbers above and below branches are bootstrap percentages and decay indices, respectively. Bullets denote clades having the same composition of taxa in all minimum-length trees from the ITS, *trn*L-F, and combined analyses. AM = andromonoecious, D = dioecious, GD = gynodioecious, M = monoecious. Tree length = 1600, consistency index = 0.46, retention index = 0.55.

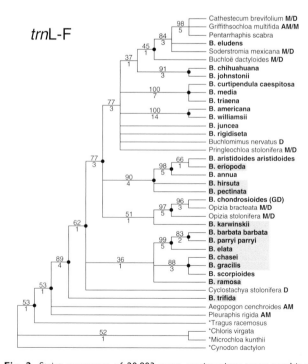

Fig. 2. Strict consensus of 20,803 most parsimonious trees resulting from an analysis of *trn*L-F sequences. Tree length = 348, consistency index = 0.76, retention index = 0.80. See Fig. 1 caption for further explanation.

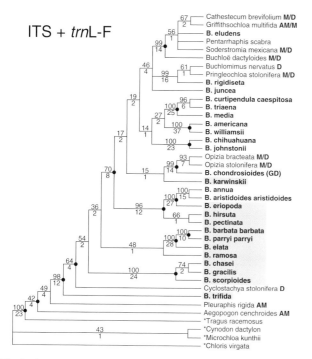

ITS + *trn*L-F

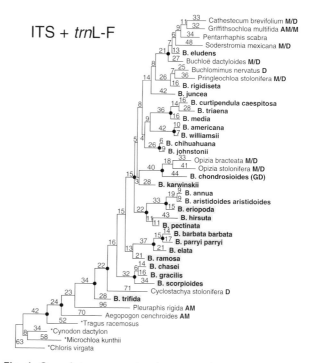

ITS + *trn*L-F

Fig. 3. Strict consensus of six most parsimonious trees resulting from a PAUP* analysis of combined ITS and *trn*L-F sequences of representatives of *Bouteloua* and related and outgroup genera. Asterisks denote outgroup taxa. In bold are members of *Bouteloua*. Shaded are members of *Bouteloua* subg. *Chondrosium*. Numbers above and below branches are bootstrap percentages and decay indices, respectively. Bullets denote clades having the same composition of taxa in all minimum-length trees from the ITS, *trn*L-F, and combined analyses. AM = andromonoecious, D = dioecious, GD = gynodioecious, M = monoecious. Tree length = 2001, consistency index = 0.51, retention index = 0.59.

Fig. 4. One of six most parsimonious trees, drawn as a phylogram, resulting from an analysis of combined ITS and *trn*L-F sequences. Numbers are branch lengths. See Fig. 3 caption for further explanation.

two occasions—in *Aegopogon* and *Pleuraphis* or a common ancestor, and in *Cathestecum* and *Griffithsochloa* or their precursor. *Soderstromia*, which has sexually dimorphic inflorescences, also has spikelets grouped in threes, but the two lateral spikelets are evident only as sterile bracts. It is possible, then, that an ancestor of *Soderstromia*, perhaps in common with *Cathestecum* and *Griffithsochloa*, had monomorphic inflorescences, functional lateral spikelets, and was andromonoecious or monoecious (like *Griffithsochloa* and most species of *Cathestecum*). Such a transition—evolution of dimorphic inflorescences and dioecism from monomorphic inflorescences and andromonoecism or monoecism—appears to have taken place more recently in *Cathestecum* (Pierce 1979). Based on the molecular data, *Buchloë* is closely related to *Cathestecum*, *Griffithsochloa*, and *Soderstromia* (Fig. 3), but there is no morphological evidence that it or any of the other taxa possessing dimorphic inflorescences ever had spikelets arranged in triads which, along with the multiple origin of inflorescence dimorphism and monoecism/dioecism, suggests that different pathways were involved.

Unlike *Cathestecum brevifolium* and *Soderstromia*, staminate branches of the other taxa having dimorphic inflorescences (*Buchloë*, *Buchlomimus*, *Cyclostachya*, *Opizia*, *Pringleochloa*) are like branches of most *Chondrosium* species, bearing numerous, spreading (pectinate) spikelets. In *Buchlomimus* and *Cyclostachya*, the carpellate branches are also *Chondrosium*-like, but those of

Buchloë, *Opizia*, and *Pringleochloa* have fewer spikelets than their staminate counterparts and differ greatly in appearance. This decoupling of the sexes affords a rare opportunity to consider inflorescence form in context of single-sex function. Pilger (1904), Reeder (1969), and Clayton and Renvoize (1986) concluded that natural selection has operated principally on the carpellate inflorescence, whereas the staminate inflorescence is relatively unmodified and similar to that of hermaphrodite relatives, a view bolstered by the rare occurrence in *Buchloë* of perfect-flowered inflorescences similar in form to staminate inflorescences (Quinn 1991). Conjecture on adaptation, therefore, has focused on the carpellate inflorescence. For instance, Reeder (1969) reasoned that the carpellate inflorescence tends to be shorter than the staminate inflorescence to enhance pollen capture and afford protection among the leaves, and carpellate spikelets tend to have better developed awns, hairs, etc., to aid in dispersal. Quinn *et al.* (1994) concluded that in *Buchloë* the position of the unawned carpellate inflorescence within the foliage is not for protection but rather to facilitate endozoochorous dispersal via large herbivores. Our interest here is the *Chondrosium*-like staminate inflorescence, which we propose may not have been the morphology of the hermaphrodite precursor but instead an architecture that has arisen repeatedly under selection for male function. It can be seen from the molecular phylogenies (Figs 1–3) that the *Chondrosium*-like staminate inflorescence appears to have evolved at least four times—in *Buchloë*, *Cyclostachya*, *Opizia* (assuming monophyly) or their precursors, and in *Buchlomimus* and *Pringleochloa* or their common ancestor. Significantly, the closest hermaphrodite relatives of these genera, except for *Cyclostachya*, do not possess the distinct *Chondrosium*-like inflorescence branches. While it is true that this inflorescence architecture is characteristic of most *Chondrosium* species and indeed may have been the morphology of the common (hermaphrodite) ancestor of the entire *Bouteloua* clade, the ITS and *trn*L-F phylogenies suggest that the remarkably similar staminate inflorescences of *Buchloë*, *Opizia*, and *Buchlomimus-Pringleochloa* do not represent the retention of an ancestral state, but instead an architecture that has evolved independently in these lineages. We submit that the staminate inflorescences of these taxa likely converged in form through selection for male function.

ACKNOWLEDGEMENTS

We thank Henry Conner and an anonymous reviewer for their comments on the manuscript. We are also grateful to Dr Conner for allowing us to incorporate his unpublished data and genetic interpretation of male sterility in *Bouteloua chondrosioides*.

REFERENCES

Beal, W. J. (1896). 'Grasses of North America.' (Henry Holt: New York.)

Beetle, A. A. (1986). Noteworthy grasses from Mexico XII. *Phytologia* **59**, 287–9.

Bentham, G. (1881). Notes on Gramineae. *The Journal of the Linnean Society. Botany* **19**, 14–134.

Bews, J. W. (1929). 'The World's Grasses: Their Differentiation, Distribution, Economics and Ecology.' (Longmans, Green and Co.: London.)

Bremer, K. (1988). The limits of amino acid sequence data in angiosperm phylogenetic reconstruction. *Evolution* **42**, 795–803.

Clayton, W. D., and Renvoize, S. A. (1986). Genera graminum: grasses of the world. *Kew Bulletin Additional Series* **13**, 1–389.

Columbus, J. T. (1996). *Bouteloua chihuahuana* (Gramineae), a new nomenclatural combination. *Aliso* **14**, 227.

Columbus, J. T. (1999). Morphology and leaf blade anatomy suggest a close relationship between *Bouteloua aristidoides* and B. (*Chondrosium*) *eriopoda* (Gramineae: Chloridoideae). *Systematic Botany* **23**, 467–478.

Columbus, J. T., Kinney, M. S., Pant, R., and Siqueiros D., M. E. (1998). Cladistic parsimony analysis of internal transcribed spacer region (nrDNA) sequences of *Bouteloua* and relatives (Gramineae: Chloridoideae). *Aliso* **17**, 99–130.

Conner, H. E., and Charlesworth, D. (1989). Genetics of male-sterility in gynodioecious *Cortaderia* (Gramineae). *Heredity* **63**, 373–82.

Donoghue, M. J., Olmstead, R. G., Smith, J. F., and Palmer, J. D. (1992). Phylogenetic relationships of Dipsacales based on *rbcL* sequences. *Annals of the Missouri Botanical Garden* **79**, 333–45.

Gould, F. W. (1980). The genus *Bouteloua* (Poaceae). *Annals of the Missouri Botanical Garden* **66**, 348–416.

Griffiths, D. (1912). The grama grasses: *Bouteloua* and related genera. *Contributions from the United States National Herbarium* **14**, 343–428.

Hackel, E. (1887–1888). Gramineae (echte Gräser). In 'Die natürlichen Pflanzenfamilien nebst ihren Gattungen und wichtigeren Arten ins-

besondere den Nutzpflanzen bearbeitet unter Mitwirkung zahlreicher hervorragender Fachgelehrten'. II. Teil. 2. Abteilung. (Eds A. Engler and K. Prantl.) pp. 1–97. (Wilhelm Engelmann: Leipzig, Germany.)

Hackel, E. (1897). Gramineae. In 'Die natürlichen Pflanzenfamilien nebst ihren Gattungen und wichtigeren Arten insbesondere den Nutzpflanzen, unter Mitwirkung zahlreicher hervorragender Fachgelehrten'. Nachträge zum II.–IV. Teil. (Ed. A. Engler.) pp. 39–47. (Wilhelm Engelmann: Leipzig, Germany.)

Hitchcock, A. S. (1913). Mexican grasses in the United States National Herbarium. *Contributions from the United States National Herbarium* **17**, 181–389.

Hitchcock, A. S. (1951). Manual of the grasses of the United States. 2nd edition, revised by A. Chase. *United States Department of Agriculture Miscellaneous Publication* **200**, 1–1051.

Lagasca, M. (1805). Memoria sobre un género nuevo de la familia de las gramas, llamado *Botelua*, y sobre otro de la misma familia que le es afine. *Variedades de Ciencias, Literatura y Artes* **2**, 129–43.

Nash, G. V. (1912). Zoysieae. In 'North American Flora'. Vol. 17 pp. 134–142. (The New York Botanical Garden: New York.)

Pierce, G. J. (1978). *Griffithsochloa*, a new genus segregated from *Cathestecum* (Gramineae). *Bulletin of the Torrey Botanical Club* **105**, 134–8.

Pierce, G. J. (1979). A biosystematic study of *Cathestecum* and *Griffithsochloa* (Gramineae). Ph. D. Thesis, University of Wyoming.

Pilger, R. (1904). Beiträge zur Kenntnis der monöcischen und diöcischen Gramineen-Gattungen. *Botanische Jahrbücher für Systematik, Pflanzengeschichte und Pflanzengeographie* **34**, 377–416.

Quinn, J. A. (1991). Evolution of dioecy in *Buchloe dactyloides* (Gramineae): tests for sex-specific vegetative characters, ecological differences, and sexual niche-partitioning. *American Journal of Botany* **78**, 481–8.

Quinn, J. A., Mowrey, D. P., Emanuele, S. M., and Whalley, R. D. B. (1994). The 'Foliage is the Fruit' hypothesis: *Buchloe dactyloides* (Poaceae) and the shortgrass prairie of North America. *American Journal of Botany* **81**, 1545–54.

Reeder, J. R. (1969). Las Gramíneas dioicas de México. *Boletín de la Sociedad Botánica de México* **30**, 121–6.

Reeder, J. R., and Reeder, C. G. (1966). Notes on Mexican grasses IV. Dioecy in *Bouteloua chondrosioides*. *Brittonia* **18**, 188–91.

Reeder, J. R., and Reeder, C. G. (1981). Systematics of *Bouteloua breviseta* and B. *ramosa* (Gramineae). *Systematic Botany* **5**, 312–21.

Roshevits, R. Yu. (1937). 'Grasses: An Introduction to the Study of Fodder and Cereal Grasses.' English translation published in 1980. (Indian National Scientific Documentation Centre: New Delhi, India.)

Swofford, D. L. (1998). 'PAUP*. Phylogenetic Analysis Using Parsimony (*and Other Methods).' Version 4.0 beta 1. (Sinauer Associates: Sunderland, Massachusetts.) (Computer software.)

Note: Since the Symposium, Columbus (1999; *Aliso* **18**, 61–5) has expanded the circumscription of *Bouteloua* to include *Buchloë, Buchlomimus, Cathestecum, Cyclostachya, Griffithsochloa, Opizia, Pentarrhaphis, Pringleochloa,* and *Soderstromia.*

Grasses: Systematics and Evolution. (2000). Eds S.W.L. Jacobs and J. Everett. (CSIRO: Melbourne)

GRASSES

SYSTEMATICS OF THE MUHLENBERGIINAE (CHLORIDOIDEAE: ERAGROSTIDEAE)

Paul M. Peterson

Department of Botany, National Museum of Natural History,
Smithsonian Institution, Washington, DC 20560-0166, U.S.A.

Abstract

An original DELTA data set of 88 primarily morphological characters was collected and then analyzed using cladistic algorithms for the New World genera (38) in the tribe Eragrostideae. The resulting phylogenetic hypotheses support the monophyly of the six genera within the Muhlenbergiinae when using *Chloris, Danthonia,* and/or *Pappophorum* as outgroups in various combinations. The six genera of the subtribe Muhlenbergiinae (*Bealia, Blepharoneuron, Chaboissaea, Lycurus, Muhlenbergia, Pereilema*) are characterized by membranous (rarely ciliate) ligules, 1 (rarely 2 or 3) floret per spikelet, 3-nerved lemmas, true caryopses, and a base chromosome number of 8 or 10. A combined analysis of chloroplast DNA restriction fragment insertions/deletions (124 characters) and morphological characters (88 characters) also supports the monophyly of the Muhlenbergiinae. The distribution of the 165 species in the subtribe is almost entirely New World; only 6 of the 151 species of *Muhlenbergia* occur in southern Asia. A historical review of the six genera currently recognized in the Muhlenbergiinae, a phylogeny and biogeographical history of the Muhlenbergiinae, a subgeneric hypothesis within *Muhlenbergia*, and a list of accepted names for all taxa within the subtribe are given.

Key words: Muhlenbergiinae, Chloridoideae, *Bealia, Blepharoneuron, Chaboissaea, Lycurus, Muhlenbergia, Pereilema*, systematics.

INTRODUCTION

The subtribe Muhlenbergiinae was first circumscribed by Pilger (1956) to include only species of *Muhlenbergia* Schreb. with narrow single-flowered spikelets, firm glumes often shorter than the awned lemmas, and cylindrical caryopses. In this same treatment Pilger recognized *Epicampes* J. Presl [sect. *Epicampes* (J. Presl) Soderstr. in *Muhlenbergia*] in subtribe Sporobolinae Ohwi, along with *Blepharoneuron* Nash, *Crypsis* Aiton, *Heleochloa* Host ex Roem., *Sphaerocaryum* Nees ex Hook. f., *Sporobolus* R. Br., and *Urochondra* C.E. Hubb. Pilger further divided *Muhlenbergia* into eight sections: *Muhlenbergia, Stenocladium* (Trin.) Bush, *Acroxis* (Trin.) Bush, *Cinnastrum* (E. Fourn.) Pilg., *Clomena* (P. Beauv.) Pilg., *Bealia* (Scribn.) Pilg., *Podosemum* (Desv.) Pilg., and *Pseudosporobolus* Parodi. Subsequent authors have agreed that Pilger's

subgeneric treatment of *Muhlenbergia* was not phylogenetically informative and not particularly useful (Soderstrom 1967; Pohl 1969; Morden 1985; Peterson and Annable 1991).

Pilger's (1956) overall classification of the grasses was very progressive and a good review of his work compared with earlier treatments appears in Soderstrom (1967). Generic limits were investigated by Phillips (1982) who concluded that the Eragrostideae Stapf are going through a period of adaptive radiation resulting in many small but distinctive genera. Pilger's treatment of tribe Eragrosteae Benth. (=Eragrostideae) included six subtribes [Eragrostinae Ohwi, Garnotiinae Pilg., Lycurinae Pilg. (*Lycurus* Kunth and *Pereilema* J. Presl), Muhlenbergiinae, Scleropogoninae Pilg., and Sporobolinae]. A comparison with more modern classifications, such as Clayton and Renvoize (1986) and Watson and

Dallwitz (1992), is found in Peterson *et al.* (1995). Peterson *et al.* (1995, 1997) placed the New World genera (including the autochthonous genera) of Eragrostideae into seven subtribes: Elusininae Dumort. (*Dactyloctenium* Willd. and *Eleusine* Gaertn.), Eragrostidinae J. Presl (*Eragrostis* Wolf, *Gouinia* E. Fourn. ex Benth., *Leptochloa* P. Beauv., *Neeragrostis* Bush, *Neesiochloa* Pilg., *Neyraudia* Hook., *Redfieldia* Vasey, *Scleropogon* Phil., *Sohnsia* Airy Shaw, *Steirachne* Ekman, *Trichoneura* Andersson, *Tridens* Roem. & Schult., *Triplasis* P. Beauv., *Tripogon* Roem. & Schult., and *Vaseyochloa* Hitchc.), Monanthochloinae Pilg. ex Potztal (*Allolepis* Soderstr. & H.F. Decker, *Distichlis* Raf., *Jouvea* E. Fourn, *Monanthochloë* Engelm., *Reederochloa* Soderstr. & H.F. Decker, and *Swallenia* Soderstr. & H.F. Decker), Muhlenbergiinae (*Bealia* Scribn., *Blepharoneuron*, *Chaboissaea* E. Fourn., *Lycurus*, *Muhlenbergia*, and *Pereilema*), Munroinae Pilg. ex P.M. Peterson (*Blepharidachne* Hack., *Dasyochloa* Willd. ex Rydb., *Erioneuron* Nash, and *Munroa* Torr.), Sporobolinae [*Calamovilfa* (A. Gray) Hack., *Crypsis*, and *Sporobolus*], and Uniolinae Clayton (*Teterachne* Nees and *Uniola* L.).

The most recent treatment of the Eragrostideae appears in a classification of the entire Chloridoideae that was based on 120 morphological and leaf anatomical characters (Van den Borre and Watson 1997). Their 'informal classification of the Chloridoideae' is quite surprising since it consists of eight main tribes or groups, these sometimes subdivided, and splits the genus *Eragrostis* into two groups, each aligned with other genera. For example, members of Muhlenbergiinae (Peterson *et al.* 1995, 1997) are found in the Zoysiinae (*Lycurus*), *Chloris* group (*Pereilema*), and *Muhlenbergia* group (*Bealia, Blepharoneuron, Chaboissaea,* and *Muhlenbergia*).

In this paper I will give a historical review of the six genera currently recognized in the Muhlenbergiinae, discuss their biogeographic trends, present cladograms of the New World Eragrostideae and Muhlenbergiinae, summarize a subgeneric hypothesis within *Muhlenbergia*, and provide a list of accepted names for all taxa within the subtribe.

HISTORICAL REVIEW

Muhlenbergia

The genus was proposed by Schreber (1789) in honor of G. H. E. Muhlenberg, a Lutheran Minister and pioneer botanist of Pennsylvannia. Many agrostologists have erected segregate genera to emphasize critical features of this large and diverse genus. Desvaux (1810) recognized the genus *Podosemum*, based on the caespitose, open-panicled, and long-awned *M. capillaris*. Palisot de Beauvois (1812) described the genus *Clomena* based on the annual, *M. peruviana*, and Presl (1830) described *Epicampes*, based on *M. robusta*. Two relatives of *M. schreberi* [*M. glomerata* and *M. andina*], were given generic status by Link (1833) as *Dactylogramma* and by Thurber (1863) as *Vaseya*, respectively. Nuttall (1848) described the genus *Calycodon* based on the widespread and often important range grass, *M. montana*. The only other generic name given to a species presently placed in *Muhlenbergia* is *Crypsinna*, described by Fournier (1886) and based on *M. macroura*. Hitchcock's (1935) transfer of many of these segregate genera to *Muhlenbergia* has been followed by most American and European botanists. The morphological

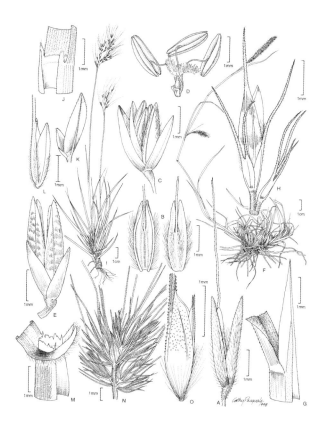

Fig. 1. The Muhlenbergiinae. *Bealia mexicana* (*Peterson & Annable 5800*): A. Spikelet. B. Deeply bilobed lemma, ventral and dorsal view. *Blepharoneuron tricholepis* (*Peterson & Annable 5567*): C. Spikelet. D. Pistil, stamens and lodicules. *Chaboissaea ligulata* (*Peterson & Annable 6198*): E. Spikelet with two florets. *Lycurus setosus* (*Peterson & Annable 11724*): F. Whole plant. G. Ligule. H. Inflorescence branch with two spikelets, upper perfect and fertile, lower sterile. *Muhlenbergia flaviseta* (*Peterson & King 8224*): I. Whole plant. J. Ligule. K. Glumes. L. Floret. *Pereilema crinitum* (*Hitchcock 9050*): M. Ligule and blade auricles. N. Inflorescence branch with spikelets subtended by sterile bristles. O. Floret.

characters that delimit the genus are spikelets with single perfect florets and hyaline or membranous lemmas with three usually prominent nerves (Fig. 1, I-L). These characters are however, not at all unique within the Eragrostideae and seem to be portrayed by about half of the genera in the tribe.

The distribution of *Muhlenbergia* is almost entirely New World, only six of the 151 species are known to occur in southern Asia. These southern Asian species of *Muhlenbergia* appear to be closely related to those species of the genus that inhabit the eastern portion of the United States and were referred to by Pohl (1969) as belonging to the subgenus *Muhlenbergia*. On the basis of anatomy, morphology, and cytology, Soderstrom (1967) distinguished two subgenera (*Muhlenbergia* and *Podosemum*) and divided subg. *Podosemum* into two sections, sect. *Podosemum* and sect. *Epicampes*. Soderstrom placed 46 species of *Muhlenbergia*, which have partially sclerosed phloem and caps of sclerenchyma associated with the primary vascular bundles, into subg. *Podosemum*. Two years later Pohl (1969) completed a revision of 12 closely related species that he believed represented the entire subg. *Muhlenbergia* in North America. Using characteristics of the rhizome (possession of very short internodes with imbricate scales) and leaf blade (thin, flat blades with low length/width ratios), Pohl distinguished these species from others in the genus. However, these same characteristics are seen in *M. californica*, a species of the mountains and valleys of southern California. Morden (1985) and Morden and Hatch (1987, 1996) have investigated the *M. repens* complex, which consists of six species in North and South America. Based on morphology, anatomy, cytology and, in part, flavonoid chemistry, a revision of 29 species of *Muhlenbergia* has been completed (Peterson 1988, 1989; Peterson and Rieseberg 1987; Peterson *et al.* 1989; Peterson and Annable 1991). More recently a biosystematic study investigating the morphology, anatomy, and flavonoid chemistry of the *M. montana* complex (consisting of 15 species) has been completed (Herrera-Arrieta and Bain 1991; Herrera-Arrieta and Grant 1993, 1994; Herrera-Arrieta 1998).

Pilger (1956) divided *Muhlenbergia* into eight sections within his monogeneric subtribe Muhlenbergiinae. The placement of *Muhlenbergia* within the Eragrostideae has long been problematical. With many of the characteristics of the entire tribe, alliances with many other genera have been suggested, most recently placed in the Sporobolinae by Clayton and Renvoize (1986) along with *Calamovilfa, Crypsis, Hubbardochloa* Auquier, *Lycurus, Pereilema, Sporobolus,* and *Urochondra* C. E. Hubb. *Hubbardochloa* and *Urochondra* are restricted to Africa and will not be discussed here whereas *Calamovilfa, Crypsis,* and *Sporobolus* seem to represent a closely allied group based on the possession of free pericarps and 1-nerved lemmas. That leaves us with two, smaller segregate genera, *Lycurus* and *Pereilema*. Based on similar leaf anatomical characters and preliminary analysis of restriction site variation of the chloroplast genome, *M. diversiglumis* has been suggested to be more closely related to species of *Pereilema* than to other species of *Muhlenbergia* (Peterson *et al.* 1989). Likewise, *M. brevis* and *M. depauperata* exhibit morphological features that suggest a close relationship with *Lycurus*. Mez (1921) indicated a relationship with *Lycurus* when he transferred *M. shaffneri* E. Fourn., considered as a synonym of *M. depauperata*, to *Lycurus*. *Muhlenbergia brevis* and *M. depauperata* share many morphological features with *Lycurus*, most importantly: spikelets borne in pairs; first glumes that are 2-nerved and 2-awned; second glumes that are 1-nerved and awned; 3-nerved, acuminate, awned lemmas with short pubescence along the margins; and pubescent paleas (Fig. 1, F-G). Perhaps these two species are intermediate between *Muhlenbergia* and *Pereilema*. Based on the lack of cork cells on the lemmatal surface *Blepharoneuron, Chaboissaea, Crypsis, Lycurus, Muhlenbergia,* and *Sporobolus* have been suggested to form an allied group (Valdes-Reyna and Hatch 1991). Recent data from restriction fragment variation of chloroplast DNA supports inclusion of *Blepharoneuron, Chaboissaea,* and *Muhlenbergia* along with *Bealia, Lycurus,* and *Pereilema* in an expanded Muhlenbergiinae (Duvall *et al.* 1994).

Lycurus

The genus was first described by Kunth (1816) as containing two species, *L. phalaroides* and *L. phleoides*. Based on the presence of perfect, paired spikelets, Nuttall (1848) described a monotypic genus, *Pleopogon setosum* Nutt. Recent studies by Reeder (1985) and Sanchez and Rugolo de Agrasar (1986) recognize three species including *L. setosus*. *Lycurus* can be separated from other

members of the Eragrostideae by possessing paired spikelets with the lower one usually being sterile and the upper one being perfect, a narrow, spiciform panicle, and 2-or 3-nerved and 2-or 3-awned first glume (Fig. 1, F-G). Based on anatomical and morphological similarities, particularly the paired spikelets, *Lycurus* was suggested to be closely related to *Aegopogon* Humb. & Bonpl. ex Willd. and *Tragus* Haller f. in the Zoysieae Benth. (Sanchez and Rugolo de Agrasar, 1986). Sanchez and Rugolo de Agrasar (1986) also point out the anatomical similarity between *Erioneuron* and *Lycurus*. Pilger (1956) erected the subtribe, Lycurinae, which included *Pereilema* as well. Hilu and Wright (1982) in a phenetic study of the Poaceae place *Lycurus* near *Hilaria*. Clayton and Renvoize (1986) take the traditional view and place *Lycurus* in the subtribe Sporobolinae along with *Calamovilfa, Crypsis, Hubbardochloa, Muhlenbergia, Pereilema, Sporobolus,* and *Urochondra*. Based on restriction site variation of the chloroplast genome, *Lycurus* appears to be most closely allied with *Muhlenbergia* and the following members of the Muhlenbergiinae: *Bealia, Blepharoneuron, Chaboissaea,* and *Pereilema* (Duvall *et al.* 1994).

Pereilema

Presl (1830) recognized *Pereilema* as a distinct genus and his concept has been narrowly interpreted by later authors. Kunth (1933) described a species of *Muhlenbergia* from Brazil that Hitchcock (1927) later transferred to *P. beyrichianum*. *Pereilema ciliatum* is native to Mexico and *P. crinitum* occurs throughout the range of the genus. A new species, *P. diandrum*, apparently restricted to Costa Rica (Davidse and Pohl 1992), brings the total to four in the genus. The genus can be separated from other members of the Eragrostideae by possessing sterile, bristle-like spikelets that subtend the fertile spikelets and prominent blade auricles that are usually ciliate (Fig. 1, M-O). The species occupy seasonally moist sites along disturbed slopes and in partial openings in the adjacent forest between 500-2100 m. *Pereilema* has been included in the Lycurinae by Pilger (1956) suggesting a relationship with *Lycurus* and was included more recently by Clayton and Renvoize (1986) in the Sporobolinae along with *Calamovilfa, Crypsis, Hubbardochloa, Lycurus, Muhlenbergia, Sporobolus, and Urochondra*. Based on restriction site variation of the chloroplast genome, *Pereilema* seems allied with *Bealia, Blepharoneuron, Chaboissaea, Lycurus,* and *Muhlenbergia* (Duvall *et al.* 1994).

Blepharoneuron

Blepharoneuron, s. str., contained a single, perennial species, *B. tricholepis*, that ranges from Colorado to Central Mexico. This species was originally placed in *Vilfa* Beauv. by Torrey (1857) and transferred to *Sporobolus* by Coulter (1885). Nash (1898) finally disagreed with Coulter's treatment and proposed the generic name, *Blepharoneuron*, which emphasizes the hairy and prominently nerved lemma and palea. The only other species included in the genus, *B. shepherdii*, was first placed in *Sporobolus* by Vasey (1887) and later transferred to *Muhlenbergia* by Swallen (1947). Peterson (1988) found *B. shepherdii* to possess a base chromosome number of x = 8 and presented a revision (Peterson and Annable 1990). The two species of *Blepharoneuron* differ from other eragrostoid genera by having a single floret per spikelet with densely appressed to spreading silky whitish hairs on the midnerve and margins of the lemma, minutely granular and capillary, wiry, flexuous, usually nodding and reflexed pedicels, and hairy paleas (Fig. 1, C, D). Reeder (1971) suggested that *Blepharoneuron* may form an alliance with *Erioneuron* because it shares certain morphological features with that genus, and shares a base chromosome number of x = 8. On the basis of cork and silica cell distribution on the surface of the lemma, Valdes-Reyna and Hatch (1991) suggested that *Blepharoneuron* is allied with *Chaboissaea, Crypsis, Lycurus, Muhlenbergia* and *Sporobolus*. Cladistic analysis of chloroplast DNA restriction fragment variation suggests that *Blepharoneuron* and *Bealia* share a most recent common ancestor within the monophyletic subtribe, Muhlenbergiinae (Duvall *et al.* 1994).

Chaboissaea

Historically, *Chaboissaea, s. str.*, contained a single species, *C. ligulata*, that ranges from northern Chihuahua to Distrito Federal, Mexico (Fournier 1886). *Chaboissaea,* although easily separated from *Muhlenbergia*, has traditionally been included in the latter genus by Hitchcock (1913), Bews (1929), Conzatti (1946), and more recently by Watson *et al.* (1985), Clayton and Renvoize (1986), and Renvoize (1998). Sohns (1953) emended the description of *C. ligulata* and suggested the genus belonged in the Festuceae rather than the Agrostideae, the latter an unnatural assemblage of mostly single-flowered genera. Reeder and Reeder (1988) transferred two Mexican annual species, *C. decumbens* and *C. subbiflora*, from *Muhlenbergia* into *Chaboissaea*. Based on morphological, anatomical, and cytological similarities, Peterson and Annable (1992) included a South American annual, *Muhlenbergia atacamensis* Parodi, in *Chaboissaea*. A revision of the genus and a hypothesized phylogeny based on morphological attributes is given in Peterson and Annable (1992). The four species of *Chaboissaea* differ from other eragrostoid genera in having gray to grayish-yellow spikelets with one or two (occasionally three) florets per spikelet, the lower floret perfect and the upper floret staminate or sterile, and a base chromosome number of x = 8 (Fig. 1, E). All four species occur in dark, clay soils in seasonally wet marshes, drainage ditches, and margins of ephemeral pools. Data from restriction site variation of chloroplast genomes suggest that *Chaboissaea* is most closely related to *Muhlenbergia*, and is best placed in the subtribe Muhlenbergiinae along with *Bealia, Blepharoneuron, Lycurus,* and *Pereilema* (Duvall *et al.* 1994). Our evidence from morphology, biogeography, soluble enzymes, and cpDNA restriction site variation suggests that the genus arose in north-central Mexico where three species still exist and that the progenitor of *C. atacamensis* migrated to South America recently (Peterson and Annable 1992; Peterson and Herrera A. 1996; Sykes *et al.* 1997). A well supported phylogeny derived from molecular data is found in Sykes *et al.* (1997) with the following topology: (*C. subbiflora* (*C. decumbens* (*C. atacamensis, C. ligulata*))).

Bealia

Vasey (1889) and Scribner first recognized the distinctive morphological features that distinguish *Bealia mexicana* from other members of *Muhlenbergia*. It has deeply bilobed lemmas with rounded to obtuse lobes, pilose or villous glumes that are 1-nerved and longer than the lemma, a crisped-curled to flexuous awn borne between the lobes, and minutely glandular pedicels (Fig. 1, A, B). Besides its morphological differences, *Bealia* has a base chromosome number of x = 8 and relatively large chromosomes when compared to *Muhlenbergia* (Peterson 1988; 1989). Although *Blepharoneuron, Dasyochloa, Erioneuron,* and *Munroa* share a base chromosome number of x = 8, only *Blepharoneuron* appears to be closely related to *Bealia*. Based on restriction fragment analysis of chloroplast DNA, *Bealia* and *Blepharoneuron* always form a clade within the monophyletic subtribe Muhlenbergiinae (Duvall *et al.* 1994). A recent study also indicated that *Bealia* is allozymically very similar to *Muhlenbergia argentea* (Peterson *et al.* 1993).

BIOGEOGRAPHY

The Muhlenbergiinae includes 165 species with three infraspecific taxa; 144 of these species occur in North America where the center of diversity is in northern Mexico (Appendix 3). There are 103 species with a distribution restricted to North America. Central America contains 39 species of Muhlenbergiinae, 24 with a North and Central American distribution pattern, 12 with a North, Central and South American distribution, and only two endemics (*Muhlenbergia xanthodas* and *Pereilema diandrum*). South America has 30 species of Muhlenbergiinae; 12 of these are endemic; 12 with a North, Central, and South American distribution; five with an amphitropical distribution (North/South America); and one with a Central and South American distribution. Asia contains six species with one infraspecific taxon primarily centered in southeastern China and Japan. One obvious hypothesis might be that this subtribe arose where it is most speciose today, i.e., northern Mexico, and has since radiated. For a dispersal event, the longer the distance from the origin, would in theory, lessen the chance of a successful introduction. Therefore, there are many species of Muhlenbergiinae in North America, less in Central America, even less in South America, although quite a few endemics, and finally very few in Asia. To address elemental patterns of past biogeographical histories among members of this tribe, along with many collaborators, I have investigated the allozyme diversity of five genera encompassing 10 species (Peterson and Columbus 1997; Peterson and Herrera A. 1996; Peterson and Morrone 1998; Peterson and Ortíz-Diaz 1998; Peterson *et al.* 1993; Sykes *et al.* 1997).

The distribution of plants on our planet has remained a subject of great interest, led by early explorers and plant geographers such as Humboldt and Bonpland (1805). The relationship between the temperate floras of North and South America was first discussed by Gray and Hooker (1880) who listed some 80 genera that shared infraspecific or congeneric species between the continents. Fifteen species in the Eragrostideae are disjunct between North and South America, and as many as 16 genera are known to have vicariant sister species (Allred 1981; Peterson *et al.* 1995). Speculation on the origin and migrational direction of these so called 'amphitropical disjuncts' has been the subject of at least four of my studies. Raven (1963), Thorne (1972), and Solbrig (1972) agree that the floristic similarity between North and South America is a result of the effects of long-distance dispersal, former continuous distributions, and parallel convergent evolution.

A classic example of a dominant North American grass species that is thought to have migrated to South America is burro grass, *Scleropogon brevifolius* Phil. A comparison of genetic identity values among populations from North and South America indicates that the genetic variation is much greater ($I = 0.88$) in North America than in South America ($I = 0.98$), and populations from South America lack 19 alleles found in the North American populations (Peterson and Columbus 1997). There are no unique alleles in the South American populations, therefore, it seems likely that *Scleropogon brevifolius* has recently dispersed to the Southern hemisphere.

Studies investigating the allozyme variation and cpDNA restriction site analysis in the amphitropical disjunct *Chaboissaea*, revealed the genetic structure and most-likely origin of this primarily Mexican genus (Peterson and Herrera A. 1996; Sykes *et al.* 1997). *Chaboissaea* includes four species, three occurring in Mexico, and one, *C. atacamensis*, in Argentina and Bolivia. *Chaboissaea atacamensis* had the highest genetic identity values ($I = 0.94$) among populations within each species suggesting recent migration to South America, probably from its closest sister *C. ligulata* (Sykes *et al.* 1997). The only tetraploid in the genus, *C. decumbens*, had no unique alleles and exhibited eight non-segregating loci, as would be expected. This north to south migration pattern is not always the case, and the reverse (south to north) is exhibited in *Bothriochloa* Kuntze (Andropogoneae; Allred 1981) and *Erioneuron* (Peterson in prep.). *Erioneuron*, like other members of subtribe Munroinae (including *Blepharidachne* and *Munroa*) has a center of diversity in South America.

Allozyme studies were useful in determining the genetic diversity and migrational direction of the amphitropical *Lycurus setosus* as well (Peterson and Morrone 1998). *Lycurus setosus* occurs in the southwestern U. S. A., northern Mexico, and again in northwestern Argentina, and Bolivia. A comparison of genetic identity values among populations from North and South America indicates that the genetic variation is greater ($I = 0.89$) in North America than in South America ($I = 0.94$), and populations from South America lack six alleles found in the North American populations. It seems likely that *Lycurus setosus* has recently dispersed to South America because the populations there contain less genetic variation.

Allozyme data were used to evaluate genetic diversity within and among populations of the amphitropical disjunct *Muhlenbergia torreyi* (Peterson and Ortíz-Diaz 1998). A comparison of genetic identity values among populations from North and South America indicates that the genetic variation is slightly greater ($I = 0.93$) in North America than in South America ($I = 0.96$). A total of 51 alleles were shared among all populations, and four unique alleles were detected; two from North American populations and two from South America. It seems likely that *Muhlenbergia torreyi* has recently dispersed to South America because the populations there are less variable. In a review on plant species disjunctions, Crawford *et al.* (1992) used Nei's (1978) formula for time ($t = D/_{2a}$), where D is the genetic distance and *a* is the substitution rate per locus per year (10^{-7} is the commonly employed rate), to calculate time of divergence. The genetic distance (D) for populations of *M. torreyi* from North America when compared to South American populations is 0.08 (+ 0.05). If we apply these

assumptions to our single species with disjunct populations, the time of divergence is calculated to be 400,000 years (±250,000 years). We hypothesize that the actual time of the colonizing event was much less than this calculated value.

Similar studies of allozyme variation in the Mexican sympatric endemics, *Bealia mexicana*, *Muhlenbergia argentea*, and *Muhlenbergia lucida*, revealed that all three species have relatively high intraspecific genetic variability (H ranging from 0.19 to 0.26) and high levels of genetic diversity (F ranging from 0.073 to 1.000) indicative of mixed mating and/or outcrossing plants (Peterson *et al.* 1993). Mean genetic identity values for pairwise comparisons of these three taxa indicated that *Bealia* was more similar to *M. argentea* than were the two species of *Muhlenbergia*. These data suggest that *M. argentea* should be placed in *Bealia* or that the recent reinstatement of *Bealia* is unwarranted. Results from a cpDNA restriction site survey indicates that *Bealia* is more closely aligned with *Blepharoneuron* by sharing three parallel site losses than with *Muhlenbergia*. It is evident that more data are needed for a thorough assessment of relationships among these taxa.

Clearly, there are two major themes that all of these disjunctions seem to have in common: 1) recent migration, since little morphological differences and very little genetic variation exist outside of North America, and 2) the putative dispersed species or taxon has migrated away from a center of diversity. Northern Mexico appears to be the center of diversity for the subtribe where there are over 70 species in Chihuahua or Durango. These general conclusions seem to illustrate the biogeographical history of the entire Muhlenbergiinae. That is, the subtribe quite possibly arose where it is most speciose (144 species found in North America) and has since radiated to other continents.

CLADISTIC ANALYSIS

Methods

A character list and their states, derived and modified from the analysis of the 38 genera of New World Eragrostideae, was used in all ensuing analyses (Peterson *et al.* 1997; see Appendix 1 for list of taxa, characters, and their states). Of these 88 characters, 21 describe the habit, 17 describe the inflorescence and other than distribution and chromosome number the remaining characters describe the spikelets (glumes, rachilla, lemmas, paleas, and caryopses). In the combined morphological and molecular analysis of relationships in the Muhlenbergiinae 124 characters were taken from a cpDNA restriction fragment analysis (Duvall *et al.* 1994). The following six restriction enzymes were scored for presence or absence: *Hpa* I, *Kpn* I, *Pst* I, *Pvu* II, *Sal* I, and *Sma* I. Molecular data for *Blepharoneuron* (*B. shepherdii and B. tricholepis*), *Chaboissaea* (*C. decumbens* and *C. ligulata*), *Muhlenbergia* (*M. andina* and *M. tenuifolia*), and *Pereilema* (*P. ciliatum* and *P. crinitum*) were pooled whenever states from the two representative species differed. Molecular data from *Dasyochloa pulchella* (Kunth) Willd. ex Rydb., *Eragrostis mexicana* (Hornem.) Link, *Leptochloa dubia* (Kunth) Nees, and *Sporobolus flexuosus* (Thurb. ex Vasey) Rydb.were used for the combined analysis. The combined data set includes a total of 212 characters. However, distribution characters (84-88) were removed from the analysis.

Phylogenetic analysis using parsimony (PAUP) test version 4d64 (issued 21 May 1998; Swofford, 1993) was used for all analyses. Files were input in nexus format using the heuristic search option with tree bisection-reconnection (TBR), mulpars (multiple parsimonious trees) option, and a simple addition sequence. Strict consensus trees were calculated for the overall analysis. Although this technique has been widely criticized (West and Faith 1990), it does provide an easy assessment for particular clades within a topology. Only parsimony-informative characters were used to analyze the six genera within the Muhlenbergiinae. Bootstrap values were calculated using a MaxTrees setting of 100 or more added sequentially and are only reported where values are greater than 50% majority rule.

New World Eragrostideae

For the overall analysis of 38 genera of New World Eragrostideae three outgroups (*Chloris* Sw., *Danthonia* DC., *and Pappophorum* Schreb.) were used simultaneously and in all possible combinations. A strict consensus of six trees using *Danthonia* as the outgroup and eliminating distribution characters (84-88) is shown in Figure 2. Three clades (A, B, C) and two sets of species pairs (D, E) correspond, in part to previously hypothesized subtribal relationships within the Eragrostideae (Peterson *et al.* 1995, 1997).

The Muhlenbergiinae (*Bealia, Blepharoneuron, Chaboissaea, Lycurus, Muhlenbergia,* and *Pereilema*), consistently occurs as a clade (Fig. 2, A) when using all outgroup combinations. The six genera of the Muhlenbergiinae are characterized by membranous ligules; 1-3-nerved first glumes; 1-4-nerved second glumes; 1(-3) florets per spikelet; 3-nerved lemmas that are awned, mucronate or unawned; truncate lodicules; true caryopses; and a base chromosome number of 8 or 10. Evidence from a chloroplast DNA restriction site survey (Duvall *et al.* 1994) and *ndh*F sequences (Clark *et al.* pers. comm.) support the monophyly of this subtribe.

Clade B (Fig. 2) containing *Blepharidachne, Dasyochloa, Erioneuron*, and *Munroa* is consistent with the subtribal designation in Peterson *et al.* (1995, 1997). These four genera form a clade when using six out of the seven outgroup combinations. Diagnostic characteristics of the subtribe include lemmas with emarginate to cleft apices and a base chromosome number of x = 7 or 8. Molecular studies support the conclusion that the Munroinae evolved from a common ancestor (Duvall *et al.* 1994). Members of the Munroinae are not always easily distinguished from the Muhlenbergiinae. However, the Munroinae always has 2-12 florets per spikelet and the lemma apex is usually emarginate to cleft, whereas, the Muhlenbergiinae usually has a single floret per spikelet and the lemma apex is mostly entire, rarely emarginate or lobed.

Clade C (Fig. 2), containing *Allolepis, Distichlis, Jouvea, Monanthochloë, Reederochloa,* and *Uniola*, is at least partially consistent with the Monanthochloinae as outlined by Clayton and Renvoize (1986) and Peterson *et al.* (1995, 1997). *Swallenia*, a monotypic genus [*S. alexandrae* (Swallen) Soderstr. & H. F. Decker] known only from the Eureka Dunes in southeastern California, where it occurs in extensive masses deeply embedded in siliceous sand, is an enigmatic taxon that never forms a clade

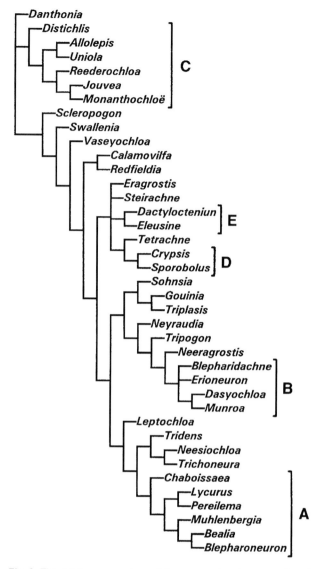

Fig. 2. The strict consensus tree of six most parsimonious trees (length = 292 steps; CI excluding uninformative characters = 0.266; RI = 0.514) analyzing 38 New World genera of Eragrostideae based on morphological data. Letters represent clades discussed in the text and partially correspond to subtribal designations as follows: A-Muhlenbergiinae, B-Munroinae, C-Monanthochloinae, D-Sporobolinae, E-Eleusininae.

with other members of the Monanthochloinae in this analysis. It is distinctive in the Monanthochloinae by possessing hermaphroditic flowers and caryopses with short and blunt style bases. Presence of distichous leaf arrangement, thick textured lemmas, and the dioecious habit distinguishes the Monanthochloinae. Secondary diagnostic characters include florets 2-25 per spikelet, fruit with an adnate pericarp, and lemmas entire at the apex. Numerous morphological and anatomical adaptions, i.e., distinctly distichous leaf arrangement and bicellular microhairs with enlarged bases, are found in some members of this subtribe. These adaptations are in direct response to the environment, since most of the species grow in saline habitats. The Monanthochloinae is primarily New World in distribution, only *Aeluropus* Trin. is restricted to the Mediterranean, northern China, Ethiopia, and Sri Lanka.

Since *Uniola* is usually a member of this clade (Fig. 2, C) even with outgroup shuffling, and *Tetrachne*, the other member of Uniolinae, is never clearly derived from a common ancestor with *Uniola*, then perhaps the Uniolinae is not really a closely related group and/or the Monanthochloinae is paraphyletic. Since only two of the four genera within the Uniolinae were part of this analysis it is not appropriate to speculate further on the validity this subtribe. The Uniolinae is a curious small subtribe that includes only four genera worldwide (Clayton and Renvoize 1986), three distributed in Africa and *Uniola* centered along the subtropical and tropical coastal regions of North and Central America. Presence of 5-20 florets per spikelet, fruit with a free pericarp, and ligule a line of hairs are characters that, when used in combination, serve to distinguish the Uniolinae from the other subtribes. Characters of secondary significance include: a 3-9-nerved lemma, primary inflorescence branches racemose or paniculate, and lemma entire at the apex. The Uniolinae seem to exhibit some affinities with the Eleusininae, since both subtribes contain species with free pericarps, laterally flattened spikelets, and spikelets that are two-ranked along primary inflorescence branches.

Crypsis and *Sporobolus* (Fig. 2, D) always form a species pair when using all possible outgroup combinations. Members of the Sporobolinae are exclusively defined by the presence of a 1-nerved lemma. Secondary diagnostic characters include one floret per spikelet and a line of hairs for a ligule. The Sporobolinae are closely linked to the Eleusininae and Uniolinae, as all three subtribes possess fruits with free pericarps. Our analysis does not indicate a close affinity between the Muhlenbergiinae and the Sporobolinae even though both have a one-flowered spikelet. Origin of the Sporobolinae probably lies in Africa or the Eastern Mediterranean where species diversity is greatest. There are at least 39 indigenous species of *Sporobolus* from southern Africa (Gibbs-Russell *et al.* 1991). *Calamovilfa*, the other member of the New World Sporobolinae usually forms a species pair with *Redfieldia*, a monotypic genus. Results from restriction site variation of the chloroplast genome (Duvall *et al.* 1994) suggest *Redfieldia* is closely related to members of the Muhlenbergiinae. Cytologically, *R. flexuosa* (Thurb. ex A. Gray) Vasey has been reported to have irregular meiosis at diakinesis and metaphase I, with 12 bivalents and one univalent or 11 bivalents and three univalents (Reeder 1971, 1976). It has been suggested by Reeder (1976) that aneuploidy is occurring and the base number for the genus is x = 10. Perhaps *Redfieldia* arose from past hybridizations among members or extinct ancestors of the Muhlenbergiinae, the maternal parent, and *Calamovilfa*.

Dactyloctenium and *Eleusine* sometimes form a pair when *Danthonia* and all the outgroups are used in the analysis (Fig. 2, E). The diagnostic character for the subtribe Eleusininae, as summarized for the New World, is an inflorescence with digitate primary branches. Additional significant characters that define the Eleusininae include the presence of free pericarps and membranous ligules. There are at least four Old World genera with caryopses with free pericarps and inflorescences consisting of digitate racemes (*Acrachne* Wight & Arn. ex Chiov., *Odyssea* Stapf, *Ochthochloa* Edgew., and *Sclerodactylon* Stapf). The Eleusininae obviously has arisen in the Old World tropics, quite possibly in East

Africa where species diversity of *Dactyloctenium* and *Eleusine* is greatest. The few species represented in the New World of these two genera suggest migration from the Old World has occurred recently. It is rather premature to assess the validity of this subtribe since the data are limited to the New World and this tribe is centered in the Old World.

The 15 genera of the New World Eragrostidinae (Peterson *et al.* 1995, 1997) do not form any recognizable clade in this analysis. I believe this subtribe is, at best, an unnatural grouping of convenience, since the relationships among these 15 genera are poorly understood. However, based on the full range of characters used here, these genera are presently excluded from the other subtribes. More than one floret per spikelet, fruit with an adnate pericarp, and base chromosome number of 10, 20 or 30 distinguishes the Eragrostidinae from the Eleusininae, Munroinae, Sporobolinae, and the Uniolinae. Important characters that can be used to separate the Eragrostidinae from most of the Monanthochloinae and Muhlenbergiinae include leaf arrangement (distichous in Monanthochloinae), sexuality (dioecious in Monanthochloinae, except *Swallenia*), lemma texture (relatively thick in Monanthochloinae), and number of lemmatal nerves (3-nerved in Muhlenbergiinae). No characters exclusively distinguish the Eragrostidinae from the Muhlenbergiinae. Eventually, as new data become available I suspect that many of these 15 taxa will be aligned within other current subtribes or placed within new, smaller monophyletic assemblages.

Muhlenbergiinae

Eragrostis, Sporobolus, Dasyochloa, and *Leptochloa* were used as outgroups for the cladistic analysis of the six genera within the Muhlenbergiinae (Figs. 3, 4, 5). When *Eragrostis* and *Sporobolus* were used as outgroups two of the four trees supported the monophyly of the ingroup one of these is shown in Figure 3. The two trees that cannot be rooted such that the ingroup is monophyletic include one tree with *Sporobolus, Eragrostis, Muhlenbergia* and a clade of the remaining five genera in a tetrachotomy and the second tree includes *Muhlenbergia* as sister to *Eragrostis* and the remaining five genera. The only other tree not shown includes a topology where *Sporobolus* and *Eragrostis* form a trichotomy with a clade with the following topology: (*Blepharoneuron* (*Bealia* (*Chaboissaea* (*Muhlenbergia* (*Lycurus, Pereilema*)))))).

The tree seen in Figure 3 is only an approximation of the phylogeny within the Muhlenbergiinae. Since there are many multiple character states scored for *Muhlenbergia*, it seems best to hypothesize that it lies at the base of the Muhlenbergiinae clade. The derived clade of *Lycurus* and *Pereilema* (bootstrap value of 63%) is supported by four apomorphies in which three are perhaps important: ciliate sheaths (16), spikelets in pairs or clusters (39), and the possession of yellow anthers (77). Pilger's (1956) original suggestion that *Lycurus* and *Pereilema* are closely related is supported by this analysis of morphological characters. *Chaboissaea*, the sister to *Lycurus-Pereilema* pair, is supported by eight apomorphies. Important character states along this clade appear to be sheaths shorter than the internodes (13), primary branches appressed to the main axis (28), second glume awned (51), the presence of sterile florets (56), and scabrous paleas (72). In a cpDNA study of *Chaboissaea* (Sykes *et al.* 1997) the sister to the

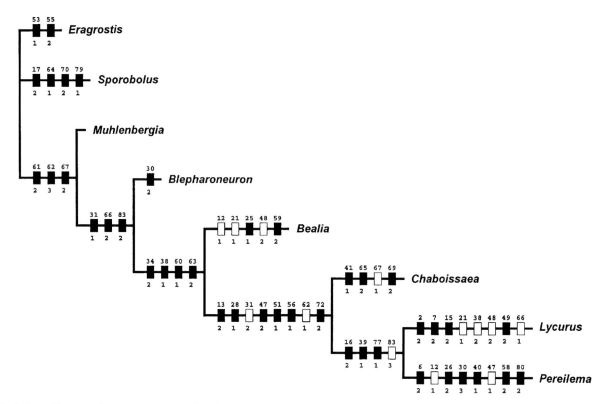

Fig. 3. One of four equally parsimonious trees (length = 54 steps, CI excluding uninformative characters = 0.643; RI = 0.546) analyzing six genera of Muhlenbergiinae based on morphological data. Solid bars = unique origin of states (synapomorphies) and hollow bars = reversals or parallel states.

four species of *Chaboissaea* was found to be *Lycurus* when *Muhlenbergia montana* and *Bealia mexicana* were also used as outgroups. The entire Muhlenbergiinae clade is supported by three apomorphies, two of these reverse later in the tree (62-lemma villous to pubescent and 67-palea hairy to glabrous), and one seems important (61-lemma glabrous to hairy).

When *Dasyochloa* and *Leptochloa* were used as outgroups 14 trees with considerable topological variation were retained. Eight trees cannot be rooted such that the ingroup is monophyletic because *Leptochloa* in seven of the eight trees, occurs on a clade with *Bealia, Blepharoneuron, Chaboissaea,* and *Muhlenbergia*. However, six trees clearly differentiate a Muhlenbergiinae clade forming a trichotomy with *Dasyochloa* and *Leptochloa*. In five of the six trees, *Muhlenbergia* and *Chaboissaea* are sisters to either *Lycurus-Pereilema* (bootstrap value of 52%) and *Bealia-Blepharoneuron* (bootstrap value of 58%) clades. The cladogram shown in Figure 4 was randomly chosen to illustrate one possible phylogeny where *Chaboissaea* is sister to the *Lycurus-Pereilema* clade and *Muhlenbergia* is sister to the *Bealia-Blepharoneuron* clade.

Morphological and molecular characters are combined in an analysis of the six genera in the Muhlenbergiinae (Figs. 5 & 6). When using the same four outgroup genera (*Eragrostis, Sporobolus, Dasyochloa,* and *Leptochloa*) in a single analysis and with any two of these genera designated as outgroups the Muhlenbergiinae form a clade with varying bootstrap values between 83-100%. No constraints were placed on the character polarization of this expanded data set. Topological variation among these seven consensus trees can perhaps shed some light on the phylog-

eny of this subtribe. All seven trees support a *Bealia/Blepharoneuron* clade (bootstraps vary from 71-98%) and a *Lycurus/Pereilema* clade (bootstraps vary from 51-82%). Four of these consensus trees, rooted with all four outgroup genera, *Dasyochloa/Sporobolus, Dasyochloa/Leptochloa,* and *Eragrostis/Leptochloa,* depict the following tetrachotomy: ((*Chaboissaea*) (*Muhlenbergia*) (*Bealia, Blepharoneuron*) (*Lycurus, Pereilema*)). Two of the six consensus trees, rooted with *Leptochloa/Sporobolus* (Fig. 5) and *Eragrostis/Sporobolus,* depict a trichotomy: ((*Chaboissaea*) (*Muhlenbergia*) (*Lycurus, Pereilema*)) with bootstrap values of 76% and 56%, respectively. The consensus tree (Fig. 5) using *Sporobolus/Leptochloa* as outgroups indicates that the Muhlenbergiinae are monophyletic with a bootstrap value of 100% and that a clade containing *Bealia* and *Blepharoneuron* (79% bootstrap) is basal to a trichotomy containing the remaining four genera (bootstrap of 76%). Another topology is seen in the consensus tree when *Dasyochloa/Eragrostis* is used as an outgroup (Fig. 6). This strict consensus of two trees again indicates that the Muhlenbergiinae are monophyletic (100% bootstrap) and that a clade containing *Lycurus/Pereilema* is basal (bootstrap value of 51%) to the remaining members of the subtribe with the following topology: ((*Muhlenbergia* (*Chaboissaea* (*Bealia, Blepharoneuron*)))). Therefore, at this time relationships among the members of the Muhlenbergiinae are highly speculative. However, a few generalizations can be made: 1) members of the subtribe shared a common ancestor in their evolutionary history, 2) *Bealia* and *Blepharoneuron* appear to be sister taxa, and 3) *Lycurus* and *Pereilema* also appear to be sister taxa.

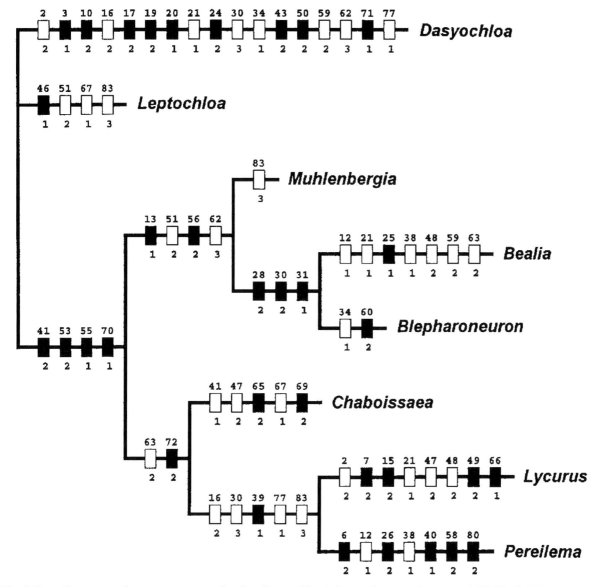

Fig. 4. One of fourteen equally parsimonious trees (length = 69 steps, CI excluding uninformative characters = 0.578; RI = 0.441) analyzing six genera of Muhlenbergiinae based on morphological data. Solid bars = unique origin of states (synapomorphies) and hollow bars = reversals or parallel states.

SUBGENERIC CLASSIFICATION OF MUHLENBERGIA

Based on a thorough study of the leaf blade anatomy as viewed in cross section (Peterson and Herrera-Arrieta, in review) the genus appears to be divisible into three subgenera. Subgenus *Muhlenbergia* is characterized as having loosely arranged chlorenchyma [phosphoenolpyruvate carboxykinase or classical PCK type, defined as centrifugal/evenly distributed photosynthetic carbon reduction (PCR) cell chloroplasts (with grana), XyMS+ and presence of PCR cell wall suberized lamella, in Hattersley and Watson's (1992) sense], shield-shaped (narrower than deep) central bulliform cells, and primary vascular bundles with non-sclerosed phloem. Thirty-six species have these characteristics; 13 of these were treated by Pohl (1969), who includes *M. andina, M. bushii, M. xcurtisetosa, M. frondosa, M. glabriflora, M. glomerata, M. mexicana, M. racemosa, M. schreberi, M. setarioides, M. sobolifera, M. sylvatica,* and *M. tenuiflora,* as consisting of a 'group of usually rhizomatous, leafy, mesic grasses confined

largely to eastern North America, the western Cordillera, and eastern Asia.' In addition to those treated by Pohl, there are eight annuals (*M. appressa, M. brandegei, M. ciliata, M. diversiglumis, M. microsperma, M. pectinata, M. tenella,* and *M. tenuifolia*) six rhizomatous perennials from Asia *(M. curviaristata, M. hakonensis, M. himalayensis, M. huegelii, M. japonica,* and *M. ramosa),* and nine more short-rhizomatous perennials from the southwestern United States and Mexico (*M. alamosae, M. arsenei, M. californica, M. dumosa, M. glauca, M. spiciformis, M. pauciflora, M. polycaulis,* and *M. thurberi*) that are included in subg. *Muhlenbergia* (see Appendix 3).

Subgenus *Trichochloa* includes 39 species that are characterized by the following anatomical features: tightly radiate chlorenchyma, primary vascular bundles with sclerosed phloem, an adaxial crown of colorless cells over the primary vascular bundles, and generally one or two continuous layers of abaxial sclerenchyma over the primary vascular bundles (Appendix 3).

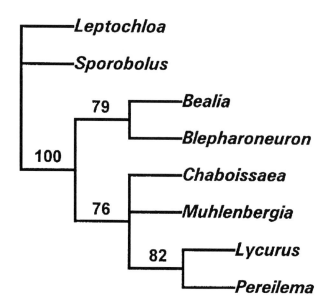

Fig. 5. The strict consensus tree of three most parsimonious trees (length = 101 steps, CI excluding uninformative characters = 0.684 ; RI = 0.633) analyzing six genera of Muhlenbergiinae based on morphological and molecular data. Numbers above branches are bootstrap values.

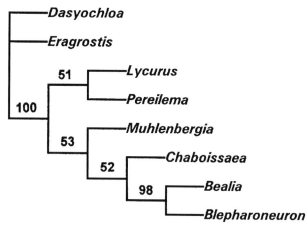

Fig. 6. The strict consensus of two most parsimonious trees (length = 128 steps, CI excluding uninformative characters = 0.658, RI = 0.576) analyzing six genera of Muhlenbergiinae based on morphological and molecular data. Numbers above branches are bootstrap values.

Soderstrom (1967) characterized subg. *Trichochloa* as having primary vascular bundles with sclerosed phloem and large caps of sclerenchyma present in the primary units (corresponds to one or two layers of abaxial sclerenchyma over the primary vascular bundles). These rather stout to robust, caespitose and non-rhizomatous perennials were divided by Soderstrom into two sections, *Podosemum* and *Epicampes*.

Anatomically, Sect. *Podosemum* is quite distinct, by having deep adaxial furrows (1/2 or more as deep as the blade width), obovate primary vascular bundle shape, vascular bundles at two or three levels in the blade, and blades with primary, secondary, and tertiary vascular bundles. Fourteen species (*M. angustata, M. capillaris, M. dubia, M. expansa, M. filipes, M. gypsophila, M. jaliscana, M. lucida, M. macroura, M. mucronata, M. nigra, M. palmeri, M. rigens,* and *M. rigida*) posses these anatomical characteristics besides having rounded basal sheaths and primarily firm ligules as pointed out by Soderstrom.

Section *Epicampes* contains 25 species (*M. aurea, M. breviligula, M. distans, M. distichophylla, M. emersleyi, M. gigantea, M. grandis, M. ×involuta, M. iridifolia, M. lehmanniana, M. lindheimeri, M. longiglumis, M. longiligula, M. mutica, M. pilosa, M. pubescens, M. pubigluma, M. reederorum, M. robusta, M. scoparia, M. speciosa, M. torreyana, M. versicolor, M. virletii, M. xanthodas*) that have compound keels containing primary and secondary vascular bundles in various combinations joined by aerenchyma tissue. In addition, the primary vascular bundles are generally rectangular in shape and the secondary and tertiary vascular bundles are about the same size as the primary bundles. Soderstrom (1967) characterized these species as having compressed-keeled basal sheaths and generally membranous ligules above.

The remaining 81 species within *Muhlenbergia* exhibit the classical NAD-ME [nicotinamide adenine dinucleotide phosphate cofactor to malic enzyme, centripetally PCR cell chloroplasts,

XyMS+ and presence of PCR cell outlines are even in transverse section, see Hattersley and Watson 1992] characteristics of many chloridoid grasses (see Appendix 3.). They appear to form a separate clade in our anatomical analysis (Peterson and Herrera-Arrieta, in review) and are placed in a third subgenus, '*Clomena*'. In general, this subgenus has primary vascular bundles that are non-sclerosed, rounded primary vascular bundles, tightly radiate chlorenchyma, and circular to fan shaped bulliform cells. In aspect these 81 species are primarily caespitose. Within '*Clomena*' there exists some resolution, for instance, the annuals (Peterson and Annable 1991; Peterson *et al.* 1989; which includes, *M. annua, M. brevis, M. crispiseta, M. depauperata, M. eludens, M. filiformis, M. flavida, M. fragilis, M. implicata, M. majalcensis, M. minutissima, M. peruviana, M. ramulosa, M. schmitzii, M. sinuosa, M. strictior, M. tenuissima,* and *M. texana*) that lack secondary vascular bundles generally form a clade, excluding *M. capillipes, M. ligularis,* and *M. vaginata*. The *Muhlenbergia repens* complex (Morden 1985, 1995; Morden and Hatch 1987, 1996; which includes, *M. fastigiata, M. plumbea, M. repens, M. richardsonis, M. utilis* and *M. villiflora*) also forms a clade in our anatomical analysis with the addition of the central Mexican endemic, *M. seatonii*. The *Muhlenbergia montana* complex (Herrera-Arrieta 1998; Herrera-Arrieta and De la Cerda-Lemus 1995; Herrera-Arrieta and Grant 1993, 1994; Herrera-Arrieta and Peterson 1992; which includes, *M. aguascalientensis, M. argentea, M. cualensis, M. curvula, M. durangensis, M. eriophylla, M. flabellata, M. flaviseta, M. michisensis, M. quadridentata,* and *M. watsoniana*) sometimes forms a partial clade, excluding *M. filiculmis, M. jonesii, M. montana,* and *M. virescens*.

ACKNOWLEDGEMENTS

I would like to thank the Smithsonian Institution's Fellowships and Grants Scholarly Studies Program, the National Museum of Natural History's Research Opportunities Fund, and the Department of Botany for providing financial support for the many collecting trips associated with this research. Special thanks are given to Rosemary Blackburn for preparing slides for the oral presentation, Yolanda Herrera-Arrieta for providing many of the anatomical sections, Robert J. Soreng for discussions and reviewing the initial version of the manuscript, Robert D. Webster for help with DELTA, Jesus Valdes-Reyna for initial morphological scoring of the taxa, and An Van den Borre for providing a critical review.

REFERENCES

Allred, K.W. (1981). Cousins to the south: Amphitropical disjunctions in the southwestern grasses. *Desert Plants* **3**, 98-106.

Beauvois, P.B. (1812). *Essai d'une nouvelle agrostographie; ou nouveaux genres des Graminées, avec figures représentant les caractères de tous les genres.* Paris.

Bews, J.W. (1929). *The Worlds's grasses; Their differentiation, distribution, economics, and ecology.* (Longmans, Green, and Company: New York.)

Clayton, W.D., and Renvoize, S.A. (1986). Genera Graminum, Grasses of the World. Royal Botanic Gardens, *Kew Bulletin Additional*, Series XIII, 389 pages.

Conzatti, C. (1946). Flora taxonomica Mexicana, vol. 1. (*Sociedad Mexicana de Historia Natural:* Distrito Federal Mexico.)

Coulter, J.M. (1885). *Manual of Botany of the Rocky Mountain Region.* (Ivison, Blakeman, Taylor, and Company: New York.)

Crawford, D.J., Lee, N.S., and Stuessy, T.F. (1992). Plant species disjunctions: Perspectives from molecular data. *Aliso* **13**, 395-409.

Davidse, G., and Pohl, R.W. (1992). New taxa and nomenclatural combinations of Mesoamerican grasses (Poaceae). *Novon* **2**, 81-110.

Desvaux, A.N. (1810). Extrait d'un mémoire sur quelques nouveau genres de la famille des Graminées. *Nouveau Bulletin des Sciences,* publié par la Société Philomatique de Paris. **2**, 187-190.

Duvall, M.R., Peterson, P.M., and Christenesen, A.H. (1994). Alliances of *Muhlenbergia* (Poaceae) within New World Eragrostideae are identified by phylogenetic analysis of mapped restriction sites from plastid DNAS. *American Journal of Botany* **81**, 622-629.

Fournier, E. (1886). Gramineae in *Mexicanas Plantas* **2**, 1-160.

Gibbs-Russell, G.E., Watson, L., Koekemoer, M., Smook, L., Barker, N.P., Anderson, H.M., and Dallwitz, M.J. (1991). Grasses of Southern Africa. *Memoirs of the Botanical Survey of South Africa* **58**, 1-437.

Gray, A., and Hooker, J.D. (1880). The vegetation of the Rocky Mountain region and a comparison with that of other parts of the world. *Bulletin of the United States Geological Survey* **6**, 1-77.

Hattersley, P.W., and Watson, L. (1992). Diversity of photosynthesis. *In* G.P. Chapman, Ed., *Grass Evolution and Domestication,* 38-116 pp. (Cambridge University Press: Cambridge.)

Herrera-Arrieta, Y. (1998). A revision of the *Muhlenbergia montana* (Nutt.) Hitchc. complex (Poaceae: Chloridoideae). *Brittonia* **50**, 23-50.

Herrera-Arrieta, Y., and Bain, J. (1991). Flavonoids of the *Muhlenbergia montana* complex. *Biochemical Systematics and Ecology* **19**, 665-672.

Herrera-Arrieta, Y., and De la Cerda-Lemus, M. (1995). *Muhlenbergia aguascalientesis,* a new species from Mexico. *Novon* **5**, 278-280.

Herrera-Arrieta, Y., and Grant, W.F. (1993). Correlation between generated morphological character data and flavonoid content in the *Muhlenbergia montana* complex. *Canadian Journal of Botany* **71**, 816-826.

Herrera-Arrieta, Y., and Grant, W.F. (1994). Anatomy of the *Muhlenbergia montana* (Poaceae) complex. *American Journal of Botany* **81**, 1038-1044.

Herrera-Arrieta, Y., and Peterson, P.M. (1992). *Muhlenbergia cualensis and M. michisensis* (Poaceae: Eragrostideae): two new species from Mexico. *Novon* **2**, 114-118.

Hilu, K.W., and Wright, K. (1982). Systematics of Gramineae: a cluster analysis study. *Taxon* **31**, 9-36.

Hitchcock, A.S. (1913). Mexican grasses in the United States National Herbarium. *Contributions from the United States National Herbarium* **17**, 181-189.

Hitchcock, A.S. (1927). The grasses of Ecuador, Peru, and Bolivia. *Contributions from the United States National Herbarium* **24**, 291-556.

Hitchcock, A.S. (1935). *Muhlenbergia.* (Poales) Poaceae (pars). *North American Flora* **17**, 431-476.

Humboldt, F.H. A. von, and Bonpland, A.J.A. (1805). Essai sur la geogr. pl. (Levrault, Schoell, and Compagnie: Paris.)

Kunth, K.S. (1816). *In* Humboldt, Bonpland, and Kunth, *Nova Genera et Species Plantarum* **I**, 141-142. pl 45.

Kunth, K.S. (1833). *Enumeratio Plantarum.* Five vol. Stutgardiae et Tubingae **I**, 173.

Link, G.M. (1833). *Hortus Regius Botanicus Berolinensis* **2**, 248. (G. Reimer: Berlin.)

Mez, C. (1921). *Gramineae novae vel minus cognitae.* Repertorium Specierum Novarum Regni Vegetablis **17**, 203-214.

Morden, C.W. (1985). A biosystematic study of the *Muhlenbergia repens* complex (Poaceae). (Ph.D. dissertation. Texas A&M University: College Station.)

Morden, C.W. (1995). A new combination in *Muhlenbergia* (Poaceae). *Phytologia* **79**, 28-30.

Morden, C.W., and Hatch, S.L. (1987). Anatomical study of the *Muhlenbergia repens* complex (Poaceae: Chloridoideae:Eragrostideae). *Sida* **12**, 347-359.

Morden, C.W., and Hatch, S.L. (1996). Morphological variation and synopsis of the *Muhlenbergia repens* complex (Poaceae). *Sida* **17**, 349-365.

Nash, G.V. (1898). New or noteworthy American grasses VIII. *Bulletin of the Torrey Botanical Club* **25**, 83-89.

Nei, M. (1978). Molecular evolution genetics. (Columbia University Press: New York.)

Nuttall, T. (1848). Descriptions of plants collected by Mr. William Gambel in the Rocky Mountains and Upper California. *Proceedings of the National Academy of Sciences in Philadelphia* **4**, 23.

Peterson, P.M. (1988). Chromosome numbers in the annual *Muhlenbergia* (Poaceae). *Madroño* **35**, 320-324.

Peterson, P.M. (1989). A re-evaluation of *Bealia mexicana* (Poaceae: Eragrostideae). *Madroño* **36**, 260-265.

Peterson, P.M., and Annable, C.R. (1990). A revision of *Blepharoneuron* (Poaceae: Eragrostideae). *Systematic Botany* **15**, 515-525.

Peterson, P.M., and Annable, C.R. (1991). Systematics of the annual species of *Muhlenbergia* (Poaceae-Eragrostideae). *Systematic Botany Monographs* **31**, 1-109.

Peterson, P.M., and Annable, C.R. (1992). A revision of *Chaboissaea* (Poaceae:Eragrostideae). *Madroño* **39**, 8-30.

Peterson, P.M., Annable, C.R., and Franceschi, V.R. (1989). Comparative leaf anatomy of the annual *Muhlenbergia* (Poaceae). *Nordic Journal of Botany* **8**, 575-583.

Peterson, P.M., and Columbus, J.T. (1997). Allelic variation in the amphitropical disjunct *Scleropogon brevifolius* (Poaceae: Eragrostideae). *BioLlannia, Edición Especial* **6**, 473-490.

Peterson, P.M., Duvall, M.R., and Christensen, A.H. (1993). Allozyme differentiation among *Bealia mexicana, Muhlenbergia argentea,* and *M. lucida* (Poaceae: Eragrostideae). *Madroño* **40**, 148-160.

Peterson, P.M., and Herrera A., Y. (1996). Allozyme variation in the amphitropical disjunct, *Chaboissaea* (Poaceae: Eragrostideae). *Madroño* **42**, 427-449.

Peterson, P.M., and Herrera-Arrieta, Y. (in review). An anatomical survey of *Muhlenbergia* (Poaceae: Muhlenbergiinae). *Systematic Botany.*

Peterson, P.M., and Morrone, O. (1998). Allelic variation in the amphitropical disjunct *Lycurus setosus* (Poaceae: Muhlenbergiinae). *Madroño* **44**, 334-346.

Peterson, P.M., and Ortíz-Diaz, J.J. (1998). Allelic variation in the amphitropical disjunct *Muhlenbergia torreyi* (Poaceae: Muhlenbergiinae). *Brittonia* **50**, 381-391.

Peterson, P.M., and Rieseberg, L.H. (1987). Flavonoids of the the annual *Muhlenbergia*. *Biochemical Systematics and Ecology* **15**, 647-652.

Peterson, P.M., Webster, R.D., and Valdes-Reyna, J. (1995). Subtribal classification of the New World Eragrostideae (Poaceae: Chloridoideae). *Sida* **16**, 529-544.

Peterson, P.M., Webster, R.D., and Valdes-Reyna, J. (1997). Genera of New World Eragrostideae (Poaceae: Chloridoideae). *Smithsonian Contributions to Botany* **87**, 1-50.

Phillips, S.M. (1982). A numerical analysis of the Eragrostideae (Gramineae). *Kew Bulletin* **37**, 133-162.

Pilger, R. (1956). Gramineae II. Unterfamilien: Micraioideae, Eragrostideae, Oryzoideae, Olyroideae. *In, Die naturalichen Pflanzenfamilien 2d ed.*, ed H. Melchoir and E. Werdermann **14**, 1-168.

Pohl, R.W. (1969). *Muhlenbergia*, subgenus *Muhlenbergia* (Gramineae) in North America. *American Midland Naturalist* **82**, 512-542.

Presl, K.B. (1830). *Reliquiae Haenkeanae* **1**, 207-356.

Raven, P.H. (1963). Amphitropical relationships in the floras of North and South America. *Quaternary Review of Biology* **38**, 151-177.

Reeder, C.G. (1985). The genus *Lycurus* in North America. *Phytologia* **57**, 283-291.

Reeder, J.R. (1971). Notes on Mexican grasses IX. Miscellaneous chromosome numbers-3. *Brittonia* **23**, 105-117.

Reeder, J.R. (1976). Systematic position of *Redfieldia* (Gramineae). *Madroño* **23**, 434-438.

Reeder, J.R., and Reeder, C.G. (1988). Aneuploidy in the *Muhlenbergia subbiflora* complex (Gramineae). *Phytologia* **65**, 155-157.

Renvoize, S.A. (1998). Gramineas de Bolivia. (The Royal Botanic Gardens: Kew.)

Sanchez, E., and Rugolo de Agrasar, Z.E. (1986). Estudio taxonomico sobre el genero *Lycurus* (Gramineae). *Parodiana* **4**, 267-310.

Schreber, J.C.D. (1789). *Genera Plantarum*. Ed. 8, 1:44-45. (Varrentrapp & Wenner: Frankfurt A.M.)

Soderstrom, T.R. (1967). Taxonomic study of the subgenus *Podosemum* and section *Epicampes* of *Muhlenbergia* (Gramineae). *Contributions from the United States National Herbarium* **34**, 75-189.

Sohns, E.R. (1953). *Chaboissaea ligulata* Fourn.: A Mexican grass. *Journal of the Washington Academy of Sciences* **43**, 405-407.

Solbrig, O.T. (1972). The floristic disjunctions between the 'Monte' in Argentina and the 'Sonoran Desert' in Mexico and the United States. *Annals of the Missouri Botany Garden* **59**, 218-223.

Swallen, J.R. (1947). The awnless annual species of *Muhlenbergia*. *Contributions from the United States National Herbarium* **29**, 203-208.

Swofford, D.L. (1993). PAUP: phylogenetic analysis using parsimony, version, 3.1.1. (Illinois Natural History Survey: Champaign, Illinois.)

Sykes, G.R., Christensen, A. H., and Peterson, P. M. (1997). A chloroplast DNA analysis of *Chaboissaea* (Poaceae: Eragrostideae). *Systematic Botany* **22**, 291-302.

Thorne, R.F. (1972). Major plant disjunctions in the geographic ranges of seed plants. *Quaternary Review of Biology* **47**, 365-411.

Thurber, G. (1863). *Gramineae*. In Enumeration of the species of plants collected by Dr. C.C. Parry, and Messrs. Elihu Hall and J. P. Harbour, during the summer and autumn of 1862, on and near the Rocky Mountains, in Colorado Territory, lat. 39^0-41^0, by Asa Gray. *Proceedings of the Academy of Natural Sciences of Philadelphia* **15**.

Torrey, J. (1857). Descriptions of the general botanical collections. *Explorations and Surveys for a Railroad Route from the Mississippi River to the Pacific Ocean: Report on the Botany of the Expedition* **4**, 1-167.

Valdes-Reyna, J., and Hatch, S.L. (1991). Lemma micromorphology in the Eragrostideae (Poaceae). *Sida* **14**, 531-549.

Van den Borre, A. and Watson, L. 1997. On the classification of the Chloridoideae (Poaceae). *Australian Systematic Botany* **10**, 491-531.

Vasey, G. (1887). New species of Mexican grasses. *Bulletin of the Torrey Botanical Club* **14**, 8-10.

Vasey, G. (1889). Gramineae. Pp. 210-214 in T. S. Brandegee, A collection of plants from Baja California. *Proceedings of the California Academy of Science* **2**, 117-216.

Watson, L., Clifford, H.T., and Dallwitz, M.J. (1985). The classification of Poaceae: Subfamilies and supertribes. *Australian Journal of Botany* **33**, 433-484.

Watson, L., and Dallwitz, M.J. (1992). The grass genera of the World. (C.A.B. International: Wallingford, U.K.)

West, J.G., and Faith, D.P. (1990). Data, methods and assumptions in phylogenetic inference. *Australian Systematic Botany* **3**, 9-20.

APPENDIX I.

Characters and their states of the New World genera of the Eragrostideae and their outgroups.

1. *Allolepis*	14. *Gouinia*	27. *Scleropogon*
2. *Bealia*	15. *Jouvea*	28. *Sohnsia*
3. *Blepharidachne*	16. *Leptochloa*	29. *Sporobolus*
4. *Blepharoneuron*	17. *Lycurus*	30. *Steirachne*
5. *Calamovilfa*	18. *Monanthochloë*	31. *Swallenia*
6. *Chaboissaea*	19. *Muhlenbergia*	32. *Tetrachne*
7. *Crypsis*	20. *Munroa*	33. *Trichoneura*
8. *Dactyloctenium*	21. *Neeragrostis*	34. *Tridens*
9. *Dasyochloa*	22. *Neesiochloa*	35. *Triplasis*
10. *Distichlis*	23. *Neyraudia*	36. *Tripogon*
11. *Eleusine*	24. *Pereilema*	37. *Uniola*
12. *Eragrostis*	25. *Redfieldia*	38. *Vaseyochloa*
13. *Erioneuron*	26. *Reederochloa*	

Outgroups: 39. *Chloris* 40. *Danthonia*   41. *Pappophorum*

1. Plants
    1. hermaphroditic = 2-9 11-14 16 17 19 20 22-25 29-41
    2. dioecious = 1 10 15 18 21 26-28 40
    2. monoecious = 3 27
    2. gynomonoecious = 20
    2. andromonecious = 6
2. Plants
    1. annual = 2-4 6-8 11 12 16 19-22 24 29 33 35 39
    2. perennial = 1 3-6 8-19 23 25-41
3. Plants
    1. stoloniferous = 1 8 9 12 13 15 18 20 26 27 29 37 39
    2. lacking stolons = 2-7 10-14 16 17 19 21-25 28-36 38-41
4. Plants
    1. rhizomatous = 1 3 5 10-12 14 15 18 19 23 25 28-32 34-40
    2. lacking rhizomes = 1 2 4 6-9 12-14 16 17 19-22 24 26 27 29 33 34 39-41
5. Flowering culms
    1. erect = 1 2 4-14 16 17 19 22-41
    2. decumbent = 3 4 6 7 9 12 15 19-21 24 39 40
    2. mat forming = 3 7 8 10 12 15 18-21 27 29 40
6. Flowering culms
    1. caespitose = 1 2 4-6 8-17 19 22 23 25-38
    2. not caespitose = 3 4 6 7 12 13 18-21 24
7. Flowering culms
    1. glabrous = 1 2 4-16 18 19 21-27 29 30 32-41
    2. hairy = 2-4 17 19-21 28 31 35 39 40
8. Flowering culms
    1. pubescent = 2-4 17 19-21 28 35 40
    2. pilose = 19 21 31 40
    2. villous = 19 21 40
9. Flowering culms
    1. with viscid internodes = 12
    1. with glaucous internodes = 19
    2. with neither viscid nor glaucous internodes = 1-41

10. Leaves
    1. cauline = 1-8 10-12 14-19 21-25 29-35 37-41
    2. mostly basal = 3 4 9 12 13 19 20 26-28 32 36
11. Leaves
    1. distinctly distichous = 10 18
    2. not distinctly distichous = 1-9 11-17 19-41
12. Leaves
    1. with sheath auricles = 2 16 19
    1. with blade auricles = 24
    2. without auricles = 1 3-23 25-38
13. Sheaths
    1. longer than the internodes = 2 4 5 10-12 14-16 18 19 24-26 28 29 31 37-41
    2. shorter than the internodes = 1 3 4 6-9 11-17 19-24 27-30 32-36 39-41
14. Sheaths
    1. glandular = 12 21
    2. not glandular = 1-20 22-41
15. Sheaths
    1. with smooth margins = 1-7 9-13 15 16 18 19 22-27 29-34 36-41
    2. with scabrous margins = 2 4 8 12 14 17 19-21 24 28 35 39 40
16. Sheaths
    1. not ciliate = 1 2 4-8 10-12 14-16 18 19 21-23 25 27-30 32-41
    2. ciliate = 3 9 10 12-14 17 19-21 24 26 29 31 39 40
17. Ligule
    1. a membrane = 2 4 6 8 14 16 17 19 24 26 33 39
    1. a ciliate membrane = 1 3 10-14 16 18 19 22 34 36 38 39
    2. a line of hairs = 5 7 9 12 15 20 21 23 25 27-32 34 35 37 39-41
18. Leaf blades
    1. filiform = 4 5 12 19 25 29 36 39-41
    1. linear = 1-11 13-20 22-24 26-38
    3. triangular = 3 12 21 39
19. Leaf blades
    1. flat = 1-4 6-8 10-12 14 16 17 19-24 27-35 37-41
    2. involute = 2-7 9 10 12 14 15 19 25 26 28 29 32 34 35 36 38-41
    3. conduplicate = 6 12 13 17-21 27 37 39 40
    4. terete = 29 40
20. Leaf blades
    1. pungent = 3 9 10 12 13 15 18-20 26 27 31 40
    2. not pungent = 1 2 4-8 11-14 16 17 19 21-25 28-30 32-41
21. Leaf blades
    1. with thickened margins = 2 9 13 17 2 39
    2. without thickened margins = 1 3-8 10-12 14-16 18 19 21 23-41
22. Inflorescence
    1. consisting of a single spikelet = 15 18 40
    1. consisting of two or three spikelets = 15 26 40
    2. consisting of more than three spikelets = 1-17 19-25 27-41
23. Inflorescence
    1. a raceme = 9 22 26 27 36 40
    2. a panicle = 1-17 19-25 27-35 37-41

24. Inflorescence
    1. exserted = 1-8 10-17 19 21-41
    2. partially included in upper sheath = 3 6 7 9 12 18 19 21 24 26 29 31 35 39-41
    2. fully included = 10 15 20
25. Main axis
    1. glandular = 2 12 21 22
    2. not glandular = 1 3-17 19 20 23-41
26. Main axis
    1. smooth = 1 2 4-9 11-13 15 19 21-23 25 26 29 32 34 37 38-41
    1. scabrous = 2 4 6 10 12 14 16 17 19 20 27 30 31 33-36 38-41
    2. hairy = 3 12 19 21 24 28 39 40
27. Primary branches
    1. digitate = 8 11 39
    2. not digitate = 1-7 9 10 12-17 19-38 40 41
28. Primary branches
    1. appressed to the main axis = 1 3 5-7 9 10 12 13 15-17 19-21 24 27-29 31 32 34 36 37 40 41
    2. spreading from the main axis = 2 4-6 8 11-14 16 19 22-24 26 28-30 33-35 37 38-40
    3. divaricate = 8 12 14 19 25 29 34 39
    3. reflexed = 8 19 29
29. Primary branches
    1. terminating in a spikelet = 1-7 9-17 19-41
    2. terminating in a bare point = 8
30. Primary branches
    1. with appressed secondary branches = 1-3 5 6 10 12 13 15 16 19 21 23 27-31 35 37-41
    2. with spreading secondary branches = 2 4 5 12 19 22 25 29 34 37-39 41
    3. reduced to a fascicle of spikelets = 9 12 24
31. Pedicels
    1. glandular = 2 4 21 22
    2. not glandular = 1 3 5-7 9 10 12-14 16 17 19-21 23-41
32. Pedicels
    1. glabrous = 1 2 4-6 9 10 12-14 16 17 19-27 29 30 32-35 37-41
    2. hairy = 3 19 21 24 28 31 39 40
33. Pedicels
    1. pubescent = 3 19 21 28 39 40
    2. pilose = 19 21 24 31 39 40
    3. villous = 19 21 24 39 40
34. Pedicels
    1. smooth = 1 3-5 9 12 13 19 21 22 24-26 29 32 34 37 39-41
    2. scabrous = 2 6 10 12 14 16 17 19 20 23 24 27 28 30 31 33-35 38-41
35. Cleistogamous spikelets
    1. present = 12 16 19 29 30 35 39-41
    2. absent = 1-29 31-34 36-41
36. Disarticulation
    1. below the glumes = 7 19 29 37
    2. above the glumes = 1-36 38-41
37. Disarticulation
    1. with the lemma and palea falling as a unit = 1-20 22 24-36 38-41

2. with the lemma and palea falling separately = 7 12 21 23
38. Callus
    1. hairy = 2 3 5 6 13 14 19 22 24 25 27 28 31 35 38-41
    2. glabrous = 1 4 6-12 15-21 23 26 27 29 30 32-34 36 37
39. Spikelets
    1. in clusters = 12 15 20 21 24 39 40
    1. paired = 17
    2. solitary = 1-14 16 18 19 22 23 25-38 40 41
40. Spikelets
    1. subtended by sterile bristle-like spikelets = 24
    2. not subtended by sterile bristle-like spikelets = 1-23 25-41
41. Spikelets
    1. laterally compressed = 2 3 5-16 18-23 25-32 34-41
    2. terete = 1 2 4 12 16 17 19 24 27 29 33 41
    2. dorsiventrally compressed = 19
42. Spikelets
    1. sessile = 7 8 11 12 15 18 21 24 39
    2. subsessile = 3 6 7 12 14 16 19-21 24 29 32 33 36 39-41
    2. pedicellate = 1-6 9 10 12-14 16 17 19-31 33-35 37 38-41
43. Spikelets
    1. up to 4.9 mm long = 2 4 6-8 12 16 17 19 24 29 32 34 36 39-41
    2. greater than 4.9 mm long = 1 3 5 8-16 18-23 25-41
44. Glumes
    1. present = 1-17 19-41
    2. absent = 15 18 20
45. Glumes
    1. shorter than the spikelets = 1 4 6-8 10-17 19-26 28-30 32 34-39 41
    1. equalling the spikelets = 2-7 13 19 27 28 31 33 34 39 41
    2. exceeding the spikelets = 2 19 39 40
46. Glumes
    1. unequal = 4 5 8 11-16 19-21 23-26 29 34-36 38 39
    2. more or less equal = 1-4 6 7 9 10 12 13 17 19 20 22 24 27 28 30 31-34 37 40 41
47. Glumes
    1. smooth = 1-5 8-10 12 13 15 16 19-32 34 37 38-41
    2. scabrous = 2 6 7 11 12 14 17 19 33 35 36 39
48. Glumes
    1. glabrous = 1 3-16 19-41
    2. hairy = 2 17 19 21
49. First glume
    1. up to 1-nerved = 1-9 11-16 19-25 27-30 32-36 39 41
    2. greater than 1-nerved = 10 14 17 19 26 27 31 37 38 40
50. Second glume
    1. shorter than lower lemma = 1 4 6 8 10-17 19-21 23-26 28 30 32 34 37-39
    1 subequal to the lower lemma = 2-5 7 19 22 27-29 31 33-36 39 41
    2 longer than lower lemma = 2 9 13 19 40
51. Second glume
    1. awned = 6 8 9 17 19 22 24 27 33 41
    2. unawned = 1-7 10-16 19-21 23 25-32 34-40
52. Second glume
    1. up to 1-nerved = 2-9 12 13 15-17 19-25 28-30 32-36 39 41
    2. greater than 1-nerved = 1 10 11 14 16 17 19 21 26 27 31 34 37-39

53. Rachilla
    1. pronounced between the florets = 1 3 6 8-16 18 20-23 25-28 30-41
    2. not pronounced between the florets = 1 2 4-6 17 19 24 29
54. Rachilla
    1. hairy = 3 22 23 28 30 31 38 40 41
    2. glabrous = 1 6 8-16 18 20 21 25-27 32-37 39 40
55. Florets
    1. up to 1 per spikelet - 2 4-7 17 19 24 29 40
    2. 2-4 per spikelet - 3 6 8 12 14 16 18 20 23 25-28 31 34 35 39-41
    2 more than 4 per spikelet - 1 8-16 18 20-23 25-28 30-34 36-38 40 41
56. Sterile florets
    1. present = 1 3 6 8-18 20-28 30-41
    2. absent = 2 4-7 12 19 29 40
57. Sterile florets
    1. above the fertile florets = 1 6 8-16 18 20-22 25-28 30 31 33-41
    2. below the fertile florets = 3 23 32 37
58. Sterile florets
    1. homomorphic = 1 3 6 8-18 21-23 25-28 30-38 40
    2. heteromorphic = 20 24 39 41
59. Lemma
    1. entire = 1 4-8 10-12 14-19 21 24-27 29-32 37-39
    2. emarginate = 13 16 19 20 22 33 34 36 40 41
    2. lobed = 2 9 13 16 20 35 40 41
    2. cleft = 3 9 23 28 40 41
60. Lemma
    1. awned = 2 3 6 8 9 13 14 16 17 19-24 27 28 33 35 36 39-41
    2. mucronate = 6 8 12 16 19 21 25 30 34 39
    2. unawned = 1 4-7 10-12 15 16 18 19 26 27 29 31 32 37-39
61. Lemma
    1. glabrous = 1 5 7 8 10-12 15 16 18 19 21 22 26-30 32 33 36 37 39 40
    2. hairy = 2-6 9 13 14 16 17 19-21 23-25 28 29 31 34 35 38-41
62. Lemma
    1. pubescent = 6 16 17 19 21 24 25 29 34 38-40
    2. pilose = 13 14 19 21 24 28 29 38-41
    3. villous = 2-4 9 14 19 21 23 31 35 39-41
63. Lemma
    1. smooth = 1 3-5 8-10 12 13 15 16 18 19 21-23 25-29 31 32 34 35 37-41
    2. scabrous = 2 6 7 11 12 14 17 19 20 24 30 33 36 39-41
64. Lemma
    1. up to 1-nerved - 5 7 29
    2. with more than 1 nerve - 1-4 6 8-28 30-41
65. Lemma
    1. hyaline = 7 12 19 21 34 39
    1. membranous = 2-4 7-9 11-14 16 17 19 20 22-24 28-30 32-36 39
    2. chartaceous = 5 6 12 15 17 19 25 31 38-41
    2. coriaceous = 1 10 12 18 20 26 27 37 40
    2. cartilaginous = no taxa
    2. indurate = 12 18

66. Lemma
    1. with glabrous nerves = 1 5 7 8 10-12 14-19 21 25-30 32 36-40
    2. with hairy nerves = 2-6 9 13 16 19 20 22-24 28 31 33-35 39-41
67. Palea
    1. glabrous = 1 6-8 10-12 14-16 18-21 23 25-30 32-34 37-41
    2. hairy = 2-5 9 12 13 17 19 21 22 24 31 34 35 40
68. Palea
    1. extending into awns = 6 14 19
    2. not extending into awns = 1-13 15-41
69. Palea
    1. hyaline = 7 12 19 21 34
    1. membranous = 2-4 8-14 16 17 19 20 22-24 27-30 32-36 39-41
    2. chartaceous = 5 6 10 12 15 19 25-27 31 37 38 40 41
    2. coriaceous = 1 18
    2. cartilaginous = no taxa
    2. indurate = 18
70. Palea
    1. margins enfolding the fruit = 1 2 4 6 10 12 17-19 24 26 40 41
    2. margins not enfolding the fruit = 3 5 7-9 11 13-16 20-23 25 27-39
71. Palea
    1. ciliate = 1 3 5 9 12-14 21 22 31 37 40 41
    2. not ciliate = 2 4 6-8 10-12 14-20 23-30 32-36 38-40
72. Palea
    1. smooth = 1-5 7-10 12 13 15 16 18-23 25-29 31-40
    2. scabrous = 6 11 12 14 17 19 24 30 39-41
73. Lodicules
    1. present = 1 2 4-6 8-17 19-41
    2. absent = 3 7 18 20 27 29 40
74. Lodicules
    1. adnate to the palea = 13 34
    2. not adnate to the palea = 1 2 4-6 8-12 14-17 19-41
75. Lodicules
    1. truncate = 2 4-6 8 10 12 14-17 19-21 24 27-32 35-38 41
    2. rounded = 25 33 34
    2. cuneate = 1 13 40
    2. acuminate = 11 22 23
76. Stamens
    1. up to 2 - 3 7 12 13 16 20 24 29 30 36
    2. greater than 2 - 1-29 31-41
77. Anther
    1. yellow = 1 3 5 8-13 15-21 23-25 27 29 32 36 37 39-41
    2. reddish-purple = 2 4 6 7 12 14-17 19 25 28-35 38-41
    3. olivaceous-plumbeous = 6 19 22 26 29
78. Stigmas
    1. up to 2 - 1-25 27-41
    2. greater than 2 - 20
79. Fruit
    1. with a free pericarp = 5 7 8 11 29 32 37
    2. with an adnate pericarp = 1-4 6 9 10 12-28 30 31 33-36 38-41
80. Caryopsis
    1. terete = 2 4-15 17-19 21 23 25 27 29-32 36-40

2. laterally compressed = 3 12 16 19 24 26
3. dorsiventrally compressed = 12 16 20 22 33-35 40 41
81. Caryopsis
  1. with persistent styles bases = 31 38
  2. lacking persistent style bases = 1-30 32-37 39-41
82. Caryopsis
  1. with pronounced sharp-pointed style bases = 38
  2. with relatively short and blunt style bases = 31
83. Base chromosome number, x=
  1. 7 = 3 20
  2. 8 = 2 4 6 7 9 13 20
  3. 9 or more = 1 5 7 8 10-12 14-17 19 21 24-27 29 31 33-37 39-41
84. Distribution by continent
  1. North America = 1-13 16-21 23-29 31-41
  2. Central America = 7 8 10-12 14-21 23 24 29 34-37 39-41
  3. South America = 3 6 8 10-14 16-20 22 24 27 29 30 32-34 36 37 39-41
  4. Africa = 7 8 11 12 16 29 32-34 36 39 40
  5. Europe = 7 8 11 12 16 29 39 40
  6. Asia = 7 8 11 12 16 19 29 33 36 39
  7. Australia = 8 10-12 16 29 39 40
  8. Pacific Islands = 11 12 16 19 29 33 39
85. Distribution for North America and Mesoamerica
  1. Greenland = no taxa
  2. Alaska = 10 19 40
  3. Canada = 10 12 19 20 29 35 40
  4. United States = 1 3-5 7-13 16-21 23 25 27 29 31-41
  5. Mexico = 1-4 6-21 23 24 26-29 33-37 39-41
  6. Central America = 8 11 12 14-17 19 24 29 35 37 39 41
  7. West Indies = 8 11 12 14 16 19 29 39-41
86. Regions of the United States divided by states
  1. Northern Pacific Region = 7 8 10-12 16 19 29 40
  2. California Region = 7 8 10-12 16 18-20 27 29 31 39 40
  3. Southwestern Region = 4 5 8-13 16 17 19 20 25 27 29 34 35 39-41
  4. Northern Plains Region = 8-13 16 19-21 25 29 35 40
  5. Central Plains Region = 3 5 8-13 16 19-21 25 29 34 35 39-41
  6. Texan Region = 1 3-5 8-13 16 18 19-21 25 27 29 33-41
  7. Southeastern Region = 8 10-12 16-19 21 23 29 34 35 37 39-41
  8. Mid-south Region = 8 11 12 16 19 21 29 37 39-41
  9. Great Lakes Region = 7 8 11 12 16 19 21 29 40
  10. North Atlantic Region = 8 10 11 12 16 19 29 39 40
  11. Central Atlantic Region = 8 10 11 12 16 19 29 32 37 39 40
  12. Alaskan Region = 11 12 19 40
87. Distribution for South America
  1. Colombia = 11 12 17 19 24 29 37 39 41
  2. Venezuela = 11 12 16 19 24 29 30 39 41
  3. Guiana, Surinan, or French Guiana = 11 12 24 29 30 39 41
  4. Ecuador = 10 11 12 17 19 24 29 33 37 39-41
  5. Peru = 11 12 13 19 24 29 33 39-41
  6. Brazil = 11 12 14 16 19 22 24 29 30 39-41
  7. Bolivia = 6 10 11 12 13 14 16 17 19 20 29 34 39-41
  8. Paraguay = 11 12 14 16 29 39-41

  9. Uruguay = 8 11 12 16 29 39-41
  10. Chile = 8 10 11 12 19 20 27 29 36 40 41
  11. Argentina = 3 6 8 10 11 12 13 14 16 17 18 19 20 27 29 32 34 36 39-41
88. Whether native or introduced
  1. Native to the New World = 1-6 9-22 24-31 33-41
  2. Introduced to the New World = 7 8 11 12 16 23 29 32

APPENDIX 2.

Description of subtribe Muhlenbergiinae based on the 88 characters in Appendix 1, taken from Peterson *et al.* (1995).

Plants hermaphroditic or andromonecious. Ligule a membrane or ciliate membrane. Leaf blades filiform or linear. Primary branches of the inflorescence not digitate. Spikelets 0.5-8 mm long; laterally compressed, terete, or dorsiventrally compressed. First glume 1-3-nerved. Second glume shorter than, about the same length as, or longer than the lower lemma; 1-4-nerved. Rachilla pronounced or not pronounced between the florets. Florets 1-3 per spikelet. Sterile florets present or absent. Lemma entire, emarginate, or lobed; awned, mucronate, or unawned; glabrous or hairy; 3-nerved; hyaline, membranous, or chartaceous. Palea hyaline, membranous, or chartaceous. Lodicules truncate. Fruit with an adnate pericarp. Base chromosome number, x = 8 or 10.

APPENDIX 3.

List of accepted species in the Muhlenbergiinae. Following some of the *Muhlenbergia* entries is a subgenus and/or section affiliation, indicated by E=sect. *Epicampes*, M= subg. *Muhlenbergia*, and P=sect. *Podosemum*; all taxa without an entry belong in 'Clomena' unless indicated with a question mark. All taxa occur in North America unless otherwise indicated by: a=Asia; c=Central America; cs=Central & South America; nc=North & Central America; ns=North and South America; ncs=North, Central & South America.

Bealia Scribn.
B. mexicana Scribn. ex Beal

Blepharoneuron Nash
B. shepherdii (Vasey) P.M. Peterson & Annable
B. tricholepis (Torr.) Nash

Chaboissaea E. Fourn.
C. atacamensis (Parodi) P.M. Peterson & Annable (s)
C. decumbens (Swallen) Reeder & C. Reeder
C. ligulata E. Fourn.
C. subbiflora (Hitchc.) Reeder & C. Reeder

Lycurus Kunth
L. phalaroides Kunth (ncs)
L. phleoides Kunth
L. setosus (Nutt.) C. Reeder (ns)

Muhlenbergia Schreb.
M. aguascalientensis Y.Herrera & De la Cerda-Lemus
M. alamosae Vasey (M)
M. andina (Nutt.) Hitchc. (M)
M. angustata (J. Presl) Kunth (P-s)

M. annua (Vasey) Swallen
M. appressa C.O. Goodd. (M)
M. arenacea (Buckley) Hitchc.
M. arenicola Buckley (ns)
M. argentea Vasey
M. arizonica Scribn.
M. arsenei Hitchc. (M)
M. articulata Scribn. (P)
M. asperifolia (Nees & Meyen ex Trin.) Parodi (ncs)
M. aurea Swallen (E-nc)
M. brandegei C. Reeder (M)
M. breviaristata (Hack.) Parodi (s)
M. brevifolia Scribn. ex Beal
M. breviligula Hitchc. (E-nc)
M. brevis C.O. Goodd.
M. breviseta Griseb. ex E. Fourn.
M. brevivaginata Swallen
M. bushii R.W. Pohl (M)
M. californica Vasey (M)
M. capillaris (Lam.) Trin. (P-nc)
M. capillipes (M.E. Jones) P.M. Peterson & Annable
M. caxamarcensis Laegaard & Sánchez Vega (s)
M. ciliata (Kunth) Trin. (M-ncs)
M. cleefii Laegaard (s)
M. crispiseta Hitchc.
M. cualensis Y. Herrera & P.M. Peterson
M. curtifolia Scribn. (M)
M. xcurtisetosa (Scribn.) Bush (M)
M. curviaristata (Ohwi) Ohwi (M-a)
M. curviaristata var. *nipponica* Ohwi (M-a)
M. curvula Swallen
M. cuspidata (Torr.) Rydb.
M. depauperata Scribn.
M. distans Swallen (E-nc)
M. distichophylla (J. Presl) Kunth (E-nc)
M. diversiglumis Trin. (M-ncs)
M. dubia E. Fourn. (P)
M. dumosa Scribn. ex Vasey (M)
M. durangensis Y. Herrera
M. elongata Scribn. ex Beal (E)
M. eludens C. Reeder
M. emersleyi Vasey (E)
M. eriophylla Swallen
M. expansa (Poir.) Trin. (E)
M. fastigiata (J. Presl) Henrard (s)
M. filiculmis Vasey
M. filiformis (Thurb. ex S. Watson) Rydb.
M. filipes M.A. Curtis
M. flabellata Mez (nc)
M. flavida Vasey
M. flaviseta Scribn.
M. flexuosa Hitchc. (s)
M. fragilis Swallen (nc)
M. frondosa(Poir.) Fernald (M)
M. gigantea (E. Fourn.) Hitchc. (E-nc)
M. glabriflora Scribn. (M)
M. glauca (Nees) B.D. Jacks. (M)
M. glomerata (Willd.) Trin. (M)

M. grandis Vasey (E)
M. gypsophila Reeder & C. Reeder (P)
M. hakonensis (Hack. ex Matsum.) Makino (M-a)
M. himalayensis Hack. ex Hook. (M-a)
M. huegelii Trin. (M-a)
M. hintonii Swallen
M. implicata (Kunth) Trin. (ncs)
M. inaequalis Soderstr. (S)
M. xinvoluta Swallen (E)
M. iridifolia Soderstr. (E)
M. jaliscana Swallen
M. japonica Steud. (M-a)
M. jonesii (Vasey) Hitchc.
M. laxa Hitchc.
M. lehmanniana Henrard (E-ncs)
M. ligularis (Hack.) Hitchc. (cs)
M. lindheimeri Hitchc. (E)
M. longiglumis Vasey (E)
M. longiligula Hitchc. (E)
M. lucida Swallen (P)
M. macroura (Kunth) Hitchc. (P-nc)
M. majalcensis P.M. Peterson
M. maxima Laegaard & Sánchez Vega (s)
M. mexicana (L.) Trin. (M)
M. mexicana var. *filiformis* (Torr.) Scribn. (M)
M. michisensis Y. Herrera & P.M. Peterson
M. microsperma (DC.) Trin. (M-ncs)
M. minutissima (Steud.) Swallen (nc)
M. montana (Nutt.) Hitchc. (nc)
M. mucronata (Kunth) Kunth (P)
M. mutica (Rupr. ex E. Fourn.) Hitchc. (E-nc)
M. nigra Hitchc. (P-nc)
M. orophila Swallen (nc)
M. palmeri Vasey (P)
M. palmirensis Grignon & Laegaard (s)
M. pauciflora Buckley (M)
M. pectinata C.O. Goodd. (M)
M. peruviana (P. Beauv.) Steud. (ncs)
M. pilosa P.M. Peterson, Wipff & S.D. Jones (E)
M. plumbea (Trin.) Hitchc. (nc)
M. polycaulis Scribn. (M)
M. porteri Scribn. ex Beal
M. pubescens (Kunth) Hitchc. (E)
M. pubigluma Swallen (E)
M. pungens Thurb. ex A. Gray
M. purpusii Mez
M. quadridentata (Kunth) Trin. (nc)
M. racemosa (Michx.) Britton, Sterns & Poggenb. (M)
M. ramosa (Hack. ex Matsum.) Makino (M-a)
M. ramulosa (Kunth) Swallen (ncs)
M. reederorum Soderstr. (E)
M. repens (J. Presl) Hitchc.
M. reverchonii Vasey & Scribn. ?
M. richardsonis (Trin.) Rydb.
M. rigens (Benth.) Hitchc. (P)
M. rigida (Kunth) Trin. (P-ncs)
M. robusta (E. Fourn.) Hitchc. (E-nc)
M. schmitzii Hack.

M. schreberi J.F. Gmel. (M-ns)
M. scoparia Vasey (E)
M. seatoni Scribn.
M. setarioides E. Fourn. (M-nc)
M. setifolia Vasey
M. sinuosa Swallen
M. sobolifera (Muhl. ex Willd.) Trin. (M)
M. speciosa Vasey (E)
M. spiciformis Trin. (M-nc)
M. straminea Hitchc.
M. stricta (J. Presl) Kunth
M. strictior Scribn. ex Beal
M. subaristata Swallen
M. sylvatica (Torr.) Torr. ex A. Gray (M)
M. tenella (Kunth) Trin. (M-ncs)
M. tenuiflora (Willd.) Britton, Sterns & Poggenb. (M)
M. tenuifolia (Kunth) Trin. (M-ns)
M. tenuissima (J. Presl) Kunth (nc)
M. texana Buckley
M. thurberi (Scribn.) Rydb. (M)

M. torreyana (Schult.) Hitchc. (E)
M. torreyi (Kunth) Hitchc. ex Bush (ns)
M. uniflora (Muhl.) Fernald
M. utilis (Torr.) Hitchc. (nc)
M. vaginata Swallen (nc)
M. venezuelae Luces (s)
M. versicolor Swallen (E-nc)
M. villiflora Hitchc.
M. villiflora var. villosa (Swallen) Morden
M. virescens (Kunth) Trin.
M. virletii (E. Fourn.) Soderstr. (E)
M. watsoniana Hitchc.
M. wrightii Vasey ex J.M. Coult.
M. xanthodas Soderstr. (E-c)

Pereilema J. Presl
P. beyrichianum (Kunth) Hitchc. (s)
P. ciliatum E. Fourn. (nc)
P. crinitum J. Presl (ncs)
P. diandrum R.W. Pohl (c)

Grasses: Systematics and Evolution. (2000). Eds S.W.L. Jacobs and J. Everett. (CSIRO: Melbourne)

G R A S S E S

A PHYLOGENY OF TRIODIEAE (POACEAE: CHLORIDOIDEAE) BASED ON THE ITS REGION OF NRDNA: TESTING CONFLICT BETWEEN ANATOMICAL AND INFLORESCENCE CHARACTERS.

J.G. Mant[A,B], R.J. Bayer[B], M.D. Crisp[A], and J.W.H. Trueman[C]

[A] Division of Botany and Zoology, Australian National University, ACT, 0200, Australia;
  E-mail: Jim.Mant@anu.edu.au
[B] Centre for Plant Biodiversity Research, CSIRO Plant Industry, PO Box 1600, ACT, 0200, Australia.
[C] Research School of Biological Sciences, Australian National University, ACT 0200, Australia.

Abstract

A phylogeny of the predominantly arid Australian grass tribe Triodieae ('Spinifex') is presented based the internal transcribed spacer (ITS) of nuclear ribosomal DNA. The main genus in the tribe is *Triodia* with 64 species (now also including the 16 species of *Plectrachne*). Of the two smaller genera, *Symplectrodia* has two species and *Monodia* is monotypic. Previous systematic treatments of this tribe have maintained a marked inconsistency in patterns between anatomical and inflorescence characters. This paper, which comprises part of a wider morphological and molecular phylogenetic study of Triodieae, tests whether convergence may have occurred in either one or both of these two suites of characters. Molecular results on 27 representative taxa and three outgroups strongly support the recognition of two lineages based on leaf blade anatomy contrasting with previous classifications based on glume and lemma characters. A 'soft' spinifex clade with centro-abaxial stomata and a paraphyletic 'hard' group with ambito-abaxial stomata are suggested. The previous circumscriptions of both *Plectrachne sens. str.* and *Triodia sens. str.* were based on floral morphology which this study finds strongly convergent. Similarly, five out of the nine informal infrageneric groups in *Triodia sens. lat.* are based on convergent morphological characters. *Symplectrodia* has a 'hard' leaf anatomy and is nested within that group, while the position of the monotypic 'soft' genus *Monodia* remains uncertain under the nrDNA data.

Key words: ITS, Poaceae, *Triodia*, spinifex.

INTRODUCTION

The taxonomically isolated tribe Triodieae is one of the most characteristic of all Australian arid zone plant groups. Vast tracts of central and north-western Australia are dominated by 'spinifex' grasses from the genera *Triodia, Monodia* and *Symplectrodia* (Poaceae: Chloridoideae). Spinifex forms a dominant vegetation community known as 'hummock grassland' on over 22% of the continent and is of key importance to the ecology of arid and semi-arid Australia (Griffin 1992).

Spinifex is abundant in all the major Australian deserts but has a much wider distribution extending into temperate and seasonally arid (monsoonal) areas, generally on poor soils such as deep siliceous sands, sandy red-earths, and skeletal soils of various origin

(Lazarides 1997). While there are several very widespread taxa, most of the species are restricted to the Hammersley, Kimberley, MacDonnell, and Musgrave mountain ranges of the arid centre and north-west of Australia (Jacobs 1982; Lazarides 1997). Despite the clear ecological significance of this tribe of grasses it has received comparatively little attention from the fields of ecology or systematics (Jacobs 1992).

The tribe is recognised by a strong synapomorphy in its C_4 Kranz tissue arrangement, which differs from that typically found in the subfamily Chloridoideae (Craig and Goodchild 1977; Renvoize 1983; Van den Borre and Watson 1997). Triodieae has been confirmed as possessing a (NAD-ME) C_4 photosynthetic pathway (McWilliam and Mison 1974; Prendergast *et al.* 1987) but has an unusual mesophyll and outer bundle sheath structure. The stomata of all species are sunken in deep, narrow grooves which contain dense interlocking papillae (McWilliam and Mison 1974). A tightly packed mesophyll is concentrated around these stomatal grooves in a disjunct position from the associated vascular bundles. Rather than being arranged concentrically around the vascular tissue, the outer bundle sheath extends laterally to connect the displaced mesophyll with the mestome and vascular tissues (McWilliam and Mison 1974; Craig and Goodchild 1977).

There are currently three genera recognised in Triodieae of which *Triodia* R.Br. is the main genus with c. 64 species (Lazarides 1997). *Monodia* S.W.L. Jacobs is monotypic (Jacobs 1985) and *Symplectrodia* Lazarides includes only two species (Lazarides 1984). A fourth genus, *Plectrachne* Henrard (with 16 species) was synonymised into *Triodia sens. lat.* in a recent revision by Lazarides (1997).

Spikelet and other macro-morphological features have formed the basis of generic and infrageneric distinctions in Triodieae despite concerns about high levels of homoplasy in these characters (Burbidge 1946b, 1953; Jacobs 1971). The previous generic distinction between *Plectrachne sens. str.* and *Triodia sens. str.* was based on glume and lemma features, *Plectrachne* being distinguished by its long linear glumes and deeply three-lobed or three-awned lemma (Lazarides 1997). *Triodia sens. str.* typically has comparatively short glumes and emarginate to shortly three-lobed lemmas, with several other variants. In the recent revision (Lazarides 1997) nine informal infrageneric groups were erected for *Triodia sens. lat.* based on general features of the inflorescence with some contribution from geographical distribution.

Overall, the plants of these (four) genera exhibit comparatively uniform floral and gross vegetative features. Yet, previous classifications have failed to account for the presence of two quite different types of leaves in the tribe. These have been recognised previously (Burbidge 1946b; Jacobs 1971) and retain the widely used common names, *'hard'* spinifex (40 species) versus *'soft'* (27 species) (Gardner 1952; Mant 1998). *'Soft'* species generally have a softer appearing foliage, a weakly pungent leaf apex, strongly curled senescent leaf, and a less-hummocked habit. Most, but not all, *'soft'* species produce a characteristic viscous resin which is used extensively as an adhesive by indigenous Australians. Rare reports of resin in some *'hard'* specimens (Burbidge 1953; Lazarides 1997) refer to droplets of a dark red exudate of a different nature to the aromatic, sticky resin of most *'soft'* species (Mant 1998). Ecologically, *'soft'* species are restricted to the monsoonal north of the continent, with the exception of *T. melvillei* (Lazarides 1997), whereas *'hard'* species are found in both monsoonal and temperate areas.

The precise nature of the distinction between *'hard'* and *'soft'*, however, involves the distribution of stomata and photosynthetic tissues on the outer (abaxial) face of the leaf (Burbidge 1946b). In *'hard'* species the stomatal grooves are spread uniformly over the abaxial leaf face, whereas in *'soft'* species the stomatal grooves and associated photosynthetic tissues are lacking from the abaxial surface except for a few that cluster around the mid-rib.

Both types of leaf occur in *Plectrachne sens. str.* and *Triodia sens. str.* and five of the nine infrageneric groups are polymorphic for this vegetative feature (Lazarides 1997). *Monodia* has a *'soft'* leaf anatomy, whereas *Symplectrodia* is *'hard'* (Mant 1998). In essence then, these classification systems reflect a weighting scheme towards inflorescence and against anatomical characters. Conflict between characters or data sets is best examined by investigating a range of different data sources, both molecular and morphological (Doyle 1997). In the present paper a molecular phylogeny of the tribe based on ITS sequence data is presented to provide an independent test of these competing claims of character importance.

METHODS

Twenty-seven representative taxa (Table 1) were chosen to encompass the morphological diversity found throughout the tribe. All nine informal infrageneric groups are represented, as are the two smaller genera, *Monodia* and *Symplectrodia*. Three outgroups were selected following a preliminary analysis of taxa from among the Chloridoideae and Arundinoideae available on GenBank (Hsaio *et al.* 1998).

Total genomic DNA was isolated either from seedling material, fresh field samples, or herbarium specimens. DNA from fresh or CTAB preserved samples was isolated following a variation from standard procedures for small quantities of tissue (Doyle and Doyle 1990; Bayer *et al.* 1996). Herbarium material was isolated then purified using the diatomite method of Gilmore *et al.* (1993).

PCR Amplification and Sequencing

The ITS region was amplified using primers ITSL (Hsaio *et al.* 1994) and ITS4 (White *et al.* 1990). Herbarium specimens were amplified in two fragments using the internal primers ITS2 and ITS3 (White *et al.* 1990) in conjunction with ITSL and ITS4 respectively. Standard polymerase chain reaction (PCR) techniques were employed as described elsewhere (Bayer *et al.* 1996). Double stranded PCR products were sequenced in both directions using the dideoxy chain termination method (Sanger *et al.* 1977) with the use of the Big Dye Terminator RR Kit ® (Perkin-Elmer Applied Biosystems) and the ABI automated sequencer at the Division of Plant Industry, CSIRO.

Phylogenetic Analysis

The analysis was conducted on PAUP* test version 4.d64 (D. Swofford unpublished). Sequences were aligned and edited using Sequencher version 3.0 and refined by eye. Minor indels of less than five base pairs were inferred in the alignment. Parsimony analyses were implemented using heuristic searches with TBR branch swapping and ten random addition sequence starting trees, with gaps treated as missing. Bootstrapping (BS) was performed with 100 replicates. T-PTP tests (Faith 1991) on individual clades were implemented. Decay indices (DI) (Bremer 1988) were calculated for the ingroup using PAUP* constraint trees and batch searches.

RESULTS

ITS Sequence Length and Nucleotide Composition

The length of ITS in Triodieae is within the range found in other studies of Poaceae (Hsaio *et al.* 1994; Buckler and Holtsford

Table 1. Voucher details for molecular accessions. All vouchers at CANB.

Taxon	Voucher/GenBank Accesssion No.	Source
M. stipoides Jacobs	A.A. Mitchell PRP3	h
S. gracilis Lazarides	C. Dunlop 4410	h
S. lanosa Lazarides	I.R. Telford 8029	h
T. angusta Burbidge	P.J. Davidson 2038A	s
T. basedowii E. Pritzel	P.J. Davidson 2054	s
T. bitextura Lazarides	G. Wells s.n.	s
T. bromoides (F. Muell.) Lazarides	Keighery 11, 171	h
T. bynoei (Hubbard) Lazarides	B.J. Carter 765	f
T. danthonioides (F.Muell.) Lazarides	S. Donaldson 1451	f
T. epactia Jacobs	P.J. Davidson 2043	s
T. fitzgeraldii Burbidge	M. Lazarides 3169	h
T. hubbardii Burbidge	G. Griffin s.n.	f
T. lanigera Domin	A.A. Mitchell PRP290	h
T. longiceps J.M. Black	S. van Leeuwen 3381	f
T. melvillei (Hubbard) Lazarides	P.J. Davidson 2035	s
T. molesta Burbidge	M. Lazarides 4393	h
T. plectrachnoides Burbidge	I. Cowie 5703	h
T. plurinervata Burbidge	D.C.F. Rentz s.n.	h
T. procera R.Br.	Waddy 794	h
T. racemigera C.A. Gardner	A. Craig s.n.	f
T. rigidissima (Pilger) Lazarides	P.J. Davidson s.n.	h
T. scariosa Burbidge	J. Mant 25	f
T. schinzii (Henr.) Lazarides	P.J. Davidson 2050	s
T. spicata Burbidge	P.J. Davidson 2057	s
T. stenostachya Domin	B. Rice 4388	h
T. tomentosa Jacobs	G. Flowers 107	f
T. wiseana C.A. Gardner	P.J. Davidson 2037	s
Eragrostis dielsii Pilger	AF019834	g
Muhlenbergia richardsonis (Trinius) Rydberg	AF019837	g
Chloris truncata R.Br.	AF019840	g

Source: (f) fresh sample; (s) seedling material; (h) herbarium specimen; (g) GenBank.

1996; Ainouche and Bayer 1997). The entire ITS region varies in length from 591 to 598 base pairs (bp) for the ingroup. Outgroup taxa vary from 596-597 bp in length. The 5.8S subunit is 164 bp in all taxa, with two variable sites in the ingroup. The ingroup (27 taxa) has 505 constant sites and 111 variable sites, 53 of these being potentially parsimony informative. The variation between ITS1 and ITS2 is comparable, with ITS2 marginally more variable (56 versus 53 variable sites).

(G+C) bias averages 59.2 % throughout the ingroup and 55.2% in the three outgroup taxa, which is comparable to other studies (Hsaio *et al.* 1998). Pairwise nucleotide divergence (Kimura two-parameter distance) among Triodieae species ranges from 1.2% to 6.5%. Intraspecific variation was less than that found between species in the five taxa where this was tested (unpublished data).

Phylogenetic Results

The maximum parsimony tree shown in Fig. 1 is a strict consensus of 22 most parsimonious trees of length 303 (CI: 0.54; RI: 0.62). There is strong ITS support shown for the monophyly of the tribe Triodieae (100% BS). This provides molecular support for the taxonomically isolated position of Triodieae that was suggested in previous morphological analyses of the chloridoids (Van den Borre and Watson 1997).

Within the ingroup the major pattern to note in the ITS trees is the split between *'hard'* and *'soft'* taxa. All those taxa with a *'soft'* leaf anatomy, except *Monodia stipoides*, are in a strongly supported clade (BS 91%; DI 5). All other ingroup taxa have a *'hard'* anatomy. *Monodia* falls outside the *'soft'* clade, although not significantly so (t-ptp test for non-monophyly p=0.68; +3 steps). *Plectrachne* is clearly not monophyletic (t-ptp test for non-monophyly p=0.01; +21 steps).

Within the *'soft'* clade there is poor resolution. Similarly, relationships among *'hard'* taxa are poorly supported. However, some groups are found consistently. Of these, a 'southern temperate' clade is found although with poor support (BS 59%; DI 1). This clade comprises representative species from south-western Western Australia (*T. danthonioides, T. bromoides*) and southern Australia (*T. scariosa* and *T. tomentosa*). However,

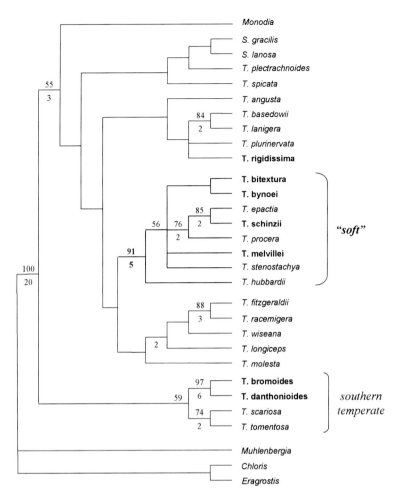

Fig. 1. ITS strict consensus of 22 most parsimonious trees of length 303 (CI: 0.54; RI: 0.62). Bootstrap (>50%) above the line, decay indicies (>1) below the line. All species with a 'soft' leaf anatomy except *Monodia stipoides* form a strongly supported clade. All other ingroup species have a 'hard' leaf anatomy. Taxa in bold are former *Plectrachne* species.

T. rigidissima, which appears to be a close morphological relative of *T. danthonioides* and *T. bromoides*, is consistently found outside the southern temperate clade (t-ptp test for non-monophyly p=0.01; +13 steps).

Outgrouping estimates the southern temperate clade as sister to the rest of the tribe. As such, the root suggests the 'hards' are paraphyletic and the 'softs' are a monophyletic group nested within the 'hards'. Given the poorly supported topologies within these two major groupings the closest sister group to the 'soft' clade remains uncertain under the ITS data.

DISCUSSION

An outstanding problem in the systematics of Triodieae has been a marked inconsistency between inflorescence and anatomical characters. This has persisted in the literature at least since the anatomical studies of Burbidge (1946a, b, 1953) in part because of an implied weighting scheme adopted towards the inflorescence characters traditionally used for taxonomic studies among closely related species of Poaceae. However, the application of molecular sequence data to this problem has provided strong evidence for a pattern of convergence in key spikelet characters (e.g. glume length and shape; lemma shape, lobing and nervation)

while supporting the hypothesis that there are two homologous leaf types in the tribe.

The molecular evidence presented strongly supports the presence of two groups within Triodieae that cut across the previous classification. A 'hard' group with ambito-abaxial stomata and a 'soft' group with centro-abaxial stomata are found. This 'hard' - 'soft' topology is strongly supported by the ITS nrDNA data and is corroborated by morphological and two cpDNA sequence data sets (Mant 1998). The monotypic *Monodia*, which has a highly apomorphic spikelet morphology, is an exception to this pattern for reasons that are not yet clear, although an origin through intergeneric hybridisation could be investigated. Overall, the floral morphology that previously circumscribed both *Plectrachne sens. str.* and *Triodia sens. str.* and which contributes to the make-up of infrageneric groups is convergent and unsuitable for generic delimitation.

While the ITS data have failed to strongly resolve relationships within these two groups, outgroup comparison suggests the 'hards' are a paraphyletic group. By extension, the root supports the suggestion of a derived 'soft' leaf anatomy. This is an intuitively appealing pattern as a loss of photosynthetic capacity from the outer side of the leaf in the *'soft'* lineage is most likely. A change from *'hard'* to *'soft'* may represent a specialisation to the more predictable rainfall of the monsoonal arid north of the continent to where species from the *'soft'* lineage are restricted. The presence of resin among most *'soft'* species, which is a relatively rare state in grasses (Burbidge 1946b), is further suggestive of a derived *'soft'* condition in the tribe.

Finally, there is a suggestion of a southern temperate clade linking the south-west of Western Australia and southern Australia. This contrasts with previous interpretations of the tribe's biogeography which were based on the division between *Triodia* and *Plectrachne* (Jacobs 1982). The south-western W.A. taxa, previously *'hard'* *Plectrachne* species, are not closely related to the *'soft'* *Plectrachne* of the monsoonal north of Western Australia. Instead, species from the south-west of W.A. appear to share a close connection with the other southern temperate species in the tribe (*T. scariosa*, *T. irritans* and allies) despite strong dissimilarities in certain spikelet characters. The position of *T. rigidissima*, a species from the Western Desert which has morphological affinities with *T. danthonioides*, conflicts with this biogeographic pattern and requires further investigation.

ACKNOWLEDGEMENTS

The funding and assistance given by the Centre for Plant Biodiversity Research, CSIRO Plant Industry is gratefully acknowledged. The project received support in part from an Hansjorg Eichler Research Grant from the Australian Systematic Botany Society. Thanks in particular to Phillip Davidson (University of Queensland) for supplying *Triodia* seed, Mike Lazarides, and to those who collected samples: Stuart Donaldson, Brian J. Carter, Andrew Craig, Graham Griffin, Stephen van Leeuwen and Helen Thompson.

REFERENCES

Ainouche, M. L., and Bayer, R. J. (1997). On the origins of the tetraploid *Bromus* species (section *Bromus*, Poaceae): insights from internal transcribed spacer sequences of nuclear ribosomal DNA. *Genome* **40**, 730-743.

Bayer, R. J., Soltis, D. E., and Soltis, P. S. (1996). Phylogenetic Inferences in *Antennaria* (Asteraceae: Gnaphalieae: Cassiniiae) based on sequences from nuclear ribosomal DNA internal transcribed spacers (ITS). *American Journal of Botany* **83**, 516-527.

Bremer, K. (1988). The limits of amino acid sequence data in angiosperm phylogenetic reconstruction. *Evolution* **42**, 795-803.

Buckler, E. S., and Holtsford, T. P. (1996). *Zea* systematics: ribosomal ITS evidence. *Molecular Biology and Evolution* **13**, 612-622.

Burbidge, N. T. (1946a). Foliar anatomy and the delimitation of the genus *Triodia* R.Br. *Blumea* Supplement **III**, 83-89.

Burbidge, N. T. (1946b). Morphology and anatomy of the Western Australian species of *Triodia* R.Br. II. Internal anatomy of leaves. *Transactions of the Royal Society of S.A.* **70**, 221-237.

Burbidge, N. T. (1953). The genus *Triodia* R.Br. *Australian Journal of Botany* **1**, 121-184.

Craig, S., and Goodchild, D. J. (1977). Leaf ultrastructure of *Triodia irritans*: a C_4 grass possessing an unusual arrangement of photosynthetic tissues. *Australian Journal of Botany* **25**, 277-290.

Doyle, J. J. (1997). Trees within trees: genes and species, molecules and morphology. *Systematic Biology* **46**, 537.

Doyle, J. J., and Doyle, J. L. (1990). Isolation of plant DNA from fresh tissue. *Focus* **12**, 13-15.

Faith, D. P. (1991). Cladistic permutation tests for monophyly and non-monophyly. *Systematic Biology* **40**, 366-375.

Gardner, C. A. (1952). Flora of Western Australia; Gramineae. (Government Printer: Perth.)

Gilmore, S., Weston, P. H., and Thompson, J. A. (1993). A simple, rapid, inexpensive and widely applicable technique for purifying plant DNA. *Australain Systematic Botany* **6**, 139-148.

Griffin, G. F. (1992). Will it burn - should it burn?: management of the spinifex grasslands of inland Australia. In 'Desertified Grasslands: Their Biology and Management'. (Ed. G. P. Chapman.) pp. 63-76. (Academic Press: London.)

Hsaio, C., Chatterson, N. J., Asay, K. H., and Jensen, K. B. (1994). Phylogenetic relationships of 10 grass species: an assessment of the phylogenetic utility of the internal transcribed spacer region in nuclear ribosomal DNA in monocots. *Genome* **37**, 112-120.

Hsaio, C., Jacobs, S. W. L., Barker, N. P., and Chatterton, N. J. (1998). A molecular phylogeny of the subfamily Arundinoideae (Poaceae) based on sequences of rDNA. *Australian Systematic Botany* **11**, 41-52.

Jacobs, S. W. L. (1971). Systematic position of the genera *Triodia* R.Br. and *Plectrachne* Henr. (Gramineae). *Proceedings of the Linnean Society of New South Wales* **96**, 175-185.

Jacobs, S. W. L. (1982). Relationships, distribution and evolution of *Triodia* and *Plectrachne*. In 'Evolution of the Flora and Fauna of Arid Australia.' (Eds W. R. Barker and P. J. M. Greenslade.) pp. 278-290. (Peacock: Adelaide.)

Jacobs, S. W. L. (1985). A new grass genus from Australia. *Kew Bulletin* **40**, 659-661.

Jacobs, S. W. L. (1992). Spinifex (*Triodia, Plectrachne, Symplectrodia* and *Monodia*: Poaceae) in Australia. In 'Desertified Grasslands: Their Biology and Management.' (Ed. G. P. Chapman.) pp. 45-62. (Academic Press: London.)

Lazarides, M. (1984). New taxa of tropical Australian grasses (Poaceae). *Nuytsia* **5**, 273-303.

Lazarides, M. (1997). A revision of *Triodia*, including *Plectrachne* (Poaceae, Eragrostideae, Triodiinae). *Australian Systematic Botany* **10**, 381-489.

Mant, J. G. (1998). 'A Phylogeny of *Triodia* and Related Genera (Poaceae: Chloridoideae).' Honours Thesis. Division of Botany and Zoology, Australian National University: Canberra.

McWilliam, J. R., and Mison, K. (1974). Significance of the C_4 pathway in *Triodia irritans* (Spinifex), a grass adapted to arid environments. *Australian Journal of Plant Physiology* **1**, 171-175.

Prendergast, H. D. V., Hattersley, P. W., and Stone, N. E. (1987). New structural / biochemical associations in leaf blades of C_4 grasses (Poaceae). *Australian Journal of Plant Physiology* **14**, 403-420.

Renvoize, S. A. (1983). A survey of leaf-blade anatomy in grasses, IV - Eragrostideae. *Kew Bulletin* **38**, 469-478.

Sanger, F., Nicklen, S., and Coulson, A. R. (1977). DNA sequencing with chain-terminating inhibitors. *Proceedings of the National Academy of Science U.S.A.* **74**, 5463-5467.

Van den Borre, A., and Watson, L. (1997). On the classification of Chloridoideae. *Australian Systematic Botany* **10**, 491-531.

White, T. J., Bruns, T., Lee, S., and Taylor, J. (1990). Amplification and direct sequencing of fungal ribosomal RNA genes for phylogenetics. In 'PCR Protocols: a guide to methods and applications.' (Eds M. Innis, D. Gelfand, J. Sninsky and T. J. White.) pp. 315-322. (Academic Press: San Diego, CA.)

ARUNDINOIDS

Austrodanthonia monticola, native to south-eastern Australia.

Grasses: Systematics and Evolution. (2000). Eds S.W.L. Jacobs and J. Everett. (CSIRO: Melbourne)

THE DANTHONIEAE: GENERIC COMPOSITION AND RELATIONSHIPS

Nigel P. Barker[A], Cynthia M. Morton[B], and H. Peter Linder[C]

[A] Department of Botany, Rhodes University, Grahamstown, 6140, South Africa.
[B] Department of Botany, University of Reading, Reading, RG6 2AS, UK.
    Present address: Director of the Herbarium, Botany and Microbiology, 101 Life Science Building,
    Auburn University, Alabama AL 36849, USA.
[C] Department of Botany, University of Cape Town, P. Bag, Rondebosch, 7700, South Africa.

Abstract

Molecular studies have shown the Arundinoideae to comprise three lineages, considered here to be the tribes Aristideae, Arundineae and Danthonieae. The Arundineae includes the reedy genera traditionally associated with this tribe as well as several additional genera previously considered to have danthonioid affinities, while the Danthonieae is now a more narrowly circumscribed group, comprising 19 genera. While there is still no clarity on the relationships of these three tribes and the Chloridoideae, the generic relationships within the Danthonieae are becoming clear and are examined in detail here. Data from three molecular data sets and a morphological data set are analysed separately and in combination. These analyses indicate that there are seven informal groups in the tribe. All analyses resolve a basal assemblage of several species of *Merxmuellera*. The other groups are: the *Pentaschistis* clade (*Pentaschistis, Prionanthium, Pentameris*), the *Pseudopentameris* clade (*Pseudopentameris, Chaetobromus*), the *Chionochloa* clade (*Chionochloa* only), the *Cortaderia* clade (*Lamprothyrsus*, South American *Cortaderia* species), the *Rytidosperma* clade *(Rytidosperma, Karroochloa, Austrodanthonia, Schismus, Notodanthonia, Tribolium, Joycea,* some species of *Merxmuellera*) and the *Danthonia* clade (*Danthonia, Plinthanthesis, Notochloe* and New Zealand species of *Cortaderia*). Almost all the analyses are in agreement over the composition of these groups, but there is limited agreement as to the relationships of these groups. The morphological data appear to be very homoplasious, and affect the resolution of the topologies obtained when this set is analysed separately and in combination. Additional sampling across the data sets is needed before these relationships can be clearly resolved.

Key words: Danthonieae, phylogeny, ITS, *rpo*C2, *rbc*L, morphology, DNA sequence data.

INTRODUCTION

Of the six subfamilies of the grasses, the Bambusoideae and Arundinoideae have been considered to be taxonomically problematic (Campbell 1985; Ellis 1987; Kellogg and Campbell 1987; Kellogg and Watson 1993). The Arundinoideae is a diverse assemblage of genera, which are considered by some to be primitive (Renvoize 1986; Kalliola and Renvoize 1994). Opinions on the generic composition of the subfamily differ, and the number of genera included in it varies from 45 (Clayton and Renvoize 1986) to more than 70 (Watson and Dallwitz 1992). Tribal level classifications vary in both the number and composition of the recognised tribes (Clayton and Renvoize 1986; Conert 1987; Watson and Dallwitz 1992; tabulated in Barker *et al.* 1995).

Early efforts to address the issue of a possibly polyphyletic Arundinoideae were based on morphological and anatomical data (Campbell 1985; Renvoize 1986; Ellis 1987; Kellogg and Campbell 1987; Kellogg and Watson 1993). Although these studies were prone to problems of analysis associated with large data sets and missing data from the more poorly known taxa, reservations about the composition, relationships and doubtful monophyly of the Arundinoideae were frequent (Campbell 1985; Ellis 1987; Kellogg and Campbell 1987). Recent molecular studies based on chloroplast gene sequence data have corroborated the taxonomic problems in the Arundinoideae. All such studies suggest that the Arundinoideae as delimited by Clayton and Renvoize (1986) is a polyphyletic assemblage (Davis and Soreng 1993; Barker *et al. 1995*; Clark *et al. 1995*; Duvall and Morton 1996; Liang and Hilu 1996; Mathews and Sharrock 1996; Barker 1997; Linder *et al.*1997; Soreng and Davis 1998; Barker *et al.* 1999). In contrast, recent studies based on sequence data from the nuclear-encoded ribosomal Internal Transcribed Spacer (ITS) support a monophyletic Arundinoideae (Hsiao *et al.* 1998, 1999). However, all these studies are open to criticisms: the *rbc*L and *ndh*F studies are poorly sampled in certain lineages, the *rpo*C2 study is subject to problems of positional homology assessment, and thus alignment, while the ITS study is flawed by the use of an inappropriate outgroup and an absence of taxa representing other major clades

Polyphyly of a major group (such as the Arundinoideae) can be the result of the erroneous inclusion of "misfit taxa" in the group and/or the presence of two or more major lineages that are not immediately related to each other (or, conversely, by the inclusion of a lineage not previously thought to be part of the group). To date, molecular studies based on *rbc*L (Barker *et al.* 1995; Barker 1997), *ndh*F (Clark *et al.* 1995) and *rpo*C2 (Barker *et al.* 1999) sequence data indicate that the Arundinoideae, as presently delimited, is polyphyletic, as all these studies demonstrate the presence (and re-aligned positions) of "misfit" taxa. Such misfit genera include *Thysanolaena*, *Spartochloa*, *Cyperochloa* and *Anisopogon*.

The removal of *Thysanolaena*, *Spartochloa*, *Cyperochloa* and *Anisopogon* still leaves the question of whether the remaining taxa form a monophyletic subfamily, now comprised of three tribes; Aristideae, Arundineae and Micraireae (all sensu Clayton and Renvoize 1986) unresolved. This can only be answered by extensive molecular and morphological studies based on as many taxa as possible, and, ideally, sequence data from both nuclear and plastid genes.

The *rpo*C2 study (Barker *et al.* 1999) has a wide sampling range of both arundinoid and other lineages. However, while able to elucidate the composition of the major lineages (subfamilies and tribes) these data are unable to elucidate relationships between these lineages. In summary, however, there appear to be four clades: one corresponding to the tribe Danthonieae sensu Watson and Dallwitz (1992), a second corresponding to the tribe Arundineae sensu stricto, a third corresponding to the Micraireae and the fourth being the tribe Aristideae. What these data have enabled is the logical sampling of a range of taxa for inclusion in studies based on other, more conserved, genes such as *rbc*L. Sequences of this gene were thus obtained from representatives of all these lineages except the Micraireae, but despite this more selective approach to sampling, no well supported resolution of the relationships of the major lineages of the Arundinoideae was obtained (Barker *et al.* 1995; Duvall and Morton 1996; Barker 1997). At best, the *rbc*L data suggest that the subfamily Chloridoideae and the remaining tribes of the Arundinoideae are a monophyletic assemblage. It appears that there is

insufficient phylogenetic signal in the *rbc*L data set to resolve these relationships, possibly as a result of a rapid radiation of these groups from a common ancestor, a sentiment echoed by Duvall and Morton (1996) and Doebley *et al.* (1990). A Chloridoideae – Arundinoideae link is also supported by the earlier chloroplast restriction site study of Davis and Soreng (1993).

The study based on *ndh*F sequence data does not sample all these lineages either, and representatives of the large Danthonieae clade are not included. Nonetheless, *ndh*F data once again imply a paraphyletic Arundinoideae, parts of which are related to the Chloridoideae (through the Micraireae), with the Arundineae basal to the remainder of the PACC clade (Clark *et al.* 1995).

There are at present fewer data sets comprising sequence data from nuclear genes. The ITS study (Hsiao *et al.*1998, 1999) is well sampled, but because of the flaws outlined above, cannot be used to effectively test questions of monophyly of the Arundinoideae. Another nuclear gene family that has been successfully used to determine phylogenetic relationships is the phytochrome gene family. Mathews and Sharrock (1996) used sequences from three paralogues of this gene (*PHYA*, *PHYB* and *PHYC*) to elucidate relationships among the major lineages of the grasses. Phylogenetic analysis of these data (combined but especially *PHYA*) showed the Arundinoideae to be paraphyletic (Fig. 5 in Mathews and Sharrock 1996). However, not all of the major clades of the Arundinoideae, or even other subfamilies were sampled in this study, so these relationships cannot be given serious consideration. Until a series of data sets all sampling the same range of taxa becomes available, the goal of determining the relationships of the major clades of the Arundinoideae will not be realised.

What can be addressed, however, are the composition and relationships within the largest of the Arundinoid lineages: the clade corresponding to the Danthonieae sensu Watson and Dallwitz (1992). Apart from the sparsely sampled *rbc*L data set, there are two molecular data sets that cover a wide range of taxa within this tribe. These data sets are the chloroplast *rpo*C2 data set, and the nuclear ITS data set. The purpose of this paper is to 1) detail the composition of this tribe, 2) provide an indication of the relationships of genera in this tribe, 3) delimit groups of genera that are well supported and 4) highlight areas where relationships are obscure and need further investigation. It must be emphasised that the analyses presented here are preliminary, as additional data are still being gathered. More detailed and in-depth analyses will be published at a later date.

This study adopts the recent generic-level classifications for the genus *Rytidosperma* sensu lato (Linder and Verboom 1996, emended by Linder 1997) and *Tribolium* (Linder and Davidse 1997). The use of these names is thus not in the context presented by Clayton and Renvoize (1986).

MATERIALS AND METHODS

Details of methods used in the DNA extraction, amplification, sequencing and sequence alignment have been previously published (Barker *et al.* 1995, 1999; Hsiao *et al.*1998). Morphological and anatomical characters were coded from herbarium specimens. A total of 67 characters was obtained (Appendix), these were largely considered to be unordered.

These data sets were analysed individually and in combination using PAUP 3.1.1 (Swofford 1993). In all analyses, *Centropodia glauca* was used as the outgroup. *Centropodia* is considered to be an arundinoid by Clayton and Renvoize (1986), but both the *rbc*L and *rpo*C2 studies place it at the base of the chloridoid lineage (Barker *et al.* 1995, 1999). This position is supported by Ellis's (1984) anatomical observations that this genus possesses the C_4 metabolic pathway and associated leaf anatomy, features typical of the general chloridoid condition. Furthermore, *Centropodia* lacks the haustorial synergid cells in the embryo that are characteristic of the danthonioid lineage (Philipson and Connor 1984; Verboom *et al.* 1994).

The alignment of the sequences is an important issue for both the *rpo*C2 and ITS data sets. For the analysis presented here, the alignment of the *rpo*C2 sequences used by Barker *et al.* (1999) is retained. The ITS sequences were re-aligned by eye, and the data set is thus not identical to that used by Hsiao *et al.* (1998).

In an attempt to find all islands of most parsimonious trees (Maddison 1991), 200 random entry input analyses (TBR swapping, MULPARS OFF) were run. The shortest trees found by this step were then subjected to a TBR branch swapping with MULPARS ON to find the full set of minimal length trees. Where more than one tree was obtained, the strict consensus was calculated. Bootstrap figures (Felsenstein 1985) were obtained from 200 replicates, each calculated from a simple addition, MAXTREES set to 200, TBR swapping with MULPARS on. All *rpo*C2 sequences and some of the ITS sequence data is lodged in GenBank (see Barker *et al.* 1999 and Hsiao *et al.* 1998 for GenBank numbers). Sequence data and voucher details of the unpublished sequence data will be made available once these molecular studies are closer to completion.

RESULTS AND DISCUSSION

The alignment of the *rpo*C2 and ITS sequence data are somewhat problematic, especially in the context of the whole grass family. However, within the Danthonieae, only a few insertion – deletion events need be hypothesised to ensure that the positional homology of the nucleotides is maintained. The nature and structure of the *rpo*C2 gene has been discussed in detail elsewhere (Cummings *et al.*1994; Barker *et al.* 1999), as have issues surrounding the alignment of these sequences.

Although the ITS data set is nuclear and the *rpo*C2 plastid based, there is some degree of agreement between the trees derived from these two data sets, both in terms of the composition of the groups within the tribe, and in terms of the relationships of these groups. In separate analyses of these molecular data sets, at least some of seven clades are well supported (bootstrap values above 70%), but the relationships of these clades are not well supported. Thus areas of conflict between the two data sets are not well supported, and the data sets can be combined (De Quieroz 1993). This combination was done in an additive manner – i.e. all taxa from both data sets are represented, not just those taxa in common. Where data from one of the data sets was lacking for a taxon, it was considered to be "missing" and coded as "?". This "ambiguity coding" was adopted as there is a very low proportion

of species common to both data sets, and we were reluctant to utilise a "splice and merge" approach (Nixon and Carpenter 1996).

The strict consensus trees derived from the analyses of these data sets, both individually and in combination, are presented in Figs 1-9. Table 1 lists the statistics from each of the various analyses conducted on the data sets, both separately and in combination. Fig. 10 provides a summary of each the analyses, showing the relationships of these groups. Upon examination of the results of most analyses, it is apparent that seven groups of taxa were usually resolved, the composition of which is generally consistent. However, the relationships between these groups are data set dependent, and are further complicated by the fact that sampling across the different data sets is not uniform. The composition and relationships of each of the clades is discussed below. These data therefore give a signal as to the groups of genera in the Danthonieae (at least the molecular data sets do – the morphological data set appears less capable of resolving the generic groups), but do not provide a robust signal on the relationships among the groups of genera.

The Basal *Merxmuellera* Assemblage (BMA).

As a genus, *Merxmuellera* is problematic, as data set resolves it as monophyletic. The anomalous *Merxmuellera rangei* is basal to the remainder of the danthonioid lineage, and is shown in other studies to have closer affinities with the Chloridoideae (Barker *et al.* 1999). It is not discussed further here. The weighted total evidence analysis (Fig. 9) shows *M. stricta, M. dura, M. guillarmodiae* and *M. disticha* to be basal to the *Rytidosperma* clade, as does the ITS (Fig. 3) and *rpo*C2 (Fig. 2) data sets. *M. setacea, M. arundinacea, M. cincta, M. decora, M. davyi* and *M. macowanii* are placed as a paraphyletic assemblage basal to the Danthonieae by both the weighted combined data set, as well as the *rpo*C2 data sets (Figs 9 and 2). Only two species were sampled for ITS, and these were placed widely separated in the cladogram, thus not supporting the paraphyletic relationship suggested by the *rpo*C2 data. The morphological data (Fig. 4) also does not recognise these as a monophyletic group, but places them in a paraphylum basal to the *Rytidosperma* clade, rather than basal to the Danthonieae. The total data analysis is thus congruent with the molecular data sets, in particular *rpo*C2. This basal paraphyletic group of species is termed here the **Basal *Merxmuellera* Assemblage** (hereafter referred to as the **BMA**).

The species considered to be included in this assemblage are *M. davyii, M. macowanii, M. cincta, M. arundinacea, M. decora* and *M. setacea*. All of these taxa are sampled for *rpo*C2 and morphology data sets, but their representation in the ITS data set is poor (two species). There are several unsampled species of *Merxmuellera* in both the *rpo*C2 and ITS data sets, and it is possible to predict, on the basis of close anatomical and morphological similarity (Ellis 1983), that species such as *M. lupulina* and *M. decora* (close relatives of *M. rufa*) will also be placed in the **BMA.** The addition of ITS sequence data from more **BMA** taxa may well modify the relationships within this assemblage. At present the composition and arrangement of this assemblage, at least as presented here, is considered to be uncertain.

223

Table 1. The tree statistics for the various data sets for the Danthonieae. MPT = Most Parsimonious Tree, ci = consistency index, ri = retention index

Data set	No. taxa (incl. outgroup)	No. MPT's	Tree length	ci	ri
ITS	37	3379	427	0.520	0.708
*rpo*C2	34	72	143	0.727	0.886
*rbc*L	7	2	61	0.803	0.821
Morphology	55	1476	318	0.261	0.630
ITS + morphology	37	750	887	0.505	0.655
ITS + *rpo*C2	52	32700	587	0.566	0.756
Morphology + *rpo*C2	34	20	884	0.502	0.607
ITS + *rbc*L + *rpo*C2 + morphology	51	396	988	0.464	0.666

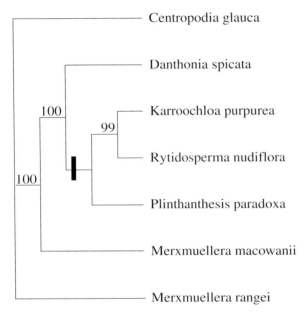

Fig. 1. One of two equally parsimonious trees obtained from the *rbc*L data set. Length = 61 steps, ci = 0.803, ri = 0.821. The bar indicates the node that collapses in the strict consensus tree. Numbers above the lines are bootstrap support values.

The *Pentaschistis* Clade.

The ***Pentaschistis* Clade** comprises the genera *Pentaschistis, Pentameris* and *Prionanthium*, and received high (99-100%) bootstrap support in the analyses of both the *rpo*C2 and ITS data sets (Figs 2 and 3). This clade has previously been thought of as a natural unit (Barker 1993; Linder and Ellis 1990; Linder *et al.* 1990) and it is readily distinguishable on morphological grounds (e.g. presence of 2-flowered spikelets, but this character is homoplasious), but was not retrieved by the morphological analysis. Of these genera, *Pentaschistis* is by far the largest, and it is possible that *Pentameris* and *Prionanthium* are specialised lineages embedded within a paraphyletic *Pentaschistis*.

The different data sets place this clade in a variety of positions. Several analyses (such as the *rpo*C2, *rpo*C2+morphology, weighted total data set) show it to be related to (basal to, sister to, or terminal to) the ***Chionochloa* clade**, but some analyses (especially those involving the morphological data) relate it to the ***Pseudopen-***

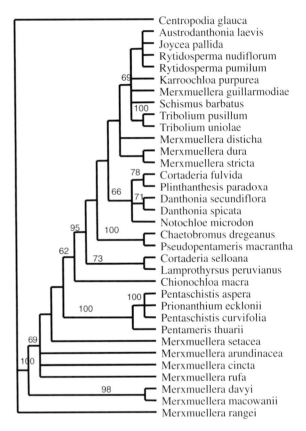

Fig. 2. Strict consensus of 72 equally parsimonious trees obtained from the *rpo*C2 data set. Length = 143 steps, ci = 0.727, ri = 0.886. Numbers above the lines are bootstrap support values.

***tameris* clade**. More detailed sampling of other clades, especially for the ITS data set, may stabilise the position of this clade.

The *Chionochloa* Clade.

All analyses retrieve a monophyletic *Chionochloa*, generally with strong bootstrap support, but the affinities of the genus remain obscure. In many analyses (e.g. *rpo*C2, ITS+*rpo*C2, *rpo*C2+morphology, total data set), *Chionochloa* is resolved as the most basal non-African genus in the danthonioid clade, and is placed above the basal African clades. Clayton and Renvoize (1986) and Conert (1987) suggest *Chionochloa* may be embedded in *Cortaderia*, but no analysis indicates a sister-group or paraphyletic relation-

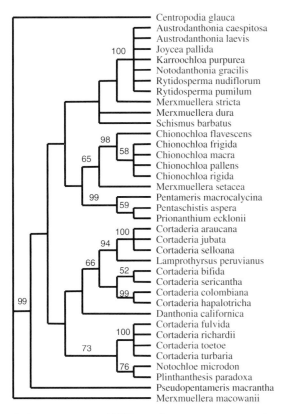

Fig. 3. Strict consensus of 3379 equally parsimonious trees obtained from the ITS data set. Length = 427 steps, c_i = 0.520, r_i = 0.708. Numbers above the lines are bootstrap support values.

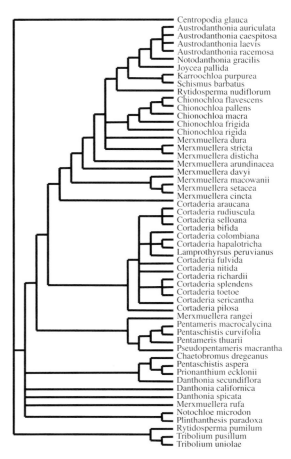

Fig. 4. Strict consensus of 1476 equally parsimonious trees obtained from the morphological data set. Length = 318 steps, c_i = 0.261, r_i = 0.630.

ship between the two genera. This supports the contention of Linder and Verboom (1996) that the genus is distinct, and only distantly related to other Australasian danthonioid grasses.

The *Cortaderia* "A" Clade.

The ***Cortaderia* "A" clade** comprises two South American genera; a species of *Cortaderia and Lamprothyrsus*. The ITS and *rpo*C2 data sets, which contain data from two or more species of *Cortaderia*, resolve this genus as polyphyletic along continental lines. The New Zealand species of *Cortaderia* are not included in this clade, but instead are considered to be part of the ***Danthonia* clade**, although the morphological data resolves them along with the South American taxa as a single genus, united by the synapomorphy of gynodioecy.

The ITS data set resolves *Lamprothyrsus* as embedded within the seven South American species of *Cortaderia*, and this clade as a whole receives moderate bootstrap support (66%; Fig. 3). The *rpo*C2 data set, which sampled just one South American species of *Cortaderia*, shows this support to be higher, at 73% (Fig. 2). The composition and relationships of *Cortaderia* are to be presented in detail elsewhere (Barker *et al.* in prep.).

The *Pseudopentameris* Clade.

This clade is another two-taxon clade comprising the southern African genera *Pseudopentameris* and *Chaetobromus*. Both of these genera are small (four and two species respectively). Although this clade is not always represented by both these genera in all the analyses, it is supported by several synapomor-

phic insertion – deletion events in the *rpo*C2 data set, as well as morphological features of the caryopsis (Barker 1994). The analysis of the morphological data does not retrieve this clade, and instead the representatives of the ***Pentaschistis* clade** and the ***Pseudopentameris* clade** are intermixed, rendering both clades paraphyletic.

The *Rytidosperma* Clade.

The ***Rytidosperma* clade** comprises the Australasian *Rytidosperma, Notodanthonia, Austrodanthonia, Joycea* as well as the African *Karroochloa, Schismus, Tribolium* and additional species of *Merxmuellera*. This clade is retrieved by most analyses, although in some cases (e.g., morphology) other clades are associated with it. However, in both the ITS and *rpo*C2 analyses, bootstrap support for this clade is weak (<50%). The morphological data set splits this clade into 3 smaller groups, ***Rytidosperma* "A"** (*Tribolium* and *Rytidosperma pumilum*), ***Rytidosperma* "B"** (*Merxmuellera disticha, M. dura, M. guillarmodiae* and *M. stricta*) and ***Rytidosperma* "C"** (*Rytidosperma nudiflora, Karroochloa, Schismus, Joycea, Notodanthonia* and *Austrodanthonia*). This is the only data set that resolves a polyphyletic ***Rytidosperma* clade**, and it is possible that these groups represent lineages within this larger clade.

At both the genus and species level, this clade is large. Although the alpha-taxonomy of many of the genera in this clade has been

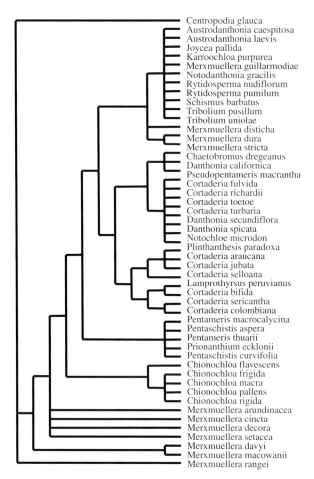

Fig. 5. Strict consensus of 32700 equally parsimonious trees obtained from the combined ITS plus *rpo*C2 data set. Length = 587 steps, ci = 0.566, ri = 0.756.

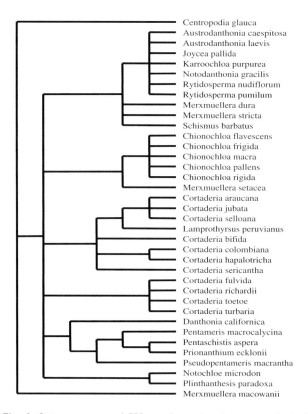

Fig. 6. Strict consensus of 750 equally parsimonious trees obtained from the combined ITS plus morphology data set. Length = 887 steps, ci = 0.505, ri = 0.655.

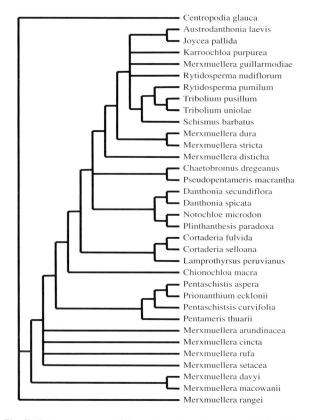

Fig. 7. Strict consensus of 20 equally parsimonious trees obtained from the combined *rpo*C2 plus morphology data set. Length = 884 steps, ci = 0.502, ri = 0.607.

completed (e.g. Linder and Verboom 1996; Linder 1997; Linder and Davidse 1997; Linder in press a; Linder in press b), sampling of this clade for molecular studies needs to be increased before any finality about generic relationships can be achieved.

The *Danthonia* Clade.

This clade comprises the genus *Danthonia*, represented by a number of different species in the different data sets, and the Australasian genera *Notochloe, Plinthanthesis* and the New Zealand species of *Cortaderia* (called *Cortaderia* "B" in the summary trees in Fig. 10). These genera are not always retrieved as a clade, and the combined ITS and morphological data set in particular shows this clade to be fragmented into three lineages (*Danthonia* "A", *Danthonia* "B" and *Cortaderia* "B", Fig. 10E). However, the majority of its constituent genera are retrieved as a clade in the *rpo*C2, ITS, ITS+*rpo*C2 and weighted total evidence analyses. Bootstrap support for this clade in the *rpo*C2 analysis is moderate (66%, Fig. 2), and slightly higher in the analysis of the ITS data (73%, Fig. 3). However, in the latter analysis, the single representative of *Danthonia* is not included in this clade. *Notochloe* and *Plinthanthesis* share an unusual leaf anatomy (Linder and Verboom 1996), and several analyses resolve these two genera as sister taxa.

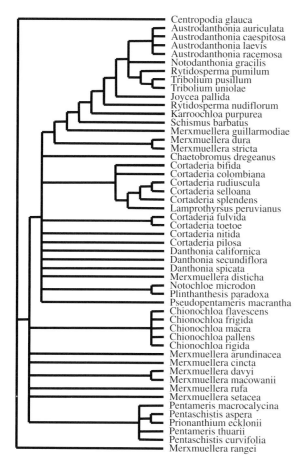

Fig. 8. Strict consensus of 396 equally parsimonious trees obtained from the combined *rbc*L, ITS, *rpo*C2 and morphology data set. Length = 988, steps, ci = 0.464, ri = 0.666.

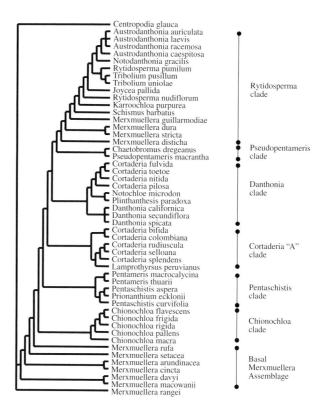

Fig. 9. Strict consensus of the 5 equally parsimonious trees obtained from the combined *rbc*L, ITS, *rpo*C2 and morphology data set, successively weighted based on character retention index values. The seven informal groups recognised in this study are shown on the right-hand side of the cladogram.

In the *rpo*C2 data set, *Danthonia* (represented by *Danthonia spicata* and its South American congener *D. secundiflora*) is monophyletic and receives 71% bootstrap support. In contrast, the morphological data does not resolve *Danthonia* as monophyletic, two of the three species forming part of a polytomy, while the third is basal to elements of the **Pentaschistis** and **Pseudopentameris clades**. *Danthonia* is a large, variable and geographically widely distributed genus, and it is possible that, like *Merxmuellera*, it is not a monophyletic lineage. However, the weighted total evidence analysis resolves *Danthonia* as a monophyletic genus, but only a study with suitably wide sample range will properly resolve this issue.

In the light of the above, it is clear that the molecular data sets resolve groups of similar composition, but the morphological data set is only partially congruent with the molecular data sets. The consistency index of the trees derived from the morphological data is very low (ci=0.261; Table 1), indicative of a high degree of homoplasy. This data set also has the lowest retention index, implying that only limited synapomorphies are present in the data. In other words, the morphological data (or elements of it) are noisy – a point that has been made previously in cladistic studies using morphological data for the grasses (Kellogg and Campbell 1986; Kellogg and Watson 1993). This discordant topology could be a result of inadequate sampling of species and/or substantial within genus variation.

There are thus several areas where further work is required. These include the acquisition of additional ITS sequence data in order to get increased overlap with the plastid *rpo*C2 data set, concentrating especially on getting data for all the species of *Merxmuellera*. Furthermore, additional samples of the widespread *Danthonia* are needed for both morphological and all molecular data sets in order to resolve relationships within the **Danthonia** clade. Only once this has been achieved can the composition and naming of these (at present) sometimes poorly supported informal clades be finalised and formalised.

CONCLUSIONS

Of the three tribes of the Arundinoideae, the generic composition of the Danthonieae is now well established. There is growing consensus on the composition of several clades within this tribe, and seven informal groups can be recognised. These are the **Basal *Merxmuellera* Assemblage**, the **Pentaschistis clade**, the **Chionochloa clade**, the **Pseudopentameris clade**, the **Cortaderia clade**, the **Rytidosperma clade** and the **Danthonia clade**.

The different data sets vary in terms of the relationships of these groups, but as there is only limited overlap in sampling, these relationships must be viewed as preliminary. Of the four data sets, the morphological data set provides particularly inconclusive results. This may be a result of under-sampling, allowing homoplasious characters to over-ride the real data. Greater overlap of sampling and additional sampling is required before a

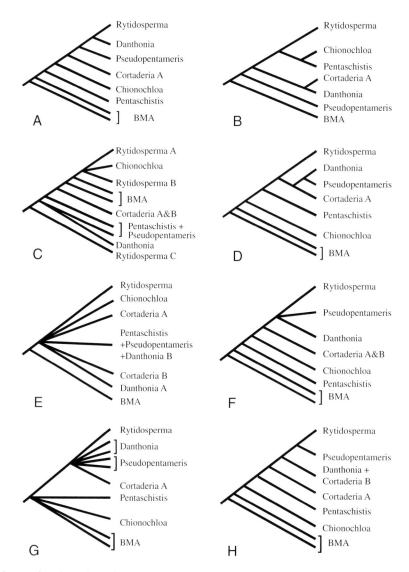

Fig. 10. Summary trees showing the relationships of the seven informal groups of genera. A) strict consensus tree from *rpo*C2 data set, B) strict consensus tree from ITS data set, C) strict consensus tree from morphological data set, D) strict consensus tree from ITS plus *rpo*C2 combined data set E) strict consensus tree from ITS plus morphology combined data set F) strict consensus tree from *rpo*C2 plus morphology data set G) strict consensus tree from all data sets combined, H) strict consensus tree from the combined data set weighted using retention index values.

convincing phylogeny of these informal groups can be achieved. For the time being these names serve a purpose in that they refer to (probably) monophyletic groups of genera that can be further studied as parts of the monophyletic Danthonieae.

ACKNOWLEDGEMENTS

NPB and CMM thank the Royal Society, London, for financial support and assistance. This research, carried out in part at the University of Reading, was made facilitated by a short term visiting research grant to NPB. NPB further acknowledges Rhodes University Joint Research Committee and the FRD for providing financial assistance to support his attendance at the Second International Monocot Conference held in Sydney, 1998.

REFERENCES

Barker, N.P. (1993). A biosystematic study of *Pentameris* (Arundineae, Poaceae). *Bothalia* **23**, 25-47.

Barker, N.P. (1994). External fruit morphology of southern African Arundineae (Arundinoideae: Poaceae). *Bothalia* **14**, 55-66.

Barker, N.P. (1997). The relationships of *Amphipogon*, *Elytrophorus* and *Cyperochloa* (Poaceae) suggested by *rbc*L sequence data. *Telopea* **7**, 205-213.

Barker, N.P., Linder, H.P., and Harley, E.H. (1995). Polyphyly of Arundinoideae (Poaceae): evidence from *rbc*L sequence data. *Systematic Botany* **20**, 423-435.

Barker, N.P., Linder, H.P., and Harley, E.H. (1999). Sequences of the grass-specific insert in the chloroplast *rpo*C2 gene elucidate generic relationships of the Arundinoideae (Poaceae). *Systematic Botany* **23**, 327-350.

Campbell, C. S. (1985). The subfamilies and tribes of Gramineae (Poaceae) in the southeastern United States. *Journal of the Arnold Arboretum* **66**, 123-199.

Clark, L. G., Zhang, W., and J. F. Wendel, J.F. (1995). A phylogeny of the grass family (Poaceae) based on *ndh*F sequence data. *Systematic Botany* **20**, 436-460.

Clayton, D.W., and Renvoize, S.A. (1986). Genera Graminum. Kew Additional Series XIII. (Her Majesty's Stationery Office: London.)

Conert, H.J. (1987). Current concepts in the systematics of the Arundinoideae. In 'Grass Systematics and Evolution'. (Eds T.R. Soderstrom, K.W. Hilu, C.S. Campbell and M.E. Barkworth.) pp. 239-250. (Smithsonian Institution Press: Washington DC.)

Cummings, M. P., King, L.M., and Kellogg, E.A. (1994). Slipped strand mispairing in a plastid gene: *rpoC2 in grasses (Poaceae). Molecular Biology and Evolution* **11**, 1-8.

Davis, J. I., and Soreng, R.J. (1993). Phylogenetic structure in the grass family (Poaceae) as inferred form chloroplast DNA restriction site variation. *American Journal of Botany* **80**, 1444-1454.

De Queiroz, A. (1993). For consensus (sometimes). *Systematic Biology* **42**, 368-372.

Doebley, J., Durbin, M., Golenberg, E.M., Clegg, M.T., and Pow Ma, D. (1990). Evolutionary analysis of the large subunit of carboxylase (*rbcL*) nucelotide sequence among the grasses (Gramineae). *Evolution* **44**, 1097-1108

Duvall, M.R., and Morton, B.R. (1996). Molecular phylogenetics of Poaceae: An expanded analysis of *rbcL* sequence data. *Molecular Phylogenetics and Evolution* **5**, 352-358.

Ellis, R.P. (1983). Leaf anatomy of the South African Danthonieae (Poaceae). VII. *Merxmuellera decora, M. lupulina* and *M. rufa. Bothalia* **14**, 197-203.

Ellis, R.P. (1984). Leaf anatomy of the South African Danthonieae (Poaceae). IX *Asthenatherum glaucum. Bothalia* **15**, 153-159.

Ellis, R.P. (1987). A review of comparative leaf blade anatomy in the systematics of the Poaceae: the past twenty-five years. In 'Grass Systematics and Evolution'. (Eds T.R. Soderstrom, K.W. Hilu, C.S. Campbell and M.E. Barkworth.) pp. 3-10. (Smithsonian Institution Press: Washington DC.)

Felsenstein, J. (1985). Confidence limits on phylogenies: an approach using the bootstrap. *Evolution* **39**, 783–791.

Hsiao, C., Jacobs, S.W.L., Barker, N.P., and Chatterton, N.J. (1998). A molecular phylogeny of the subfamily Arundinoideae (Poaceae) based on sequences of rDNA. *Australian Systematic Botany* **11**, 41-52.

Hsiao, C., Jacobs, S.W.L., Chatterton, N.J., and Assaay, K.H. (1999). A molecular phylogeny of the grass family (Poaceae) based on the sequences of nuclear ribosomal DNA (ITS). *Australian Systematic Botany* **11**, 667-688.

Kalliola, R., and Renvoize, S.A. (1994). One or more species of *Gynerium? (Poaceae). Kew Bulletin* **49**, 305-320.

Kellogg, E.A., and Campbell, C.S. (1987). Phylogenetic analysis of the Gramineae. In 'Grass Systematics and Evolution'. (Eds T.R. Soderstrom, K.W. Hilu, C.S. Campbell and M.E. Barkworth.) pp. 310-322. (Smithsonian Institution Press: Washington DC.)

Kellogg, E.A., and Watson, L. (1993). Phylogenetic studies of a large data set. I. Bambusoideae, Andropogonoideae, and Pooideae (Gramineae). *Botanical Review* **59**, 273-343.

Liang, H., and Hilu, K.W. (1996). Application of the *matK* gene sequences to grass systematics. *Canadian Journal of Botany* **74**, 125-134.

Linder, H.P. (1997). Nomenclatural corrections in the *Rytidosperma* complex (Danthonieae, Poaceae). *Telopea* **7**, 269-274.

Linder, H.P. (in press a) *Rytidosperma vickeryae* - a new danthonioid grass from Kosciuszko (New South Wales, Australia): morphology, phylogeny and biogeography. *Australian Systematic Botany* [accepted 10 Feb . 1999, scheduled for vol 12, no. 6]

Linder, H.P. (in press, b) Danthonieae. In 'Flora of Australia' (Ed. A. E. Orchard), pp. 321-367. (ABRS/CSIRO Australia, Melbourne.)

Linder, H.P., and Ellis, R.P. (1990). A revision of *Pentaschistis* (Arundineae: Poaceae). *Contributions from the Bolus Herbarium* **12**, 1-124.

Linder, H.P., Thompson, J.F., Ellis, R.P., and Perold, S.M. (1990). The occurrence, anatomy and systematic implications of the glands in *Pentaschists* and *Prionanthium* (Poaceae, Arundinoideae, Arundineae). *Botanical Gazette* **151**, 221-233.

Linder, H.P, and Verboom, G.A. (1996). Generic limits in the *Rytidosperma (Danthonieae, Poaceae) complex. Telopea* **6**, 597-627.

Linder, H.P, Verboom, G.A., and Barker, N.P. (1997). Phylogeny and evolution in the *Crinipes group of grasses (Arundinoideae: Poaceae). Kew Bulletin* **52**, 91-110.

Linder, H.P., and Davidse, G. (1997). The systematics of *Tribolium* Desv. (Danthonieae: Poaceae). *Botanischer Jahrbücher Systematiek* **119**, 445-507.

Maddison, D.R. (1991). The discovery and importance of multiple islands of most-parsimonious trees. *Systematic Zoology* **40**, 315-328.

Mathews, S., and Sharrock, R.A. (1996). The phytochrome gene family in grasses (Poaceae): a phylogeny and evidence that grasses have a subset of loci found in dicot angiosperms. *Molecular Biology and Evolution* **13**, 1141-150.

Nixon, K.C., and Carpenter, J.M. (1996). On simultaneous analysis. *Cladistics* **12**, 221-241.

Philipson, M.N., and Connor, H.E. (1984). Haustorial synergids in Danthonioid grasses. *Botanical Gazette* **145**, 78-82.

Renvoize, S.A. (1986). A survey of leaf-blade anatomy in grasses VIII. Arundinoideae. *Kew Bulletin* **41**, 323-338.

Soreng, R.J., and Davis, J.I. (1998). Phylogenetics and character evolution in the grass family (Poaceae): simultaneous analysis of morphological and chloroplast restriction site character sets. *The Botanical Review* **64**, 1-85.

Swofford, D. L. (1993). PAUP: Phylogenetic Analysis Using Parsimony. Version 3.1.1. (Smithsonian Institute: Washington DC.)

Verboom, G. A., Linder, H.P., and Barker, N.P. (1994). Haustorial synergids: An important character in the systematics of danthonioid grasses (Arundinoideae: Poaceae). *American Journal of Botany* **81**, 1601-1610.

Watson, L., and Dallwitz, M.J. (1992). The grass genera of the world. (C.A.B. International: Cambridge.)

APPENDIX

Morphological Characters used in phylogenetic analysis. Unless otherwise indicated by a "*", all characters are considered to be unordered.

1   Tiller base: slender / swollen & tuberous
2   Plant sexuality: bisexual / gynodioecious
3   Basal sheaths: entire / fragmenting transversely / fragmenting lengthwise
4   Old leaves: caducuous / persisting as a peg
5   Basal sheaths: straight / spiralling / twisted
6   Basal leaf margins: glabrous / villous
7   Basal sheaths: pale / dark
8   Basal sheaths: persisting as tangled fibres / not persisting
9   Leaves: soft-tipped / pungent
10 Inflorescence: paniculate / capitate / spicate / distichous
11 Inflorescence: huge, > 30cm / smaller, <30cm
12 Inflorescence branches: glabrous / pubescent
13 Axillary cleistogamous flowers: absent / present
14 Upper cleistogamous florets: absent / present
15 Pedicel indumentum: glabrous or scabrous / basal tuft
16 *Glume length: taller than florets / equalling florets / shorter than florets
17 Glume veins: 1 / more than 2
18 Callus: oblique / horizontal
19 Callus size: smaller or equal to rachilla / larger than rachilla
20 Number of florets per spikelet: 3-8 / 2
21 Lemma shape: linear-lanceolate / narrowly ovate
22 Lemma indumentum: absent / longitudinal lines / transverse tufts / scattered at base / two long clumps / scattered tufts
23 Lemma hairs: tapering / clubshaped
24 *Lemma veins: 3 / 5-7 / 9
25 Lemma apex: entire / obscure bilobed / deeply bilobed
26 *Setae: substantial / minute / absent
27 Setae position: terminal / sinus
28 Palea-margin hairs: short or absent / long and dense
29 Palea-keels: folded / flat
30 Palea keels: scabrid / glabrous
31 *Palea-lemma ratio: palea > lemma / palea ca.1 lemma / palea 0.5-1 lemma / palea 0.3-0.5 lemma
32 Lodicule shape: cuneate / rhomboid
33 Lodicule microhairs: absent / present
34 Lodicule macrohairs: absent / present
35 Caryopsis shape: linear / obovate
36 Caryopsis apex: glabrous / villous
37 Hilum: linear / punctate
38 Clear cells in chlorenchyma: absent / present
39 Clear cells under abaxial epidermis: absent / sparse (1-2 layers) / thick (+3 layers)
40 *Bulliform cells: absent / only flanking midrib / on most furrows
41 Adaxial ridging: absent / present
42 Abaxial ridging: absent / present
43 Adaxial papillae: absent / present
44 Adaxial papillae: All over / basaly on ridges only
45 Adaxial prickle hairs: absent / present
46 Adaxial zonation: absent / present
47 Leaf-symmetry: symmetrical / asymmetrical
48 Phloem: entire / divided / doubly-divided
49 Extension cells: absent / present
50 Vascular Bundles (Sheath?): longer than wide / wider than long
51 Outer Bundle Sheath (secondary): not lignified / lignified
52 Midrib: like other ribs / raised, massive / smaller
53 Multicellular glands: absent / present
54 Adaxial microhairs: not overlapping / overlapping in furrows
55 Sclerenchyma cap: anchor-shaped / lens-shaped / pyramidal /massive-square
56 Outer Bundle Sheath (primary vbs): parenchymatous /thickened
57 Scattered large cells: absent / present
58 Chlorenchyma spaces: absent / present
59 Cushion-based macrohairs: absent / present
60 Intercostal silica body shape: round / dumbbell / absent
61 Costal silica body shape: round / dumbbell / absent
62 Chromosome base number 6 / 7
63 Chromosome base 6 ploidy level: diploid (2) / tetraploid (4) / hexaploid (6) / octoploid (8) / decaploid (10) / duodecaploid (12)
64 Adaxial microhairs: absent / present
65 Abaxial epidermis: ca. =adaxial epidermis / +x2 adaxial epidermis
66 Erect microhairs in grooves: absent / present
67 Abaxial silica: common / very sparse / absent

Grasses: Systematics and Evolution. (2000). Eds S.W.L. Jacobs and J. Everett. (CSIRO: Melbourne)

BIOGEOGRAPHY OF THE DANTHONIEAE

H.P. Linder[A] and N.P. Barker[B]

[A] Bolus Herbarium, University of Cape Town, Rondebosch 7701, South Africa.
[B] Department of Botany, Rhodes University, Grahamstown, South Africa.

Abstract

The Danthonieae are temperate grasses, largely restricted to the southern hemisphere, and are common on all the southern continents. Only the type genus *Danthonia* is found on the boreal continents, and even *Danthonia* has some species in the south. Although the phylogeny of the Danthonieae is still imperfectly known, it is possible to postulate generic groups, and to analyse the biogeographical patterns in each of these groups. The *Danthonia* group has a distribution similar to that of *Nothofagus*, and shows a biogeographical pattern equivalent to *Nothofagus*, which is interpreted as reflecting vicariance across the South Pacific. The *Rytidosperma* clade is less well known phylogenetically. It is disjunct across all southern oceans, but the phylogenetic relationships among the species from South America, Africa and Australasia remains unclear. Within Australasia, analyses allowing for both dispersal and vicariance suggest an ancestral area on Tasmania, with subsequent dispersal to New Zealand, Kosciusko and New Guinea. The *Pentaschistis* clade is currently the least well known. The clade is restricted to Africa, unless the indications from one molecular data set of a relationship to the New Zealand snowgrasses, *Chionochloa*, are taken seriously. Both these clades have radiated extensively in their respective areas.

'The Arundinoideae . . . has failed to make a spectacular impact on the environment and remains a subfamily of rather loosely related, mostly mediocre genera, as would be expected in a waning group' (Renvoize 1981).

Key words: Danthonieae, biogeography.

INTRODUCTION

The Arundinoideae as defined by Renvoize (1977) are probably polyphyletic (Kellogg and Campbell 1987; Barker *et al.* 1995; Barker *et al.* this volume; GPWG this volume; but see Hsiao *et al.* 1997), including many disparate elements that belong to other lineages, and so obviously make no biogeographical sense. It is then not surprising that the members of the assemblage were regarded as relicts of a once more successful group, and the description is correct in the sense that several groups basal to other clades were included in Arundinoideae. However, with the emerging evidence that one subset of the subfamily, the Dantho-nieae as circumscribed by Barker *et al.* (1995, 1999) and others (e.g., Verboom *et al.* 1994; Barker *et al.* this volume) is monophyletic, questions about the geographical history of this austral group become tractable, and we can search for an explanatory narrative for the tribe.

The Danthonieae clade is found predominantly in the southern hemisphere, and may be regarded as a south-temperate group. All genera and 280 of the 292 species used in this study occur on the southern continents (South America, Africa, Australasia and Malesia). The only genus with species in North America and Eurasia is the type genus *Danthonia*. The tribe is well represented

Table 1. Genera recognised in the Danthonieae, with the number of species, and the most recent revision. Some genera have not had adequate attention, so no revisions are cited for them.

Genus	Clade: *Rytidosperma, Danthonia* Pentaschistis	Continent / area	Species	Revision
Austrodanthonia	Ryt.	Australia, New Zealand	28	(Linder and Verboom 1996; Linder 1997)
Chaetobromus	Pent.	Africa	1	(Verboom and Linder 1998)
Chionochloa	Pent.	New Zealand, Australia	24	(Connor 1991)
Cortaderia s.l.	Dan.	South America, New Zealand	26	(Connor 1983)
Danthonia	Dan.	Both Americas, Eurasia	22	
Joycea	Ryt.	Australia	3	(Linder and Verboom 1996)
Karroochloa	Ryt.	Africa	4	(Conert and Türpe 1969)
Lamprothyrsus	Dan.	South America	2	
Merxmuellera	?	Africa	19	
Notochloë	Dan.	Australia	1	
Notodanthonia	Ryt.	Australia, New Zealand	5	(Linder and Verboom 1996; Linder 1997)
Pentameris	Pent.	Africa	9	(Barker 1993)
Pentaschistis	Pent.	Africa	c. 66	(Linder and Ellis 1990; Phillips 1994, 1995)
Plinthanthesis	Dan.	Australia	3	
Prionanthium	Pent.	Africa	3	(Davidse 1988)
Pseudopentameris	Pent.	Africa	1	(Barker 1995)
Rytidosperma	Ryt.	Australia, New Zealand, South America	36	(Linder and Verboom 1996)
Schismus	Ryt.	Africa / Eurasia	5	(Conert and Türpe 1974)
Tribolium	Ryt.	Africa	10	(Linder and Davidse 1997)

on all four austral terrains (Table 1 and Fig. 1), and in the Cape Floristic Region in Africa it forms the dominant temperate grass group (e.g. Southern Africa – see Linder 1989, 1994) In Australia the temperate grasslands on the Southern Tablelands are dominated by the danthonioid genus *Austrodanthonia*, while the montane slopes of the mountains on the South Island, New Zealand, are dominated by the danthonioid genus *Chionochloa*.

The distribution pattern of the tribe is typical of a 'Gondwanan' pattern, thus suggesting that the biogeography could be interesting in terms of understanding the distributional disjunctions across the large southern oceans: the Indian, Atlantic and Pacific. Of further interest is the relationship between the continuous distributions of the species in the southern temperate grasslands, and the more scattered, disjunct distributions at lower latitudes in montane or subalpine environments. This paper is an attempt to develop a narrative or story about the biogeographical history of the Danthonieae.

METHODS

Phylogeny

No new phylogenetic analyses were performed for this paper, and all phylogenies are published, or were presented at this conference. Despite substantial activity, there are still a host of phylogenetic problems within the tribe. Although many of the genera have recently been redelimited in an attempt to establish monophyletic units (e.g., Linder and Verboom (1996) for the Australian Danthonieae, and Linder and Davidse (1997) for the *Tribolium/*

Urochlaena problem), the generic limits between *Pentaschistis, Pentameris* and *Prionanthium* (e.g., Barker 1993) remain unsatisfactory, it is possible that *Cortaderia* might be paraphyletic, and Barker *et al.* (1995) have shown that earlier worries about the monophyly of *Merxmuellera* might be well-founded.

There are several data sets that inform partially on the phylogenetic patterns in the clade, they are reviewed at this meeting by Barker *et al.* (this volume). Although there is substantial congruence between the phylogenies based on *rpo*C2 and ITS (Barker *et al.* this volume, figs 2 and 3 respectively), the morphological data (Barker *et al.* this volume, fig. 4) suggest a bizarre phylogeny. In order to simplify the biogeographical interpretation, we largely base this analysis on the ITS-based phylogeny (Barker *et al.* this volume, fig. 3). On the basis of this phylogeny, we can detect three clades: the *Danthonia* clade, *Rytidosperma* clade, and the *Pentaschistis* clade. These clades are larger than the ones delimited by Barker *et al.* (this volume), but this was done deliberately to maximise the biogeographical interpretations possible. The relationships among these clades are not clear.

However, the ITS sampling and resolution is not currently adequate to resolve the relationships within the *Rytidosperma* and *Pentaschistis* clades, but it does provide resolution in the *Danthonia* clade. In the *Rytidosperma* clade two partial morphological phylogenies are available (Linder and Verboom 1996; Linder and Davidse 1997), and these have also been used in the interpretations.

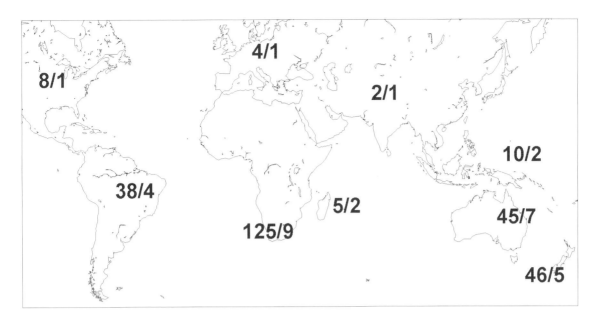

Fig. 1. Numbers of species and genera of Danthonieae from each continental area.

Area Relationships

There is a plethora of methods for analysing biogeographical data. We used only two methods in this study. The 'default' method is a cladistic biogeographical method that searches for paralogy-free subtrees, and then searches for the best-fit general pattern that is supported by these paralogy-free subtrees (Nelson and Ladiges 1996; Ladiges *et al.* 1997). This method searches for information indicating that two areas are more closely related to each other than either is to a third area. It provides no indication of whether this relationship is the result of splitting of a once continuous distribution range (vicariance) or of long-distance dispersal across pre-existing barriers. The taxon names on a taxon cladogram are replaced by area names, and the area cladogram is inspected to find three-area components that indicate two areas are sister-areas relative to a third area, while ignoring any duplications of the areas. In order to avoid the possibility that areas may have been established by dispersal, widespread areas of taxa on the taxon cladogram are decomposed into their constituent parts, and these are all analysed separately. This method should locate instances of vicariance (if they exist).

The second method used locates both dispersal and vicariance explanations, while giving priority to vicariance explanations. Areas are mapped on the internal nodes of the tree in such a way that additive mapping carries no cost, but extinctions and dispersal events each carry a cost of one additional step (Ronquist 1997). This operation is performed by the software DIVA ver. 1.1. (Ronquist 1996). The resultant output indicates at which nodes the most parsimonious explanation is dispersal, and at which it would be vicariance. Athough this method is particularly suitable for the study of single taxa, it may be less suitable for entire floras, as it is not clear how the individual narratives should be summed to a general statement.

Searching for paralogy-free subtrees is done by hand, and therefor does not need a fully resolved cladogram. DIVA does require a fully resolved tree – this was achieved by only including those clades for which adequate resolution was available, and those for which any resolution would give the same result (e.g., in the *Danthonia* clade, for the New Zealand species of *Cortaderia*).

RESULTS

Danthonia Clade

The *Danthonia* clade includes *Cortaderia, Danthonia, Plinthanthesis* and *Notochloë*. With the exception of *Danthonia* this clade has been well sampled for ITS. The largest radiation in this clade is in South America, with *Danthonia, Cortaderia* and *Lamprothyrsus*: 28 species out of the total of 54 species. The Australian members are *Plinthanthesis* (3 species) and *Notochloë* (1 species), both largely restricted to the ancient sandstones of the eastern seaboard, which could be regarded as relictual. This applies particularly to *Notochloë*, growing along streams in the Blue Mountains (Fig. 2).

The clade contains both soft herbacous grasses (*Danthonia*), small sclerophyllous tussocks (*Plinthanthesis*) and large sclerophyllous ever-green tussocks (*Cortaderia*). It is not possible to determine what the ancestral growth-form was, as the sister-clade would need to be confidently established. The plants also grow in a diversity of habitats: heaths (*Plinthanthesis*), stream-margins in light shade (*Notochloë*), temperate grasslands (*Danthonia*) and disturbed habitats and cliffs (at least the New Zealand species of *Cortaderia*). A detailed resolution of this clade could be of interest in reconstructing its ecological history.

The phylogenetic analysis presented at this conference by Barker *et al.* (this volume) indicates that *Cortaderia* is paraphyletic, with one segregate of *Cortaderia* restricted to New Zealand, sister to the Australian genera *Plinthanthesis* and *Notochloë,* while the South America species of *Cortaderia* are sister to the American genus *Danthonia*. This relationship indicates that New Zealand is the sister area to Australia, rather than South America. This areal relationship was also demonstrated for *Nothofagus* (Linder and Crisp 1995), but is contrary to the (Australia (New Zealand,

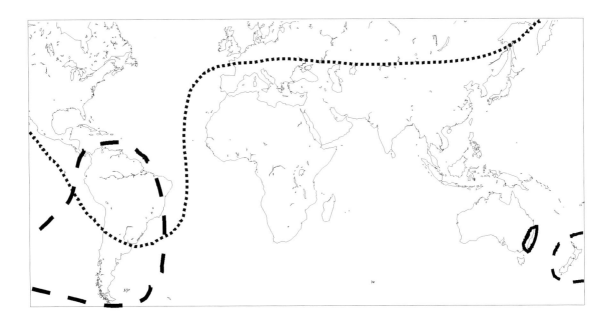

Fig. 2. Distribution patterns of the *Danthonia* clade. *Danthonia* distribution is outlined by the dotted line, *Cortaderia* by the dashed line, *Plinthanthesis* and *Notochloë* by a solid line

South America)) pattern demonstrated for insects (Brundin 1965; 1966; Edmunds 1981; Cranston and Naumann 1991). Analysis using DIVA also produced a pattern of vicariance (Fig. 3) with dispersal from South America northwards to North America.

Some authors regard these patterns as the result of long-distance dispersal (Hill 1992; Pole 1993). An alternative explanation is that this pattern is the result of the breakup of Gondwana, and therefore of vicariance rather than dispersal (Hooker 1860; Humphries 1983; Humphries *et al.* 1986; Humphries and Parenti 1986; Crisci *et al.* 1991; Seberg 1991; Linder and Crisp 1995). If this is a vicariance pattern, then it suggests that the Danthonieae co-occurred with *Nothofagus* in the southern temperate forests at the end of the Cretaceous, some 80 mya. This would be significantly earlier than the first fossil grass pollen from the southern Hemisphere, which dates from the Paleocene (Muller 1981), with the first grass pollen from New Zealand being recorded from the Oligocene (Mildenhall 1980). It is possible that the grass pollen simply failed to make it into the fossil record, but this appears to be unlikely, as wind-pollinated plants like grasses produce substantial quantities of pollen, and could therefore, like *Nothofagus*, be expected to be common in the fossil record. However, currently the Danthonieae occur only rarely with *Nothofagus* – most of the Danthonieae are found in substantially drier habitats, which might not have as many suitable habitats for fossilisation. It is reasonable to assume that even in the Cretaceous there must have been disturbed or arid habitats, unsuitable for forests, which must have carried a more ephemeral, early-succession flora, that might have included the Danthonieae.

These early dates (late Cretaceous) are based on the reconstructions which suggest that New Zealand has been isolated for the last 80 million years (Crook 1981). However, the biogeographical history of New Zealand is complex, and there is a rich diversity of hypotheses about it. Maybe more interesting is that Australia and South America have been in contact via Antarctica

until ca. 30 mya, when Australia separated from Antarctica, and when the Drake Passage between South America and the Antarctic Peninsula opened (Barker and Burrell 1977; Coleman 1980; Crook 1981). Thus the vicariance pattern in the *Danthonia* clade (excluding its New Zealand occurrence) might have been established as recently as 30 mya, well within the known age of the family. The trans-Tasman disjunction should then be considered as a separate issue: it may be dispersal (as argued below for the *Rytidosperma* clade) or vicariance.

Rytidosperma Clade

The *Rytidosperma* clade includes *Rytidosperma, Austrodanthonia, Notodanthonia, Joycea, Karroochloa, Schismus* and *Tribolium*. We do not have a robust, well-resolved phylogeny for this clade, and the molecular trees offer little resolution. The *rpo*C2 data indicate a clade of *Rytidosperma, Austrodanthonia* and *Joycea* (Barker *et al.* 1999). Morphological data indicate that *Karroochloa* might be nested in this clade (Linder and Verboom 1996), but this is contrary to cytological information which was difficult to code adequately for the morphological analysis. Furthermore, morphological data indicate that *Schismus* and *Tribolium* are sisters (Linder and Davidse 1997). We postulate that the best summary of the data is: ((*Tribolium, Schismus*), (*Karroochloa,* (*Rytidosperma, Notodanthonia, Joycea, Austrodanthonia*))), and this was used for the biogeographical analysis. This clade would benefit from a detailed molecular analysis, as the cladistic relationships postulated above have not been adequately corroborated. Both ITS and *rpo*C2 sequence data indicate that one or more species of polyphyletic *Merxmuellera* might also belong to this clade.

The *Rytidosperma* clade is large, containing 86 species (excluding an unknown number of *Merxmuellera* species). With the exception of *Merxmuellera*, the species are all orthophyllous, herbaceous tufted grasses. The major radiation of the clade has been in

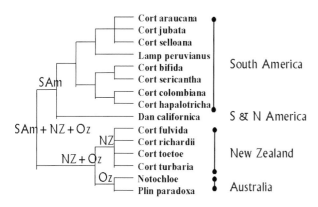

Fig. 3. Area cladogram of the *Danthonia* clade, with the ancestral areas optimized onto the nodes, using DIVA. "Cort" = *Cortaderia*, "Lamp" = *Lamprothyrsus*, "Dan" = *Danthonia*, "Plin" = *Plinthanthesis*. SAm = South America, NZ = New Zealand, Oz = Australia.

Australasia, where *Austrodanthonia* is very common on the Southern and Central Tablelands and Western Slopes, with some species reaching Western Australia. In addition, *Rytidosperma* is common in subalpine or montane habitats (nine species in Australia), and is common on South Island, New Zealand (13 species). There are also six species in South America (Baeza P. 1996), and nine in the Malesian mountains (Veldkamp 1979, 1993). The 15 African species in *Schismus* and *Tribolium* are weakly perennial or annuals (Conert and Türpe 1974; Linder and Davidse 1997) (Fig. 4).

If it is assumed that *Karroochloa* is not embedded in the *Rytidosperma, Austrodanthonia, Joycea, Notodanthonia* clade, then the areal relationships are (Africa, (Australia, New Zealand, South America)). *Rytidosperma* is found in all the areas excluding Africa, while *Notodanthonia* and *Austrodanthonia* are endemic to Australia and New Zealand, and *Joycea* is endemic only to Australia. If we assume that widespread distributions indicate that all the areas in which the widespread taxon occur are each others closest relatives (Assumption 0) then these areas are grouped together. If, however, we assume that widespread distribution may have been established by dispersal, and may therefore not be informative about the relationships among the areas (Assumption 2), then several solutions are possible: Australia is sister to either New Zealand or South America. Combining these options leads to the areal relationships suggested above.

There are two possible explanations for this trans-Indian Ocean disjunction. The first would be that it is an ancient disjunction, dating back to a period when the Indian Ocean was much smaller: the Indian – African break has been dated to 118 mya, but suggestions are that the ocean only opened up properly 96 mya (Powell *et al.* 1988). This is an explanation that has been applied to the Restionaceae (Linder 1987) and the Proteaceae (Johnson and Briggs 1975). Whether this is a 'pure' vicariance explanation, with the original distribution range without any breaks, or whether there were narrow waterways separating the subsequent continents, is academic. However, there are indications that the African and Australasian members of the *Rytidosperma* clade may have separated relatively recently (at least more recently than the end of the Cretaceous, by which time the Indian Ocean was already well established). Morphologically it is

difficult to distinguish *Karroochloa* from *Rytidosperma*, and there is very little ITS sequence divergence between these two groups. This suggests that it is very unlikely that these clades have been separated for 96 million years. The alternative explanation to vicariance would be long-distance dispersal across the Indian Ocean. Grass caryopses, enclosed in awned and bearded lemmas, might be capable of long distance dispersal (Davidse 1987), and indeed, one species of *Pentaschistis* reached Amsterdam Island, a tiny rocky outcrop in the middle of the Indian Ocean, halfway between Africa (where the rest of the genus occurs) and West Australia (Linder and Ellis 1990). The *Rytidosperma* clade also contains other surprises, such as the disjunct distribution in *Schismus*: there are four species in South Africa, restricted to the drier south-western areas, and two species around the Mediterranean, in similar habitats. *Schismus arabicus* from the Mediterranean is clearly the sister species to *S. barbatus*, which occurs both around the Mediterranean, and in Namaqualand in South Africa (Linder and Davidse 1997). This might, however, be a relict of the 'arid corridor' which spans Africa from the south-west to the north-east (Verdcourt 1969; Freitag and Robinson 1993; Jürgens 1997), and so a dispersal route appears possible. Several recent analyses of trans Indian Ocean disjunctions have used molecular clocks to argue for long-distance dispersal from Africa to Australia (e.g. Bakker *et al.* 1998; Baum *et al.* 1998).

Analysis of the geographical patterns for *Rytidosperma* in Australasia (Linder 1999) indicates that New Zealand and Australia form the sister area to New Guinea, and that the *Rytidosperma* floras of Tasmania and the Australian Alps are so similar that they should probably be regarded as one centre of endemism. This pattern does not match the geological history, as the New Guinean mountains are substantially younger than the Tasman sea, thus suggesting that this pattern might reflect recency of dispersal. This would indicate that it is easier for these grasses to cross the Tasman Sea than it is to cross Queensland and the Torres Strait. If dispersal is a major feature in determining the geographical patterns in Australasian *Rytidosperma*, then an analysis using a method that may locate the 'ancestral area' would be appropriate.

The 'dispersal – vicariance ' method of Ronquist (1997), implemented in DIVA 1.1 (Ronquist 1996), leads to the pattern in Fig. 5, which has a total of 10 dispersal events and no vicariance events. The ancestral area is optimised to Tasmania, but with a dispersal event to New Zealand and a subsequent dispersal back to Tasmania. The Ancestral Area method of Bremer (1992) also locates Tasmania as the ancestral area. Kosciusko and New Guinea are optimised to be colonised twice: once from New Zealand and once from Tasmania. The close link between the northern alpine areas (New Guinea and Kosciusko) with New Zealand rather than Tasmania is intriguing, and might require an explanation. However, with a small change in the optimisations on the internal nodes, it could be postulated that Tasmania is the ancestral area, and that both New Zealand and the northern areas were occupied from an ancestral Tasmanian population, possibly with some subsequent dispersal across the Tasman from New Zealand to Kosciusko, as has been postulated for *Hebe* (Garnock-Jones 1993). Ollier (1986) suggests that the New Zealand alpine areas are relatively youthful, with mountain formation only since the

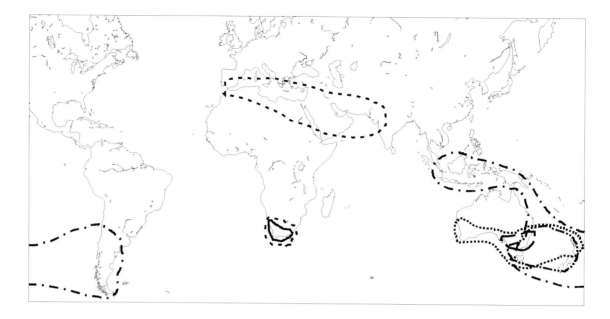

Fig. 4. Distribution pastterns of the *Rytidosperma* clade. Solid lines = *Karroochloa, Tribolium*; dotted line = *Austrodanthonia*; dashes = *Schismus*; long dashes = *Joycea*; dash and one dot = *Rytidosperma*; dashes and two dots = *Notodanthonia*.

Miocene. It is therefore possible that the requisite environments may not have existed in New Zealand before this date, making the above reconstruction ecologically sensible.

Pentaschistis Clade

The *Pentaschistis* clade is retrieved by all data sets, and includes the genera *Pentaschistis, Pentameris* and *Prionanthium*. This clade is entirely African. There are some morphological indications that *Pseudopentameris* and *Chaetobromus* may also belong to this clade – however, this is not supported by the molecular data. The inclusion or not of these two genera in the *Pentaschistis* clade would not affect the biogeographical arguments, as these two genera are also African.

The bulk of the radiation is in South Africa, in fact in the Western Cape, but with species common north through the Drakensberg to Ethiopia. There is as yet no phylogeny for this group, but the pattern is typical for many other 'Cape' clades, where the bulk of the species are in the south, with a reduction in species numbers to the north .

The relationship between the great species richness in the south, and the northern species isolated on the central African and East African peaks, would be interesting to explore. It appears to parallel the patterns observed in *Rytidosperma* in Australasia. It is tempting to draw a further parallel to *Danthonia* in the Americas, where the connection extends all the way through to the north-temperate regions. Detailed phylogenies of these clades might show parallel northwards migration as the equatorial mountains formed during the Neogene or later, thus opening a temperate migration route from the south-temperate to the north-temperate regions.

The relationships of the *Pentaschistis* clade are unclear. The *rpo*C2 data set places the clade directly on the stem of the Danthonieae, but ITS provides weak indications of a sister-group relationship to *Chionochloa*. The ITS results are followed here, although they might well be incorrect, as they disappear in the combined analysis of *rpo*C2 and ITS. It is equally difficult to determine the phylogenetic affinities of *Chionochloa*. The sampling of the genus was too poor in the *rpo*C2 analysis to provide a confident solution, and ITS places it weakly next to the *Pentaschistis* clade. On morphology, *Chionochloa* approaches *Cortaderia*, and *Chionochloa / Cortaderia archboldii* from New Guinea is intermediate between the two genera (Conert 1975; Clayton and Renvoize 1986).

Chionochloa has radiated in New Zealand, with some 24 poorly defined species. The basal members of the genus are found on North Island and in the lowlands, as well as one on Lord Howe Island, and one on Mt Kosciusko (Linder, unpublished information). The major radiation in the genus occurs on the subalpine regions of the mountains of South Island. If indeed *Chionochloa* is the sister of the *Pentaschistis* clade, this would be further evidence for a trans-Indian Ocean connection, with radiations both in Africa and New Zealand.

SUMMARY

These data indicate a relationship of (Africa, (South America, (Australia, New Zealand))). This is consistent with the global patterns suggested by Linder and Crisp (1995). It is missing the eastern South American area which they found to be sister to Africa – if this area is deleted, the relationship is identical. The Linder and Crisp hypothesis was based on a large group of families, including Haemodoraceae, Iridaceae, Restionaceae, Winteraceae, some Proteaceae and *Nothofagus*. The South America – Australasian relationships appear to be quite strong, and mirror the patterns of *Nothofagus*. This relationship could be corroborated by the fossil record, but to date sufficiently early fossils of grasses are still missing.

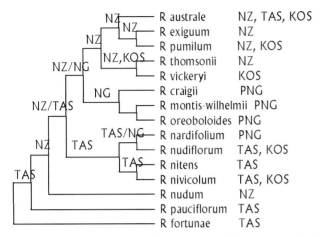

Fig. 5. Cladogram of a portion of *Rytidosperma*, with the areas optimized on using DIVA, thus mapping the ancestral area to Tasmania. NZ = New Zealand, TAS = Tasmania, KOS = Kosciuzko; PNG = New Guinea.

However, the relationships to Africa remain enigmatic – both in terms of the phylogenetic affinities, and in finding a reasonable mechanism by which this pattern could be established. This relationship is confounded by our ignorance of the phylogenetic status of the African genus *Merxmuellera* and its segregates, the isolated position of the *Pentaschistis* clade, and the close relationship between *Karroochloa* and *Rytidosperma*. Only when these phylogenetic problems have been resolved will it become possible to suggest a biogeographical relationship between Africa and the Australo-South American areas, and to determine whether this distribution is the result of dispersal or vicariance, or a combination of these two mechanisms.

Within-continent patterns might well be the result of dispersal. This area would benefit from further exploration, as it might be informative on global changes since the mid-Tertiary, allowing the northward spread of south-temperate groups.

The results available here are not adequate for a complete narrative of the evolution of the Danthonieae, and there are several outstanding phylogenetic problems that need to be solved first. But a pattern of vicariance, dispersal and radiation in the southern hemisphere is emerging.

ACKNOWLEDGEMENTS

HPL would like to acknowledge funding from the Foundation of Research Development (Pretoria) and the University of Cape Town. NPB thanks the Royal Society, London, for financial assistance in the form of a of short term visiting research grant (to carry out part of this research at the University of Reading) Rhodes University and the FRD for providing subsistence and travel funding to support his attendance at the Second International Monocot Conference held in Sydney, 1998.

REFERENCES

Baeza P., C. M. (1996). Los géneros *Danthonia* D.C. y *Rytidosperma* Steud. (Poaceae) en América - una revisión. *Sendtnera* **3**, 11-93.

Bakker, F. T., Hellbrügge, D., Culham, A., and Gibby, M. (1998). Phylogenetic relationships within *Pelargonium* sect. *Peristera (Gera-*

niaceae), inferred from nrDNA and cpDNA sequence comparisons. *Plant Systematics and Evolution* **211**, 273-287.

Barker, N. P. (1993). A biosystematic study of *Pentameris* (Arundineae, Poaceae). *Bothalia* **23**, 25-47.

Barker, N. P. (1995). A systematic study of the genus *Pseudopentameris* (Arundinoideae: Poaceae). *Bothalia* **25**, 141-148.

Barker, P. F., and Burrell, J. (1977). The opening of the Drake Passage. *Marine Geology* **25**, 15-34.

Barker, N. P., Linder, H. P., and Harley, E. H. (1995). Polyphyly of Arundinoideae (Poaceae): evidence from *rbc*L sequence data. *Systematic Botany* **20**, 423-435.

Barker, N. P., Linder, H. P., and Harley, E. H. (1999). Sequences of the grass-specific insert in the chloroplast *rpo*C2 gene elucidate generic relationships of the Arundinoideae (Poaceae). *Systematic Botany* **23**, 327-350.

Barker, N.P., Morton, C.M., and Linder, H.P. The Danthonieae: generic composition and relationships, this volume.

Baum, D. A., Small, R. L., and Wendel, J. F. (1998). Biogeography and floral evolution of baobabs (*Adansonia*, Bombaceae) as inferred from multiple data sets. *Systematic Biology* **47**, 181-207.

Bremer, K. (1992). Ancestral areas: a cladistic reinterpretation of the center of origin concept. *Systematic Biology* **41**, 436-445.

Brundin, L. (1965). On the real nature of transantarctic relationships. *Evolution* **19**, 496-505.

Brundin, L. (1966). Transatlantic relationships and their significance. *Kungliga Svenska Vetenskapsakademiens Handlingar, ser 4* **11**, 1-472.

Clayton, W. D., and Renvoize, S. A. (1986). 'Genera Graminum. Grasses of the World' (Her Majesty's Stationary Office: London.)

Coleman, P. J. (1980). Plate tectonics background to biogeographic development in the soutwest Pacific over the last 100 million years. *Palaeogeography, Palaeoclimatology, Palaeoecology* **31**, 105-121.

Conert, H. J. (1975). Die *Chionochloa*-Arten von Australien und Neuguinea. *Senckenbergiana Biologica* **56**, 153-164.

Conert, H. J., and Türpe, A. M. (1969). *Karroochloa*, eine neue Gattung de Gramineen (Poaceae, Arundinoideae: Danthonieae). *Senckenbergiana Biologica* **50**, 289-318.

Conert, H. J., and Türpe, A. M. (1974). Revision der Gattung *Schismus* (Poaceae: Arundiniodeae: Danthonieae). *Abhandlungen der Senckenbergischen Naturforschenden Gesellschaft* **532**, 1-81.

Connor, H. E. (1983). Names and types in *Cortaderia* Stapf (Gramineae) II. *Taxon* **32**, 633-634.

Connor, H. E. (1991). *Chionochloa* Zotov (Gramineae) in New Zealand. *New Zealand Journal of Botany* **29**, 219-282.

Cranston, P. S., and Naumann, I. D. (1991). Biogeography. In 'The insects of Australia' (Ed. I. D. Naumann.) pp. 180-197. (Melbourne University Press: Melbourne.)

Crisci, J. V., Cigliano, M. M., Morrone, J. J., and Roig-Junent, S. (1991). A comparative review of cladistic approaches to historical biogeography of southern South America. *Australian Systematic Botany* **4**, 117-126.

Crook, K. A. W. (1981). The break-up of the Australian-Antarctic segment of Gondwanaland. In 'Ecological Biogeography of Australia' (Ed. A. Keast.) pp. 3-14. (W. Junk: The Hague.)

Davidse, G. (1987). Fruit dispersal in the Poaceae. In 'Grass systematics and evolution' (Eds T. R. Soderstrom, K. W. Hilu, C. S. Campbell and M. E. Barkworth.) pp. 143-155. (Smithsonian Institution Press: Washington.)

Davidse, G. (1988). A revision of the genus *Prionanthium* (Poaceae:Arundinae). *Bothalia* **18**, 143-153.

Edmunds, G. F. (1981). Comments. In 'Vicariance biogeography: a critique' (Eds G. Nelson and D. Rosen.) pp. 287–297 (Columbia University Press: New York.)

Freitag, S., and Robinson, T. J. (1993). Phylogeographic patterns in mitochondrial DNA of the ostrich (*Struthio camelus*). *Auk* **110**, 614-622.

Garnock-Jones, P. J. (1993). Phylogeny of the *Hebe* complex (Scrophulariaceae: Veroniceae). *Australian Systematic Botany* **6**, 457-479.

GPWG. A phylogeny of the grass family (Poaceae), as inferred from eight character sets. This volume.

Hill, R. S. (1992). *Nothofagus*: evolution from a southern perspective. *Trends in Ecology and Systematics* **7**, 190-194.

Hooker, J. D. (1860). 'Introductory Essay.In 'The botany of the Antarctic Voyage" (Lovell Reeve: London.)

Hsiao, C., Jacobs, S. W. L., Barker, N. P., and Chatterton, N. J. (1997). A molecular phylogeny of the subfamily Arundinoideae (Poaceae) based on sequences of rDNA (ITS). *Australian Systematic Botany* **11**, 41-52.

Humphries, C. J. (1983). Biogeographical explanations and the southern beeches. In 'Evolution, Time and Space: the emergence ofthe Biosphere' (Eds R. W. Sims, J. H. Price and P. E. S. Whalley.) pp. 335-365. (Academic Press: London and New York.)

Humphries, C. J., Cox, J. M., and Nielsen, E. S. (1986). *Nothofagus* and its parasites: a cladistic approach to coevolution. In 'Coevolution and systematics' (Eds A. R. Stone and D. L. Hawksworth.) pp. 55-76. (Clarendon Press: Oxford.)

Humphries, C. J., and Parenti, L. E. (1986). 'Cladistic biogeography' (Oxford monographs in biogeography: Oxford.)

Johnson, L. A. S., and Briggs, B. G. (1975). On the Proteaceae - the evolution and classification of a southern family. *Botanical Journal of the Linnean Society* **70**, 83-182.

Jürgens, N. (1997). Floristic biodiversity and history of African arid regions. *Biodiversity and Conservation* **6**, 495-514.

Kellogg, E. A., and Campbell, C. S. (1987). Phylogenetic analyses of the Gramineae. In 'Grass Systematics and Evolution' (Eds T. R. Soderstrom, K. W. Hilu, C. S. Campbell and M. E. Barkworth.) pp. 310-322. (Smithsonian Institution Press: Washington.)

Ladiges, P. Y., Nelson, G., and Grimes, J. (1997). Subtree analysis, *Nothofagus* and Pacific biogeography. *Cladistics* **13**, 125-129.

Linder, H. P. (1987). The evolutionary history of the Poales/Restionales - a hypothesis. *Kew Bulletin* **42**, 297-318.

Linder, H. P. (1989). Grasses in the Cape Floristic Region: phytogeographical implications. *South African Journal of Science* **85**, 502-505.

Linder, H. P. (1994). Afrotemperate phytogeography: implications of cladistic biogeographical analysis. In 'Proceedings of the XIIIth Plenary Meeting AETFAT, Malawi.' (Eds J. H. Seyani and A. C. Chikuni.) pp. 913-930. (National Herbarium and Botanic Gardens, Malawi: Zomba.)

Linder, H. P. (1997). Nomenclatural corrections in the *Rytidosperma* complex (Danthonieae, Poaceae). *Telopea* **7**, 269-274.

Linder, H. P. (1999). *Rytidosperma vickeryae* - a new danthonioid grass from Kosciuszko (New South Wales, Australia): morphology, phylogeny and biogeography. *Australian Systematic Botany* **12**, 743-755.

Linder, H. P., and Crisp, M. D. (1995). *Nothofagus* and Pacific biogeography. *Cladistics* **11**, 5-32.

Linder, H. P., and Davidse, G. (1997). The systematics of *Tribolium* Desv. (Danthonieae: Poaceae). *Botanische Jahrbuecher fuer Systematic* **119**, 445-507.

Linder, H. P., and Ellis, R. P. (1990). A revision of *Pentaschistis* (Arundineae: Poaceae). *Contributions from the Bolus Herbarium* **12**, 1-124.

Linder, H. P., and Verboom, G. A. (1996). Generic limits in the *Rytidosperma* (Danthonieae, Poaceae) complex. *Telopea* **6**, 597-627.

Mildenhall, D. C. (1980). New Zealand Late Cretaceous and Cenozoic plant biogeography: a contribution. *Palaeogeography, palaeoclimatology, palaeoecology* **31**, 197-233.

Muller, J. (1981). Fossil pollen records of extant angiosperms. *Botanical Review* **47**, 1-142.

Nelson, G., and Ladiges, P. Y. (1996). Paralogy in cladistic biogeography and analysis of paralogy-free subtrees. *American Museum Novitates* **3167**, 1-44.

Ollier, C. D. (1986). The origin of alpine landforms in Australasia. In 'Flora and Fauna of Alpine Australasia' (Ed. B. A. Barlow.) pp. 3-26. (CSIRO: Canberra.)

Phillips, S. M. (1994). Variation in the *Pentaschistis pictigluma* complex (Gramineae). In 'Proceedings of the XIII Plenary Meeting of AETFAT, Zomba, Malawi' (Eds J. H. Seyani and A. C. Chikuni.) pp. 359-372. (National Herbarium and Botanic Gardens of Malawi: Zomba.)

Phillips, S. M. (1995). A new species of *Pentaschistis* (Gramineae) from Ethiopia. *Kew Bulletin* **50**, 615-617.

Pole, M. S. (1993). Keeping in touch: vegetation prehistory on both sides of the Tasman. *Australian Systematic Botany* **6**, 387-397.

Powell, C. M., Roots, S. R., and Veevers, J. J. (1988). Pre-breakup continental extension in East Gondwanaland and the early opening of the eastern Indian Ocean. *Tectonophysics* **155**, 261-283.

Renvoize, S. (1981). The subfamily Arundinoideae and its position in relation to a general classification of the Gramineae. *Kew Bulletin* **36**, 85-102.

Renvoize, S. A. (1977). The anatomy and systematics of the Gramineae sub-family Arundinoideae. Thesis for M. Phil., University of Reading.

Ronquist, F. (1996). DIVA (Uppsala University: Uppsala.)

Ronquist, F. (1997). Dispersal - vicariance analysis: a new approach to the quantification of historical biogeography. *Systematic Biology* **46**, 195-203.

Seberg, O. (1991). Biogeographic congruence in the South Pacific. *Australian Systematic Botany* **4**, 127-136.

Veldkamp, J. F. (1979). Poaceae. In 'The Alpine Flora of New Guinea' (Ed. P. van Royen.) pp. 1035-1224. (J. Cramer: Vaduz.)

Veldkamp, J. F. (1993). Miscellaneous notes on southeast Asian Gramineae. VIII. New species in *Danthonia*. *Blumea* **38**, 217-219.

Verboom, G. A., and Linder, H. P. (1998). A re-evaluation of species limits in *Chaetobromus* (Danthonieae: Poaceae). *Nordic Journal of Botany* **18**, 57-77.

Verboom, G. A., Linder, H. P., and Barker, N. P. (1994). Haustorial synergids: an important character in the systematics of danthonoid grasses (Arundinoideae: Poaceae) *American Journal of Botany* **81**, 1601-1610.

Verdcourt, B. (1969). The arid corridor between the northeast and southwest areas of Africa. *Palaeoecology of Africa* **4**, 140-144.

PHYSIOLOGY/ECOLOGY

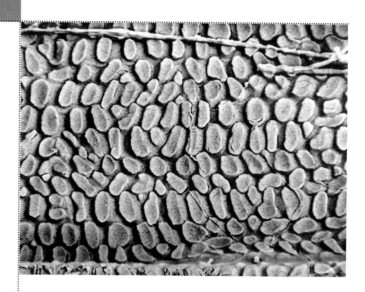

A Scanning Electron Micrograph of the lemma surface of *Agrostis clavata* showing the 'Trichodium Net', the pattern of cells with silicified walls characteristic of lemmas belonging to true *Agrostis* species.

Grasses: Systematics and Evolution. (2000). Eds S.W.L. Jacobs and J. Everett. (CSIRO: Melbourne)

CONTRIBUTIONS OF PROLAMIN SIZE DIVERSITY AND STRUCTURE TO THE SYSTEMATICS OF THE POACEAE

K.W. Hilu

Department of Biology, Virginia Polytechnic Institute and State University, Blacksburg VA 24061, USA.
e-mail: hilukw@vt.edu

Abstract

Prolamin is the principle storage protein in the grain of most grasses and is encoded by several multigene families. Studies on prolamin size, polypeptide heterogeneity and immunological similarities were conducted for over 120 species from 83 genera representing all seven major grass subfamilies and numerous tribes and subtribes. Prolamins range in structure from small molecular weight polypeptides consisting only of conserved domains to larger ones comprised partially or completely of repeat motifs. Major polypeptide components fall into three size classes, 10-17 kDa, 20-35 kDa, and 35-100 kDa, that characterize three grass groups: bambusoid-oryzoid, PACC group (Panicoideae, Arundinoideae, Centothecoideae, Chloridoideae) and pooids, respectively. Immunological studies have confirmed the structural divergence of the size classes and corroborated the correlations with the grass lineages. Immunology-based phylogeny is congruent with current views on grass systematics. Sequence information from prolamin at the amino acid and nucleic acid levels shows promise for further contributions to grass systematics.

Key words: Prolamin, protein, Poaceae, grasses, systematics, evolution, phylogeny.

INTRODUCTION

Molecular information from proteins and nucleic acids have contributed considerably to our understanding of systematic relationships and evolution in the Poaceae. Although contributions of nucleic acid studies to grass systematics have been consistently stressed, the impact of protein data has not. This article focuses on the utility of prolamin data in grass systematics. Prolamin, an alcohol-soluble component of seed storage protein, is a prominent feature of the Poaceae. Seed storage proteins function primarily as stores of nitrogen and sulfur for use during germination (Raghavan 1997). It is believed to be unique to the Poaceae with some structural similarity to the 2S albumin of dicots (Shewry *et al.* 1995). In this article, I review previous work on prolamins, provide new data, and focus on actual and potential contributions of these macromolecules to grass systematics above the

genus level. Contributions of prolamin data will be addressed both at the protein and gene levels.

OVERVIEW OF PROLAMIN STRUCTURE

Seed storage proteins are unique in their high content of certain amino acids (e.g., glutamine, asparagine, and/or arginine), tissue-specific expression (confined to seeds), temporal expression, and unusual abundance (Raghavan 1997). Globulins, prolamins, and glutelins are considered true storage proteins. Globulins are the predominant proteins in dicots while prolamins and glutelins are dominant in monocots. Prolamins are unique to the Poaceae and constitute the principal storage protein in most grasses, accounting for approximately half of the nitrogen in the grain (Shewry *et al.* 1995). The exceptions are rice and oat where the major storage proteins are 11S globulin-like proteins; however, the contribution of prolamin to total protein content in rice has been underestimated (Krishhnan and White 1995). Prolamin is found in membrane-bound protein bodies confined to the endosperm (Raghavan 1997). Characteristically, prolamin is rich in hydrophobic and uncharged amino acids and poor in charged ones, and is soluble in 50-70% aqueous alcohol. Cereal prolamin has high glutamine and proline content (Shewry *et al.* 1995) but is very low in lysine and tryptophan. In spite of its apparent non-specific function, prolamin tends to be under selection to maintain certain levels of charge and hydrophobicity (Hilu and Esen 1988). These selectional constraints are enforced by structural requirements for processing, sequestering, deposition, and mobilization of prolamins in the endosperm.

Genomic and cDNA sequences of prolamins have provided valuable information on the structure and diversity of the prolamin proteins and the number of genes and gene families that encode them. Shewry *et al.* (1995) presented a critical comparative analysis of prolamin structure in a review of seed storage protein. Prolamin structure was first described in the pooids barley, wheat and rye, and later classified by Miflin *et al.* (1983) into three groups: S-rich, S-poor, and high molecular weight (HMW). The S-rich prolamins include B- and γ-hordeins of barley, two types of γ-secalins of rye, and α- and γ-gliadins, and low molecular weight glutenin of wheat. Their structure is represented by a N-terminal domain composed of repeated sequences and a non-repetitive C-terminal domain (Shewry *et al.* 1995). The S-poor prolamins include C-hordein of barley, the ω-secalins of rye, and the ω-gliadins of wheat (Kasarda *at al.* 1983; Hull *et al.* 1991). These proteins are comprised almost completely of repeat sequences. The HMW prolamin is typified by subunits of wheat glutelin. Extensive repeated sequences are present in the HMW prolamin, flanked by non-repetitive N- and C-terminal domains. Differences in the number of repeated peptides are largely responsible for variation in HMW subunit size (Shewry *et al.* 1995).

In the Panicoideae, maize zeins are among the well-studied prolamins. Zeins fall into four principal subclasses - 10 kDa, 15 kDa, 19 kDa, and 22 kDa proteins (Llaca and Messing 1998). The central regions of the polypeptides in zein sequences contain tandem repeats of 20 amino acids (Pedersen *et al.* 1986). Prolamins of *Oryza* (Oryzoideae) and *Phyllostachys* (Bambusoideae) were examined for the low molecular weight class (LMW; see Hilu and Sharova 1998). These prolamins contain the S-rich 10 kDa protein as well as the 13 kDa and 16 kDa proteins (Kim and Okita 1988a,b; Masumura *et al.* 1989,1990; Barbier and Ishihama 1990; Feng 1990; Wen *et al.* 1993; Hilu and Sharova 1998).

Studies of prolamin molecular masses and structure have underscored the diversity existing in the Poaceae. Prolamins display intriguing patterns of complexity from small molecular weight polypeptides consisting of conserved domains and devoid of repeat motifs, to larger ones comprised partially or completely of repeat domains (Shewry *et al.* 1995). Differences in structure and organization among prolamin classes might be construed as evidence of possible multiple origins for these genes and their translated products. In a comparative study of seed storage proteins, Shewry *et al.* (1995) contrasted amino acid composition of various grass prolamins and proposed the presence of three conserved and some repetitive domains. They proposed the evolution of prolamin gene families by duplication in a single conserved domain to produce the three conserved domains followed by insertion of repetitive sequences among these domains. This hypothesis stresses the unique origin of prolamin genes in Poaceae.

PROLAMIN DIVERSITY IN POACEAE AND ITS TAXONOMIC IMPLICATIONS

Methods and Systematic Representation

In a series of publications, prolamin size, polypeptide heterogeneity and immunological similarities have been studied in over 120 species from 83 genera representing all seven major grass subfamilies and numerous tribes and subtribes (Hilu and Esen 1988, 1990, 1993; Esen and Hilu 1989, 1991,1993). In addition, new reciprocal immunological information is presented here for prolamins of species representing 23 genera from seven subfamilies (Table 1). Methods for protein extraction and immunological reactions of this new study followed Hilu and Esen (1993). Antisera were raised to total prolamin fractions, cross reactions were measured with ELISA (enzyme-linked immunosorbent assay) as described in Esen and Hilu (1989). The immunological similarity index (ISI) between a pair of taxa was calculated as: (Heterologous reactions / Heterologous followed by homologous) X 100. Heterologous reaction gives a measure of shared epitopes, while the heterologous followed by homologous give a measure of total antigenic epitopes. Both heterologous and homologous reactions were conducted at antibody excess to insure the participation of all potential reactive epitopes in the reaction.

A distance matrix (Sneath and Sokal 1973) was generated and analyzed phylogenetically with the Fitch and Margoliash (1967) method using the bambusoid *Dendrocalamus* as an outgroup. This choice of outgroup is necessitated by the unavailability of prolamin from the sister family of the grasses, Joinvilleaceae, and is supported by the strong evidence in favor of the Bambusoideae as a basal lineage in Poaceae (Barker *et al.* 1995; Clark *et al.* 1995; Duvall and Morton 1996; Liang and Hilu 1996; Soreng and Davis 1998).

In all prolamin studies focusing on the systematics of Poaceae (cited above), prolamin was extracted from grains using 60% isopropanol, and polypeptide components were resolved by SDS-PAGE (sodium dodecyl sulphate polyacrylamide gel electrophoresis). Immunological cross-reactivities were measured by ELISA. In the immunoblotting experiments, proteins were electrophoresed on 12% SDS-PAGE, transferred electrophoretically to nitrocellulose sheets following Towbin *et al.* (1979), and blots were probed with antisera to whole prolamins or specific polypeptides. The methods for SDS-PAGE, ELISA and immunoblotting are detailed in Hilu and Esen (1988) and Esen and Hilu (1991). ELISA data were analyzed phenetically with the unweighted pair-group method (UPGMA; Sneath and Sokal 1973) and phylogenetically using the Fitch-Margoliash algorithm for genetic distances (Fitch and Margoliash 1967) as described in Hilu and Esen (1993). The NTSYS-pc package (Rohlf 1993) was used for UPGMA and PHYLIP (Felsenstein 1988) for the Fitch-Margoliash method.

Table 1. Species used in the immunological study along with their respective higher categories and the sources of the material.

Subfamilies & tribes	Species	Sources of material
Arundinoideae		
Aristideae	*Aristida adscensionis* L.	USDA, PI 269867
	Aristida congesta Roem. &Schult.	USDA, PI 364388
Arundineae	*Danthonia californica* Boland	USDA, PI 232247
	Danthonia pilosa R. Br.	USDA, PI 202162
	Cortaderia selloana (Schult. &Schult. f.) Asch. & Graebn.	Ana Anton de Triquell
	Phragmites australis (Cav.) Trin. ex Steud.	Gary P. Fienming, Assatague Island, VA
Chloridoideae		
Sporoboleae	*Sporobolus indicus* (L.) R. Br.	USDA, PI 31039
Chlorideae	*Bouteloua gracilis* (H.B.K.) Log. ex Steud.	Native Plant Inc., BOGR 9392
	Chloris distichophylla Lag.	USDA, PI 404297
	Spartina cynosoroides (L.) Roth.	Hilu & Esen, KH 2528
	Vaseyochloa multinervosa (Vasey.) Hitchc.	USDA, PI 216663
Pappophoreae	*Enneapogon glaber* Burbidge	USDA, PI 257721
Eragrosteae	*Eragrostis capensis* (Thunb.) Trin.	USDA, PI 208127
Panicoideae		
Andropogoneae	*Sorghum halepense* (L.) Pers.	USDA, PI 271615
	Coix lacryma-job L.	USDA, PI 326342
Paniceae	*Setaria faberi* Herrm.	Hilu, KH 2529
	Pennisetum orientale Rich.	USDA, PI 271596
Arundinelleae	*Danthoniopsis dinteri* (Pilger) C. E. Hubb.	USDA, PI 207548
Pooideae		
Triticeae	*Hordeum vulgare* L.	Hilu, KH3
Aveneae	*Avena sativa* L.	Hilu, KH9406
Poeae	*Festuca arundinacea* Scherb.	Hilu, KH2
Bambusoideae		
Bambuseae	*Dendrocalamus strictus* (Roxb.) Nees	Thomas Soderstrom
Oryzoideae		
Ehrharteae	*Ehrharta erecta* Lam.	Hilu, KH 9417

Prolamins as Molecular Characters in Poaceae

Prolamin polypeptide size in Poaceae ranges from about 10-100 kDa (Hilu and Esen 1988). The major polypeptide components of the prolamin generally fall in three size classes: 10-17 kDa, 20-35 kDa, and 35-100 kDa. The first size class characterizes the Bambusoideae and Oryzoideae, the second the PACC subfamilies (Panicoideae, Arundinoideae, Centothecoideae, Chloridoideae), and the third the Pooideae (Fig. 1). The first two lineages have a relatively narrow range of polypeptide size components and are quite distinct as size classes. In contrast, some members of the Aveneae and Poeae of the Pooideae (Fig. 1, lanes 4,5) displayed major polypeptide components that are within the PACC group size class (also see Hilu and Esen 1988). Several pooid, panicoid, and all *Aristida* species examined possess minor polypeptide components 8-16 kDa in size, corresponding to the bambusoid-oryzoid size class. The major prolamin component of the pooid *Stipa* is ca. 14 kDa. The systematic implications of variation in both major and minor prolamin polypeptide components are discussed below. Although some overlap in polypeptide components exists, Esen and Hilu's (1989) immunological study and this one (Fig. 2) demonstrate pronounced structural differences among the three major size classes as members of those classes display higher immunological similarities among themselves than with members of the other classes. Detailed immunological studies within the Arundinoideae, Panicoideae, Chloridoideae, and Aristideae (Hilu and Esen 1990, 1993; Esen and Hilu 1991, 1993), in which outgroup taxa from related subfamilies were used further corroborated the clear distinction between the size classes and the subfamilies that possess them.

The importance of prolamin polypeptide sizes as useful characters in the systematics of the Poaceae stems from the inherent nature of the SDS-PAGE method used in resolving the protein components. This technique resolves polypeptide components primarily on the basis of size differences. Size-related mutations represent major evolutionary steps and, consequently, signify important taxonomic characters. Diversity in polypeptide size is the product of corresponding diversity in gene families. Multigene families are believed to originate by gene duplication followed by further differentiation and divergence of the new genes (Ohta 1991). Premature termination of the open reading frame (ORF) due to point mutations or insertion/deletion events (indels) that are not in multiples of three can result in a truncated product (smaller in size). Indels of this type can also produce a frame shift that may extend the ORF, producing larger polypeptides. The development of

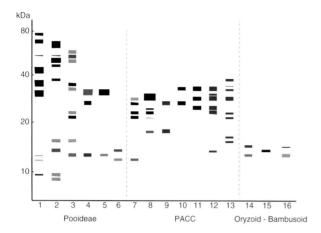

Fig. 1. Schematic representation of prolamin patterns in the Poaceae summarized from the studies of Hilu and Esen (1988, 1990, 1993) and Esen and Hilu (1989, 1991, 1993). The genera represented are 1. *Hordeum*, 2. *Bromus*, 3. *Festuca*, 4. *Dactylis*, 5. *Phleum*, 6. *Stipa*, 7. *Aristida*, 8. *Danthonia*, 9. *Phragmites*, 10. *Cynodon*, 11. *Digitaria*, 12. *Panicum*, 13. *Danthoniopsis*, 14. *Ehrharta*, 15. *Oryza*, 16. *Dendrocalamus*.

repeat domains further increases size and structural diversity in prolamin (Shewry *et al.* 1995). Knowing the constraints imposed on prolamin structure, such mutations are likely to occur under stringent selection conditions and to be fixed in populations at low rates. Consequently, these mutations are detectable, and useful, primarily at higher taxonomic levels.

Systematic Implications at and above the Subfamily Level

The phylogeny based on prolamin immunology demonstrates the sister relationship between the oryzoid *Ehrharta* and the bambusoid *Dendrocalamus* (Fig. 2), which is supported by the presence in both taxa of 10–17 kDa polypeptides (Hilu and Esen 1988). The two subfamilies grouped together on the basis of immunological data, but with low similarity (Esen and Hilu 1989). This type of association highlights the disputable taxonomic treatment of the oryzoid grasses. The oryzoid grasses have either been placed as a tribe in the Bambusoideae or as a distinct subfamily closely allied to the Bambusoideae (Hilu and Wright 1982; Soreng and Davis 1998). Prolamin data, thus, suggest a distinct subfamilial level for the Oryzoideae and support a close relationship with the Bambusoideae, a relationship gaining support from nucleic acid information (Davis and Soreng 1993; Barker *et al.* 1995; Clark *et al.* 1995; Duvall and Morton 1996; Liang and Hilu 1996).

The Pooideae form a monophyletic group diverging after *Ehrharta* (Fig. 2). The immense degree of heterogeneity in prolamin size components is quite apparent in the Pooideae, but polypeptide components with large molecular masses are unique to this subfamily (Fig 1). The wide occurrence in the Pooideae of low molecular weight polypeptides characteristic of the bambusoid-oryzoid group is a striking feature (Fig. 1). The divergence of the pooid genera in this immunology-based phylogeny directly after the oryzoid *Ehrharta* may reflect this phenomenon. The phylogenetic position for the Pooideae in this study (Fig. 2)

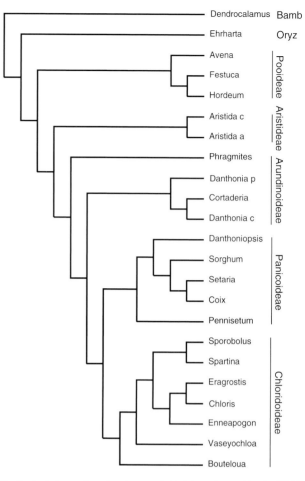

Fig. 2. A phylogeny based on immunological data obtained by the ELISA technique and analyzed with the Fitch-Margoliash distance method. The tree is rooted with the bambusoid *Dendrocalamus* (see reasoning for choice of outgroup in text).

is supported by other molecular data (Barker *et al.* 1995; Clark *et al.* 1995; Liang and Hilu 1996; Hilu *et al.* in press).

The monophyly of the PACC group is supported in this immunological study (Fig. 2). Within the PACC group, *Aristida* is the most basal lineage followed by the paraphyletic Arundinoideae. The Panicoideae and Chloridoideae form two monophyletic lineages. The paraphyly of the Arundinoideae is due to the position of *Phragmites*. In a detailed immunological study of this subfamily (Hilu and Esen 1990), *Phragmites* appeared isolated and basal within a monophyletic Arundinoideae. The prolamin-based association of *Aristida* with the PACC clade (Esen and Hilu 1989) is systematically sound. The tribe has been placed either in the Chloridoideae or Arundinoideae (Hilu and Wright 1982). Based on a more detailed prolamin study that included representative species of *Aristida* and *Stipagrostis*, Esen and Hilu (1991) showed marked differences in prolamin components of the two genera. All *Aristida* species examined by Esen and Hilu (1991) contained a pronounced prolamin component ca. 15 kDa in size resembling the oryzoid-bambusoid prolamin type. *Stipagrostis* lacks the characteristic prolamin profile of *Aristida* and displays low ELISA immunological affinities to the latter, grouping with the Chloridoideae. Immunoblot experiments confirmed these results as antiserum to *Stipagrostis* prolamin reacted weakly with prolamin of *Aristida* but strongly with those of the chloridoid taxa *Bouteloua* and *Spartina*. *Stipagrostis* has a normal C_4 anatomy in contrast with the unique double Kranz bundle sheaths of *Aristida* (deWinter 1965). Molecular data from nucleic acids would be most valuable in assessing the monophyly of the Aristideae and determining its phylogenetic position in the PACC clade.

The 20-35 kDa size class of prolamin characterizes the PACC subfamilies and the immunological data strongly support structural similarities in this size class (Hilu and Esen 1988, 1993; Esen and Hilu 1989, 1993). The expanded study presented in this paper (Fig. 2) corroborates the findings of previous studies and provides additional support for the coherence of the PACC assemblage. The affinities among the Panicoideae, Arundinoideae, Centothecoideae, and Chloridoideae were first reported by Hilu and Wright (1982) based on a UPGMA analysis of 85 morphological-anatomical characters in 215 grass genera. This association also emerged in a chloroplast DNA reassociation study in the Poaceae (Hilu and Johnson 1991) and in later molecular investigations (Davis and Soreng 1993; Barker *et al.* 1995; Clark *et al.* 1995; Duvall and Morton 1996; Liang and Hilu 1996).

Variation in prolamin size in the Poaceae is summarized in Fig. 3 and is superimposed on a representation of the immunology-based phylogeny of Fig. 2. The 10-17 size class is the proposed ancestral form. Based on prolamin size and immunological similarities, one can envision the bambusoid and oryzoid lineage radiating from an ancestral group possessing a 10-17 kDa or smaller prolamin size class. I have detected genes with the potential of encoding 10 kDa prolamin in the Joinvilleaceae (Hilu, unpublished data). The Pooideae could have diverged either directly from the ancestral stock or possibly from an oryzoid-bambusoid line (Fig. 3). *Stipa s. l.*, a genus with a demonstrated basal position in the Pooideae (Barker *et al.* 1995; Hilu *et al.* in press) and sister to other pooid grasses possesses the ancestral prolamin (15-17 kDa) as a major component. Higher molecular weight prolamins were evolved in other pooid tribes via gene duplication and insertions of single copy and repeat domains in their genes. The larger molecular weight prolamin is correlated with higher prolamin content in the grain (Hilu and Esen 1988), and may possibly be acquired as an adaptation for faster development at the seedling stages in less mesic environments than those of the bambusoid-oryzoid grasses.

Members of the Pooideae also have prolamin components within the size range of the PACC group (Fig. 1, 3). *Aristida*, one of the most basal lineages in the PACC clade (Soreng and Davis 1998; Hilu *et al.* in press) possesses both the ancestral prolamin class and the derived intermediate size group characteristic of the PACC clade (Hilu and Esen 1991) and which is also found in some Pooideae taxa (Fig. 1, lanes 4, 5, 7). Although the Arundinoideae do not commonly display the ancestral type, several taxa in the subfamily possess prominent prolamin fractions that are ca. 17 kDa in size (Hilu and Esen 1990). The ancestral prolamin type is still expressed in the Panicoideae but is not detectable in most of the Chloridoideae species examined (Esen and Hilu 1993, Hilu and Esen 1993). Thus, in the PACC group, there appears to be a gradual shift in prolamin size from *Aristida-*

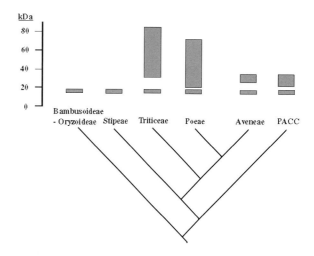

Fig. 3. Schematic representation of variation in prolamin size in the Poaceae summarized from the studies of Hilu and Esen (1988, 1990, 1993) and Esen and Hilu (1989) and superimposed on a representation of the immunology-based phylogeny of Fig. 2.

Arundinoideae with the smaller prolamin fractions to the more derived lineages of the Panicoideae and Chloridoideae. The adaptation of the Chloridoideae to mostly xeric environments may imply low advantages for the LMW prolamin.

Systematic Implications below the Subfamily Level

Although the major contributions of prolamin data to grass systematics have been at and above the subfamily level, useful information at the tribal level has been obtained. The most significant is the distinctness of the Aristideae as a tribe and its phylogenetic position apart from the Chloridoideae and Arundinoideae (Esen and Hilu 1991; this study, Fig. 2). Within the Pooideae, the Aveneae appears as a distinct tribe with the Triticeae and Poeae grouping together (Esen and Hilu 1989; Fig. 2). This is in contrast to Davis and Soreng's (1993) chloroplast restriction site study where the Triticeae formed a basal clade sister to taxa in the Poeae and Aveneae. Among the Arundinoideae genera examined by Hilu and Esen (1990), *Phragmites* emerged as a distinct taxon, having the least heterogeneous and most distinct prolamin profile. The divergence in its prolamin is well demonstrated in the immunoblot experiment (Hilu and Esen 1990). These data explain the phylogenetic position of *Phragmites* and the consequent paraphyly of the Arundinoideae in this study (Fig. 2). Renvoize (1986) excluded *Phragmites* from the core genera of the Arundinoideae, placing it among his 'peripheral genera' on the basis of distinctive anatomical characteristics. Of the six *Danthonia* species studied by Hilu and Esen (1990), the two North American species, *D. spicata* and *D. californica*, have prolamin patterns distinct from the Old World species and display high immunological affinities to South American *Cortaderia*. Immunoblot data provide strong support for the structural differences between prolamin of the New and Old World *Danthonia* species examined in that study and revealed high similarities between *D. spicata* and *D. californica* and *Cortaderia*. Close relationship between *Cortaderia* and the danthonoid group has been suggested by Clifford and Watson (1977).

In the Panicoideae, prolamin size and immunological data support the monophyly of the Panicoideae, the distinct boundaries of its major tribes Paniceae and Andropogoneae, and the ancestral status of the Arundinelleae (Esen and Hilu 1993). Tribal segregation in this study (Fig. 2) is not evident probably due to the smaller sample size compared with Esen and Hilu (1993). In the latter study, seven of the 11 Andropogoneae subtribes recognized by Clayton and Renvoize (1986) were represented. In the Paniceae, *Cenchrus* and *Pennisetum* (Cenchrinae) formed a single group linked to a lineage containing *Brachiaria*, *Setaria*, and *Panicum* (Setariinae). Immunoblot reactions between *Cenchrus* and *Pennisetum* prolamins were quite high compared to other panicoid grasses. The data support previously suspected taxonomic affinities between the two genera on the basis of morphology (Clayton and Renvoize 1986) and chloroplast DNA data (Clegg *et al.* 1984). *Chionachne* (Chionachninae) emerged as a taxon distinct from the subfamily in terms of immunological affinities, reacting only to prolamins of representative species of *Digitaria* and *Chloris*. *Coix* (Coicinae) appeared in a terminal lineage in the Andropogoneae. This information is consistent with some specialization in the inflorescence and floret structure in the monotypic Coicinae, especially the hard cup-shaped spatheole subtending the florets and the presence of a prophyll between male and female racemes (Clayton and Renvoize 1986).

The examples cited in this section highlight the contribution of prolamin data to grass systematics at the tribal and, in some cases, at the genus level.

PROLAMIN CONTRIBUTIONS AT THE PROTEIN AND GENE LEVELS: PROBLEMS AND PROSPECTS

Prolamin Size and Immunology

Contributions of prolamin protein variation to our understanding of grass systematics and evolution are quite evident. Prolamin size classes have provided molecular markers that characterize major lineages in the Poaceae. The immunological data (ELISA and immunoblots) provide valuable insight into our understanding of evolution of major lineages, monophyly of the currently recognized subfamilies (possibly except Arundinoideae), and placement of genera with disputable taxonomic positions. Prolamin studies have also been useful at the tribal level, although resolution in some cases has not been adequate, possibly due to the conserved nature of prolamins and/or the quantitative nature of the immunological procedure. The only subfamily that has not yet been examined in detail is the Pooideae, an intriguing group of grasses with the highest prolamin polypeptide heterogeneity and largest molecular weight units.

Availability of seed material constitutes the major limitation in prolamin size studies. Fortunately, the SDS-PAGE method does not require a large amount of seed tissue: 0.2 to 0.5 g of material is usually sufficient to resolve major and minor polypeptide components. Equipment and supplies for such studies are reasonably inexpensive. The ELISA immunological study, in contrast, represents a major limitation, particularly when reciprocal reactions are to be emphasized. Reciprocal immunological reactions are recommended because one-way reaction between a pair of taxa could be biased by overdominance of certain epitopes in the antisera of the species chosen for generating the antibodies. In those cases, performing reciprocal reactions and averaging the data help minimize the effect of epitope overdominance. The generation of antisera is a lengthy procedure, relatively expensive and requires the use of rabbits. Another limitation of the method is, ironically, its high sensitivity, making it inadvisable to combine data from different sets of experiments. Immunological reactions have to be performed in the same experiment in order to compare taxa. Therefore, representative taxa from previous studies have to be included in current immunological experiments for comparisons to be made. This situation is similar to DNA reassociation and, to a certain degree, chloroplast restriction fragment length studies.

Genes Encoding Prolamins

Studies on prolamin polypeptide structure and size diversity have paved the way for a prospective focus on the genes that encode this class of proteins. Prolamins are encoded by a group of multigene families that vary in gene number and degree of conservatism. The subject has been critically reviewed by Shewry *et al.* (1995) and is discussed earlier in this chapter. Although, prolamin encoding genes have been sequenced from across the Poaceae, the focus has been on a limited number of economically important cereal crops. Those studies laid down foundations for some intriguing questions regarding the origin, genetic diversity and evolution of such a complex, yet well structured group of multigene families. In our laboratory, we have been focusing for the past few years on the utility of sequence information from genes encoding the LMW prolamins in understanding divergence of basal lineages in the Poaceae and in the systematics of the Oryzoideae. Our identification of prolamin genes in the sister family Joinvilleaceae will help in rooting phylogenies. The issue that remains to be critically addressed when multigene families such as these are used in molecular systematics is paralogy *vs.* orthology and its impact on phylogenetic reconstruction (Doyle 1992). The homogenization via concerted evolution of prolamin genes present in multiple copies, such as those encoding 10 kDa and zein prolamins, renders phylogenetic reconstruction more reliable. In addition, the different rates of substitutions in the different gene families make possible the utility of the genes at different taxonomic levels. Our comparative analyses of genes encoding prolamins in oryzoid and bambusoid grasses show genes encoding 10 kDa prolamins to be very highly conserved compared with the genes encoding the 13 and 16 kDa prolamins (Hilu and Sharova 1998; Boyle and Hilu unpublished data).

CONCLUSIONS

Patterns of variation in prolamin polypeptide size and structural similarity as measured by immunological methods have provided important insights into grass systematics at various taxonomic levels. Present data on amino acids and nucleic acid sequences show that this approach holds promise for further contributions to grass systematics and evolution that will augment previous studies based on prolamin size and immunological similarities. With the major contributions in grass molecular systematics being from the chloroplast genome, information from the nuclear genome such as the prolamin gene are most needed.

ACKNOWLEDGEMENTS

The author thanks the U.S. Department of Agriculture, Agricultural Research Service-National Plant Germplasm System for providing seed material, and L. A. Alice, T. Borsch and I. M. Boyle for their comments on the manuscript. This work is supported by NSF grant No. BSR 8707959.

REFERENCES

Barbier, P., and Ishihama A. (1990). Variation on the nucleotide sequence of a prolamin gene family in wild rice. *Plant Molecular Biology* **15**, 191-195.

Barker, N. P., Linder, H. P, and Harley. E. H. (1995). Polyphyly of Arundinoideae (Poaceae): evidence from *rbc*L sequence data. *Systematic Botany* **20**, 423-435.

Clark, L. G., Zhang, W., and Wendel, J. F. (1995). A phylogeny of the grass family (Poaceae) based on *ndh*F sequence data. *Systematic Botany* **20**, 436-460.

Clayton, W. D., and Renvoize, S. A. (1986). 'Genera graminum.' (HMSO publications: London.)

Clegg, M. T., Rawson, J. R. Y., and Thomas, K. (1984). Chloroplast DNA variation in pearl millet and related species. *Genetics* **106**, 449-461.

Clifford, H. T., and Watson, L. (1977). 'Identifying grasses: data, methods, and illustrations.' (Univ. Queensland Press: St. Lucia, Australia.)

Davis, J. I., and Soreng R. J. (1993). Phylogenetic structure in the grass family (Poaceae) as inferred from chloroplast DNA restriction site variation. *American Journal of Botany* **80**, 1444-1454.

de Winter, B. (1965). The South African Stipeae and Aristideae (Gramineae). An anatomical, cytological and taxonomic study. *Bothalia* **8**, 201-404.

Doyle, J. J. (1992). Gene trees and species trees: molecular systematics as one-character taxonomy. *Systematic Botany* **17**, 144-163.

Duvall, M. R., and B. R. Morton. (1996). Molecular phylogenetics of Poaceae: an expanded analysis of *rbc*L sequence data. *Molecular Phylogenetics and Evolution* **5**, 352-358.

Esen, A., and Hilu, K. W. (1989). Immunological affinities among subfamilies of the Poaceae. *American Journal of Botany* **76**, 196-203.

Esen, A., and Hilu, K. W. (1991). Electrophoretic and immunological studies of prolamins in the Poaceae: II. Phylogenetic affinities of the Aristideae. *Taxon* **40**, 5-17.

Esen, A. and K. W. Hilu. (1993). Prolamin and immunological studies in the Poaceae: IV. Subfamily Panicoideae. *Canadian Journal of Botany* **71**, 315-322.

Felsenstein, J. (1988). 'PHYLIP (Phylogeny Inference Package), version 3.1.' (The Univ. of Washington: Seattle, Washington.)

Feng G., Wen L., Huang J. K., Shorrosh B. S., Mythukrishnan S., and Reeck G. R. 1990 Nucleotide sequence of a cloned rice genomic DNA fragment that encodes a 10 kDa prolamin polypeptide. *Nucleic Acids Research* **18**, 683.

Fitch, W. M., and Margoliash, E. (1967). Construction of phylogenetic trees. *Science* **155**, 279-284.

Hilu, K. W., Alice, L. A., and Liang, H. (in press). Phylogeny of Poaceae inferred from matK sequences. *Annals of the Missouri Botanic Garden*.

Hilu, K. W., and Esen, A. (1988). Prolamin size diversity in the Poaceae. *Biochemical Systematics and Ecology* **16**, 457-465.

Hilu, K. W., and Esen, A. (1990). Prolamin and immunological studies in Poaceae. I. Subfamily Arundinoideae. *Plant Systematics and Evolution* **173**, 57-70.

Hilu, K. W., and Esen, A. (1993). Prolamin and immunological studies in the Poaceae: III. Subfamily Chloridoideae. *American Journal of Botany* **80**, 104-113.

Hilu, K. W., and Johnson, J. L. (1991). Chloroplast DNA sequence variation in the Poaceae. *Plant Systematics and Evolution* **176**, 21-31.

Hilu, K. W., and Sharova, L. (1998). Characterization of 10 kDa prolamin genes in *Phyllostachys aurea* Riv. (Bambusoideae, Poaceae). *American Journal of Botany* **85**, 1033-1037.

Hilu, K. W., and Wright, K. (1982). Systematics of Gramineae: A cluster analysis study. *Taxon* **31**, 9-36.

Hull G. A., Halford N. G., Kreis M., and Shewry P. R. (1991). Isolation and characterization of genes encoding rye prolamins containing a highly repetitive motif . *Plant Molecular Biology* **17**, 1111-1115

Kasarda D. D., Autran J. -C., Lew E.J.-L., Nimmo C. C., and Sherry P. R. (1983). N-terminal amino acid sequences of ω-gliadins and ω-secalins implications for the evolution of prolamin genes. *Biochemical Biophysics Acta* **747**, 138-150.

Kim W. T., and Okita T. W. (1988a). Structure, expression, and heterogeneity of the rice seed prolamins. *Plant Physiology* **88**, 649-655.

Kim W. T., and Okita T. W. (1988b). Nucleotide and primary sequence of a major rice prolamin. *FEBS Letters* **231**, 308-310.

Krishhnan, H. B., and White, J. A. (1995). Morphometric analysis of rice seed protein bodies. Implication for a significant contribution of prolamin to the total protein content of rice endosperm. *Plant Physiology* **109**, 1491-1495.

Liang, H., and Hilu K. W. (1996). Application of *mat*K gene sequences to grass systematics and evolution. *Canadian Journal of Botany* **74**, 125-134.

Llaca, V, and Messing, J. (1998). Amplicons of maize genes are conserved within genic but expanded and constricted in intergenic regions. *The Plant Journal* **15**, 211-220.

Masumura T., Shibata D., Hibino T., Kato T., Kawabe k., Takeba G., Tanaka K., and Fujii S. (1989). cDNA cloning of an mRNA encoding a sulfur-rich 10 kDa prolamin polypeptide in rice seeds. *Plant Molecular Biology* **12**, 123-130.

Masumura T., Hibino T., Kidzu K., Mitsukawa N., Tanaka K., and Fujii S. (1990). Cloning and characterization of a cDNA encoding a rice 13 kDa prolamin . *Molecular Genral Genetics* **221**, 1-7.

Miflin, B. J., Field, J. M., and Shewry, P. R. (1983). Cereal storage proteins and their effects on technological properties. In 'Seed Proteins'. (Eds. J. Daussant, J. Mosse, and J. Vaughan.) pp. 255-319. (Academic Press: London.)

Pedersen k., Argos P., Naravana S. V. L., and Larkins B. A. (1986). Sequence analysis and characterization of a maize gene encoding a high-sulfur zein protein of 15,000. *Journal of Biological Chemistry* **261**, 6279-6284.

Ohta, T (1991). Multigene families and the evolution of complexity. *Journal of Molecular Evolution* **33**, 34-41.

Raghavan, V. (1997). ' Molecular Embryology of Higher Plants.' (Cambridge University Press: Cambridge.)

Renvoize, S. A. (1986). A survey of leaf blade anatomy in grasses: VIII. Arundinoideae. *Kew Bulletin* **41**, 323-342.

Shewry P. R., Napier J. A., and Tatham A. S. (1995). Seed storage proteins: Structures and biosynthesis. *Plant Cell* **7**, 945-956.

Rohlf, F. J. (1993). 'NTSYS-pc, numerical taxonomy and multivariate analysis system, version 1.8.' (Exeter Publishing, LTD: Setauket, New York.)

Sneath, P. H. A., and Sokal, R. R. (1973). 'Numerical taxonomy.' (W. H. Freeman: San Francisco.)

Soreng, R. J., and J. I. Davis. (1998). Phylogenetic and character evolution in the grass family (Poaceae): simultaneous analysis of morphological and chloroplast DNA restriction site character sets. *The Botanical Review* **64**, 1-67.

Towbin, H., Staehelin, T., and Gordon, J. (1979). Electrophoretic transfer of protein from polyacrylamide gels to nitrocellulose sheets. *Proceedings of the National Academy of Sciences, USA* **76**, 4350-4354.

Wen T. N., Shyur L. F., Su J. Ch., and Chen Ch. S. (1993). Nucleotide sequence of a rice (*Oryza sativa*) prolamin storage protein gene, RP6. *Plant Physiology* **101**, 1115-1116.

GRASSES

Grasses: Systematics and Evolution. (2000). Eds S.W.L. Jacobs and J. Everett. (CSIRO: Melbourne)

ECOLOGICAL SIGNIFICANCE OF SOUTH-WEST AFRICAN GRASS LEAF PHYTOLITHS: A CLIMATIC RESPONSE OF VEGETATION BIOMES TO MODERN ARIDIFICATION TRENDS

Christian Mulder [A] *and Roger P. Ellis* [B]

[A] Laboratory of Palaeobotany and Palynology, Utrecht University, Budapestlaan 4, 3584 CD Utrecht, the Netherlands; E-mail: C.Mulder@bio.uu.nl

[B] Grassland Research Centre, Department of Agricultural Development, P.B. X05, Lynn East, Pretoria 0039, South Africa; E-mail: mariana@IGS1.agric.za

Abstract

Recent advances in autecology and plant physiology on the effects of aridity-stress conditions on the micromorphology of plant epidermal short-cell silica bodies (phytoliths) provide a new tool for evaluating subtle change in precipitation. An on-going, integrated project concerning the modern phytoliths/vegetation/climate relationships of the Namibian area is outlined. The photosynthetic pathway appears to be the major factor determining how successful plant species are under different climates. Contemporary grass phytoliths clearly illustrate how the chorology of species with C_3 or C_4 photosynthetic pathways depends on the mean annual rainfall. The photosynthetic pathway of grass species can be determined by examining the shape of the silica bodies formed in the short-cells of the grass leaf epidermis. The differences in these shapes can be used to show how the relative abundance of species varies with the annual rainfall. Therefore modern grass leaf phytoliths are physiologically informative, and can easily be distinguished from those of other parts of the plant (and from non-grasses). It is demonstrated that African grass phytoliths have to be used as a proxy for aridity to improve the resolution of fossil records, in that changes in the total phytolith assemblages provide information about phytoclimatic shifts of the deciduous open vegetation in response to past rainfall events. Phytoliths offer a powerful addition to conventional palynology in arid areas. Since their biogenic signal can also be recovered from sediments, this approach: (1) can contribute to explain the functional patterns of the past vegetation, and (2) can help to predict future trends of the biomes of arid lands.

Key words: C_4 (four-carbon chain fixation), K (Kranz anatomy), me (malate enzyme), NAD (nicotinamide-adeninedinucleotide), NADP (nicotinamideadeninedinucleotidephosphate), PCK (phosphoenolpyruvate-carboxykinase), Okavango, silica bodies, annual rainfall, vegetation biomes.

INTRODUCTION

The sensitivity of present-day plant species and plant communities to climate can be assessed in different ways, at a taxonomic level (biogeography of specific taxa) and at the level of the plant formation (Box 1981), but the real behaviour of an ecosystem cannot be predicted from lower levels (Weiner 1996). Since extrapolation from short-term observation often fails to predict the functional patterns of the vegetation (Prentice *et al.* 1992), long-term field experiments on the response of plant specimens, plant populations and plant communities are necessary in predicting the ecological effects of climate.

Prognostic models of the response of the vegetation to climatic changes are mainly focused on the ecophysiological constraints of the plant functional types (Claussen 1994; Haxeltine and Prentice 1996). However, present-day field measurements and modern data sets often do not agree with the simulation, for instance, failing to distinguish between natural and potential vegetation. According to Prentice *et al.* (1998), the vegetation-atmosphere interaction has been disregarded in most prognostic models. The arid scenario of South-West Africa provides evidence of the weakness of any implementation of the biome reconstruction models based only on the (scarce) pollen records so far available. Especially in Namibia, an interpretation of still supposed Quaternary biome shifts cannot rely on any terrestrial pollen-based evidence (Heine 1998). Tarasov *et al.* (1998) suggested that an objective assessment of the biophysical feedback of arid landscapes must rely on the statistical weight of the herbaceous taxa. The ecological information supplied by herbs and grasses growing under extreme environmental conditions is critical in the prediction of the aridity effects (which will otherwise be overestimated in most simulations).

In using a realistic global vegetation-climate coupling to evaluate the vegetation distribution in arid and semi-arid zones of Africa, we have to take the carbon storage into account. In southern Africa (an area still critical for all the BIOME-models mentioned before), the present-day desert is an important source of carbon to the atmosphere (cf. Lioubimtseva *et al.* 1998). The Kalahari and the Namib-Naukluft deserts are contrasted offshore by the Benguela Upwelling System. The Benguela Current in the SE Atlantic Ocean is regarded as a carbon sink and a major component in the equatorial circulation across the South and the North Atlantic (Wefer *et al.* 1996). This land-sea coupling contributed during the Quaternary to both short- and long-term desertification trends. However, the magnitude of simulated changes in Africa during the Late Quaternary shows some persistent mismatches with the fossil pollen records (Jolly *et al.* 1998). A much larger modern data set is necessary to test correctly the lateral expansion of the distribution of the xerophytic plant associations.

The photosynthetic pathway has been chosen as the main factor in vegetation response to climatic changes. In the study of marine or lacustrine Quaternary sediments, the commonly used tool to unravel physiologic trends of the vegetation is the signal of the C_3/C_4 ratio derived from land plants (recovered offshore as terrigenous organic matter), as this terrestrial ratio provides an evaluation of the stable isotope signal from bulk organic material (Kelly *et al.* 1991). However, even subtle differences in grass physiology can be inferred from morphological evidence and quantified by LM (light microscope) observations of the epidermal cells (Ellis 1979, 1988).

Cumulative interactions between CO_2, temperature and precipitation in the past have been recognised in several parts of the world, but palynological sequences and independent sedimentological records are difficult to correlate in time with temperature and rainfall trends. According to Ehleringer *et al.* (1997), Cerling *et al.* (1998), Collatz *et al.* (1998) and Mulder (1999), the low atmospheric CO_2 content and the low temperatures during the last glacial maximum (18,000 yr B.P.) enabled C_4-dominated biomes to expand their global distribution, but actually to what extent? Although it is known that the plant available moisture is important in determining the distribution of present-day plants (Ellery *et al.* 1991; Rutherford and Westfall 1994), the temporal changes along the NE-SW African rainfall gradient (related to the upwelling variations of the Benguela Current) remain unclear because of the lack of an unambiguous aridity/humidity signal on land (cf. Partridge *et al.* 1990). Stokes *et al.* (1997) suggested that several significant short-lived (5-20,000 yr long) episodes of aridity in the Kalahari are related to changes in the sea-surface temperature (SST) gradient across southern Africa. However, the low resolution of most proxy data available in Africa

makes the timing of the Holocene Climatic Optimum (hypsithermal) uncertain (Cohen and Tyson 1995).

This pilot study is designed to assess whether present-day leaf phytoliths (mainly epidermal short-cell silica bodies) can be used to quantify the degree of aridity and its rate of change in the past. Phytoliths are produced by passive silica up-take and deposition in the epidermal cells. Present-day phytoliths were studied under different environmental conditions throughout the modern vegetation patterns of southern Africa. In tracing such present-day patterns, the study of taxonomically identified phytoliths provides an assessment of the biotic response of the flora. Spatial ecological constraints change the degree of occurrence of individual species (competitivity), which may be recognised in temporal trends of the total percentage of the phytolith types occurring in the sediments (fossil phytolith assemblage).

Fossil plant phytoliths have already been recovered from continental (Fredlund and Tieszen 1997a; Lu *et al.* 1998) and from marine sediments (Jansen and van Iperen 1991; Romero *et al.*, 1998). The phytolith dispersal is linked to vegetation (canopy cover, clearings, and fires), autecology (plant tissue desiccation, water stress), and geomorphology (surface run-off, soil erosion, wind threshold velocity). This study focuses on the grass phytoliths because the distribution of many species in the flora of Namibia is ecologically dependent on the precipitation thresholds (Ellis *et al.* 1980).

STUDY AREA

The study area of about 1,000,000 km^2 is south of the Cunene River (18°S) and ranges from the coastal Namib Desert up to the fringes of the Kalahari desert, where the Okavango River discharges into a seismically active extension of the East African Rift Valley System (Fig. 1). Even though some geological records are available (Rust and Vogel 1988; Thomas and Shaw 1991; Heine 1992), the palaeobiological record of the study area is still hardly explored, mainly because of the lack of knowledge of the ecology of the desert plant species involved and of the dynamics of the plant populations (Jürgens *et al.* 1997).

Throughout Namibia only a small variation in the biogenic signal of $\delta^{13}C$ values has been measured in spite of a steep SW-NE-rainfall gradient (50 up to 600 mm/yr in Schulze *et al.* 1996) and a dominant ecological gradient running in a WSW-ENE direction (Jürgens *et al.* 1997). The regional rainfall variations can be easily recognised (Fig. 1). In most of Namibia complicating factors such as soils, herbivore and/or human disturbance, and surface fire regime do not obscure the biogenic signal. The local physiological competition (Werger and Ellis 1981) and the lack of secondary variation in air humidity make the chosen area ideal for temporal aridification studies.

Up to now, the phases of aridity can be judged merely from gaps in time in lacustrine and fluvial deposits (Thomas *et al.* 1997), or from randomised pollen recovered from fossil middens (Scott 1996). Phytoliths are often preserved in large amounts in several sediments of the study area, such as in Okavango peats (where up to 70% biogenic silica occurs according to McCarthy *et al.* 1989), in the Ncamasere Valley (Meadows, pers. comm.), and in some calcretes and river terraces of the Kuiseb and the Tsondab

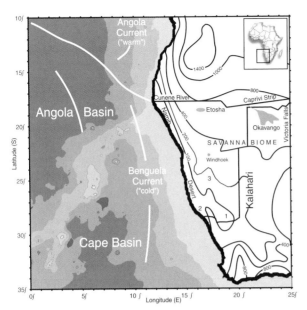

Fig. 1. The boundaries of the selected biomes in the study area depend on the mean annual precipitation (the rainfall is indicated in mm/yr by dotted lines). Grass samples were collected in the Okavango Fan (Botswana), along the Caprivi Strip (Namibian border) and near the Victoria Falls (Zambezi River, Zimbabwe). The vegetation of the savanna is the most variable of the biomes. This is evident in NW Botswana, where the Okavango sedimentation processes strongly affect the local vegetation succession. Relevès were taken for the dominant grass species in: the Namib Desert (**1.**), the Succulent Karoo Biome (**2.**), and the Nama-Karoo (**3.**), and compared with minimum -, mean -, and maximum annual rainfall, and the distribution area of the investigated herbaria specimens (reference vouchers in the Appendix).

drainage basins related to past erosive events (van der Wateren, pers. comm.). We assume that these deposits contain also less severe dry phases (apart from the gaps), that can be checked by the analyses of phytolith assemblages.

MATERIALS AND METHODS

Laboratory Treatment

Grass samples of 68 locally common species belonging to 37 genera were collected during two expeditions (September-October 1997, and July-August 1998). For each taxon at least 10 leaves were sampled under different environmental conditions. The leaves from field specimens (mainly from the Okavango area) and from the Namibian herbarium vouchers (J, PRE) were ashed in an oven at 280°C for 1 hour to extract the phytoliths. At the same time, a reference collection of Light Microscopy (LM) and Scanning Electron Microscopy (SEM) photographs according to the method suggested by Theunissen (1994) was established in order to facilitate identification of phytoliths from other sources. Phytoliths were described and identified respectively at a 1,000 x or 500 x magnification with an Olympus BH-2 microscope (objectives: SPlan100 1.25 Oil and DPlan50 0.90 Oil).

In addition, some surface samples ('litter' from monospecific populations) were prepared for palynological treatment by flotation and decanted supernatant (Kondo *et al.* 1994). HCl and HNO_3 were added to remove carbonates and organic matter. The phytoliths were transferred to test tubes by centrifugation after heavy liquid flotation ($ZnCl_2$ at a specific gravity of 2.2).

Phytoliths and cuticula remains were mounted in silicone oil 2000CS or in glycerine jelly after drying by acetone.

Statistical Approaches

The reliability of palaeoenvironmental reconstructions depends on a complete assessment of the modern analogues. The high-resolution study of present-day flora and vegetation patterns is always a contribution towards a better interpretation of past environmental trends. Several grass-phytolith keys are available for archaeobotanical purposes (Mulholland and Rapp Jr 1992; Piperno and Pearsall 1998), but the nomenclature followed in Watson and Dallwitz (1994) has been chosen for our purpose.

This pilot study is the first assessment of the response to climate of modern grass leaves involving their physiology and the phytolith morphology. To avoid possible spurious relationships in the resulting ordination, the taxonomic species-matrix deserves close scrutiny. The non-diagnostic phytoliths are excluded from Table 1, as they have been downweighted at this stage. The total variance of all the phytolith types taken into account is set to 100%.

PCA (indirect gradient approach)

The Principal Component Analysis (PCA) of the grass phytolith types (based on SEM observations of leaves from the herbarium reference vouchers given in the Appendix) has been performed with the PrinComp program (SYN-TAX 5.0 package of Podani 1993). The statistics of the phytolith variables are given in Table 1, while in the Appendix the percentages of variance of objects accounted for by each component are given. A graphical synthesis of the principal components of the investigated species (Appendix) weighted by the phytolith types taken into account (Table 1) is shown in Fig. 2.

CCA (direct gradient approach)

The Canonical Correspondence Analysis (CCA) developed by ter Braak (1987) was carried out to integrate the indirect approach of the PrinComp program. CCA incorporates both correlation and regression between floristic data and environmental factors; in this way, an integrated direct ordination is obtained (Kent and Coker 1992). The input parameters have to be of high quality and restricted in number. A too large number of environmental variables would support any pattern and would give misleading results in the direct gradient analysis (McCune 1997). The important underlying variables have been measured or taken from the literature. Each grass species was plotted against its $\delta^{13}C$ value, the minimum and maximum annual rainfall of the growing-area (Schulze *et al.* 1996), the tribe, the Kranz anatomy of the leaves (as defined in Ellis 1988), and the biomes within which it occurs viz. Savanna, Nama-Karoo, Succulent Karoo, and Desert (Gibbs Russell *et al.* 1991). The latter chorological variables were weighted against the relevès and the mean annual precipitation. The floristic (principal) input was also defined by plotting the same grass species against the phytolith types (based on LM observations of specimens growing under different environmental conditions). The biplot was performed with the CANOCO 3.10 program (ter Braak 1991) for a canonical correspondence between the local environmental variables and the biogenic silica forms occurring in the leaf sections of the dominant grasses of each biome (Fig. 3c).

Table 1. Principal Component Analysis of the Namibian grass phytolith types. Data set of the observed frequencies of the different phytolith types, established for each phytolith (sub) type of each grass species as an average value of the J or PRE herbarium vouchers, the specimens collected on the field and/or the samples from the plots where the taxa have been relevéd. Pooled variance = 90.2607

PHYTOLITH STANDARDIZED CORRELATION (PrinComp)						
Phytolith type	Mean	Standard Deviation	Variance	Variance as Percentage	Eigenvalues	Eigenvalues as Percentage
Pooid type	0.3857	1.8360	3.3708	3.735 %	3.896	32.47 %
Smooth type	0.6286	1.9424	3.7731	4.180 %	1.650	13.75 %
Round type	1.1000	2.6438	6.9899	7.744 %	1.108	9.23 %
Saddle type	1.8714	3.3705	11.3600	12.586 %	1.023	8.52 %
Tall-and-Narrow type	0.6429	1.8730	3.5083	3.887 %	0.954	7.95 %
Crenate type	1.0714	2.4573	6.0383	6.690 %	0.894	7.45 %
Oryzoid type	0.3429	1.5778	2.4894	2.758 %	0.827	6.89 %
Panicoid type	5.4857	4.3330	18.7752	20.801 %	0.599	4.99 %
Cross-Shaped subtype	2.1143	3.2372	10.4795	11.610 %	0.475	3.96 %
Butterfly-Shaped subtype	1.7143	2.9740	8.8447	9.799 %	0.323	2.69 %
Dumb-bell subtype	2.2143	2.9971	8.9824	9.952 %	0.161	1.34 %
Nodular subtype	1.2143	2.3768	5.6491	6.259 %	0.091	0.76 %

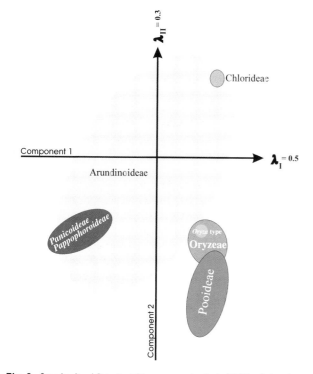

Fig. 2. Standardized Principal Component Analysis (PCA) of the phytolith types found in the herbarium vouchers weighted by the plant available moisture. In a PCA the eigenvalues measure the importance of each axis. λ_I and λ_{II} are the eigenvalues of axes I and II, respectively. These eigenvalues indicate that the axes account for 49.78% and 32.03%, respectively, of the total variation in the data set. Axis III is not shown (λ_{III} = .18). The total variance of the phytoliths is set to 1 (100%). To avoid crowding on the plot the position of the 68 Namibian grass species is not shown on this diagram, but their PCA values are given in Appendix.

RESULTS AND DISCUSSION

A strong linear relation between the leaf characters and a combination of climate variables has been known for decades (Raunkiaer 1934; Dolph and Dilcher 1979), and even the seasonality in the rainfall is reflected in the foliar physiognomy of perennial plants (Werger and Morris 1991; Jacobs 1999). According to Tieszen *et al.* (1979) and Twiss (1992), the prediction capacity of the response to climate of the functional types occurring within a flora in Africa suggested a different environmental scenario from that occurring in the mid-latitudes, where the productivity of arido-active plants is controlled by a combination of temperature and rainfall (Epstein *et al.* 1997; Fredlund and Tieszen 1997b). Since the rainfall is the dominant controlling factor in arid lands, the climatic balance between C_3 grasses (the usual metabolism of plants growing in temperate and moist areas) and C_4 grasses (the dominant metabolism of plants growing in semi-desertic biomes) provides more information if the C_4 metabolism is split up in the various physiotypes (Fig. 3a). Our preliminary results in SW Africa show to what extent this can actually be achieved.

The occurrence of phytoliths infilling the cell lumina of the leaves of most taxa points to strong links between the morphology of biogenic silica bodies and the Kranz anatomy typical for the C_4 metabolism (small substomatal spaces, dense mesophyll, and large bundle sheath cells). Short-cell phytoliths clearly reflect the grass physiology in response to climate. Table 1 shows that the highest eigenvalue of the phytolith types comes from the C_3 Pooid phytolith type (3.896, covering 32.47 % of the total ecological significance of grass phytoliths), while the typical nodular subtype, only present in C_4 grasses, has the lowest eigenvalue of the whole set. Since all the panicoid-like phytolith (sub) types score together only 14.01 % of the total variance, it becomes evident that variously shaped elongate phytoliths (like those occurring in intercostal long cells, in cell walls or in extracellular spaces) are even less significant than the latter subtype due to the lack of a distinctive micromorphology (i.e., taxonomically or physiologically not diagnostic).

A clear link between taxonomy and phytolith morphology is not evident in Fig. 2. In fact, the principal component analysis (PCA) shows that a potentially misleading noise is due to overlap

Table 2. Canonical Correspondence Analysis of the Namibian grasses. Autecological intraset of $\delta^{13}C$ (reflecting the actual physiology, from Schulze *et al.* 1996), minimum and maximum annual rainfall from the plots where the specimens were collected, where the J or PRE vouchers have been sampled, or where the taxa were relevéd (according to Müller 1982), tribe (according to Watson and Dallwitz, 1994), and leaf transectional anatomy (the anatomical types of Ellis 1988) vs. Synecological intraset, defined by the biomes where the selected species occur (according to Gibbs Russell *et al.* 1991). Wilks λ for the full set = .24233.

Roots Removed	Canonical R	R–Squared	χ^2	Degrees of Freedom	λ Prime
Full set	.7748	.600	90.01	25	.2423
1 removed	.5583	.312	31.78	16	.6063
2 removed	.2810	.079	8.06	9	.8808
3 removed	.2087	.044	2.83	4	.9564
4 removed	.0080	.000	.00	1	.9999

of the phytolith types belonging to the Oryzeae (Bambusoideae) with the physiologically variable Arundinoideae. In the Appendix the statistically significant values of the first two components are marked in bold, and show taxa of high ecological significance for future applications. For instance, the occurrence and the physiology of facultative perennials like *Stipagrostis* are put into evidence by a high significance of both the 1st and 2nd component values, since a particular kind of C_4 NAD-me metabolism enables this genus to respond opportunistically to minimal rainfall (Seely 1990). We would also assume a high ecological significance of the Oryzeae, since all C_3 species show an extremely low 2nd component value. However, if we compare the Appendix with Table 1, we can see that even if the phytoliths produced by *Oryza* belong always to the Oryzoid type and the species occurring in NW Botswana is statistically very significant (–2.983), the phytolith morphology scores only 0.827 (Table 1). Also the ecology of *Leersia* is statistically significant (-3.039), but no diagnostic phytolith types mark the occurrence of this species.

The Chloridoideae include species with either C_4 NAD-me or C_4 PCK metabolism in various degrees. However, the Chlorideae even include one species of *Eragrostis* described as a C_3 plant by Ellis (1984) which is the only non-Kranz species where also the 1st component value is statistically significant. Furthermore, *Eragrostis walteri* shows the lowest 1st component value of the whole Appendix (-5.46, surely due to extreme environmental conditions!). Within the Panicoideae the same C_3 metabolism occurs in *Sacciolepis* (2nd component value always significant throughout the genus), while otherwise most Panicoideae species have a C_4 NADP-me or C_4 PCK metabolism.

The correlation between the photosynthetic pathway and the annual precipitation (grouped into 6 clusters) is demonstrated in Fig. 3b, where the percentage of the total physiological types recognised in the transectional anatomy of the collected leaves and/or the herbarium vouchers is plotted against the mean annual precipitation of each study site where the samples were collected. Moisture availability during the growth season was obtained from an assessment of the climate at a biome-scale. Among the C_4 species, the pyruvate-formers show a clear negative correlation with the annual rainfall only within the C_4 NAD-me metabolic type, while the malate formers (C_4 NADP-me) show a positive correlation with annual rainfall. The C_4 PCK grasses were not taken into further account, but their per-

centage can be evinced from Fig. 3b as the difference [C_4 PCK % = (100 – total shown percentages)]. The canonical correspondence analysis (CCA) of the phytoliths in the biplot of Fig. 3c provides evidence for an implementation of the prognostic models mentioned before, and enhances the temporal interpretation of environmental records in showing phytoclimatic shifts of biomes. As explained by ter Braak (1987), this kind of output clarifies the (present-day) floristic response to environmental conditions.

A multiple regression for each phytolith type on each axis selected the independent variable that best explains its variation of occurrence in any biome. The component scores in any R analysis are different from the eigenvector scores on the first two axes (phytolith loadings, compare Table 2). The biome scores and the phytolith type scores are computed at different scales on the same graph (Fig. 3c). Direction and length of the arrows indicate the rate of spatial change of each biome: the short arrow of the Desert biome indicates an extremely rapid change, and the long arrow of the Nama-Karoo biome indicates a gradual change. Furthermore, the position of each arrow indicates its degree of correlation with each CANOCO axis.

This implies that the boundaries of the Nama-Karoo biome are the most closely rainfall-related (its arrow is almost parallel to the second axis), while the Succulent Karoo and the Desert are almost equally related to both axes. The Savanna biome is even more related to the first axis, corresponding to an increase of temperature (and CO_2, as suggested from a C_4 grass enrichment in the hyrax diet during the Namibian Holocene by Scott, 1996). This different vegetation response pointed out by the grass phytolith morphology of modern southern African grasses is a reliable reflection of their physiology.

According to Low and Rebelo (1996) the Succulent Karoo has few C_3 grasses, the Nama-Karoo is dominated by C_4 grasses, and the Savanna grass layer consists mostly of C_4 grasses in areas where the growing season is hot, but in the region with winter rainfall C_3 grasses are dominant. The biome arrows in the biplot of Fig. 3c show this climatic threshold, since the winter rainfall boundary can be related to the minimum annual rainfall. Both the Savanna and the Succulent Karoo are linked to an area of (very) low winter rainfall, where the (primary) moisture availability is increased by the continuous effects of the coastal fog in the latter biome.

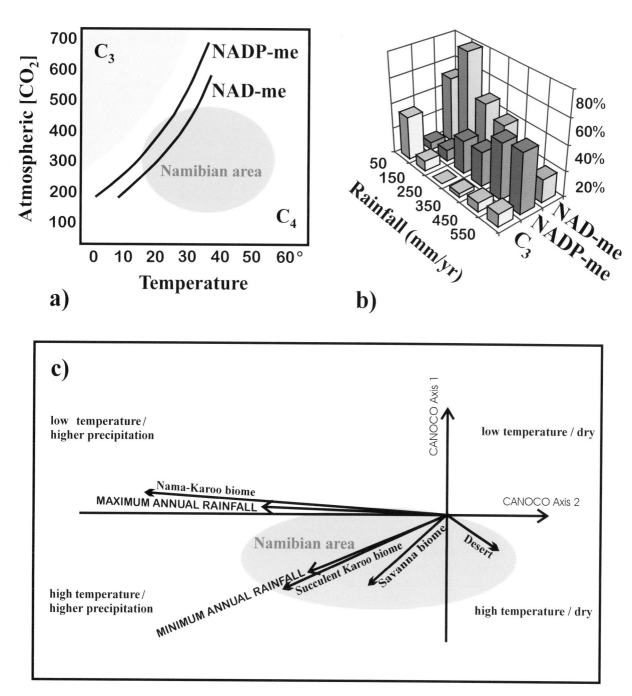

Fig. 3. *a)*. Crossover daytime temperatures during the growing-season of the correspondent CO_2-uptake quantum yield as a function of atmospheric CO_2 (concentration in ppmv). Changes in the CO_2-uptake quantum yield mainly due to photorespiration; average temperature in Celsius degrees of the different CO_2-uptakes equivalent for both the C_3 and C_4 monocots (modified from Ehleringer *et al.* 1997). Estimated (palaeo) climatic boundaries of the study area given by the shaded area (cf. Müller 1982; Lioubimtseva *et al.* 1998). *b)*. Relationships between the annual precipitation and the share of metabolic types of the Namibian grass species [%C_3 + %C_4 PCK + %C_4 NAD-me + %C_4 NADP-me = 100 %]. Specimens were collected from the Okavango, throughout Caprivi and Etosha, up to the littoral. Please see the text for further explanations. *c)*. Environmental co-ordinates for a CANOCO t-value biplot (no transformation). The vertical Axis I corresponds to a temperature diminishing during the wet season. The horizontal Axis 2 corresponds to a fall in precipitation, reflecting the gradient of plant moisture availability (with the moist biomes on the left of the CCA). Estimated vegetation boundaries of the study area given by the shaded area. The superimposed length of the arrows representing each environmental variable (vegetation biome and rainfall) indicates the direction to which the variability of grass species (magnitude of change) increases more rapidly. The longer arrows (maximal rainfall and Nama-Karoo) are more closely related to the ordinates since they indicate a gradual magnitude of spatial change, while the short arrow (Desert) represents abrupt spatial changes. The rate of change is bidimensional (spatial patterns), as no time factor was taken into account.

Some C_3-type phytolith assemblages recovered from the Okavango Savanna ecotones suggest a higher amount of water, available at the seasonal-scale. In most regional time sequences in the Okavango channels the development of the cross-sectional area (with its decline in current velocity and the debris blockages) plays an important role (Ellery *et al.* 1992). At the scale of a biome the relative abundance of C_3-type phytoliths points to changes in the rate of precipitation, useful for long-term correlations (especially the statistically significant Pooid phytolith type of Table 1).

The zonal distribution of the C_4 grasses, and the relative abundance of their phytoliths, reflect at the regional scale the water stress variation of the plant community. Aridification trends in southern Africa can be inferred at the biome-scale from C_4 grass phytoliths (especially in the karroid vegetation assemblages). The use of this tool after additional investigations will help us in the future to integrate the marine records offshore Namibia.

ACKNOWLEDGEMENTS

The stimulating discussions and hospitality offered by K. Balkwill (J, Witwatersrand), W.N. Ellery (UNATAL, Durban), and E.M. Veenendaal (HOORC, Maun) are gratefully acknowledged. We thank A. Cadman (UWits, Johannesburg), L.M. Dupont (GeoB, Bremen), L. Fish (PRE, Pretoria), G.G. Fredlund (UWI, Milwaukee), H. Hooghiemstra (UvA, Amsterdam), C.R. Janssen (UU, Utrecht), M.A. Prins (VU, Amsterdam), L. Ramberg (UBOT, Gaborone), J.-B.W. Stuut (NIOZ, Texel), H. Visscher (UU, Utrecht), and M.J.A. Werger (UU, Utrecht) for valuable comments. Mary Barkworth (Utah SU, Logan) is gratefully acknowledged for precious notes and detailed advice. The C.E. Moss Herbarium of Witwatersrand (J) and the National Herbarium (PRE) are thanked for the kind loan of their vouchers. The research started as a pilot study within the framework of the NSG-NIOZ-GeoB project on high-resolution land-sea correlations between the Benguela Upwelling System and the Kalahari Desert (1996-1998). The fieldwork and the participation at MONOCOTS II of the first author were financially supported by the Netherlands Research School of Sedimentary Geology.

REFERENCES

Box, E.O. (1981). Macroclimate and plant forms: An introduction to predictive modeling in phytogeography. *Tasks for vegetation science* **1**, 1-258.

Cerling, T.E., Ehleringer, J.R., and Harris, J.M. (1998). Carbon dioxide starvation, the development of C_4 ecosystems, and mammalian evolution. *Philosophical Transactions of the Royal Society of London*, Series B, **353**, 159-171.

Claussen, M. (1994). On coupling global biome models with climate models. *Climate Research* **4**, 203-221.

Cohen, A.L., and Tyson, P.D. (1995). Sea-surface fluctuations during the Holocene off the south coast of Africa: implications for terrestrial climate and rainfall. *The Holocene* **5**, 304-312.

Collatz, G.J., Berry, J.A., and Clark, J.S. (1998). Effects of climate and atmospheric CO_2 partial pressure on the global distribution of C_4 grasses: present, past and future. *Oecologia* **114**, 441-454.

Dolph, G.E., and Dilcher, D.L. (1979). Foliar physiognomy as an aid in determining paleoclimate. *Palaeontographica, Abt.* **B 170**, 151-172.

Ehleringer, J.R., Cerling, T.E., and Helliker, B.R. (1997). C_4 photosynthesis, atmospheric CO_2, and climate. *Oecologia* **112**, 285-299.

Ellery, K., Ellery, W.N., and Verhagen, B.Th. (1992). The distribution of C_3 and C_4 plants in a successional sequence in the Okavango Delta. *Suid-Afrikaanse tydskrif vir plantkunde* **58**, 400-402.

Ellery, W.N., Scholes, R.J., and Mentis, M.T. (1991). An initial approach to predicting the sensitivity of the South African grassland biome to climate change. *South African Journal of Science* **87**, 499-503.

Ellis, R.P. (1979). A procedure for standardizing comparative leaf anatomy in the Poaceae. II. The epidermis as seen in surface view. *Bothalia* **12**, 641-671.

Ellis, R.P. (1984). *Eragrostis walteri* - a first record of non-Kranz leaf anatomy in the sub-family Chloridoideae (Poaceae). *Suid-Afrikaanse tydskrif vir plantkunde* **3**, 380-386.

Ellis, R.P. (1988). Leaf anatomy and systematics in *Panicum* (Poaceae: Panicoideae) in southern Africa. Modern Systematic Studies in African Botany. *Monographs in Systematic Botany* **25**, 129-156.

Ellis, R.P., Vogel, J.C., and Fuls, A. (1980). Photosynthetic pathways and the geographical distribution of grasses in South West Africa/Namibia. *South African Journal of Science* **76**, 307-314.

Epstein, H.E., Lauenroth, W.K., Burke, I.C., and Coffin, D.P. (1997). Productivity patterns of C_3 and C_4 functional types in the U.S. Great Plains. *Ecology* **78**, 722-731.

Fredlund, G.G., and Tieszen, L.L. (1997a). Calibrating grass phytolith assemblages in climatic terms: Application to late Pleistocene assemblages from Kansas and Nebraska. *Palaeogeography, Palaeoclimatology, Palaeoecology* **136**, 199-211.

Fredlund, G.G., and Tieszen, L.L. (1997b). Phytolith and Carbon Isotope Evidence for Late Quaternary Vegetation and Climate Change in the Southern Black Hills, South Dakota. *Quaternary Research* **47**, 206-217.

Gibbs Russell, G.E., Watson, L., Koekmoer, M., Smook, L., Barker, N.P., Anderson, H.M., and Dallwitz, M.J. (1991). Grasses of southern Africa. *Memoirs van die botaniese opname van Suid-Afrika* **58**, 1-437.

Haxeltine, A., and Prentice, I.C. (1996). BIOME3: An equilibrium terrestrial biosphere model based on ecophysiological constraints, resource availability, and competition among plant functional types. *Global Biogeochemical Cycles* **10**, 693-709.

Heine, K. (1992). On the ages of humid Late Quaternary phases in southern African areas (Namibia, Botswana). *Palaeoecology of Africa* **23**, 149-164.

Heine, K. (1998). Climate change over the past 135,000 years in the Namib Desert (Namibia) derived from proxy data. *Palaeoecology of Africa* **25**, 171-198.

Jacobs, B.F. (1999). Estimation of rainfall variables from leaf characters in tropical Africa. *Palaeogeography, Palaeoclimatology, Palaeoecology* **145**, 231-250.

Jansen, J.H.F., and van Iperen, J.M. (1991). A 220,000-year climatic record for the east-equatorial Atlantic Ocean and equatorial Africa: Evidence from diatoms and opal phytoliths in the Zaire (Congo) deep-sea fan. *Paleoceanography* **6**, 573-591.

Jolly, D., Harrison, S.P., Damnati, B., and Bonnefille, R. (1998). Simulated climate and biomes of Africa during the Late Quaternary: Comparison with pollen and lake status data. *Quaternary Science Reviews* **17**, 629-657.

Jürgens, N., Burke, A., Seely, M.K., and Jacobson, K.M. (1997). Desert. In: 'Vegetation of Southern Africa.' (Eds R.M. Cowling, D.M. Richardson, and S.M. Pierce) pp: 189-214. (Cambridge University Press: Cambridge).

Kelly, E.F., Amundson, R.G., Marino, B.D., and DeNiro, M.J. (1991). Stable isotope ratios of carbon in phytoliths as a quantitative method of monitoring vegetation and climate change. *Quaternary Research* **35**, 222-233.

Kent, M., and Coker, P. (1992). 'Vegetation Description and Analysis. A Practical Approach.' (Belhaven Press: London).

Kondo, R., Childs, C., and Atkinson, I. (1994). 'Opal Phytoliths of New Zealand.' (Manaaki Whenua Press: Lincoln).

Lioubimtseva, E., Simon, B., Faure, H., Faure-Denard, L., and Adams, J.M. (1998). Impacts of climatic change on carbon storage in the Sahara-Gobi desert belt since the Last Glacial Maximum. *Global and Planetary Change* **16-17**, 95-105.

Low, A.B., and Rebelo, A.G. Eds. (1996). 'Vegetation of South Africa, Lesotho and Swaziland. A companion to the Vegetation Map of South Africa, Lesotho and Swaziland.' (Department of Environmental Affairs & Tourism: Pretoria).

Lu, H., Wu, N., Liu, T., Han, J., Qin, X., Sun, X., and Wang, Y. (1998). Seasonal Climatic Variation recorded by Phytolith Assemblages from the Baoji Loess Sequence in Central China over the last 150 ka. *PAGES* **6**, 4-5.

McCarthy, T.S., McIver, J.R., Cairncross, B., Ellery, W.N., and Ellery, K. (1989). The inorganic chemistry of peat from the Maunachira channel-swamp system, Okavango Delta, Botswana. *Geochimica Cosmochimica Acta* **53**, 1077-1089.

McCune, B. (1997). Influence of noisy environmental data on canonical correspondence analysis. *Ecology* **78**, 2617-2623.

Mulder, Ch. (1999). Biogeographic re-appraisal of the Chenopodiaceae of Mediterranean drylands: A quantitative outline of their general ecological significance in the Holocene. *Palaeoecology of Africa* **26**, 161-188.

Mulholland, S.C., and Rapp, G. jr (1992). A morphological classification of grass silica-bodies. In: 'Phytolith Systematics' (Eds. G. Rapp jr., and S.C. Mulholland). *Emerging Issues, Advances in Archaeological and Museum Science* **1**, 65-89.

Müller, M.J. (1982). Selected climatic data for a global set of standard stations for vegetation science. *Tasks for vegetation science* **5**, 1-306.

Partridge, T.C., Avery, D.M., Botha, G.A., Brink, J.A., Deacon, J., Herbert, R.S., Maud, R.R., Scott, L., Talma, A.S., and Vogel, J.C. (1990). Late Pleistocene and Holocene climatic change in southern Africa. *South African Journal of Science* **86**, 302-306.

Piperno, D.R., and Pearsall, D.M. (1998). The silica bodies of tropical American grasses: Morphology, taxonomy, and implications for grass systematics and fossil phytolith identification. *Smithsonian Contributions to Botany* **85**, 1-40.

Podani, J. (1993).' SYN-TAXpc version 5.0: Computer Programs for Multivariate Data analysis in Ecology and systematics. User's Guide.' (Scientia Publishing: Budapest).

Prentice, I.C., Cramer, W., Harrison, S.P., Leemans, R., Monserud, R.A., and Solomon, A.M. (1992). A global biome model based on plant physiology and dominance, soil properties and climate. *Journal of Biogeography* **19**, 117-134.

Prentice, I.C., Harrison, S.P., Jolly, D., and Guiot, J. (1998). The climate and biomes of Europe at 6000 yr BP: Comparison of model simulations and pollen-based reconstructions. *Quaternary Science Reviews* **17**, 659-668.

Raunkiaer, Ch. (1934). The use of leaf size in biological plant geography. In 'The life forms of plants and statistical plant geography.' (Coll. Papers of Ch. Raunkiaer) pp. 368-378. (Clarendon Press: Oxford).

Romero, O., Lange, C., Swap, R., and Wefer, G. (1998). Eolian-transported freshwater diatoms and phytoliths across the equatorial Atlantic record temporal changes in Saharan dust transport patterns. *Journal of Geophysical Research – Oceans* **104 C2**, 3211-3222.

Rust, U., and Vogel, J.C. (1988). Late Quaternary environmental changes in the northern Namib Desert as evidenced by fluvial landforms. *Palaeoecology of Africa* **19**, 127-137.

Rutherford, M.C., and Westfall, R.H. (1994). 'Biomes of southern Africa: an objective categorization.' *Memoirs van die botaniese opname van Suid-Afrika* **63**, 1-94.

Schulze, E.-D., Ellis, R.P., Schulze, W., Trimborn, P., and Ziegler, H. (1996). Diversity, metabolic types and $\delta^{13}C$ carbon isotope ratios in the grass flora of Namibia in relation to growth form, precipitation and habitat conditions. *Oecologia* **106**, 352-369.

Scott, L. (1996). Palynology of hyrax middens: 2000 years of palaeoenvironmental history in Namibia. *Quaternary International* **33**, 73-79.

Seely, M.K. (1990). Patterns of plant establishment on a linear desert dune. *Israel Journal of Botany* **39**, 443-451.

Stokes, S., Thomas, D.S.G., and Washington, R. (1997). Multiple episodes of aridity in southern Africa since the last interglacial period. *Nature* **388**, 154-158.

Tarasov, P.E., Cheddadi, R., Guiot, J., Bottema, S., Peyron, O., Belmonte, J., Ruiz-Sanchez, V., Saadi, F., and Brewer, S. (1998). A method to determine warm and cool steppe biomes from pollen data; application to the Mediterranean and Kazakhstan regions. *Journal of Quaternary Science* **13**, 335-344.

ter Braak, C.J.F. (1987). Ordination. In: 'Data analysis in community and landscape ecology' (Eds R.H.G. Jongman, C.J.F. ter Braak, and O.F.R. van Tongeren) pp. 91-173 (Pudoc: Wageningen).

ter Braak, C.J.F. (1991). 'Update Notes: CANOCO Version 3.12.' (Agricultural Mathematics Group DLO, Wageningen).

Theunissen, J.D. (1994). A Method for Isolating and Preparing Silica Bodies in Grasses for Scanning Electron Microscopy. *Biotechnic & Histochemistry* **69**, 291-294.

Thomas, D.S.G., and Shaw, P.A. (1991). 'The Kalahari Environment.' (Cambridge University Press: Cambridge).

Thomas, D.S.G., Stokes, S., and Shaw, P.A. (1997). Holocene aeolian activity in the southwestern Kalahari Desert, southern Africa: significance and relationships to late-Pleistocene dune-building events. *The Holocene* **7**, 273-281.

Tieszen, L.L., Senyimba, M.M., Imbamba, S.K., and Troughton, J.H. (1979). The distribution of C_3 and C_4 grasses and carbon isotope discrimination along an altitudinal and moisture gradient in Kenya. *Oecologia* **37**, 337-350.

Twiss, P.C. (1992). Predicted world distribution of C_3 and C_4 grass phytoliths. In 'Phytolith Systematics' (Eds. G. Rapp jr., and S.C. Mulholland). *Emerging Issues, Advances in Archaeological and Museum Science* **1**, 113-128.

Watson, L., and Dallwitz, M.J. (1994). 'The Grass Genera of the World.' 2nd edition. (CAB International: Wallingford).

Wefer, G., Berger, W.H., Siedler, G., and Webb, D.J. Eds. (1996). 'The South Atlantic: Present and Past Circulation.' (Springer Verlag: Berlin, Heidelberg).

Weiner, J. (1996). Problems in Predicting the Ecological Effects of Elevated CO_2. In 'Carbon Dioxide, Populations, and Communities' (Eds C. Körner, and F.A. Bazzaz) pp. 431-441 (Academic Press: San Diego).

Werger, M.J.A., and Ellis, R.P. (1981). Photosynthetic pathways in the arid regions of South Africa. *Flora* **171**, 64-75.

Werger, M.J.A., and Morris, J.W. (1991). Climatic control of vegetation structure and leaf characteristics along an aridity gradient. *Annali di Botanica* **49**, 203-215.

Christian Mulder and Roger P. Ellis

Appendix. Percentage of variance of objects (phytoliths) accounted for by each component (grass species). Autecological intraset of δ^{13}C (Schulze *et al.* 1996), minimum and maximum annual rainfall (according to Müller 1982), tribe (according to Watson and Dallwitz, 1994), and leaf transectional anatomy (types in Ellis 1988). Reference vouchers from the herbaria of J and PRE. Statistical significant values are marked in **bold**, and the relevant species (highest or lowest scores) in *CAPITALS*. *(Continued over)*

Reference Vouchers		Min. mm/yr	Max. mm/yr	δ^{13}C	First Perc. Var.	Second Perc. Var.	Third Perc. Var.	Mean	First Comp.	Second Comp.	Third Comp.	Phytochem
Andropogon eucomus Ellery 53904, J; Ellis 1357, PRE	A N	400	600	-11.55	81.646	14.925	3.429	600	0.684	0.292	-0.140	NADP-me
Andropogon schirensis Ellis 970, PRE	A N	400	600	-11.41	79.422	17.046	3.532	600	0.694	0.322	-0.146	NADP-me
Aristida congesta Dye 60148, J	A R	200	500	-12.91	62.163	35.236	2.601	400	-1.065	0.802	-0.218	NADP-me
Aristida stipoides Smith 1954, PRE; s.n. 1, J	A R	500	600	-12.49	78.406	1.385	20.21	600	1.027	-0.137	-0.522	NADP-me
Brachiaria deflexa Smook 1138, PRE	P A	200	500	-13.98	90.778	8.262	0.960	300	-1.216	0.367	-0.125	PCK
Brachiaria humidicola Ellery 57600, J; Ellis 343, PRE	P A	400	600	-11.69	83.878	12.808	3.314	600	0.674	0.263	-0.134	PCK
Cenchrus ciliaris Balkwill 85016, J; Ellis 4603, PRE	P A	100	600	-12.26	14.925	29.465	55.610	500	-0.601	0.845	1.161	NADP-me
Chloris gayana Ellery 331, J; Ellis 3746, PRE	C H	500	600	-11.58	78.951	0.184	20.864	600	1.093	0.053	-0.562	PCK
Cymbopogon excavatus Ellis 99, PRE; Maguire 67135, J	A N	300	600	-11.70	17.266	61.613	21.122	600	0.262	0.495	0.290	NADP-me
Cynodon dactylon Dye 20, J; Ellis 332, PRE	C H	100	600	-15.60	29.049	0.919	70.032	450	-0.843	0.150	1.309	NAD-me
Dactyloctenium giganteum Baines&Parry 74962, J; Ellis 3847, PRE	C H	500	600	-11.63	79.024	0.120	20.857	600	1.090	0.042	-0.560	PCK
Digitaria brazzae Ellis 1338, PRE	P A	500	600	-10.41	74.910	4.739	20.351	600	1.178	0.296	-0.614	NADP-me
Digitaria debilis Dye 9, J; Ellis 1545, PRE	P A	400	600	-11.35	78.476	17.951	3.574	600	0.698	0.334	-0.149	NADP-me
Digitaria eylesii Ellis 2081, PRE	P A	500	600	-10.21	73.902	5.934	20.163	600	1.192	0.338	-0.623	NADP-me
Digitaria milanjiana Voster 2779, PRE	P A	400	600	-11.25	76.914	19.448	3.638	600	0.706	0.355	-0.153	NADP-me
Echinochloa colona Ellis 2908, PRE	P A	300	600	-12.58	16.068	39.720	44.212	600	0.198	0.312	0.329	NADP-me
Echinochloa crus-galli Smook 5871, PRE	P A	400	600	-11.08	74.312	21.950	3.738	600	0.718	0.390	-0.161	NADP-me
Echinochloa pyramidalis Ellery 54479, J; Ellis 1892, PRE	P A	300	600	-13.41	11.086	11.211	77.703	600	0.138	0.139	0.366	NADP-me
Echinochloa stagnina Cohen 23353, J; Jacobsen 2978, PRE	P A	300	600	-10.63	16.727	74.756	8.517	600	0.339	0.717	0.242	NADP-me
Enteropogon macrostachyus Bainer&Parry 75022, J; Ellis 2078, PRE	C H	400	600	-13.60	93.364	5.850	0.786	600	0.536	-0.134	-0.049	NAD-me
Eragrostis aspera Killick 1714, PRE	C H	300	500	-12.41	44.599	19.766	35.635	400	-0.691	0.460	-0.618	NAD-me
Eragrostis cilianensis Ellery 58261, J; Ellis 888, PRE	C H	100	500	-14.32	87.822	9.023	3.156	400	-1.652	0.530	0.313	NAD-me
Eragrostis echinochloidea De Winter&Wiss 4427, Retief 1530, PRE	C H	100	600	-12.61	16.553	25.126	58.321	300	-0.627	0.772	1.176	NAD-me
Eragrostis lappula Ellis 2014, PRE; Henning 77512, J	C H	500	600	-12.65	77.853	2.176	19.972	600	1.016	-0.170	-0.514	PCK
Eragrostis pilosa Smook 1765, PRE	C H	500	600	-15.04	53.725	33.662	12.613	600	0.843	-0.667	-0.408	NAD-me

Appendix. Percentage of variance of objects (phytoliths) accounted for by each component (grass species). Autecological intraset of $\delta^{13}C$ (Schulze *et al.* 1996), minimum and maximum annual rainfall (according to Müller 1982), tribe (according to Watson and Dallwitz, 1994), and leaf transectional anatomy (types in Ellis 1988). Reference vouchers from the herbaria of J and PRE. Statistical significant values are marked in **bold**, and the relevant species (highest or lowest scores) in *CAPITALS*. *(Continued over)*

Reference Vouchers		Min. mm/yr	Max. mm/yr	$\delta^{13}C$	First Perc. Var.	Second Perc. Var.	Third Perc. Var.	Mean	First Comp.	Second Comp.	Third Comp.	Phytochem
Eragrostis porosa De Winter&Codd 284, PRE	C H	100	500	-12.64	73.844	24.362	1.794	400	-1.531	0.879	0.239	NAD-me
Eragrostis superba Ellery 52841, J	C H	300	600	-12.47	16.392	43.154	40.455	600	0.206	0.334	0.324	NAD-me
Eragrostis viscosa Ellis 530, PRE; Witkowski s.n., J	C H	300	600	-13.94	6.132	0.504	93.364	500	0.100	0.029	0.389	NAD-me
ERAGROSTIS WALTERI Ellis 4344, 4346, PRE; Giess 8104, PRE	C H	50	200	-26.71	84.258	7.163	8.579	100	**-5.460**	**-1.592**	-1.742	n-K
Eriochrysis pallida Ellery 52833, Killick 251, PRE	P A	400	400	-11.84	23.507	3.803	72.691	400	-1.140	0.459	-2.005	NADP-me
Heteropogon contortus Ellery&Ellery 78945, J; Ellis 52, PRE	A N	200	600	-12.10	3.245	42.372	54.383	600	-0.178	0.645	0.730	NADP-me
Hyparrhenia filipendula Codd 6880, PRE	A N	500	600	-12.00	79.214	0.084	20.703	600	1.063	-0.035	-0.543	NADP-me
Hyparrhenia hirta Ellis 4772, PRE	A N	100	600	-13.03	18.542	20.134	61.323	500	-0.657	0.685	1.195	NADP-me
Hyparrhenia rufa Ellis 3715, PRE; Williamson&Payet 84896, J	A N	600	600	-11.18	69.846	0.281	29.873	600	1.533	-0.097	-1.003	NADP-me
Hyperthelia dissoluta Ellis 340, PRE	A N	500	600	-11.44	78.695	0.434	20.871	600	1.103	0.082	-0.568	NADP-me
Imperata cylindrica Ellis 320, PRE; Henning 321, J	A N	300	600	-12.19	16.947	50.993	32.06	600	0.226	0.393	0.311	NADP-me
LEERSIA HEXANDRA Ellery 269, J; Ellis 3713, PRE	B A	500	600	-26.44	0.003	99.893	0.104	500	0.018	**-3.039**	0.098	n-K
Miscanthus junceus Ellery 54480, J; Ellis 317, PRE	A N	400	600	-12.76	97.498	0.454	2.048	600	0.596	0.041	-0.086	NADP-me
ORYZA LONGISTAMINATA Ellery 300, J; Ellis 3677, PRE	B A	500	600	-26.17	0.015	99.902	0.083	600	0.037	**-2.983**	0.086	n-K
Panicum maximum Ellery 257, J; Ellis 3850, PRE	P A	300	600	-12.83	15.063	31.294	53.643	600	0.180	0.260	0.340	PCK
Panicum repens Ellery 53905, J; Smook 4211, PRE	P A	300	600	-12.50	16.310	42.236	41.454	600	0.204	0.328	0.325	PCK
Paspalidium geminatum Giess 3132, PRE	P A	200	600	-12.68	5.427	30.686	63.887	600	-0.220	0.524	0.756	NADP-me
Paspalum scrobiculatum Smook 1006, PRE	P A	300	600	-12.31	16.751	47.796	35.453	600	0.218	0.368	0.317	NADP-me
PASPALUM VAGINATUM Ellis 276, PRE	P A	100	300	-12.82	74.656	7.605	17.739	100	**-3.347**	**1.068**	-1.631	NADP-me
Pennisetum glaucocladum Ellis 2922, PRE; Maguire 75405, J	P A	500	600	-10.94	77.215	2.057	20.728	600	1.139	0.186	-0.590	NADP-me
Perotis patens Balkwill 80940, J; Ellis 1926, PRE	C H	400	600	-13.03	98.292	0.071	1.637	600	0.577	-0.015	-0.074	NAD-me
PHRAGMITES AUSTRALIS Ellis 187, PRE; Maguire 68446, J	A R	100	600	-25.62	26.506	40.342	33.152	600	-1.568	**-1.935**	1.754	n-K
Pogonarthria squarrosa Baines&Parry 75285, J; Ellis 2089, PRE	C H	300	600	-13.04	13.892	23.811	62.298	600	0.165	0.216	0.349	NAD-me
SACCIOLEPIS AFRICANA De Winter&Marais 4528, PRE	P A	400	600	-26.92	2.058	94.644	3.297	600	-0.428	**-2.905**	0.542	n-K
SACCIOLEPIS HUILLENSIS Johnstone 356, PRE	P A	400	600	-27.71	2.360	94.304	3.336	600	-0.486	**-3.070**	0.577	n-K

Appendix. Percentage of variance of objects (phytoliths) accounted for by each component (grass species). Autecological intraset of $\delta^{13}C$ (Schulze *et al.* 1996), minimum and maximum annual rainfall (according to Müller 1982), tribe (according to Watson and Dallwitz, 1994), and leaf transectional anatomy (types in Ellis 1988). Reference vouchers from the herbaria of J and PRE. Statistical significant values are marked in **bold**, and the relevant species (highest or lowest scores) in *CAPITALS*.

Reference Vouchers		Min. mm/yr	Max. mm/yr	$\delta^{13}C$	First Perc. Var.	Second Perc. Var.	Third Perc. Var.	Mean	First Comp.	Second Comp.	Third Comp.	Phytochem
SACCIOLEPIS TYPHURA Ellery 58278, J; Ellis 1527, PRE	P A	500	600	-26.08	0.022	99.902	0.076	600	0.044	**-2.964**	0.082	n-K
Schmidtia pappophoroides Dye 12, J; Ellis 929, PRE	C H	100	600	-12.57	16.365	25.616	58.019	500	-0.624	0.780	1.174	PCK
Setaria sagittifolia Ellery 241, J; Ellis 2085, PRE	P A	400	600	-12.07	89.761	7.294	2.945	600	0.646	0.184	-0.117	NADP-me
Setaria sphacelata Ellery 312, J; Ellis 329, PRE	P A	500	600	-11.92	79.227	0.022	20.751	600	1.069	-0.018	-0.547	NADP-me
SETARIA VERTICILLATA Ellery 58288, J; Ellis 764, PRE	P A	100	600	-11.63	12.137	37.435	50.427	500	-0.556	**0.976**	1.133	NADP-me
SORGHASTRUM FRIESII Ellery 334, J; Ellis 3695, PRE	A N	600	600	-12.54	66.597	4.677	28.726	600	**1.435**	-0.380	-0.942	NADP-me
SPOROBOLUS AFRICANUS Smook 5456, PRE	C H	600	600	-13.05	64.355	7.790	27.855	600	**1.398**	-0.486	-0.920	PCK
Sporobolus fimbriatus Dye 19, J; Ellis 4074, PRE	C H	300	600	-13.35	11.599	13.096	75.304	600	0.142	0.151	0.363	PCK
Sporobolus ioclados De Winter&Codd 340, PRE; Traill 61140, J	C H	200	600	-13.23	8.013	19.863	72.124	500	-0.260	0.410	0.781	NAD-me
Sporobolus pyramidalis Smook 5043, PRE	C H	300	600	-11.78	17.254	60.128	22.618	600	0.256	0.478	0.293	PCK
Sporobolus spicatus Ellery 244, J; Ellis 3706, PRE	C H	200	600	-13.15	7.612	21.373	71.015	600	-0.254	0.426	0.777	NAD-me
STIPAGROSTIS CILIATA Maguire 74002, J	A R	50	300	-13.54	81.254	6.700	12.046	300	**-3.605**	1.035	-1.388	NAD-me
Themeda triandra Ellis 295, 371, 2023, PRE	A N	300	600	-11.59	17.263	63.513	19.225	600	0.270	0.518	0.285	NADP-me
Trachypogon spicatus Ellery 58304, J; Ellis 143, PRE	A N	500	600	-13.15	75.149	5.873	18.978	600	0.980	-0.274	-0.492	NADP-me
Tragus berteronianus Ellery 243, J; Ellis 2021, PRE	C H	100	500	-15.07	91.816	4.399	3.785	500	-1.706	0.374	0.346	NAD-me
Tragus racemosus Smook 2774, PRE	C H	200	600	-12.32	3.999	37.988	58.013	500	-0.194	0.599	0.740	NAD-me
Vetiveria nigritana Davidson 41756, J; Ellis 342, PRE	A N	400	600	-12.14	90.777	6.356	2.867	600	0.641	0.170	-0.114	NADP-me
VOSSIA CUSPIDATA Ellis 3708, PRE; Henning 77475, J	A N	600	600	-10.56	70.097	0.028	29.875	600	**1.578**	0.032	-1.030	NADP-me

Grasses: Systematics and Evolution. (2000). Eds S.W.L. Jacobs and J. Everett. (CSIRO: Melbourne)

GRASSES

EFFECTS OF ELEVATED ATMOSPHERIC [CO$_2$] IN *PANICUM* SPECIES OF DIFFERENT PHOTOSYNTHETIC MODES (POACEAE: PANICOIDEAE)

Claudia Tipping and David R. Murray

Faculty of Science and Technology, University of Western Sydney - Hawkesbury Campus, Locked Bag No. 1, Richmond NSW 2753.

Abstract

Panicum tricanthum Nees, *P. antidotale* Retz. and *P. decipiens* Nees ex Trin., respectively, were selected to represent C$_3$, C$_4$ and C$_3$/C$_4$ intermediate perennial species of *Panicum*. Plants grown from seed with 900 ppm [CO$_2$] under natural sunlight and controlled temperatures (30°/22°C) were compared with plants grown at ambient [CO$_2$]. The anatomy of the last fully expanded leaf of the main tiller was studied by light microscopy with computerized graphic image analysis, and by transmission electron microscopy. The leaf anatomy did not change qualitatively in response to elevated [CO$_2$], but there were changes in leaf thickness, and in the proportions of total trans-sectional area occupied by mesophyll, bundle sheath cells, vascular elements, and sclerenchyma, depending on species. The abaxial stomatal frequency decreased by 22% for *P. tricanthum*, but increased by around 30% for the others. With 900 ppm CO$_2$, all three species showed a massive increase in leaf starch content (to>30% of dry matter). Starch granules accumulated in the chloroplasts of the mesophyll as well as the bundle sheath cells. Increased leaf glaucousness in response to elevated [CO$_2$] was correlated with increased or modified deposition of epicuticular wax on both leaf surfaces. The patterns were studied by scanning electron microscopy. This response to elevated [CO$_2$] has not previously been recorded for Monocotyledons. *Panicum decipiens* rarely responded to elevated [CO$_2$] in a truly intermediate fashion, but its responses resembled those of *either* the C$_3$ *or* the C$_4$ species. C$_3$/C$_4$ intermediates may thus be interpreted as developmental chimeras, and not as species in transition between C$_3$ and C$_4$ modes in an evolutionary sense.

Key words: Carbon dioxide, leaf anatomy, *Panicum*, stomatal distribution, wax deposition.

INTRODUCTION

Panicum comprises more than 300 annual or perennial species found in tropical or subtropical regions of both hemispheres, extending into warm temperate areas. There are more than 30 species in Australia, mostly indigenous, and considerable research on their potential as tropical forage crops has taken place over the past 100 years (Breakwell 1918; Wilson 1976, 1991; Akin *et al.* 1983; Wilson *et al.* 1983). Several species of *Panicum* featured prominently in the first biochemical and anatomical characterizations of C$_4$ tropical grasses: *P. obseptum*, *P. pygmaeum* and *P. simile* (Carolin *et al.* 1973); *P. miliaceum* and *P.*

maximum (Edwards and Gutierrez 1972; Hatch *et al.* 1975). A remarkable feature of the genus is that its C$_4$ species are distributed amongst all three known metabolic sub-groups (Hatch *et al.* 1975; Dengler *et al.* 1986; Prendergast *et al.* 1987). *Panicum* also includes C$_3$/C$_4$ intermediate species, such as *P. milioides* (Brown and Brown 1975).

The panicoid pattern of leaf anatomy is distinctive, with either a single radial file of chlorenchyma cells encircling each vascular bundle (Esau 1960; Carolin *et al.* 1973), or an inner mestome sheath of parenchyma separating the photosynthetic sheath from the vascular tissues of the bundle (Dengler *et al.* 1985, 1986;

Prendergast *et al.* 1987). The bundle sheath chlorenchyma of C_4 species of *Panicum* are well developed, with somewhat thicker cell walls than the background mesophyll cells, whereas the parenchyma surrounding the veins in C_3 species are simply enlarged.

Global average atmospheric CO_2 concentration has increased from 290 ppm mid-19th century to about 360 ppm at present, and may increase to 660 ppm before the end of the 21st century (reviewed by Rogers and Dahlman 1993; Murray 1995, 1997). A continued increase will eventually have an impact on the growth of crop plants, irrespective of any direct contribution of CO_2 to climate change. One aim of the present study was to ascertain whether the growth responses to elevated [CO_2] of species representing C_3, C_4 and C_3/C_4 intermediate species of *Panicum* include the changes in leaf morphology most often observed in dicotyledonous C_3 species, such as leaf thickening and starch accumulation (Thomas and Harvey 1983; Yelle *et al.* 1989; Radoglou and Jarvis 1990; Tipping and Murray 1997). Changes in specific leaf area, internal leaf anatomy, leaf starch content and distribution, the abundance of stomata, and in epicuticular wax patterns are described.

MATERIALS AND METHODS

The species chosen for this investigation were: *Panicum tricanthum* Nees (C_3), *P. antidotale* Retz. (C_4 - NADP-ME group) and *P. decipiens* Nees ex Trin.(a C_3/C_4 intermediate species with Kranz anatomy). These species are amongst those showing the greatest extent of leaf dry matter digestibility when assessed as potential forage crops (Akin *et al.* 1983; Wilson *et al.* 1983). The seeds planted were of the same accessions as used in these earlier studies (J. Wilson, personal communication).

Growth Conditions

Seeds were planted in a 60:40 sand:peat mixture in seed trays and after 3 weeks seedlings were transferred individually to pots (350 mL or 1 L) filled with potting mix. Pots were watered daily. Mineral nutrients were supplied as Osmocote (Sierra Chemicals) slow-release fertiliser in accordance with recommendations in the literature (Thomas and Harvey 1983). The plants were grown with controlled temperature (30°/22°C, day/night) and relative humidity range 80%/50% in transparent microclimate chambers held inside a glasshouse. They were exposed either to CO_2 augmented from a concentrated pure source (Commonwealth Industrial Gases; 900 +/- 50 ppm), or to ambient CO_2 (350 ppm), using circulation and monitoring systems similar to those described by Conroy *et al.* (1986). The CO_2 concentration 900 ppm was chosen according to the results of preliminary experiments in which [CO_2] was varied; 900 ppm was the lowest that gave obvious changes in epicuticular wax deposition. The observations presented are for plants grown in 1994 and 1995, at times that avoided the shortest winter days or the longest summer days.

Light Microscopy and Graphic Image Analysis

Sampling and specimen preparation procedures were based on methods described by Carolin *et al.* (1973, 1975, 1977). A 3 mm-wide piece was cut from the mid-point of two or three fully expanded final leaves from main tillers (sampled at 9.30 a.m.) and fixed in 3% (v/v) glutaraldehyde in 0.025M phosphate buffer, pH 6.8. These pieces were then dehydrated in a graded ethanol or acetone series and embedded in Spurr's embedding medium (Spurr 1969). Sections (1 μm) were cut using a glass knife and microtome. They were stained in 0.6% toluidine blue and photographed with a Zeiss Orthomat light microscope. Magnifications were calibrated using a stage micrometer. Leaf blade thickness was determined as the mean of measurements made at three locations along the section.

Image analysis was performed essentially according to Wilson (1976) but with the advantage of using a Noran TN/8502 Dedicated Image Analyzer and Macintosh computer. The area of each type of tissue was obtained from a portion of the transverse section including three major vascular bundles (veins) on each side of the midrib, and expressed as a proportion of the total cross-sectional area. The areas of individual bulliform cells or mesophyll cells were determined from sets of five cells selected within a field of view, repeated five times.

Transmission Electron Microscopy (TEM)

The last fully expanded leaf of the main tiller was harvested soon after the onset of the light period and the embedding procedure described above was carried out, including a post-fixation in 2% osmium tetroxide solution for 2 h before the dehydration series. Several variations in the fixation times at several steps were made in order to optimise resolution. Fixation was particularly difficult when the starch content had increased as a consequence of growth with elevated [CO_2]. Ultrathin sections were stained sequentially in 2% aqueous uranyl acetate for 15 min, and with lead citrate for 15 min. The ratios of appressed (grana-forming) and unappressed thylakoids for mesophyll chloroplasts were obtained by drawing ten transect lines across five representative chloroplasts in TEM images of five leaves of each species and each treatment, and scoring along each line (Tipping 1996).

Leaf Starch Content

Starch analysis was carried out for the uppermost fully expanded leaves from the main tiller for three separate trials, each subjected to four replicate assays. The leaf samples were taken soon after the onset of the light period, dried at 70°C for 2 days, then milled. Sub-samples of 50 mg dried milled material were assayed enzymatically employing a Megazyme reagent system (assay format 2; α-amylase/pullulanase/β-amylase).

Specific Leaf Area

Ten plants per treatment or control group were harvested after 9 weeks growth. Total leaf area (including sheath, which is a minor proportion of the whole leaf in these species) was measured using a leaf area meter (Delta-T Devices). The dry matter content was determined after incubation at 80°C for 24 h. Mean specific leaf area (cm^2 g^{-1}) was calculated by dividing total leaf area by total dry matter content.

Observations on Stomata

Replicas of the lower (abaxial) surface near the midpoint of a leafblade were prepared using cellulose acetate mixed with 0.1% aniline blue (North 1956). The replicas were mounted on microscope slides with double-sided tape and counts of stomatal frequency were calibrated with a grid reticule in the eyepiece.

Table 1. Changes in tissue proportions within the transsectional area of fully expanded upper leaves of *Panicum decipiens* in response to growth with 900 ppm CO_2.[A]

Mean values within a row followed by an asterisk are significantly different (p<0.05).

	At 350 ppm	At 900 ppm	Ratio 900/350
Epidermis	0.19	0.19	1.00
Mesophyll	0.42	0.41	0.98
Sclerenchyma	0.056*	0.072*	1.34
Vascular elements	0.085*	0.074*	0.87
Bundle sheath cells	0.213	0.208	0.98

[A]The proportions are relative to a total value of 1.00 and were obtained as described by computer-assisted graphic image analysis. The mean thicknesses of leaves were 0.205 mm and 0.168 mm respectively and this difference was significant (p<0.05).

Table 2. Changes in tissue proportions within the transsectional area of fully expanded upper leaves of *Panicum antidotale* in response to growth with 900 ppm CO_2.[A]

Mean values within a row followed by an asterisk are significantly different (p<0.05).

	At 350 ppm	At 900 ppm	Ratio 900/350
Epidermis	0.152	0.158	1.04
Mesophyll	0.63*	0.43*	0.68
Sclerenchyma	0.047*	0.063*	1.34
Vascular elements	0.067*	0.114*	1.70
Bundle sheath cells	0.094*	0.225*	2.39

[A]The proportions are relative to a total value of 1.00 and were obtained as described by computer-assisted graphic image analysis. The mean thicknesses of leaves were 0.26 mm and 0.21 mm respectively and this difference was significant (p<0.05).

Means were calculated from three leaves per treatment, and counts from 20 fields per leaf were added. From the same slides, measurements were made of the mean pore length of the stoma and the overall maximum width of the stomatal complex, which includes a pair of subsidiary cells (see Fig. 7.6 of Esau 1960).

SEM Procedures

Either a JEOL 35 C or Philips 503 scanning electron microscope (SEM) was employed. Freshly cut leaf pieces about 3 mm square were attached with carbon paint (dag) to custom-designed brass stubs (Romeo 1996) and plunged into liquid nitrogen for 25 s. The uncoated stub was immediately placed in the SEM. The frozen specimens were examined at 7.5 kV following sublimation of the surface ice (which took about 20 min) and before shrinkage could commence. This corresponds to a temperature of about 65°C (Anton *et al.* 1994). Several techniques were compared in order to confirm that the procedure adopted gave least disruption to the observed patterns of epicuticular wax deposition. These included: air-drying leaf samples plus gold or platinum coating; freeze-drying with gold or platinum coating; critical point drying, and the preparation of resin replicas, followed by gold or platinum coating (Tipping 1996).

Statistical Treatment

For graphical presentation, two-way analysis of variance was carried out using the JMP procedures of SAS (1994). Tests for signif-

Table 3. Changes in tissue proportions within the transsectional area of fully expanded upper leaves of *Panicum tricanthum* in response to growth with 900 ppm CO_2.[A]

Mean values within a row followed by an asterisk are significantly different (p<0.05).

	At 350 ppm	At 900 ppm	Ratio 900/350
Epidermis	0.20	0.20	1.00
Mesophyll	0.51	0.53	1.04
Sclerenchyma	0.042*	0.054*	1.29
Vascular elements	0.056	0.057	1.01
Bundle sheath cells	0.169*	0.145*	0.86

[A]The proportions are relative to a total value of 1.00 and were obtained as described by computer-assisted graphic image analysis. The mean thicknesses of leaves were 0.185 mm and 0.2051 mm respectively and this difference was significant (p<0.05).

Table 4. Effects of growth with 900 ppm CO_2 on stomatal frequency (number mm^{-2}) for the last fully expanded leaf on the main tiller of three species of *Panicum*.

Mean values within a row followed by an asterisk are significantly different (p<0.02).

	At 350 ppm	At 900 ppm	Ratio 900/350
Panicum antidotale	121.8*	156.2*	1.28
Panicum decipiens	124.0*	166.6*	1.34
Panicum tricanthum	216.0*	169.0*	0.78

icance between mean values were made using the CONTRAST statement in the JMP procedure. The standard error bars in Fig. 1 indicate pooled standard errors of the means. For tabular data, a t-test of significance was performed. A p-value of less than or equal to 0.05 was taken to indicate a significant difference except for the data of Table 4, where the p-value was less than or equal to 0.02.

RESULTS

Effects of [CO_2] on Leaf Anatomy

The C_3/C_4 species in this comparison, *P. decipiens*, resembles the C_4 species in having a well defined bundle sheath and closely packed mesophyll cells (Akin *et al.* 1983). However, the positioning of chloroplasts in the bundle sheath cells of *P. decipiens* is centripetal, whereas their arrangement in the bundle sheath cells of *P. antidotale* is centrifugal, as expected for NADP-malic enzyme C_4 species (Prendergast *et al.* 1987).

Under the influence of elevated [CO_2], there were no qualitative changes in the leaf anatomy of any of these three species. However, there were detectable differences in leaf thickness: an increase of about 10% for the C_3 species, but decreases for both the C_4 and C_3/C_4 intermediate species (Tables 1-3). There were reductions in mean specific leaf area (expressed as cm^2 g^{-1} dry matter) of 12% for the C_3 species, 22% for the C_4 species, and 34% for *P. decipiens*. Graphic image analysis also indicated some significant changes in the proportions of trans-sectional area occupied by different tissues. The sclerenchyma proportion increased by about 30% in all three species (Tables 1-3). The vascular elements increased their proportion markedly only in the C_4 species (Table 2). *Panicum antidotale* also displayed a corresponding increase in bundle sheath cell proportion, and a decrease in mesophyll (Table 2). This was accomplished by an

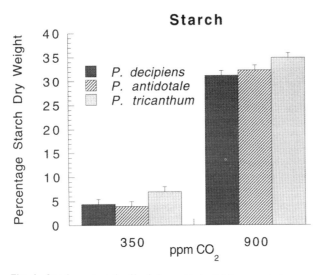

Fig. 1. Starch content (as % of dry matter) of fully expanded upper leaves of *Panicum* species growing with ambient or elevated [CO_2].

increased frequency of veins rather than substantial changes in cell size. By comparison, the mesophyll proportion did not change significantly in either *P. decipiens* or *P. tricanthum*, and the bundle sheath cell proportion either declined in *P. tricanthum* (Table 3) or remained essentially the same in *P. decipiens* (Table 1).

As judged by trans-sectional area, the mean size of individual mesophyll cells increased slightly in response to growth with 900 ppm CO_2, but these differences were statistically significant only in the case of *P. decipiens*, where an increase from 63 μm^2 to 90 μm^2 was measured. Bulliform cells within the adaxial epidermis were also affected in *P. decipiens* when plants were grown with elevated [CO_2]. There were fewer bulliform cells, and their individual trans-sectional areas approximately doubled. No such change occurred for the C_3 species, *P. tricanthum*. This comparison was restricted, since *P. antidotale* lacks bulliform cells. Nevertheless, the proportion of total trans-sectional area occupied by the epidermises was unaltered in *P. decipiens* (Table 1).

Effects of [CO_2] on Leaf Starch Content and Chloroplast Structure

Under elevated [CO_2], values of starch content for mature leaves of all three species were greater than 30% (Fig. 1), compared to values ranging between 4% and 7% with ambient CO_2. Under ambient conditions, bundle sheath chloroplasts of C_4 species generally store starch preferentially compared to mesophyll chloroplasts (Gallaher and Brown 1977; Walbot 1977), although circumstances are known where starch is also present in the mesophyll chloroplasts (Carolin *et al.* 1973; Forde *et al.* 1975).

For *P. decipiens*, starch was already evident in bundle sheath and mesophyll chloroplasts under ambient conditions, and there was enhanced accumulation of starch in chloroplasts in both tissues under elevated [CO_2]. In *P. antidotale*, chloroplasts in the bundle sheath accumulated starch under both ambient and elevated [CO_2]. Moreover, with 900 ppm CO_2, chloroplasts in the mesophyll accumulated abundant starch as well. The starch was clearly evident as a number of oval to lenticular granules per

plastid. In *P. tricanthum*, there was little obvious starch in chloroplasts under ambient conditions, but an accumulation of starch in chloroplasts occurred throughout the bundle sheath and mesophyll cells under elevated [CO_2].

The starch granules that accumulated during growth with elevated CO_2 appeared to distort the final thylakoid organization within each plastid. However, grana are clearly present in both kinds of chloroplast in all three species under ambient and elevated [CO_2], and there were no differences in the ratios of appressed to non-appressed thylakoids observed for any species (data not shown; Tipping 1996).

The peripheral reticulum observed for chloroplasts of some C_4 grasses (Carolin *et al.* 1973; Forde *et al.* 1975; Bruhl and Perry 1995) was not evident in the chloroplasts of any of these three species.

Stomatal Abundance

Following growth with elevated [CO_2], the stomatal frequency for the abaxial epidermis had decreased by about 22% in the case of the C_3 species, *P. tricanthum* (Table 4). In contrast, the stomatal frequency had increased by about 30% for both other species (Table 4). There were no changes in the dimensions of individual stomata or their subsidiary cells (Tipping 1996).

Effects of Elevated [CO_2] on Epicuticular Wax Deposition

Under ambient conditions, the leaves of *P. antidotale* showed a more pronounced wax bloom than the other two species. The abaxial epidermis (Fig, 2, A1) exhibited more wax than the adaxial epidermis (Fig. 2, B1). The wax appears as a closely textured dendritic filigree, sometimes suggestive of snowflakes. The term 'crystalline' has been applied to similar wax formations (Percy and Baker 1990). In response to growth with 900 ppm CO_2, the texture of the wax on the abaxial surface appeared to be more open, and coarser (Fig. 2, A2). The amount of wax on the adaxial epidermis increased considerably (Fig. 2, B2 vs. B1). The individual epidermal cells are more or less rectangular, and their positions are always clear beneath the wax coating. The subsidiary cells of stomata are covered to about the same extent as the other epidermal cells.

For the leaf of *P. decipiens* grown with ambient CO_2, the abaxial epidermis exhibits amorphous wax, concentrated in ridges along anticlinal walls between files of epidermal cells. The adaxial surface displays a more uniform coating of compact, dendritic wax with a few small plates. The positions of bulliform cells were pronounced. In response to growth with 900 ppm CO_2, the amorphous wax on the abaxial surface remained concentrated within the depressions between files of epidermal cells. The aggregation of dendritic wax on the adaxial epidermis became much finer in texture. The amounts of wax deposited on both abaxial and adaxial surfaces appeared to have increased.

For the leaves of *P. tricanthum* grown with ambient CO_2, the wax on the abaxial epidermis was both sparse and amorphous. In contrast, the adaxial epidermis showed a uniform coverage of wax with a dendritic texture. In response to growth with 900 ppm CO_2, the wax on the abaxial surface increased in amount, and altered in texture to become dendritic, although the coverage was still not uniform. The wax on the adaxial surface became slightly more compact and finer in texture towards the outer limits of individual filaments. A summary of these SEM observations is given in Table 5.

DISCUSSION

Changes in Leaf Anatomy

No qualitative or quantitative anatomical changes were observed following growth with elevated [CO_2] for the related C_4 species *Zea mays* (Thomas and Harvey 1983), nor for *P. laxum* (C_3) or *P. milioides* (C_3/C_4) in the study by Byrd and Brown (1989). Nevertheless, the C_4 species *P. antidotale* has responded to elevated [CO_2] with reciprocal changes in mesophyll versus bundle sheath and vascular tissues (Table 2). This response may have come about because *P. antidotale* belongs to the NADP-malic enzyme sub-group, and so the single bundle-sheath layer differentiates from the procambium, whereas the background mesophyll differentiates from the ground meristem (Dengler *et al.* 1985, 1986). In contrast, the mesophyll and the photosynthetic bundle sheath cells all differentiate from ground meristem in C_4 species of *Panicum* belonging to the NAD-malic enzyme or PEP carboxykinase sub-groups.

Starch and Photosynthetic Capacity

In leaf blades, starch accumulates only inside the chloroplast (Walker 1974). The acquisition of very high starch contents in the leaves of these three species (Fig. 1) has involved granule formation in both mesophyll chloroplasts and bundle sheath chloroplasts.

The starch content of chloroplasts normally fluctuates diurnally, in keeping with the mobilization of photosynthetic products to benefit non-photosynthetic parts of the plant. When [CO_2] is raised, both minimum and maximum starch contents increase (e.g. tomato, Yelle *et al.* 1989). It has been suggested in several studies that starch may lower photosynthetic capacity by inhibiting CO_2 transport or disrupting chloroplast internal organization (Wulff and Strain 1982). However, other observations imply that the presence of starch in chloroplasts has little detrimental impact on photosynthetic rates.

There is evidence that the extent to which starch can accumulate in chloroplasts is under genetic control, and varies within a genus. For example, the average contents of starch are higher in leaves of *Lycopersicon esculentum* than in the corresponding leaves of *L. chmielewski*, yet the former displays higher photosynthetic capacity than the latter (Yelle *et al.* 1989). Forde *et al.* (1975) studied the C_4 grass *Paspalum dilatatum*, which can accumulate up to 20% of leaf dry matter as starch under ambient [CO_2], and found that "the photosynthetic rate of cabinet-grown plants remained constant through the day as starch accumulated". They concluded that "the absence of a significant decline in photosynthetic rates during the day indicates that the starch accumulating in the leaf did not have a negative feedback effect on assimilation under these conditions".

The growth responses to elevated [CO_2] demonstrated by these species (Tipping 1996; Tipping and Murray 1998) indicate

higher photosynthetic capacity under conditions that permitted starch to accumulate. This suggests that disruption of chloroplast structure apparently caused by starch granules probably does not reflect the situation *in vivo*, and this impression could result mainly from inherent difficulties in obtaining uniform fixation, or in sectioning heterogeneous tissues.

Stomatal Abundance

The measurement and interpretation of changes in stomatal abundance have been controversial (Woodward 1987; Körner 1988; Malone *et al.* 1993; Murray 1995, 1997). The differentiation of stomata in the leaf surface commences "shortly before the main period of meristematic activity in the epidermis is completed" (Esau 1960) and may continue well into the period of general cell expansion. In monocotyledons with parallel venation, the formation of stomata in longitudinal rows proceeds from the apex downwards (Esau 1960). There are clearly opportunities for elevated [CO_2] to influence this process.

A decrease in abaxial stomatal frequency in response to elevated [CO_2] is the usual result reported for amphistomatous leaves of dicotyledons (Beerling and Kelly 1997) or for cladodes (North *et al.* 1995). For some species there may be no effect (Malone *et al.* 1993). There are fewer reports for monocotyledons, but usually there is no change in stomatal frequency, as in leaves of *Zea mays* (Thomas and Harvey 1983) and the flag leaf of *Avena sativa* (Malone *et al.* 1993).

In a comparison of leaves studied by E. J. Salisbury in 1927 with newly collected samples of what are intended to be the same species, Beerling and Kelly (1997) inferred that substantial decreases in stomatal frequency (of the order of 40%) had occurred for the abaxial leaf surfaces of the monocotyledons *Luzula sylvatica* and *Arum maculatum*. However, an increase of the order of 40% was noted for the abaxial leaf surface of blue-bell (*Hyacinthoides non-scripta*). These three species are not closely related. The present study is thus the first to indicate that changes in stomatal abundance in response to elevated [CO_2] can occur in opposite directions in species belonging to the same genus. This will remain the case even if *P. decipiens* is reclassified as *Steinchisma decipiens*.

Whether such alterations in stomatal frequency have any major impact on water-use efficiency is not known. Water-use efficiency should be improved with 900 ppm CO_2 regardless of stomatal frequency because of the narrowing effect of elevated [CO_2] on stomatal apertures (for review see Chapter 6 of Murray 1997). Whether a decrease in stomatal abundance (Table 4) is consistently correlated with C_3, or an increase with C_4 or C_3/C_4 intermediate status, is a matter for further observation. So far as *P. antidotale* is concerned, the extent of the increase in abaxial stomatal abundance conforms with the increase in vascular and bundle sheath tissues in their proportion of cross-sectional area (Table 2), since longitudinal files of cells that include the stomata tend to lie over the veins.

Epicuticular Wax Deposition and [CO_2]

The outermost cell layers of leaves and primary stems are coated with waxes, which help to prevent desiccation and also reflect some incident electromagnetic radiation (Martin and Juniper 1970; Araus *et al.* 1991; Febrero and Araus 1994). The waxes are

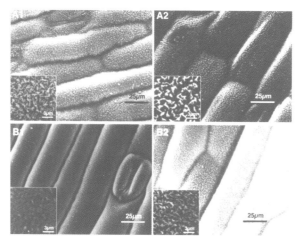

Fig. 2. SEMs of leaf surfaces of *P. antidotale* plants grown with 350 ppm CO_2 (A1,B1) or 900 ppm CO_2 (A2,B2). Abundant dendritic (crystalline) wax is normally present on the abaxial surface (A1) and tends to become coarser with 900 ppm CO_2 (A2). The adaxial epidermis shows little wax normally (B1) but this intensifies in response to growth with 900 ppm CO_2 (B2). Scales are as shown within each SEM.

Table 5. Epicuticular wax distribution on fully expanded upper leaves of three species of *Panicum*, and summary of changes in response to growth under 900 ppm CO_2 as observed by cold stage cryo-SEM procedures.

Species	Surface	At 350 ppm CO_2	At 900 ppm CO_2
P. antidotale	Abaxial	dendritic aggregate	coarser texture
	Adaxial	dendritic aggregate	increased amount
P. decipiens	Abaxial	amorphous	increase (in ridges)
	Adaxial	dendritic aggregate	more, finer texture
P. tricanthum	Abaxial	sparse, amorphous	increase, dendritic
	Adaxial	dendritic aggregate	more, finer texture

- alteration of the nature of light penetrating the epidermal cell layer from focussed to diffuse (Vogelmann *et al.* 1996);

- a reduction to some extent in photosynthetic photon flux density (PPFD) in the bundle sheath and mesophyll cells;

- a reduction in UV-B radiation penetrating guard cell chloroplasts and underlying chloroplasts, according to wax composition (content of phenolics or other screening compounds).

synthesized in underlying epidermal cells, and somehow released beyond the secondary and primary layers of the cuticle. Although wax eleboration is constitutive in some species, there is strong evidence that in other species environmental parameters such as temperature, relative humidity, soil moisture, light intensity, wind, and exposure to acid pH can influence wax micro-structure and quantity (Percy and Baker 1990).

An increase in epicuticular wax deposition due to elevated [CO_2] has seldom been recorded since its discovery by Thomas and Harvey (1983). This effect of CO_2 is strongly dependent on species as well as on concentration of CO_2. For soybean (*Glycine max*), only the abaxial leaf epidermis showed this response, and the texture of the wax remained dendritic (Fig. 3 of Thomas and Harvey 1983). The response did not occur at elevated concentrations of [CO_2] below 910 ppm (i.e. between 340 and 718 ppm). For *Zea mays*, the response did not occur even at 910 ppm (Thomas and Harvey 1983).

The three species in the present study show different characteristic patterns of epicuticular wax deposition when grown with ambient CO_2, and different kinds of response to elevated [CO_2] (Fig. 2, Table 5). A qualitative change in the texture or shape of the extruded wax was observed in several instances (Table 5). An increased deposition on the adaxial surface relative to the abaxial is similar to the normal situation for some other monocotyledonous species such as barley (*Hordeum vulgare*) at ambient CO_2 (Febrero and Araus 1994). On balance, the wax patterns and responses of the C_3/C_4 intermediate species, *P. decipiens*, resemble the C_4 species more than the C_3.

There are several adaptive advantages of increased epicuticular wax deposition on leaves, cladodes or phyllodes. They include:

- water conservation through a reduction in non-stomatal transpiration;

- enlargement of the boundary layer near the stomata;

Since enhanced growth responses were observed for these three species of *Panicum* (Tipping 1996; Tipping and Murray 1998) it is clear that the available PPFD has not been seriously impaired by any increased production of waxes or alteration in their texture. In the future, under external conditions with increased UV-B as well as elevated [CO_2], the beneficial effects of an increased wax coating may prove to be an extremely important adaptive feature, enhancing the potential value of these perennial species as forage crops.

Status of C_3/C_4 Intermediates

There is no doubt that C_4 photosynthesis is polyphyletic (Laetsch 1974). This 'syndrome' has arisen an unknown number of times in Poaceae, and at least four times in the Cyperaceae (Carolin *et al.* 1977; Bruhl and Perry 1995; Soros and Bruhl 1998). How-

Table 6. A summary of the ways in which the responses to elevated [CO_2] of the C_3/C_4 intermediate species, *P. decipiens*, resemble either the C_4 species (*P. antidotale*) or the C_3 (*P. tricanthum*).

Parameter	Resembles: C_4	C_3
Total (above-ground) dry matter[A]	+	
Stem dry matter[A]		+
Stem height[A]	+	
Internodes on main tiller[A]		+
Tiller frequencies[A]		+
Leaf specific area	+	
Leaf thickness	+	
Mesophyll proportion		+
Bundle sheath proportion		+
Stomatal frequency	+	
Leaf epicuticular wax deposition	+	

[A]Tipping 1996; Tipping and Murray 1998.

ever, the status of C_3/C_4 intermediates is unclear. Among monocotyledons they are known in *Panicum*, and also in *Eleocharis* (*E. pusilla*; Bruhl and Perry 1995). Most studies of C_3/C_4 intermediate plants have focussed on dicotyledons, especially species of *Atriplex* [Chenopodiaceae] (Björkman *et al.* 1971) and *Flaveria* [Asteraceae] (Monson *et al.* 1984; Ku *et al.* 1991).

The view that such species represent evolutionary intermediates is often expressed, however Monson *et al.* (1984) long ago questioned "whether they represent species in the process of evolution from the C_3 to the C_4 pathway, a stabilized photosynthetic alternative that will not evolve to the level of the C_4 pathway, or stabilized hybrids between closely related C_3 and C_4 species." The "evolutionary intermediate" presumption has again been questioned, by Kopriwa *et al.* (1996), and in view of the difficulty of recovering certain genotypes in hybridization experiments with *Atriplex* and *Panicum* (Bouton *et al.* 1986), the "stabilized photosynthetic alternative" hypothesis from among these suggestions appears most plausible. Extending this suggestion, we propose that in *Panicum* at least, such an intermediate represents a developmental chimera, since in so many respects the C_3/C_4 species resembles *either* the C_4 *or* the C_3 in its response (Table 6).

ACKNOWLEDGEMENTS

This study was supported by German Government Academic Exchange and Australian Federal Government Scholarships to CT. The microscopy was carried out at the Electron Microscope Unit at the University of Sydney, NSW Australia. The authors thank Tony Romeo and Dennis Dwarte for their valuable assistance; Dr John Wilson for providing *Panicum* seeds and advice; Prof. E. W. R. Barlow for the provision of growth and CO_2 monitoring facilities at the Hawkesbury Campus of the University of Western Sydney, and Dr Maret Vesk for advice during the course of the study, and helpful comments on a draft manuscript. We also thank the reviewers for their suggestions, and acknowledge helpful discussion with Dr Mirta Arriaga and Dr Jeremy Bruhl in September 1998.

REFERENCES

Akin, D. E., Wilson, J. R., and Windham, W. R. (1983). Site and rate of tissue digestion in leaves of C_3, C_4 and C_3/C_4 intermediate *Panicum* species. *Crop Science* **23**, 147-155.

Anton, L. H., Ewers, F. W., Hammerschmidt, R., and Klomparens, K. L. (1994). Mechanisms of deposition of epicuticular wax in leaves of broccoli, *Brassica oleracea* L. var. *capitata* L. *New Phytologist* **126**, 505-510.

Araus, J. L., Febrero, A., and Vendrell, P. (1991). Epidermal conductance in different parts of durum wheat grown under Mediterranean conditions: the role of epicuticular waxes and stomata. *Plant, Cell and Environment* **14**, 545-558.

Beerling, D. J., and Kelly, C. K. (1997). Stomatal density responses of temperate woodland plants over the past seven decades of CO_2 increase: a comparison of Salisbury (1927) with contemporary data. *American Journal of Botany* **84**, 1572-1583.

Björkman, O., Nobs, M., Pearcy, R., Boynton, J., and Berry, J. (1971). Characteristics of hybrids between C_3 and C_4 species of *Atriplex*. In 'Photosynthesis and Photorespiration'. (Eds M. D. Hatch, C. B. Osmond and R. O. Slatyer) pp. 105-119. (Wiley Interscience, Sydney, Australia.)

Bouton, J. H., Brown, R. H., Evans, P. T., and Jernstedt, J. A. (1986). Photosynthesis, leaf anatomy, and morphology from progeny from hybrids between C_3 and C_3/C_4 *Panicum* species. *Plant Physiology* **80**, 487-492.

Breakwell, E. (1918). Popular descriptions of grasses - the panic grasses. *Agricultural Gazette of New South Wales* **29**, 836-847.

Brown, R. H., and Brown, W. V. (1975). Photosynthetic characteristics of *Panicum milioides*, a species with reduced photorespiration. *Crop Science* **15**, 681-685.

Bruhl, J. J., and Perry, S. (1995). Photosynthetic pathway-related ultrastructure of C_3, C_4, and C_3-like C_3/C_4 intermediate sedges (Cyperaceae) with special reference to *Eleocharis*. *Australian Journal of Plant Physiology* **22**, 521-530.

Byrd, G. T., and Brown, R. H. (1989). Environmental effects on photorespiration of C_3/C_4 species. I. Influence of CO_2 and O_2 during growth on photorespiratory characteristics and leaf anatomy. *Plant Physiology* **90**, 1022-1028.

Carolin, R. C., Jacobs, S. W. L., and Vesk, M. (1973). The structure of the cells of the mesophyll and parenchymatous bundle sheath of the Gramineae. *Botanical Journal of the Linnean Society* **66**, 259-275.

Carolin, R. C., Jacobs, S. W. L., and Vesk, M. (1975). Leaf structure in Chenopodiaceae. *Botanische Jahrbuch Systematische Pflanzengesichte und Pflanzengeographie* **95**, 226-255.

Carolin, R. C., Jacobs, S. W. L., and Vesk, M. (1977). The ultrastructure of Kranz cells in the Family Cyperaceae. *Botanical Gazette* **138**, 413-419.

Conroy, J. P., Smillie, R. M., Kuppers, M., Bevege, D. I., and Barlow, E. W. R. (1986). Chlorophyll *a* fluorescence and photosynthetic and growth responses of *Pinus radiata* to phosphorus deficiency, drought stress, and high CO_2. *Plant Physiology* **81**, 423-430.

Dengler, N. G., Dengler, R. E., and Hattersley, P. W. (1985). Differing ontogenetic origins of PCR ("Kranz") sheaths in leaf blades of C_4 grasses (Poaceae). *American Journal of Botany* **72**, 284-302.

Dengler, N. G., Dengler, R. E., and Hattersley, P. W. (1986). Comparative bundle sheath and mesophyll differentiation in the leaves of the C_4 grasses *Panicum effusum* and *P. bulbosum*. *American Journal of Botany* **73**, 1431-1442.

Edwards, G. E., and Gutierrez, M. (1972). Metabolic activities in extracts of mesophyll and bundle sheath cells of *Panicum miliaceum* (L.) in relation to the C_4 dicarboxylic acid pathway of photosynthesis. *Plant Physiology* **50**, 728-732.

Esau, K. (1960). 'Anatomy of Seed Plants.' (John Wiley & Sons Inc., New York and London.)

Febrero, A., and Araus, J. L. (1994). Epicuticular wax load of near isogenic barley lines differing in glaucousness. *Scanning Microscopy* **8**, 735-748.

Forde, B. J., Whitehead, H. C. M., and Rowley, J. A. (1975). Effect of light intensity and temperature on photosynthetic rate, leaf starch content and ultrastructure of *Paspalum dilatatum*. *Australian Journal of Plant Physiology* **2**, 185-195.

Gallaher, R. N., and Brown, R. H. (1977). Starch storage in C_4 vs C_3 grass leaf cells as related to nitrogen deficiency. *Crop Science* **17**, 85-88.

Hatch, M. D., Kagawa, T., and Craig, S. (1975). Subdivision of C_4-pathway species based on differing C_4 acid decarboxylating systems and ultrastructural features. *Australian Journal of Plant Physiology* **2**, 111-128.

Kopriwa, S., Chu, C.-C., and Bauwe, H. (1996). Molecular phylogeny of *Flaveria* as deduced from the analysis of nucleotide sequences encoding the H-protein of the glycine cleavage system. *Plant, Cell and Environment* **19**, 1028-1036.

Körner, C. (1988). Does global increase of CO_2 alter stomatal density? *Flora* **181**, 253-257.

Ku, M. S., Wu, J., Dai, Z., Scott, R., Chu, C., and Edwards, G. E. (1991). Photosynthetic and photorespiratory characteristics of *Flaveria* species. *Plant Physiology* **96**, 518-528.

Laetsch, W. M. (1974). The C_4 syndrome: a structural analysis. *Annual Review of Plant Physiology* **25**, 27-52.

Malone, S. R., Mayeux, H. S., Johnson, H. B., and Polley, H. W. (1993). Stomatal density and aperture length in four plant species grown across a subambient CO_2 gradient. *American Journal of Botany* **80**, 1413-1418.

Martin, J. T., and Juniper, B. E. (1970). 'The Cuticles of Plants.' (Edward Arnold Lyd, London U.K.).

Monson, R. K., Edwards, G. E., and Ku, M. S. B. (1984). C_3-C_4 intermediate photosynthesis in plants. *Bioscience* **34**, 563-574.

Murray, D. R. (1995). Plant responses to carbon dioxide. *American Journal of Botany* **82**, 690-697.

Murray, D. R. (1997). 'Carbon Dioxide and Plant Responses.' (Research Studies Press/John Wiley, U.K.)

North, C. (1956). A technique for measuring structural features of plant epidermis using cellulose acetate films. *Nature* **178**, 1186-1187.

North, G. B., Moore, T. L., and Nobel, P. S. (1995). Cladode development for *Opuntia ficus-indica* under current and doubled CO_2 concentrations. *American Journal of Botany* **82**, 159-166.

Percy, K. E., and Baker, E. A. (1990). Effects of simulated acid rain on epicuticular wax production, morphology, chemical composition and on cuticular membrane thickness in two clones of Sitka spruce [*Picea sitchensis* (Bong.) Carr.]. *New Phytologist* **116**, 79-87.

Prendergast, H. D. V., Hattersley, P. W., and Stone, N. E. (1987). New structural/biochemical associations in leaf blades of C_4 grasses (Poaceae). *Australian Journal of Plant Physiology* **14**, 403-420.

Radoglou, K. M., and Jarvis, P. G. (1990). Effects of CO_2 enrichment on four poplar clones. I. Growth and leaf anatomy. *Annals of Botany* **65**, 617-626.

Rogers, H. H., and Dahlman, R. C. (1993). Crop responses to CO_2 enrichment. *Vegetatio* **104/105**, 117-131.

Romeo, T. (1996). Simple cold stage for the SEM. *International Conference on Electron Microscopy, Proceedings*, Abstract 14. (University of Sydney, Sydney.)

SAS. (1994). 'SAS/STAT user's guide.' (SAS Institute, Cary, NC.)

Soros, C. L., and Bruhl, J. J. (1998). Multiple evolutionary origins of C_4 photosynthesis in the Cyperaceae. *Second International Conference on the Comparative Biology of the Monocotyledons and Third International Symposium on Grass Systematics and Evolution*, Abstracts, p.52. (University of NSW, Sydney, Australia.)

Spurr, A. R. (1969). A low viscosity epoxy resis embedding medium for electron microscopy. *Journal of Ultrastructure Research* **26**, 31-43.

Thomas, J. F., and Harvey, C. N. (1983). Leaf anatomy of four species grown under continuous CO_2 enrichment. *Botanical Gazette* **144**, 303-309.

Tipping, C. (1996). 'Morphological and Structural Investigations into C_3, C_4 and C_3/C_4 Members of the Genus *Panicum* Grown at Elevated CO_2 Concentrations.' Ph.D. Thesis. (University of Western Sydney, Richmond, NSW, Australia.)

Tipping, C., and Murray, D. R. (1997). Effects of elevated atmospheric [CO_2] on leaf structure. In 'Carbon Dioxide and Plant Responses.' (Ed. D. R. Murray.) pp. 71-98. (Research Studies Press/John Wiley, U.K.)

Tipping, C., and Murray, D. R. (1998). Growth responses to elevated atmospheric [CO_2] in C_4 and C_3/C_4 Members of *Panicum* and other Monocotyledons. *Second International Conference on the Comparative Biology of the Monocotyledons and Third International Symposium on Grass Systematics and Evolution*, Abstracts, p.88. (University of NSW, Sydney, Australia.)

Vogelmann, T. C., Borneman, J. F., and Yates, D. J. (1996). Focussing of light by leaf epidermal cells. *Physiologia Plantarum* **98**, 43-56.

Walbot, V. (1977). Use of silica sol step gradients to prepare bundle sheath and mesophyll chloroplasts from *Panicum maximum*. *Plant Physiology* **60**, 102-108.

Walker, D. A. (1974). Chloroplast and cell - the movement of key substances, etc. across the chloroplast envelope. In 'Plant Biochemistry', Biochemistry Series One, Volume 11. (Ed. D. H. Northcote.) pp. 1-49. (Butterworths, London, and University Park Press, Baltimore.)

Wilson, J. R. (1976). Variation of leaf characteristics with level of insertion on a grass tiller. II Anatomy. *Australian Journal of Agricultural Research* **27**, 355-364.

Wilson, J. R. (1991). Plant structures: their digestive and physical breakdown. In 'Recent Advances on the Nutrition of Herbivores.' (Eds Y. W. Ho, H. K. Wong, N. Abdullah, and Z. A. Tajuddin.) pp. 207-216. (MSAP, Kuala Lumpur, Malaysia.)

Wilson, J. R., Brown, R. H., and Windham, W. R. (1983). Influence of C_4, C_3, and intermediate types of anatomy on the dry matter digestibility of *Panicum* grass species. *Crop Science* **23**, 141-146.

Woodward, F. I. (1987). Stomatal numbers are sensitive to increases in CO_2 from pre-industrial levels. *Nature* **327**, 617-618.

Wulff, R. D., and Strain, B. R. (1982). Effects of CO_2 enrichment on growth and photosynthesis in *Desmodium paniculatum*. *Canadian Journal of Botany* **60**, 1084-1091.

Yelle, S., Beeson, R. C., Trudel, M. J., and Grosselin, A. (1989). Acclimation of two tomato species to high atmospheric carbon dioxide. I. Sugar and starch concentrations. *Plant Physiology* **90**, 1465-1472.

Grasses: Systematics and Evolution. (2000). Eds S.W.L. Jacobs and J. Everett. (CSIRO: Melbourne)

ECOPHYSIOLOGICAL INVESTIGATIONS OF THE DISTRIBUTION OF POACEAE AND RESTIONACEAE IN THE CAPE FLORISTIC REGION, SOUTH AFRICA

T. L. Bell[A], W. D. Stock[B] and H. P. Linder[B]

[A]Department of Botany, University of Western Australia, Nedlands 6907, Western Australia, Australia.
[B]Department of Botany, University of Cape Town, Private Bag Rondebosch 7700, South Africa.

Abstract

The graminoid layer of fynbos vegetation is chiefly composed of members of the family Restionaceae rather than Poaceae, whereas the latter tend to predominate in all other South African biomes including renosterveld. To determine possible ecophysiological mechanisms for such distinct distributions, this study investigated the effects of nutrient addition, competition and soil texture on growth and nutrient uptake of glasshouse-grown plants of common fynbos species of Restionaceae and Poaceae. When grown in pots in monoculture, the majority of species did not display significant differences in above- and below-ground biomass accumulation when subjected to nutrient additions at levels matching those found in native fynbos and renosterveld habitats. However, when grown in paired species culture and supplied with nutrients at levels similar to those associated with agricultural plants a two-fold increase in root and shoot or culm biomass, and a ten-fold increase in leaf or culm photosynthetic area was observed in all species. At comparable nutrient levels, all species grew significantly better when grown in monoculture than when paired with another species. The restio, *Restio festucae-formis* Masters, appeared to be the better competitor when paired with another restio but not when associated with the grass *Pentaschistis papillosa* (Steud.) Linder. In contrast, the restioid *Ischyrolepis subverticillata* Steud. grew larger in the presence of a grass partner than with another restio. Soils of different textures promoted significant differences in biomass accumulation in only two species tested: *Restio festucaeformis* and *Pentaschistis papillosa*. It is concluded that no single variable tested here could be held responsible for explaining the distribution of grasses and restios in natural habitats; however, nutrient availability, competition and soil texture were all considered likely to play collective roles in biomass accumulation of species of both families.

Key words: Fynbos, renosterbos, Restionaceae, Poaceae, nitrogen, phosphorus, competition.

INTRODUCTION

The fynbos of the Cape Floristic Region has long been grouped physiognomically and physiologically with other mediterranean heathlands of the world, including those of the northern hemisphere, Californian chaparral, South American mattoral and perhaps the most similar analogy of fynbos — the kwongan of Western Australia (Specht 1979). All are characterised by marked dominance of evergreen sclerophyllous shrubs, oligotrophic soils and a cool, wet winter/hot, dry summer weather pattern. Fynbos has been classified in many ways (Bond and Goldblatt 1984; Moll and Jarman 1984; Cowling and Holmes 1992) but one of the simplest floristically based definitions is that of Taylor (1978). By this description, fynbos consists of three elements: (a) a restioid element that is almost always present and includes wiry, aphyllous hemicryptophytes of the Restionaceae and Cyperaceae, (b) an ericoid element that is often, but not always, present and (c) a proteoid element that is frequently, but also not always, present. In contrast, renosterveld (Boucher and Moll 1980, renosterbosveld, renoster shrubland), which generally lies adjacent to fynbos (Cowling and Holmes 1992), is readily characterised by the presence of *Elytropappus rhinocerotis,* a small grey-leafed shrub in the family Asteraceae with grasses dominating the graminoid layer between these bushes. This vegetation type is often associated with heavier shale-derived soils that have a higher nutrient content than fynbos soil (Low 1983; Stock and Allsopp 1992).

In a recent review of the South African Restionaceae it was noted that Restionaceae replace Poaceae as dominants in the graminoid layer of the Cape Floristic Region (Linder 1991a). No physiological explanation was given for this observation although certain edaphic and climatic patterns have been selected as reliable predictors of such distribution. For example, it has frequently been noted that distribution patterns of restios are closely associated with the occurrence of low nutrient, sandstone-derived soils, while grasses are more common in higher nutrient granitic- or shale-derived soils (Acocks 1975; Goldblatt 1978; Bond and Goldblatt 1984; Campbell 1985; Hoffmann *et al.* 1987; Campbell and Werger 1988). Similarly, rainfall has been suggested to be a good indicator of vegetation types; however, it is not considered a causative factor of distribution (Linder 1991b). In support of this, early studies in low nutrient environments of Australia suggest that limiting climatic influences often run parallel with those relating to nutrient-poor soils, but it is the latter which are considered to be the overriding influence on the type of vegetation which develops (Beadle 1954; Specht 1979).

Only a handful of ecophysiological studies are presently available to help unravel the complex distribution patterns of fynbos vegetation. A study by Trollope (1973) has shown that grasses eventually dominate where fynbos is frequently burnt, supposedly due to greater availability of nutrients. Similarly, Hoffmann *et al.* (1987) have found that grasses are common in low nutrient soils only in the first few years after fire and Witkowski and Mitchell (1989) have shown that cover and biomass of grasses increase dramatically where fertilisers have been added to nutrient-poor fynbos soils. In comparison to grasses, restios have the capacity to conserve nutrients by means of highly effective internal remobilisation from older culms and rhizomes (Low 1984; Stock *et al.* 1987; Meney and Dixon 1988; Meney *et al.* 1990a) and may therefore be argued to be particularly well adapted to low nutrient environments. Grasses do not generally show such frugal use of nutrients but are rather considered to be better competitors when nutrients are relatively abundant (Stock *et al.* 1987; Witkowski and Mitchell 1989).

This study was performed in two parts. The first part was aimed at characterising a number of soil properties associated with fynbos and renosterveld vegetation growing under the same climatic conditions. Secondly, a series of glasshouse experiments investigated the effects of nutrient addition, competition and soil texture on biomass production of common fynbos grasses and restios. Using an inert sand mix as basal growth medium three hypotheses were tested: (a) that grasses would respond to nutrient additions with increased growth relative to restios, (b) restios would be better competitors than grasses under low nutrient conditions, and (c) soil texture would affect root biomass production in restios but not grasses. The overall aim of this part of the study was to identify and uncover potential factors which might be contributing to the dominance of restios rather than grasses in the graminoid layer of native fynbos vegetation.

MATERIALS AND METHODS
Characterisation of Native Habitats
Two sites were selected on private farmland (Fairfield Farm) near Napier, approximately 250 km east of Cape Town. One of the sites supported typical fynbos vegetation on low nutrient, highly leached quartzitic sandstone-derived soil. The second site, located within 5 km of the fynbos site, consisted of typical renosterveld vegetation growing on relatively high nutrient, base-saturated, shale-derived soil.

Soils from both sites were sampled monthly from November 1996 to June 1997 for determination of seasonal change in percentage moisture, pH (using 2 M $CaCl_2$), conductivity, organic matter content (volatile C lost following combustion at 550 °C calculated as a percentage of initial soil weight), total N (Kjeldahl digestion for total N determination according to the method outlined by Stock and Lewis (1986)), and total P concentrations (triacid digestion for total P determination according to the methods of Grimshaw (1985) and Murphy and Riley (1962)). Each month soils were sampled from three separate locations at both sites and collected at 5 cm increments to a total depth of 20 cm. Rainfall and maximum and minimum temperature data for the same period were provided by the South African Weather Bureau.

Glasshouse Experiments
Species used in glasshouse experiments included three perennial fynbos grasses, *Pentaschistis papillosa* (Steud.) Linder, *P. malouinensis* (Steud.) Clayton, and *Tribolium brachystachyum* (Nees) Renvoize, and six fynbos restios, *Calopsis paniculatus* (Rottb.) Desv., *Chondropetalum tectorum* (L.f.) Rafin., *Ischyrolepis subverticillata* Steud., *Restio festucaeformis* Masters, *Rhodocoma gigantea* Kunth. Linder and *Thamnochortus cinereus* Linder. Details of the distribution and growth form of each species are given in Table 1.

Seed of the grasses collected in November 1993 (*P. papillosa*) and January 1995 (*P. malouinensis*, *T. brachystachyum*) were sown into seedling trays (3:1 sand:perlite mix) and their germination stimulated by application of smoke-water (A. Hitchcock, personal communication). Seed of restio species gathered from routine annual collections (1995–6) were germinated in a similar fashion but had previously been shown not to require smoke-water treatment (H. Jamison, personal communication). All seed was germinated in early June 1996 at Kirstenbosch Botanic Garden, Cape Town.

Six-week-old seedlings were transplanted into 3 L black plastic nursery plant bags containing a mix of well-watered, coarse (No. 4, Consol Glass) glass-making-grade silica sand and 'fynbos' potting mix (packed by Kirstenbosch Botanic Garden) in a 2:1 ratio. Such a homogeneous and inert growth medium was selected to prevent nutrient immobilisation likely to occur in native soils, for good drainage and for the ability to control particle size and organic matter content. Plants were established in 'monoculture' with four plants per species per bag and in 'paired-culture' with two plants each of the two selected species to allow for deaths following transplantation. Cultures were thinned after 12 weeks to half this density (i.e. a total of two plants per bag).

Equal numbers of pots representing monocultures of a single species (Experiment 1) received either 'medium' (4%) or 'low'

Table 1. Distribution and important characteristics of species of Poaceae and Restionaceae used in glasshouse experiments. All species are native to the Cape Floristic Region.

Species	Distribution and characteristics
Poaceae	
Pentaschistis papillosa	Short spreading perennial common on low altitude sandstone associated with mesic mountain fynbos vegetation [B,C], competition avoider [D], resprouter [C]
Pentaschistis malouinensis	Short tufted perennial, common in a range of vegetation types at various densities and over a wide altitudinal range [A,B,C], competition avoider [D], resprouter [C]
Tribolium brachystachyum	Prostrate to tufted perennial preferring sandy soils in mountainous and disturbed areas, common on soils of better nutrient status and higher organic matter content [B,C], fire response unknown[F]
Restionaceae	
Calopsis paniculatus	Rhizomatous perennial with culms to 2 m tall, widespread and common to riverside floras, particularly on soils with high organic matter content [B,E], resprouter [F]
Chondropetalum tectorum	Tussock-forming restio to 1 m tall, widespread and common in marshes and seeps such as ditches and vleis [B,E], seeder [F]
Ischyrolepis subverticillata	Caespitose perennial to 1.5 m tall, widespread and common along streams and riverbanks [B,E], shade-tolerant [E], likely to be a resprouter [F]
Restio festucaeformis	Short, caespitose perennial, restricted distribution in association with richer soils [B,E], seeder [F]
Rhodocoma gigantea	Large tufted perennial with culms up to 3 m tall, widespread distribution, found in high positions along river banks [B,E], seeder [F]
Thamnochortus cinereus	Medium-sized caespitose perennial, widespread and common on rocky outcrops [B,C], fire response unknown [F]

[A] Dahlgren and Clifford (1982), [B] Bond and Goldblatt (1984), [C] Gibbs Russell *et al.* (1990), [D] Linder and Ellis (1990), [E] Linder (1991a), [F] Linder (personal communication).

(1%) levels of the recommended strength for regular Hoagland's solution containing P and K as KH_2PO_4/K_2HPO_4, N as NH_4NO_3, and Mg^{+2}, Ca^{+2}, K^+, Fe-Na-EDTA, SO_4^{-2}, Cl^- and micronutrients, B, Mn, Zn, Cu, Mo and Co). Concentrations of available phosphorus and nitrogen (i.e. medium at 40 µM P and 20 µM N, low at 10 µM P and 5 µM N) were judged to be close to levels of nutrients found in renosterveld and fynbos soils respectively (see Table 2 and published information relating to nutrient status of soils, i.e., Low 1983; Mitchell *et al.* 1986; Witkowski and Mitchell 1987; Stock and Allsopp 1992). Plants grown in paired-culture combination (Experiment 2) were supplied with either 'high' (25%) or low (1%) strength Hoagland's solution. Nutrient levels of the high nutrient treatment were similar to those recommended for use with agricultural plants (i.e., 25% strength Hoagland's solution with 250 µM P and 125 µM N). Nutrient solutions were added fortnightly as 300 mL aliquots and pots given a thorough watering using deionised water a few days prior to nutrient application to flush out accumulated nutrients. Plants received deionised water between nutrient applications to maintain all cultures at or close to field capacity. Each treatment in Experiments 1 and 2 had ten replicates per species or species pair.

A further experiment (Experiment 3) investigating potential growth responses to substrate particle size involved replicate two paired-cultures of a grass and a restio (*P. papillosa* with *R. festucaeformis* and *P. malouinensis* with *I. subverticillata*) transplanted into bags of a well-watered mix of fine glass-making-grade silica sand (No. 1, Consol Glass) and fynbos potting mix (2:1). All plants received low (1%) strength nutrient solution as outlined above. This experiment had 20 replicates per species pair. All plants in all experiments were maintained at 20-25 °C under ambient sunlight and day length in glasshouse facilities of the

Botany Department, University of Cape Town between July 1996 and April 1997.

Plants were harvested after 10 months of growth with below-ground biomass (roots or roots and rhizomes) being carefully washed free of adhering sand and separated from above-ground biomass (shoots or culms). Measurements included determination of above- and below-ground dry mass increments (mean dry mass of above- or below-ground components of seedlings at the time of transplanting into pots subtracted from corresponding final dry mass values for each species), photosynthetic area of shoots or culms and shoot or culm:root dry weight ratios. Coarse roots (with secondary thickening) were separated from fine roots (with no secondary thickening) and lengths of each fraction were determined using subsamples of five individuals of each species in all treatments using the line intercept method of Giovanetti and Mosse (1980). Presence or absence of vesicular-arbuscular (VA) mycorrhizal infection and capillaroid roots was determined simultaneously using roots stained in lactic-glycerol blue solution (1.25:1:1.5 lactic acid:glycerol:water with 0.65 g trypan blue powder (1%)) for 24 hours. After examination, all root material was washed repeatedly in running water to remove residual storage solution (1:2:1 lactic acid:glycerol:water), dried and returned to the original root system for inclusion in total dry weight determinations.

RESULTS

Characterisation of Fynbos and Renosterveld Soils

Rainfall and temperature data for the field study period during 1996 and 1997 are presented in Fig. 1. Since both sites were less than 5 km apart and separated by only very low undulating land they were deemed to experience near identical climatic influences. Rainfall events were well reflected in soil moisture content

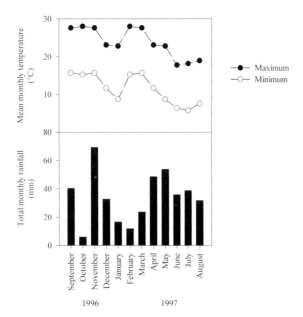

Fig. 1. Mean monthly maximum and minimum temperatures and total rainfall for Fairfield Farm near Napier, South Africa for the study period during 1996 and 1997.

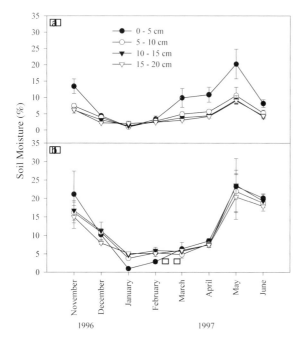

Fig. 2. Mean monthly percentages of soil moisture for four soil depths (0-5, 5-10, 10-15 and 15-20 cm) for (a) fynbos and (b) renosterveld sites at Fairfield Farm, near Napier, South Africa for the study period during 1996 and 1997. Data point and whisker represent mean ± S. E.

with lowest levels in both soil types during the driest months of January through to March (Fig. 2). A very wet November in 1996 contributed to increased soil moisture content in both soil types despite higher daytime temperatures. The renosterveld soil remained wetter for longer after the onset of more typical summer weather (December and January) than did the fynbos soil and never became quite as dry as fynbos soil, at least at depths

greater than 5 cm. The upper 5 cm layer of fynbos soil increased in moisture content at a faster rate than renosterveld soil once winter rains commenced (April) but lower soil layers never had more than 10% of weight as water. Renosterveld soil showed less variation in moisture content with depth than did fynbos soil, particularly during the wetter months of the year.

Soil properties such as pH, conductivity, organic matter content, and N and P content were analysed monthly but did not vary greatly throughout the study period (2-25%). Thus, mean values recorded are given in Table 2, with the first column pair relating to the 0-5 cm layer of soil and the second column pair to the lower 15-20 cm layer. The fynbos soil was always more acidic than the renosterveld soil and conductivity measurements were substantially different. Percentages of total organic matter in the fynbos soil were approximately half those of the renosterveld soil and amounts of total soil N and P were four to five times lower in fynbos soils. Based on the above it was concluded that the properties of the sand mixes used in glasshouse experiments (coarse and fine sand, column 3, Table 2) were much closer to soil of fynbos than of renosterveld soil. The sand mix was also rated low in total N and P prior to nutrient applications.

Experiment 1-Monoculture, Medium or Low Nutrients, Coarse Sand Mix

There was little significant difference in accumulation of above- and below-ground biomass for most species grown in monoculture at nutrient levels similar to those found in natural habitats of fynbos (i.e. 'low nutrients') or renosterveld (i.e. 'medium nutrients', Fig. 3a). Exceptions include a significant above-ground biomass increment (indicated by an asterisk, Student's t-test, $p < 0.05$) in the grass *P. papillosa* when grown under low nutrient conditions and significant below-ground biomass incremental differences in *C. paniculatus* and *R. festucaeformis* grown under low nutrient conditions and *P. malouinensis* grown under medium nutrient conditions. Overall, the grasses did not grow to a greater extent than the restios with relatively higher nutrient application, and this trend was also reflected by similarities in shoot or culm photosynthetic area and shoot or culm:root dry weight ratios of each of the species (Figs 3b and c). Exceptions to these general trends observed for photosynthetic area and dry weight ratios were found for two species of restios, *C. paniculatus* and *R. festucaeformis* (indicated by an asterisk, Student's t-test, $p < 0.05$). There were no significant differences between coarse and fine root lengths, and hence total root lengths, of all species grown in monoculture whether at medium and low nutrient levels (results not shown).

Experiment 2-Paired-culture, High or Low Nutrients, Coarse Sand Mix

Plants grown in paired-culture with greatly differing nutrient supplies showed at least a three-fold increase in above- and below-ground biomass increment when supplied with nutrients equivalent to agricultural plant requirements ('high nutrients', Figs 4a and b). Similarly, a large increase in photosynthetic area of leaves or culms was recorded for species receiving a high nutrient complement when compared to those receiving low amounts of nutrients (Figs 5a and b). Such increases were up to seven-fold in the restio species *I. subverticillata* and 70-fold in the grass

Table 2. Comparison of characteristics of soils associated with fynbos and renosterveld vegetation in the top 20 cm layer (November 1996-June 1997) and those used in glasshouse culture (July 1996) prior to nutrient application. The first column of each set of measurements from the field sites is the mean value of the top 0-5 cm of soil and the second column is the mean value measured for the lower 15-20 cm. The first column for the glasshouse sand relates to the coarse sand used in Experiments 1 and 2 and the second column relates to the finer sand used in Experiment 3.

Soil Property	Renosterveld		Fynbos		Glasshouse	
	0-5 cm	15-20 cm	0-5 cm	15-20 cm	Coarse	Fine
pH (CaCl$_2$)	5.0	4.9	3.7	3.8	4.7	4.6
Conductivity (μS)	333.5	294.0	73.5	56.7	50	89
Organic matter (%)	10.5	8.3	5.6	2.0	2.7	2.8
Total N (mg g soil^{-1})	8.4	4.1	2.7	0.6	0.1	0.1
Total P (μg g soil^{-1})	258.9	235.5	44.4	33.0	88.8	88.4

species *P. malouinensis* and *T. brachystachyum* and were significant for all species (Student's t-test, $p < 0.05$, significance not indicated on graphs). Shoot or culm:root dry weight ratios were much higher when nutrients were in good supply (0.8-6.7, Fig. 5d) than at lower levels (0.3-1.6, Fig. 5e) suggesting that feedback of assimilates into shoot growth increased predominantly under more favourable nutrient regimes. Measurements of coarse and fine root lengths showed similar significant patterns of increase for plants fed with agricultural-strength nutrients than fynbos-strength nutrients (results not shown) but relative proportions of coarse and fine roots remained similar in both treatments (i.e. 12-34% of total root length as fine roots).

When the contingent of species grown in paired-culture (Experiment 2) was compared to the same species grown in monoculture at equivalent low nutrient levels (Experiment 1), the majority of species showed significantly smaller increments in above- and below-ground biomass and photosynthetic area when grown in paired-culture than in monoculture (Table 3). An exception to this was the restio *R. festucaeformis* which did not exhibit a significant difference in culm or root biomass increment whether grown in mono- or paired-culture. Shoot or culm:root dry weight ratios generally remained unchanged whether the species was grown in monoculture or paired-culture (Table 3). Fine and coarse root lengths of all species did not differ significantly when grown in monoculture or paired-culture, with the exception of decreases in coarse root lengths of *Restio festucaeformis* when grown with the grass *Pentaschistis papillosa*, and *P. malouinensis* when grown with the restio *Ischyrolepis subverticillata* (Table 3).

Comparing the two species that were used more than once in paired-culture combinations, it appeared that *R. festucaeformis* accumulated more biomass in the presence of another restio than in the presence of a grass (*P. papillosa*) at low nutrient concentrations (Table 3). In contrast, the restio *I. subverticillata* grew better in terms of biomass increase in the presence of a grass (*P. malouensis*) than another restio.

Statistical comparisons of the growth variables measured for the various treatments investigating competition between species are presented in Table 4. Firstly, by comparing the monoculture growth of each of the species used in the paired-culture treatments, significant differences in size and weights were evident for most species, but not in relation to coarse, fine and total root lengths (Table 4a). The few exceptions included similarly-sized culms of the two restios *C. tectorum* and *R. festucaeformis* and root biomass of the grass *P. papillosa* and the restio *R. festucaeformis* after 10 months of growth in monoculture. When selected plants were grown together in paired-culture under either low (Table 4b) or high nutrient conditions (Table 4c), size and weight differences observed were similar to those observed when plants were grown in monoculture. In other words, the larger plant of the species pair remained large despite being grown in competition with another individual of the same or a different species. Interestingly, the grass *P. malouinensis* and the restio *I. subverticillata* were significantly different in size when grown in monoculture but similarly sized when cultivated together. The difference in fine root length (but not coarse root length) of the grass *T. brachystachyum*, and the restio *T. cinereus*, which was evident when both plants were grown in monoculture, was also observed when grown together under low nutrient conditions.

Experiment 3-Paired-culture, Low Nutrients, Fine or Coarse Sand Mix

Comparing the two species pairs (*P. papillosa* grown with *R. festucaeformis* and *P. malouinensis* grown with *I. subverticillata*) grown in fine sand and fed low nutrients *versus* coarse sand under the same nutrient conditions (Figs 4b and c), there were no clear patterns of growth response in terms of biomass accumulation. One species pair, *P. papillosa* and *R. festucaeformis*, produced significantly less root biomass (indicated by an asterisk, Student's t-test, $p < 0.05$, Fig. 4c) when grown in fine sand compared to coarse sand, but the same did not apply to the second experimental grass/restio pair (*P. malouensis* and *I. subverticillata*). The only significant difference in photosynthetic area was recorded for the restio *R. festucaeformis* which produced a greater photosynthetic area when grown in coarse compared to fine sand (Figs 5b and c). Shoot:root and culm:root dry weight ratios were similar for species pairs grown in coarse or fine sand at similar nutrient regimes (Fig. 5e and f). Coarse and fine root lengths were both

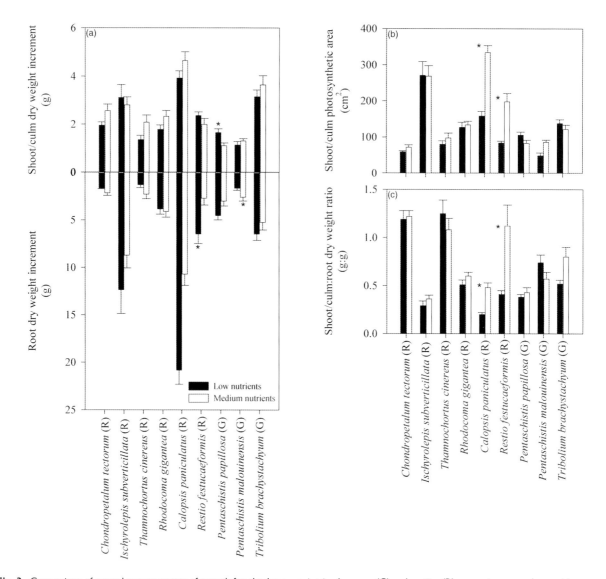

Fig. 3. Comparison of several measurements of growth for glasshouse-maintained grasses (G) and restios (R) grown in monoculture with coarse soil rooting substrates and medium or low nutrient supply: (a) shoot/culm and root dry weight increments, (b) shoot/culm photosynthetic area, and (c) shoot/culm:root dry weight ratios. Bar and whisker represent mean + S. E. Asterisks indicate significant differences (Student's t-test, $p < 0.05$) between a single species grown at medium or low nutrient additions.

significantly greater for *P. papillosa* and *R. festucaeformis* when grown in coarse sand but not for *P. malouinensis* and *I. subverticillata* (results not shown).

Despite absence of any applied soil inoculum in the glasshouse experiments, at least one representative of each grass species in each treatment showed some evidence of VA infection (Table 5). Infection was most likely from natural low levels of inoculum as spores or hyphal fragments present in the unsterilised fynbos mix initially incorporated into the sand. Capillaroid roots (*c.f.* Lamont 1982) were observed for at least 5% and up to 50% of the restio individuals, with some representation in all species.

DISCUSSION

Species richness in the Cape Floristic Region of South Africa has been correlated most frequently with edaphic factors (Goldblatt

1978; Oliver *et al.* 1983; Bond and Goldblatt 1984; Hoffmann *et al.* 1987; Campbell and Werger 1988) although rainfall has been cited as being more important in determining patterns of diversity (Linder 1991b). Beadle (1954) and Specht (1979) also suggest that climatic influences are often parallel to that of nutrient-impoverished soil but the latter has the overriding influence on species distribution. With this in mind the series of nutrient-availability experiments described in this study were established. It is evident from the results that grasses and restios representative of fynbos vegetation are capable of responding equally well in terms of biomass increase when supplied with nutrients equivalent to levels (moderate treatments) found in surrounding renosterveld habitats. In addition, species of both families have the capacity to respond to levels of nutrients well above those they would normally experience. Such flexibility in nutrient utilisation has been found for other members of fynbos (e.g. certain

Table 3. Comparison of measured variables for each grass or restio species when grown in mono- and paired-culture under low nutrient conditions. Significance was tested using Students t-test, p < 0.05. Numbers in bold represent means ± S. E. of plants grown in monoculture, numbers in normal text represent means ± S. E. of plants grown in paired-culture.

Species	Shoot/culm increment (g)	Root increment (g)	Shoot/culm photosynthet- icarea (cm²)	Shoot/culm:ro ot dry weight ratio	Coarse root length (cm)	Fine root length (cm)	Total root length (cm)
Chondropetalum tectorum (R)	**1.9 ± 0.1**	**1.7 ± 0.1**	**58.7 ± 3.3**	**1.2 ± 0.1**	**113.5 ± 6.6**	**285.0 ± 21.8**	**398.6 ± 25.8**
	0.4 ± 0.0	0.8 ± 0.2	13.2 ± 1.4	0.7 ± 0.2	97.8 ± 8.3	317.7 ± 33.0	415.5 ± 38.9
	***	***	***	*	NSD	NSD	NSD
Restio festucaeformis (R) [when grown with *Chondropetalum tectorum*(R)]	**2.4 ± 0.2**	**6.5 ± 1.0**	**83.8 ± 5.2**	**0.4 ± 0.0**	**237.9 ± 16.2**	**456.5 ± 33.3**	**694.5 ± 37.5**
	2.4 ± 0.3	5.3 ± 0.9	141.5 ± 15.6	0.5 ± 0.1	161.8 ± 11.0	570.1 ± 66.7	731.9 ± 71.7
	NSD[A]	NSD	***	NSD	NSD	NSD	NSD
Ischyrolepis subverticillata (R) [when grown with *Rhodocoma gigantea* (R)]	**3.1 ± 0.5**	**12.3 ± 2.5**	**271.2 ± 38.4**	**0.3 ± 0.0**	**215.0 ± 38.7**	**337.0 ± 26.7**	**552.0 ± 61.9**
	0.4 ± 0.1	0.9 ± 0.2	23.3 ± 4.3	0.5 ± 0.1	74.9 ± 2.8	318.9 ± 25.4	393.7 ± 25.1
	***	***	***	*	NSD	NSD	NSD
Rhodocoma gigantea (R)	**1.78 ± 0.2**	**3.8 ± 0.6**	**127.0 ± 13.9**	**0.5 ± 0.0**	**213.8 ± 27.6**	**680.0 ± 72.3**	**893.8 ± 78.7**
	0.7 ± 0.1	2.2 ± 0.4	32.7 ± 3.6	0.4 ± 0.1	115.9 ± 5.6	514.5 ± 37.9	630.5 ± 41.0
	***	*	***	NSD	NSD	NSD	NSD
Tribolium brachystachyum (G)	**3.1 ± 0.3**	**6.5 ± 0.7**	**138.1 ± 10.3**	**0.5 ± 0.0**	**152.2 ± 4.1**	**1041.1 ± 48.7**	**1193.3 ± 52.1**
	0.7 ± 0.1	1.5 ± 0.2	57.9 ± 5.5	0.5 ± 0.1	117.1 ± 7.4	888.9 ± 72.1	1006.1 ± 78.5
	***	***	***	NSD	NSD	NSD	NSD
Thamnochortus cinereus (R)	**1.4 ± 0.2**	**1.3 ± 0.3**	**80.0 ± 10.0**	**1.3 ± 0.1**	**165.5 ± 11.4**	**548.3 ± 20.8**	**713.8 ± 30.3**
	0.1 ± 0.0	0.1 ± 0.0	12.6 ± 2.8	1.4 ± 0.2	128.0 ± 6.5	373.2 ± 43.6	501.2 ± 46.0
	***	***	***	NSD	NSD	NSD	NSD
Pentaschistis papillosa (G)	**1.7 ± 0.2**	**4.5 ± 0.5**	**105.6 ± 8.7**	**0.4 ± 0.0**	**230.7 ± 11.9**	**963.8 ± 94.6**	**1194.5 ± 99.8**
	1.1 ± 0.2	4.7 ± 1.1	58.8 ± 8.1	0.3 ± 0.1	198.1 ± 6.8	752.5 ± 83.9	950.5 ± 90.5
	**	NSD	***	NSD	NSD	NSD	NSD
Restio festucaeformis (R) [when grown with *Pentaschistis papillosa* (G)]	**2.4 ± 0.2**	**6.5 ± 1.0**	**83.8 ± 5.2**	**0.4 ± 0.0**	**237.9 ± 16.2**	**456.5 ± 33.3**	**694.5 ± 37.5**
	1.3 ± 0.3	1.9 ± 0.8	81.1 ± 17.7	1.6 ± 0.3	122.3 ± 9.3	489.1 ± 56.2	611.4 ± 64.5
	***	***	NSD	***	*	NSD	NSD
Pentaschistis malouinensis (G)	**1.1 ± 0.1**	**1.7 ± 0.2**	**1.1 ± 0.1**	**0.7 ± 0.1**	**171.5 ± 14.8**	**601.5 ± 69.4**	**773.0 ± 59.7**
	0.1 ± 0.0	0.2 ± 0.1	5.2 ± 1.3	1.0 ± 0.3	71.3 ± 10.7	323.7 ± 23.1	394.9 ± 32.5
	***	***	***	NSD	*	NSD	*
Ischyrolepis subverticillata (R) [when grown with *Pentaschistis malouinensis* (G)]	**3.1 ± 0.5**	**12.3 ± 2.5**	**271.2 ± 38.4**	**0.3 ± 0.0**	**215.0 ± 38.7**	**337.0 ± 26.7**	**552.0 ± 61.9**
	0.8 ± 0.1	2.3 ± 0.5	58.5 ± 4.8	0.4 ± 0.1	134.1 ± 8.8	467.4 ± 44.5	601.5 ± 50.2
	***	***	***	NSD	NSD	NSD	NSD

[A] NSD - no significant difference, p > 0.05, * p < 0.05, ** p < 0.02, *** p < 0.001

annual, graminoid and restioid growth forms, Witkowski 1988) and in restios of Western Australian heathland (Meney *et al.* 1990a). This is contrary to the nutrient utilisation of other groups common in fynbos (e.g. proteoid and ericoid growth forms, Witkowski 1988) and for heathlands in general (Gimingham 1972; Helsper *et al.* 1983; Heil and Diemont 1983) where increased nutrient supply does not result in marked increases in biomass accumulation. Given the increase in biomass with addition of nutrients recorded in this study, it may be reasonably concluded that if nutrient availability resulting from differing soil types were the sole variable determining distribution of grasses and restios then restios would be expected to grow and survive equally as well as grasses in higher nutrient renosterveld and grasses equally as well as restios in nutrient-impoverished fynbos soils. Clearly this is not the case and other features of the environment are at play along with nutrient availability in determining distribution patterns of grasses and restios. In addition to considering the abiotic environment it must be kept in mind that adaptations to the biotic environment, such as defense against herbivory, may result in competitive advantages of certain species (Whittaker 1979).

Despite the evidence above, it must be recognised that experimental plants were grown under warm, well-watered conditions with a continuous supply of nutrients for a period of time in excess of their normal growing season in the wild. This is well substantiated considering the seasonal nature of moisture availability indicated by the field study. Under such abnormal conditions, typical growth patterns, and hence nutrient uptake, may not have been strictly adhered to. For example, when grown in native habitat the restio *Thamnochortus punctatus* Pill. displays asynchronous growth patterns, with the development of vegetative culms in spring and summer, root and rhizome growth in wetter months and reproductive events in late summer (Stock *et al.* 1987). Similar phenological patterns have been reported for three other species of restios of mediterranean Western Australia (Meney and Dixon 1988; Meney *et al.* 1990 a, b) suggesting that plasticity in growth phenology may be common in the family. Consequently, growth of restios in extended favorable conditions of the glasshouse study may have resulted in enhanced deployment of nutrients to vegetative organs than would normally be experienced during a shorter growing period in the field. The growth and allocation patterns

Table 4. Comparison of measured variables for each species pair when grown in the glasshouse in (a) monoculture under low nutrient conditions, (b) paired-culture under low nutrient conditions, and (c) in paired-culture under high nutrient conditions. Significance was tested using Students t-test, p < 0.05.

Species pair	Shoot/ culm inc.	Root inc.	Shoot/ culm area	Shoot/ culm:root dry weight ratio	Coarse root length	Fine root length	Total root length
(a) Monoculture, low nutrients							
Chondropetalum tectorum (R) & *Restio festucaeformis* (R)	NSD[A]	***	***	***	**	NSD	**
Ischyrolepis subverticillata (R) & *Rhodocoma gigantea* (R)	*	***	***	***	NSD	NSD	NSD
Tribolium brachystachyum (G) & *Thamnochortus cinereus* (R)	***	***	***	***	NSD	***	***
Pentaschistis papillosa (G) & *Restio festucaeformis* (R)	***	NSD	*	NSD	NSD	NSD	NSD
Pentaschistis malouinensis (G) & *Ischyrolepis subverticillata* (R)	***	***	***	***	NSD	NSD	NSD
(b) Paired-culture, low nutrients							
Chondropetalum tectorum (R) & *Restio festucaeformis* (R)	***	***	***	**	NSD	NSD	NSD
Ischyrolepis subverticillata (R) & *Rhodocoma gigantea* (R)	**	*	*	NSD	**	NSD	NSD
Tribolium brachystachyum (G) & *Thamnochortus cinereus* (R)	***	***	***	NSD	NSD	*	*
Pentaschistis papillosa (G) & *Restio festucaeformis* (R)	***	***	***	***	**	NSD	NSD
Pentaschistis malouinensis (G) & *Ischyrolepis subverticillata* (R)	NSD	NSD	NSD	NSD	NSD	NSD	NSD
(c) Paired-culture, high nutrients							
Chondropetalum tectorum (R) & *Restio festucaeformis* (R)	***	***	***	**	NSD	NSD	NSD
Ischyrolepis subverticillata (R) & *Rhodocoma gigantea* (R)	**	*	*	NSD	NSD	NSD	NSD
Tribolium brachystachyum (G) & *Thamnochortus cinereus* (R)	***	***	***	NSD	-	-	-
Pentaschistis papillosa (G) & *Restio festucaeformis* (R)	***	***	***	***	NSD	NSD	NSD
Pentaschistis malouinensis (G) & *Ischyrolepis subverticillata* (R)	NSD	NSD	NSD	NSD	NSD	NSD	NSD

[A]NSD no significant difference, p > 0.05, * p < 0.05, ** p < 0.02, *** p < 0.001

of nutrients in fynbos grasses is virtually unknown. However, two of the species found in more mesic habitats (*P. papillosa* and *T. brachystachum*, Table 1) retain leaves for only a single season so that new leaves must be produced annually. These species would be expected to complete their growth and reproduction within the few months of the year when water and nutrients were most abundant. *Pentaschistis malouensis*, with longer-lived leaves, may have the luxury of accommodating more asynchronous nutrient allocation patterns. In either case, the extended growth season afforded in the glasshouse may well give misleading patterns of the actual nutrient requirements likely in native habitats. In addition there is the need to consider that there are different levels of conservation of nutrients within each plant with some species displaying greater nutrient use efficiency than others (e.g. Aerts 1990; Gutschick 1993; Vazquez de Aldana and Berendse 1997). For example, recycling

of nutrients from older vegetative parts to newly forming organs of the clonal restio *Alexgeorgea nitens* (Nees) Johnson & Briggs has been estimated to be from 14% for P requirements to 58% for K (Meney *et al.* 1990a).

A second feature investigated in the glasshouse study was the effect of particle size of the rooting media or soil texture. Renosterveld soils are shale-derived and consequently are composed of very fine particles (>60% as grains less than 0.5 mm diameter) in comparison to the coarser sandstone-derived fynbos soils (<40% as grains less than 0.5 mm diameter). The effect of particle size on biomass accumulation in grasses and restios grown under glasshouse conditions with a neutral media was mixed suggesting it is a species-specific response related more to root structure, function or architecture. However, it is likely that the effects of particle size on drainage, rates of soil drying, moisture holding capacity, aeration, amounts of nutrients adsorbed onto

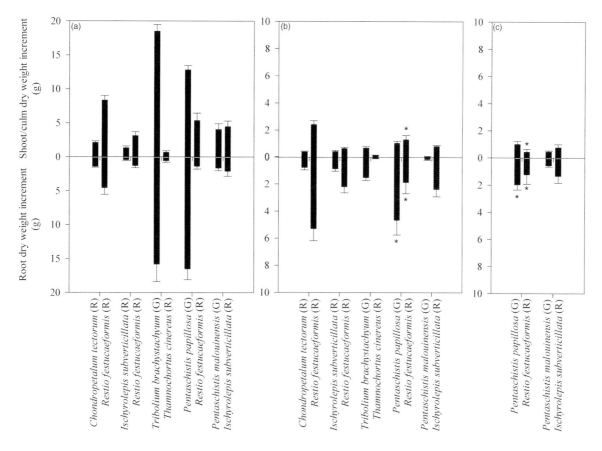

Fig. 4. Comparison of shoot or culm and root dry weight increments of five glasshouse-maintained restio (R)/restio (R) or grass (G)/restio (R) combinations grown in paired-culture with coarse soil rooting substrates and supplied with either (a) high or (b) low nutrients, and (c) two grass (G)/restio (R) combinations grown in paired-culture with fine rooting substrates and low nutrient supply. Bar and whisker represent mean + S. E. Asterisks in (b) and (c) indicate significant differences (Student's t-test, p < 0.05) between a single species when grown in coarse or fine sand mixes with similar low nutrient additions.

soil colloids and maintenance of root contact with particles will have a greater influence on plant growth when considering soils derived from different parent materials and their associated chemistry (Fitter 1985).

The effects on growth of other edaphic variables such as moisture availability, pH, conductivity and organic C could not be tested methodically in the present glasshouse study but were investigated seasonally in native habitats to determine variability over time and with soil depth. Fynbos and renosterveld soils were then shown to contrast greatly overall with respect to moisture content over time and with depth. Water availability has obvious important implications for biomass accumulation, particularly for initial seedling establishment and growth. Further rigorous testing of the effects of moisture availability on growth of grasses and restios is certainly required and potentially lends itself to glasshouse studies. The great disparity between the chemistry of the two soil types has implications for different growth responses for grasses and restios, and also offers potential variables for future manipulation in the glasshouse setting. For example, very acid fynbos soils might well foster a different process of mineral exchange than in the less acidic renosterveld soils, and this difference might in turn affect the availability of micronutrients even to the point of elevating levels of certain metal complexes to toxic levels, particularly

with respect to grasses. Availability of organic and inorganic N and P might also be totally different in the two soil types.

Competition between individuals for nutrients was also investigated in this study. It is evident that under low nutrient conditions there was a greater degree of inter- than intraspecific competition regardless of whether the competitor was a member of the same family or not. Such results are contrary to findings made by Richards (1993), who concluded that competition had an important role in determining spacing and patterns of individuals within fynbos communities but only a minor role in determining species distributions. The general pattern of greater accumulation of biomass with intraspecific competition compared to interspecific competition may be attributed in part to complementary sharing of nutrients between plants of the same species (e.g., Ritz and Newman 1984; Newman 1988; Frey and Schuepp 1993). This would be expected to give specific portions of a population a competitive advantage over more species-rich areas of the same population. Two of the grass species used in this study, *T. brachystachyum* and *P. papillosa*, had a greater growth response compared to their restio partners when nutrients were in plentiful supply. This suggests that the grasses possessed a greater ability to use nutrients when readily available whereas the growth of the restio partners was more conservative

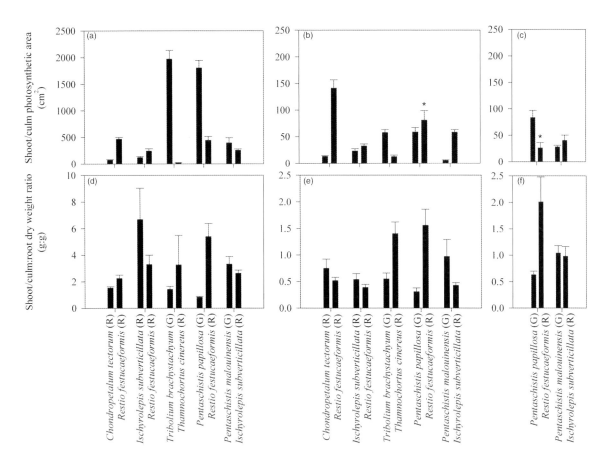

Fig. 5. Comparison of shoot or culm photosynthetic areas (a-c) and shoot or culm:root dry weight ratios (d-f) of five glasshouse-maintained restio (R)/restio (R) or grass (G)/restio (R) combinations grown in paired-culture with coarse soil rooting substrates and supplied with either high (a, d) or low (b, e) nutrients, and two grass (G)/restio (R) combinations grown in paired-culture with fine rooting substrates and low nutrient supply (c, f). Bar and whisker represent mean + S. E. Asterisks in (b) and (c) indicate significant differences (Student's t-test, p < 0.05) between a single species when grow in coarse or fine sand mixes with similar low nutrient additions.

under the same conditions. Unfortunately, because of this type of effect, it cannot be determined whether the biomass increases observed were due to better competitive ability of the grasses for nutrient uptake. On the other hand, the grass *P. malouinensis* appeared to have a poorer competitive ability than did its restio partner even when nutrients were in ready supply. This fits well with the 'competition avoider' classification of this species by Linder and Ellis (1990). Of the restio species used in this study, *R. festucaeformis* and *R. gigantea* were possibly better competitors for nutrients than were other members of their family. This may reflect the faster growth rates generally associated with the seeder fire response strategy (Bell *et al.* 1984; Bell and Pate 1993; Meney and Pate 1999, Table 1). Further glasshouse experimentation using matched pairs of seeders and resprouters would be useful in testing this hypothesis.

Both restios and grasses are generally described as having root systems that are shallow, fibrous and dense (Higgins *et al.* 1987). Low (1984) attributed much of the success of members of the Restionaceae in nutrient-impoverished habitats of fynbos to possessing such types of root systems, considering them to be well adapted for optimum nutrient and moisture uptake. It has been shown in this study that members of both families have equivalent and consistent proportions of fine (12-34%) and coarse

roots (66-88%) in their below-ground biomass. However, the three grass species had up to twice the total root length of their restio counterparts. This was evident for two of the three grass species when grown under the lowest nutrient conditions but became especially obvious in all grass species grown under higher nutrient conditions. Grasses grown in competition with restios also showed production of greater total root lengths. This suggests that even though the basic root types of grasses and restios may be considered to be broadly similar, subtle differences in amounts of roots produced may vary considerably under different nutrient regimes. This may well mean that grasses have potentially greater access to nutrient stores in the soil by possessing a larger nutrient absorbing area than restios. Such an enhanced nutrient uptake mechanism can only be increased in grasses with the presence of mycorrhizal symbionts (Harley and Smith 1983) whereas restios are generally thought to be weakly or non-mycorrhizal (Low 1980; Lamont 1984; Meney *et al.* 1993). Instead, restios have been noted to possess capillaroid roots which function mainly to increase the physical absorptive areas of roots (Lamont 1984). Such evidence suggests that nutrient enhancing mechanisms must also be taken into account when considering the distribution patterns of vegetative elements of an ecosystem.

Table 5. Frequency of observations (%) of VA mycorrhizal infection in trypan blue stained roots of grasses and presence of capillaroid roots (%) of restios for glasshouse-grown experimental plants.

Species	VA mycorrhizal infection (%)	Capillaroid roots (%)
Pentaschistis papillosa (G)	12	-
Pentaschistis malouinensis (G)	16	-
Tribolium brachystachyum (G)	5	-
Calopsis paniculatus (R)	-	20
Chondropetalum tectorum (R)	-	45
Ischyrolepis subverticillata (R)	-	51
Restio festucaeformis (R)	-	7
Rhodocoma gigantea (R)	-	28
Thamnochortus cinereus (R)	-	5

It should be concluded from this study that it is not yet possible to select any single variable from those examined to explain the distribution of grasses and restios of the Cape Floristic Region. However, the glasshouse studies show that nutrient availability, soil texture and competition do play formative roles in above- and below-ground biomass accumulation of individual species of restios or grasses, and there is every likelihood that moisture availability may also be a major contributing factor.

ACKNOWLEDGEMENTS

The authors wish to thank the Hon. P. K. van der Byl for access to Fairfield Farm, Mr. and Mrs. C. A. van Zyl for farm liaison and A. J. Smit, T. Verboom, C. Morrow, F. Ojeda and A. Lemson for help with field collections. Thanks to Prof. J. S. Pate and a reviewer for comments on an earlier draft of the manuscript. One author (TLB) was in receipt of a Postdoctoral Fellowship from the University of Cape Town, South Africa.

REFERENCES

Acocks, J. P. H. (1975). Veld types of South Africa. *Memoirs of the Botanical Survey of South Africa* **40**, 1-128.

Aerts, R. (1990). Nutrient use efficiency in evergreen and deciduous species from heathlands. *Oecologia* **78**, 115-305.

Beadle, N. C. (1954). Soil phosphate and delimitation of plant communities in eastern Australia. *Ecology* **25**, 370-374.

Bond, P., and Goldblatt, P. (1984). Plants of the Cape Flora: a descriptive catalogue. *Journal of South African Botany, Supplementary Volume* **13**, 98-127.

Bell, D. T., Hopkins, A. J. M., and Pate, J. S. (1984). Fire in the Kwongan. In 'Kwongan-Plant Life of the Sandplain'. (Eds J. S. Pate and J. S. Beard.) pp. 1-26. (University of Western Australia Press: Nedlands.)

Bell, T. L., and Pate, J. S. (1993). Morphotypic differentiation in the south-western Australian restiad *Lyginia barbata* R. Br. (Restionaceae). *Australian Journal of Botany* **41**, 91-104.

Boucher, C., and Moll, E. J. (1980). South African mediterranean shrublands. In 'Mediterranean Type Shrublands'. (Eds F. Di Castri, D. W. Goodall and R. L. Specht.) pp. 233-248. (Elsevier: Amsterdam.)

Campbell, B. M. (1985). A classification of the mountain vegetation of the fynbos biome. *Memoirs of the Botanical Survey of South Africa* **50**, 1-115.

Campbell, B. M., and Werger, M. J. A. (1988). Plant form in the mountains of the Cape, South Africa. *Vegetatio* **43**, 43-47.

Cowling, R. M., and Holmes, P. M. (1992). Flora and vegetation. In 'The Ecology of Fynbos-Nutrients, Fire and Diversity'. (Ed. R. M. Cowling.) pp. 23-61. (Oxford University Press: Cape Town.)

Dahlgren, R. M. T., and Clifford, H. T. (1982). 'The Monocotyledons. A Comparative Study'. (Academic Press: London.)

Fitter, A. H. (1985). Functional significance of root morphology and root system architecture. In 'Ecological Interactions in the Soil-Plants, Microbes and Animals'. (Ed. A. H. Fitter.) pp. 87-106. (Blackwell Scientific Publications: Oxford.)

Frey, B., and Schüepp, H. (1993). A role of vesicular-arbuscular (VA) mycorrhizal fungi in facilitating interplant nitrogen transfer. *Soil Biology and Biochemistry* **25**, 651-658.

Gibbs Russell, G. E., Watson, L., Koekemoer, M., Smook, L., Barker, N. P., Anderson, H. M., and Dallwitz, M. J. (1990). 'Grasses of Southern Africa'. *Memoirs of the Botanical Survey of South Africa* No. 58, BRI, Pretoria.

Gimingham, C. H. (1972). 'Ecology of Heathlands'. (Chapman and Hall: London.)

Giovanetti, M., and Mosse, B. (1980). An evaluation of techniques for measuring vesicular arbuscular mycorrhizal infection in roots. *New Phytologist* **84**, 489-500.

Goldblatt, P. (1978). An analysis of the flora of southern Africa: its characteristics, relationships and origins. *Annals of the Missouri Botanical Garden* **65**, 369-436.

Grimshaw, H. M. (1985). The determination of total P in soils by acid digestion. In 'Chemical Analysis in Environmental Research'. (Ed. A. P. Rowland.) pp. 92-95. (ITE Symposium No. 18, Natural Environmental Research Council: Cumbria.)

Gutschick, V. P. (1993). Nutrient-limited growth rates: roles of nutrient-use efficiency and of adaptations to increased uptake rate. *Journal of Experimental Botany* **44**, 41-51.

Harley, J. L., and Smith, S. E. (1983). 'Mycorrhizal Symbiosis'. (Academic Press: New York.)

Heil, G. W., and Diemont, W. H. (1983). Raised nutrient levels change heathland into grassland. *Vegetatio* **53**, 113-120.

Helsper, H. P. G., Glenn-Lewis, D., and Werger, M. J. A. (1983). Early regeneration of *Calluna* heathland under various fertilization treatments. *Oecologia* **58**, 208-214.

Higgins, K. B., Lamb, A. J., and van Wilgen, B. W. (1987). Root systems of selected plant species in mesic mountain fynbos in the Jonkershoek Valley, south-western Cape Province. *South African Journal of Botany* **53**, 249-257.

Hoffman, M. T., Moll, E. J., and Boucher, C. (1987). Post-fire succession at Pella, a South African lowland fynbos site. *South African Journal of Botany* **53**, 370-374.

Lamont, B. B. (1982). Mechanisms for enhancing nutrient uptake in plants, with particular reference to mediterranean South Africa and Western Australia. *Botanical Review* **48**, 597-689.

Lamont, B. B. (1984). Specialised modes of nutrition. In 'Kwongan-Plant Life of the Sandplain'. (Eds J. S. Pate and J. S. Beard.) pp. 126-145. (University of Western Australia Press: Nedlands.)

Linder, H. P. (1989). Grasses in the Cape Floristic Region: phytogeographical implications. *South African Journal of Science* **85**, 502-505.

Linder, H. P. (1991a). A Review of the Southern African Restionaceae. *Contributions form the Bolus Herbarium* **13**, 209-364.

Linder, H. P. (1991b). Environmental correlates of patterns of species richness in the south-western Cape Province of South Africa. *Journal of Biogeography* **18**, 509-518.

Linder, H. P., and Ellis, R. P. (1990). Vegetative morphology and interfire survival strategies in the Cape Fynbos grasses. *Bothalia* **20**, 91-103.

Low, A. B. (1980). Preliminary observations on specialized root morphologies in plants of the Western Cape Province. *South African Journal of Science* **76**, 513-516.

Low, A. B. (1983). Phytomass and major nutrient pools in an 11-year post-fire coastal fynbos community. *South African Journal of Botany* **2**, 98-104.

Low, A. B. (1984). The Cape restios-A modern success story? In 'Proceedings of the Fourth International Conference on Mediterranean Ecosystems'. (Ed. B. Dell.) pp. 93-94. Perth, Western Australia.

Manders, P. T. (1990). Fire and other variables as determinants of forest/fynbos boundaries in the Cape Province. *Journal of Vegetation Science* **1**, 483-490.

Meney, K. A., and Dixon, K. W. (1988). Phenology, reproductive biology and seed development in four rush and sedge species from Western Australia. *Australian Journal of Botany* **36**, 711-726.

Meney, K. A., Dixon, K. W., Scheltema, M., and Pate, J. S. (1993). Occurrence of vesicular mycorrhizal fungi in dryland species of Restionaceae and Cyperaceae from south-west Western Australia. *Australian Journal of Botany* **41**, 733-737.

Meney, K. A., and Pate, J. S. (1999). Seasonal growth and nutrition of Restionaceae. In 'Australian Rushes. Biology, Identification and Conservation of Restionaceae and Allied Families'. (Eds K. A. Meney and J. S. Pate.) In press. (University of Western Australia Press, Nedlands.)

Meney, K. A., Pate, J. S., and Dixon, K. W. (1990a). Phenology of growth and resource deployment in *Alexgeorgea nitens* (Nees) Johnson & Briggs (Restionaceae), a clonal species from south-western Western Australia. *Australian Journal of Botany* **38**, 543-557.

Meney, K. A., Pate, J. S., and Dixon, K. W. (1990b). Comparative morphology, anatomy, phenology and reproductive biology of *Alexgeorgea* spp. (Restionaceae) from south-western Western Australia. *Australian Journal of Botany* **38**, 523-541.

Mitchell, D. T., Coley, P. G. F., Webb, S., and Allsopp, N. (1986). Litterfall and decomposition processes in the coastal fynbos vegetation, south-western Cape, South Africa. *Journal of Ecology* **74**, 977-993.

Moll, E. J., and Jarman, M. L. (1984). Clarification of the term fynbos. *South African Journal of Science* **80**, 351-352.

Murphy, J., and Riley, J. P. (1962). A modified single-solution method for the determination of phosphate in natural waters. *Analytical Chimica Acta* **27**, 31-36.

Newman, E. I. (1988). Mycorrhizal links between plants: their functioning and ecological significance. *Advances in Ecological Research* **18**, 243-270.

Oliver, E. G. H., Linder, H. P., and Rourke, J. P. (1983). Geographical distribution of present day taxa and their phytogeographical significance. *Bothalia* **14**, 427-440.

Richards, M. B. (1993). Soil factors and competition as determinants of fynbos plant species distributions in the south-western Cape, South Africa. PhD thesis, University of Cape Town, South Africa.

Ritz, K., and Newman, E. I. (1984). Movement of ^{32}P between intact grassland plants of the same age. *Oikos* **43**, 138-142.

Specht, R. L. (1979). Heathlands and related shrublands of the world. In 'Ecosystems of the World. Heathlands and Related Shrublands. Descriptive Studies. Volume 9A'. (Ed. R. L. Specht.) pp. 1-18. (Elsevier Scientific Publishing Company: Amsterdam.)

Stock, W. D., and Allsopp, N. (1992). Functional perspective of ecosystems. In 'The Ecology of Fynbos-Nutrients, Fire and Diversity'. (Ed. R. M. Cowling.) pp. 241-259. (Oxford University Press: Cape Town.)

Stock, W. D., and Lewis, O. A. M. (1986). Soil nitrogen and the role of fire as a mineralizing agent in a South African coastal fynbos ecosystem. *Journal of Ecology* **74**, 317-328.

Stock, W. D., Sommerville, J. E. M., and Lewis, O. A. M. (1987). Seasonal allocation of dry mass and nitrogen in a fynbos endemic Restionaceae species *Thamnochortus punctatus* Pill. *Oecologia* **72**, 315-320.

Taylor, H. C. (1978). Capensis. In 'Biogeography and Ecology of Southern Africa'. (Ed. M. J. A. Werger.) pp. 171-229. (Junk: The Hague.)

Trollope, W. S. W. (1973). Fire as a method of controlling macchia (fynbos) vegetation on the Amatole Mountains of the Eastern Cape. *Proceedings of the Grasslands Society of South Africa* **8**, 35-41.

Vazquez de Aldana, B. R., and Berendse, F. (1997). Nitrogen-use efficiency in six perennial grasses from contrasting habitats. *Functional Ecology* **11**, 619-626.

Witkowski, E. T. F. (1988). Response of a low-land fynbos ecosystem to nutrient additions. PhD thesis, University of Cape Town, South Africa.

Witkowski, E. T. F., and Mitchell, D. T. (1987). Variations in soil phosphorus in the fynbos biome, South Africa. *Journal of Ecology* **75**, 1159-1171.

Witkowski, E. T. F., and Mitchell, D. T. (1989). The effects of nutrient additions on aboveground and phytomass and its phosphorus and nitrogen contents of sand-plain lowland fynbos. *South African Journal of Botany* **55**, 243-249.

Whittaker, J. B. (1979). Invertebrate grazing: Competition and plant dynamics. In 'Population Dynamics'. (Ed. O. T. Solbrig.) pp. 207-222. (Blackwell Scientific Publications: Oxford.)

BREEDING SYSTEMS

Female inflorescence of the dioecious tropical Australian strand species, *Spinifex longifolius.*

Grasses: Systematics and Evolution. (2000). Eds S.W.L. Jacobs and J. Everett. (CSIRO: Melbourne)

ADAPTIVE PLASTICITY IN REPRODUCTION AND REPRODUCTIVE SYSTEMS OF GRASSES

James A. Quinn

Department of Ecology, Evolution, and Natural Resources, Rutgers University, New Brunswick, NJ 08901-1582
  USA.

Abstract

Genetically-determined plasticity plays a major role in the fine-tuning of reproduction and reproductive systems of grasses in response to temporal or spatial environmental variation. In the hermaphroditic *Notodanthonia caespitosa* (Gaud.) Zotov, some populations in Australia show genetic differentiation in maximizing local environmental control (and plasticity) of flowering phenology, and others exhibit genetic differentiation in the direction of seasonal timing of flowering. North American populations of *Sporobolus cryptandrus* (Torr.) A. Gray show a highly significant relation between local environmental variability/unpredictability and the relative importance of plasticity in reproduction. In addition, a population's breeding system, and its environmentally-governed versatility, can be shaped by its ecological history. In *Buchloe dactyloides* (Nutt.) Engelm., genotypes vary markedly in type and lability of sex expression, and populations vary in the numbers of individuals that are male, female, or 'monoecious' (inconstant with M:F ratios affected by environment) in relation to prior or current environmental conditions. Considering the 'individualistic' population concept, the study of the breeding system of a species should not only compare the differential responses of genotypes and populations to a range of environments, but also equally emphasize the amounts and patterns of plasticity in reproductive strategies exhibited across this range of environments.

Key words: Breeding systems, genetic differentiation, phenology, phenotypic plasticity, reproduction, reproductive versatility.

INTRODUCTION

Grasses display an extraordinary diversity of breeding systems – hermaphroditism to most forms of monoecism, dioecism, and apomixis (Connor 1979, 1981, 1987), and genetic differentiation in reproductive strategies and breeding systems has been frequently demonstrated among populations of wide-ranging species (Hodgkinson and Quinn 1978; Quinn 1998). Equally important is the role of genetically determined plasticity in the fine-tuning of reproductive allocation or of the breeding system in response to temporal or spatial environmental variation (Quinn 1987, 1998). 'Phenotypic plasticity' in this paper will refer to 'the amount by which the expressions of individual char-

acteristics of a genotype are changed by different environments' (Bradshaw 1965). Bradshaw (1965, 1974) documented the existence and importance of genetically based phenotypic plasticity in plant populations, and subsequent research has demonstrated that the phenotypic plasticity of a trait can be genetically determined and affected by selection, and suggested that populations of a species may therefore vary in plasticity in relation to local environmental variability and predictability (Quinn 1987).

ADAPTIVE VARIATION IN FLOWERING PHENOLOGY

The proper timing of flowering is essential for maximizing the production of progeny, and genetic differentiation in flowering phenology has been frequently demonstrated among populations of wide-ranging grass species (McMillan 1959; Evans and Knox 1969; Quinn 1969; Quinn and Ward 1969; Rotsettis *et al.* 1972; Groves 1975). In my opinion, no example is more striking, or better illustrates the interaction of genetic differentiation and phenotypic plasticity, than *Notodanthonia caespitosa*. This species is a perennial bunchgrass that occurs across the southern half of Australia, and latitudinally ranges from 31° to 42° S , an 11° latititudinal range (Hodgkinson and Quinn 1978). Its climatic range extends from a hot arid environment at its northern limit, where the mean annual rainfall may be as low as 220 mm, to a cool and moist temperate environment at its southern limit in Tasmania (>1250 mm). In the field the northernmost populations flower up to 4 months earlier than Tasmanian populations, but it was not known how much of this difference was due to genetic differentiation and how much was due to a developmental response to differing environmental conditions. Five populations (closed circles in Fig. 1) were selected along a N-S transect which spanned the species' latitudinal range. Plants were transplanted to a garden at Deniliquin, and grown from seeds in the CERES Phytotron in Canberra to determine the effect of daylength, temperature, and vernalization on floral initiation and inflorescence development. Key results from Hodgkinson and Quinn (1978) are as follows:

(*i*) *N. caespitosa* was consistently found to be a long-day plant;

(*ii*) northern populations only required daylengths longer than 9.5 hr for floral induction, compared with 11 hr or longer (up to 12.5+) for southern populations;

(*iii*) the time in the proper inductive conditions required for this floral initiation was 5-7 days for the three northernmost populations and 21-25 days for the two southern populations; and

(*iv*) inflorescence development (initiation to flag leaf stage) was considerably slower in southern populations (4 times longer for Heathcote than for Nyngan), and appeared to be important in the timing of reproduction in the field.

Of far greater ecological/evolutionary significance than the latitudinal pattern in daylength requirements and inflorescence development was the difference among populations in the environmental factor(s) controlling the timing of reproduction! When critical daylengths for floral initiation were related to daylengths in natural habitats (Hodgkinson and Quinn 1978), it became apparent that daylength *per se* was unlikely to have a significant effect on the timing of reproduction in Nyngan and Lake Cargelligo populations. For these two northern populations, tillers arising at any time of the year would be exposed to daylengths that exceed their critical value for floral initiation. This would mean that once a tiller forms, it could soon become reproductive, and that temperature and soil moisture would largely determine the rate of its reproductive development. This characteristic would seem to be of adaptive significance, since it would enable opportunistic reproduction to occur in an environment characterized by erratic, unpredictable rainfall (Hodgkinson and Quinn 1978). Conversely, in southern locations, typified by Heathcote and Conara Junction, daylength appears to play an important role in the timing of reproduction. Here, soil moisture availability rarely limits the period of

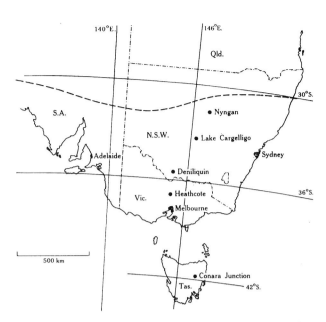

Fig. 1. Locations (•) of the five sites in southeastern Australia from which seeds and plants of *Notodanthonia caespitosa* were collected. The dashed line represents the northern limit for the distribution of the species as derived from herbarium records listed by Vickery (1956). (From Hodgkinson and Quinn 1976).

plant growth, and the growing season is predictable (Hodgkinson and Quinn 1978).

In summary, in cool and moist temperate environments the reproduction of *N. caespitosa* is programmed via daylength responses and inherent rate of inflorescence development to coincide with the termination of a relatively predictable growing season, whereas in hot semi-arid and arid environments this control is relaxed, which permits opportunistic reproduction whenever soil moisture and temperature permit growth.

Thus, in the populations of *N. caespitosa* there has been genetic differentiation in the direction of maximizing local environmental control (and plasticity) of flowering phenology (i.e. permitting 'opportunistic reproduction') and genetic differentiation in the direction of a seasonal timing of flowering.

FINE-TUNING OF FLOWERING PHENOLOGY IN RESPONSE TO YEARLY FLUCTUATIONS

Even in those populations in which the flowering phenology is programmed through daylength responses, yearly fluctuation in temperatures and soil moisture availability can affect the timing of flowering through effects on growth rates and inflorescence development. This fine-tuning of flowering phenology, i.e. the responses to yearly fluctuations in the local environment, could also be adaptive, and should be particularly important in day-neutral species.

Demonstration of such adaptive plasticity requires a clear separation of the variance among genotypes, the variance among environments, and the variance attributable to the interaction of genotypes with environments, and this can be best accomplished by the clonal replication of genotypes over a range of environments. An additional requirement would be a series of populations of different ecological histories of environmental variability and predictability. All these requirements are met by a uniform garden study with replicated blocks of cloned genotypes initiated in 1963 on the day-neutral *Sporobolus cryptandrus* (Quinn and Ward 1969). *Sporobolus* is an inbreeding, perennial bunchgrass, which is successional and relatively short-lived. Three years of data on flowering phenology were available for treatment as characters measured in three environments (year-to-year variation in a local environment); only the data from the second year had been published, and indicative of the late 1960s, attention was focused primarily on the differences in mean population responses to a common environment.

Thus, a new analysis of these uniform garden data (J. A. Quinn and J. D. Wetherington, in prep.) provides a unique opportunity to test the null hypothesis that a population's ecological history has no effect on the relative importance of phenotypic plasticity in the timing of reproduction.

Key elements of the experimental design and data collection (Quinn and Ward 1969) were:

(*i*) genotypes were collected from 16 populations of *Sporobolus cryptandrus* in Colorado and adjacent areas in the USA;

(*ii*) genotypes were divided into ramets, and a randomized block design was established in the summer of 1963 near Ft. Collins, CO (six genotypes per population, three blocks); and

(*iii*) weekly phenology readings were taken in 1964, 1965, and 1966. Flowering phenology was best represented by 'Flag Leaf Apparent' (the collar of the leaf that will subtend the inflorescence has emerged from the sheath of the preceding leaf).

In our reanalysis, the 3 years of data were treated as a character measured in three environments as a way of quantifying 'environmental sensitivity' and partitioning the variance (Falconer 1989). The variance component for clones (V_c) indicated genetic variation among clones, the variance component for years (V_y) indicated the amount of phenotypic plasticity, and the clone x year component (V_{cy}) indicated the genotype x environment interaction, i.e., the differences among clones in their 'pattern of phenotypic plasticity.'

It should be emphasized that there were pronounced differences in the environments of the three years. For example, the differences between the low and high years in annual precipitation and in March-July precipitation were 200% and 300%, respectively; evaporation in the highest year was 37% greater; and the last date of spring frost varied by 3 months (U. S. Department of Commerce 1965, 1966, 1967).

Results of ANOVA of the selected phenological stage indicated that the variation among populations and the variation among clones were both statistically significant at the 0.01 level in each of the three years. Figure 2 is a plot of the 16 *Sporobolus* population means, showing the timing of the first appearance of a flag leaf across the three years. The populations not only vary within a year but also vary both in the amount of their year-to-year variation (or plasticity) and their pattern of plasticity (shape of reaction norm). ANOVAs of the individual populations produced a range of results in which either no variance component or 1, 2,

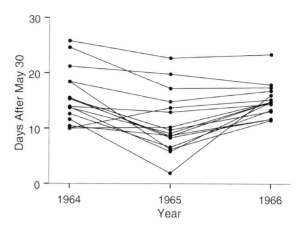

Fig. 2. Mean dates for the first appearance of a flag leaf for the 16 *Sporobolus cryptandrus* populations at the Fort Collins transplant garden in 1964, 1965, and 1966.

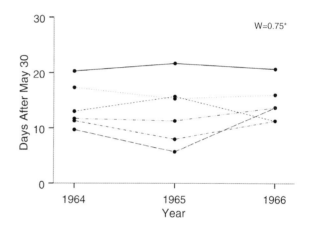

Fig. 3. Mean dates for the first appearance of a flag leaf for the six clones of the Cortez, Colorado population of *Sporobolus cryptandrus* at the Fort Collins transplant garden in 1964, 1965, and 1966. W = Kendall's coefficient of concordance.

or 3 of the clone, year, and clone x year components were significant (or highly significant, $P < 0.01$). Illustrating a portion of this variation, Figs 3 and 4 each show the values for the individual clones of a population plotted across the three years. Figure 3 shows a population that had a significant clone variance component but a lack of a significant year-to-year variation or of Genotype-Environment interaction (lines crossing). Verifying this lack of a G-E interaction, Kendall's coefficient of concordance (W) was significant at the .05 level. Figure 4 shows a population whose only significant variance component was the clone x year; in this case the W was very low, indicating a lack of a similar pattern shown by the clones and the importance of the G-E interaction component.

Levene's test (Levene 1960) verified that the populations varied significantly in their amount of plasticity (V_y component) and highly significantly in their pattern of plasticity (V_{cy} component).

In order to test the hypothesis that phenotypic plasticity should be greatest for populations in unpredictable environments and/or under short growing seasons, an environmental index was developed to indicate 'the potential of a population site to select

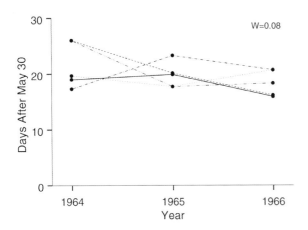

Fig. 4. Mean dates for the first appearance of a flag leaf for the six clones of the Grand Junction, Colorado population of *Sporobolus cryptandrus* at the Fort Collins transplant garden in 1964, 1965, and 1966. The solid black line represents two clones which had the same mean values for all three years. W = Kendall's coefficient of concordance.

Fig. 5. Sex expression in a single trimonoecious genotype of *Buchloe dactyloides*, ranging from inflorescences 100% male in morphology and sex expression (at left) through intermediates in both morphology and sex expression (including perfect flowers) to the normal female coalesced inflorescence (at right). (From Yin and Quinn 1994).

for plasticity in phenology' (J. A. Quinn and J. D. Wetherinton, in prep.). Using the prior 16 years of data at each site, this index was developed from the equal weighting of:

(*i*) mean growing season length (number of days between the last spring and first fall dates of temperatures 28°F or below);

(*ii*) mean annual precipitation (interacts with temperature to determine effective growing season length);

(*iii*) the coefficient of variation (CV) of the growing season length; and

(*iv*) the CV of the annual precipitation.

In order to pool the above four values for a site so that the highest total value would represent the greatest potential for selection for plasticity, the lowest value for the first two and the highest value for the last two were given a maximum value of 100, with the rest of the values scaled accordingly.

Pearson correlation coefficients of the environmental index with the variance components for the populations of *Sporobolus* showed highly significant positive relationships between the index and the 'year' (0.77, $P < 0.01$) and the 'clone x year' (0.63, $P < 0.01$) components, i.e., a highly significant relationship between the ecological history of a population and the relative importance of plasticity in the timing of reproduction. Because *Sporobolus* is found over a wider range of climatic and soil conditions than any other grass in the Southwest USA, such a relationship is remarkable, if one considers the many factors working against a perfect fit: founder effects, genetic drift, soils, interspecific competition, herbivory, past disturbance events, etc.

ADAPTIVE VARIATION AND PLASTICITY IN THE BREEDING SYSTEM

A population's breeding system, and its environmentally-governed versatility, can be shaped by its ecological history, e.g. past levels of *r-* *vs* *K*-selection and DI *vs* DD mortality (Law *et al.* 1977; Roos and Quinn 1977; Morishima 1985; Richards 1990),

frequency and intensity of past disturbance events (Heslop-Harrison 1961, 1966; McNamara and Quinn 1977; Cheplick and Quinn 1982), heterogeneity of habitat in time and space (McMillan 1965; Campbell *et al.* 1983; Bell and Quinn 1987; Richards 1990), and biome-level differences in growing season, resources, and biotic interactions (Edwards 1974; Tallowin 1977; Soreng and Van Devender 1989; Elmquist and Cox 1996). Of the many examples available, I will use the results of our recent and ongoing research on buffalograss (*Buchloe dactyloides*) (Yin and Quinn 1994, 1995). Buffalograss is usually considered a 'dioecious' species because most plants are either male or female, and extremely stable in sex expression, but some populations contain a significant number of individuals that are labile or monoecious in their sex expression (Yin and Quinn 1994). For certain of these *Buchloe* genotypes, sex expression can be readily affected by environmental conditions or experimental manipulation of the hormone level. Indeed, we found that the relative numbers of female and male flowers in monoecious plants, and the relative numbers of female, perfect, and male flowers in trimonoecious plants, could be dramatically affected through the use of gibberellin and its inhibitor, paclobutrazol or PAC (Yin and Quinn 1995). Figure 5 shows the range of sex expression in a single trimonoecious genotype, ranging from inflorescences 100% male in morphology and sex expression (at left) through intermediates in both morphology and sex expression (including perfect flowers) to the normal female coalesced inflorescence (at right). A high concentration of gibberellin applied over a sufficient length of time will lead to a plant with all male inflorescences (similar to the one on the left), and a high concentration of PAC will produce plants with all female inflorescences (similar to the one on the right).

In natural populations of *Buchloe dactyloides*, there is evidence that population densities may affect both the gender allocation of individual monoecious plants, and the frequency of sex forms. Looking first at gender allocation of individual plants, certain monoecious plants produce predominantly male inflorescences after cloning and transplanting to new soil (J. A. Quinn, pers. obser.). There have also been reports that male inflorescences

Table 1. Sex form distribution of vegetative plant samples and seed samples collected from eight natural populations of buffalograss Sex forms are denoted as male (M), inconstant male (IM), inconstant female (IF), and female (F). The percentage of inconstant (or monoecious) sex forms within each sample is labeled as %I. Chi-square was used to test the independence between sex form distribution and population location. ***Significant at the 0.001 probability level. Table by permission from Huff and Wu (1992)

Population	Vegetative sample			Seed sample		
	Sex form distribution			**Sex form distribution**		
	N	M : IM : IF : F	%I	N	M : IM : IF : F	%I
		(No. of individuals)			(No. of individuals)	
Hennessey	15	6 : 1 : 1 : 7	13	–	– – – –	–
Chillicothe	26	9 : 2 : 5 : 10	27	37	6 : 9 : 16 : 6	68
Guymon	20	10 : 0 : 0 : 10	0	82	36 : 2 : 3 : 41	6
Wilderado	21	11 : 2 : 0 : 8	10	25	11 : 0 : 3 : 11	12
Clayton	19	9 : 0 : 3 : 7	16	24	10 : 2 : 2 : 10	17
Glenrio	18	7 : 0 : 3 : 8	17	45	12 : 4 : 6 : 23	22
Springer	20	9 : 1 : 3 : 7	20	–	– – – –	–
Apache	17	6 : 3 : 2 : 6	29	–	– – – –	–
Pooled	156	67 : 9 : 17 : 63	17	213	75 : 17 : 31 : 91	22
χ^2 independence	10.1			60.2***		

predominate at higher nitrogen levels (Huff and Wu 1987), on plants with longer stolons (Plank 1892; Arber 1934), in mowed fields (Wu *et al.* 1984), and in open/low density sites (Shaw *et al.* 1987; J. A. Quinn, pers. obser.). Although this male inflorescence production may be mostly a response to increased light and resources, it does obviate a pollen limitation problem for *Buchloe* at low densities. A Nebraska study (Jones and Newell 1946) documented the markedly limited dispersal of pollen from the short inflorescence culms, and my personal observations indicate that the number of seeds within diaspores is affected by the percentage cover of buffalograss and/or distance from male plants (Quinn 1985).

Considering the frequency of sex forms within populations in relation to density, Huff and Wu (1992) reported a relationship between percentage cover of *Buchloe* and the frequency of monoecious plants in eight populations in two E-W transects across the shortgrass prairie of central North America (Table 1). As can be seen in their table, the number of monoecious plants (%I) in the vegetative sample on the left ranged from 0% at Guymon, Oklahoma (in the central part of the range with 91% *Buchloe* cover) to 29% at Apache Springs, New Mexico (a population on the periphery of *Buchloe*'s range with 17% cover of *Buchloe*). Markedly more monoecious plants were found in seed samples (right side of table) and, in general, in peripheral populations where the *Buchloe* sod is sparse and discontinuous and potential mates are rare, monoecious genotypes may have an advantage over unisexual individuals by possessing the ability to self-fertilize. However, at vegetatively more dense locations, they suggested that the self-fertility of monoecious genotypes might lead to inbreeding depression and less competitive and less variable progenies than those arising from cross-fertilization. There are alternative hypotheses (Huff and Wu 1992), and much additional research will be required to determine the genetic and ecological significance of such population variation in sex form frequencies in grasses (Yin and Quinn 1994).

SUMMARY AND CONCLUSIONS

Grasses not only display an extraordinary diversity of breeding systems–hermaphroditism to dioecism, but many also have marked environmentally-governed reproductive versatility, combining a mixture of outcrossing, selfing, and asexual strategies.

Excluding phylogenetic and/or genetic and morphological constraints, the breeding system, and its phenotypic plasticity, of a population should be influenced by the levels of *r*- vs *K*-selection and DI vs DD mortality, frequency and intensity of past disturbance events, heterogeneity of habitat in time and space, and biome-level differences in growing season, resources, and biotic interactions (Quinn 1998).

As a result, 'No two plant populations have exactly similar breeding systems and exactly similar patterns of variation.' (Richards 1986).

Considering this 'individualistic' population concept (Quinn, 1987), we must avoid breeding system studies that utilize only one population of a species or only two or three individuals each from a few populations. If we are to appreciate the range of adaptive variation in sex expression in grasses, we must not only compare the differential responses of individuals and populations to a range of environments, but also equally emphasize the amounts and patterns of plasticity in reproductive strategies exhibited across this range of environments.

REFERENCES

Arber, A. (1934). 'The Gramineae. A Study of Cereal, Bamboo, and Grass.' (Cambridge University Press: Cambridge.)

Bell, T. J., and Quinn, J. A. (1987). Effects of soil moisture and light intensity on the chasmogamous and cleistogamous components of reproductive effort of *Dichanthelium clandestinum* populations. *Canadian Journal of Botany* **65**, 2243-2249.

Bradshaw, A. D. (1965). Evolutionary significance of phenotypic plasticity in plants. *Advances in Genetics* **13**, 115-155.

Bradshaw, A. D. (1974). Environment and plasticity. *Brookhaven Symposia in Biology* **25**, 75-94.

Campbell, C. S., Quinn, J. A., Cheplick, G. P., and Bell, T. J. (1983). Cleistogamy in grasses. *Annual Review of Ecology and Systematics* **14**, 411-441.

Cheplick, G. P., and Quinn, J. A. (1982). *Amphicarpum purshii* and the 'pessimistic strategy' in amphicarpic annuals. *Oecologia* **52**, 327-332.

Connor, H. E. (1979). Breeding systems in the grasses: a survey. *New Zealand Journal of Botany* **17**, 547-574.

Connor, H. E. (1981). Evolution of reproductive systems in the Gramineae. *Annals of the Missouri Botanical Garden* **68**, 48-74.

Connor, H. E. (1987). Reproductive biology in the grasses. In 'Grass Systematics and Evolution'. (Eds T. R. Soderstrom, K. W. Hilu, C. S. Campbell, and M. E. Barkworth.) pp. 117-132. (Smithsonian Institution Press: Washington, DC.)

Edwards, J. A. (1974). Studies in *Colobanthus quitensis* (Kunth) Barl. and *Deschampsia antarctica* Desv. VI. Reproductive performance on Signy Island. *British Antarctic Survey Bulletin* **39**, 67-86.

Elmqvist, T., and Cox, P. A. (1996). The evolution of vivipary in flowering plants. *Oikos* **77**, 3-9.

Evans, L. T., and Knox, R. B. (1969). Environmental control of reproduction in *Themeda australis*. *Australian Journal of Botany* **17**, 375-389.

Falconer, D. S. (1989). 'Introduction to Quantitative Genetics.' (Longman: Essex, England.)

Groves, R. H. (1975). Growth and development of five populations of *Themeda australis* in response to temperature. *Australian Journal of Botany* **23**, 951-963.

Heslop-Harrison, J. (1961). The function of the glume pit and the control of cleistogamy in *Bothriochloa decipiens* (Hack.) C. E. Hubbard. *Phytomorphology* **11**, 378-383.

Heslop-Harrison, J. (1966). Reflections on the role of environmentally-governed reproductive versatility in the adaptation of plant populations. *Transactions of the Botanical Society of Edinburgh* **40**, 159-168.

Hodgkinson, K. C., and Quinn, J. A. (1976). Adaptive variability in the growth of *Danthonia caespitosa* Gaud. populations at different temperatures. *Australian Journal of Botany* **24**, 381-396.

Hodgkinson, K. C., and Quinn, J. A. (1978). Environmental and genetic control of reproduction in *Danthonia caespitosa* populations. *Australian Journal of Botany* **26**, 351-364.

Huff, D. R., and Wu, L. (1987). Sex expression in buffalograss under different environments. *Crop Science* **27**, 623-626.

Huff, D. R., and Wu, L. (1992). Distribution and inheritance of inconstant sex forms in natural populations of dioecious buffalograss (*Buchloe dactyloides*). *American Journal of Botany* **79**, 207-215.

Jones, M. D., and Newell, L. C. (1946). Pollination cycles and pollen dispersal in relation to grass improvement. Nebraska Agricultural Experiment Station, Research Bulletin No. 148, Lincoln.

Law, R., Bradshaw, A. D., and Putwain, P. D. (1977). Life-history variation in *Poa annua*. *Evolution* **31**, 233-246.

Levene, H. (1960). Robust tests for equality of variances. In 'Contributions to Probability and Statistics'. (Eds I. Olkin, S. G. Ghurye, W. Hoeffding, W. G. Madow, and H. B. Mann.) pp. 278-292. (Stanford University Press: Stanford, CA.)

McMillan, C. (1959). The role of ecotypic variation in the distribution of the central grassland of North America. *Ecological Monographs* **29**, 285-308.

McMillan, C. (1965). Grassland community fractions from central North America under simulated climates. *American Journal of Botany* **52**, 109-116.

McNamara, J., and Quinn, J. A. (1977). Resource allocation and reproduction in populations of *Amphicarpum purshii* (Gramineae). *American Journal of Botany* **64**, 17-23.

Morishima, H. (1985). Habitat, genetic structure and dynamics of perennial and annual populations of the Asian wild rice *Oryza perennis*. In 'Genetic Differentiation and Dispersal in Plants'. (Eds P. Jacquard, G. Heim, and J. Antonovics.) pp. 179-190. (Springer-Verlag: Berlin.)

Plank, E. N. (1892). *Buchloe dactyloides*, Engelm., not a dioecious grass. *Bulletin of the Torrey Botanical Club* **19**, 303-306.

Quinn, J. A. (1969). Variability among High Plains populations of *Panicum virgatum*. *Bulletin of the Torrey Botanical Club* **96**, 20-41.

Quinn, J. A. (1985). Validity of breeding for a female bias in the dioecious buffalograss (*Buchloe dactyloides*). In 'Proceedings of the XV International Grassland Congress'. (Eds T. Okubo and M. Shiyomi.) pp. 297-299. (The Science Council of Japan and the Japanese Society of Grassland Science: Nishi-nasuno, Tochigi-ken.)

Quinn, J. A. (1987). Complex patterns of genetic differentiation and phenotypic plasticity versus an outmoded ecotype terminology. In 'Differentiation Patterns in Higher Plants'. (Ed. K. M. Urbanska.) pp. 95-113. (Academic Press: London.)

Quinn, J. A. (1998). Ecological aspects of sex expression in grasses. In 'Population Biology of Grasses'. (Ed. G. P. Cheplick.) pp. 136-154. (Cambridge University Press: Cambridge.)

Quinn, J. A., and Ward, R. T. (1969). Ecological differentiation in sand dropseed (*Sporobolus cryptandrus*). *Ecological Monographs* **39**, 61-78.

Richards, A. J. (1986). 'Plant Breeding Systems.' (George Allen and Unwin: London.)

Richards, A. J. (1990). The implications of reproductive versatility for the structure of grass populations. In 'Reproductive Versatility in the Grasses'. (Ed. G. P. Chapman.) pp. 131-153. (Cambridge University Press: Cambridge.)

Roos, F. H., and Quinn, J. A. (1977). Phenology and reproductive allocation in *Andropogon scoparius* (Gramineae) populations in communities of different successional stages. *American Journal of Botany* **64**, 535-540.

Rotsettis, J., Quinn, J. A., and Fairbrothers, D. E. (1972). Growth and flowering of *Danthonia sericea* populations. *Ecology* **53**, 227-234.

Shaw, R. B., Bem, C. M., and Winkler, G. L. (1987). Sex ratios of *Buchloe dactyloides* (Nutt.) Engelm. along catenas on the shortgrass steppe. *Botanical Gazette* **148**, 85-89.

Soreng, R. J., and Van Devender, T. R. (1989). Late Quaternary fossils of *Poa fendleriana* (muttongrass): Holocene expansions of apomicts. *Southwestern Naturalist* **34**, 35-45.

Tallowin, J. R. B. (1977). The reproductive strategies of a subantarctic grass, *Festuca contracta* T. Kirk. In 'Adaptations within Antarctic Ecosystems'. (Ed. G. A. Llano.) pp. 967-980. (Smithsonian Institution Press: Washington, DC.)

U. S. Department of Commerce. (1965). Climatological data, Colorado. Government Printing Office, Volume 69 No. 13, Washington, DC.

U. S. Department of Commerce. (1966). Climatological data, Colorado. Government Printing Office, Volume 70 No. 13, Washington, DC.

U. S. Department of Commerce. (1967). Climatological data, Colorado. Government Printing Office, Volume 71 No. 13, Washington, DC.

Vickery, J. W. (1956). A revision of the Australian species of *Danthonia* D.C. *Contributions from the New South Wales National Herbarium* **2**, 249-325.

Wu, L., Harvindi, A. H, and Gibeault, V. A. (1984). Observations on buffalograss sexual characteristics and potential for seed production improvement. *HortScience* **19**, 505-506.

Yin, T., and Quinn, J. A. (1994). Effects of exogenous growth regulators and a gibberellin inhibitor on sex expression and growth form in buffalograss (*Buchloe dactyloides*), and their ecological significance. *Bulletin of the Torrey Botanical Club* **121**, 170-179.

Yin, T., and Quinn, J. A. (1965). Tests of a mechanistic model of one hormone regulating both sexes in *Buchloe dactyloides* (Poaceae). *American Journal of Botany* **82**, 745-751.

Grasses: Systematics and Evolution. (2000). Eds S.W.L. Jacobs and J. Everett. (CSIRO: Melbourne)

DIOECISM IN GRASSES IN ARGENTINA

H.E. Connor[A], Ana M. Anton[B] and Marta E. Astegiano[C]

[A]Department of Geography, University of Canterbury, Christchurch, New Zealand.
[B]Instituto Multidisciplinario de Biología Vegetal, Universidad Nacional de Córdoba, Córdoba, Argentina.
[C]Facultad de Ciencias Agropecuarias, Universidad Nacional de Córdoba, Córdoba, Argentina.

Abstract

Dioecism is a low frequency phenomenon in the grasses, and is primarily a New World feature centred in Mexico. Of the array of about 20 dioecious genera, four occur in Argentina, together with three danthonioids where apomixis in female plants has replaced the sexual state. In Argentina most dioecism occurs in *Poa,* and in genetic experiments orthodox 1M:1F sex ratios are obtained, but there are no field data. Field sampling in halophytic *Monanthochloe acerosa* indicated a 1:1 ratio. Dioecism is incomplete in *Scleropogon brevifolius,* and although F/n = 0.51, the balance of the population consists, in descending order of frequency, of andromonoecids, males, and an array of monoecious variants. *Scleropogon* exemplifies the ambivalence between the results of a detailed examination of the floral biology of a taxon and a simple flora entry "dioecious". The genetic base for sex form segregation in *Cortaderia,* and for departures from sexuality for it and *Lamprothyrsus,* is broader than for most Argentinian taxa.

Key words: Grasses, dioecism, sex ratios, autonomous apomixis, *Cortaderia, Lamprothyrsus, Monanthochloe, Poa, Scleropogon, Gynerium.*

INTRODUCTION

Dioecism, the presence in natural populations of plants of two sex forms, one pistillate and seed bearing, and the other staminate and pollen producing, is an allogamous system (Cruden and Lloyd 1995; Sakai and Weller 1999); all pollinations are assumed to be compatible.

The frequencies of the sex forms are well studied in those countries with significant dioecious taxa, New Zealand for example, and models for the evolution of dioecism such as those of Charlesworth and Charlesworth (1978 a,b), Lloyd (1980), and Ross (1982) are becoming verified by empirical studies on many genera especially those suited to experiment (Webb 1999). Dioecism in the grasses is at a very low frequency - in c. 3% of the genera, and in c. 100 species, <1% of species.

Data presented here on some dioecious Argentine grasses are to show the extent to which Argentinian grasses conform to the expectations of dioecious taxa, and their role in what is predominantly a phenomenon of New World grasses.

DIOECISM AND GRASS GENERA

The incidence of dioecism in flowering plants is summarised by Sakai and Weller (1999); the highest frequencies are in insular floras, all of them in the Pacific Ocean except Mauritius. Dioecious species frequencies as in Sakai and Weller are, for example, Mauritius 11%, New Zealand 13%, Hawaii 15%, and Samoa 17%. Within this perspective, dioecious grasses are discordant because their highest frequency is in the continental New World, and in Mesoamerica in particular.

The dioecious grass genera spread among five tribes are (number of species in each genus in parentheses): Arundineae: *Gynerium* (1); Chlorideae: *Buchloe* (1), *Buchlomimus* (1), *Cyclostachya* (1), *Opizia* (2 or 3), *Pringleochloa* (1), *Soderstromia* (1); Eragrostideae:

Allolepis (1), *Distichlis* (13), *Jouvea* (2), *Monanthochloe (3)*, *Neeragrostis* (2), *Reederochloa* (1), *Scleropogon* (1), *Sohnsia* (1); Paniceae: *Pseudochaetochloa* (1), *Spinifex* (4), *Zygochloa* (1), Poeae: *Festuca* (400), *Poa* (500) (Reeder 1969; Tzvelev 1976; Connor 1979; Clayton and Renvoize 1986; Watson and Dallwitz 1992; Peterson *et al.* 1997). Monotypy is frequent. Exceptions to New World dominance are the austral panicoids, disjunct *Distichlis*, and cosmopolitan *Festuca* and *Poa*.

The array of dioecious grasses in Argentina is few in genera and numerically small except *Poa* c. 35 spp.; to these should be added *Monanthochloe* 1 sp., *Distichlis* 4 spp., *Gynerium* 1 sp., and the genera in which dioecism is incomplete in its expression: *Cortaderia* 6 spp.; *Lamprothyrsus* 1 sp.; *Scleropogon* 1 sp. This may not be the widest base on which to make predictions or generalisations on dioecism, but data are available for all except *Distichlis* spp.

The primary sex ratio M/F=1 was recognized by Fisher (1930) and given precision by Shaw and Mohler (1953). Sex ratio data allow us to detect if grasses correspond to ratios in other groups and conform to theoretical propositions (Connor 1981, 1987). The data presented here are simple counts of M and F in natural and in experimental populations. To these primary data are added the development of sexual dimorphisms in florets and spikelets, and of resulting dispersal mechanisms (Davidse 1987), the adjustment of floral sex ratios, and the presence of the colony-forming habit.

RESULTS

Poa (Poeae)

Anton and Connor (1995) estimated that there are c. 35 dioecious species of *Poa* in Argentina, most of them in Patagonia and the Andes. There are no data on sex form frequencies in natural populations, but results from experimental progenies (Table 1) show 1M:1F segregation ratios, that the genetic system of control of sex is effective between different levels of ploidy, and that there is no conflict with the general hypothesis of male heterogamety.

Floret dimorphism in Argentinian *Poa* is described by Nicora (1978); dimorphism does not occur in dioecious species in North America except in *P. arachnifera*, or in New Zealand (Anton and Connor 1995).

Monanthochloe (Eragrostideae)

This genus of halophytic sites is an amphitropical disjunct; *M. acerosa* occurs in central and western Argentina (Parodi 1954; Villamil 1969). At four localities sex form frequency counts were made on 13 November 1993 (Table 2); anthesis occurred from 2.30 p.m. until 6 p.m. on a day of full sun. There is a good fit to 1M:1F and no heterogeneity among the samples. Colonies may be up to 10 m in diameter, and isolated from each other by about the same distance, on the broad, flat, saline plain near the boundary between the provinces of Córdoba and Catamarca. At San José de las Salinas, Salinas Grandes, in low scrub at the margin of the great salt plain, male and female colonies are often smaller and the two may be adjacent or 0.5 - 1 m apart. There are no ecological preferences for the sex forms.

An inflorescence has a single spikelet; in males there are 3-10 florets/spikelet, and in females 2-4-(7) (Villamil 1969). There is no marked floret or spikelet dimorphism, but the floral sex ratio is adjusted in favour of pollen production.

Scleropogon (Eragrostideae)

The conspicuously dimorphic, stoloniferous species, *S. brevifolius*, is a New World amphitropical disjunct. Though Philippi (1870) had described it as dioecious its sex expression had proved difficult to interpret even after Pilger (1940) realised that it could also be monoecious. Reeder and Toolin (1987) found populations in North America where there were M and F plants, and monoecious as well as gynomonoecious elements.

A population from the general area of the *locus classicus*, Uspallata, Mendoza, was sampled in 1994, and described in Anton *et al.* (1998). A broad array of sex forms was found, varying from orthodox M and F plants, to plants with elaborate trimonoecious inflorescences (Table 3).

In one salient respect this population meets the requirement for a dioecious species since F/n=0.5. In four other respects it conforms to dioecism in the grasses: the inflorescence has a pronounced sexual dimorphism; the floral sex ratio between male and female spikelets and inflorescences is adjusted in favour of pollen production; it has evolved a chamaechorous habit; and it is stoloniferous. The major discrepancy lies in the low frequency of male plants (0.149) which is offset by the high frequency of andromonoecious plants (0.186); between them they account for most of the pollen production in the population. Andromonoecious plants and the other monoecious variants, together with a modest contribution from hermaphrodites, contribute about half of the ovules available for fertilization. There are no data on the compatibility reaction of any of the morphs.

This population shows every evidence of an evolving dioecism. Many simple intermediate and penultimate steps between hermaphroditism and dioecism can be identified. Femaleness, assumed to be controlled by homozygous genes for male sterility, stabilized relatively easily and perhaps relatively quickly; maleness which would require female sterility to be dominant to male sterility, and a linkage (Richards 1986), is incomplete in its evolution. Reasons for this genetic tardiness are unknown, but a high genetic value for ovules may possibly prevent the final steps to M/n=0.5.

The Mendozan population agrees in broad terms with the sexual display of plants in Arizona as described by Reeder and Toolin (1987). Despite the similar floral diversity in the disjunct populations, there is a high level of genetic variation in allozyme alleles in North America which is absent from South American plants (Peterson and Columbus 1997). That low allozyme variability in Argentina persuaded Peterson and Columbus that South American populations have only recently arrived from the Northern Hemisphere, and that very few individuals, perhaps even in a single event, formed the initial colony. The two data sets are discordant in Mendoza - high sexual diversity, and low allozyme variability.

Dimorphism for sex form in floret and spikelet characters is developed to a high degree of differentiation especially in the anthoecium where the robust lemma of the female has three awns c. 35 times longer than the awn of the smaller thinner male lemma. Flowers are simple: a tristaminate androecium in the lodiculate male, and a bistigmatic ovary in the elodiculate female.

Table 1. Sex form Segregation in *Poa* (after Anton and Connor 1995)

	M	F	$1:1\chi^2$
A. Open pollination:			
P. barrosiana	43	42	0.012
P. pilcomayensis	23	20	0.012
B. Interspecific hybrids			
Tetraploids			
P. iridifolia x *P. lanigera*	6	7	0.077
x *P. ligularis*	16	12	0.571
x *P. phalaroides*	7	5	0.333
P. ligularis x *P. phalaroides*	10	20	3.333
Tetraploids x Octoploids			
P. arachnifera x *P. iridifolia*	7	9	0.25
x *P. lanigera*	8	8	0
Octoploids			
P. arachnifera x *P. bonariensis*	14	6	3.200

Heterogeneity χ^2_6 = 7.574 n.s.

Table 2. Sex form frequencies in *Monanthochloe acerosa*, central Argentina (November 1993)

Locality	M	F	$1:1\chi^2$
San José de Las Salinas	22	24	0.087
Las Salinas - Las Cañas	6	4	0.400
Córdoba-Catamarca boundary	46	45	0.110
Catamarca Province	22	18	0.400
Heterogeneity χ^2_3 0.764 n.s.			

Cortaderia (Danthonieae)

Cortaderia in Argentina is either basically subdioecious or comprises exclusively apomictic females (Astegiano *et al.* 1995, 1996). Traditionally the genus is interpreted as dioecious, but morphologically and genetically it is gynodioecious with plants that bear either pistillate flowers (F) or perfect flowers (H). Both sex forms set seeds. In those species where all plants are female, reproduction is by precocious, autonomous, aposporous apomixis (Connor 1974; Philipson 1978).

The only count of sex form frequency in Argentina indicated 1H:1F in octoploid *C. selloana* (Astegiano *et al.* 1995); this is in accord with counts made in New Zealand (Connor 1965). There are no data for the southern sexual species tetraploid *C. pilosa* and octoploid *C. araucana*.

In New Zealand, where *C. selloana* is now widely naturalised, hermaphrodite plants set few seeds under self-pollination but are cross-compatible. Their seed-producing role is usually doubted or ignored, but under cross-pollination in New Zealand seed set was 47.3% in hermaphrodites compared with 98.7% in female plants. Germination of seeds from hermaphrodites is 5-6%, and together with the lower seed set ensures that little is contributed

to the next generation by seeds from that sex form (Connor 1974). Wind dispersal of seeds from female plants is aided by floret dimorphism.

For all practical and genetic purposes *C. selloana* should be regarded as functionally dioecious, and the 1:1 primary sex ratio occurs in the wild in Argentina, and in New Zealand where it is ecologically successful.

Male sterility in *C. selloana* is under recessive genic control, the double recessive at any of up to three complementary loci causing male sterility (Connor and Charlesworth 1989). There is no evidence of dominant genetic male sterility or of nucleo-cytoplasmic inheritance considered by Couvet *et al.* (1986), Maurice *et al.* (1993, 1994), and Schultz (1994) to be more common than genetic control and important in the evolution of dioecism.

Just more than 13,000 plants were grown in experiments in New Zealand from controlled cross-pollinations of females; in many families segregation was 1H:1F, but in others ratios of 3H:5F and 3H:13F could be fitted (Connor and Charlesworth 1989). In progenies from females pollinated by homozygous dominant males the ratio is 1H:0F; such pollen donors can only be the descendants of hermaphrodite plants.

Table 3. Sex forms and frequencies in *Scleropogon brevifolius* at Uspallata, Mendoza (after Anton *et al.* 1998) (n = 483 plants; 600 inflorescences; c.850 spikelets)

	Frequency	Spikelets/Inflorescence	Florets/Spikelets	Contribution	
				Pollen	Ovules
Male (M)	0.149	2.17	6.76	0.393	-
Andromonoecious (MH)	0.186	2.28	2.93	0.444	0.321
Monoecious (MF)	0.023	1.50	2.65	0.022	0.015
Trimonoecious (MHF)	0.079	1.16	2.36	0.089	0.103
Hermaphrodite (H)	0.025	1.91	4.82	0.041	0.065
Gynomonoecious (HF)	0.029	1.11	2.10	0.013	0.041
Female (F)	0.509	1.50	2.14	-	0.465

Data from three generations of the experimental *C. araucana* (2n = 8*x* = 72) x *C. selloana* (2n = 8*x* = 72) indicate that control of male sterility is identical in both species (Connor 1983). Our expectation is that *C. araucana* populations will show 1:1 ratios in nature.

Three highly polyploid species, *C. jubata* (2n = 12*x* = 108), *C. rudiuscula* (2n = 12*x* = 108), and *C. speciosa* (2n = 8*x* = 72), have substituted autonomous apospory for dioecism (Connor 1974; Philipson 1978). All plants are female, and all plants set seeds; all are ecologically successful.

Lamprothyrsus (Danthonieae)

The South American genus *Lamprothyrsus* is ditypic with *L. hieronymii* in Argentina and Bolivia, which Parodi (1949) had noted as seed-producing females in his garden, and *L. peruvianus* of Ecuador and Peru, which Connor (1979) had noted as seed-producing females in his experimental garden. In both species all plants examined are female, except for two male specimens of *L. hieronymii* recognised by Bernadello (1979). Polliniferous plants of *L. peruvianus* are unreported. All plants of *L. hieronymii* seen at La Caldera, Salta, in the forest and on the hillside in November 1993, were female.

Lamprothyrsus, in the opinion of Connor and Dawson (1993), is the contemporary, high polyploid (2n = c.136), asexual residue of an austral, palaeo-allopolyploid genus that had become dioecious but in which autonomous apospory has later been substituted for dioecism as the result of the invasion of genes for apomixis. The genus is ecologically successful.

Even if there were pollen fertile plants in populations of *L. hieronymii* they could not contribute to the maintenance of dioecism because precocious embryogenesis and endosperm formation at anthesis prevent opportunities for cross-pollination, leaving it to depend on an exclusively maternal fitness based on gametophyic apomixis.

Gynerium (Danthonieae)

Monotypic, rhizomatous *Gynerium sagittatum* with markedly dimorphic, large inflorescences, and extremely tall culms, is known in Argentina only in the northwestern Province of Salta and in the northeastern provinces of Chaco, Misiones and Corrientes, at the southern limits for the species.

In Ecuador the sex ratio is approximately 1M:1F (S.A. Renvoize in Astegiano *et al.* 1996). All specimens of Argentinian origin are seed-bearing females; no plants with staminate flowers have been collected in the wild.

Rhizomes of *G. sagittatum* were probably brought to northern Argentinian river banks and riparian islands by flood waters from Bolivia and Brazil, where autonomous apomictic reproduction must have already replaced dioecism.

Discussion

Not all genera described as dioecious are in fact so, and morphological or genetic uncertainty about sex forms, as for example, in *Scleropogon* or *Cortaderia,* demonstrate the point which was similarly made about some Mexican grasses by Reeder and Reeder (1963, 1966), Reeder *et al.* (1965), and Columbus (this volume). The need for precise floral detail remains unfulfilled.

Sex Ratios in Grasses

The sex ratio data for natural populations of dioecious grasses are sparse (Table 4). There is an excess of females in apomixis-prone *Poa cusickii,* (Soreng 1991); for other ratios in *Poa* spp. see Soreng (this volume). An excess of males in psammophilous *Spinifex hirsutus* (Kirby 1988 and *in litt.*) is equalled by a similar excess in *S. sericeus* on Valla Beach, New South Wales, Australia, in an October 1984 sample (Maze and Whalley 1990); and another excess of males is reported for *Festuca kingii* (Fox and Harrison 1981). In the other dioecious grasses, 1M:1F generally holds true; there is no habitat selection or ecological differentiation between sex forms in *Buchloe* (Shaw *et al.* 1987, Quinn 1991), *Buchlomimus* (Reeder *et al.* 1965), *Distichlis* (Connor and Jacobs 1991), *Spinifex* (Connor 1984; Maze and Whalley 1990; Connor and Jacobs 1991); *Pseudochaetochloa* (Connor and Jacobs 1991), *Scleropogon* (Anton *et al.* 1998), or *Zygochloa* (Connor and Jacobs 1991). The two species where positive ecological differentiation is recorded are *Distichlis spicata* for salinity (Freeman *et al.* 1976), and *Festuca kingii* for soil moisture (Fox and Harrison 1981). There is no rigorous support for the spatial segregation of sex forms as outlined by Bierzychudek and Eckhart (1988) and Freeman *et al.* (1997).

In Argentina, conformity to the primary sex ratio 1M:1F and to features associated with dioecism are present in natural populations of *Monanthochloe acerosa,* a species which forms large colonies, and has adjusted floral sex ratios. The 1M:1F primary sex ratio obtains in families of *Poa* spp. raised from seeds gathered in the wild from open-pollinated females, and in artificial F_1 inter-

Table 4. Primary Sex Ratios of Gramineae in Natural Populations

Poeae			
Festuca kingii	0.58M	0.42F	(Fox & Harrison 1981)
Poa cusickii			
ssp. *cusickii*	0.35M	0.65F	(Soreng 1991)
ssp. *pallida*	0.46M	0.54F	(Soreng 1991)
P. pringlei	0.50M	0.50F	(Soreng 1991)
Eragrostideae			
Distichlis distichophylla	0M	1F[1]	(Connor & Jacobs 1991)
	0.5M	0.5F	(Connor & Jacobs 1991)
	0.9M	0.1F	(Connor & Jacobs 1991)
	1M[1]	0F	(Connor & Jacobs 1991)
D. spicata	0.49M	0.51F	(Freeman *et al.* 1976)
Chlorideae			
Buchloe dactyloides	0.52M	0.48F	(Schaffner 1920)
	0.52M	0.48F	(Quinn & Engel 1986)
	0.48M	0.52F	(Shaw *et al.* 1987)
	0.49M[2]	0.51F	(Huff and Wu 1992)
Paniceae			
Pseudochaetochloa australiensis	0.47M	0.53F	(Connor & Jacobs 1991)
Spinifex hirsutus	0.57M	0.43F	(Kirby 1988 and *in litt.*)
S. longifolius[3]	0.49M	0.51F	(Connor & Jacobs 1991)
S. sericeus	0.52M	0.48F	(Connor 1984)
	0.51M	0.49F	(McDonald 1983)
	0.52M	0.48F	(Kirby 1988 and *in litt.*)
	0.51M	0.49F	(Maze & Whalley 1990 for Stuart Point 1983)
	0.61M	0.39F	(Maze & Whalley 1990, for Valla Beach Oct 1984)
Danthonieae			
Gynerium sagittatum	0.5M	0.5F	(S.A. Renvoize *in litt.*)

1 Unisexual colonies.

2 Inconstant M and F occurred in 7 of 8 populations, in total 26 plants among 156..

3 The frequency 0.9M 0.1F unwisely reported by Connor (1984) for plants near Darwin, Northern Territory, Australia, should be disregarded; that population was examined early in the season when male inflorescences had emerged in their characteristic manner , before those on female plants.

specific hybrids; the genetic control of sex form persists in high polyploids; anthoecium dimorphism is present.

Argentine taxa reveal some exceptions. *Scleropogon brevifolius* meets the sex ratio for the female component and F/n = 0.5, but fails with M/n = 0.15; floral sex ratios are adjusted so that spikelets/inflorescence and florets/spikelets are more numerous in males than in females; sexual dimorphism is maximised in floret and spikelet characters; the habit is stoloniferous. Subdioecious *Cortaderia selloana* meets the primary sex ratio F/n = 0.5 because, although genetically gynodioecious, seeds from hermaphrodites make little contribution to the next generation; male sterility is under recessive gene control, and no equivalent female sterility system is as yet recognised, so that the ratio is not exactly determined by the genetics of dioecism. Sex form

dimorphism occurs in florets. Control of male sterility in *C. araucana* is identical.

There is no support for spatial or ecological separation of the sex forms in these taxa, nor in species of stoloniferous or rhizomatous habit is there any risk that colonies may become so large that pollination distances are exceeded; see also Quinn (this volume) and Soreng (this volume) for North American species.

Departures from Dioecism to Apomixis

A switch from hermaphroditism to dioecism is expected to be associated with some genetic benefit which is frequently interpreted as avoidance of strict inbreeding and its consequences (Charlesworth and Charlesworth 1978a, b; Freeman *et al.* 1997; Sakai and Weller 1999) or to differential resource allocation.

The further switch to apomixis in female plants is towards dependence on maternal selection in populations consisting of female plants.

Because many of the dioecious grasses are monotypic there is little opportunity to determine any association between the hermaphroditism-dioecism switch and a response to a compatibility reaction. Sporadic monoecism as described by Reeder (1969) and Columbus (this volume) does offer an opportunity to assess compatibility reactions, but in the only species examined, *Buchloe dactyloides,* they are equivocally reported: self-compatible (Quinn and Engel 1986, Huff in Huff and Wu 1992), and self-incompatible (Wu *et al.* 1984). *Poa* is fundamentally a self-incompatible genus with perfect flowers, but self-compatibility is at least present in cleistogamous species (Campbell *et al.* 1983; Anton and Connor 1995).

Though the compatibility reaction prior to the evolution of dioecism is usually unknown, a subsequent shift to precocious, autonomous apospory in females is now well documented in *Lamprothyrsus* (Bernadello 1979; Connor and Dawson 1993), *Cortaderia* (Connor 1974), and *Gynerium sagittatum* (Astegiano *et al.* 1996). For none of these Argentinian taxa can causality for the departure be established. They correspond to similar movements in dicotyledonous dioecious taxa where causality is equally obscure. So far as is known dioecious species of *Poa* in Argentina have not adopted this apomictic pathway, one already well established in North America (Soreng 1991); apomixis in Argentine *Poa* is known in two gynomonoecious species.

These substitutions are effected by the invasion of gene(s) for apomixis which can spread through females in 1M:1F population leading to fixation, and to the production of families of female plants (Maynard Smith 1971, Charlesworth 1980, Marshall and Brown 1981, Mogie 1992). This substitution and the generation of exclusively female plants was recognised by Gustafsson (1947); Charlesworth (1980) and Mogie (1992) independently offer genetic support to his conclusion.

So far as one can judge, no dioecious species in North American Chlorideae, e.g., *Buchloe, Cyclostachya, Soderstromia* has departed from sexuality, nor any in the Eragrostideae of which *Monanthochloe, Scleropogon* and *Distichlis* are in Argentina. *Cortaderia, Lamprothyrsus* and *Gynerium,* which lie taxonomically outside the main chloridoid concentration of dioecism, have evolved asexuality. *Poa* is archetypically apomictic (Anton and Connor 1995; Soreng 1991; Soreng and Van Devender 1989).

CONCLUSIONS

The reproductive biology of dioecious grasses has been little studied compared with dicotyledonous groups. This examination of dioecious Argentine grasses indicates that: (i) dioecious genera of predominantly austral distribution are more frequent defectors to apomixis than those in other regions of the New World; (ii) 1M:1F in natural populations is in agreement with many other field estimates, but there are exceptions; (iii) segregations in experimental families of *Poa* and *Cortaderia* fit 1:1 ratios. The development of sexual floral dimorphism is of frequent but varied nature and extent, and although associated both with seed

dispersal and the adjustment of floral sex ratios, it fits as a common correlative of dioecism. The association of dioecism and the colony-forming habit occurs in *Monanthochloe* and *Scleropogon,* but other taxa are usually caespitose.

The main theoretical contribution to dioecism by Argentine grasses is in data from experiments in *Poa* and *Cortaderia;* a secondary contribution lies in the details of an actively evolving dioecism in *Scleropogon.*

REFERENCES

Anton, A.M., and Connor, H.E. (1995). Floral biology and reproduction in *Poa* (Poeae:Gramineae). *Australian Journal of Botany* **43**, 577-599.

Anton, A.M., Connor, H.E., and Astegiano, M.E. (1998). Taxonomy and floral biology in *Scleropogon* (Eragrostideae:Gramineae). *Plant Species Biology* **13**, 35-50.

Astegiano, M.E., Anton, A.M., and Connor, H.E. (1995). Sinopsis del género *Cortaderia* (Poaceae) en Argentina. *Darwiniana* **33**, 43-51.

Astegiano, M.E., Anton, A.M., and Connor, H.E. (1996). Arundineae: In 'Flora Fanerogámica Argentina'. Dir. A.T. Hunziker. *Fasciculo* **22**, 1-21.

Bernardello, L.M. (1979). Sobre, el género *Lamprothyrsus* (Poaceae) en Argentina. *Kurtziana* **12-13**, 119-132.

Bierzychudek, P., and Eckhart, V. (1988). Spatial segregation of the sexes of dioecious plants. *American Naturalist* **132**, 34-43.

Campbell, C.S., Quinn, J.A., Cheplick, G.P., and Bell, T.J. (1983). Cleistogamy in Grasses. *Annual Review of Ecology and Systematics* **14**, 411-441.

Charlesworth, B. (1980). The cost of sex in relation to mating system. *Journal of Theoretical Biology* **84**, 655-671.

Charlesworth, B., and Charlesworth, D. (1978a). A model for the evolution of dioecy and gynodioecy. *American Naturalist* **112**, 975-997.

Charlesworth, D., and Charlesworth B. (1978b). Population genetics of partial male sterility and the evolution of monoecy and dioecy. *Heredity* **41**, 136-153.

Clayton, W.D., and Renvoize, S.A. (1986). 'Genera Graminum: Grasses of the World'. *Kew Bulletin Additional Series* **13**.

Connor, H.E. (1965). Breeding systems in New Zealand grasses. V. Naturalised species of *Cortaderia. New Zealand Journal of Botany* **3**, 17-23.

Connor, H.E. (1974) Breeding systems in *Cortaderia* (Gramineae). *Evolution* **27**, 663-678.

Connor, H.E. (1979). Breeding systems in the grasses: a survey. *New Zealand Journal of Botany* **17**, 547-574.

Connor, H.E. (1981). Evolution of reproductive systems in the Gramineae. *Annals of the Missouri Botanical Garden* **68**, 48-74.

Connor, H.E. (1983). *Cortaderia* (Gramineae): Interspecific hybrids and the breeding system. *Heredity* **51**, 395-403.

Connor, H.E. (1984). Breeding systems in New Zealand grasses. XI. Sex ratios in dioecious *Spinifex sericeus. New Zealand Journal of Botany* **22**, 569-574.

Connor, H.E. (1987). Reproductive biology in the grasses. In 'Grass systematics and evolution'. (Eds T.R. Soderstrom, K.W. Hilu, C.S. Campbell, and M.E. Barkworth) pp.117-132. (Smithsonian Institution Press: Washington, D.C.)

Connor, H.E., and Charlesworth, D. (1989). Genetics of male sterility in gynodioecious *Cortaderia* (Gramineae). *Heredity* **63**, 373-382.

Connor, H.E., and Dawson, M.I. (1993). Evolution of reproduction in *Lamprothyrsus* (Arundineae:Gramineae). *Annals of the Missouri Botanical Garden* **80**, 512-517.

Connor, H.E., and Jacobs, S.W.L. (1991). Sex ratios in dioecious Australian grasses: a preliminary assessment. *Cunninghamia* **2**, 385-390.

Couvet, D., Bonnemaison, F., and Gouyon, P-H. (1986). The maintenance of females among hermaphrodites: The importance of nuclear cytoplasmic interactions. *Heredity* **57**, 325-330.

Cruden, R.W., and Lloyd, R.M. (1995). Embryophytes have equivalent sexual phenotypes and breeding systems: why not a common terminology to describe them. *American Journal of Botany* **82**, 816-825.

Davidse, G. (1987). Fruit dispersal in the Poaceae. In 'Grass systematics and evolution' (Eds T.R. Soderstrom, K.W. Hilu, C.S. Campbell, and M.E. Barkworth) pp.143-165. (Smithsonian Institution Press: Washington, D.C.)

Fisher, R.A. (1930). 'The genetical theory of natural selection'. (Clarendon Press: Oxford.)

Fox, J.F., and Harrison, A.T. (1981). Habitat assortment of sexes and water balance in a dioecious grass. *Oecologia (Berlin)* **49**, 233-235.

Freeman, D.C., Klikoff, L.G., and Harper, K.T. (1976). Differential resource utilization by the sexes of dioecious plants. *Science* **21**, 72-77.

Freeman, D.C., Lovett Doust, J., El-Keblawy, A., Miglia, K.J., and McArthur, E.D. (1997). Sexual specialization and inbreeding avoidance in the evolution of dioecy. *Botanical Review* **63**, 65-92.

Gustafsson, A. (1947). 'Apomixis in higher plants. II. The causal aspects of apomixis'. (G.W.K. Gleerup: Lund.)

Huff, D.R. and Wu, L. (1992). Distribution and inheritance of inconstant sex forms in natural populations of dioecious buffalograss *(Buchloe dactyloides)*. *American Journal of Botany* **79**, 207-215.

Kirby, G.C. (1988). The population biology of a smut fungus, *Ustilago spinificus* Ludw. I. Geographic distribution and abundance. *Australian Journal of Botany* **36**, 339-346.

Lloyd, D.G. (1980). The distribution of gender in four angiosperm species illustrating two evolutionary pathways to dioecy. *Evolution* **34**, 123-134.

McDonald, T.J. (1983). Life cycle studies on sand spinifex grass (*Spinifex hirsutus*). Report no. D02.12. pp.75-99 in Research Reports Beach Protection Authority, Brisbane.

Marshall, D.R., and Brown, A.H.D. (1981). The evolution of apomixis. *Heredity* **47**, 1-15.

Maurice, S., Charlesworth, D., Desfeux, C., Couvet, D., and Gouyon, P-H. (1993). The evolution of gender in hermaphrodites of gynodioecious populations with nucleo-cytoplasmic male sterility. *Proceedings of the Royal Society of London. Series B. Biological Sciences* **251**, 253-261.

Maurice, S., Belhassen, E., Couvet, D., and Gouyon, P.H. (1994). Evolution of dioecy; Can nuclear-cytoplasmic interactions select for maleness? *Heredity* **73**, 346-354.

Maynard Smith, J. (1971). The origin and maintenance of sex. In 'Group selection'. (Ed. G.C. Williams) pp.163-175. (Aldine Atherton: Chicago.)

Maze, K.M., and Whalley, R.D.B. (1990). Sex ratios and related characteristics in *Spinifex sericeus* R.Br. (Poaceae). *Australian Journal of Botany* **38**, 153-160.

Mogie, M. (1992). 'The evolution of asexual reproduction in plants'. (Chapman & Hall, London.)

Nicora, E. (1978). 'Flora Patagónica. Parte III. Gramineae'. (Colección Cientifica del Instituto Nacional de Tecnología Agropecuaria Buenos Aires.)

Parodi, L.R. (1949). Los géneros de Avenéas de la flora Argentina. *Revista Argentina de Agronomía* **16**, 205-223.

Parodi, L.R. (1954). Nota preliminar sobre el género *Monanthochloe* (Gramineae) en Argentina. *Physis* **20**, 1-3.

Peterson, P.M., and Columbus, J.T. (1997). Allelic variation in the amphitropical disjunct *Scleropogon brevifolius* (Poaceae:Eragrostideae). *Biolliana Edicion Especial* **6**, 473-490.

Peterson, P.M., Webster, R.D., and Valdes-Reyna, J. (1997). Genera of New World Eragrostideae (Poaceae:Chloridoideae). *Smithsonian Contributions to Botany* **87**, 50pp.

Philippi, R.A. (1870). Sertum mendocinum. *Anales Universidad de Chile* **36**, 159-171.

Philipson, M.N. (1978). Apomixis in *Cortaderia jubata* (Gramineae). *New Zealand Journal of Botany* **16**, 45-59.

Pilger, R. (1940). Über die Gattung *Scleropogon* Phil. *Notizblatt des Botanischen Gartens und Museums zu Berlin-Dahlem* **15**, 15-22.

Quinn, J.A. (1991). Evolution of dioecy in *Buchloe dactyloides* (Gramineae): Tests for sex-specific vegetative characters, ecological differences, and sexual niche-partitioning. *American Journal of Botany* **78**, 481-488.

Quinn, J.A., and Engel, J.L. (1986). Life-history strategies and sex ratios for a cultivar and a wild population of *Buchloe dactyloides* (Gramineae). *American Journal of Botany* **73**, 874-881.

Reeder, J.R. (1969). Las gramíneas dioicas de México. *Bóletin de la Sociedad Botánica de México* **30**, 121-126.

Reeder, J.C., and Reeder, C.G. (1963). Notes on Mexican grasses. II. *Cyclostachya* a new dioecious genus. *Bulletin of the Torrey Botanical Club* **90**, 193-201.

Reeder, J.R., and Reeder, C.G. (1966). Notes on Mexican grasses. IV. Dioecy in *Bouteloua chondrosioides*. *Brittonia* **18**, 188-191.

Reeder, J.C., Reeder, C.G., and Rzedowski, J. (1965). Notes on Mexican grasses. III. *Buchlomimus* another dioecious genus. *Brittonia* **17**, 26-33.

Reeder, J.C., and Toolin, L.J. (1987). *Scleropogon* (Gramineae), a monotypic genus with disjunct distribution. *Phytologia* **62**, 267-274.

Richards, A.J. (1986). 'Plant breeding systems'. (George Allen & Unwin: London.)

Ross, M.D. (1982). Five evolutionary pathways to subdioecy. *American Naturalist* **119**, 297-318.

Sakai, A.K.. and Weller, S.G. (1999). Gender and sexual dimorphism in flowering plants: a review of terminology, biogeographic patterns, ecological correlates, and phylogenetic approaches. In 'Sexual and gender dimorphism in flowering plants'. (Eds M.A. Geber, T.E. Dawson, and L.F Delph.) pp.1-31. (Springer-Verlag: Berlin Heidelberg.)

Schaffner, J.H. (1920). The dioecious nature of buffalo-grass. *Bulletin Torrey Botanical Club* **47**, 119-124.

Schultz, S.T. (1994). Nucleo-cytoplasmic male sterility and alternative routes to dioecy. *Evolution* **48**, 1933-1945.

Shaw, R.B., Bern, C.M., and Winkler, G.L. (1987). Sex ratios in *Buchloe dactyloides* (Nutt.) Engelm. *Botanical Gazette* **148**, 85-89.

Shaw, R.F., and Mohler, J.D. (1953). The selective advantage of the sex ratio. *American Naturalist* **87**, 337-342.

Soreng, R.J. (1991). Systematics of the 'Epiles' group of *Poa* (Poaceae). *Systematic Botany* **16**, 507-528.

Soreng, R.J., and Van Devender, T.R. (1989). Late Quaternary fossils of *Poa fendleriana* (Muttongrass): Holocene expansions of apomicts. *Southwestern Naturalist* **34**, 35-45.

Tzvelev, N.N. (1976). 'Zlaki SSSR.' Nauka Leningrad. ('Grasses of the Soviet Union. Part 1'. Amerind Publishing. New Delhi).

Villamil, C.B. (1969). El género *Monanthochloe* (Gramineae). Estudios morfológicos taxonomicos con especial referencia a la especie Argentina. *Kurtziana* **5**, 369-391.

Watson, L. and Dallwitz, M.J. (1992). 'The grass genera of the world'. (C.A.B. International, Wallingford.)

Webb, C.J. (1999). Empirical studies: evolution and maintenance of dimorphic breeding systems. In 'Sexual and gender dimorphism in flowering plants'. (Eds M.A. Geber, T.E. Dawson, and L.F. Delph.) pp.61-95. (Springer-Verlag: Berlin Heidelberg.)

Wu, L., Harvindi, A.H., and Gibeault, V.A. (1984). Observations on buffalo grass sexual characteristics and potential for seed production improvement. *Hortscience* **19**, 504-506.

GRASSES

Grasses: Systematics and Evolution. (2000). Eds S.W.L. Jacobs and J. Everett. (CSIRO: Melbourne)

APOMIXIS AND AMPHIMIXIS COMPARATIVE BIOGEOGRAPHY: A STUDY IN POA (POACEAE)

Robert J. Soreng[AB]

[A]Biology Dept., New Mexico State University, Las Cruces New Mexico, USA.
[B]Current Address: Natural History Museum, Smithsonian Institution, Washington DC , USA 20560-0166.

Abstract

The distributions of diclinous *Poa* of western North America were studied to compare patterns of occurrence of sexual and asexual reproduction. The distribution of nine apomicts and their nearest sexual relatives are compared over geographical ranges and climatic gradients. Sex-ratios determined from more than 180 population samples and 4,500 herbarium specimens, and the distribution of sexes over geographic ranges, were used to map the distribution of predominantly asexual and sexual modes of reproduction. Elevational lapse rates were used to examine the distribution of amphimicts and apomicts over thermal and moisture gradients. In this group of *Poa* apomicts all taxa are more successful than amphimicts in cool environments with short frost-free seasons, but frequently are also as successful as amphimicts in warm environments with long frost-free seasons. Apomicts occur over broader ranges of most variables considered and are more numerous than amphimicts. They occur in continuous diverse habitats as well as in more island-like habitats. Although apomicts are frequently parapatric to amphimicts, they appear to be mostly excluded from environments in which amphimicts are common. Amphimicts appear to be most successful in mild mesic environments with long frost-free seasons, environments in which shade, pest pressures (but not necessarily species richness or competitor abundance) are likely to be relatively high.

Key words: Apomixis, biogeography, dioecy, parthenogenesis, *Poa*, Poaceae.

INTRODUCTION

In a complex environment, where would you expect to find apomicts and their sexual relatives? This question has theoretical implications for understanding the evolution of sex (Maynard Smith 1978), systematic implications for classification and understanding of variation patterns within apomictic groups (Stebbins 1950; Lynch 1984), and practical implications for plant breeding and conservation of genetic resources in apomictic species. (In the present paper apomixis refers to agamospermy, asexual reproduction by seed.)

In a broad survey of the distribution of apomixis in plant taxa, Bierzychudek (1985) concluded that apomicts exhibit broader geographic ranges and tend to occur at higher elevations and lat-

itudes than their sexual counterparts. Similar patterns have been found among animal parthenogens, where there are also positive correlations between asexual reproduction and habitat disturbance, island-like distribution, and aridity (Bell 1982). There were too few examples in the literature for Bierzychudek to confirm these additional differences in plants.

Although trends in latitude and elevation reveal something about the significance of apomictic reproduction, they do little to predict where sexual relatives of apomicts are most likely to be found. Neither do they explain why the ranges of apomicts and their sexual relatives are frequently disjunct (Lynch 1984). It remains an enigma to students of apomixis why, given density-independent-reproduction and release from the two-fold cost-of-sex (Maynard Smith 1978), apomicts have not completely over-run their sexual progenitors (Glesener and Tilman 1978).

One way to refine our knowledge of the abiotic and biotic factors influencing the distribution and relative abundance of apomicts and amphimicts (sexually reproducing individuals) is to develop detailed knowledge of the distribution patterns of apomicts and their sexual progenitors. Dioecious herbaceous plants provide ideal organisms for such studies because the distribution of asexual reproduction can be described in detail from the geographic and population absence of male plants (Bayer and Stebbins 1983). The present paper presents a study of the distribution of reproductive modes among diclinous bluegrasses, genus *Poa*, in North America, a group in which the occurrence of apomixis is well known (Stebbins 1950; Marsh 1952; Grun 1955; Soreng 1986, 1991; Soreng and Van Devender 1989). Little is known of the type of apomixis in these species except that in *P. wheeleri* embryosac development is diplosporous (Grun 1955; as apomictic *P. nervosa*).

The objectives of the paper are to 1) map the geographic distribution of apomictic and sexual reproduction, 2) compare distributions within and among pairs of apomictic and related amphimictic taxa over several major environmental gradients, 3) compare these findings to those of other authors, and 4) use this information to draw general conclusions about habitat characteristics useful for predicting where sexual counterparts of apomicts are most likely to be found.

METHODS

The nomenclature and taxonomic relationships of the study species (Table 1) follow Soreng (1985, 1990, 1991, 1994, 1998). More than 4500 herbarium specimens and 180 populations of diclinous *Poa* and their close relatives from western North America were examined. Specimens were borrowed from ARIZ, ASU, CAS, DS, GH, ID, MICH, MSC, MO, NY, ORE, OSC, TAES, TEX, UC, US, WIS, WSU, WTU, or studied at CSU, MONT, NMC, RSA, UNM, UTP (herbarium acronyms after Holmgren *et al.* 1990). Complete lists of exsiccatae are on file at MO and US. Compilations of collections of *P. nervosa* and *P. wheeleri* rely in part on D.D. Keck's annotation notes (on file at US and WIS). All duplicates of a distinct collection were treated as one accession, or two accessions if both sexes were represented.

The sample includes 22 taxa, including 17 species and seven subspecies. Apomixis is evident in nine of these taxa in four species, amphimixis is apparently absent in three of these nine, and both apomixis and amphimixis are apparent in the other six. Apomixis appears to be absent in the remaining 13 species.

The ecological distribution of each of the nine taxa in which apomixis occurs is compared with that of its most closely related sexual (paired) progenitor (Table 1). In six comparisons the paired amphimicts are delimited as regions of sexual reproduction within the same species or subspecies as the apomicts they are paired with (Figs 1-6). In these cases it was not possible to distinguish the apomicts morphologically from sexual plants of the same taxon other than by the occurrence of staminate individuals in populations. In the three other comparisons the asexual taxa are morphologically distinct from any populations in which staminate plants are present (Figs 7-9). These apomicts are

compared with sexual taxa that are believed by the author to have been involved in their origins. The 13 strictly sexual taxa employed in this study (Table 1) are divided into two groups, 'unpaired' and 'paired' with an apomict. This allows comparisons within and between nine apomict/sexual pairs, and further comparison with the set of ten unpaired sexual taxa from the same taxonomic subgroups (Soreng 1990, 1991) which have not given rise to any apomictic lineages or been a parent of any apomictic taxon.

The distribution of different sex plants was recorded for each taxon and the distribution of apomixis and amphimixis mapped (Figs 1-9) from the following evidence: 1) sex-ratios – a high female bias within population and herbarium samples; 2) geography – a low frequency or absence of staminate plants in herbarium samples from regions where pistillate plants are common; 3) seed-set – production of seed in individuals or populations isolated from pollen sources. There is no obvious external sexual dimorphism in these species (except in flower), and both sexes are assumed to be equally likely to be collected.

The occurrence of apomixis, and thus the suitability of this approach for studying geographic parthenogenesis (as the subject is often known) in these *Poa*, is indicated by the production of ample seed in individuals isolated from pollen and in populations that are entirely pistillate, and by the absence of plants with developed stamens over most of the geographic ranges of these taxa (Soreng 1986). Furthermore, it has been demonstrated that significantly more seed is set in populations with few or no males than in those with from less than 1:1 to 3:1 females per males, a situation that is best explained (population density aside) by the presence of a high percentage of apomictic females (Soreng 1986). Population samples were collected in the field from 1980 to 1985. In most cases population samples represent at least 30 individuals, but smaller samples were included if all or nearly all individuals in a stand were sampled. Sex-ratios were used to group populations into three classes 1) predominantly sexual – up to three females per male; 2) intermediate – between three and eight females per male, and; 3) predominantly asexual – more than eight females per male.

To map the distribution of amphimixis and apomixis, individual collections were classified as being from predominantly sexual or predominantly asexual geographic regions. If the average sex-ratio in sexual populations is one female to one male, and this appears to be the case in most of these species (estimated from population samples, Soreng 1986), it is reasonable to assume that populations with a high proportion of females are predominantly asexual. The extent of sexual and predominantly asexual regions were first approximated by tabulation of the ratios of female to male specimens collected within counties. If there were one or more males per five females within a county, or no males but less than four females in a county adjacent to a county otherwise considered to be in the amphimictic zone, then that county was considered to be in the amphimictic zone. All other counties, those with females only or less than one male per five females, were considered to be in the asexual zone. As delimited, little or no sexual reproduction occurs within the delimited asexual zones: males are very rare in the latter regions. This procedure slightly biases the results toward larger amphimictic zones but, in view of the paucity of collections in some regions, it reduced the

Table 1. Pairs of apomicts and their closest sexual relatives, and unpaired strictly sexual taxa. Apomicts are boldfaced.

Sexual	(compared with)	Apomict
	Poa sect. *Dioicopoa* E. Desv.	
Poa arachnifera Torrey		none
	Poa sect. *Madropoa* Soreng subsect. *Madropoa* Soreng	
Poa atropurpurea Scribn.		none
Poa confinis Vasey		none
Poa cusickii		Vasey
subsp. *cusickii*		**self**
subsp. *pallida* Soreng		**self**
subspp. *cusickii* and *pallida* /		**P. cusickii subsp. epilis (Scribn.) W.A. Webber**
Poa douglasii Nees		none
Poa fendleriana Steud.		
subsp. *albescens* (Hitchc.) Soreng		**self**
subsp. *fendleriana*		**self**
subsp. *longiligula* (Scribn. & T. A. Williams) Soreng		**self**
Poa leibergii Vasey		none
Poa macrantha Vasey		none
Poa piperi Hitchc.		none
Poa porsildii Gjaerevoll		none
Poa pringlei Scribn.		**self**
Poa stebbinsii Soreng		none
	Poa sect. *Homalopoa* Dumort. 'P. nervosa group'	
Poa sierrae J. T. Howell		none
Poa chambersii Soreng and *Poa rhizomata* Hitchc. /		**Poa cusickii subsp. purpurascens (Vasey) Soreng**
Poa nervosa (Hook) Vasey		**Poa wheeleri Vasey**

possibility of underestimating the size of amphimictic zones. Where asexual zones surround all or part of the amphimictic zones, the extent of zone overlap was estimated by sampling populations within the amphimictic zones.

A second data set includes the elevation, geographic coordinates, and climatic region of each collection site. Baker (1944) divided the western United States into 28 regions with relatively homogeneous climates, and calculated environmental 'lapse rates' within each region for temperature and precipitation. These rates of change over elevation were used to estimate temperature and precipitation conditions for each collection site. The set of collections occurring within each of Baker's regions was divided into three subsets for each taxon pair: 1) asexual – females occurring in regions where males are absent; 2) sexual – all collections of males (except in the rare *Poa nervosa*, *P. rhizomata,* and *P. chambersii,* where the amphimicts are morphologically distinguishable from the apomicts and all plants are counted as sexual; male plants are taken as de-facto proxies of the occurrence of sexual reproduction in populations of all of the taxa, regardless of sex-ratios); 3) mixed sexual and asexual – females from within regions where males occurred. The latter set was used to check for deviations in abundance and elevational range of females from the males in the climate regions where males occurred. The subsets of data are presented as curves for each climate region and, except for elevation (where statistical range is portrayed),

represent one standard-deviation on either side of the mean (Figs 12-16). This organization of the data allows visualization of the ecological distribution of plants within individual regions as well as across the whole ranges of the taxa.

A Chi-square statistic derived from 2×2 contingency tables is used to test the null hypothesis of no pattern of ecological difference between the apomictic and sexual members of pairs. If eight of nine or seven of eight apomicts deviate in the same direction from corresponding sexual relatives, then the pattern is considered significant at the 90% confidence level.

RESULTS

Abundance in Nature

In Table 2 the number of collections of sexual and asexual plants is estimated from herbarium samples. In estimating the proportion of collections that are asexual it is assumed that the average ratio of sexes within sexual taxa is one to one. In eight of nine cases asexual plants are more commonly collected. The ratios of asexual collections to sexual collections within the different pairs are 13.2, 5.7, 3.4, 3.2, 3.2, 1.9, 1.2, 1.0 and 0.5; apomicts outnumbering amphimicts in six cases, nearly equalling them in two, and being outnumbered in one. However, for determining the overall sex-ratio for each taxon, individual collections with a plant of each sex were counted as two plants, whereas pistillate

Figs 1-9. Distribution of zones of predominantly sexual and asexual reproduction in diclinous *Poa* of western North America. Sexual Zone – hatching upper left to lower right. Asexual Zone – hatching upper right to lower left.

collections are counted as only one plant. Correcting for this sampling bias (i.e., counting mixed sex collections as one sample), there are 19, 6, 6, 4, 3.8, 3, 1.3, 1.3, and 0.8 times more apomict collections than amphimict collections (Table 2). Viewed in this way the apomicts outnumber the amphimicts in all but one case.

Geographic Ranges

Distributional ranges are assessed from the standpoint of areal differences (Fig. 10), latitude (Fig. 11) and longitude extensions, extent of island-like occurrences, and geographical overlap. The

areal ranges of the asexual zones (apomicts) are significantly larger than those of amphimictic zones (paired amphimicts) in eight of nine cases. Only within *Poa fendleriana* subsp. *albescens* is there little difference (Figs 3 and 10). The average areal range of the paired amphimicts is larger than that of the unpaired amphimicts, and provides no evidence that the apomicts might have caused contractions in the ranges of paired amphimicts. The influences of different dispersal capabilities on range size are not known. However, the taxa with lemma pubescence, which aids in dispersal of *Poa* seed, do not appear to have greater ranges than taxa with glabrous lemmas (Fig. 10).

Table 2. Frequency of collection of sexual and asexual *Poa*. Columns represent the ratio of apomictic to amphimictic collections (corrected); percentage of the collections estimated to be apomictic (corrected (*a*) and uncorrected (*b*), see text for explanation); total numbers of collections overall, and totals within sexual and asexual zones. Within the sexual zone the sex-ratio among collections (as pistillate to staminate, unless otherwise indicated) is given, along with an estimated number of collections of amphimictic and possibly apomictic (surplus pistillate) individuals assuming a 1:1 sex ratio. Within the asexual zone the number of collections of pistillate (P) and staminate (S) individuals is given.

	Ratio Apomicts/Amphimicts	Overall % Apomicts		Total Coll.	Sexual Zone				Asexual Zome	
		(a)	(b)		Total Colls.	Sex Ratio	Total Sexual	Surplus Pistillate	P	S
Poa arachnifera				90	90	1:1				
Poa atropurpurea				19	19	0.9:1				
Poa chambersii				9	9	0.8:1				
Poa confinis (^A g/d)				90	90	1.4:1				
Poa cusickii										
subsp. *cusickii*	1.3:1	56	55	190	110	1.6:1	86	24	80	5
subsp. *pallida*	4:1	80	77	253	52	1.1:1	50	2	194	7
subsp. *epilis*	3.8:1	79	76	384	compared with subspp. *cusickii* & *pallida*				384	
subsp. *purpurascens*	6:1	86	85	132	compared with *P. chambersii* & *rhizomata*				132	
Poa douglasii				68	68	1.2:1				
Poa fendleriana										
subsp. *albescens*	0.8:1	45	34	65	65	2:1	43	22	B	
subsp. *fendleriana*	3:1	75	66	688	333	1.9:1	229	103	349	6
subsp. *longiligula*	6:1	86	76	699	212	3.2:1	100	112	431	26
Poa leibergii (g)				87	87	0.9:1 (perfect)^C				
Poa macrantha				134	134	1.4:1				
Poa nervosa (pg)				15	15	1.3:1 (perfect & mixed perfect/pistillate)^C				
Poa piperi				61	61	1:1				
Poa porsildii										
Poa pringlei	1.3:1	58	51	68	33	0.9:1	33	35		
Poa rhizomata				15	15	0.9:1				
Poa sierrae				21	21	0.6:1				
Poa stebbinsii (pg)				58	47	0.4:1 (staminate and perfect)^C				
Poa wheeleri	19:1	94	92	284 compared with *P. nervosa*					278	6

^A Reproductive systems: pg = partially gynodioecious, g = gynodioecious, d = dioecious, unmarked taxa are dioecious only.

^B It was not possible to clearly delimit a geographically distinct asexual zone in this subspecies.

^C Ratio of pistillate plants to the set of other plants specified in parentheses.

The latitudinal ranges of apomicts are broader in seven cases and no different in two cases (Fig. 11), and the average latitudinal range of the apomicts is significantly broader (p < 0.01). However, total extensions of ranges are only slightly more northward than southward (average 3.5°N versus 2.5°S). Apomicts occur north of paired amphimicts in six cases and south of them in seven cases. The above analysis does not account for different abundances of the apomicts to the north or south, but it is not apparent from specimen distribution, except in a few cases, that apomicts are more common in the northern than the southern halves of their latitudinal ranges, or that densities are higher in northern versus southern extensions beyond the amphimictic zones.

Apomicts also have broader longitudinal ranges in seven cases. Only weak directional trends emerged among the apomict/amphimict pairs. However, apomicts are uncommon west of the Cascade – Sierra Nevada axis. In contrast eight of the 19 sexual taxa are centered there. This strongly suggests that there are factors encouraging the success of apomixis in the interior continental climatic regions of North America and discouraging its evolution or success in regions of more maritime, in this case Mediterranean climates.

Island-like distributions are found in all the taxa, depending on how population scale is defined. For the purposes of this analysis islands are considered as disjunctions of over 100 km between nearest collections. Considered in this way eight of nine apomicts exhibited island-like distributions and most occur on several islands. In contrast both the paired and unpaired amphimicts have more continuous distributions. However, apomicts also occupy equally as broad or broader areas of continuous distribution.

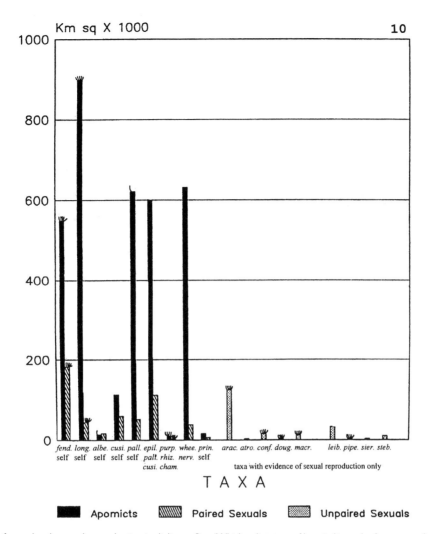

Fig. 10. Areal extent of sexual and asexual reproduction in diclinous *Poa*. 'Whiskers' at tops of bars indicate the frequency of plants in that taxon with pubescent lemmas: five whiskers = all; one = infrequent or rare; none = zero.

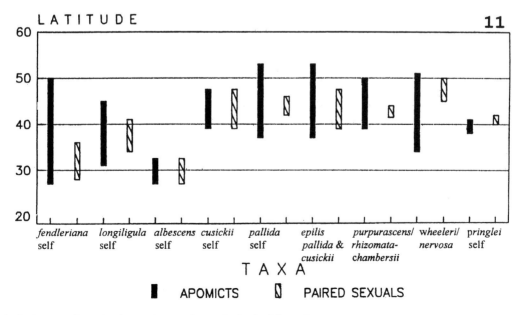

Fig. 11. Latitudinal ranges of sexual and asexual zones of reproduction in diclinous *Poa*.

Apomicts are clearly allopatric from paired amphimicts in only two cases: *P. wheeleri*, and *P. cusickii* subsp. *purpurascens* (Figs 8-9). In addition, sex-ratio data indicate that in *P. pringlei* (Fig. 6) apomicts are rare or absent in the amphimictic zone; and elevational data (Fig. 12) indicate *P. cusickii* subsp. *epilis* occurs at consistently higher elevations than amphimicts of *P. cusickii s.l.* Within the remaining five pairs apomicts and amphimicts occur sympatrically or parapatrically over some portion of their ranges. Although partially apomictic populations are not uncommon, population sex-ratios indicate sexual populations often seemingly exist in isolation from apomicts at the level of local distribution (Soreng 1986). In these cases the apomicts surround the amphimicts, extending through them in latitude, longitude and elevation, but occur at a lower frequency within the ecological center of the amphimictic zone than in peripheral environments.

Gradients

Elevation is primarily of interest as a modifier of temperature and precipitation. These variables have different ranges in each climate region, and although they both change with elevation their relative rates differ among climate regions. Therefore figures of elevation and elevationally modified gradients (e.g., Figs 12-16) are presented so that the distribution of taxa within each climatic region can be examined.

In eight of nine cases apomicts exhibit broader elevational ranges than their paired amphimicts (Fig. 12). The unpaired amphimicts exhibit comparatively much narrower elevational limits than paired amphimicts. The apomicts occur from 305 to 3050 m higher than paired amphimicts. The upper elevation limits of both apomicts and amphimicts in *P. fendleriana* subsp. *albescens* are terminated by the limited height of the Sierra Madre peaks, not by climate constraints, and that case is disqualified from effected analyses. With regard to low elevation, apomicts and paired amphimicts are equally likely to occur at the lowest sites: generally, there was little difference between the terminal lowest points.

The distribution of apomicts was examined along three thermal gradients; mean July temperature (Fig. 13), mean minimum January temperature (Fig. 14), and the length of the frost-free season (Fig. 15). In only three cases did apomicts occur over broader temperature ranges, and in the remaining cases the members of the pairs occur over equally broad ranges. Although seven of eight apomicts are more likely to be found where summer high temperatures are cooler, the average extreme high temperature limits are nearly the same for paired apomicts and amphimicts (Fig. 13). There was more constancy in the high temperature limit (at one standard deviation above the mean for each group) among the sampled taxa and within the pairs than in any other gradient average. The high temperature limit is 21°C.

In all cases apomicts occur over broader ranges of mean minimum January temperature than paired amphimicts, and in eight cases they reach much colder situations (Fig. 14).

Apomicts occur over broader ranges of frost-free season lengths in seven of nine cases (Fig. 15). The mean length of amphimicts'

frost-free season is shorter than that of apomicts in eight cases. Although three paired amphimicts reach the zero day frost-free season, they extend into that climate zone little if at all. In contrast, in eight cases apomicts occur occasionally within that zone or are centered near it.

Apomicts occupy broader ranges of mean annual precipitation than paired amphimicts in eight of nine cases (Fig. 16). Apomicts occur where precipitation is greater by 50–500 mm (mean difference 240 mm) than paired amphimicts in six cases, as well as where precipitation is less in seven cases (mean difference 150 mm). Among these taxa the apomicts do not consistently reach wetter or drier zones than paired amphimicts. Average total snow is greater where apomicts occur than where paired amphimicts occur in eight of nine cases, but there is little difference in low snow amounts.

With regard to major climatic regions (Good 1974; Mitchell 1976), five paired and 10 unpaired amphimicts (15 of 19 amphimicts surveyed) occur in a Mediterranean type climate, as do seven of nine apomicts. Apomicts, however, tend to occur in more than one climatic regime, and amphimicts in only one. Apomicts slightly exceed the range of major climate types tolerated by the paired amphimicts collectively. With regard to tolerance of climatic heterogeneity throughout the western United States (Baker 1944), apomicts occur in more of these minor climatic regions than paired amphimicts in six cases and in an equal number in three cases.

The results are summarized in Table 3.

DISCUSSION

The present analysis of the ecological distribution of apomicts and their sexual counterparts has several advantages over most previous attempts at similar analyses in plants. First, the species considered are diclinous, either gynodioecious or dioecious. This allows us to describe the geographical ranges of sexual and apomictic reproduction based on the frequency of staminate and pistillate plants in herbarium and population samples (Bayer and Stebbins 1983). Usually apomixis is inferred by less direct means and the data are more difficult and time consuming to achieve, and scant (Bierzychudek 1985). Most apomictic plants are pseudogamous and facultatively apomictic, factors which interfere with comparison of the extent of sexual and asexual reproduction (Kellogg 1987). The apomicts in this study, however, do not require pollen for stimulation of seed set, as confirmed by study of plants in isolation or bagging of inflorescences (Soreng 1986). As such they may be considered obligately apomictic over large portions of their geographic ranges. Second, studying a closely related group of apomicts and their sexual relatives in which there are also sexual taxa that have not given rise to asexual taxa, allows us to make comparisons between the two types of sexual populations.

Third, differences in ploidy levels are not a confounding factor in at least three of the pairs assessed. Amphimicts and apomicts of *Poa cusickii* subsp. *pallida*, *P. fendleriana* subsp. *fendleriana*, and subsp. *longiligula*, are octoploid ($2n=56$). Ploidy differences are known between *P. nervosa* ($2n=28$) and its asexual counterpart *P.*

Table 3. Comparison of amphimictic and apomictic *Poa* distributions.

Tests determined significant at the 90% level
Amphimicts occur over narrower geographical ranges (8 of 9 cases).
Amphimicts occur over narrower elevational ranges (8 of 9 cases).
Amphimicts occur over narrower ranges of mean minimum winter temperatures (all cases).
Amphimicts occur over narrower ranges of frost-free season lengths (8 of 9 cases).
Amphimicts occur over narrower ranges of mean annual precipitation (8 of 9 cases).
Apomicts extend to higher elevations (8 of 9 cases).
Apomicts predominate in colder summer climates (7 of 8 cases) and in colder winter climates (7 of 8 cases).
Apomicts predominate where growing seasons are shorter (8 of 9 cases).
Apomicts predominate where there is more total snow (8 of 9 cases).
Apomicts range into alpine habitats (8 of 9 cases).

Additional observations
Apomicts consistently outnumber amphimicts, but not greatly so in most cases.
There is no evidence of contraction of ranges of amphimicts as a possible result of interaction between apomicts and amphimicts.
There is no consistent pattern of sympatry or parapatry versus allopatry between the apomicts and amphimicts.
Apomicts occur in more island-like habitats than amphimicts, but also occur in as continuous and diverse habitats as amphimicts.
Apomicts latitudinal ranges do not extend consistently to the north, nor do apomicts as a group extend on the average much more to the north than south of the amphimicts.
Apomicts do not regularly range into drier climates than amphimicts.
Amphimicts are most abundant in the middle to high end of the temperature range of apomicts, but apomicts may be as abundant as amphimicts in that temperature range.
Apomicts do not regularly range into warmer climates than amphimicts.

wheeleri (2n=56-91) (Grun, l955). A criticism has been levied that differential environmental tolerances might result from multiplied chromatin contents rather than from mode of reproduction (Bierzychudek 1985), although in this instance Stebbins (pers. comm.) discounts the weight of polyploidy as an interfering factor as all 19 taxa with counts are polyploid (Soreng 1994; 2n=28 or higher, *P. chambersii, P. pringlei,* and *P. sierrae* remaining uncounted; *Poa x*=7).

From previous studies of the distributional patterns of apomixis in plants we would expect apomicts to cover larger geographic ranges than their sexual progenitors, and to extend beyond them into terrain that is higher latitude and higher elevation, and thus colder, more island-like, and possibly more arid and more glaciated. The one-sided nature of this picture is relatively uninformative and possibly misleading to researchers interested in where sexual populations are likely to occur.

Although we know apomicts tend to occur in certain extreme conditions, few studies have addressed where they are most abundant or what specific conditions are likely to favor sexual over asexual reproduction. Apomicts in *Poa* often extend through the geographic and climatic ranges of amphimicts, overall exhibiting broader and inclusive ranges of ecological tolerances, and frequently are abundant where their ranges adjoin those of the amphimicts. The following discussion focuses on the common denominators in the distributions of the amphimicts.

The results of the present study in *Poa* agree with the general patterns found in plants, that the ranges of sexual progenitors are smaller than those of their apomictic counterparts. In addi-

tion, the amphimicts occur in one or few local climate regions, and rarely occur in more than one climatic regime whereas apomicts usually occur in more than one climatic regime. The present results also agree that apomicts often occur at higher elevations, but add that apomicts extend throughout the elevational ranges of the amphimicts. The results differ, however, with regard to latitude.

When studying latitudinal aspects of geographic ranges, regional climate variations and geographic barriers to migration need to be considered. In the area studied, the effects of latitude (and elevation) on thermal and precipitation gradients are greatly modified by global atmospheric and oceanic circulation patterns, landmass size, shape and topography (Baker 1944; Bryson and Hare 1974; Mitchell 1976). In contrast to the existence of major geographic barriers to latitudinal migration of European apomicts (i.e., the Baltic and Mediterranean seas and east to west tending mountain ranges) the long north-south tending mountain ranges of North America present a more even terrain in which to study latitudinal migration.

In these *Poa* it appears that latitudinal range extensions of apomicts can be accounted for by larger range sizes, and that directional extensions are species specific. However, it is important to consider the dynamics of species ranges. In theory, apomicts will migrate in the direction of sustaining climate shift, and do so faster than the obligately outcrossing amphimicts. In the Holocene Epoch low latitude temperate and boreal populations are often relictual and high latitude ones established as the result of post-glacial period migrations. Fossil evidence suggests outcrossing sexual populations of

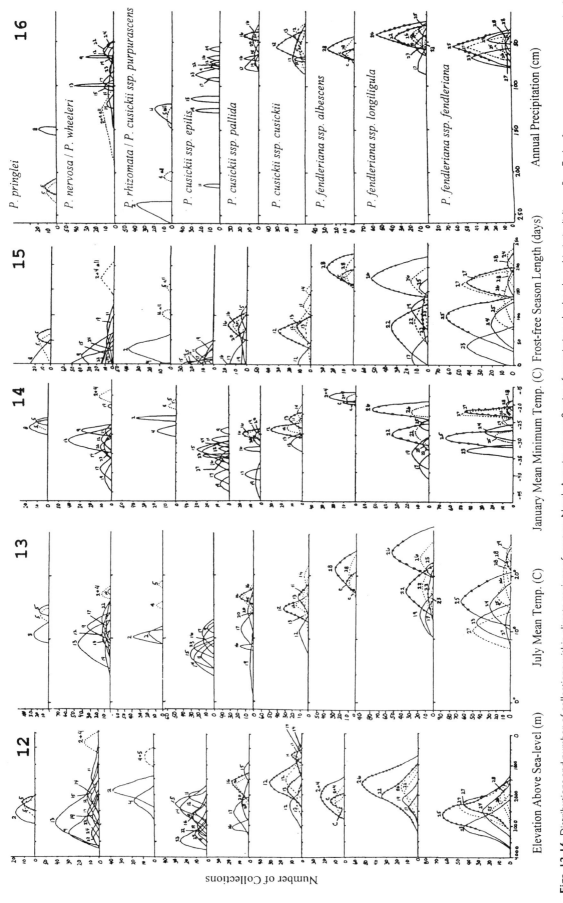

Figs. 12-16 Distributions and number of collections within climate regions of western North America among 9 pairs of apomictic and related amphimictic diclinous *Poa*. Ends of curves represent one standard deviation, or the full range for elevation, above and below the mean (unless terminated within a standard deviation as in fig. 15) within climate regions numbered according to Baker (1944). Peaks of curves correspond to the number of distinct collections of a taxon and their means within climate regions. Simple lines represent pistillate plants from zones of apomictic reproduction. Beaded lines represent pistillate plants from zones and elevations where staminate plants occur. Dashed lines represent only staminate plants except in the cases where apomicts and amphimicts are morphologically distinct and then they represent all plants of the amphimict taxon. 12. Elevation. 13. July mean temperature. 14. January mean minimum temperature. 15. Frost-free season length. 16. Annual precipitation.

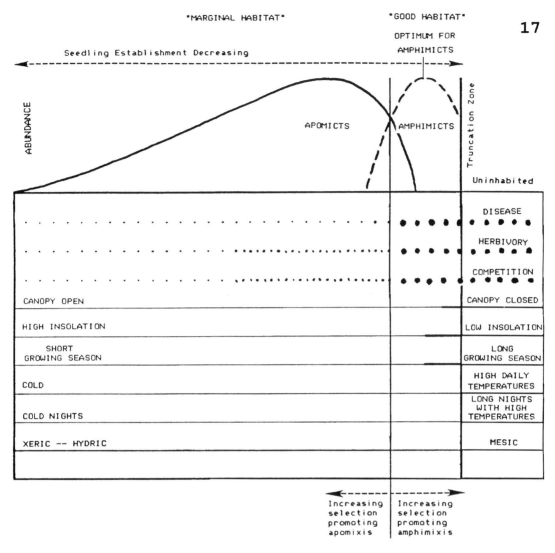

RANGE OF ECOLOGICAL OCCURRENCE

Fig. 17. A model of environmental factors that may exclude apomicts from habitat that is colonizable by amphimicts, and visa-versa for *Poa*. The model accounts for differences in density dependent reproduction, genetic adaptability, and fecundity between amphimicts and apomicts within the framework of abiotic and biotic environmental gradients. Individual factors are expected to differ among taxonomic groups, and to a lesser extent among species within them.

P. fendleriana were much slower than apomictic populations to migrate or expand their ranges (Soreng and Van Devender 1989). This is consistent with the expectation that obligately outcrossing taxa have more capacity to adapt, but, because of lower seed set, have less opportunity to migrate. The degree of separation of climate from latitude in western North America, along with contingencies of history, make extension to higher latitudes a weak characteristic for predicting where sexual reproduction is likely to be found. In addition, the apomicts may have extended to much lower latitudes than amphimicts during long peri-glacial cycles.

It has been suggested that apomicts are more likely to colonize habitats with uniform but island-like distributions, and that amphimicts are more likely to occur in more continuous but variable habitats such as highly dissected mountain ranges or

climatically unstable or unpredictable regions (Bayer and Stebbins 1983). The findings of the present study agree that apomicts are more likely than amphimicts to colonize island-like habitats. However, it appears these *Poa* apomicts are equally as likely as amphimicts to colonize continuous, variable habitats.

It has been suggested for plants (Bierzychudek 1985) and has been demonstrated for animals (Bell 1982) that apomicts and parthenogens occur in more arid regions than their sexual counterparts. This is predicted by the theory that apomicts will be less successful than sexual organisms where competition and pest/predator pressures are high (Van Valen 1973; Levin 1975; Bell 1982). However, when temperature and precipitation gradients are compared, there is no apparent trend toward occurrence in arid environments among diclinous *Poa* apomicts.

Diclinous *Poa* apomicts consistently predominate at higher elevations, and thus cooler environments, than do their sexual counterparts. In addition, apomicts and amphimicts exhibited fairly uniform reductions in abundance with respect to high temperatures: all of the taxa dropped out near the 21°C July mean. This value agrees well with the general distributional limit of the 24°C July mean for the occurrence of species of the genus *Poa* worldwide (Hartley 1961; Hiesey and Nobs 1982). These patterns suggest that in general the apomicts are under less heat stress than amphimicts.

In consort with temperature, precipitation values for the taxa do not agree with the aridity expectation. Orographic and interior continental influences are significant here. Apomicts predominate where there is greater precipitation more often than where there is less precipitation. In addition, apomicts predominate where total snow accumulation was greater than for amphimicts. This also suggests that the apomicts are under less moisture stress.

The frequent occurrence of the apomicts in situations with short or zero frost-free seasons is not surprising in *Poa*, a genus of cool season grasses in which the majority of species occur in montane temperate and boreal habitats. However, that these amphimicts appear to be consistently restricted from this zone, yet often approach it, is of interest. The factors preventing the amphimicts from establishing populations in the year-round frost zone (high elevations and latitudes) need examination. Some evidence has been gathered (Soreng 1986) suggesting that a high degree of herbaceous competition in lower alpine habitats may reduce the likelihood of obligately outcrossed species establishing reproductively stable populations. Specifically, associated herbaceous plant cover, especially grass cover, was frequently as high or higher for apomictic than sexual populations. Furthermore, as attested to by records of habitat variables (Soreng 1986), amphimicts of *P. fendleriana*, *P. nervosa*, *P. piperi*, *P. rhizomata*, and *P. sierrae* (and to a lesser degree in *P. cusickii*), occur most frequently in relatively protected forest openings, where herbaceous cover is low, species richness is moderate, and sunlight is limited. The apomicts are most frequently found in more open and exposed sites, and infrequently occur in abundance in semi-closed, closed canopy, or otherwise shady habitats. Apomicts' freedom from mate density-dependence would be advantageous in environments crowded by other species, especially if pest pressures are reduced by frequent freezing or aridity. It needs to be considered that selection against high fecundity in shady habitats, which characteristically have low energy budgets (Salisbury 1942), may reduce the survival rate of apomicts relative to amphimicts there.

SUMMARY AND CONCLUSIONS

Within the limits of resolution of the present study *Poa* amphimicts may be seen as reproducing best in situations where, on the average, temperatures are warmer, growing seasons are longer, and precipitation is lower than for apomicts. However, using elevational lapse rates (Baker 1944) to estimate the position of plants along gradients of temperature and precipitation does not account for mitigating, site specific factors. Thus moisture conditions may not actually be more stringent.

More detailed studies of population distribution carried out for *Poa fendleriana* subsp. *fendleriana* (Soreng 1986) suggested mean minimum January temperature, length of frost-free season, and total snow are more important determinants of the distribution of amphimicts and apomicts than are slope and aspect over the 23° latitudinal range of this taxon. However, within regions of overlapping zones, sexual populations relative to apomictic populations, were concentrated in habitats with lower solar irradiation (adjusted for slope, aspect, and latitude), moderate diversity of herbaceous cover, high percentage tree cover, and protected topographic position. As a result the present discussion must be tempered, but even if apomicts occur on more extreme exposures their predominance at higher elevations would compensate to some degree for increased evapotranspiration.

One of the simplest and most explanatory hypotheses concerning the different ranges of apomicts and amphimicts is that amphimicts have reduced migration and colonization capacities due to limitations in their reproductive capacity. The apomicts may migrate in the direction of climate change while the amphimicts are constrained to adapt in place. However, this hypothesis does not explain why apomicts, in regions where they overlap and surround amphimicts, appear to be excluded from habitats in which the amphimicts are abundant.

Although in these *Poa* the apomicts are more numerous than the amphimicts, in most cases apomicts were only between two and six times more abundant. In *Poa fendleriana* apomicts are only twice as abundant as amphimicts, yet seed-set in apomictic populations of *Poa fendleriana* averaged four times greater than in sexual populations (Soreng 1986). Given that all apomictic progeny are female, there is an eight-fold divergence in their yearly reproductive capacities. Evidently, low survivorship in these apomicts is a general pattern among apomicts (Glesener and Tilman 1978; Maynard Smith 1978), and there are a variety of plausible explanations for why this may be so. Differential survival rates may stem from genetic deterioration of the apomicts (Muller 1964), differences in resource allocation (Salisbury 1942; Harper 1977), and inability of apomictic lineages to produce new genotypes in response to selective pressures such as intra- and interspecific competition (Bell 1982), microparasite (Bremermann 1985) predator pressures, and climatic variability (Williams 1975). The relative abundance of apomicts is also contingent on the life history, age, and origin of the apomicts (Stebbins 1950 p. 414, 1985; Soreng 1986).

The present study provided no evidence to support the hypothesis (Lynch 1984) that interaction (e.g., competition or reproductive interference) between apomicts and amphimicts caused reductions in the ranges of the amphimicts. By comparing the paired amphimicts to related sexual species without apomictic counterparts, it was apparent that the abundance, range size, and other distributional characteristics of the paired amphimicts were greater than for unpaired amphimicts in most taxa for all variables tested. In the majority of cases the apomicts were observed to occur all around the amphimicts but appeared to be excluded from the habitat in which the amphimicts were most abundant. Reduced abundances of apomicts where they are parapatric or sympatric with amphimicts, appears to be better explained by differential abiotic or biotic tolerances, than by interaction.

Relative to the amphimicts, it appears that these *Poa* apomicts are more successful in more exposed terrain, in colder environments with shorter growing seasons, but they also may reproduce as well as amphimicts in warmer environments with long growing seasons. Amphimicts, in contrast, appear to be most successful in protected terrain, in relatively mesic environments with long growing seasons and mild temperatures, environments in which disease and pest pressures are expected to be high, and selection to be 'K' directed.

To a certain extent the response of apomicts are expected to be group specific and predictable based on latent selection potentials (i.e., limitations imposed by the evolutionary history of the group). *Poa*, for example, is well adapted to cold, and intolerant of sustained high temperatures (Hartley 1961; Hiesey and Nobs 1982). Thus, these apomicts have not broken any heat barriers, but are capable of migrating into cold regions. In contrast, *Parthenium* species are relatively intolerant of cold, but have adaptations for surviving in hot dry climates, and apomicts in that genus appear to respond to this potential (Rollins 1949).

Beyond genetic background, the advantages of apomixis are in broadly tolerant genotypes and high reproductive rates and thus the ability to migrate as climate changes. Amphimicts occur predominantly in environments with abiotic factors promoting K selection and with biotic pressures favoring genetic flexibility. Thus, if biologists want to find sexual populations of other apomicts in which sexual reproduction is more difficult to detect (e.g., *Poa pratensis*, and *P. secunda*), they should look in areas of the species ranges with long growing seasons, with abiotic factors selecting against high expenditures on reproduction by seed, and with conditions mesic enough to promote abundance and diversity of predator and disease organisms (Fig. 17).

The habitat parameters favorable to sexual reproduction outlined here should have general application to other groups. It is hoped that the examples and ideas presented here will be useful in efforts to model the distribution of sexual reproduction in other apomictic groups and will aid in the discovery of new sexual populations within the ranges of known apomicts.

REFERENCES

Baker, F. S. (1944). Mountain climates of the western Unites Sates. *Ecological Monographs* **14**, 223-254.

Bayer, R. J., and Stebbins, G. L. (1983). Distribution of sexual and apomictic populations of *Antennaria parlinii*. *Evolution* **37**, 555-561.

Bell, G. (1982). 'The Masterpiece of Nature, the Evolution and Genetics of Sexuality'. (University of California Press: Berkeley.)

Bierzychudek, P. (1985). Patterns in plant parthenogenesis. *Experientia* **41**, 1255-1264.

Bremermann, H. H. (1985). The adaptive significance of sexuality. *Experientia* **41**, 1245-1254.

Bryson, R. A., and Hare, F. K. (1974). The climates of North America. In 'World Survey of Climatology' (Ed. H. E. Landsburg) (Elsevier Scientific Publishing Co.: New York.)

Glesener, R. R., and Tilman, D. (1978). Sexuality and the components of environmental uncertainty: Clues from geographic parthenogenesis in terrestrial animals. *American Naturalist* **112**, 659-673.

Good, R. (1974). 'The Geography of the Flowering Plants' (4th ed.) (John Wiley and Sons Inc.: New York.)

Grun. P. (1955). Cytogenetic studies in *Poa*: III. Variation within *P. nervosa*, an obligate apomict. *American Journal of Botany* **42**, 778-784.

Harper, J. L. (1977). 'Population Biology of Plants'. (Academic Press: New York.)

Hartley, W. (1961). Studies on the origin, evolution, and distribution of the Gramineae IV: The genus *Poa*. *Australian Journal of Botany* **9**, 152-161.

Hiesey, W. M., and Nobs, M. A. (1982). Experimental studies on the nature of species VI. Interspecific hybrid derivatives between facultatively apomictic species of bluegrasses and their responses to contrasting environments. Carnegie Institute of Washington, Publication No. 636.

Holmgren, P. K., Holmgren, N. H., and Barnett, L. C. (1990). 'Index Herbariorum, Part I: The Herbaria of the World' (8th ed.) (New York Botanical Garden: Bronx.)

Kellogg, E. A. (1987). Apomixis in the *Poa secunda* complex. *American Journal of Botany* **74**, 1431-1437.

Levin, D. A. (1975). Pest pressure and recombination systems in plants. *American Naturalist* **109**, 437-451.

Lynch, M. (1984). Destabilizing hybridization, general purpose genotypes and geographic parthenogenesis. *Quarterly Review of Biology* **59**, 257-290.

Marsh, V. L. (1952). 'A Taxonomic Revision of the Genus *Poa* of the United States and Southern Canada'. (Ph.D. dissertation University of Washington: Seattle.)

Maynard Smith, J. (1978). 'The Evolution of Sex'. (Cambridge University Press: London.)

Mitchell, V. L. (1976). The regionalization of climate in the western United States. *Journal of Applied Meteorology* 15, 920-927.

Muller, H. J. (1964). The relation of recombination to mutational advance. *Mutation Research* **1**, 2-9.

Rollins, R. C. (1949). Sources of genetic variation in *Parthenium argentatum* Gray. *Evolution* **3**, 358-368.

Salisbury, E. J. (1942). 'The Reproductive Capacity of Plants, Studies in Quantitative Biology'. (G. Bell and Sons Ltd.: London.)

Soreng, R. J. (1985). *Poa* in new Mexico, with a key to Middle and Southern Rocky Mountain species (*Poaceae*). *Great Basin Naturalist* **45**, 395-422.

Soreng, R. J. (1986). 'Distribution and Evolutionary Significance of Apomixis in Diclinous *Poa* of Western North America'. (Ph.D. dissertation New Mexico State University, Las Cruces.)

Soreng, R. J. (1990). Chloroplast-DNA phylogenetics and biogeography in a reticulating group: Study in *Poa* (Poaceae). *American Journal of Botany* **77**, 1383-1400.

Soreng, R. J. (1991). Systematics of the 'Epiles' group of *Poa* (*Poaceae*). *Systematic Botany* **16**, 507-528.

Soreng, R. J. (1994). *Poa* L. In 'The Jepson Manual, Higher Plants of California'. (Ed. J. C. Hickman) pp. 1284-1291. (University of California Press: Berkeley.)

Soreng, R. J., (1998). An infrageneric classification of *Poa* in North America, and other notes on sections, species, and subspecies of *Poa*, *Puccinellia*, and *Dissanthelium* (*Poa*ceae). *Novon* **8**,187-202.

Soreng, R. J., and Van Devender, T. R. (1989). Late quaternary fossils of *Poa fendleriana* (Muttongrass): Holocene expansions of apomicts. *The Southwestern Naturalist* **34**, 35-45.

Stebbins, G. L. (1950). 'Variation and Evolution in Plants'. (Columbia University Press: New York.)

Stebbins, G. L. (1985). Polyploidy, hybridization, and the invasion of new habitats. *Annals of the Missouri Botanical Garden* **72**, 824-832.

Van Valen, L. (1973). A new evolutionary law. *Evolutionary Theory* **1**, 1-30.

Williams, G. C. (1975). 'Sex and Evolution'. (Princeton University Press: New Jersey.)

Grasses: Systematics and Evolution. (2000). Eds S.W.L. Jacobs and J. Everett. (CSIRO: Melbourne)

COMPARATIVE REPRODUCTIVE BIOLOGY OF THE VULNERABLE AND COMMON GRASSES IN *BOTHRIOCHLOA* AND *DICHANTHIUM*

Ping Yu, N. Prakash and R. D. B. Whalley

Division of Botany, School of Rural Science and Natural Resources,
University of New England, Armidale NSW 2351, Australia.

Abstract

Apomixis is found in the vulnerable grasses *Bothriochloa biloba* S. T. Blake and *Dichanthium setosum* S. T. Blake and it is of the pseudogamous apospory and adventive embryony types. Apomixis reduces seed production in *B. biloba* and *D. setosum*. The low seed production is mostly caused by competition for resources among multiple embryo sacs with pseudogamous apospory, adventive embryony and polyembryony accompanying apomixis and a shortage of endosperm for supplying nutrition in adventive embryony. High ratio of embryo sac abortion is also an important factor causing sterility in *D. setosum*. A comparison between obligate sexual reproduction and facultative apomixis indicates that obligate sexual reproduction leads to high seedset in the common grasses, *Bothriochloa macra* S. T. Blake and *Dichanthium sericeum* A. Camus. The resource limitation caused by a high pollen/ovule ratio in *D. setosum* may be an additional cause for the extremely low seed set in this species.

Key words: Apomixis, sexual reproduction, embryological development, pollen/ovule ratio, rare species, Poaceae.

INTRODUCTION

Which factors cause rare plant species to be rare and common species to be common? By reviewing the published literature on rare-common species comparisons, Kunin and Gaston (1993) concluded that the general traits of rare species appear to be that 'the reproductive characters of rare species tend to be biased away from outcrossing and sexual reproduction'. Rare taxa show a tendency towards asexual reproduction, lower levels of self-incompatibility, lower overall reproductive effort and poorer seed dispersal (Kunin and Gaston 1993). For example, the rare species *Eupatorium resinosum* had lower levels of sexual reproduction compared with its common congener *E. perfoliatum* (Byers and Meagher 1997). One of the main factors blamed for the rarity of the endangered aquatic plant *Howellia aquatilis* was the lack of genetic variability caused by obligate self-fertilization (Lesica *et al.* 1988). Longton (1992) suggested that if self-fertilization becomes obligate in species of British mosses, the species tend to become rare. A low level of self-incompatibility is expressed in these two cases of rarity. A trend towards asexual reproduction and poorer dispersal exists in the endangered shrub *Haloragodendron lucasii* (Sydes and Peakall 1998) with extensive clonality as its main reproductive mode.

Seed production may be an important factor limiting population size of a vulnerable plant species. There are many factors that limit the production of viable seeds. Intrinsic limitations include genetically programmed breeding systems, embryonic abnormalities, pollination mechanisms and self-incompatibility (Pavlik *et al.* 1993; Wiens *et al.* 1987, 1989). Extrinsic limitations include abiotic resource levels, pollinator availability and predation on seeds, fruits or portions of the whole plant (Pavlik *et al.* 1993).

The breeding system is a major determinant of the genetic structure of a population. Most grasses have flexible breeding systems, which are a mixture of asexual, autogamous, and xenogamous strategies (Richards 1990). The *Bothriochloa-Dichanthium* complex as a whole consists of diploid sexual races and polyploids that vary from almost completely sexual to obligate apomicts (Celarier and Harlan 1957; Harlan *et al.* 1964; Saran and de Wet 1970).

Genetically mediated embryonic abnormalities caused sterility in *Hilaria belangeri* (Steud.) Nash and *Hilaira mutica* (Buckl.) Benth. (Brown and Coe 1951). It is also responsible for the relic nature of *Dedeckera eurekensis* and might ultimately result in its extinction (Wiens *et al.* 1989).

Pollination in grasses is mainly by wind, therefore production of enough viable pollen is critical for effective pollination. Sex allocation theory explains the influence of pollen production (male investment) on seed production (female investment) as follows: on the one hand if the pollen/ovule ratio (P/O) is below the threshold for effective pollination, the amount of viable pollen reaching stigmas is insufficient for fertilization of all the ovules that a plant could produce, fecundity decreases (Cruden 1977; Totland 1997). On the other hand, the total resource allocation for reproduction must be divided between male and female functions. Therefore, if pollen production is above the threshold for effective pollination, a trade-off takes place between paternal investment (as P/O) and maternal investment in seed production (Charnov 1979, 1982, 1987; Charlesworth and Charlesworth 1981; Garnier *et al.* 1993).

Little work has been done to compare the reproductive biology of rare and common grasses and to assess intrinsic limitations on seed production in vulnerable grasses. The aim of the present study was to compare the reproductive biology of the vulnerable grasses *B. biloba* and *D. setosum* with two closely related and co-occurring common congeners *B. macra* and *D. sericeum*, and to investigate the reproductive factors limiting seed production of the two vulnerable grasses.

MATERIALS AND METHODS

Study Species

Both *Bothriochloa* and *Dichanthium* belong to the Poaceae, Subfamily Panicoideae, Tribe Andropogoneae. Species of *Bothriochloa*, *Dichanthium* and *Capillipedium* are called Old World Bluestem (OWB) and they are united across the bridging compilospecies *B. intermedia* (Harlan and de Wet 1963; De Wet and Harlan 1966). De Wet and Harlan (1970) suggested that *Bothriochloa*, *Capillipedium* and *Dichanthium* should be combined under the generic name *Dichanthium*.

Bothriochloa biloba and *D. setosum* are listed as vulnerable in 'The Threatened Species Conservation Act 1995', New South Wales, Australia. *Bothriochloa biloba* is also listed as 3V in 'Rare and Threatened Australian Plants' (Briggs and Leigh 1996), i.e. its geographical range in Australia is greater than 100 km, and it is at risk in a longer period (20-50 years) by continued depletion. *D. setosum* was listed as 3VC in the 1988 edition of 'Rare and Threatened Australian Plants' (Briggs and Leigh), it was removed from the list in the 1996 edition. The distribution range of *B.*

biloba is 35°17'S-21°S, 148°E-152°30'E and *D. setosum* is distributed in the range from 42°S-20°36'S, 147°50'E-152°20'E.

Bothriochloa macra and *D. sericeum* are widely distributed in Australia and *D. sericeum* is in New Guinea as well. All four species are often found in grasslands or woodlands to which fertilizer has not been added. At our experimental sites, *B. biloba* and *D. sericeum* co-existed in grasslands and *B. macra* and *D. setosum* co-occurred in grassland or woodland. The four grasses commence growth in spring, flower in summer and become dormant in late autumn to winter, so they are warm season perennial grasses.

Flowering Phenology and Embryological Studies

Flowering phenology observations were conducted once per day on eight inflorescences each from different plants of each species in the glasshouse commencing in December 1997. Flowering phenology of *B. macra* and *D. setosum* was monitored from two populations in the field near Armidale (30°27'S, 151°40'E; 30°30'S, 151°43'E) and that of *D. sericeum* and *B. biloba* was observed from one population near Inverell (29°50'S, 150°59'E) and one population at Warialda (29°25'S, 150°28'E) once per month. After walking in a straight line from one point in the field, one inflorescence was selected from the closest plant every 20 steps. Every bisexual floret on the inflorescence was dissected to check for chasmogamy or cleistogamy.

For embryological studies, fresh inflorescences of *B. biloba* were collected from Inverell, Warialda and from glasshouse plants transplanted from Attunga (30°56'S, 150°35'E) and Inverell. Inflorescences of *B. macra* and *D. setosum* were collected from two sites near Armidale or from glasshouse plants grown from seeds collected at Armidale. Inflorescences of *D. sericeum* were collected from Armidale and Inverell or glasshouse plants grown from seeds collected at Inverell. The collecting period was from December to February 1996. Pistils and anthers were fixed in FPA (five parts formalin, five parts propionic acid and 90 parts 70% ethanol), embedded in paraplast, sectioned at 10-12 μm and stained in safranin /fast green. In order to decide whether the ovules were fertilized, ovules dissected before anthesis and after anthesis were carefully separated.

Microscopic images of anthers and ovules were captured by a DAGE-MTI B-chip video camera connected to a Zeiss Axioskop Microscope, digitized by a Scion Image frame-grabber in a Macintosh 9600 computer. NIH-Image software was used for frame grabbing and Adobe photoshop software was used to adjust contrast and brightness of the image. Embryological pictures were composed in Adobe PageMaker software.

Cytological Studies

Seeds from each population were germinated in an incubator. The seedlings were transplanted into pots after germination, and grown in the glasshouse for one month. Then root tips were collected, squashed with 2% aceto-orcein and chromosomes were counted (Mujeeb-Kazi and Miranda 1985).

Pollen Production

Species of *Bothriochloa* and *Dichanthium* have paired spikelets, one sessile, fertile and awned, and the other pedicellate, unawned. Pedicellate spikelets in *B. biloba, B. macra* and

D. sericeum are sterile while those in *D. setosum* are male (Wheeler *et al.* 1982). One pair of spikelets just below those where the florets of the sessile spikelets had already opened, were collected from each of eight plants of each species growing in the glasshouse. All the anthers from one floret were placed in a vial, squashed using a glass rod, then 20 μl Aniline blue solution was added and mixed thoroughly. After staining for ten minutes, one drop of the solution with evenly dispersed pollen was placed on the grid of a hemacytometer, and viable and non-viable pollen grains were counted using a microscope (Kearns and Inouye 1993). As the hemacytometer has two big gratings each with a volume of 0.9 mm³, one count in each grating of one hemacytometer was conducted for each sampling. Dark blue-stained grains were recorded as viable, pale blue-stained and colorless grains were decided as non-viable. The number of viable pollen grains and total pollen grains per floret were estimated by multiplying the mean number of pollen grains in the two gratings of one hemacytometer by the appropriate dilution factors. The anthers in the male, pedicellate spikelets of *D. setosum* were examined and viable and non-viable pollen grains recorded as above.

Seed Production per Inflorescence

Twenty inflorescences at the first sign of anthesis, each from different plants of the four species in the glasshouse were bagged and the bagged inflorescences were collected four weeks later. Empty spikelets and seeds in each inflorescence were recorded. The same treatment was conducted in the field to compare the seed production in the glasshouse and in the field.

Germination

In order to check the occurrence of seeds with multiple embryos, germination experiments were conducted in petri dishes containing germination pads and Kimpak absorbent paper (Seedburo Equipment, Chicago, Illinois). Seed samples of *B. macra* and *D. setosum* were collected from Armidale and those of *B. biloba* and *D. setosum* were collected from Inverell and stored at room temperature for 12 months. Before placing the seeds in the petri dishes, the germination pads were moistened with tap water to the point of run-off. Four replicates of 50 seeds from each species were tested for germination in an incubator set at 30°C/20°C in 12 hourly cycles, lit by 3 x 30 W Cool White fluorescent light tubes during the high temperature phase. Germinated seeds were examined for the presence of multiple embryos two weeks later.

Statistical Analysis

Pollen production and seed production were analyzed using analysis of variance and, when necessary, data were arc sin transformed to conform to the assumptions for ANOVA. Paired t-test was conducted to test the difference in number of viable pollen, total pollen and percentage viable pollen per floret between paired sessile bisexual floret and pedicellate male floret in *D. setosum*.

RESULTS

Flowering Phenology of the Four Species

Bothriochloa macra, *D. sericeum* and *D. setosum* started flowering in early November whereas *B. biloba* started flowering in Decem-

Table 1. Occurrence of chasmogamy only, cleistogamy only or co-occurrence of both on individual racemes.

Species	Chasmogamy only	Cleistogamy only	Both co-occur
B. biloba	+	+	+
B. macra	+		+
D. sericeum	+	+	+
D. setosum	+		

ber. Flowering finished at the end of March in *B. macra* and *D. setosum*, at the end of April in *D. sericeum* and seedhead fall had finished in late May for *B. biloba*. The sequence of anthesis varied. In all four species anthesis started from the apex and then progressed downwards or started from a floret close to the apex and then progressed upwards and downwards simultaneously.

Stigmas appeared earlier than anthers in chasmogamous florets, indicating that all four species are protogynous. Only chasmogamy was recorded in *D. setosum* while in the other three species chasmogamous or cleistogamous or both kinds of florets were recorded on individual racemes (Table 1). The ratio of chasmogamy and cleistogamy varied among inflorescences of the same plant, different plants in the same population, different populations of the same species at the same growing period and the same population at different growing periods. Paleas and lemmas of cleistogamous florets remained closed all the way through to seed maturity. Anthesis occurred late in the afternoon in *D. setosum* and early in the morning in the other three species. Anthesis did not occur in the pedicellate male florets in *D. setosum* until all the bisexual florets on the same inflorescence had finished. Consequently, pollen of the male florets would not pollinate pistils of the same inflorescence.

Chromosome Numbers

Both diploid (2n = 20) and tetraploid (2n = 40) plants were found in the Inverell population of *B. biloba*, while in the other three species only diploids (2n = 20) were recorded.

Reproduction

Apomixis commonly occurred in *B. biloba* and *D. setosum*, as well as normal sexual reproduction. However, only sexual reproduction was found in *B. macra* and *D. sericeum*.

Embryological Development in Sexual Reproduction

Microsporogenesis is normal in all four species, even in *B. biloba*, which is predominately apomictic. The anther tapetum is of the secretory type, with successive cytokineses in pollen mother cells and three-celled mature pollen grains (Johri *et al.* 1992).

Megasporogenesis and embryo sac development in the sexual development pathway in all four species followed the common developmental pattern for grasses described by Johri *et al.* (1992): a *Polygonum* type of embryo sac generated from a single archesporial cell; anatropous, tenui- or crassinucellar, bitegmic ovules; secondary multiplication of antipodal cells; and the nucellar cells close to the micropyle enlarging to form a nucellar cap (Fig. 1A).

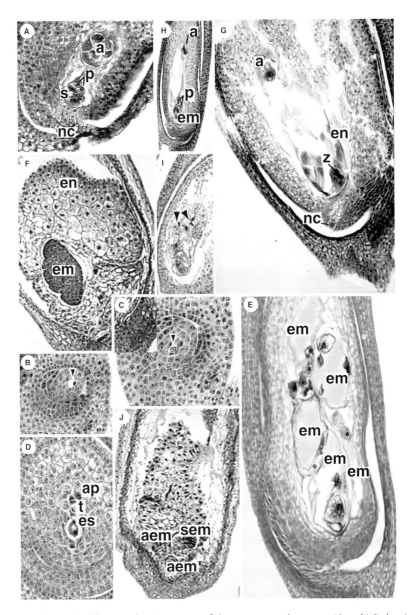

Fig. 1. (A) Longitudinal section of an ovule of *B. macra* showing a mature *Polygonum* type embryo sac with multiplied antipodal cells and nucellar cap (Bar, 10μm). (B-J) Longitudinal sections of ovules of *B. biloba*. (B-C) Successive sections of one ovule (Bars, 10μm): arrow in Fig. 1B shows a degenerating archesporial cell; arrow in Fig. 1C shows a healthy aposporial embryo sac initial. (D) Three lines of development co-exist in this ovule: a 2-nucleate embryo sac, a triad and an aposporial embryo sac initial (Bar, 10μm). (E) Five aposporial embryo sacs in one ovule (Bar, 60μm). (F) A pseudogamous aposporial embryo with cellular endosperm (Bar, 80μm). (G) A fertilized embryo sac with degenerating zygote; one antipodal cell and one nucellar cap cell started dividing (Bar, 30μm). (H) An unfertilized embryo sac with an adventive embryo; one antipodal cell started dividing (Bar, 20μm). (I) Arrows show three adventive embryos in one unfertilized ovule (Bar, 20μm). (J) Fertilized ovule with a sexual embryo and two apomictic embryos (Bar, 60μm).

(*a*: antipodal cell, *aem*: apomictic embryo, *ap*: aposporial embryo sac initial, *em*: embryo, *en*: endosperm, *es*, embryo sac; *nc*: nucellar cap, *p*: polar nuclei, *sem*: sexual embryo, *t*: triad)

Apomixis

Apomixis in *B. biloba* and *D. setosum* was either pseudogamous apospory or adventive embryony or both co-occurred.

Pseudogamous Apospory

Apomictic embryo sac initials usually originated from nucellar cells after the degeneration of the megaspore mother cell (MMC) or occurred parallel to the development of the MMC (Fig. 1B-C). They can be distinguished from sexual structures by having mul-

tiple archesporial cells and the degeneration of some archesporial cells. The co-existence of sexual and apomictic embryo sacs in *B. biloba* was quite common. Fig. 1D shows that in one ovule three pathways of development co-existed: (1) a 2-nucleate embryo sac, (2) a triad and (3) an aposporial embryo sac initial cell. The mature aposporious embryo sac is 2-nucleate, 3-nucleate, 4-nucleate or 8-nucleate with irregular distribution of nuclei. Up to five embryo sacs were found in one ovule (Fig. 1E). Fig. 1F shows one aposporous embryo surrounded by cellular endosperm.

Because fertilization is needed for the formation of endosperm, this kind of apomixis is called pseudogamous apospory.

Adventive Embryony

Adventive embryos were found to initiate from nucellar cells, cells inside apomictic embryo sacs, egg, synergid or antipodal cell inside a sexual embryo sac. In Fig. 1G an antipodal cell and one nucellar cap cell have restarted division and both have conspicuous nuclei and dark-stained cytoplasm. The ovule was from a floret where anthesis had been completed and a nuclear endosperm and a degenerating zygote were found in the sexual embryo sac. Fig. 1H shows an adventive embryo formed before anthesis in an ovule of *B. biloba.* In this ovule two polar nuclei still existed while an antipodal cell showed signs of dividing by its dark-stained cytoplasm. Arrows in Fig. 1I show three adventive embryos in one ovule of *B. biloba*. Fig. 1J shows three embryos side by side in one embryo sac with cellular endosperm. From the position of the three embryos in the embryo sac, it could be determined that one embryo developed from a fertilized zygote while one of the other two might have come from an unfertilized synergid and the third from the adventive embryony of a nucellar cell. As the embryo that developed from the synergid was not a result of fertilization, its chromosome number should be half that of the maternal parent. This perhaps explains why there were chromosome counts of 20 and 40 in one population of *B. biloba*.

When several embryos co-existed in one ovule, the sexual embryo could normally be distinguished from the apomictic ones by its smaller, dark-stained cells, its position close to the micropyle, and by its accompanying endosperm. However, as apomictic embryos could develop by adventive embryony from cells inside a sexual or apomictic embryo sac or nucellar cells before or after anthesis or pseudogamous apospory after pollination, it is difficult to distinguish adventive embryos from pseudogamous aposporous embryos.

Comparative Seed Production in Sexual and Apomictic Reproduction

The proportions of apomixis and sexual reproduction of each species in the glasshouse and the field were compared. In *B. macra* and *D. sericeum* only sexual reproduction was recorded, in *D. setosum* a few apomictic ovules were found in both situations and the frequency of apomixis is very similar between florets collected from the glasshouse (62.96%, 166 ovules) and florets collected from Inverell (60.46%, 193 ovules) in *B. biloba*. Data from all locations for every species were pooled for interspecies comparison.

The proportion of ovules that produced seeds in the glasshouse and the field was compared for each species. No significant difference was found between the glasshouse condition (mean = 0.503) and the field (mean = 0.602) in *B. biloba*, *B. macra* and *D. sericeum* (all p > 0.05). However, significantly higher seed production was recorded in the glasshouse (mean = 0.268) than in the field (mean = 0.091) in *D. setosum* (p < 0.002). The proportion of seed set was significantly lower in *D. setosum* than in *B. macra* and *D. sericeum* in the glasshouse (Fig. 2) as well as in the field. As it is impossible to find all four species at any one experimental site,

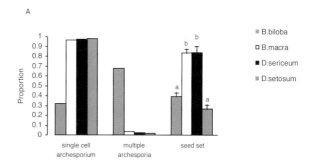

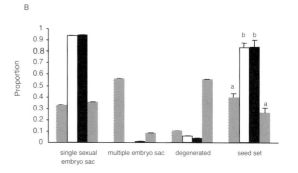

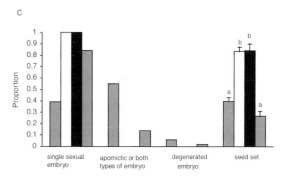

Fig. 2. The relationship between proportion of different structures at various embryological development stages and proportion of seed set in four species (+ s.e.). (A) The relationship between archesporium type and seed production. (B) The relationship between mature embryo sac type and seed production. (C) The relationship between embryo type and seed production. Columns with the same letter are not significantly different (p>0.05).

seed production data recorded in the glasshouse were used for comparing seed production of the four species.

Multiple archesporia, multiple embryo sacs and multiple embryos accompanied apomixis in *B. biloba* and *D. setosum* (Fig. 2). *Bothriochloa biloba* had the lowest proportion of single cell archesporium and the highest proportion of multiple archesporial cells compared with the other three species, yet its proportion of seedset was the second lowest in the four species (Fig. 2A) (29 - 72 ovules). *D. setosum*, with the highest ratio of embryo sac abortion, produced the lowest amount of seeds (Fig. 2B). *Bothriochloa macra* and *D. sericeum* had the highest proportion of sexual embryo sac and seed production, while *B. biloba* and *D. setosum,* with fewer single sexual embryo sacs, more apomictic and degenerated embryo sacs, produced much less seeds (Fig. 2B) (68 - 201 ovules). Only one sexual embryo was found in

Fig. 3. Shows one embryo seedling (left) and twin embryo seedling (right).

each ovule of *B. macra* and *D. sericeum* at the embryo stage, while numerous apomictic embryos of both sexual and apomictic type were found in *B. biloba* and *D. setosum* and these species had much lower seed production (Fig. 2C) (44 - 100 ovules).

Examination of the proportion of sexual and apomictic embryo sac combinations in ovules of *B. biloba* showed that half of the ovules contained two or multiple embryo sacs. The occurrence of twin seedlings (Fig. 3) was 0.75% (1000 seeds germinated) in *B. biloba*. This indicates that most of the multiple embryo sacs or multiple embryos were partly or totally degenerated during some stage of development. No twin seedlings or seedlings with multiple embryo were found in the other three species (1000 seeds for each species), which suggests that only one mature embryo was formed in these species.

Pollen/ovule Ratio and Seed Production

The results of the paired t-test on pollen production between sessile bisexual and pedicellate male florets of each pair of spikelets in *D. setosum* indicated that there was no significant difference in viable pollen production, pollen production and percentage viable pollen production between them (all p>0.05). When estimating viable pollen and total pollen per ovule for *D. setosum*, data from the bisexual floret and male floret of the paired spikelets were added, respectively; whereas in the other three species, because pedicellate spikelets are sterile, only data from the sessile fertile florets were taken into account.

Pollen production in contrast with seed production in the four species was compared in Fig. 4. *Dichanthium setosum* produced much more viable pollen and had a significantly higher pollen/ovule ratio and percentage viable pollen than the other three species (Fig. 4A-C). There were no significant differences among the other species with respect to these variables. *Dichanthium setosum* produced a significantly lower number of seeds per inflorescence than the other three species whereas *D. sericeum* produced the highest number of seeds (p<0.05). There were no significant differences in seeds/inflorescence between *B. biloba* and *B. macra* (Fig. 4D).

Bothriochloa biloba had the significantly highest number of spikelets/inflorescence and *D. setosum* had the significantly lowest number of spikelets/inflorescence (Fig. 4E). Percentage seed-

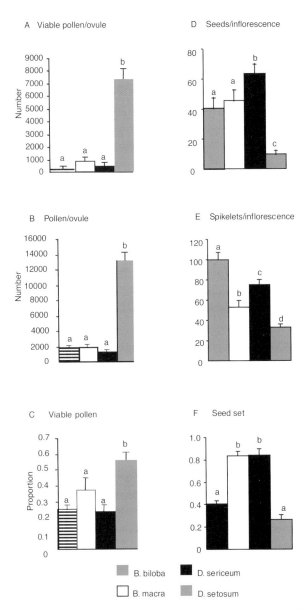

Fig. 4. The comparison between pollen and seed production in the four species (+ s.e.). Columns with the same letter are not significantly different (p>0.05).

set was not significantly different in *B. biloba* and *D. setosum*, the same as in *B. macra* and *D. sericeum*. However, it was significantly higher in *B. macra* and *D. sericeum* than in *B. biloba* or *D. setosum* (Fig. 4F).

Discussion

Ploidy Level and Reproductive Mode

In Old World Bluestem (OWB) grasses, diploid species with 2n = 20 are believed to be sexual and tetraploids are believed to be facultative apomicts (Celariar and Harlan 1957; Harlan *et al.* 1961). However, in diploid *D. setosum*, some facultative apomixis was encountered.

The co-existence of diploids and tetraploids in one population of *B. biloba* could be explained by adventive embryony of egg, synergids, or antipodals inside the sexual embryo sac which has only half the number of chromosomes of its tetraploid parent. Autonomous parthenogenetic adventive embryony in reduced sexual embryo sacs in OWB grasses has already been discovered by Harlan *et al.* (1964) who even suggested that when apomictic and sexual embryo sacs co-occur, apomixis tends to induce precocious adventive embryos in sexual embryo sacs.

Apomixis and Seed Production

The occurrence of apomixis in *B. biloba* and *D. setosum* follows the rare species traits described by Kunin and Gaston (1993). A detailed comparison of the number of sexual and apomictic structures at the different developmental stages among the four species indicates that competition among multiple archesporial cells, multiple embryo sacs and multiple embryos associated with apomixis can be the main reason for the low seed set in the vulnerable grasses *B. biloba* and *D. setosum*. As no endosperm forms to match the development of adventive embryos and there is often more than one embryo formed in one ovule, the simultaneous development of all embryos inevitably leads to a nutritional shortage and eventually collapse of part or the whole ovule.

The number of seeds per inflorescence was significantly lower in *B. biloba* than in *D. sericeum* (Fig. 4D), and percentage seed set in *B. biloba* was significantly lower than that of *B. macra* and *D. sericeum* (Fig. 4F). This may be caused by the high frequency of facultative apomixis in *B. biloba*, and competition among multiple developmental pathways associated with apomixis, which leads to embryo sac degeneration and embryo abortion (Fig 2).

In the obligate sexual reproduction of *B. macra* and *D. sericeum*, only one archesporial cell develops into one Polygonum type of embryo sac; after double fertilization, only one embryo will develop with its endosperm providing nutrition. With all the resources of the ovule focused on this single developmental pathway, a good seed set is guaranteed compared with the multiple embryos that often lack endosperm in the two vulnerable species with apomictic embryo development. Our findings agree with Ellstrand and Antonovics's (1985) suggestion that the fecundity of the populations of sexually derived individuals was higher than that of the populations of asexually derived individuals.

Ecological Implications of Facultative Apomixis

Seeds are the bearers of genetic information and the agents of dispersal. The apomictically produced seeds carry genetic information identical to their mother, so they are highly efficient dispersal agents for the original genotype. In a stable environment where the risk of death is negligible, apomictic reproduction, which avoids variation, might be an advantage (Drury 1974).

However, in a variable environment the properties of the original plant may not be compatible with different conditions, so genetic variation is more successful. In such a case, sexual reproduction, which promotes genetic variation, is an advantage (Asker and Jerling 1992).

Sexual reproduction promotes genetic variation, but the cost of males for the production of enough viable pollen and pollination processes reduces the advantage; loss of genes in meiotic division gives sexual reproduction a half disadvantage in relation to asex-

ual reproduction and there is a risk of producing unfit offspring. These are called 'the cost of meiosis' and 'the cost of sex' (Williams 1975; Richards 1986). There are other costs related to morphological differentiation, which makes sexual reproduction possible, and the cost of mating, such as the resources for male competition and female choice. Outcrossing sexuals will be less reproductively efficient than asexuals in conditions of limited cross pollination (Richards 1986). All these costs make sexual reproduction a costly and time-consuming process (Asker and Jerling 1992).

In facultative apomixis, the production of new genotypes by amphimixis could be followed by fecund production of the best genotypes by apomixis. The relative proportions of the two types of reproduction can be fine-tuned by selection (Asker and Jerling 1992). Facultative apomixis and sexual reproduction therefore form a well-balanced genetic system with an improved balance between constancy and change. Apomicts avoid meiosis, which conserves heterozygosity of hybrid polyploids, thus reducing or eliminating inbreeding depression. Apomixis also slows down gene flow by avoiding cross-pollination so that a high level of local adaptation can be preserved (Asker and Jerling 1992). Therefore, although facultative apomixis reduces seed production in *B. biloba* and *D. setosum*, it may play an important role in preserving gene heterozygosity, so that these species are able to live in a variety of environments because of the diversity of their genotypes.

Pollen/ovule Ratios and Seed Production

The number of viable pollen grains per ovule and pollen grains per ovule is not significantly different in *B. biloba*, *B. macra* and *D. sericeum* (Fig. 4A-B), and both of them are highly significantly lower than that for *D. setosum*. The good seed set per inflorescence in *B. biloba*, *B. macra* and *D. sericeum* compared with extremely poor seed set in *D. setosum* indicates that enough pollen is produced for effective pollination and no resources are wasted on male investment in these species (Fig. 4D).

Dichanthium setosum produced the lowest number of spikelets per inflorescence and its seed production is also the lowest (Fig. 4D-F). The occurrence of apomixis and the accompanying multiple developmental pathways and embryo abortion could account for this low seed production, but the frequency of the above happenings is much lower in *D. setosum* compared with that in *B. biloba*. The very high rate of embryo sac abortion (Fig. 2) and pollen production in *D. setosum* also account for its very low seed set.

In comparison with the other three species in which chasmogamy and cleistogamy co-occur, completely chasmogamous *D. setosum* may have a much higher level of outcrossing and be less efficient in pollen transport and, as a compensation, more investment is placed on male function. *Dichanthium setosum* produced seven-fold more viable pollen, had six-fold higher pollen/ovule ratios and much higher percentage of viable pollen than the other three species (Fig. 4A-C). However, it produces the lowest number of seeds per inflorescence and has the lowest percentage seedset. According to the sex allocation theory, at the same resource level, if more resources are allocated to the male function, the female investment, such as the seed production, will be negatively affected (Charlesworth and Charlesworth 1981). The resource limitation caused by the remarkably high pollen/ovule ratio may be an additional reason for the low seed production in *D. setosum*.

Conclusions

Competition among multiple developmental pathways accompanying apomixis and a lack of endosperm leads to a low seed set in *B. biloba* and *D. setosum*. As apomixis commonly occurs in *B. biloba,* it is the main cause of low seed production in this species. Comparative pollen production among the four grasses indicates that pollen production does not limit seed production in *B. biloba*.

Apomixis only occurs occasionally in *D. setosum*, so it may only play a minor role in low seed set in this species. In *D. setosum* the extremely high ratio of embryo sac abortion and resource allocation to male function may be the main reason for low seed production.

Theoretically, facultative apomixis and sexual reproduction form a well-balanced genetic system. Sexual reproduction promotes genetic variation, whereas apomixis preserves gene heterozygosity. Such a balanced genetic system should enable *B. biloba* and *D. setosum* to live in a variety of environments because of the diversity of their genotypes. Therefore, the authors believe that although facultative apomixis reduces seed production, it could benefit the species in the long run. However, if the cost is dramatically reduced seed production, then the species could be facing extinction.

Enough viable pollen is produced for effective pollination, and no resources are wasted on production of surplus pollen in *B. macra* and *D. sericeum*. Thus a good maternal investment is achieved by economic allocation of resources. All the resources that an ovule could require focus on the single sexual developmental pathway, and a good nutrition supply certainly leads to high seed production in these common species.

Acknowledgements

We would like to thank Dr. Glenda Vaughton for her valuable comments on the manuscript, Associate Prof. Paul Whitington for providing the access to the digital camera and computer, Mr. Douglas Beckers for collecting plants.

References

Asker, S. E., and Jerling, L. (1992). 'Apomixis in plants.' (CRC Press: Boca Raton, Ann Arbor, London and Tokyo.)

Briggs, J. and Leigh, J. (1988). 'Rare or threatened Australian plants,' revised edition. Australian National Parks and Wildlife Service, speacial Publication no. 14.

Briggs, J. D., and Leigh, J. H. (1996). 'Rare or threatened Australian Plants,' revised edition. (CSIRO Australia: Collingwood, Victoria.)

Brown, W. V., and Coe, G. E. (1951). A study of sterility in *Hilaria belangeri* (Sreud.) Nash and *Hilaria mutica* (Buckl.) Benth. *American Journal of Botany* **38**, 823-829.

Byers, D. L., and Meagher, T. R. (1997). Comparison of demographic characteristics in a rare and a common species of *Eupatorium. Ecological Applications* **7**, 519-530.

Celarier, R. P., and Harlan, J. R. (1957). Apomixis in *Bothriochloa, Dichanthium*, and *Capillipedium. Phytomorphology* **7**, 93-102.

Charlesworth, D., and Charlesworth, B. (1981). Allocation of resources to male and female functions in hermaprodites. *Biological Journal of Linnean Society* **15**, 57-74.

Charnov, E. L. (1979). Simultaneous hermaphroditism and sexual selection. *Proceedings of National Academy of Science. U.S. A.* **76**, 2480-2484.

Charnov, E. L. (1982). Parent-offspring conflict over reproductive effort. *American Naturalist* **119**, 736-737.

Charnov, E. L. (1987). On sex allocation and selfing in higher plants. *Evolutionary Ecology* **1**, 30-36.

Cruden, R. W. (1977). Pollen-ovule ratios: a conservative indicator of breeding systems in flowering plants. *Evolution* **31**, 31-46.

De Wet, J. M. J., and Harlan, J. R. (1966). Morphology of the compilospecies *Bothriochloa intermedia. American Journal of Botany* **53,** 94-98.

De Wet, J. M. J., and Harlan, J. R. (1970). *Bothriochloa intermedia* - a taxonomic dilemma. *Taxon* **19**, 339-340.

Drury, W. H. (1974). Rare species. *Biological Conservation* **6**, 162-169.

Ellstrand, N. C., and Antonovics, J. A. (1985). Experimental studies of the evolutionary significance of sexual reproduction II. A test of the density dependent selection hypothesis. *Evolution* **39**, 657-66.

Garnier, P., Maurice, S., and Olivieri, I. (1993). Costly pollen in maize. *Evolution* **47**, 946-949.

Harlan, J. R., and de Wet, J. M. J. (1963). Role of apomixis in the evolution of the *Bothriochloa-Dichanthium* complex. *Crop Science* **3**, 314-316.

Harlan, J. R., de Wet, J. M. J., Richardson, W. L., and Cheda, H. R. (1961). Studies on Old World Bluestems III. *Technical bulletin, Oklahoma Agricultural Experiment Station* **T-91**, 1-30.

Harlan, J.R., Brooks, M.H., Borgaonkar, D.S., and de Wet, J. M. J.(1964). Nature and inheritance of apomixis in *Bothriochloa* and *Dichanthium. Botanical Gazette* **125**, 41-46.

Johri, B. M., Ambegaokar, K. B., and Srivastava, P. S. (1992). 'Comparative embryology of Angiosperms.' (Springer-Verlag: Berlin, Heidelberg.)

Kearns, C. A., and Inouye, D. W. (1993). 'Techniques for pollination biologists.' (University Press of Colorado: Ninot, Colorado.)

Kunin, W. E., and Gaston, K. J. (1993). The biology of rarity: patterns, causes and consequences. *Trends in Ecology and Evolution* **8**, 298-301.

Lesica, P., Leary, R. F., Allendorf, F. W., and Bilderback, D. E. (1988). Lack of genic diversity within and among populations of an endangered plant, *Howellia aquatilis. Conservation Biology* **2**, 275-282.

Longton, R. E. (1992). Reproduction and rarity in British mosses. *Biological Conservation* **59**, 89-98.

Mujeeb-Kazi, A., and Miranda, J. L. (1985). Enhanced resolution of somatic chromosome constrictions as an aid to identifying intergeneric hybrids among some Triticeae. *Cytologia* **50**, 701-709.

Pavlik, B. M., Ferguson, N., and Nelson, M. (1993). Assessing limitations on the growth of endangered plant populations, II. seed production and seed bank dynamics of *Erysimum capitatum* ssp. *angustatum* and *Oenothera deltoides* ssp. *howellii. Biological conservation* **65**, 267-278.

Richards, A. J. (1986). 'Plant breeding systems.' (George Allen & Unwin: London.)

Richards, A. J. (1990). The implications of reproductive versatility for the structure of grass populations. In *Reproductive versatility in the grasses*, edited by G. P. Chapman, pp. 131-153. (Cambridge University Press: Cambridge.)

Saran, S., and de Wet, J. M. J. (1970). The mode of reproduction in *Dichanthium intermedium* (Gramineae). *Bulletin of the Torrey Botanical Club* **97**, 6-13.

Sydes, M. A., and Peakall, R. (1998). Extensive clonality in the endangered shrub *Haloragodendron lucasii* (Haloragaceae) revealed by allozymes and RAPDs. *Molecular Ecology* **7**, 87-93.

Totland, Ø. (1997). Limitations on reproduction in alpine *Ranunculus acris*. *Canadian Journal of Botany* **75**, 137-44.

Wheeler, D. J. B., Jacobs, S.W. L., and Norton, B. E. (1982). 'Grasses of New South Wales.' (The University of New England: Armidale, NSW, Australia.)

Wiens, D., Calvin, C. L., Wilson, C. A., Davern, C. I., Frank, D., and Seavey, S. R. (1987). Reproductive success, spontaneous embryo abortion, and genetic load in flowering plants. *Oecologia* **71**, 501-9.

Wiens, D., Nickrent, D. L., Davern, C. I., Calvin, C. L., and Vivrette, N. J. (1989). Developmental failure and loss of reproductive capacity in the rare paleoendemic shrub *Dedeckera eurekensis*. *Nature* **338**, 65-7.

Williams, G. C. (1975). 'Sex and evolution.' (Princeton University Press: Princeton, New Jersey.)

GRASSES

Grasses: Systematics and Evolution. (2000). Eds S.W.L. Jacobs and J. Everett. (CSIRO: Melbourne)

APOMIXIS IN MONOCOTYLEDONS

R. Czapik

Institute of Botany, Department of Plant Cytology and Embryology, Jagellonian University, Grodzka 52, 31-044 Krakow, Poland. E-mail:ubczapik@cyf-kr.edu.pl

Abstract

Studies in Monocotyledons confirm the unity of morphological processes and problems of agamospermous Angiosperms, the occurrence of recurrent and non-recurrent apomixis and the prevalence of facultative apomixis and pseudogamy. Apospory and diplospory, the processes of unreduced embryo sac development, are combined with one or more modes of embryo formation (parthenogenesis, apogamety, hemigamy, androgenesis, adventitious embryony) in a complete apomictic cycle. Either complete apomixis, or elementary apomictic processes, have been noted in c. 110 genera of Monocotyledons. Sixty-two of those genera belong to Gramineae, the family whose contribution to apomictic studies is the most conspicuous. The important results of these studies concern genetic and phenotypic control of apomixis. Recently, new techniques have opened wider perspectives for research and investigations on the molecular basis of the phenomena, focusing on theoretical and practical objectives, seem to be most promising.

Key words: Monocotyledoneae, apomixis, agamospermy, apospory, diplospory, parthenogenesis, apogamety, hemigamy, androgenesis, adventitious embryony.

INTRODUCTION

Apomictic studies in Angiosperms began in 1841 when the first plant developing seeds without the action of pollen, *Alchornea ilicifolia* (= *Coelebogyne ilicifolia*, Euphorbiaceae), was described by John Smith at Kew. During the following decades students of apomixis focused their investigations on individual plants and their progenies. The stress was laid on morphology and genetics as much as on taxonomic, populational, and ecological problems of apomicts. The formation of new forms within apomictic groups was described and the opinion spread that facultative apomixis was a common phenomenon, and that obligatory apomictic species or specimens were very rare if found at all. Recently some new techniques have been introduced to studies carried out both *in vivo* and *in vitro*. Investigations on the molecular basis of apomixis have commenced and are increasing in number.

Successful manipulation of apomixis with practical objectives for plant breeding and selection is a starting point of many new projects. Leading programmes concern the problem of transferring apomictic factors to cultivated plants from their wild relatives. The stress placed on the practical aspects of apomictic reproduction has contributed to the idea of an 'Asexual Revolution' in plant breeding and international agriculture in the coming century (Gillman 1994). Gramineae, along with other members of Monocotyledoneae, which include many economically important plants, are at the centre of this scientific movement.

Generally, the end of the 20th century is characterised in botany by an increasing interest in the principal problems of apomixis. Simultaneously, the search for new apomictic taxa seems to have attracted less attention. Examination of the qualitative and quantitative state of the research after the lapse of c. 150 years of studies is worthwhile for the group in which lies the Gramineae, the family with the largest number of apomictic records.

Forms of Apomixis and their Embryology in Monocotyledons

The general distribution of apomixis in Monocotyledons is typical of Angiosperms. Yet, it is not only the number of taxa with apomictic records that is important but also the contribution to science brought by research within those taxonomic groups.

Apomixis in Monocotyledons is realised as agamospermy, the apomictic mode of reproduction by seeds, in which meiosis and fertilization are constantly lacking. Vegetative apomixis, recognized by some authors (Gustafsson 1946) and denied its apomictic status by others (Battaglia 1963; Solntseva 1967; Nogler 1984; Asker and Jerling 1992), is represented chiefly by viviparous forms in some monocotyledoneous taxa. This and related phenomena are a special theoretical problem for plant reproduction but are not considered in the present review.

Apomixis, a complex system of processes, is genetically regulated in its recurring form. The opposite possibility, so called non-recurrent apomixis, is usually connected with occasional formation of haploids in the process of haploid parthenogenesis or apogamety. Examples of non-recurrent apomixis in Monocotyledons are included in Table 1 but not distinguished from other forms. That decision was supported by some broad definitions proposed for apomixis (Battaglia 1963; Khokhlov 1967) which emphasise one condition only — the lack of fertilization. Such a view indicates that the term apomixis or agamospermy has become synonymous with parthenogenesis, an opinion expressed by zoologists (Soumalainen *et al.* 1987).

In the view accepted in the present paper parthenogenesis is important, but is only one of the seven elementary apomictic processes involved in embryo and embryo sac formation. All of them occur in Monocotyledons, either mutually combined or as a single apomictic process complemented with a sexual one. The occurrence of a single elementary process in the cycle of reproduction usually means a tendency towards apomixis (Gustafsson 1946), but it may also reflect the incomplete state of knowledge.

The elementary apomictic processes are involved in embryo sac and embryo formation. Apomeiotic embryo sacs with unreduced chromosome numbers develop either 1) from archesporial cells (diplospory) or 2) from somatic cells of the ovule (apospory). An embryo may develop without fertilization either 1) from an egg cell (parthenogenesis), 2) from any cell of the embryo sac except the egg (apogamety), 3) from an egg cell after disturbed fusion of female and male nuclei which divide independently and thus form a mosaic embryo (hemigamy), 4) from an egg cell with a sperm nucleus (androgenesis), or 5) from any cell of the ovule outside the embryo sac (adventitious embryony).

Apospory and diplospory are usually separated in examined species. As far as published records are concerned it is apospory that seems to prevail in Monocotyledons. In some species, not only in genera, e.g. in *Pennisetum ciliare* both processes occur, and in *Paspalum minus* aposporous and diploporous embryo sacs develop in the same plant and even the same ovule. Such a

Table 1. List of Genera with Apomictic Records.

Genera with records published for at least one species in which complete apomixis or elementary apomictic processes were found (taxa according to Brummitt (1992)). The list was checked with original papers and compared with Khokhlov (1967), Watson (1997) and *Apomixis Newsletter* vol 1 - 10. Sources not referenced are available from the author.

GRAMINEAE - total 657 genera, apomixis in 62 genera (c. 9.3%)

AEGILOPS hapl
x AEGILOTRITICUM exp
AGROPYRON D, P, hapl
AGROSTIS exp, hapl
ALOPECURUS hapl
ANDROPOGON A, P hapl
ANTHEPHORA A P
APLUDA (apomixis: Watson 1990)
ARRHENATHERUM hapl
AVENA hapl

BOTHRIOCHLOA A, P, AE, Ps
BOUTELOUA A, P
BRACHIARIA A, P, hapl
BROMUS AE, hapl
BUCHLOE A, P

CALAMAGROSTIS D, P, aut
CENCHRUS A, P, AE, aut
CHLORIS A, P
COIX H, hapl
CORTADERIA A

DACTYLIS A, hapl
DESCHAMPSIA hapl
DICHANTHIUM A, D, P, Ps

ECHINOCHLOA A
ELYMUS D, exp
ERAGROSTIS D, P
EREMOPOGON A, exp
ERIOCHLOA A, P
EUCHLAENA exp, hapl, androg
EUSTACHYS A

FESTUCA hapl, embryological studies
FINGERHUTHIA A, P

HARPOCHLOA A
HETEROPOGON A, P,
HIEROCHLOE A, P Ps, hapl
HILARIA (apomixis: Watson 1990)
HOLCUS hapl
HORDEUM exp, hapl, androg
HYPARRHENIA A, P

LAMPROTHYRSUS A, aut

NARDUS D, P, aut

ORYZA AE, hapl

PANICUM A, P, Ps
PASPALUM A, D, P, Ps
PENNISETUM A, D, P, AE, Ps
PHLEUM D, hapzl
POA A , D, P, AE, Ps, hapl

RENDLIA A

Abbreviations: A - apospory; D - diplospory; AE - adventitious embryony; H - hemigamy; Ag - apogamety; androg - androgenesis; Ps - pseudogamy; aut - autonomous apomixis; exp - apomixis established in experiment (the morphological form unknown); ind - artificially induced apomixis; hapl - haploid plants or embryos formed sporadically in addition to recurrent apomixis.

Table 1. List of Genera with Apomictic Records. *(Continued)*

SACCHARUM D, Ps
SCHIZACHYRIUM A
SCHMIDTIA A
SECALE D, exp, ind, hapl
SETARIA A, P, Ps
SORGHUM A, D, AE, hapl
THEMEDA A, P
TRIBOLIUM A
TRICHOLAENA A, P
TRIPSACUM D, P, Ag, androg
x TRITICOSECALE exp
TRITICUM exp, ind, hapl
UROCHLOA A, P
ZEA hapl

ORCHIDACEAE - total 835 genera
apomixis in 19 genera (c. 2.7%)

BLETILLA hapl, Ag
CEPHALANTHERA hapl, Ag
CYNORCHIS AE
CYNOSORCHIS AE
DACTYLORCHIS A, P, Ps, hapl
EPIDENDRUM AE
EPIPACTIS hapl
GASTRODIA AE
GYMNADENIA AE
HABENARIA exp
LISTERA hapl
MAXILLARIA AE
NIGRITELLA AE, aut
ORCHIS Ag, hapl
PLATANTHERA Ag, hapl
SPIRANTHES apomeiosis, AE, aut
VANDA exp
ZEUXINE AE, aut
ZYGOPETALUM A, AE, hapl, Ag

18 OTHER FAMILIES - total 559 genera, **apomixis in 29 genera**
(c. 5.4%)

AGLAONEMA A
ALLIUM A, D, P, AE, Ps, hapl, Ag
AMARYLLIS exp
AMORPHOPHALLUS exp
ANTHURIUM exp
ASPARAGUS hapl
BURMANNIA D
CALOSTEMMA AE
COCOS exp, hapl
COLCHICUM AE
CYANELLA AE
DIANELLA exp.
FRITILLARIA exp

Abbreviations: A - apospory; D - diplospory; AE - adventitious embryony; H - hemigamy; Ag - apogamety; androg - androgenesis; Ps - pseudogamy; aut - autonomous apomixis; exp - apomixis established in experiment (the morphological form unknown); ind - artificially induced apomixis; hapl - haploid plants or embryos formed sporadically in addition to recurrent apomixis.

Table 1. List of Genera with Apomictic Records. *(Continued)*

HABRANTHUS D
HEDYCHIUM hapl, Ag
HEMEROCALLIS AE
HOSTA AE
LILIUM D, P, hapl, Ag
MUSA exp
NERINE Ag
NOTHOSCORDUM AE
ORNITHOGALUM A
POTAMOGETON D, P, aut
SMILACINA AE
SPATHIPHYLLUM AE
TACCA A
TRILLIUM AE, Ag
TULIPA AE
ZEPHYRANTHES D, P, H

Abbreviations: A - apospory; D - diplospory; AE - adventitious embryony; H - hemigamy; Ag - apogamety; androg - androgenesis; Ps - pseudogamy; aut - autonomous apomixis; exp - apomixis established in experiment (the morphological form unknown); ind - artificially induced apomixis; hapl - haploid plants or embryos formed sporadically in addition to recurrent apomixis.

situation must be considered in studies on the genetic problems of apomixis.

Neither an aposporous or diplosporous origin influences the organization of a functional embryo sac if the early stages proceed undisturbed. One type of embryo sac is four-nucleate, the others are eight-nucleate in the normal pattern of their structure. However, they differ in some details of the developmental mechanisms. These differences, together with the cytologically unreduced state of embryo sacs, are reflected in the terminology which is different for apomictic and sexual plants. Three types of unreduced embryo sacs recognized in Monocotyledons were first described in Dicotyledons, and two in Monocotyledons. Among eight-nucleate embryo sacs the aposporous one is of the *Hieracium* type, two others are diplosporous: *Antennaria* and *Taraxacum* types. The fourth type of embryo sac, *Panicum* type, is specific for some groups of Gramineae; it was described as an aposporous, four-nucleate embryo sac in the subfamily Panicoideae. The lack of antipodals in its simple structure is a good morphological marker that facilitates the identification of an apomictic mode of reproduction. The absence of four-nucleate embryo sacs in diplosporous plants is an argument supporting the view by Reddy (1977) that diplospory should be evolutionarily more primitive than apospory.

The fifth type is a peculiar embryo sac - *Allium* type - described for the first time in Monocotyledons by Håkansson and Levan (1957). It has the reduced chromosome number of the meiotic embryo sac but it develops from an archesporial cell with a doubled chromosome number. Thus, the embryo sac and the egg cell have an unreduced chromosome number in relation to the surrounding somatic cells of the ovule.

Apogamety, hemigamy and androgenesis are rare phenomena in the Angiospermae. Nevertheless, in the evaluation by Solntseva (1998) apogamety was noted in 77 species from 17 families of

Monocotyledons. Apogamety is, as a rule, connected with polyembryony and occurs irregularly but more often than the two other processes. Synergidal apogamety was described, e.g., in *Allium, Oryza* and *Trillium.* Antipodal, haploid or diploid apogamety was noted in *Allium, Fritillaria, Gagea, Hordeum, Paspalum, Poa* and *Tripsacum* (Solntseva 1995). Genetically determined hemigamy occurs in five genera of the Angiospermae, two of them monocotyledonous: *Coix* (Gramineae) and *Zephyranthes* (Amaryllidaceae). Androgenesis was found in single cases, e.g., in the genera *Euchlaena, Hordeum, Tripsacum.*

Adventitious embryony prevails in the investigated taxa of Orchidaceae, the family with the highest number of recognized genera in the Monocotyledons (835 according to Brummitt, 1992). Adventitious embryony was noted in 10 out of 19 genera with apomictic records (Table 1). Such a low frequency of records does not allow, however, the formulation of any general conclusions on the frequency of the phenomenon because one must take into consideration the relationship between published data and the number of genera still uninvestigated, and the level of interest in embryological research within the family.

The embryology of apomicts in Monocotyledons is comparatively well documented at the level of light microscopy, but that does not mean that nothing more could be done, e.g., with the problems of endosperm, or morphology of the mechanisms in pseudogamy. The examination of apomicts by electron microscopy began in the early 1980s but it is progressing slowly. Important for the Monocotyledons are comparative investigations of the egg and synergid structure in connection with apogamety in *Trillium camtschatcense* (Naumova 1991), the ultrastructure of apospory studied in *Poa nemoralis* and *P. palustris* (Osadtchiy and Naumova 1996), and of diplospory in *Panicum maximum* (Naumova and Willemse 1995), as well as the comparative studies of apomictic and meiotic embryo sacs in *Pennisetum* (Chapman and Busri 1994).

It is well known that irregularities of embryo sac formation and structure are more often noted in apomictic than in sexual plants. In Monocotyledons irregular embryo sacs were described showing retardation of development or disturbed organization of non-functional embryo sacs. Their abnormalities are so characteristic of apomicts, that among others they were the indirect argument in the recognition of the apomictic nature of four species of *Festuca* (Table 1) by Shishkinskaya (1983).

The pseudogamous form of agamospermy prevails in Monocotyledons. This means that fertilization of the central cell is necessary for the complete development of the embryo and for formation of endosperm in a functional seed, even in those species in which embryos start development precociously before anthesis. Autonomous apomixis is rare. It is noted in e.g. *Calamagrostis, Cenchrus, Nardus* and *Potamogeton.*

According to the rule established for agamospermy, facultative apomixis prevails in Monocotyledons, too, and obligatory apomixis is rare, or non-existent. In a few species some authors describe occurrences of obligatory apomictic plants and researchers always look for such plants as valuable material for studies on genetic problems.

In several of the investigated agamic complexes (mostly of Gramineae) sexually reproducing plants were noted, e.g. in *Panicum maximum* and *Cenchrus ciliaris.*

Agamospermy is rare in diploid taxa, but apomictic diploids are known in *Hierochloe australis, Panicum antidotale, Pennisetum hohenackeri* and in *P. ramosum.* Recently apospory was discovered in diploid species of *Tribolium:* two from section *Tribolium* - *T. echinatum* and *T. utriculosum* and one from section *Uniolae* - *T. uniolea* (Visser and Spiess 1994). The possibility of complete apomixis awaits investigations of embryo development in the taxa mentioned above.

CONTRIBUTION OF MONOCOTYLEDONS TO APOMIXIOLOGY

Studies on Monocotyledons have conspicuously improved the state of knowledge of the apomictic mode of reproduction mostly because of the interest paid to economically important plants such as grasses. Essential to the solution of the fundamental problems of apomixis are: collection of new data on the occurrence of apomictic plants within Angiosperms; investigations on the mechanisms and morphological forms of the processes; and studies on genetic and phenotypic control of apomixis.

The number of published records of apomictic processes in Monocotyledons include 110 genera from 20 families (Table 1). Two families account for 73% of the records: c. 56% of the data (62 genera) belong to Gramineae and c. 17% (19 genera) to Orchidaceae. In the remaining 18 families the percentage of genera with apomictic records amounts to 26% of all the data. Of these, five genera belong to Amaryllidaceae, four to Araceae and Liliaceae, two to Alliaceae; for 13 families only one record each has been published.

Such a comparative evaluation points only to the present state of research. It means that apomictic processes in Gramineae have been noted in 60 genera and two intergeneric artificial hybrids, and in Orchidaceae in 19 genera. Conclusions could not be drawn about either the general frequency of apomixis or its distribution among genera. Nevertheless, Gramineae hold the position of the most explored family, and the highest number of apomictic records. Numbers in the two families next highest to Gramineae and Orchidaceae are: (1) Compositae 45 genera, c. 3% of recognized genera, and (2) Rosaceae 18 genera, 18.5 % of genera in the family (Czapik 1996).

The data compiled in Table 1 allow us to state that more should be done before it will be possible to formulate any general conclusions about the frequency of apomixis in Monocotyledons.

Well documented discoveries of apomixis are always associated with embryological studies and supported by experimental crosses and progeny tests. The results so far in Monocotyledons confirm the similarity of morphological processes of the apomictic mode of reproduction in Angiosperms, with their regularity and deviation. Nevertheless, the embryological data have shown up two types of embryo sacs described for the first time in Monocotyledons. One of them is typical for some grasses, i.e., the four-nucleate *Panicum* type; the second, the *Allium* type, is an example of a complicated mechanism of development that leads to the unreduced state of the egg formed after undisturbed meiosis.

Studies on the genetic control of apomixis began in Dicotyledons, but recently more extensive investigations have been carried out on grasses. The results are often controversial as far as the evaluation of the number of genes involved in the hereditary process is concerned, and the recessiveness or dominance of these genes. Thus several models of genetic control have been proposed and efforts made towards isolating the gene(s) for apomixis, their location in chromosomes, and the mapping of the loci responsible for the main elementary apomictic processes. Some results were achieved from the studies on the genetic regulation of apospory in *Pennisetum, Panicum* and *Tripsacum* (Asker and Jerling 1992; see also the *Apomixis Newsletter* vol. 1-10) but much more is still to be done.

PRESENT STATE AND PERSPECTIVES OF APOMICTIC STUDIES IN MONOCOTYLEDONS

In contemporary apomictic studies one can observe a decreasing interest in the search for new records beyond the classic apomictic complexes. Thus, the results do not allow us to draw any conclusions about the general distribution of agamospermy, or some of its forms, within taxa above the level of genus. However, investigations at the specific level in genera renowned for apomixis have been progressing quite well. This trend is characteristic for grasses, especially in the economically important groups, - one searches for wild apomictic relatives of cultivated plants.

The gaps of unexamined taxa could be covered in close collaboration with embryological and experimental research focused on the mode of reproduction (Czapik and Koscinska-Pajǎk 1998). This effort could be increased. It is expected that the results obtained with the light microscope will be enriched by details observed in electron microscopy, particularly studies on ultrastructure of generative cells and tissues.

The genetic control of apomixis, and the transfer of 'apomictic genes' to sexually reproducing plants are currently in focus. Trials towards the induction of parthenogenesis, and the classic methods of crossing *in vivo,* have been recently assisted by *in vitro* cultures, and some results are expected from cell hybridisation. Molecular methods should soon help to isolate genes controlling apomixis and these studies will have a direct practical meaning for plant breeding and selection (Koltunow *et al.* 1995) when the easy manipulation of apomixis will be an important tool.

Special emphasis in research projects on Monocotyledons is given to experiments with sexually reproducing *Zea mays* and its wild relatives – diplosporous species of *Tripsacum* - and other economically important grasses, e.g., *Oryza sativa,* species of *Bothriochloa, Pennisetum.* At the Apomixis Workshop in Atlanta (Elgin and Miksche 1992) 16 speakers presented apomixis research programmes in the U.S.A, their future plans and possibility of studies. The objects were grasses important in plant breeding, among other Monocotyledons - *Zephryanthes* was mentioned. Two contributions connected with Dicotyledons: citrus and cotton were only a small addition there. It is also significant that the comprehensive article on use of apomixis in cultivar development by Hanna (1995) deals almost exclusively

with studies on grasses. The above examples highlight the importance of monocotyledonous taxa in applied apomixiology.

Generally, the trends of experimental research, supported by observational data, seem to dominate in Monocotyledons. Molecular techniques (e.g., DNA fingerprinting, isozyme methods) support the classic morphological and cytological methods of evaluating modes of reproduction in experimental progenies of apomictic plants, or in sympatric populations in the wild that comprise asexual and sexual plants. Biochemical analyses, e.g., in *Poa* and *Panicum,* have helped to identify their apomictic progenies, and to estimate the degree of apomixis in *Allium tuberosum* (Kojima and Nagato 1992).

Monocotyledons and especially grasses have contributed a great deal to our knowledge of phenotypic control of apomixis. Such external factors as light regime, temperature, and chemicals were examined in natural and experimental conditions. The most popular have been the studies carried out on the influence of photoperiod on the degree of apomixis. Apomictic species tested included *Dichanthium aristatum* (several times), *Themeda australis, Paspalum cromyorrhizon,* and *Pennisetum ciliare* (Knox and Heslop-Harrison 1963; Evans and Knox 1967; Quarín 1986; Hussey *et al.* 1991).

Other general problems of apomixis have been discussed in several papers using examples from Monocotyledons, e.g. the origin and evolution of apomixis in Gramineae by Reddy (1977). Links between apomixis and taxonomy have also been considered by several authors. There are two main obstacles in the discussion on this and other problems in the whole of the Angiospermae and not only in Monocotyledoneae: the lack of a sufficient number of facts based on observations, and a persistent controversy over the definition and range of apomixis because of the complexity of its mechanisms.

REFERENCES

Asker, S., and Jerling, L. (1992). 'Apomixis in Plants'. (CRC Press: Boca Raton.)

Bashaw, E.C., and Hanna, W.W. (1990). Apomictic reproduction. In 'Reproductive Versatility in the Grasses' (Ed. G. P. Chapman) pp. 100-130. (Cambridge University Press: Cambridge.)

Battaglia, E. (1963). Apomixis. In 'Recent Advances in Embryology of the Angiosperms' (Ed. P. Maheshwari) pp. 222-264. (Catholic Pres: Ranchi, India.)

Bonilla, J. R., and Quarín, C. L. (1997). Diplosporous and aposporous apomixis in a pentaploid race of *Paspalum minus. Plant Science* **127**, 97-104.

Brummitt, R.K. (1992). 'Vascular Plant Families and Genera'. (Royal Botanic Gardens: Kew.)

Chapman, G. P., and Busri, N. (1994). Apomixis in *Pennisetum*: An ultrastructural study. *International Journal of Plant Sciences* **155**, 492-497.

Czapik, R. (1996). Problems of apomictic reproduction in the families Compositae and Rosaceae. *Folia Geobotanica et Phytotaxonomica* **3**, 381-387.

Czapik, R., and Koscinska-Pajǎk, M. (1998). Frequency of apomixis in seed plants of Poland. *Folia Morphologica* **57**, p. 16.

Elgin, J. H. Jr., and Miksche, J. B. (Eds) (1992). Proceedings of the Apomixis Workshop, February 11-12, 1992, Atlanta, Georgia. *Agricultural Research Service* **104**, 1-66.

Evans, L. T., and Knox, R. B. (1969). Environmental control of reproduction in *Themeda australis*. *Australian Journal of Botany* **17**, 375-389.

Gillman, H. (1994). The small farmer's friend. *Ceres* No. **149 (vol. 26, No. 5),** 17-22.

Gustafsson, A. (1946). Apomixis in higher plants. Part I. *Lunds Universitets Årsskrift* N. F. 2, **42**, 1-67.

Hanna, W.W. (1995). Use of apomixis in cultivar development. *Advances in Agronomy* **59**, 333-350.

Håkansson, A., and Levan, A. (1957). Endo-duplicational meiosis in *Allium odorum*. *Hereditas*, **43,** 179-200.

Hussey, M. A., Bashaw, E. C., Hignight, K. W., and Dahmer, M. L. (1991). Influence of photoperiod on the frequency of sexual embryo sacs in facultative apomictic buffelgrass. *Euphytica* **54**, 141-145.

Khokhlov, S. S. (1967). Apomixis: classification and distribution (in Russian). In 'Uspiekhi sovremennoy genetiki' vol 1 (Ed. H. P. Dubinin) pp 43 - 105. (Nauka: Moskva.)

Knox, R. B, and Heslop-Harrison, J. (1963). Experimental control of aposporous apomixis in a grass of the *Andropogoneae*. *Botaniska Notiser* **116**, 127-11.

Kojima, A. and Nagato, Y. (1992). Pseudogamous embryogenesis and the degree of parthenogenesis in *Allium tuberosum*. *Sexual Plant Reproduction* **5**, 79-85.

Koltunow, A.M., Bicknell, R.A., Chaudhury A.M. (1995). Apomixis: molecular strategies for the generation of genetically identical seed without fertilization. *Plant Physiology* **108**, 1345-1353.

Naumova, T. N. (1991). Apogamety in *Trillium camschatcense*: ultrastructural aspects. *Apomixis Newsletter* **3**, 16-17. [Cited with permission of the author].

Naumova, T.N. (1993). 'Apomixis in Angiosperms. Nucellar and Integemantary Embryony.' (CRC Press: Boca Raton.)

Naumova, T. N., and Willemse M. T. M. (1995). Ultrastructural characterization of apospory in *Panicum maximum*. *Sexual Plant Reproduction* **8**, 197-204.

Nogler, G. A. (1984). Gametophytic apomixis. In 'Embryology of Angiosperms' (Ed. B.M. Johri) pp. 475-518 (Springer: Berlin, Heidelberg.)

Quarín, C. L. (1986). Seasonal changes in the incidence of apomixis of diploid, triploid and tetraploid plants of *Paspalum cromyrrhizon*. *Euphytica* **35**, 515-522.

Osadtchiy, J., and Naumova, T. N. (1996). Diplospory in *Poa nemoralis* and *P. palustris*: ultrastructural aspects. *Apomixis Newsletter* **9**, 9-6. [Cited with permission of the authors.]

Reddy, P.S. **(**1977). Evolution of apomictic mechanisms in Gramineae - a concept. *Phytomorphology* **27**, 45-50.

Shishkinskaya, N. (1983). Cytoembryological investigation of apomixis in weed grass species. In 'Fertilization and Embryogenesis in Ovulated Plants' (Ed. O. Erdelska) pp. 363-365. (Veda: Bratislava.)

Smith, J. (1841). Notice of a plant which produces perfect seeds without any apparent action of pollen. *Transactions of the Linnean Society* **18**, 509-512.

Solntseva, M. P. (1969). Principles of embryological classification of apomixis in Angiosperms. *Revue de Cytologie et de Biologie Végétales*. **32**, 371-377.

Solntseva, M. P. (1995). Letters. *Apomixis Newsletter* **8**, 55-57 [Cited with permission of the author.]

Solntseva, M. P. (1998). On the application of double names for characterizing apomictic plants. *Botnicheskij Zhurnal* **82,** 49-58.

Suomalainen, E., Saura, A., and Lokki, J. (1987). 'Cytology and Evolution in Parthenogenesis'. (CRC Press: Boca Raton.)

Visser, N. C., and Spies, J. J. (1994). Cytogenetic studies in the genus *Tribolium* (*Poaceae: Danthonieae*). II. A report on embryo sac development, with special reference to the occurrence of apomixis in diploid specimens. *South African Journal of Botany* **60**, 22-26.

Watson, L. (1990). The grass family, Poaceae. In 'Reproductive versatility in the Grasses' (Ed. G. P. Chapman ') pp. 1 - 31. (Cambridge University Press: Cambridge.)

BIOGEOGRAPHY

Triodia scariosa subsp. *bunicola* in the Flinders Ranges, South Australia

Grasses: Systematics and Evolution. (2000). Eds S.W.L. Jacobs and J. Everett. (CSIRO: Melbourne)

Chorology of Grasses – a Review

T.A. Cope

Royal Botanic Gardens, Kew, Richmond, Surrey TW9 3AE, UK.

Abstract.

Chorology is the study of distribution and before the word became popular for several different aspects of phytogeography it had been given a special meaning for the grasses. The ecological background to chorology is summarised in this paper and the role of Kew in the study of grass chorology is explained. Most of the world's major phytogeographical units are based on limited numbers of selected species, occupy mutually exclusive territories (within the limits of mapping scale) and are mapped with contiguous boundaries. Phytosociology has played a large part in delimiting these units, but species distributions do not always conform to those of the communities in which they participate. Grasses from particular ecosystems are prone to invade other nearby ecological niches when allowed to do so with the result that their distribution does not always conform to established patterns; thus the traditional approach to the study of world vegetation does not adequately account for the observed distribution of grasses. A chorological study is, therefore, best performed by removing species from their communities and granting them an existence in their own right. Endemics, and more widespread species that transgress traditional boundaries, have presented particular problems to phytogeographers in the past, but chorology, as practised at Kew, can account for them. Studies of grasses have shown that: species distributions tend to be concentric rather than coincident; phytochoria are based on formations and are under climatic control; phytochoria fade away at the edges with adjacent ones overlapping at their margins; and that phytochoria lend themselves extremely well to a nested hierarchy of units of decreasing geographical range.

Key words: Biogeography, chorology, endemism, phytochoria, phytogeography, *Poaceae*.

Introduction

The Oxford English Dictionary defines chorology as 'the scientific study of the geographical extent or limits of anything.' The earliest usage of the word cited by the OED is that of Lingard (1858) who employed it in a non-biological context (specifically, the distribution of Anglo-Saxon bishoprics); in a biological context it dates from Haeckel (1866). Since nothing more precise is implicit in the word, chorology is used today for a number of different aspects within the realms of plant and animal geography, and indeed it has acquired so many shades of meaning that it can now scarcely be used without first being defined. Browicz (1994) used the term for his collection of dot-maps of Near and Middle Eastern tree species; and Zohary's (1973, p.81) 'phytogeographical' map of the Middle East was followed by Léonard's (1989, p.81) very similar 'chorological' map. When first used by Clayton for his new method of analysing grass distributions (Clayton and Hepper 1974) the word had been neglected for a considerable time, but it has now come back into vogue and is widely used in place of the more accessible words 'biogeography' and 'distribution.' Any special meaning that Clayton tried to attach to it has now more or less been lost.

The ecological background to chorology and the need for a new method of analysing grass distributions were discussed at length by

Clayton and Hepper (1974) and summarised by Clayton and Cope (1980a) and Cope and Simon (1995). The history of chorology, its broader concepts and its relationships to the general field of phytogeography were critically reviewed by Friis (1992). Numerous attempts to produce global vegetation maps were discussed by Friis and they seem to fall into two schools. On the one hand is the view of Schouw (1822) that phytogeographical units can be arranged in a nested hierarchy, while on the other is the view of White (1971, 1976) that the hierarchy should be discarded. Both schools base their classifications on the concept of 'endemics' and both recognise that there are troublesome species that transgress the boundaries between phytochoria. Schouw recommended that phytochoria should be defined in terms of their characteristic vegetation, thus ignoring the problem species, but White introduced the concept of transition zones to accommodate them. Whatever the differences between these two approaches they share the common ground of defining their phytochoria in terms of limited numbers of selected species. Friis did not dwell on chorology as practised at Kew, nor did he mention any of the interesting findings that have been achieved from our studies of the grasses. The work begun by Clayton and Hepper (1974), and developed by Clayton (1975), Clayton and Panigrahi (1975), Cope (1977a), Clayton and Cope (1980a, 1980b), Cope and Simon (1995) and Cope (ined.), continues at Kew and that it should continue can be justified in two ways: first, analyses with such detail as those for Africa (White 1976, 1988) and Southwest Asia (Zohary 1973; Léonard 1989) are not available for most of the rest of the world; and secondly, the maps presented by these authors do not adequately account for the observed distribution of grasses. It is well understood that grasses can be rather imprecise in their ecological preferences, many of them being exceedingly difficult to describe (for example, *Dichanthium micranthum* Cope occurs in southern Arabia in a wide range of habitats from rocks in the splash zone above high water mark via monsoon grassland and open woodland to harsh rocky desert), and consequently their adherence to the phytogeographical and chorological patterns laid out by the cited authors is equally imprecise. An alternative strategy, specifically to examine grass chorology (since this is where both the taxonomic expertise and phytogeographic interest currently reside at Kew) is needed and the method devised at Kew more than two decades ago has proved to be both informative and statistically robust. It is reviewed here.

BACKGROUND

The spatial sorting of species into communities by the action of local climatic, edaphic and biotic factors is the realm of ecology. The discipline aims to describe and map these communities, to explain their relationships with the environment, to examine the interactions of their components, and to organise them into a hierarchical classification according to the degree of similarity between them. Phytogeography extends the community classification to a global study of vegetation by shifting the similarity criteria from species-composition to physiognomy. The vegetation types at the summit of the hierarchy ('formations') are thus essentially demarcated by differences in their dominant life-form, and because environmental factors conform to a nested hierarchy of geographical scale (climate being the most wide-

spread; biotic the most fragmented and local) classifying these vegetation types into a similar nested hierarchy is feasible. The edaphic and biotic environments are largely climatically dependent and seldom give rise to distributions that transgress climatic boundaries; thus the world's vegetation types are mostly determined by climate.

Both communities and formations are conventionally defined in terms of limited numbers of selected species. While the distribution of species and communities are clearly inter-related they are by no means equivalent and it is a mistake to assume that all species' distributions correspond with those of their characteristic communities. This is particularly true of the grasses and forms the basis of a new approach to understanding their distribution. Chorology, as defined at Kew, sets out to examine the properties of species' distributions when they are granted an existence in their own right, irrespective of the communities in which they participate, and takes no account of their ecology.

While local climatic, edaphic and biotic factors act together, or separately, to limit the occurrence of individual species and the communities into which they are sorted, they do not themselves constitute effective barriers to the free dispersal of seed. Since seed is of wider occurrence than the plants that generate it we can envisage these plants being surrounded by a halo of seed extending for a lesser or greater distance. Much of this seed will be wasted, of course, because local factors may not allow its germination or the survival of any seedlings it does produce, but with a patchwork of communities in close proximity it can be appreciated that the participants in each of these communities have been recruited from a common seed pool. In a sense, therefore, chorology is the study of the distribution of these seed pools and thus the study of impediments to migration within the range of habitats ecologically suitable for occupation by a species. The discipline may be approached in two ways: first, by plotting present day distribution patterns and exploring the inferences about past migration that may be drawn from them; and secondly, by investigating the actual means by which dispersal was, or at least could have been, effected, and examining the consequences (Clayton 1976; Wickens 1976). Plant dispersal is often treated as if it were a deterministic process, with the underlying assumption that some set of biological laws involving autecology, target habitat and means of transport determine whether a species will migrate or not. However, it seems more realistic to regard dispersal as a statistical process in which migration depends upon the favourable coincidence of several fluctuating factors, so that its outcome is not predictable with certainty but can be expressed only in terms of probability (Simpson 1952). Seen in this light, the distribution of individual species assumes less importance, because all of them are subject to random variation, and a small proportion of quite improbable disjunctions is to be expected. It is, rather, the average distribution of species that is significant.

TECHNIQUES

The method of chorology, as practiced at Kew, has been fully described in previous papers but is summarised here for convenience.

Step 1: The species list

All species from the area under study are listed. Hitherto, we have looked at local floras in isolation but the present author is currently working towards a world overview that will involve something in excess of 8500 species. Not all species will be appropriate and a selection will have to be discarded. The latter includes all cereals and other species known only from cultivation (or now so widely cultivated that their native distribution is obscure). The analysis relies on sound taxonomy on a global scale and any species whose taxonomic status is in any doubt must also be discarded.

Step 2: The recording units

The study area must now be divided into units for recording purposes. There is no set size – or indeed shape – for these units but they must be small enough to produce maps of sufficient detail and large enough to be sure of capturing the appropriate information. The following factors should be taken into account:

a) if of sufficient size the recording units will sample climatically dependent species-groupings but will be largely insensitive to the multitude of species-associations produced by edaphic and biotic factors;

b) each recording unit should contain a comprehensive mosaic of communities in order to avoid mapping at a purely ecological scale and to ensure that every species in the seed pool has an equal chance of being recorded. Since it is unlikely that every unit would contain a stretch of coast any purely maritime species will be unrepresented in most units and they should therefore also be eliminated from the species list;

c) the recording units should ideally conform to the world's major phytogeographical boundaries, but since the exercise is to identify these boundaries the argument here is somewhat circular;

d) the recording units should, as far as is practicable, be of equal size since units of unequal size are statistically objectionable;

e) the size of a recording unit will to some extent be determined by the intensity of botanical collecting. Since this is very uneven, account must be taken of parts of the world where it is particularly low, and an undercollected area should be amalgamated with a neighbouring well-collected one to eliminate unnecessary gaps in the data matrix;

f) since most of the records are to be compiled from herbarium specimens, augmented where appropriate by reliable local Floras, it is important that the boundaries of the recording units be readily apparent.

With these factors in mind, we are virtually forced to use country-sized recording units employing current political boundaries. Large countries can be subdivided along existing administrative boundaries, and small neighbouring countries amalgamated.

Step 3: The data matrix

The occurrence of species within recording units is now recorded. There is much debate on the suitability of simple presence or absence records and Friis (1992) has expressed the view that a measure of abundance as well as presence would be desirable. However, since we believe that species distributions are probabilistic rather than deterministic (see **Background** above) frequency of occurrence within any particular recording area is unimportant, and because the data matrix is compiled from herbarium specimens it is, in any case, often quite impossible to ascertain. An illustration of this problem concerns a new species of grass described from Dhofar Province, Oman (Cope 1977b). Between the time the original collections were made (1943) and the time the species (*Dichanthium micranthum*) was described, there were just two known specimens. Since numerous collectors had been to this area over this period and none had ever gathered any more of the species the natural assumption was that it was probably extremely rare. Since its publication, however, collectors have become much more aware of it and there have been 27 subsequent specimens added to the Kew herbarium. A visit to the area by the author in 1991 revealed, against all expectation, that the species is probably the single most abundant and widespread grass in Dhofar. The sparsity of herbarium records seems, paradoxically, to be a measure of the species' abundance in the field, and reflects an aspect of the psychology of collectors that urges them to deem something so common not to be worth gathering. Whatever the case, the abundance of a species in the herbarium is not a reliable measure of its abundance in the field and Friis's desire cannot be fulfilled (except in rare cases of adequate local knowledge).

At this point in the exercise the species list should be reconsidered and all known or suspected introductions eliminated.

Step 4: The analysis

When this form of chorology was being developed at Kew in the 1970s analysis proceeded by computation. The software developed by W.D. Clayton and C.J. Cousins (unpublished) was, however, written for a now-obsolete machine and in a now-obsolete language. The machine itself was not very powerful by today's standards and was capable of handling only a limited data-set. As time and experience progressed it became possible to dispense with the computer altogether (the software now being more or less redundant and not worth rewriting for a modern machine) and analysis could proceed by manual manipulation of the data. The practitioners of Kew chorology have become very well versed in the patterns that the computer would be likely to find and are now confident of detecting them without its aid (indeed, many of them become apparent even while the data-matrix is being compiled). The original task that the computer was asked to perform was to compare distributions by calculating Jaccard's Coefficient of Similarity (number of recording units in common divided by the total number occupied) for every pair of species; from the output a minimum spanning tree could be constructed and phytochoria identified by inspection. Inspection begins at an arbitrary point on the tree, usually at the centre of a tight cluster, and adjacent species' distributions compared. Those that are broadly similar are accumulated and when the overall shape of the distribution pattern changes between adjacent species in the tree it can be assumed that the limit of a species-group has been transgressed. The computer does not identify all the phytochoria and some of them may be fragmented within the minimum spanning tree; up to a point, therefore, identifying them becomes an intuitive process.

Step 5: Drawing the maps

Once the content of a species-group (which we can now call a phytochorion) has been determined a map can be drawn. Total species counts for each recording unit involved in the phytochorion are entered on an outline map and lines (which are termed isochores) are drawn around those units with certain proportions of the total species count. Our normal practice was to draw the 10%, 25%, 50% and 75% isochores. Local knowledge of the vegetation and topography often helps in accurate positioning of the isochores, but in general they are lines that serve only to give an impression of the shape of the phytochorion and do not reflect the complex interdigitation of neighbouring phytochoria that to some extent exists on the ground.

RESULTS

Once our analyses were completed, and the phytochoria identified and mapped, we were able to make the following observations about grasses:

a) species distributions are not random, forming a disordered mosaic, but are superimposed to such an extent that a limited number of generalised patterns – phytochoria – suffice to describe them;

b) phytochoria, at least at one level, are centred on formations as comparison between our maps and various regional vegetation maps reveals. We can detect these formations by their grass floras even though there are occasions when, paradoxically, relatively few of the participating grasses actually grow within them. It is possible that grasses are unusual in the degree to which they show this effect since they have a remarkable capacity for colonising the seral stages of a vegetation type whose climax state would otherwise exclude them (particularly so in forest ecosystems where grasses are not particularly prominent under a closed canopy). It is therefore to our advantage that a large proportion of the territory assigned to some formations is still – or again – in the early seral stages of succession, or indeed at a disclimax, for this allows the expansion of a considerable grass flora, either from adjacent areas or from a historical seed-bank, into areas that would otherwise be quite unsuitable. In fact, it is not the formations themselves that we have detected but rather the extent of those factors that limit and maintain them. Grasses are responding to the local climatic factors and not to any component of the vegetation itself;

c) species vary widely in the amplitude of their climatic tolerance. The specialists, which make stringent demands of their environment, are confined to the heart of a phytochorion where conditions are at their most stable. The opportunists, whose adaptability enables them to invade environmental outliers wherever local competition is sufficiently relaxed, can make their way to the periphery. Thus, distributions tend to be concentric rather than coincident, an important difference between formations and communities on one hand and our concept of phytochoria on the other. Formations and communities occupy mutually exclusive territories and can be mapped with clear-cut, contiguous boundaries; phytochoria, however, overlap, fading away imperceptibly at the edges and are best mapped by the described method of contouring.

d) once the phytochoria have been mapped and inspected, it can be observed that there is an obvious tendency for the lesser to nest wholly within the geographical territory of the greater. There are, therefore, natural hierarchies that can be used as the basis of a formal classification, a property noted by Schouw (1822) more than a century and a half ago.

The main levels in the hierarchy are:

Kingdom. This comprises species of more or less continent-wide distribution limited north and south by the world's major latitudinal zones. Since the grass sub-families are confined more or less to latitudinal belts (Cross 1980) and approximately 2/3 of genera are confined to single continents (Clayton 1975), the various kingdoms largely differ from one another in their generic components.

Region. Regions are composed of species whose distributions correspond with the world's main vegetation types (formations) and are under climatic control.

Domain. On the whole domains comprise species of limited range within a region reflecting partial fragmentation of the species-pool by recent changes in climate accompanied by the introduction into the system of newly evolved species in centres now isolated by the imposition of geographical or climatological obstacles to their dispersal.

Where necessary, intermediate ranks (sub-kingdom, sub-region etc.) can be interposed. The grass species in any one particular community could have been drawn from any or all of the locally represented ranks in the chorological hierarchy.

ENDEMICS

Endemic species often attract considerable attention because their presence is thought to indicate centres of particular interest. However, simplistic interpretations for endemism are discouraged by the disconcertingly high numbers that are revealed (Clayton and Cope 1980a). They present a number of methodological problems, not the least of which is that of defining them. Daydon Jackson (1928) defines endemic as 'confined to a given region, as an island or country'; thus the quality of endemism – island endemics aside as they are a special case – results from the imposition of arbitrary boundaries. As an example of the scale of this problem, a recent analysis of the grasses of Australasia (i.e., Australia plus New Zealand, but excluding off-shore territories; Cope and Simon 1995), was based on an initial list of 1054 grass species that were recorded from the area so defined. Of these, 976 were 'endemic' to Australasia (i.e. absent from all other parts of the world); of these, 747 were endemic to Australia (i.e., absent from New Zealand); of these, 66 were endemic to the State of Western Australia (i.e., absent from the rest of the country); of these, 27 were endemic to the southwest corner of the State (i.e., absent from elsewhere in the State); and so on. To White (1971), endemics were the only species that mattered in chorology; species not endemic to a phytochorion did not contribute to the circumscription of that phytochorion. Those that did transgress the boundary had, nevertheless, to be accounted for. In his latest map of Africa (White 1976) several phytochoria no longer have contiguous boundaries, but are separated one from another by 'transition zones'. These comprise species that occur in the adjacent phytochoria, but are not endemic to them. The transition zone itself is characterised by having few or no endemic species of its own, all or most of the species being found in the adjacent phytochoria. The zone only exists because the adjacent areas of endemism have already been defined in terms of other, selected species. Endemism is not therefore an innate quality of the species themselves but rather the product of the philosophy that defined the phytochoria in the first place. This would seem to be a needlessly complicated way of resolving the problem of species that transgress predetermined boundaries. Schouw (1822) would have taken the pragmatic view and simply ignored them. In Kew's concept of chorology the problem does not arise; evidence from the grasses – to which our studies have thus far been confined – indicates that phytochoria fade away at the edges and those species at the periphery are no less legitimate members of the appropriate phytochorion than those at its heart. That phytochoria cannot be mapped with single lines delimiting mutually exclusive territories is therefore not viewed as a problem either.

There are, however, certain species that are a problem and these are the 'one-record endemics.' Because of the tendency for phytochoria to overlap, one-record endemics cannot always be ascribed to a particular phytochorion with confidence, especially in the area of overlap. If those that correspond with the heart of a phytochorion are added to the species count for that phytochorion they tend to distort the percentages upon which its isochores are calculated and the isochores will contract about its centre. By contrast, species occurring in two or three adjacent recording units do not greatly disturb the relative patterns of either endemism or phytochoria, whichever way they are placed, but they raise the awkward question of how widely distributed a species may be before it ceases to be considered endemic. As a reasonable compromise, therefore, the wider endemics have been regarded as species of narrow geographical amplitude at the heart of a phytochorion, but the one-record endemics are listed separately and not used in the construction of isochores. This is a departure from treatments in our earlier papers where the concept of discrete endemic centres was firmly entrenched. Wickens (1979) urged that a critical treatment of endemism should delve deeply into taxonomic relationships probing for causal factors, and would eventually, no doubt, reveal several different kinds of endemics, but this is beyond the scope of what we have been doing at Kew. What we have found, however, at a more superficial level is that the notion of compact discrete centres of endemism is largely untenable. Endemics occur in a number of broad geographical patterns, among them being the following:

1. *Southern extremities.* The greatest concentrations of endemic grasses are found at the southern tips of the continents (South Africa, Madagascar, India, Southeast Asia and South America). Climatic isolation from the continuous land mass to the north is an obvious contributory factor, but this does not altogether explain why speciation in South Africa, for example, should have been quite so prolific (Goldblatt 1978).

2. *Mountains.* There is an obvious relationship between endemism and mountains which, in the major orogenic zones, has produced a high level of local endemism over very extensive areas. Such terrain affords a rich variety of ecological niches, so that new species (and displaced refuges) have a better chance of persisting for a time than they would under the stringent competition induced by a more demanding and uniform habitat. It also affords a favourable degree of environmental stability because climatic change, which may have a catastrophic ecological effect upon the plains, merely causes a shift of altitudinal zonation in the uplands.

3. *Background.* A modest level of endemicity is almost ubiquitous, though the species concerned are sometimes little better than local ecotypes. However, this does suggest that speciation is a widespread phenomenon and that, in assessing the evolutionary significance of the process, we should not be overawed by the numerical superiority of upland endemics.

No distinctions were attempted, in our analyses, between refugia and centres of origin, for the differences are difficult to quantify, but many of the endemic areas do contain one or two genera that show a marked proliferation of species, and in this sense they are centres of speciation. However, the scanty evidence available suggests that successful new species spread rather rapidly; for example, the new, fertile, hybridogenous species *Spartina anglica* C.E. Hubb. has virtually displaced its parents (*S. maritima* (Curtis) Fern. and *S. alterniflora* Loisel.) and its sterile counterpart (*S. x townsendii* Groves & J. Groves) in a period of little more than 50 years. The implication from our work is that endemics are often rather unsuccessful species confined to special environments, and it seems likely that regions of endemism are more often sanatoria for the weak than nurseries for the strong, but they may constitute reservoirs of colonisers in the event of climatic change.

CONCLUSIONS

Our work on chorology at Kew has revealed a remarkable degree of order in the distribution of grasses at species level; earlier work we did revealed similar order at sub-familial, tribal and generic level. Our interest has thus far been with the one family that we know well and in which we can be confident that the standard of taxonomy is at a reasonably high level on a global scale. We have deliberately looked at species in isolation from the communities in which they grow in an attempt to discover the underlying patterns of distribution of grasses that are not always evident from other vegetational studies. The future of chorology is obvious and should clearly concentrate on two lines of investigation. Firstly, we need to know whether the patterns of distribution among grasses are peculiar to them or whether they represent more general rules applicable to all angiosperms. Those of us who have practiced chorology on the grasses lack the taxonomic expertise to carry the studies across into other large angiosperm families, so the future of this line of investigation depends upon the willingness of others to pursue our lines of thought. The possibility of this is something we await with interest. Secondly, we need to add a cladistic component to the exercise in order to learn what this may reveal about more general questions of the evolution of the grasses. The compilation, at Kew, of a DELTA database for the world's grasses will be an important contributory factor; only the man-power is currently lacking.

ACKNOWLEDGEMENTS

I am especially grateful to Dr Derek Clayton, Royal Botanic Gardens, Kew, for carefully guiding me through his concept of chorology and for entrusting the completion of his project to my hands. His previous work, scattered over a number of papers, has formed the basis of this review paper and I have unashamedly helped myself to large segments of his original texts on the understanding that little of what he has written can be improved upon.

REFERENCES

Browicz, K. (1994). 'Chorology of trees and shrubs in South-west Asia and adjacent regions.' (Polish Academy of Sciences: Poznan.)

Clayton, W.D. (1975). Chorology of the genera of Gramineae. *Kew Bulletin* **30**, 111–132.

Clayton, W.D. (1976). The chorology of African mountain grasses. *Kew Bulletin* **31**, 273–288

Clayton, W.D., and Cope, T.A. (1980a). The chorology of Old World species of Gramineae. *Kew Bulletin* **35**, 135–171.

Clayton, W.D., and Cope, T.A. (1980b). The chorology of North American species of Gramineae. *Kew Bulletin* **35**, 567–576.

Clayton, W.D., and Hepper, F.N. (1974). Computer-aided chorology of West African grasses. *Kew Bulletin* **29**, 213–234.

Clayton, W.D., and Panigrahi, G. (1975). Computer-aided chorology of Indian grasses. *Kew Bulletin* **29**, 669–686.

Cope, T.A. (1977a). Computer-aided chorology of Middle Eastern grasses. *Kew Bulletin* **31**, 819–828.

Cope, T.A. (1977b). A new species of *Dichanthium* from Southern Arabia. *Publications from Cairo University Herbarium* **7** & **8**, 325.

Cope, T.A. (ined.). The chorology of Central and South American grasses. *Kew Bulletin* **00**, 000–000.

Cope, T.A., and Simon, B.K. (1995). The chorology of Australasian grasses. *Kew Bulletin* **50**, 367–378.

Cross, R.A. (1980). Distribution of subfamilies of Gramineae in the Old World. *Kew Bulletin* **35**, 279–289.

Daydon Jackson, B. (1928). 'A glossary of botanic terms', ed. 4. (Duckworth: London & New York.)

Friis, I. (1992). Forests and forest trees of Northeast tropical Africa. *Kew Bulletin, additional series* **XV**.

Goldblatt, P. (1978). An analysis of the flora of southern Africa. *Annals of the Missouri Botanical Garden* **65**, 369–436.

Haeckel, E. (1866). 'Generelle Morphologie der Organismen.' Vol. 1–2. (Reimer: Berlin.)

Léonard, J. (1989). 'Contribution à l'étude de la flore et la végétation des déserts d'Iran. Fascicule 9, Considérations phytogéographiques sur les phytochories irano-turanienne, saharo-sindienne et de la Somalie-pays Masai.' (Jardin botanique national de Belgique: Meise.)

Lingard, J. (1858). 'The history and antiquities of the Anglo-Saxon church' edn 3, 1, app. F, 349.

Schouw, J.F. (1822). 'Grundtræk af den almindelige Plantegeografi.' (Copenhagen.)

Simpson, G.G. (1952). Probabilities of dispersal in geologic time. *Bulletin of the American Museum of Natural History* **99**, 163–176.

White, F. (1971). The taxonomic and ecological basis of chorology. *Mitteilungen aus der Botanischen Staatssammlung München* **10**, 91–112.

White, F. (1976). The vegetation map of Africa. The history of a completed project. *Boissiera* **24**, 659–666.

White, F. (1988). Taxonomy, ecology and chorology of African Ebenaceae. II. The non-Guineo-Congolian species of *Diospyros* (excluding sect. *Royena*). *Bulletin du Jardin Botanique National de Belgique* **58**, 325–448.

Wickens, G.E. (1976). Speculations on long distance dispersal and the flora of Jebel Marra, Sudan Republic. *Kew Bulletin* **31**, 105–150.

Wickens, G.E. (1979). Speculations on seed dispersal and the flora of the Aldabra archipelago. *Philosophical Transactions of the Royal Society of London, B*, **286**, 85–87.

Zohary, M. (1973). 'Geobotanical Foundations of the Middle East', vol 1. (Gustav Fischer Verlag: Stuttgart; Swets & Zeitlinger: Amsterdam.)

Grasses: Systematics and Evolution. (2000). Eds S.W.L. Jacobs and J. Everett. (CSIRO: Melbourne)

GRASSES

GRASSES IN NORTH AMERICA: A GEOGRAPHIC PERSPECTIVE

Mary E. Barkworth and Kathleen M. Capels,

Intermountain Herbarium, Biology Department,
 Utah State University, Logan, Utah 84322-5305, U.S.A.

Abstract

We have databased information on grass distributions in North America from a variety of sources in connection with the *Manual of Grasses for the Continental United States and Canada* project. These data, and treatments submitted by the contributors, indicate that the *Manual* region (= continental United States and Canada) has 896 native grass species and 331 established introductions. Of the native species, 609 occur only in the U.S. portion of the region and 11 only in the Canadian portion. There are 410 species endemic to the *Manual* region, with 262 restricted to the United States. Eighty-four species are endemic to a single State or Province, with the largest number (37) being in California. There are 209 native species whose range extends into northern Mexico, 74 that extend into Central America or the Caribbean, and 70 that extend into South America. There are also 22 native species that have a disjunct distribution in South America. There are 93 native species whose range includes temperate or arctic regions outside of the *Manual* region, and 17 whose extra-regional distribution is tropical.

The amount of data for the counties in the contiguous U.S. varies substantially. This must be considered in evaluating the distribution maps presented, but the data are sufficiently robust to support some general statements about grass distributions in the region. The region of highest grass diversity is the southwest. The Orcuttieae and Hainardeae are restricted to the west coast, primarily California. The Cynodonteae, Eragrostideae, and Pappophoreae are southwestern tribes. The Cynodonteae differs from the Eragrostideae in having its highest relative abundance more narrowly concentrated on the eastern slope of the Rocky Mountains. The Stipeae is richest in the southwest, but its relative abundance is greatest in the west. The Aristideae is rather evenly distributed across the southern U.S., but never forms a high proportion of the grass flora. The Diarrheneae and Brachyelytreae have a northeastern distribution in the U.S. but are also present in Asia. The Phareae, Bambuseae, Oryzeae, Centotheceae, Paniceae, and Andropogoneae have their greatest diversity in the southeastern U.S. The Andropogoneae's area of high relative abundance extends farther north and west than that of the Paniceae. The Aveneae, Poeae, Meliceae, and Danthonieae have similar, northern distributions. The Bromeae and Triticeae have their highest diversity towards the southwest, but their highest relative abundance is towards the northern part of the contiguous U.S. The Arundineae, represented only by *Phragmites australis* (Cav.) Steudel, occurs in two disjunct regions, one restricted to coastal areas around the Gulf of Mexico and the other extending northeast of a line extending from western Texas to North Carolina.

Key words: Biogeography, grasses, grass distributions, North America, photosynthetic pathways.

INTRODUCTION

The *Manual of Grasses for the Continental United States and Canada* (Barkworth *et al.* in prep.) will include keys, descriptions, illustrations, distribution maps, and nomenclatural information for the grasses that have been found in the continental U.S. or Canada. Preparation of the *Manual* is a cooperative effort, involving over 70 contributors (Appendix A) located at a wide range of institutions. Its funding (U.S. Department of Agriculture-

Agricultural Research Service US$173,000; Utah Agricultural Experiment Station US$85,300) has been earmarked for editorial assistance, illustrations, and supplies. Contributors received no funding from the project despite being expected to provide all information for their genera, including distribution maps for each species recognized.

For many contributors this last requirement proved impossible. They hold positions that allow little or no time for such research and provide no facilities or support for handling loans of specimens. Even contributors with significant research responsibilities find it hard to devote time to floristic research when tenure and promotion criteria demand that they obtain competitive outside grants. An additional problem was that many genera had no designated contributor, and hence no one to provide distribution maps.

To address these problems, we began databasing distributional information for grasses from a wide range of resources (Appendix B) in addition to specimens. This database is primarily a default resource. If a contributor has provided maps, we compare the distributions shown by our database with his or her maps. If there are significant discrepancies, or even just some questionable records, we look for an explanation. In most cases the variations are slight or easily explained by differences in taxonomic opinion. In a few instances, further investigation is required. In this paper we present information on the geographic distribution of grass tribes in North America and share the insights we have gained from developing the database.

MATERIALS AND METHODS

Sources of Information
The information in our database comes from many resources (Appendix B). Before entering any data in the files, we convert the nomenclature used to that employed in the *Manual*, using the synonymy provided by the appropriate contributor. We have not attempted to impose a uniform species concept on *Manual* contributors. Data are recorded at the lowest taxonomic level used in a resource but, because many of our sources provide data only at the species level, we summarized the data at the species level before constructing the tribal summaries presented in this paper.

Geographic Recording Units
For the contiguous 48 States of the U.S. (those south of Canada), our primary recording unit is the county (called a 'parish' in Louisiana). Each State is divided into counties, and these usually appear on herbarium specimens from these States. Counties are also frequently used as the recording unit for atlases and checklists.

Counties are convenient, but not ideal, recording units. They range in size from 1.97 km² to 23,470 km², with the 30 smallest counties being in Virginia, whereas the 30 largest counties are in five western States: Arizona, California, Nevada, Oregon, and Wyoming. In addition, although the borders of eastern counties tend to follow geographic features, those of western counties often follow lines of latitude or longitude. There is, however, no other geographic unit for which one could obtain a comparable amount of information without major funding.

Specimens from the contiguous U.S. rarely bear latitude and longitude data. With the advent of inexpensive GPS (Global Positioning System) units, latitude and longitude data or UTM coordinates are found with increasing frequency on specimens collected in this portion of the *Manual* region.

Alaska and parts of Canada are divided into counties, but county information from these regions rarely appears on herbarium labels and has never been used in developing checklists or floristic atlases. Data from these two regions are entered in the database as point data, using latitude and longitude. Such data are often included on the labels of specimens from these areas, particularly specimens collected at high latitudes.

We use several programs for determining the latitude and longitude at which a specimen was collected. Many specimens from the western U.S. (but not Alaska) have their source recorded in terms of township, range, and section. A section is a unit of land that is, theoretically, a square one mile [1.6 km] long on each side. For these specimens, latitude and longitude were determined using a conversion program written by Wefald (1995). It is accurate to the nearest section. For other U.S. specimens, the latitude and longitude data were obtained from Street Atlas USA (DeLorme 1997), TripMaker (Rand McNally 1997), or the Geographic Names Information System compact disk (U.S. Geological Survey 1993). In the last instance, the accuracy used was usually 20 miles [32 km], locations up to 20 miles from a reference point being interpreted as having the same latitude and longitude as the reference point. Greater accuracy was provided for specimens from Utah. For Canadian specimens, we have generally obtained the geographic coordinates from provincial and territorial gazetteers but TripMaker (Rand McNally 1997), a recent acquisition, can be used for this purpose.

While the new software makes it much easier to obtain latitude and longitude data for a location than a topographic map, it is still a time-consuming procedure and the accuracy is often limited by the imprecision of the original collection data. For these reasons, we have deduced the latitude and longitude data only for specimens from the contiguous 48 states that are of particular interest.

We use the Geographic Transformer program (Blue Marble Geographics 1997) to obtain latitude and longitude data for localities shown on published maps. This program calculates the projection of a map from eight reference points. These data are used to construct a quadratic function for determining the latitude and longitude of other locations on the map, and an estimate of the error associated with each determination.

The Data Files
The data are recorded in tribal files, with each tribe being represented by both a datapoint file of latitude-longitude data and a county file for county level data. We use Alpha 5 (Alpha Software 1996) to manipulate the database files outside of the mapping program.

The datapoint files store information about the source of each record, e.g., the herbarium code and accession number, the collector's name and collection number, the citation for a flora or other publication that contained the original information, or the name of the individual sending us information and the year in which it was received. The structure of the county files precludes storing such information with each record, so we use a code to indicate the class of data source and record information about the source in Papyrus (Research Software Design 1992). The classes used are contributor, State or Provincial atlas or manual, regional atlas or manual, reliable observation, other publication (e.g., journal article, county publication, thesis, ecological publication), herbarium database, cultivated plant, and specimen (usually in the Intermountain Herbarium, UTC). When we have two sources for a particular record, the order of priority is contributor's data, specimen data, published data, reliable observation. So far as we are aware, contributors who have provided county level data have done so on the basis of specimens or careful observation.

A species with infraspecific taxa is considered to be present in a county if any of its infraspecific taxa are present. It is also treated as present if there is a specimen for which we have latitude and longitude data for a location within a particular county, perhaps from a published distribution map that did not show counties. For this paper, we excluded records that we knew were based on cultivated plants.

Data Analysis

To obtain summary statistics, we used Excel (Microsoft 1997) to record, for each tribe, which species occurred in which counties and whether the species was native or introduced. Species that are native in some part of the *Manual* region were counted as being native throughout their range in the region. This is usually, but not always, true. The two species groups (native and introduced) were then summarized separately to obtain the total number of each present within a county. These county totals were used in developing the maps.

We also created a database to store information about the overall distribution of each species. This indicates whether a species is introduced or, if native, endemic to a single State or Province, Canada, the continental U.S., or the *Manual* region (i.e. the continental U.S. plus Canada). We used several categories to indicate the range of a species beyond the *Manual* region: northern Mexico (north of Guerrrero, Oaxaca, and Tabasco); Central America and the Caribbean islands; South America; America; Europe; Eurasia; Asia; Africa; India; Australia; transberingian; circumboreal; tropical; or cosmopolitan. American species are those that have a continuous distribution from the *Manual* region to South America; species with a disjunct population in South America are recorded as having part of their distribution in South America. These categories were, in general, mutually exclusive, except that a species could extend across the Bering Strait and into northern Mexico. This range information was obtained from a variety of references, notably Clayton and Renvoize (1968), Cope (1982), COTECOCA (1983, 1987, 1991, 1995), Davidse *et al.* (1994), Douglas *et al.* (1994), Gibbs-Russell *et al.* (1991), Gould (1975), Gould and Moran (1981), Hickman (1993), Hitchcock (1951), Judziewicz (1990), Koyama (1987), Scoggan (1978), Tolmachev (1995), Tsvelev (1983), and Tutin *et al.* (1980).

Mapping Species Diversity

For the contiguous 48 States, we summarized by tribe the total number of species, number of native species, and number of introduced species present within a county. We report these

numbers as species counts. For native species, we calculated two additional measures of diversity, species density and relative abundance. Species density was obtained by dividing the number of species in a county by the county's area, as provided in the mapping program, Atlas GIS (ESRI 1998). The relative abundance of a tribe in a county is the number of native species of that tribe in the county as a percentage of the total number of native species in the county. This measure differs from Hartley's percent frequency (1950, 1954, 1958a, 1958b; Hartley and Slater 1960) in its exclusion of introduced taxa.

We used Atlas GIS (ESRI 1998) to portray the geographic distribution of the various diversity measures. For tribes that have more than five species in the *Manual* region, the range of species counts was divided into five equal ranges, with a count of zero being treated as a sixth range. For species density, the scores were divided into five ranges in such a way that each range was represented by the same number of counties. Relative abundance was plotted using both approaches, dividing the total range into five equal intervals (equal range approach) and dividing the total range in such a way as to have an equal number of counties in each range. In a few instances, the lowest two ranges both showed zero abundance, to the first four significant figures. In such instances, the two ranges were treated as a single range.

Our intent was to divide Canada and Alaska into artificial 'counties' of approximately the same size as the average county size in the contiguous 48 States, so that we could look at variations in grass diversity throughout the *Manual* region. This could be accomplished with Atlas GIS (ESRI 1998), but we could not afford the time to do so.

The summaries presented here reflect the data that we had entered through December 31, 1998. A project such as this is never complete, but we shall continue to incorporate additional data, to the extent that our resources permit, until the *Manual* is published. We are also pursuing avenues that would make the development of well-documented, locality-based distribution maps an ongoing collaborative project in which any herbarium can participate to the extent that its director deems appropriate. This is obviously a desirable objective, but one that requires consideration of the costs and benefits for all involved, particularly for the most important individuals, those who collect and identify specimens. The software for integrating, on demand, information from multiple institutions exists (Kaiser 1999). Many herbaria are already databasing their information in an object-oriented database, the only major requirement of the existing software. The major problem to integrating data from herbaria is the lack of agreement that exists concerning the taxonomic treatment of many taxa and, of course, the large number of existing specimens that do not have geographic coordinate data. The existence of inexpensive GPS units and the integrating software described by Kaiser should encourage future collectors to include geographic coordinates on all their labels.

RESULTS

Summary Data

The *Manual* currently includes information on 1390 species. Of these, 896 are native to the *Manual* region, and 331 are established introductions. The remainder are cultivated or transient species. There are probably less than 10 species whose status, whether native or introduced, is doubtful. The cultivated species include both ornamentals and species commonly grown for research purposes.

There are 276 species in both the U.S. and Canada, 609 only in the U.S., and 11 that are restricted to Canada. Unfortunately, we did not keep a separate count for Alaskan species. Had we done so, the number of species in both the U.S. and Canada would have been lower because many of the species included in that count occur only in Alaska and Canada, not in the contiguous 48 States. We would also have been able to determine how many species occur only north of the forty-ninth parallel (Canada's southern boundary), as well as those that are restricted to Alaska. We could extract the data from our current files, but unfortunately it would require more time than we can afford at present.

The *Manual* region has 410 endemic species, 262 of which are restricted to the U.S. and four to Canada. One tribe, the Orcuttieae, is restricted to California and Baja California, Mexico. Eighty-four species are endemic to a single State or Province. California, with 37, has the largest number of such endemics. Florida has 14, Texas 11, Oregon 3, Washington 2, and Arizona, Colorado, Montana, New York, British Columbia, and Saskatchewan 1 each. Florida is much smaller (192,666 km^2) than either California (464,304 km^2) or Texas (849,095 km^2). Adding Georgia and South Carolina to Florida would create a region comparable in size to California (471,699 km^2) having 23 endemic species.

There are 209 species whose range extends from the *Manual* region into northern Mexico and an additional 15 with a similar range plus a disjunct range in South America. The corresponding figures for extending into Central America (in which we included southern Mexico and the Caribbean) are 74 and 7. Seventy species extend from the U.S. into South America.

It was harder to characterize the extra-American distribution of native northern species. As they are currently categorized, there are 33 circumboreal species, 25 transberingian, 18 Eurasian, 17 tropical, 11 Asian, 4 cosmopolitan [*Deschampsia cespitosa* (L.) P. Beauv., *Distichlis spicata* (L.) Greene, *Phragmites australis* (Cav.) Trin., and *Trisetum spicatum* (L.) K. Richter], and 2 amphi-Atlantic [*Puccinellia vaginata* (Lange) Fernald & Weath. and *P. vahliana* (Liebm.) Scribner & Merr.].

Diversity Data for the Contiguous 48 States

There are 3111 counties in the contiguous U.S., and the completeness of our data varies from county to county. We have very few data for major portions of the country (Figs 1-3): Washington and Montana in the northwest; Maryland, New Jersey, and Rhode Island in the northeast; Mississippi and Alabama in the southeast; Oklahoma and central Texas in the south; and parts of California. There are two counties for which we have no records (Wahkiakum County, Washington, and Claiborne County, Mississippi) and 1213 with only 1-44 species. Some of these counties are in States with a State atlas, e.g. Ohio (Braun 1967) and Georgia (Jones and Coile 1988). This suggests that the problem is sometimes lack of collecting activity, not the difficulty of acquiring unpublished data. Thirty-one counties have 177-220 species, with the largest number, 220, being found in Doña Ana County, New Mexico.

The uneven data density in our database means that interpretations based on it should be treated with caution. Nevertheless, the overall picture of greatest species counts in the southwest is probably true. Acquisition of additional data from Mississipi and Alabama might suggest that the southeast is about as rich as the southwest, but this may not necessarily be the case, as we had a good source of data for Georgia (Jones and Coile 1988). The ten counties with the most native grass species are Pima Co., Arizona (161), Brewster Co., Texas (157), Coconino Co., Arizona (156), Doña Ana Co., New Mexico (149), Larimer Co., Colorado (148), Mobile Co., Alabama (147), Cochise Co., Arizona (146), Harris Co., Texas (145), and Franklin Co., Florida (141). Coconino, Pima, Cochise, Doña Ana, and Brewster Counties are all southwestern counties with considerable topographic variation. Harris County (which includes the city of Houston) is highly ecologically diverse, ranging all the way from dry uplands to saltwater marshes. Franklin County, in the Florida panhandle, includes both the Apalachicola River watershed and part of the coastal region. Larimer County is on the east slope of the Rockies and adjacent to the Wyoming-Colorado border. It has considerable topographic variation, but it is also home to Colorado State University, a Land Grant institution. Such institutions usually have a major herbarium and a history of floristic activity. Doña Ana County is also home to such an institution, New Mexico State University. Mobile County houses the University of South Alabama which, while it does not have a strong botanical emphasis, has Dr. Michel Lelong, an agrostologist, as a member of its Biology Department. There is no Land Grant institution in Franklin County, but there is one, Florida State University, in nearby Tallahassee.

The important role an individual or institution plays in determining the number of species recorded for a county is more evident in the map of species density for native grasses (Fig. 4). Most of the counties with high species density are in the eastern portion of the U.S. because of the small size of the counties there. Almost all the western counties with high species density are associated with a taxonomist or institution having a strong floristic research program. Eastern States have such institutions, and they have, in many instances, been in existence much longer than those in the west. Nevertheless, the rather uniformly high species densities of counties in the eastern States suggests that the primary factor involved is the small size of the counties in these States.

The importance of field-oriented taxonomists and institutions with a strong floristic research program in documenting the diversity of a region comes as no surprise to taxonomists, but it needs to be emphasized. Making data from all existing specimens widely available would improve the situation, but there are many areas of the country that are still poorly collected.

Of the ten counties with the most introduced species, nine are in California. The tenth is Philadelphia, Pennsylvania. All except Butte County, California, are coastal counties with significant ports. The presence of so many Californian counties in this top ten came as a surprise, but California has a long history of commerce with many other parts of the world, and a great diversity of habitats. It may also, however, reflect the appreciation of those living in California for the richness of their native flora and a concern for protecting it from further erosion by the invasion of aggressive weedy species. This appreciation is also evident in the number of county checklists (8) available for California, the support that was raised for publishing the *Jepson Manual* (Hickman 1993), and state legislation that mandates use of native species in all revegetation projects on public land. The high count for Butte County reflects, to a substantial extent, the interest shown by one of the ranchers in the county, Larry Ahern, who started by wanting to know the plants on his ranch and then became interested in the flora of the whole county.

The data for introduced species are less reliable than those for native species, but it is difficult to assess how much less. Many collectors tend to ignore introduced species, particularly weedy ones, so herbaria provide a poor record of the distribution of such species. Some of the references we used also failed to show the distribution of such species, in part because of their reliance on herbarium specimens.

Tribal Distribution Patterns
We identified six different distribution patterns among native North American grasses. The Orcuttieae (Fig. 5) and Hainardieae (Fig. 6) are confined to the extreme west. Members of the Orcuttieae are associated with vernal pools and are found only in California and Baja California, Mexico. The Hainardieae is a small tribe, most of whose members are Mediterranean or Eurasian. Only one genus, *Scribneria* Hackel, occurs elsewhere. Its only species is most abundant in California. It also occurs in southern Washington but has not been found in Oregon, the intervening state. The closest other member of the Hainardieae is *Pholiurus pannonicis* (Host) Trin., which grows in eastern Europe and central Asia (Clayton and Renvoize 1986).

Counts for four tribes, the Pappophoreae (Fig. 7), Eragrostideae (Figs 8, 9), Cynodonteae (Figs 10, 11), and Stipeae (Figs 16, 17), suggest a primarily southwestern distribution, although the map for the Stipeae has a more western and northern distribution than the other three tribes. The Stipeae is most abundant in counties around the Great Basin, whereas the other tribes have their highest species counts at the southern border of the U.S. The Aristideae (Figs 12, 13, 14) also has a strong southwestern presence, but it appears to be better represented in the southeast than the Cynodonteae and Eragrostideae.

The Pappophoreae is a small tribe, represented in the Americas by two genera, both of which occur in the *Manual* region. The three other genera occur in Africa, one being restricted to that continent, the second extending into Pakistan, and the largest being widespread in the tropics and subtropics, but particularly speciose in Australia (Clayton and Renvoize 1986).

The other two tribes in this southwestern group show rather different patterns in their relative abundance data. The region of relatively high abundance for the Eragostideae (Fig. 9) extends from the U.S.-Mexican border to southern Nevada and eastern California and, east of the Rocky Mountains, toward southern Manitoba and Ontario, whereas the Cynodonteae (Fig. 11) has a strong presence only in southern Arizona and on the eastern slope of the Rocky Mountains. Whether this reflects a climatic restriction or failure to compete with other established species on the eastern portion of the Central Plains is not evident. Consideration of the species involved, which is beyond the scope of this paper, might be informative. The eastern slope of the Rockies

has lower summer temperatures than the areas to the east or the south, but why this should favor the Cynodonteae is not clear.

The Aristideae is not a major component of the native grass flora over any part of its range. It forms a relatively high proportion of the native grass flora only in those counties for which we have few data (Fig. 13). When its relative abundance is divided so that there is an equal number of counties in all ranges used (Fig. 14), the tribe is shown to be rather evenly distributed across the southern portion of the region, with lobes of high frequency extending north across the Central Plains and the east coast.

Hartley and Slater (1960) reported that the Eragrostoideae [= Chloridoideae], in which they included the Aristideae, is associated with warm, arid climates with mean mid-winter temperatures above 10°C and non-seasonal or summer precipitation. Their map (Hartley and Slater 1960, [their] Fig. 3), which is based on State level data (Hitchcock 1951), is similar to ours (Fig. 15), but our percentages for the Chloridoideae are higher than theirs. This may be because we calculated relative abundance on the basis of native species whereas Hartley and Slater considered all species present.

The Stipeae is the only one of the southwestern tribes to have C_3 photosynthesis. Its predominantly western and generally more northern distribution compared to the other tribes is more evident in the relative abundance map (Fig. 17). The large number of counties in which it forms a rather high proportion of the native flora is probably an artifact of Barkworth's interest in the tribe. We have incorporated data she obtained from examining over 20,000 North American specimens, plus the results of a survey of over 200 U.S. herbaria for two species in the tribe, *Piptochaetium avenaceum* (L.) Parodi and *Achnatherum hymenoides* (Roemer & Schultes) Barkworth (Barkworth and Pierson 1994).

Two small tribes, the Brachyelytreae (Fig. 18) and Diarrheneae (Fig. 19), have a primarily northeastern distribution. Neither extends far into Canada, but both are also represented in eastern Asia, suggesting that their present distribution in North America is a relict of an earlier, more widespread flora (Koyama and Kawano 1964).

Six tribes are primarily southeastern in their distribution. Four of these, the Phareae (1 species; Fig. 20), Bambuseae (1 species; Fig. 21), Oryzeae (12 species; Fig. 22), and Centotheceae (4 species; Fig. 24) are considered to be old tribes. Their current representation in North America is probably a fragment of a once more widespread distribution, but possibly one that did not extend west of what are now the Central Plains. Only one species from these four tribes, *Leersia oryzoides* (L.) Sw., is native in western North America.

The other two southeastern tribes, the Paniceae and Andropogoneae, are considered to be relatively recently derived tribes. They have probably expanded their range in North America since the last glaciation. Their distribution patterns in the contiguous U.S. are very similar if one looks simply at their species counts (Figs 24, 26), although the Paniceae is more widespread, and speciose, than the Andropogoneae. The relative abundance of the Paniceae is, however, concentrated in the southeastern portion of the region (Fig. 25), whereas the Andropogoneae maintains a more

335

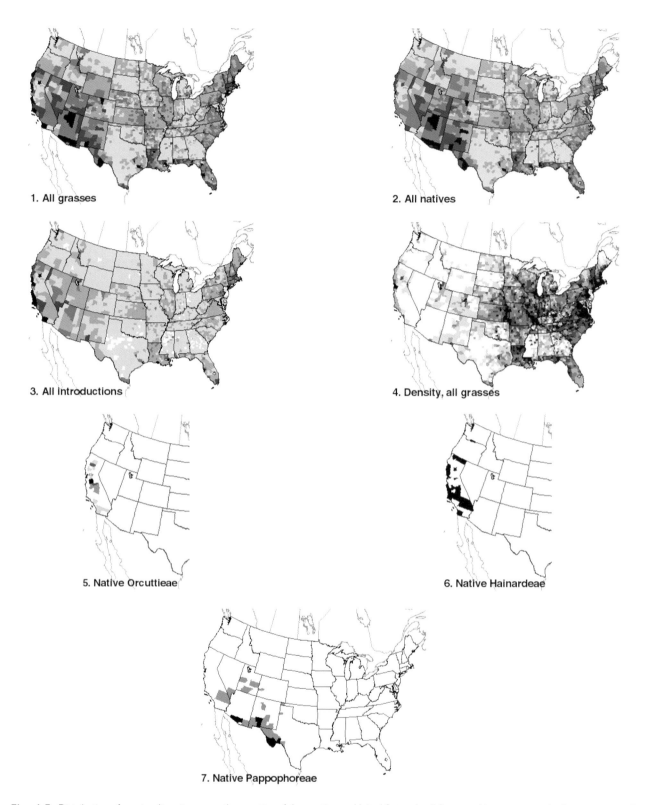

Figs. 1-7. Distribution of species diversity among the counties of the contiguous United States. In all figures, white represents the lowest range and the darkest color the highest range. The ranges used for each figure are given with the individual figure legends. The number in parentheses indicate how many counties are in each range. Figure 1. Total number of grass species. 0 (2), 1-44 (1213), 45-88 (1209), 89-132 (525), 133-176 (131), 177-220 (31). Figure 2. Number of native grass species. 0 (3), 1-32 (1139), 33-64 (1270), 65-96 (558), 97-128 (124), 129-161 (17). Figure 3. Number of introduced grass species. 0 (120), 1-20 (2054), 21-40 (778), 41-61 (124), 62-81 (25), 82-102 (10). Figure 4. Number of native grass species per square kilometer. 0-0.00800 (518), 0.00800-0.01580 (518), 0.01581-0.02675 (519), 0.02675-0.04022 (518), 0.04022-0.05944 (519), 0.05944-0.88989 (518). Figure 5. Number of native Orcuttieae. 0 (3095), 1 (8), 2 (4), 3 (2), 4 (2). Figure 6. Number of native Hainardeae. 0 (3091), 1 (20). Figure 7. Number of native Pappophoreae. 0 (3077), 1 (27), 2 (7).

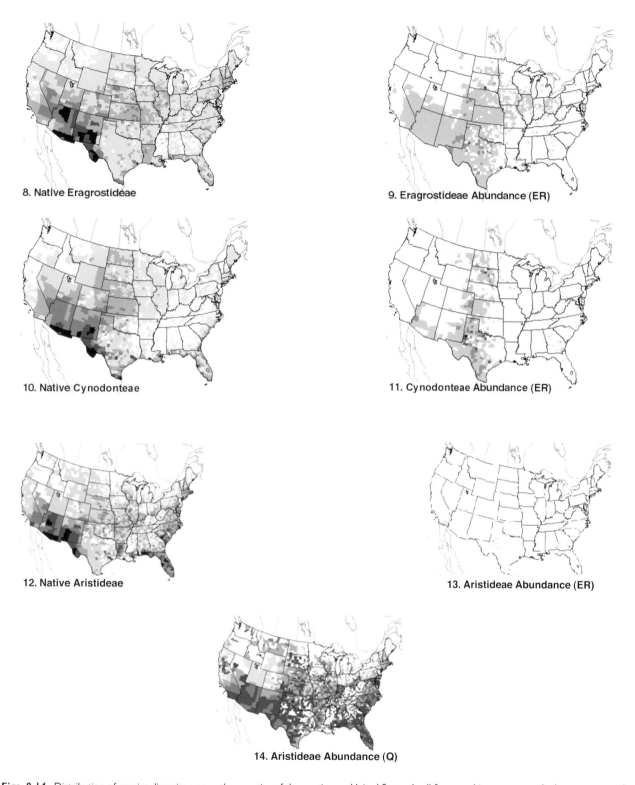

8. Native Eragrostideae

9. Eragrostideae Abundance (ER)

10. Native Cynodonteae

11. Cynodonteae Abundance (ER)

12. Native Aristideae

13. Aristideae Abundance (ER)

14. Aristideae Abundance (Q)

Figs. 8-14. Distribution of species diversity among the counties of the contiguous United States. In all figures, white represents the lowest range and the darkest color the highest range. The ranges used for each figure are given with the individual figure legends. The number in parentheses indicate how many counties are in each range. Figure 8. Number of native Eragrostideae. 0 (344), 1-10 (1953), 11-21 (719), 22-31 (74), 32-42 (10), 43-53 (11). Figure 9. Relative abundance of native Eragrostideae (Equal Ranges). 0-20% (2249), 20-40% (820), 40-60% (35), 60-80% (3), 80-100% (1). Figure 10. Number of native Cynodonteae. 0 (1289), 1-3 (1327), 4-6 (391), 7-10 (82), 11-13 (14), 14-17 (8). Figure 11. Relative abundance of native Cynodonteae (Equal Ranges). 0-10% (2728), 10-20% (314), 20-30% (51), 30-40% (14), 40-50% (1). Figure 12. Number of native Aristideae. 0 (948), 1-2 (1339), 3-4 (562), 5-6 (181), 7-8 (61), 9-10 (19). Figure 13. Relative abundance of native Aristideae (Equal Ranges). 0-20% (3071), 20-40% (32), 40-60% (3), 60-80% (0), 80-100% (2). Figure 14. Relative abundance of native Aristideae (Quantiles). 0 (948), 0.01-2.17% (305), 2.17-4.17% (632), 4.17-6.35% (608), 6.35-100% (615) [The lowest two ranges are combined].

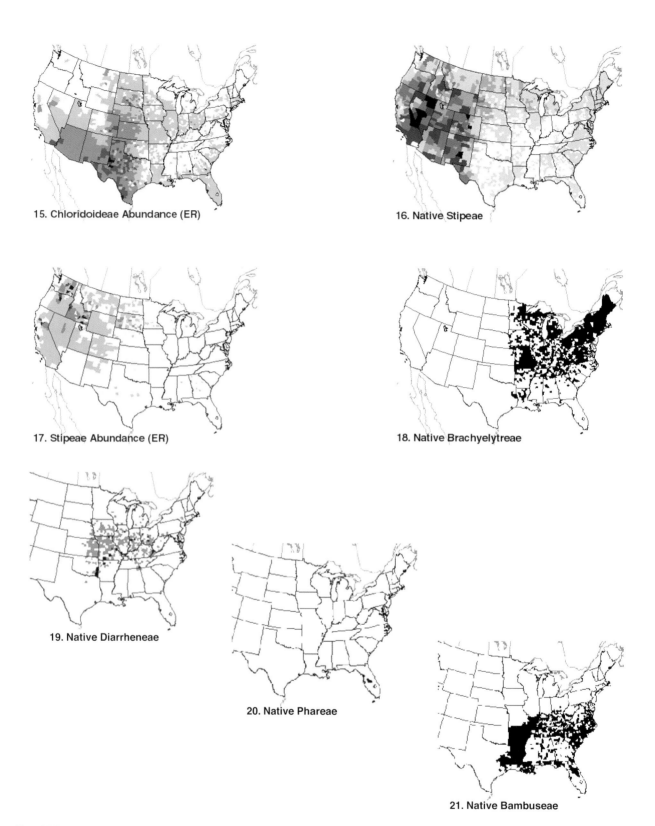

Figs. 15-21. Distribution of species diversity among the counties of the contiguous United States. In all figures, white represents the lowest range and the darkest color the highest range. The ranges used for each figure are given with the individual figure legends. The number in parentheses indicate how many counties are in each range. Figure 15. Relative abundance of Chloridoideae (Equal Ranges). 0-20 (1422), 20-40 (1287), 40-60 (322), 60-80 (68), 80-100 (9). Figure 16. Number of native Stipeae. 0 (1278), 1-3 (1480), 4-6 (214), 7-9 (90), 10-12 (42), 13-15 (7). Figure 17. Relative abundance of Stipeae (Equal Ranges). 0% (1278), 0.001-2.48% (587), 2.48-6.06% (626), 6.06-50% (617). Figure 18. Number of native Brachyelytreae. 0 (2186), 1 (925). Figure 19. Number of native Diarrheneae. 0 (2739), 1 (338), 2 (34). Figure 20. Number of native Phareae. 0 (3109), 1 (2). Figure 21. Number of native Bambuseae. 0 (2550), 1 (561).

22. Native Oryzeae

23. Native Centotheceae

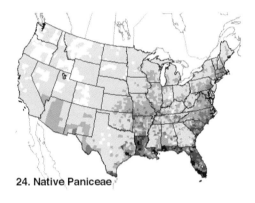

24. Native Paniceae

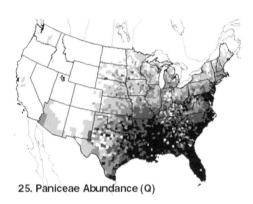

25. Paniceae Abundance (Q)

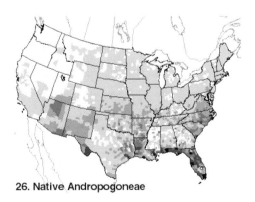

26. Native Andropogoneae

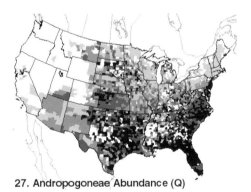

27. Andropogoneae Abundance (Q)

Figs. 22-27. Distribution of species diversity among the counties of the contiguous United States. In all figures, white represents the lowest range and the darkest color the highest range. The ranges used for each figure are given with the individual figure legends. The number in parentheses indicate how many counties are in each range. Figure 22. Number of native Oryzeae. 0 (1319), 1 (566), 2 (754), 3 (429) 5 (30) 6-7 (13). Figure 23. Number of native Centothecae. 0 (2155), 1 (668), 2 (277), 3 (10), 4 (1). Figure 24. Number of native Paniceae. 0 (181), 1-12 (1795), 13-24 (823), 25-37 (232), 38-49 (64), 50-62 (16). Figure 25. Relative abundance of native Paniceae (Quantiles). 0.001-12.73% (622), 12.73-20.63% (622), 20.63-28.57% (638), 28.57-38.71% (606), 38.71-100% (620). Figure 26. Number of native Andropogoneae. 0 (564), 1-5 (1924), 6-10 (484), 11-15 (101), 16-20 (30), 21-26 (8). Figure 27. Relative abundance of native Andropogoneae (Quantiles). 0.001-2.22% (623), 2.22-5.66% (622), 5.66-8.33% (625), 8.33-12.5% (660), 12.5-100% (578).

339

even relative abundance from the southeast to the western Central Plains (Fig. 27).

Hartley (1958a, 1958b) found that the Andropogoneae was better represented in areas with a bimodal precipitation pattern, whereas the Paniceae was more abundant in tropical climates with a unimodal pattern. Our data suggest that North American Andropogoneae also have a greater tolerance for cold winters than the Paniceae. Our data also agree with another of Hartley's observations, that dominance in a region and species richness are not necessarily correlated. The Andropogoneae was the dominant grass over the Central Plains before these were converted to agricultural use, but the Appalachian Mountains and eastern coastal plains have more species per county.

The Aveneae (Figs 28, 29), Poeae (Figs 30, 31), Meliceae (Figs 32, 33), Bromeae (Figs 34, 35), Triticeae (Figs 36, 37), and Danthonieae (Figs 38, 39) have a primarily northern distribution, extending south along the mountains on both coasts. This statement is based in part on knowledge of their Canadian distributions, but it is also indicated by maps of their relative abundance. The Bromeae and Danthonieae are less abundant in the north, but otherwise conform to the northern distribution pattern. Of these northern tribes, all but the Danthonieae are members of the Pooideae.

The Arundineae is represented in North America by a single species, *Phragmites australis* (Cav.) Steudel, but the distribution of this species in the U.S. came as a surprise. Overall, it extends from the southwestern to northeastern U.S., and across the southeastern U.S., but there is a conspicuous gap between the two portions of its distribution (Fig. 40). This is not simply an artifact of our data, for we have good records from some of the States in this gap, e.g. Tennessee (Chester *et al.* 1993) and Kentucky (J.J.N. Campbell *in litt.*). We have no explanation for the disjunction.

DISCUSSION

Recent Changes in Distribution

We have not included collection year in our distribution file and cannot, therefore, provide information on changes in distribution during the last 250 years, the period during which taxonomic botanists have been active on the continent. Even if we had, the collecting density for most species is probably not sufficient to provide much information. Many of the records for native species in our database are undoubtedly purely historical, for they are based on collections made in what are now urban or agricultural areas.

If the collection date were included, it would be possible to track the spread of species that have been introduced or become weedy during this century, at least in some instances. Thus Mack (1981) traced the early and rapid spread of *Bromus tectorum* L. in the western U.S. from herbarium specimens and Barkworth (unpubl.) was able to follow the spread of *Ranunculus testiculatus* Crantz (Ranunculaceae). It may be harder to trace the spread of species deliberately introduced for forage improvement or soil stabilization, but expansion of the North American range of *Elymus elongatus* subsp. *ponticus* (Podp.) Melderis [= *Thinopyrum ponticum* (Podp.) Z.W. Liu & R.R.-C. Wang] has been traced (Darbyshire 1997 and references therein). This species was origi-

nally introduced for stabilizing the alkaline soils that prevail in the arid west, but its range is increasing spontaneously along roads and highways that are salted in winter in order to prevent the buildup of snow and ice on their surface.

Ancient and Future Changes in Distribution

The primary determinant of the large-scale distribution of grasses is their photosynthetic pathway. Teeri and Stowe (1976) showed a correlation of 0.97 between the percentage of C_4 species in an area and the minimum growing season temperature. In a review of data based on studies in many different countries, Ehleringer *et al.* (1997) noted that the transition temperature (the temperature associated with the switch from C_3 dominance to C_4 dominance) ranges from 20°-28°C. Species with the NAD-me C_4 pathway are more abundant in drier regions, whereas those with the NADP-me pathway prevail in areas with higher precipitation. Our data are consistent with these observations for members of the Cynodonteae, Eragrostideae, and Pappophoreae, which tend to have the NAD-me pathway, whereas C_4 members of the Paniceae, Andropogoneae, and Aristideae have the NADP-me pathway.

Ehleringer *et al.* (1997) also showed that, at any given temperature, changes in carbon dioxide concentration would lead to changes in the relative abundance of species with different pathways. Low concentrations would favor NAD-me grasses; increasing concentrations would favor NADP-me grasses, but at the highest concentrations C_3 grasses would be favored over all C_4 grasses. They suggested C_4 dominated ecosystems expanded during glacial maxima because of the low atmospheric CO_2 concentrations prevailing during these periods. They also noted that the atmospheric CO_2 concentration prior to industrialization, 280 ppmV, suggests that C_4 ecosystems, such as those found in the Great Plains, would have been more extensive at that time than they are now (current atmospheric CO_2 concentration is 350 ppmV). They cite several soil studies that support this observation.

Ehleringer *et al.*'s (1997) observations imply we can expect a further shrinking of C_4 ecosystems as atmospheric CO_2 concentrations continue to increase. The effect will be slowed by increasing temperatures, but not necessarily stopped. The implications for human welfare are scary. The central and southern portion of the Great Plains, one of the most agriculturally productive areas of North America, is currently dominated by C_4 ecosystems. Much of the arid southwestern U.S. is now dominated by C_3 ecosystems, but these are not areas noted for their food production. The northern plains are also dominated by C_3 ecosystems, but these are ecosystems adapted to relatively low temperatures. Whether new cultivars of crop plants can be developed sufficiently fast to compensate both for climate changes and the rapidly increasing human population is doubtful.

Distribution Patterns below the Tribal Level

Our county level data on tribal distributions of grasses in the contiguous U.S. have revealed some interesting patterns, despite the limitations inherent in the uneven data density resulting from our approach. Examination of generic and species maps reveals additional patterns. For instance, the Stipeae is a temper-

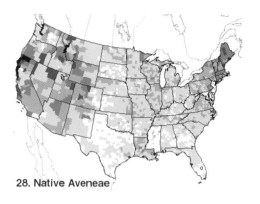

28. Native Aveneae

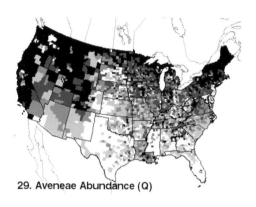

29. Aveneae Abundance (Q)

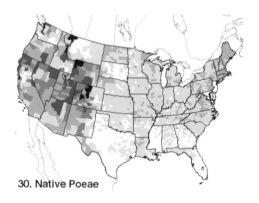

30. Native Poeae

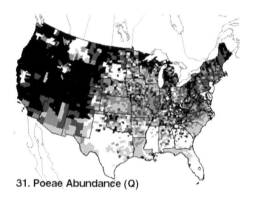

31. Poeae Abundance (Q)

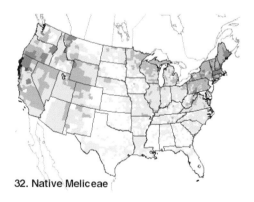

32. Native Meliceae

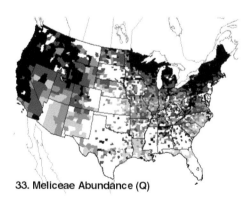

33. Meliceae Abundance (Q)

Figs. 28-33. Distribution of species diversity among the counties of the contiguous United States. In all figures, white represents the lowest range and the darkest color the highest range. The ranges used for each figure are given with the individual figure legends. The number in parentheses indicate how many counties are in each range. Figure 28. Number of native Aveneae. 0 (533), 1-6 (1802), 7-12 (630), 13-18 (117), 19-24 (21), 25-31 (8). Figure 29. Relative abundance of native Aveneae (Quantiles). 0.001-2.7% (623), 2.7-7.41 (626), 7.41-10.34% (618), 10.34-14.29% (629), 14.29-55.56% (612). Figure 30. Number of native Poeae. 0 (697), 1-5 (1861), 6-11 (446), 12-16 (61), 17-22 (34), 23-31 (12). Figure 31. Relative abundance of native Poeae (Quantiles). 0 (697), 0.001-5.08% (548), 5.08-7.27% (625), 7.27-10.53% (627), 10.53-100% (611). Figure 32. Number of native Meliceae. 0 (1003), 1-3 (1641), 4-6 (394), 7-9 (68), 10-12 (2), 13-15 (3). Figure 33. Relative abundance of native Meliceae (Quantiles). 0 (1003), 0.001-1.94% (242), 1.94-3.64% (624), 3.64-6.67% (641), 6.67-100% (598).

34. Native Bromeae

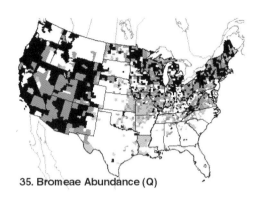

35. Bromeae Abundance (Q)

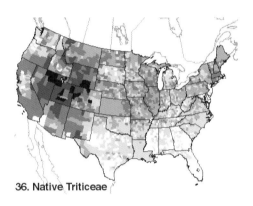

36. Native Triticeae

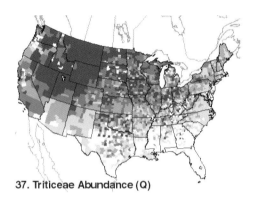

37. Triticeae Abundance (Q)

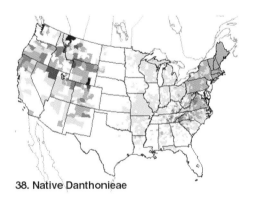

38. Native Danthonieae

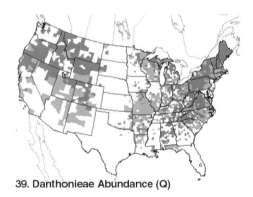

39. Danthonieae Abundance (Q)

Figs. 34- 39. Distribution of species diversity among the counties of the contiguous United States. In all figures, white represents the lowest range and the darkest color the highest range. The ranges used for each figure are given with the individual figure legends. The number in parentheses indicate how many counties are in each range. Figure 34. Number of native Bromeae. 0 (1051), 1-2 (1186), 3-4 (331), 5-6 (77), 7-8 (13), 9-10 (3). Figure 35. Relative abundance of native Bromeae (Quantiles). 0 (1051), 0.001-2.17% (371), 2.17-4.26% (623), 4.26-20% (613). Figure 36. Number of native Triticeae. 0 (503), 1-3 (1068), 4-6 (1016), 7-10 (449), 11-13 (62), 14-17 (13). Figure 37. Relative abundance of native Triticeae (Quantiles). 0.001-2.63% (622), 2.63-6.94% (622), 6.94-10% (652), 10-13.73% (592), 13.73-100% (620). Figure 38. Number of native Danthonieae. 0 (1665), 1 (998), 2 (367), 3 (72), 4 (6), 5 (3). Figure 39. Relative abundance of Danthonieae (Quantiles). 0 (1665), 0.001-2.35% (670), 2.35-50% (776).

ate tribe, but the five native North American genera have different distributions: *Piptochaetium* J. Presl and *Nassella* E. Desv. being primarily South American genera that extend into the *Manual* region from the south; *Hesperostipa* (Elias) Barkworth being primarily a genus of the Central Plains; and *Piptatherum* P. Beauv. and *Oryzopsis* Michx. being primarily northern genera. *Achnatherum* P. Beauv., as currently circumscribed, extends into both Eurasia and South America, but its limits need examination. Within the Meliceae, *Glyceria* R. Br. and *Schizachne* Hackel extend across Canada and southward along the mountains on either side of the continent, but *Melica* L. is more southern in its distribution. Moreover, it is represented by two morphologically and, in North America, geographically distinct subgenera: subg. *Melica* in the west and subg. *Bromelica* Thurber in the east. These and other patterns are best explored by taxonomists specializing in the tribes and genera concerned, for they are in a better position to discuss the relationships between current distributions and the evolutionary history of the taxa.

There are, of course, many limitations to developing distribution maps in this fashion. First there is the question of data reliability. Were the specimens identified correctly and in accordance with the taxonomic concepts adopted by the *Manual* contributors? We have not distributed the generic keys, so there are bound to be some differences in taxonomic concepts between those used in the *Manual* and those used by our data sources. If species in the *Manual* have been split compared to most existing treatments, the distribution of the segregates will be poorly represented in the literature. Lumping presents much less of a problem. The most severely affected genera are probably *Bromus* L., *Elymus* L., *Panicum* L., and *Poa* L., but we were given county level data by the contributors of *Elymus* and *Panicum*.

There are also many more analyses that could be performed on our data, including an examination of the tribal distribution patterns in terms of various climatic factors, such as those used by Hartley (1954, 1958a, 1958b, 1961; Hartley and Slater 1960) and the distribution of counties with similar agrostological indices (Hartley 1950). We do not have the time to do so, and our mapping program, Atlas GIS (ESRI 1998), has limited capabilities in this regard. It does, however, make it easy to generate a variety of different distribution maps. It is also simple to learn and relatively inexpensive. Hartley would have welcomed it.

The primary means of developing better, more reliable distribution maps is to persuade funding sources at all levels that floristic research, including that conducted in response to legal requirements, should be documented by herbarium specimens and that facilities for curating such specimens need to be maintained by qualified taxonomists. Another major source of improvement would be in expediting the incorporation of individual specimens into databases that could be accessed by software such as that described by Kaiser (1999). As Dr. Meredith Lane, a science advisor to President Clinton, noted, 'Australia is far and away ahead of the rest of the world in expertise and knowledge in this area' [quoted by T. Thwaites in *Ecos* (Melbourne) **97**:32 (1998)]. In the United States, many would like to see something similar. Unfortunately, national funding agencies have seemed to attach more importance to developing a 'perfect' herbarium database system than to creating systems with limited abilities that will

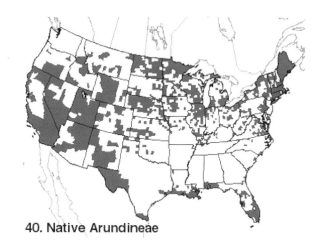

40. Native Arundineae

Fig. 40. Number of native Arundineae. 0 (2306), 1 (805).

provide immediate, or almost immediate, benefits to the most important individuals in developing our knowledge of plant diversity within a country, well-trained floristic botanists. Without such individuals, our understanding of a region's biodiversity, a biodiversity that we are rapidly eroding, can never improve.

ACKNOWLEDGEMENTS

We offer our profound thanks to the individuals and institutions whose generosity in making their data freely available to the *Manual* project, and in sending us news of new discoveries, made this paper feasible. We also applaud and thank those who have published checklists and atlases for portions of the *Manual* region. We have listed in Appendix B the sources of the geographic data that we have used in preparing this paper. We would welcome copies or information about additional checklists. We also thank the contributors to the *Manual* whose treatments we used in interpreting the data, the people at ESRI, particularly Sam Berg, for replying clearly and promptly to the questions we sent him about Atlas GIS, and Nicole DeBloois and Marianne Harris for helping us acquire locality data from published maps.

REFERENCES

Alpha Software (1996). Alpha Five, Version 3. (Alpha Software: Burlington, Massachusetts.)

Barkworth. M.E., and Pierson, K. (1994). Preparing distribution maps: from how many herbaria should one borrow, and which? *American Journal of Botany* **81** [Suppl.], 142.

Blue Marble Geographics (1997). The Geographic Transformer, Version 3.07. (Blue Marble Geographics: Gardiner, Maine.)

Braun, E.L. (1967). 'The Monocotyledonea: Cat-tails to Orchids.' The Vascular Flora of Ohio, Volume 1. (Ohio State University Press: Columbus.)

Chester, E.W., Wofford, B.E., DeSelm, H.R., and Evans, A.M. (1993). 'Atlas of Tennessee Vascular Plants. Pteridophytes, Gymnosperms, Angiosperms: Monocots, Vol. 1.' Miscellaneous Publication No. 9. (The Center for Field Biology, Austin Peay State University: Clarksville, Tennessee.)

Clayton, W.D., and Renvoize, S.A. (1986). 'Genera Graminum: Grasses of the World.' Kew Bulletin Additional Series XIII. (Her Majesty's Stationery Office: London.)

Cope, T.A. (1982). 'Flora of Pakistan 143. Poaceae.' E. Nasir and S.I. Ali (Eds.). (Department of Botany, University of Karachi and National Herbarium (Stewart Collection), Agricultural Research Council: Islamabad.)

COTECOA/Comisión Ténico Concultiva para la Determinación Regional de los Coeficientes de Agostadero (1983). 'Las Gramíneas de México, Tomo I.' (Secretaría de Agricultura y Recursos Hidráulicos: México, D.F.)

COTECOA/Comisión Ténico Concultiva para la Determinación Regional de los Coeficientes de Agostadero (1987). 'Las Gramíneas de México, Tomo II.' (Secretaría de Agricultura y Recursos Hidráulicos: México, D.F.)

COTECOA/Comisión Ténico Concultiva para la Determinación Regional de los Coeficientes de Agostadero (1991). 'Las Gramíneas de México, Tomo III.' (Secretaría de Agricultura y Recursos Hidráulicos: México, D.F.)

COTECOA/Comisión Ténico Concultiva para la Determinación Regional de los Coeficientes de Agostadero (1995). 'Las Gramíneas de México, Tomo IV.' (Secretaría de Agricultura, Ganadería y Desarrollo Rural: México, D.F.)

Darbyshire, S.J. (1997). Tall wheatgrass, *Elymus elongatus* subsp. *ponticus*, in Nova Scotia. *Rhodora* **99**, 161-165.

Davidse, G., Sousa S., M., and Chater, A.O. (Eds.) (1994). 'Flora Mesoamericana, Volume 6.' (Universidad Nacional Autónoma de México, Instituto de Biología: México, D.F.)

DeLorme (1997). Street Atlas USA, Version 5.0. (DeLorme: Yarmouth, Maine.)

Douglas, G.W., Straley, G.B., and Meidinger, D. (Eds.) (1994). 'The Vascular Plants of British Columbia. Part 4 – Monocotyledons.' Ministry of Forests Research Program. (Crown Publications, Inc.: Victoria.)

Ehleringer, J.R., Cerling, T.E., and Helliker, B.R. (1997). C_4 photosynthesis, atmospheric CO_2, and climate. *Oecologia* **112**, 285-299.

ESRI (1998). Atlas GIS, Version 4.0. (Environmental Systems Research Institute, Inc.: Redlands, California.)

Gibbs-Russell, G.E., Watson, L., Koekemoer, M., Smook, L., Barker, N.P., Anderson, H.M., and Dallwitz, M.J. (1991). 'Grasses of Southern Africa.' O.A. Leistner (Ed.). (National Botanic Gardens/Botanical Research Institute: Pretoria.)

Gould, F.W. (1975). 'The Grasses of Texas.' (Texas A&M University Press: College Station.)

Gould, F.W., and Moran, R. (1981). 'The Grasses of Baja California, Mexico.' Memoir 12. (San Diego Society of Natural History: San Diego, California.)

Hartley, W. (1950). The global distribution of tribes of the Gramineae in relation to historical and environmental factors. *Australian Journal of Agricultural Research* **1**, 355-373.

Hartley, W. (1954). The agrostological index: A phytogeographical approach to the problems of pasture plant introduction. *Australian Journal of Botany* **2**, 1-21.

Hartley, W. (1958a). Studies on the origin, evolution, and distribution of the Gramineae. The Tribe Andropogoneae. *Australian Journal of Botany* **6**, 115-128.

Hartley, W. (1958b). Studies on the origin, evolution, and distribution of the Gramineae. The Tribe Paniceae. *Australian Journal of Botany* **6**, 343-357.

Hartley, W. (1961). Studies on the origin, evolution, and distribution of the Gramineae. IV. The genus *Poa* L. *Australian Journal of Botany* **9**, 152-161.

Hartley, W., and Slater, C. (1960). Studies on the origin, evolution, and distribution of the Gramineae. III. The tribes of the subfamily Eragrostideae. *Australian Journal of Botany* **8**, 256-276.

Hickman, J.C. (Ed.) (1993). 'The Jepson Manual: Higher Plants of California.' (University of California Press: Berkeley and Los Angeles, and London.)

Hitchcock, A.S. (1951). 'Manual of Grasses of the United States.' Edition 2, revised by A. Chase. U.S. Department of Agriculture Miscellaneous Publication 200. (U.S. Government Printing Office: Washington, D.C.)

Jones, S.B., Jr., and Coile, N.C. (1988). 'The Distribution of the Vascular Flora of Georgia.' (Department of Botany, University of Georgia: Athens, Georgia.)

Judziewicz, E.J. (1990). 'Flora of the Guianas 187. Poaceae (Gramineae)'. A.R.A. Görts-Van Rijn (Ed.). (Koeltz Scientific Books: Koenigstein, Germany.)

Kaiser, J. (1999). Searching museums from your desktop. *Science* **284**, 888.

Koyama, T. (1987). 'Grasses of Japan and Its Neighboring Regions. An Identification Manual.' (Kodansha Ltd.: Tokyo.)

Koyama, T., and Kawano, S. (1964). Critical taxa of grasses with North American and eastern Asiatic distribution. *Canadian Journal of Botany* **42**, 859-867.

Mack, R.N. (1981). Invasion of *Bromus tectorum* L. into western North America: An ecological chronicle. *Agro-Ecosystems* **7**, 145-165.

Microsoft (1997). Excel 97. (Microsoft Corporation: Bellevue, Washington.)

Rand McNally (1997). TripMaker, 1997 Edition. (Rand McNally: Skokie, Illinois.)

Research Software Design (1992). Papyrus, Version 7.0.0. (Research Software Design, Portland, Oregon.)

Scoggan, H.J. (1978). 'The Flora of Canada, Part 2 – Pteridophyta, Gymnospermae, Monocotyledoneae.' National Museum of Natural Sciences Publications in Botany No. 7(2). (National Museums of Canada: Ottawa, Ontario.)

Teeri, J.A., and Stowe, L.G. (1976). Climatic patterns and the distribution of C_4 grasses in North America. *Oecologia* **23**, 1-12.

Tolmachev, A.I. (1995). 'Flora of the Russian Arctic: A Critical Review of the Vascular Plants Occurring in the Arctic Region of the Former Soviet Union.' (Trans. C.D.G. Griffiths. J.G. Packer, Ed., English edition). (University of Alberta Press: Edmonton.)

Tsvelev, N.N. (1983). 'Grasses of the Soviet Union.' (Amerind Publishing Co.: New Delhi, India.) [English translation, published for the Smithsonian Institution Libraries and the National Science Foundation, Washington, D.C.]

Tutin, T.G., Heywood, V.H., Burges, N.A., Moore, D.M., Valentine, D.H., Walters, S.M., and Webb, D.A. (Eds) (1980). 'Flora Europaea, Volume 5.' (Cambridge University Press: Cambridge.)

U.S. Geological Survey, National Mapping Division (1993). US GeoData: GNIS/Geographic Names Information System. (Earth Science Information Center: Reston, Virginia.)

Wefald, M. (1995). TRS2LL.EXE, a computer program for converting between Township, Range, Section data and latitude and longitude in the western United States. (Martin Wefald: 784-48th Avenue, San Francisco, California.)

APPENDIX A.

Contributors of Generic Treatments for the Manual of Grassses for the Continental United States and Canada

Susan G. Aiken, Charles M. Allen, Kelly W. Allred, Carol R. Annable, Cynthia Aulbach-Smith, Claus Baden, Mary E. Barkworth, Bernard R. Baum, Alan A. Beetle, Christine M. Bern, Roland von Bothmer, David M. Brandenburg, Paul P.H. But, Christopher S. Campbell, Julian J.N. Campbell, Jack R. Carlson, Jacques Cayouette, Lynn G. Clark, J. Travis Columbus, Laurie L. Consaul, William J. Crins, Thomas F. Daniel, Stephen J. Darbyshire, Gerrit Davidse, Patricia Dâvila Aranda, Jerrold I. Davis, Robert W. Freckmann, Mark Gabel, Craig W. Greene, Gerald F. Guala, II, David W. Hall, Barry Hammel, LeRoy H. Harvey, M.J. Harvey, Stephan L. Hatch, Khidir W. Hilu, Hugh H. Iltis, Surrey W. Jacobs, Niels Jacobsen, Emmet J. Judziewicz, Elizabeth A. Kellogg, Michel G. Lelong, Robert I. Lonard, Laura A. Morrison, Leon E. Pavlick, Paul M. Peterson, Grant L. Pyrah, Charlotte O. Reeder, John R. Reeder, James M. Rominger, Jack Rumely, Evangelina A. Sánchez, Sandra Saufferer, Robert B. Shaw, James P. Smith, Jr., Neil Snow, Robert J. Soreng, Lisa A. Standley, Steven N. Stephenson, Michael T. Stieber, Edward E. Terrell, John W. Thieret, Rahmona A. Thompson, Gordon C. Tucker, Jesús Valdés-Reyna, Alan S. Weakley, Robert D. Webster, J.K. Wipff, Thomas Worley, and H. Oliver Yates.

APPENDIX B.

Sources of Mapping Data

1. Individuals.

Charles M. Allen, Kelly W. Allred, Brock Benson, Julian J.N. Campbell, Adolf Ceska, Kenton L. Chambers, Laurie L. Consaul, Michael L. Curto, Stephen J. Darbyshire, Steven Dewey, Barbara Ertter, Myrna Fleming, Donna Ford, Ben Franklin, Gerald F. Guala, II, Vernon L. Harms, Alton M. Harvill, Douglass M. Henderson, Harold Hinds, Judy Hoy, Tom John, Michael Johnson, John Kartesz, R.Q. Landers, Frank Lomer, Francis E. Northam, Jeanette C. Oliver, Steve J. Popovich, H. Roemer, Lynn Rubright, Bruce A. Sorrie, Peter F. Stickney, Edward E. Terrell, Gordon C. Tucker, Margriet Weatherwax, Patrick Whitmarsh, Barbara L. Wilson, J.K. Wipff, and Vern Yadon.

2. Institutional Databases and Internet Resources.

Humboldt State University herbarium (HSC), plus some data from JEPS collections; Missouri Botanical Garden specimen database; Montana Natural History Program (Web site); Montana State University (Web site); Morton Peck's herbarium (WILLU, housed at Oregon State University, but not integrated into the OSC collections); Mount Rainier National Park, Washington; Nevada State Museum herbarium (NSMC); Nevada Test Site herbarium (NTS), Mercury, Nevada; Northern Prairie Science Center Herbarium (NPWRC), Jamestown, North Dakota; Oregon State University herbarium (OSC); Pacific Lutheran University herbarium, Tacoma, Washington; Royal British Columbia Museum herbarium (V); San Jose State University herbarium (SJSU); SERFIS (SouthEastern Regional Floristic Information System) Web site; Spring Branch Science Center herbarium (SBSC); Texas A&M University (TAMU) Web site, with data from four Texas herbaria; University of Alaska Museum, Fairbanks herbarium (ALA); University of Colorado, Boulder herbarium (COLO); University of Idaho herbarium (ID); University of Kansas herbarium (KANU); University of Manitoba herbarium (WIN); University of Nebraska, Omaha herbarium (OMA); University of Nevada, Las Vegas herbarium (UNLV); University of Wisconsin-Madison herbarium (WIS); West Virginia University herbarium (WVA); Western Washington University herbarium (WWB).

3. Source Publications.

Aiken, S.G., W.G. Dore, L.P. Lefkovitch and K.C. Armstrong. 1989. *Calamagrostis epigejos* (Poaceae) in North America, especially Ontario. Canad. J. Bot. 67:3205-3218.

Albee, B.J., L.M. Shultz, and S. Goodrich. 1988. Atlas of the Vascular Plants of Utah. Utah Museum of Natural History Occasional Publication No. 7, The Utah Museum of Natural History, Salt Lake City, Utah.

Alex, J.F. 1987. Quackgrass: origin, distribution, description and taxonomy. Pp. 1-16 in J.L. Glick (Ed.), Quackgrass Action Committee Workshop Technical Proceedings: March 10 and 11, 1987, Westin Hotel, Winnipeg, Manitoba. Monsanto Canada Ltd., Winnipeg, Manitoba.

Alinot, S.F. 1973. The vascular flora of Glen Helen, Clifton Gorge, and John Bryan State Park. Ohio Biological Survey, Biological Notes No. 5, The Ohio State University, Columbus, Ohio.

Allen, C.M. 1992. Grasses of Louisiana. 2nd ed. Cajun Prairie Habitat Preservation Society, Eunice, Louisiana.

Allen, L. and M. Curto. 1996. Noteworthy collections. Madroño 43:337-338.

Allred, K.W. 1993. A Field Guide to the Grasses of New Mexico. New Mexico Agricultural Experiment Station, Department of Agricultural Communications, New Mexico State University, Las Cruces, New Mexico.

Amoroso, J.L. and W.S. Judd. 1995. A floristic study of the Cedar Key Scrub State Reserve, Levy County, Florida. Castanea 60:210-232.

Anderson, D.E. 1961. Taxonomy and distribution of the genus *Phalaris*. Iowa State Coll. J. Sci. 36:1-96.

Anderson, L.C. 1995. Noteworthy plants from north Florida: VI. Sida 16:581-587.

Angelo, R. and D.E. Boufford. 1998. Atlas of the flora of New England: Poaceae. Rhodora 100:101-233.

Anonymous. 1973. Check List of Vermont Plants, Including all Vascular Plants Growing Without Cultivation. Vermont Botanical and Bird Club, Vermont.

Anonymous. 1990: Weed alert! Be on the lookout for serrated tussock, a Federal noxious weed (*Nassella trichotoma*).

Anonymous. 1995. Checklist of Harney County, Oregon.

Anonymous. 1996. Mount Revelstoke & Glacier National Parks: Vascular Plant Checklist.

Anonymous. 1998. A Checklist of Butte County, Idaho plants [based on R. Wunner's A Flora of Craters of the Moon National Monument].

Banks, D.L. and S. Boyd. 1998. Noteworthy collections. Madroño 45:85-86.

Barkley, T.M. (Ed.). 1977. Atlas of the Flora of the Great Plains. The Iowa State University Press, Ames, Iowa.

Barrett, S.C.H. and D.E. Seaman. 1980. The weed flora of Californian rice fields. Aquatic Bot. 9:351-376.

Basinger, M.A. and P. Rogertson. 1996. Vascular flora and ecological survey of an old-growth forest remnant in the Ozark Hills of southern Illinois. Phytologia 80:352-367.

Beatley, J.C. 1969. Vascular plants of the Nevada Test Site, Nellis Air Force Range, and Ash Meadows. University of California Laboratory of Nuclear Medicine and Radiation Biology Atomic Energy Commission contract report.

Beatley, J.C. 1970. Additions to vascular plants of the Nevada Test Site, Nellis Air Force Range, and Ash Meadows. University of California Laboratory of Nuclear Medicine and Radiation Biology Atomic Energy Commission contract report.

Beatley, J.C. 1971. Vascular plants of Ash Meadows, Nevada. University of California Laboratory of Nuclear Medicine and Radiation Biology Atomic Energy Commission contract report.

Beatley, J.C. 1973. Check list of vascular plants of the Nevada Test Site and central-southern Nevada. Department of Biological Sciences, University of Cincinnati Atomic Energy Commission contract report.

Best, C., J.T. Howell, W. Knight, I. Knight, and M. Wells. 1996. A Flora of Sonoma County: Manual of the Flowering Plants and Ferns of Sonoma County, California. California Native Plant Society, Sacramento, California.

Best, K.F., J.D. Banting, and G.G. Bowes. 1978. The biology of Canadian weeds. 31. *Hordeum jubatum* L. Canad. J. Pl. Sci. 58:699-708.

Borowski, M. and W.C. Holmes. 1996. *Phyllostachys aurea* Riv. (Gramineae: Bambuseae) in Texas. Phytologia 80:30-34.

Bowcutt, F. 1996.: A floristic study of Delta Meadows River Park, Sacramento County, California. Madroño 43:417-431.

Boyle, W.S. 1945. A Cyto-taxonomic study of the North American species of *Melica*. Madroño 8:1-26.

Braun, E.L. 1967. The Monocotyledoneae: Cat-tails to Orchids. (The Vascular Flora of Ohio, Volume One.) Ohio State University Press, Columbus, Ohio.

Brown, L.E. 1993. The deletion of *Sporobolus heterolepis* (Poaceae) from the Texas and Louisiana floras, and the addition of *Sporobolus silveanus* to the Oklahoma flora. Phytologia 74:371-381.

Brown, L.E. and S.J. Marcus. 1998. Notes on the flora of Texas with additions and other significant records. Sida 18:315-324.

Brown, L.E. and J. Schultz. 1991. *Arthraxon hispidus* (Poaceae), new to Texas. Phytologia 71:379-381.

Bruner, J.L. 1987. Systematics of the *Schizachyrium scoparium* (Poaceae) complex in North America. Ph.D. Dissertation, Ohio State University, Columbus, Ohio.

Bryson, C.T. 1993. *Sacciolepis indica* (Poaceae) new to Mississippi. Sida 15:555.

Calder, J.A. and R.L. Taylor. 1968. Flora of the Queen Charlotte Islands, Part I: Systematics of the Vascular Plants. (Monograph No. 4, Part 1) Research Branch, Canada Department of Agriculture, Ottawa, Ontario.

Callihan, R.H. 1987. Eradication technology of matgrass (*Nardus strictus* L.) in the Clearwater National Forest. Report to the U.S. Forest Supervisor, Clearwater National Forest, Idaho.

Callihan, R.H. and T.W. Miller. 1998. Noxious weed identification guide. [info. from WorldWide Web].

Callihan, R.H. and D. Pavek. 1988. May, 1988 *Milium vernale* survey in Idaho County, Idaho. Report to the Idaho County Commissioners, in cooperation with the Idaho County Weed Control Supervisor.

Catling, P.M., D.S. Erskine and R.B. MacLaren. 1985. The Plants of Prince Edward Island, with New Records, Nomenclatural Changes, and Corrections and Deletions. Revised ed. (Research Branch, Agriculture Canada, Publication 1798) Canadian Government Publishing Centre, Ottawa, Ontario.

Catling, P.M., A.A. Reznicek, and J.L. Riley. 1977. Some new and interesting grass records from southern Ontario. Canad. Field-Naturalist 91:350-359.

Chester, E.W. 1996. Rare and noteworthy vascular plants from the Fort Campbell Military Reservation, Kentucky and Tennessee. Sida 17:269-274.

Chester, E.W., B.E. Wofford, H.R. DeSelm, and A.M. Evans. 1993. Atlas of Tennessee Vascular Plants. Pteridophytes, Gymnosperms, Angiosperms: Monocots. Vol. 1. (Miscellaneous Publication No. 9) The Center for Field Biology, Austin Peay State University, Clarksville, Tennessee.

Clark, C. 1990. Vascular plants of the underdeveloped areas of California State Polytechnic University, Pomona. Crossosoma 16:1-7.

Clark, D.A. 1996. Natural History Inventory of Colorado: A Floristic Survey of the Mesa de Maya Region, Las Animas County, Colorado. Vol. 17. University of Colorado Museum, Boulder, Colorado.

Clay, K. 1995. Noteworthy collections. Castanea 60:84-85.

Clokey, I.W. 1951. Flora of the Charleston Mountains, Clark County, Nevada. Univ. Calif. Publ. Bot. 24:1-274.

Cchrane, T.S., M.M. Rice, M. Marion, and W.E. Rice. 1984. The flora of Rock County, Wisconsin: Supplement I. Michigan Bot. 23:121-133.

Cody, W.J. 1988. Plants of Riding Mountain National Park, Manitoba. (Research Branch, Agriculture Canada, Publication 1818/E) Canadian Government Publishing Centre, Ottawa, Ontario.

Cody, W.J. 1996. Flora of the Yukon Territory. National Resource Council of Canada, Research Press, Ottawa, Ontario.

Cody, W.J., S.J. Darbyshire, and C.E. Kennedy. 1990. A bluegrass, *Poa pseudoabbreviata* Roshev., new to the flora of Canada, and some additional records from Alaska. Canad. Field-Naturalist 104: 589-591.

Cody, W.J., C.E. Kennedy, and B. Bennett. 1998. New records of vascular plants in the Yukon Territory. Canad. Field-Naturalist 112:289-328.

Cope, E.A. 1994. Further notes on beachgrasses (*Ammophila*) in northeastern North America. Newsletter New York Fl. Assoc. 5:5-7.

Crampton, B. 1955. *Scribneria* in California. Leafl. W. Bot. 7:219-220.

Crouch, V.E and M.S. Golden. 1997. Floristics of a bottomland forest and adjacent uplands near the Tombigbee River, Choctaw County, Alabama. Castanea 62:219-238.

Curto, M. and D.M. Henderson. 1998. A new *Stipa* (Poaceae: Stipeae) from Idaho and Nevada. Madroño 45:57-63.

Curto, M. and L. Allen. 1992. California Native Plant Society grass walk. Cuyamaca Rancho State Park [San Diego County, California].

Cutler, H.C. and E. Anderson. 1941. A preliminary survey of the genus *Tripsacum*. Ann. Missouri Bot. Gard. 28:249-269.

Darbyshire, S.J. 1997. Tall wheatgrass, *Elymus elongatus* subsp. *ponticus*, in Nova Scotia. Rhodora 99:161-165.

Davis, J.I. 1990. *Puccinellia howellii* (Poaceae), a new species from California. Madroño 37:55-58.

Deam, C.C. 1940. Flora of Indiana. Department of Conservation, Division of Forestry, State of Indiana, Indianapolis, Indiana.

DeSelm, H.R. 1975. *Schizachyrium stoloniferum* Nash var. *wolfei* DeSelm. Sida 6:114-115.

Deshaye, J. and J. Cayouette. 1988. La flore vasculaire des îles et de la presqu'île de Manitounuk, Baie d'Hudson: Structure phytogéographique et interprétation bioclimatique. Provancheria 21:1-74.

Diamond, A.R., Jr. and J.D. Freeman. 1993. Vascular flora of Conecuh Co., Alabama. Sida 15:623-638.

Dix, W.L. 1945. Will the stowaway, *Molinia caerulea*, become naturalized? Bartonia 23:41-42.

Dore, W.G. and J. McNeill. 1980. Grasses of Ontario. (Research Branch, Agriculture Canada Monograph 26) Canadian Government Publishing Centre, Hull, Québec.

Douglas, G.W., G.B. Straley, and D. Meidinger (Eds.). 1994. The Vascular Plants of British Columbia. Part 4 - Monocotyledons. Research Branch, Ministry of Forests, Victoria, British Columbia.

Douglas, G.W. and R.J. Taylor. 1970. Contributions to the flora of Washington. Rhodora 72:496-501.

Drew, M.B., and L.K. Kirkman, and A.K. Gholson, Jr. 1998. The vascular flora of Ichauway, Baker County, Georgia: A remnant longleaf pine/wiregrass ecosystem. Castanea 63:1-24.

Dubé, M. 1983. Addition de *Festuca gigantea* à la flore du Canada. Ludoviciana 14:213-215.

Eddy, T. and N.A. Harriman. 1992. *Muhlenbergia richardsonis* in Wisconsin. Michigan Bot. 31:39-40.

Eddy, T.L. 1983. A vascular flora of Green Lake County, Wisconsin. M.S. Thesis, University of Wisconsin-Oshkosh, Oshkosh, Wisconsin.

Eilers, L.J. and D.M. Roosa. 1994. The Vascular Plants of Iowa. University of Iowa Press, Iowa City, Iowa.

Ertter, B. 1997. Annotated checklist of the East Bay Flora: Native and Naturalized Vascular Plants of Alameda and Contra Costa Counties, California. California Native Plant Society, East Bay Chapter, Berkeley, California.

Fassett, N.C. 1951. Grasses of Wisconsin: The Taxonomy, Ecology, and Distribution of the Gramineae Growing in the State without Cultivation. The University of Wisconsin Press, Madison, Wisconsin.

Ferlatte, W.J. 1974. A Flora of the Trinity Alps of Northern California. University of California Press, Berkeley, California.

Fishbein, M. 1995. Noteworthy collections. Madroño 42:83.

Fleming, G.P. and J.C. Ludwig. 1996. Noteworthy collections. Castanea 61:89-95.

Fleming, P. and R. Kanal. 1995. Annotated checklist of vascular plants of Rock Creek Park, National Park Service, Washington, D.C. Castanea 60:283-316.

Fox, W.E. and S.L. Hatch. 1996. *Brachiaria eruciformis* and *Urochloa brizantha* (Poaceae: Paniceae) new to Texas. Sida 17:287-288.

Franklin, J.F. and C. Wiberg. 1979. Goat Marsh Research Natural Area. (Supplement No. 10) (Federal Research Natural Areas in Oregon and Washington: A Guidebook for Scientists and Educators) U.S. Department of Agriculture Forest Service, Pacific Northwest Forest and Range Experiment Station, Corvallis, Oregon.

Freckmann, R.W. 1972. Grasses of Central Wisconsin. (Reports on the Fauna and Flora of Wisconsin, Report No. 6) The Museum of Natural History, Stevens Point, Wisconsin.

Freeman, C.C., R.L. McGregor, and C.A. Morse. 1998. Vascular plants new to Kansas. Sida 18:593-604.

Frelich, L. 1979. Vascular Plants of Newport State Park, Wisconsin. (Department of Natural Resources Research Report No. 100) Department of Natural Resources, Madison, Wisconsin.

Gandhi, K.N. 1989. A biosystematic study of the *Schizachyrium scoparium* complex. Ph.D. Dissertation, Texas A&M University, College Station, Texas.

Garlitz, R. 1989. Rare and interesting grass records from the northeastern Lower Peninsula. Michigan Bot. 28:67-71.

Garlitz, R. and D. Garlitz. 1986. *Bouteloua gracilis*, a new grass to Michigan. Michigan Bot. 25:123-124.

Gillett, G.W., J.T. Howell, and H. Leschke. 1995. A Flora of Lassen Volcanic National Park, California. Revised ed. (rev. V.H. Oswald, D.W. Showers, and M.A. Showers). California Native Plant Society, Sacramento, California.

Gilman, A.V. 1993. Four recent additions to the vascular flora of Vermont. Maine Naturalist 1:31-32.

Gould, F.W. 1969. Taxonomy of the *Bouteloua repens* complex. Brittonia 21:261-274.

Gould, F.W. 1979. The genus *Bouteloua* (Poaceae). Ann. Missouri Bot. Gard. 66:348-416.

Gould, F.W., M.A. Ali, and D.E. Fairbrothers. 1972. A revision of *Echinochloa* in the United States. Amer. Midl. Naturalist 87:36-59.

Gould, F.W. and Z.J. Kapadia. 1964. Biosystematic studies in the *Bouteloua curtipendula* complex. II. Taxonomy. Brittonia 16:182-207.

Griffin, J.R. 1990. Flora of Hastings Reservation, Carmel Valley, California. 3rd ed. Hastings Natural History Reservation, University of California-Berkeley, Berkeley, California.

Guala, G.F. II. 1988. *Poa bulbosa* L. (Poaceae) in Michigan. Michigan Bot. 27:13-14.

Hall, D.W. 1982. *Sorghastrum* (Poaceae) in Florida. Sida 9:302-308.

Hall, H.M. 1902. A botanical survey of San Jacinto Mountain. Univ. Calif. Publ. Bot. 1:1-140.

Hallsten, G.P., Q.D. Skinner, and A.A. Beetle. 1987. Grasses of Wyoming. 3rd ed. (Research Journal 202.) Agricultural Experiment Station, University of Wyoming, Laramie, Wyoming.

Hämet-Ahti, L. 1965. Vascular plants of Wells Gray Provincial Park and its vicinity, in eastern British Columbia. Ann. Bot. Fenn. 2:138-164.

Hansen, B.F. and R.P. Wunderlin. 1996. Grasses of Florida: A Checklist of the Poaceae of Florida Along with Their Distribution by County [database update of 1992 checklist]. Institute for Systematic Botany, University of South Florida, Tampa, Florida.

Harris, S.K. 1975. The Flora of Essex County, Masachusetts. Peabody Museum, Salem, Massachusetts.

Harvill, A.M. Jr., T.R. Bradley, C.E.Stevens, T.F. Wieboldt, D.E. Ware, D.W. Ogle, G.W. Ramsey, and G.P. Fleming. 1992. Atlas of the Virginia Flora. 3rd ed. Virginia Botanical Associates, Burkeville, Virginia.

Hatch, S.L. 1975. A biosystematic study of the *Schizachyrium cirratum - Schizachyrium sanguineum* complex (Poaceae). Ph.D. Dissertation, Texas A&M University, College Station, Texas.

Hatch, S.L., W.E. Fox, III, and J.E. Dawson, III. 1998. *Triraphis mollis* (Poaceae: Arundineae) a species reported new to the United States. Sida 18:365-368.

Hatch, S.L., D.J. Rosen, J.A. Thomas, and J.E. Dawson, III. 1998. *Luziola peruviana* (Poaceae: Oryzeae) previously unreported from Texas and a key to Texas species. Sida 18:611-614.

Hawkins, T.K. and E.L. Richards. 1995. A floristic study of two bogs on Crowley's Ridge in Greene County, Arkansas. Castanea 60:233-244.

Hays, J. 1995. A floristic survey of Falls Hollow sandstone glades, Pulaski County, Missouri. Phytologia 78:264-276.

Heidel, B. 1996. Noteworthy collections. Madroño 43:436-440.

Hellquist, C.E. and G.E. Crow. 1997. The bryophyte and vascular flora of Little Dollar Lake peatland, Mackinac County, Michigan. Rhodora 99:195-222.

Henry, L.K. 1978. Vascular Flora of Bedford County, Pennsylvania: An Annotated Checklist. Carnegie Museum of Natural History, Pittsburgh, Pennsylvania.

Hinds, H.R. 1986. The Flora of New Brunswick. Primrose Press, Fredericton, New Brunswick.

Hodgdon, A.R., G.E. Crow, and F.L. Steele. 1979. Grasses of New Hampshire: I. Tribes Poeae (Festuceae) and Triticeae (Hordeae). New Hampshire Agric. Exp. Sta. Bull. 512:1-53.

Hounsell, R.W. and E.C. Smith. 1966. Contributions of the flora of Nova Scotia. VIII. Distribution of arctic-alpine and boreal disjuncts. Rhodora 68:409-419.

Howell, J.T. 1970. Marin Flora: Manual of the Flowering Plants and Ferns of Marin County, California. 2nd ed. University of California Press, Berkeley, California.

Howell, J.T. 1979. A reconsideration of *Trisetum projectum* (Gramineae). Wasmann J. Biol. 37:21-23.

Howell, J.T., G.H. True, and C. Best. 1981. Notes on Marin County plants 1970-1980. Four Seasons 6:3-6.

Hunt, D.M. and R.E. Zaremba. 1992. The northeastward spread of *Microstegium vimineum* (Poaceae) into New York and adjacent states. Rhodora 94:167-170.

Jones, S.B., Jr and N.C. Coile. 1988. The Distribution of the Vascular Flora of Georgia. University of Georgia, Athens, Georgia.

Jones, S.D. and G. Jones. 1992. *Cynodon nlemfuensis* (Poaceae: Chlorideae) previously unreported in Texas. Phytologia 72:93-95.

Jones, S.D and J.K. Wipff. 1992. *Eustachys retusa* (Poaceae), the first report in Florida and a key to *Eustachys* in Florida. Phytologia 73:274-276.

Judziewicz, E.J. and R.G. Koch. 1993. Flora and vegetation of the Apostle Islands National Lakeshore and Madeline Island, Ashland and Bayfield Counties, Wisconsin. Michigan Bot. 32:43-193.

Junak, S., T. Ayers, R. Scott, D. Wilken, and D. Young. 1995. A Flora of Santa Cruz Island. Santa Barbara Botanic Garden, Santa Barbara, California.

Kearney, T.H and R.H. Peebles. 1951. Arizona Flora. University of California Press, Berkeley and Los Angeles, California.

Kiger, R.W. 1971. *Arthraxon hispidus* (Gramineae) in the United States: Taxonomic and floristic status. Rhodora 73:39-46.

Lackschewitz, K. 1991. Vascular Plants of West-Central Montana – Identification Guidebook. (General Technical Report INT-277) U.S. Department of Agriculture, Forest Service, Intermountain Research Station, Ogden, Utah.

Landry, G.P. 1996. Noteworthy collections. Castanea 61:197.

Lange, K.I. 1998. Flora of Sauk County and Caledonia Township, Columbia County, South Central Wisconsin. (Technical Bulletin No. 190) Department of Natural Resources, Madison, Wisconsin.

Lawrence, D.L and J.T. Romo. 1995. Tree and shrub communities of wooded draws near the Matador Research Station in southern Saskatchewan. Canad. Field-Naturalist 108:397-409.

Layser, E.F., Jr. 1969. A floristic study of Pend Oreille County, Washington. M.S. Thesis, State University College of Forestry, Syracuse University, Syracuse, New York.

Lelong, M.G. 1977. Annotated list of vascular plants in Mobile, Alabama. Sida 7:118-146.

Lesica, P. 1985. Checklist of the Vascular Plants of Glacier National Park. (Proceedings of the Montana Academy of Sciences Monograph No. 4) Montana Academy of Sciences, Missoula, Montana.

Lesica, P. 1994. Noteworthy collections. Madroño 41:231.

Lewis, M.E. 1971. Flora and major plant communities of the Ruby-East Humboldt Mountains, with special emphasis on Lamoille Canyon. Report to the U.S. Forest Service, Humboldt National Forest, Nevada.

Lloyd, R.M. and R.S. Mitchell. 1965. Plants of the White Mountains, California and Nevada. Revised ed. University of California White Mountain Research Station, Berkeley, California.

Lonard, R.I. 1993. Guide to the Grasses of the Lower Rio Grande Valley, Texas. University of Texas-Pan American Press, Edinburg, Texas.

MacDonald, J. 1996. A survey of the flora of Monroe County, Mississippi. M.S. Thesis, Mississippi State University, Mississippi State, Mississippi.

MacRoberts, B.R. and M.H. MacRoberts. 1995. Floristics of xeric sandhills in northwestern Louisiana. Phytologia 79:123-131.

MacRoberts, B.R. and M.H. MacRoberts. 1995. Vascular flora of two calcareous prairie remnants on the Kisatchie National Forest. Phytologia 78:18-27.

MacRoberts, B.R. and M.H. MacRoberts. 1996. The floristics of calcareous prairies on the Kisatchie National Forest, Louisiana. Phytologia 81:35-43.

MacRoberts, B.R. and M.H. MacRoberts. 1996. Floristics of xeric sandhills in east Texas. Phytologia 80:1-7.

MacRoberts, B.R. and M.H. MacRoberts. 1997. Floristics of beech-hardwood forest in east Texas. Phytologia 82:20-29.

MacRoberts, M.H. and B.R. MacRoberts. 1993. Vascular flora of sandstone outcrop communities in western Louisiana, with notes on rare and noteworthy species. Phytologia 75:463-480.

MacRoberts, M.H. and B.R. MacRoberts. 1995. Noteworthy vascular plant collections on the Kisatchie National Forest, Louisiana. Phytologia 78:291-313.

MacRoberts, M.H. and B.R. MacRoberts. 1998. Noteworthy vascular plant collections on the Angelina and Sabine National Forests, Texas. Phytologia 84:1-27.

Magee, D. 1993. Manual of the Vascular Flora of New England and Adjacent New York. Normandeau Associates, Inc., Bedford, New Hampshire.

Maley, A. 1994. Natural History Inventory of Colorado: A Floristic Survey of the Black Forest of the Colorado Front Range. Vol. 14. University of Colorado Museum, Boulder, Colorado.

Matthews, M.A. 1997. An Illustrated Field Key to the Flowering Plants of Monterey County and Ferns, Fern Allies, and Conifers. California Native Plant Society, Sacramento, California.

Maze, J. and K.A. Robson. 1996. A new species of *Achnatherum (Oryzopsis)* from Oregon. Madroño 43:393-403.

McClintock, E., P. Reeberg, and W. Knight. 1990. A Flora of the San Bruno Mountains, San Mateo County, California. (Special Publication No. 8) California Native Plant Society, Sacramento, California.

McNeill, J. 1981. *Apera*, silky-bent or windgrass, an important weed genus recently discovered in Ontario, Canada. Canad. J. Pl. Sci. 61:479-485.

Meeks, D.N. 1984. A floristic study of northern Tippah County, Mississippi. M.S. Thesis, Mississippi State University, Mississippi State, Mississippi.

Mitchell, R.S. and G.C. Tucker. 1997. Revised Checklist of New York State Plants. (New York State Museum Bulletin No. 490) (Series Ed: Mitchell, Richard S. Contributions to a Flora of New York State, Checklist IV) State Education Department, University of the State of New York, Albany, New York.

Mitchell, W.W. and H.J. Hodgson. 1965. The status of hybridization between *Agropyron sericeum* and *Elymus sibiricus* in Alaska. Canad. J. Bot. 43:855-859.

Moe, L.M. and E.C. Twisselmann. 1995. A Key to Vascular Plant Species of Kern County, California & A Flora of Kern County, California. California Native Plant Society, Sacramento, California.

Mohlenbrock, R.H. and D.M. Ladd. 1978. Distribution of Illinois Vascular Plants. Southern Illinois University Press, Carbondale, Illinois.

Morris, M.W. 1987. The vascular flora of Grenada County, Mississippi. M.S. Thesis, Mississippi State University, Mississippi State, Mississippi.

Morris, M.W. 1997. Contributions to the flora and ecology of the northern longleaf pine belt in Rankin County, Mississippi. Sida 17:615-626.

Morrone, O. and F.O. Zuloaga. 1992. Revisión de las especies sudamericanas nativas e introducidas de los géneros *Brachiaria* y *Urochloa* (Poaceae: Panicoideae: Paniceae). Darwiniana 31:43-109.

Morrone, O. and F.O. Zuloaga. 1993. Synopsis del género *Urochloa* (Poaceae: Panicoideae: Paniceae) para México y América Central. Darwiniana 32:59-75.

Morrone, O., A.S. Vega, and F.O. Zuloaga. 1996. Revisión de las especies del género *Paspalum* L. (Poaceae: Panicoideae: Paniceae), grupo Dissecta (*s. str.*). Candollea 51:103-138.

Morton, J.K. and J.M. Venn. 1984. The Flora of Manitoulin Island. 2nd ed. Dept. of Biology, University of Waterloo, Waterloo, Ontario.

Moseley, R.K. 1996. Vascular flora of subalpine parks in the Coeur d'Alene River drainage, northern Idaho. Madroño 43:479-492.

Moss, E.H. 1983. Flora of Alberta: A Manual of Flowering Plants, Conifers, Ferns and Fern Allies Found Growing without Cultivation in the Province of Alberta, Canada. 2nd ed. (rev. John G. Packer). University of Toronto Press, Toronto, Ontario.

Musselman, L.J., T.S. Cochrane, W.E. Rice, and M. Marion. 1971. The flora of Rock County, Wisconsin. Michigan Bot. 10:147-205.

Naumann, T. 1996. Plant list, including scientific and common names, Dinosaur National Monument [Dinosaur, Colorado].

Nelson, J.B. and K.B. Kelly. 1997. Noteworthy collections. Castanea 62:283-287.

New York Flora Association. 1990. Preliminary Vouchered Atlas of New York State Flora. New York State Museum Institute, Albany, New York.

New York Flora Association. 1990-1998. Miscellaneous articles and species lists [additions to Preliminary Vouchered Atlas]. Newslett. New York Fl. Assoc. 1-9.

Northam, F.E. and R.H. Callihan. 1992. Morphology and phenology of interrupted windgrass in northern Idaho. J. Idaho Acad. Sci. 28:15-19.

Northam, F.E., R.H. Callihan, and R.R. Old. 1989. *Sporobolus vaginiflorus* (Torrey ex Gray) Wood: Biology and pest implications of an alien grass recorded in Idaho. J. Idaho Acad. Sci. 25:49-55.

Northam, F.E., R.H. Callihan and R.R. Old. 1991. Range extensions of four introduced grasses in Idaho. J. Idaho Acad. Sci. 27:19-21.

Northam, F.E., R.R. Old and R.H. Callihan. 1993. Little lovegrass (*Eragrostis minor*) distribution in Idaho and Washington. Weed Technol. 7:771-775.

Northam, F.E., and P.W. Stahlman. 1995. Range extension of southwestern cupgrass (*Eriochloa acuminata*) into Kansas. Trans. Kansas Acad. Sci. 98:68-71.

Oldham, M.J., S.J. Darbyshire, D. McLeod, D.A. Sutherland, D. Tiedje, and J.M. Bowles. 1995. New and noteworthy Ontario grass (Poaceae) records. Michigan Bot. 34:105-132.

Olham, M.J. and S.J. Darbyshire. 1993. The adventive grasses, *Apera interrupta* and *Deschampsia danthonoides*, new to Maine. Maine Naturalist 1:231-232.

Oswald, V.and L. Ahart. 1994. Flora of Butte County, California. California Native Plant Society Press, Sacramento, California.

Ownbey, G.B.and T. Morley. 1991. Vascular Plants of Minnesota: A Checklist and Atlas. University of Minnesota, Minneapolis, Minnesota.

Parker, D.S. and J.M. Stucky. 1995. Southwestern cupgrass (*Eriochloa acuminata* (Presl) Kunth (Poaceae: Paniceae)) in North Carolina. Report to North Carolina State University, Raleigh, North Carolina.

Perkins, B.E. and T.S. Patrick. 1980. Status report on Tennessee populations of *Calamovilfa arcuata*. Threatened and endangered species report, University of Tennessee, Knoxville, Tennessee.

Peterson, P.M. 1986. A flora of the Cottonwood Mountains, Death Valley National Monument, California. Wasmann J. Biol. 44:73-126.

Plunkett, G.M. and G.W. Hall. 1995. The vascular flora and vegetation of western Isle of Wight County, Virginia. Castanea 60:30-59.

Pohl, R.W. 1959. Introduced weedy grasses in Iowa. Proc. Iowa Acad. Sci. 66:160-162.

Pohl, R.W. 1966. The grasses of Iowa. Iowa State Coll. J. Sci. 40:341-566.

Popovich, S.J. and D.M. Henderson. 1994. Noteworthy collections. Madroño 41:149-150.

Popovich, S.J., W.D. Shepperd, D.W. Reichert, and M.A. Cone. 1993. Flora of the Fraser Experimental Forest, Colorado. (General Technical Report RM-233) U.S. Department of Agriculture, Forest Service, Rocky Mountain Forest and Range Experiment Station, Fort Collins, Colorado.

Porsild, A.E. and W.J. Cody. 1980. Vascular Plants of the Continental Northwest Territories, Canada. National Museum of Natural Sciences, National Museum of Canada, Ottawa, Ontario.

Powell, A.M. 1994. Grasses of the Trans-Pecos and Adjacent Areas. University of Texas Press, Austin, Texas.

Radford, A.E., H.E. Ahles, and C.R. Bell.. 1965. Atlas of the Vascular Flora of the Carolinas. (Technical Bulletin No. 165) North Carolina Agricultural Experiment Station, University of North Carolina-Raleigh, Raleigh, North Carolina.

Raven, P.H., H.J. Thompson, and B.A. Prigge. 1986. Flora of the Santa Monica Mountains, California. 2nd ed. (Southern California Botanists, Special Publication No. 2) University of California-Los Angeles, Los Angeles, California.

Rawinski, T.J., M.N. Rasmussen, and S.C. Rooney. 1989. Discovery of *Sporobolus asper* (Poaceae) in Maine. Rhodora 91:220-221.

Read, J.C. and B.J. Simpson. 1992. Documented chromosome numbers 1992:3. Documentation and notes on the distribution of *Melica montezumae*. Sida 15:151-152.

Redman, D.E. 1995. Noteworthy collections. Castanea 60:82-84.

Redman, D.E. 1995. Distribution and habitat types for Nepal microstegium in Maryland and the District of Columbia. Castanea 60:270-275.

Reed, C.F. 1964. A flora of the chrome and manganese ore piles at Canton, in the Port of Baltimore, Maryland and at Newport News, Virginia, with descriptions of genera and species new to the flora of the eastern United States. Phytologia 10:321-405.

Reeder, J.R. 1995. *Stipa tenuissima* (Gramineae) in Arizona – a comedy of errors. Madroño 41:328-329.

Reeder, J.R. and C.G. Reeder. 1980. Systematics of *Bouteloua breviseta* and *B. ramosa* (Gramineae). Syst. Bot. 5:312-321.

Reeder, J.R. and C.G. Reeder. 1990. *Bouteloua eludens*: Elusive indeed, but not rare. Desert Pl. 10:19-22, 31.

Rhoads, A.F. and W.M. Klein, Jr. 1993. The Vascular Flora of Pennsylvania: Annotated Checklist and Atlas. American Philosophical Society, Philadephia, Pennsylvania.

Richards, C.D., F. Hyland, and L.M. Eastman. 1983. Check-list of the vascular plants of Maine: Second revisied edition. Bull. Josselyn Bot. Soc. Maine 11:1-73.

Riefner, R.E. Jr and D.R. Pryor. 1996. New locations and interpretation of vernal pools in southern California. Phytologia 80:296-327.

Riley, J.L. 1979. Some new and interesting vascular plant records from northern Ontario. Canad. Field-Naturalist 93:355-362.

Riley, J.LI and S.M. McKay. 1980. The Vegetation and Phytogeography of Coastal Southwestern James Bay. (Life Sciences Contributions, Royal Ontario Museum, No. 124) The Royal Ontario Museum, Toronto, Ontario.

Rill, K.D. 1983. A vascular flora of Winnebago County, Wisconsin. Trans. Wisconsin Acad. Sci. 71:155-180.

Roalson, E.H. and K.W. Allred. 1998. A floristic study in the Diamond Creek drainage area, Gila National Forest, New Mexico. Aliso 17:47-62.

Roberts, F.M., Jr. 1989. A Checklist of the Vascular Plants of Orange County, California. (Museum of Systematic Biology Research Series No. 6) University of California-Irvine, Irvine, California.

Rogers, B.S. and A. Tiehm. 1979. Vascular Plants of the Sheldon National Wildlife Refuge, with Special Reference to Possible Threatened and Endangered Species. (Department of the Interior, U.S. Fish & Wildlife Service, Region 1, Portland, Oregon.) U.S. Government Printing Office, Washington, D.C.

Rouleau, E.and G. Lamoureux. 1992. Atlas of the Vascular Plants of the Island of Newfoundland and of the Islands of Saint-Pierre-et-Miquelon. Fleurbec, Québec City, Québec.

Rousseau, C.. 1968. Histoire, habitat et distribution de 220 plantes introduites au Québec. Naturaliste Canad. 95:49-169.

Rousseau, C. 1974. Géographie Floristique du Québec-Labrador: Distribution des principales espèces vasculaires. Les Presses de l'Université Laval, Québec City, Québec.

Rubtzoff, P. 1961. Notes on fresh-water marsh and aquatic plants in California. Leafl. W. Bot. 9:165-180.

Sanders, A.C. 1996. Noteworthy collections. Madroño 43:524-532.

Scheffer, T.H. 1945. The introduction of *Spartina alternifolia* to Washington with oyster culture. Leafl. W. Bot. 4:163-164.

Sharma, M.P. and W.H. Vanden Born. 1978. The biology of Canadian weeds. 27. *Avena fatua* L. Canad. J. Pl. Sci. 58:141-157.

Simmons, M.P., D.M.E. Ware, and W.J. Hayden. 1995. The vascular flora of the Potomac River watershed of King George County, Virginia. Castanea 60:179-200.

Simpson, M.G., S.C. McMillan, C. Scott, and B.L. Stone. 1995. Checklist of the Vascular Plants of San Diego County [California]. San Diego State University Herbarium Press, San Diego, California.

Smith, C.F. 1976. A Flora of the Santa Barbara Region, California. Santa Barbara Museum of Natural History, Santa Barbara, California.

Smith, E.B. 1988. An Atlas and Annotated List of the Vascular Plants of Arkansas. 2nd ed. Edwin B. Smith, Fayetteville, Arkansas.

Smith, G.L. and C.R. Wheeler. 1990-91. A flora of the vascular plants of Mendocino County, California. Wasmann J. Biol. 48/49:63-81.

Sorrie, B.A. 1987. Notes on the rare flora of Massachusetts. Rhodora 89:113-196.

Sorrie, B.A. and P.W. Dunwiddie. 1990. *Amphicarpum purshii* (Poaceae), a genus and species new to New England. Rhodora 92:105-107.

Sorrie, B.A., B. Van Eerden, and M.J. Russo. 1997. Noteworthy plants from Fort Bragg and Camp MacKall, North Carolina. Castanea 62:239-259.

Sparks, L.H., R. Del Moral, A.F. Watson, and A.R. Kruckeberg. 1976. The distribution of vascular plant species on Sergief Island, southeast Alaska. Syesis 10:5-9.

Spicher, D. and M. Josselyn. 1985. *Spartina* (Gramineae) in northern California: Distribution and taxonomic notes. Madroño 32:158-167.

Stalter, R. and E. Lamont. 1996. Noteworthy collections. Castanea 61:396-397.

Stevenson, G.A. 1965. Notes on the more recently adventive flora of the Brandon area, Manitoba. Canad. Field-Naturalist 79:174-177.

Steyermark, J.A. 1963. Flora of Missouri. The Iowa State University Press, Ames, Iowa.

Stickney, P.F. 1961. Range of rough fescue (*Festuca scabrella* Torr.) in Montana. Proc. Montana Acad. Sci. 20:12-17.

Stiles, B.J. and C.L. Howel. 1998. Floristic survey of Rabun County, Georgia, part II. Castanea 63:154-160.

Stuart, J.D., T. Worley, and A.C. Buell. 1996. Plant associations of Castle Crags State Park, Shasta County, California. Madroño 43:273-291.

Taylor, R.J. and C.E. Taylor. 1980. Status report on *Calamovilfa arcuata*. Threatened and endangered species report, University of Tennessee, Knoxville, Tennessee.

Taylor, R.J. and C.E. Taylor. 1987. Additions to the vascular flora of Oklahoma. IV. Sida 12:233-237.

Terrell, E.E. and J.L. Reveal. 1996. Noteworthy collections. Castanea 61:95-96.

Thorne, R.F., B.A. Prigge, and J. Henrickson. 1981. A flora of the higher ranges and the Kelso Dunes of the eastern Mojave Desert in California. Aliso 10:71-186.

Tucker, G.C. 1996. The genera of Pooideae (Gramineae) in the southeastern United States. Harvard Pap. Bot. 9:11-90.

Vanderhorst, J.P. 1993. Flora of the Flat Tops, White River Plateau, and vicinity in northwestern Colorado. M.S. Thesis, University of Wyoming, Laramie, Wyoming

Villamil, C.B. 1969. El género *Monanthochloë* (Gramineae). Estudios morfológicos y taxonómicos con especial referencia a la especia Argentina. Kurtziana 5:369-391.

Voss, E.G. 1972. Michigan Flora: A Guide to the Identification and Occurrence of the Native and Naturalized Seed-Plants of the State.

Part I, Gymnosperms and Monocots. Cranbrook Institute of Science, Bloomfield Hills, Michigan.

Ward, G.H. 1948. A flora of Chelan County, Washington. M.S. Thesis, State College of Washington [Washington State University], Pullman, Washington.

Warwick, S.I. 1979. The biology of Canadian weeds. 37. *Poa annua* L. Canad. J. Pl. Sci. 59:1053-1066.

Warwick, S.I. and S.G. Aiken. 1986. Electrophoretic evidence for the recognition of two species in annual wild rice (*Zizania*, Poaceae). Syst. Bot. 11:464-473.

Warwick, S.I. and L.D. Black. 1983. The biology of Canadian weeds. 61. *Sorghum halepense*. Canad. J. Pl. Sci. 63:997-1014.

Warwick, S.I., L.D. Black, and B.F. Zilkey. 1985. The biology of Canadian weeds. 72. *Apera spica-venti*. Canad. J. Pl. Sci. 65:711-721.

Weber, W.A. 1984. A new genus of grasses from the western oil shales. Phytologia 55:1-2.

Weber, W.A. 1995. Natural History Inventory of Colorado: Checklist of Vascular Plants of Boulder County, Colorado. Vol. 16. University of Colorado Museum, Boulder, Colorado.

Werner, P.A. and R. Rioux. 1977. The biology of Canadian weeds. 24. *Agropyron repens* (L.) Beauv. Canad. J. Pl. Sci. 57:905-919.

Whitney, K.D. 1996. Noteworthy collections. Madroño 43:336-337.

Williams, A.H. 1997. Range expansion northward in Illinois and into Wisconsin of *Tridens flavus* (Poaceae). Rhodora 99:344-351.

Winstead, R. 1990. A taxonomic and ecological survey of the plant communities of Attala County, Mississippi. M.S. Thesis, Mississippi State University, Mississippi State, Mississippi.

Wipff, J.K. and S.L. Hatch. 1992. *Eustachys caribaea* (Poaceae: Chlorideae) in Texas. Sida 15:160-161.

Wipff, J.K. and S.D. Jones. 1994. *Melica subulata* (Poaceae: Meliceae) The first report for Colorado. Sida 16:210-211.

Wipff, J.K., S.D. Jones, and C.T. Bryson. 1994. *Eustachys glauca* and *E. caribaea* (Poaceae: Chlorideae) The first reports for Mississippi. Sida 16:211.

Wipff, J.K., R.I. Lonard, S.D. Jones, and S.L. Hatch. 1993. The genus *Urochloa* (Poaceae: Paniceae) in Texas, including one previously unreported species for the state. Sida 15:405-413.

Yadon, V. 1995. Checklist of the vascular plants of Monterey County, California, revised to conform with the Jepson Manual. (Unpublished manuscript.)

Zebryk, T.M. 1998. Noteworthy collections. Castanea 63:78-79.

Zika, P.F. 1989. Noteworthy collections. Madroño 36:207.

Zika, P.F. 1990. Range expansions of some grasses in Vermont. Rhodora 92:80-89.

Zika, P.F., and E.J. Marshall. 1991. Contributions to the flora of the Lake Champlain Valley, New York and Vermont, III. Bull. Torrey Bot. Club 118: 58-61.

Zika, P.F., R.J. Sternand, and H.E. Ahles. 1983. Contributions to the flora of the Lake Champlain Valley, New York and Vermont. Bull. Torrey Bot. Club 110:366-369.

Zika, P.F. and B. Wilson. 1998. Noteworthy collections. Madroño 45:86-87.

Zinck, M. 1998. Roland's Flora of Nova Scotia [revised by M. Zinck]. 3rd ed. Nimbus Publishing and Nova Scotia Museum, Province of Nova Scotia.

Zobel, D.B. and C.R. Wasem. 1979. Pyramid Lake Research Natural Area. (Supplement No. 8) (Federal Research Natural Areas in Oregon and Washington: A Guidebook for Scientists and Educators.) U.S. Department of Agriculture, Forest Service, Pacific Northwest Forest and Range Experiment Station, Corvallis, Oregon.

TROPICAL GRASSLANDS AND SAVANNAS

Robert B. Shaw

Department of Forest Sciences, Center for Ecological Management of Military Lands
Colorado State University, Fort Collins, CO 80523, USA.
e-mail: rshaw@cemml.colostate.edu

Abstract

The tropical or torrid zone lies between the Tropic of Cancer (23.5° north) and the Tropic of Capricorn (23.5° south). The area is characterized by a winterless climate with no month with a mean temperature less than 18° Celsius. Uniformly distributed precipitation in the tropics gives rise to typical evergreen tropical rainforest vegetation, while a distinct dry season of two months or more supports tropical savanna or grassland vegetation. Evaporation which equals or exceeds precipitation also is characteristic of savanna/grasslands within the tropical climatic/geographic zone. Humid tropical savanna and dry tropical-subtropical grassland (steppe) regions account for 17% and 10% of the global land surface area, respectively. Africa and South America have the largest areas dominated by this type of vegetation. Much smaller areas are found in Australia, Asia, and Central America. Members of the subfamilies Panicoideae and Arundinoideae are the predominant grasses. Paniceae genera are more common in the mesic areas, while genera belonging to the Andropogoneae are more prevalent in arid areas. Desertification caused by overgrazing by domestic as well as native ungulates and cropping of marginal lands is one of the major threats to tropical grasslands and savannas. Global climate change models predict a major shift in tropical savanna/grassland vegetation. Areas currently dominated by grass species which use the C_4 photosynthetic pathway are predicted to become desert or C_3 dominated by the year 2090. Tropical grasslands could be reduced by as much as 90% in the next one hundred years.

Key words: Grasslands, savannas, tropical, global climate change, weeds, desertification, overgrazing.

INTRODUCTION

The purpose of this paper is to summarize some of the information concerning tropical grasslands on a global scale. This should provide for a quick path into the literature for interested readers. Included will be remarks about the global and continental distribution of tropical grasslands and savannas, the climate associated with tropical grasslands and savannas, taxonomic structure of tropical grasslands and savannas, and finally, the future of these ecosystems.

The terminology concerning tropical grasslands is very convoluted and confusing (Singh *et al.* 1983). Grass-dominated plant communities most commonly are referred to as grasslands, prairies, steppes, or savannas (Singh *et al.* 1983). South African grasslands or savannas are frequently called velds (Acocks 1960). While in South America campo, pampa, llano, and cerrado have been used for savannas or grasslands (Blydenstein 1961; Burkhart 1975; Eiten 1972).

The term savanna has been used in a most confusing manner as well. Savanna, a word of Spanish or French origin, originally implied land 'without trees' [Hill (1965) in Bourliere & Hadley (1970)]. They found that savanna 'has been applied to almost any formation in any part of the world that has some grasses in the herb layer.' Huntley and Walker (1982) apply the widest possible definition for tropical savanna. They consider it to be 'those ecosystems which lie between the equatorial rainforests and the deserts and semi-deserts of Africa, Australia and South America. These savannas form a continuum of physiognomic types ranging from closed woodlands with a heliophytic grass understory, through open savanna woodland to treeless edaphic grasslands.' Obviously the woody plant layer influences the understory herbaceous grass layer in some manner (Yangambi 1956).

I find that savannas and grasslands differ structurally and functionally and represent different ecosystems. Savannas support an overstory of woody dicot species (in some cases monocot species such as palms). The density, height, and percent canopy cover of this woody component varies considerably globally; but, the woody overstory does variously influence the graminoid understory. Grasslands are void of a woody overstory component. It is nearly impossible to separate the literature of grasslands and savannas in the tropics, but, I will attempt to differentiate my remarks by savannas/grasslands where possible.

DISTRIBUTION

Grasses are the most ubiquitous of all vascular plant groups and plant communities dominated by grasses occur globally. Approximately 14 million square miles (24%) of the earth's surface is classified as some sort of grassland (Barnard and Frankel 1964). A notable portion of these grass dominated communities are found in the tropics, that area of the earth's surface that lies between 23.5°N (Tropic of Cancer) and 23.5°S (Tropic of Capricorn). This includes most of central Africa, large regions of central and northern South America, extreme southern Asia, and much of the land mass of Oceania. All of North America and Europe are excluded as is most of Asia. Huntley and Walker (1982) report that 65% of Africa, 60% of Australia, and 40% of South America are classified as tropical savanna.

The largest area of tropical grasslands occurs within Africa. The Sahelian region (Le Houerou 1993) is a vast area nearly 600 km wide and 6000 km long situated south of the Sahara Desert. Tropical African savannas have been reviewed by Menaut (1983) and Menaut and Cesar (1982). Most of tropical east Africa (Herlocker *et al.* 1993) supports complex savanna communities. 'Grasslands free from any significant growth of woody plants are extremely limited within East Africa. Where such grasslands do occur, they are the consequence of local edaphic and pyric conditions. Thus, most grasslands include a woody component and appear to have been derived from woody communities through burning and clearing' (Herlocker *et al.* 1993).

Tropical and subtropical grasslands and savannas in South America are composed of the llanos in Venezuela and Colombia, the highlands and basins of Guiana, and the campos cerrados in the tablelands of central Brazil (Bucher 1982; Coupland 1993; Eiten 1982). Burkhart (1975) also includes a narrow belt of native grassland that occurs at mid-elevation along the eastern edge of the Andes. This area runs from Venezuela and Colombia to Bolivia and northwestern Argentina (Coupland 1993). Tropical American savannas have been reviewed by Sarmiento (1983).

Approximately 53% of the Australian continent is covered by savannas (Walker and Gillison 1982; Gillison 1983). Tropical savannas and grasslands occupy the upper third of Australia.

Moore (1993) has extensively mapped and described the grasslands of the continent. He delineated the tropical grasslands into tropical tallgrass, xerophytic hummockgrass, xerophytic tussockgrass, *Acacia* shrub-shortgrass, and xerophytic midgrass.

CLIMATE (BAILEY 1996)

The major macroclimatic feature of the tropics is the lack of winter. Winterless zones where the average temperature of the coldest month is 18°C, correspond closely with the Tropic of Cancer and the Tropic of Capricorn. Similarly, precipitation, runoff and evapotranspiration form a zonal pattern like temperature. Air masses converge at or near the tropics and yield abundant precipitation, runoff and subsequent evapotranspiration. Nearer the Tropics of Cancer and Capricorn, subtropical high pressure zones exist that lower precipitation to the point where tree growth is limited. Thus, the tropical zone can be characterized as both hot and humid. Combining the temperature and moisture regimes, the climate of the region is classified as tropical humid. Furthermore, the climate can be segregated based on monthly precipitation patterns within the humid tropics. Rainforest-type vegetation is characterized by humid tropical climate with no distinct dry season. This corresponds to a tropical wet climate with all months above 18°C and no dry season (Trewartha 1968). Savanna-type vegetation is characterized by humid tropical climate with a distinct dry season. Trewartha's (1968) classification gives rise to a very predictable pattern of climatic/ecoregion distribution. Around the equator is a very large area of rainforest-type vegetation. Further from the equator, savanna-type vegetation occurs in areas with a two month distinct dry season. If evaporation exceeds precipitation within this savanna region, a steppe (i.e. grassland form of vegetation) will occur. Approximately 25% of the earth's land area falls within the humid tropical climate region. About 17% of this area is within the savannas and 8% in tropical rainforests (Bailey 1996).

TAXONOMY

Clayton and Renvoize (1986) hypothesize that grasses first arrived on the scene during the Paleocene and became abundant in the Oligocene. They felt that grasses probably evolved along the forest edges and began to dominate the vegetation as the forests retreated during times of hotter and drier paleoclimates. Evolution of grasses, grasslands/savannas, and herbivores during the Oligocene and Miocene has been well documented (Dix 1964; Stebbins 1981).

Three evolutionary lines of development can be delineated based on taxonomic evidence. The first are bambusoid types, concentrated in the tropics, which specialize in forest and aquatic habitats. The second are Chloridoid/Panicoid, mainly tropical in distribution. Finally, the third are Pooid, temperate in nature, and if occurring in the tropics only at higher altitudes. Hartley (1958a, 1958b, 1961, 1963; Hartley and Slater 1960) seems to agree with these generalizations.

If generic diversification is any indication, or can be correlated to areas of origin, there are three major global zones which show uniqueness (Clayton and Renvoize 1983). The first is in eastern and southern Africa with the greatest number of genera restricted to a very small zone of great diversity. The second is the Indian peninsula and southeast Asia. This group probably helped give rise to the diversity that is evident in Australia. Finally, the tropical regions of South America are another zone of differentiation of grass genera.

No strong patterns of taxonomic dominance of any specific grass groups in the major tropical regions of the globe exist. African grasslands and savannas tend to be dominated by species of the Panicoideae. Members of the Andropogoneae generally are more abundant. The Paniceae genera become prevalent in the more mesic environments, while Chlorideae and Eragrosteae species are more common at the arid end of the moisture gradient. South American grasslands and savannas are generally dominated by members of the Panicoideae and Arundinoideae. The tropical areas of the Australian continent contain some of the greatest mixtures of grass groups. Predictably, the more mesic areas are dominated by species of the Panicoideae/Paniceae. As the climate becomes more arid, the members of the Andropogoneae, Chloridoideae and Arundinoideae/Aristideae begin to dominate the vegetation. Perhaps the most important feature is the predominance of grasses that use the C_4 photosynthetic pathway in all the tropical grassland and savanna types. The number of C_4 grasses in the flora is positively correlated to temperature (i.e., the greater the temperature the larger the number of C_4 grass species)(Ellis *et al.* 1980; Hattersley 1983).

FUTURE

The future of tropical grasslands and savannas is related to the threats and uses made of these ecosystems by man as well as by results of other activities by man. These tropical systems face similar challenges to the temperate grasslands of the northern and southern hemispheres. The major threats are overgrazing, desertification, conversion to 'improved' pastures or crop land, invasion by alien species, and climate change. Review information is available on impacts of pastoralism (Lamprey 1983), desertification (WMO 1997), and range 'improvement' practices in the tropics (Pratt and Gwynee 1977). Reviews on alien plant invasion globally and climate change specific to tropical savannas and grasslands are not readily available.

I will use the Hawaiian Islands as an example of the impacts of alien plant invasion for three reasons. First, they are possibly the 'most extraordinary living museum of evolution on the planet' (Vitousek *et al.* 1987). Second, alien plant invasions and their ecological impacts are most severe on oceanic islands (Loope 1992) and invasion of alien plants are considered worse in Hawaii than anywhere else on the globe (Taylor 1992). Finally, my students and I have studied a specific alien plant species that has had a detrimental impact on a tropical grassland in the Hawaiian Islands (Castillo 1997).

Invasion by alien species causes displacement and/or replacement of native species (Jacobi and Scott 1985). Biological invasions not only alter community composition, structure, and function, but also cause large scale ecosystem changes (Vitousek 1990, 1992). Wester (1992) estimates that there are over 800 alien plant species reproducing without direct human assistance in the Hawaiian Islands. One of the most noxious of these alien plants

is *Pennisetum setaceum* (Fountain Grass). It is a C_4 grass native to the Saharan desert region and the African Mediterranean coast. It is a perennial bunchgrass that is apomictic but capable of wide phenotypic variation. The plant is found from sea level to nearly 3,000 m (10,000 ft) on the Island of Hawaii. The first known collection of the species in Hawaii was from the Island of Lanai in 1914 (Jacobi and Warshauer 1992), and it was introduced as an ornamental to the leeward side of the Island of Hawaii in 1917. Since that time it has become the dominant herbaceous species over most dry land areas on the west side of the island (Cuddihy and Stone 1990). Fire has not been a major evolutionary factor in the development of vegetation in Hawaii (Mueller-Dombois 1981). *Pennisetum setaceum*, however, is very fire-tolerant (like most grasses) and after repeated burning will eventually form a near monotypic stand. Native shrubs and some grasses do not recover as rapidly as *P. setaceum. Pennisetum setaceum* recovered to 92% of original basal and foliar cover in 21 months following a wildfire, while many native shrubs and grasses did not recover (Shaw *et al.* 1997). This species also invades nearly barren, very young lava flows and will eventually alter the primary successional patterns on these areas and many naturally occurring ecosystems will be lost forever (Cuddihy and Stone 1990). *Pennisetum setaceum* currently is sold as an ornamental species in the warm areas of the world. It is widely used in Australia and the United States and has already escaped from cultivation; further problems with the species may be expected in the future.

Global climate change also is expected to have a negative impact on tropical grasslands and savannas in the near future. Cannell (1997) reports on models that track changes in ecosystem structure due to patterns and trends in global climate. The impact of emissions and aerosols were modeled using the HadCM2 model (Mitchell *et al.* 1995). A 'business as usual' scenario for greenhouse gas emissions and population growth was assumed. These data were put into a global ecosystem model to determine the potential responses of natural ecosystems (based on vegetative structure) to climate change caused by the HadCM2 output. The Hydrid v4.1 model was used (Friend *et al.* 1997) to predict the global distribution of plant and vegetation types. The model determines the response of individual overstory species and the understory herbaceous layer with the environment at the intra- and interspecific competition for light, moisture, and nitrogen (Cannell 1997). Seven groups were modeled: 1. No vegetation, 2. C_3 herbaceous, 3. Tropical broadleaf evergreen, 4. Cold deciduous, 5. C_4 grasses, 6. Dry deciduous broadleaf, and 7. Needleleaf. Global vegetation change from the 1990s to the 2090s was determined as areal percent cover over the dominant plant type. The most striking change was in C_4 grass dominated areas. Tropical grasslands and savannas were predicted to change into deserts or C_3 grasslands. A reduction from 10% in the 1990s to 1% in the 2090s was calculated. Changes were expected to occur by the end of the next century and were a result of decreased precipitation and increases of 8°C in regions such as northern South America, the Sahel, northern Australia and India. The C_4 grasslands in these regions were converted to deserts because of the increased temperature and decreased precipitation predicted by the models.

Human uses of natural resources will continue to reduce these assets globally during the next century. Almost all tropical ecosystems can be expected to decrease in the future. It appears that the 'developed' nations of the temperate northern hemisphere (Europe and North America) will continue to benefit even from global climate change. Conversely, the 'underdeveloped' nations of the tropics, particularly South America, India, and the Sahel of central Africa, will continue to be areas of poverty, hunger, and resultant strife. Regional armed conflicts are expected to continue within these areas threatening global security. It perhaps will be impossible to contain these as regional problems, and global armed conflict potentially could result.

REFERENCES

Acocks, J. P. (1960). Veld types of South Africa.. *Botanical Survey of South Africa Memoir* **38**, 201-341.

Bailey, R. G. (1996). 'Ecosystem Geography'. (Springer-Verlag, Inc.: New York.)

Barnard, C. and O. H. Frankel. (1964). Grass, Grazing Animals, and Man in Historic Perspective. In "Grasses and Grasslands". (Ed. C. Barnard.) (Macmillam & Co, LTD: London.)

Bourliere, F., and Hadley, M. (1970). The ecology of tropical savannas. *Annual Review of Ecology and Systematics* **1**, 125-152.

Blydenstein, J. (1961). Tropical savanna vegetation of the llanos of Colombia. *Ecology* **48**, 1-15.

Bucher, E. H. (1982). Chaco and Caatinga - South American arid savannas, woodlands and thickets. In 'Ecology of Tropical Savannas, Ecological Studies 42'. (Eds B.J. Huntley and B. H. Walker). (Springer-Verlag: Heidelberg.)

Burkhart, A. (1975). Evolution of grasses and grasslands in South America. *Taxon* **24**, 53-66.

Cannell, M. G. R. (1997). Global change. Scientific report of the Institute of Terrestrial Ecology, Centre for Ecology and Hydrology, Natural Environmental Research Council, Monks Wood, Abbots Ripton, Huntington, Cambs.

Castillo, M. J. (1997). Control of *Pennisetum setaceum* (Forssk.) Chiov. in native Hawaiian dry upland ecosystems. MS Thesis. Department of Forest Sciences, Colorado State University, Fort Collins, CO.

Clayton, W. D., and S. A. Renvoize. (1986). Genera Graminum: Grasses of the World. Kew Bulletin Additional Series XIII. (Royal Botanic Gardens: Kew.)

Coupland, R. T. (1993). Overview of South American Grasslands. In 'Natural Ecosystems: eastern hemisphere and resume. Ecosystems of the World 8B' (Ed. R.T. Coupland) (Elsevier: Amsterdam.)

Cuddihy, L. W., and C. P. Stone. (1990). 'Alteration of native Hawaiian vegetation: effects of humans, their activities and introductions'. (University of Hawaii Press: Honolulu.)

Dix, R. L. (1964). A history of biotic and climatic changes within the North American grassland. In 'Grazing in terrestrial and marine environments'. British Ecological Society Symposium **4**, 71-89.

Eiten, G. (1972). The cerrado vegetation of Brazil. *Botanical Review* **38,** 201-341.

Eiten, G. (1982). Brazilian 'savannas'. In 'Ecology of Tropical Savannas, Ecological Studies 42'. (Eds B. J. Huntley and B.H. Walker.) (Springer-Verlag: Heidelberg.)

Ellis R. P., Vogel, J. C., and A. Fuls. (1980). Photosynthetic pathways and the geographic distribution of grasses in southwest Namibia. *South African Journal of Science* **76**, 307-314.

Friend, A. D., A. K. Stevens, R. G. Knox, and M. G. R. Cannell. (1997). A process-based, terrestrial biosphere model of ecosystem dynamics (Hydrid v3.0). *Ecological Modeling* **95**, 249-287.

Gillison, A. N. (1983). Tropical savannas of Australia and the southwest Pacific. In 'Tropical Savannas. Ecosystems of the World, 13'. (Ed. F. Bourliere.) (Elsevir: Amsterdam.)

Hartley, W. (1958a). Studies on the origin, evolution, and distribution of the Gramineae. I. The tribe Andropogoneae. *Australian Journal of Botany* **6**, 115-128.

Hartley, W. (1958b). Studies on the origin, evolution, and distribution of the Gramineae. II. The tribe Paniceae. *Australian Journal of Botany* **6**, 343-357.

Hartley, W. (1961). Studies on the origin, evolution, and distribution of the Gramineae. IV. The genus *Poa. Australian Journal of Botany* **9**, 152-161.

Hartley, W. (1963). Distribution of the grasses. In 'Grasses and Grasslands'. (Ed. C. Barnard.) (Macmillam & Co, LTD: London.)

Hartley, W. And C. Slater. (1960). Studies on the origin, evolution, and distribution of the Gramineae. III. The tribes of the sub-family Eragrostoideae. *Australian Journal of Botany* **8**, 256-276.

Hattersley, P. (1983). The distribution of C_3 and C_4 grasses in Australia in relation to climate. *Oecologia* **57**, 113-128.

Herlocker, D. J., H. J. Dirschl and G. Frame. (1993). Grasslands of East Africa. In 'Natural Ecosystems: eastern hemisphere and resume. Ecosystems of the World 8B'. (Ed. R. T. Coupland.) (Elsevier: Amsterdam.)

Hill, T. L. (1965). Savannas: a review of a major research problem in tropical geography. Savanna Res. Ser., No 3. (McGill Univ: Montreal.)

Huntley, B. J., and B. H. Walker. (1982). Introduction. In 'Ecology of Tropical Savannas, Ecological Studies 42'. (Eds B.J. Huntley and B. H. Walker.) (Springer-Verlag: Heidelberg.)

Jacobi, J. D., and Scott, J. M. (1985). An assessment of the current status of native upland habitats and associated endangered species on the island of Hawaii. In 'Hawaii's Terrestrial Ecosystems: Preservation and Management.' (Eds C. P. Stone, C. W. Smith, and J. T. Tunison.) (Cooperative National Park Resources Unit, University of Hawaii: Manoa.)

Jacobi, J. D. and R. Warshauer. (1992). Distribution of six alien palnt species in upland habitats of the Island of Hawaii. In 'Alien plant invasions in native ecosystems of Hawaii: management and research.' (Eds C. P. Stone, C. W. Smith, and J. T. Tunison.) (Cooperative National Park Resources Unit, University of Hawaii: Manoa.)

Lamprey, H. F. (1983). Pastoralism yesterday and today: the overgrazing problem. In 'Tropical Savannas. Ecosystems of the World, 13'. (Ed. F. Bourliere.) (Elsevir: Amsterdam.)

Le Houerou, H. N. (1993). Grasslands of the Sahel. In 'Natural Ecosystems: eastern hemisphere and resume. Ecosystems of the World 8B'. (Ed. C. T. Coupland.) (Elsevir: Amsterdam.)

Loope, L. L. (1992). An overview of problems with introduced plant species in national parks and biosphere reserves of the United States. In 'Alien plant invasions in native ecosystems of Hawaii: management and research.' (Eds C. P. Stone, C. W. Smith, and J. T. Tunison.) (Cooperative National Park Resources Unit, University of Hawaii: Manoa.)

Menaut, J. C. (1983). The vegetation of African savannas. In 'Tropical Savannas. Ecosystems of the World, 13'. (Ed. F. Bourliere.) (Elsevir: Amsterdam.)

Menaut J. C., and J. Cesar. (1982). The structure and dynamics of a West African savanna. In "Ecology of Tropical Savannas, Ecological Studies 42'. (Eds B. J. Huntley and B. H. Walker.) (Springer-Verlag: Heidelberg.)

Mitchell, J. F. B., T. C. Johns, J. M..Gregory and S. F. B. Tett. (1995). Climatic response to increasing levels of greenhouse gases and sulphate aerosols. *Nature* **376**, 501-504.

Moore, R. M. (1993). Grasslands of Australia. In 'Natural Ecosystems: eastern hemisphere and resume. Ecosystems of the World 8B'. (Ed. R. T. Coupland.) (Elsevier: Amsterdam).

Mueller-Dombois, D. (1981). Fires in tropical ecosystems. In 'Fire Regimes and Ecosystem Properties'. (Eds H. A. Monney, T. M. Bonnicksen, N. L. Christensen, J. E. Lotan, and W. A. Reiners.) USDA, For. Ser. Gen. Tech. Rep. WO-26: Washington, DC.

Pratt, D. J., and M.. D. Gwynne. (1977). 'Rangeland management and ecology in east Africa' (Robert E. Krieger Publ. Co.: Huntington, New York.)

Sarmiento, G. (1983). The savannas of tropical America. In 'Tropical Savannas. Ecosystems of the World, 13'. (Ed. F. Bourliere.) (Elsevier: Amsterdam.)

Shaw, R. B., J. M. Castillo, and R. D. Laven. (1997). Impacts of wildfire on vegetation and rare plants within the Kipuka Kalawamauna endangered plants habitat area, Pohakuloa Training Area, Hawaii. Proceeding - Fire Effects on Rare Plant Species and Habitats Conference. International Association of Wildfire.

Singh, J. S., W. K. Lauenroth and D. J. Milchunas. (1983). Geography of grassland ecosystems. *Progress in Physical Geography* **7**, 46-80.

Stebbins, G. L. (1981). Coevolution of grass and herbivores. *Annuals of Missouri Botanic Garden* **68**, 75-86.

Taylor, D. 1992. Controlling weeds in natural areas in Hawaii: a management perspective. In 'Alien plant invasions in native ecosystems of Hawaii: management and research'. (Eds C. P. Stone, C. W. Smith, and J. T. Tunison.) (Cooperative National Park Resources Unit, University of Hawaii: Manoa.)

Trewartha, G. T. (1968). 'An introduction to climate. 4th ed'. (McGraw-Hill: New York.)

Vitousek, P. M., L. L. Loope, and C. P. Stone. (1987). Introduced species in Hawaii: biological opportunities or ecological research. *Trends in Ecology and Evolution* **2**, 224-227.

Vitousek, P. M. (1990). Biological invasions and ecosystem processes: towards an integration of population biology and ecosystem studies. *Oikos* **57**, 7-13.

Vitousek, P. M. (1992). Effects of alien plants on native Hawaiian ecosystems. In 'Alien plant invasions in native ecosystems of Hawaii: management and research'. (Eds C. P. Stone, C. W. Smith, and J. T. Tunison.) (Cooperative National Park Resources Unit, University of Hawaii: Manoa.)

Walker, J., and A. N. Gillison. (1982). Australian savannas. In 'Ecology of Tropical Savannas, Ecological Studies 42'. (Eds B. J. Huntley and B. H. Walker.) (Springer-Verlag: Heidelberg.)

Wester, L. 1992. Origin and distribution of adventive alien flowering plants in Hawaii. In 'Alien plant invasions in native ecosystems of Hawaii: management and research'. (Eds C. P. Stone, C. W. Smith, and J. T. Tunison.) (Cooperative National Park Resources Unit, University of Hawaii: Manoa.)

World Meterorological Organization. (1997). 'Climate, drought, and desertification'. WMO Series No. 869. Geneva: Switzerland.

Yangambi, N. I. (1956). 'Phytogeography'. (Conseil Scientifique pour l'Afrique: London.)

GRASSES

Grasses: Systematics and Evolution. (2000). Eds S.W.L. Jacobs and J. Everett. (CSIRO: Melbourne)

TEMPERATE GRASSLANDS OF THE SOUTHERN HEMISPHERE

R.H. Groves

CSIRO Plant Industry & CRC for Weed Management Systems, GPO Box 1600, Canberra, ACT 2601, Australia.

Abstract

The grasslands of the temperate regions of Argentina, South Africa, southeastern Australia and New Zealand are compared and contrasted in terms of their floristics and their different histories of usage and abusage over the periods of European settlement. Exchanges of plant species between all four regions have occurred which have important implications for present and future management of these grasslands both for animal production and for their conservation; some of these implications will be explored. The grass *Nassella trichotoma* (Nees) Hack. ex Arech. is now widespread in all regions and its ecology in both its native and introduced ranges is compared in relation to limiting the impact of this species as well as of some other major invasive species.

Key words: Grasslands, Argentina, South Africa, Australia, New Zealand, Serrated Tussock, *Nassella trichotoma*.

INTRODUCTION

Not all the world's grass species and genera are confined to grasslands *sens. strict*. Grasses occur in nearly all vegetation types to some extent. In savanna systems (grasslands with a sparse tree or shrub stratum), the grassy stratum is of prime importance to the functioning of those ecosystems and the role of grazing animals in them (Tainton and Walker 1993). Some contemporary grasslands can be thought of as 'secondary' or derived, in that the tree stratum of what was once a woodland has been removed and subsequent regimes of regular burning and continuous grazing prevent re-invasion of woody plants. Whatever their original status in a botanical sense, grasslands are important to humans because they have been the source of some of the economically important grass species able to be domesticated and used as staples for human food supplies. In this sense, they have been basic to the development of human societies as we know them. Grasslands also provide habitat for many animals, and especially both indigenous and domesticated vertebrates. Most grasslands were used by indigenous peoples for grazing and hunting of native vertebrates; more recently, Euro-pean settlers have grazed domesticated vertebrates on natural or derived grasslands. In nearly all instances, the role of natural and deliberately lit fires has been linked inextricably to the role of grazing, albeit to varying extents in different regions. Although floristic and faunal composition of grasslands changes regionally, there seems to be little obvious that distinguishes management of grasslands between southern and northern hemispheres. Rather, there may be greater differences between grasslands in tropical and sub-tropical climatic regions and those in temperate regions, irrespective of the hemisphere in which they occur.

This contribution seeks to identify distinguishing characteristics of grasslands in four southern regions of temperate climate (Argentina, southern Africa, southeastern Australia and New Zealand) in terms of floristic composition, their management and their susceptibility to invasion by introduced plants. Because the tussock grass *Nassella trichotoma* (Nees) Hack. ex Arech. (syn. *Stipa trichotoma* Nees) is indigenous to southern South America (Argentina and elsewhere) and seriously invasive in the three other regions in the Southern Hemisphere, I shall use it as an

Table 1. The main grass tribes present in some seven Southern Hemisphere temperate grasslands, compared with a global coverage of 64 sites expressed as % contributed by each major tribe (from Hartley (1950) Table 1).

Locality	Percentage of Total Grass Species for each Locality*					
	Agros.#	Andr.	Aven.	Eragr.	Fest.	Panic.
Corrientes, Argentina	4.5	11.7	0.9	9.9	9.9	39.7
Santa Fe, Argentina	2.3	9.9	0.8	6.9	9.9	42.7
Riversdale, South Africa	1.8	7.3	23.8	9.2	16.5	13.8
George district, South Africa	3.8	11.3	17.3	9.8	14.3	15.8
Southern South Australia	10.4	4.5	12.3	2.0	26.0	9.7
Tasmania, Australia	21.2	3.0	14.1	2.0	27.3	4.0
SW South Island, New Zealand	25.3	0	29.3	0	30.7	0
Average(sthtemp)	**9.9**	**6.8**	**14.1**	**5.7**	**19.2**	**18.0**
Average(global)	**8.2**	**11.9**	**6.3**	**8.1**	**16.5**	**24.7**

*Full references to floras and floristic lists used as sources of data on grass distribution not cited in original
#Agros. = Agrosteae, Andr. = Andropogoneae, Aven. = Aveneae, Eragr. = Eragrostideae, Fest. = Festuceae, Panic. = Paniceae

example of the major changes currently occurring in terms of floristic composition and functioning of temperate grasslands as a result of the increasing rates of exchange between southern grassland regions. Some other invasions of southern grasslands will also be described.

FLORISTIC COMPOSITION AND FUNCTIONAL GROUPS

In this review, I wish to present floristic information at the level of tribe and in terms of functional groups, rather than at the level of individual species. Information on individual species is available for representative grasslands at localities in all four regions (Soriano 1992: Killick 1963: Groves 1965: Wardle 1991, for Argentina, South Africa, Australia and New Zealand respectively).

Hartley (1950) analysed the global distribution of the major grass tribes and showed that, although there are some local differences within regions, the main differences occurred between regions, with the southern Australian and New Zealand grasslands differing most from the global averages (Table 1). Of the six major tribes, the Aveneae and to a lesser extent the Agrosteae were under-represented and the Paniceae over-represented in the Argentinian grasslands, whereas the two South African grasslands had the Agrosteae and the Paniceae under-represented and the Aveneae over-represented. Species belonging to the tribes Andropogoneae, Eragrostideae and the Paniceae were under-represented in the southern Australian grasslands and absent from the New Zealand grasslands, whereas in the same three grasslands the other three major tribes (the Agrosteae, Aveneae and Festuceae) were over-represented. Few other generalisations are possible from Hartley's data.

Gillison (1993) followed a similar approach in outlining a regional biogeographic framework for the grasslands of Oceania and showed that the tribes Agrosteae and Aveneae were over-represented and the Paniceae under-represented in the temperate regions of Oceania compared with northern (tropical) regions. We may conclude more generally, on the basis of tribal representation, that the southern temperate grasslands differ less between themselves than with their tropical counterparts found at lower latitudes.

In southern temperate grasslands, four floristic components can usually be identified. Native perennial grasses with a 'C_3' pathway for photosynthesis, such as species of *Danthonia*, *Poa* and *Stipa sens. lat.*, grow mainly in the spring and autumn, whereas warm-season 'C_4' grasses, such as *Themeda triandra* and *Bothriochloa*, grow mainly in summer. A third component comprises a suite of colourful native forbs that occupy the inter-tussock spaces in the grasslands. These three components comprise the indigenous element in the vegetation prior to occupation of all grassland regions by Europeans. Since that time, a fourth component of various introduced grasses and herbs (having annual, biennial and perennial life cycles) now characterises all southern temperate grasslands, the floristic composition of which in any locality depends primarily on previous disturbance regimes. Each component has its own cycle of seasonal growth and development and responses to factors such as grazing, fire and addition of fertiliser (Fig. 1). In what follows, I wish to comment briefly on the first three indigenous elements in grassland floristics before describing some of the changes in floristics as a result of species invasions. The three indigenous components may be thought of as 'functional groups' and the different species invasions may arise because of inadequate functioning of one or all of these indigenous groups. Recognition of such functional groups

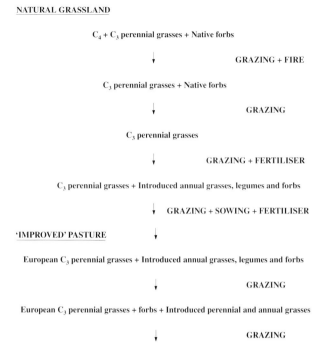

NATURAL GRASSLAND

$C_4 + C_3$ perennial grasses + Native forbs

↓   GRAZING + FIRE

C_3 perennial grasses + Native forbs

↓   GRAZING

C_3 perennial grasses

↓   GRAZING + FERTILISER

C_3 perennial grasses + Introduced annual grasses, legumes and forbs

↓   GRAZING + SOWING + FERTILISER

'IMPROVED' PASTURE   ↓

European C_3 perennial grasses + Introduced annual grasses, legumes and forbs

↓   GRAZING

European C_3 perennial grasses + forbs + Introduced perennial and annual grasses

↓   GRAZING

Fig. 1. Changes in temperate grassland functional groups as a result of 200 years of European settlement in south-eastern Australia (freely derived from Moore 1970).

in grasslands may be especially relevant to the planning of re-vegetation programs for grassland species (Groves 1990), as well as for their management for grazing or for nature conservation.

USAGE AND ABUSAGE OF SOUTHERN TEMPERATE GRASSLANDS

The patterns of management of southern temperate grasslands prior to the arrival of Europeans are largely unknown. Certainly, they were burnt and often for purposes of hunting the indigenous vertebrates that lived in such grasslands. For instance, Pyne (1991) describes the deliberate burning of some Tasmanian grasslands by aborigines for hunting. Some of the New Zealand grasslands are probably derived from forests burnt by the Maori 1000 years ago (Wardle 1991). This use of fire does not necessarily distinguish temperate grasslands from others; rather, it is the frequency of burning (irregular cf. annual) and to a lesser extent the different season of burning (late spring-early summer cf. winter) that may distinguish them. The burning led to new growth that attracted grazing animals that in turn could be hunted more readily.

With the advent of Europeans with their domesticated grazing animals, regular burning of temperate grasslands continued and *may* have increased in frequency. The feature that changed most, however, was probably the continuity of grazing associated with that regular burning. In other words, it was the interaction between fire and the grazing regime that led to profound change in the floristic composition of the grasslands. Wherever the warm season grasses had been present, they disappeared under continuous grazing, and especially in response to grazing shortly after fire. One major functional group of summer-growing, deep-rooted perennial grasses at first decreased and then disap-

peared in many cases. The inter-tussock native forbs were selectively grazed and with a consequent increase in soil nutrients came to be replaced by introduced annual grasses and legumes and rosette herbs of European origin. Moore (1970) described the sequential change in floristic composition in dominant species in a south-eastern Australian grassland following from continuous grazing. The cooler-season native grasses, such as species of *Danthonia* and *Stipa sens. lat.*, were usually the last of the three functional groups to disappear. In most southern temperate regions some vestiges of this group remain. Although individual species may change within the southern temperate region, the overall pattern from tall warm-season native grasses to shorter native grasses able to grow in spring and autumn to introduced annual grasses of Mediterranean European origin probably applies to all southern temperate regions. It was only the much later, regular additions of phosphatic fertiliser that led to the introduced annual legume component (especially *Trifolium* spp.) becoming dominant and the eventual extirpation of the spring- and autumn-growing native perennial grasses.

Large areas of Argentinian, southern Australian and New Zealand grasslands have thus been profoundly changed floristically and functionally. The effects have been less in South Africa only because the area of the winter-rainfall region is less and the 'resting' of grasslands during the summer-autumn period is a more common practice. In many cases the carrying capacity of such derived grasslands for domesticated stock has increased. But the extirpation of the deep-rooted summer-growing perennials has had a significant effect on seasonal patterns of soil moisture usage and consequent water and salt yields. Such derived grasslands may not be as stable floristically when grazed through periods of drought as the native perennial grasslands they replaced. The end result may be an increased rate of invasion by undesirable introduced plants. In addition, nature conservation is confined to fragments of the original grasslands and these may be inadequate for long-term survival of the rare and endangered animals and inter-tussock grassland forbs as well as being more prone to invasion. Over the 200 to 300 year period of European use of southern temperate grasslands, short-term economic gain has often been at the expense of wiser land and water management and of nature conservation.

INVASIVE SPECIES IN SOUTHERN TEMPERATE GRASSLANDS

The usage and abusage of land originally covered with southern temperate grassland has led in all four regions to profound changes in floristic composition of the modified vegetation. Invasions of pest animals and introduced weeds have followed, often with disastrous effects to the derived grasslands and their carrying capacities as well as for nature conservation. In New Zealand a number of species in the European forb genus *Hieracium* have spread throughout the tussock grasslands and occupied the inter-tussock spaces, often in association with rabbit grazing. In southern Australia the unpalatable perennial grass *Eragrostis curvula,* of South African origin, has spread to replace the native summer-growing perennial grasses. European thistles have invaded all four regions. In South Africa as well as in New Zealand and southern Australia, the unpalatable South American grass *Nassella trichotoma* has invaded temperate grasslands with disastrous effects to sheep-carrying capacity of land. In southern Australia remnants of original grassland are being invaded by other *Nassella* species with equally disastrous effects for conservation of those remnants. In all regions, introduced annual grasses have become dominant and reduced stocking capacity, especially through the summer months. I wish now to highlight just one such major invasion because it impinges on all four regions of southern temperate grasslands, viz. the example of *Nassella trichotoma*.

Nassella trichotoma

Nassella trichotoma is a species indigenous to South American grasslands, especially and most abundantly in an arc from Mar del Plata west to the eastern section of La Pampa Province and northward to Cordoba Province of Argentina (Connor 1960). This species of tussock grass also occurs naturally in Peru, Chile and Uruguay (Parodi 1930, as quoted by Campbell and Vere 1995). In this region, it occurs on both plains and mountainous areas, often on locally dry sites, at which it may be heavily grazed. Whether *N. trichotoma* is a weed in South America remains a moot point. Connor concluded that *N. trichotoma* was not a weed in its native Argentina, whereas Vervoorst (1967, as quoted by Campbell and Vere 1995) noted subsequently that it was invading grazing land and cultivated areas in its native range and thereby becoming weedy.

When the species was introduced accidentally to South Africa, southeastern Australia and New Zealand, there was no doubt about its weed status, however. Within two years of its identification in New South Wales in 1935, *N. trichotoma* was assessed as a potential weed (Cross 1937) and now, 60 years later, is arguably one of the worst weeds in that State. The grass had been introduced accidentally to all three regions probably much earlier but remained unidentified until 1934 in New Zealand (Allen 1935) and until 1952 in South Africa (Wells 1974). It is now widespread in all three regions of the Southern Hemisphere, as well as occurring in a minor way at several sites in Europe (introduced as a contaminant of fleece?) and more recently in some regions of the United States (as a contaminant of pasture seed from Argentina) (Campbell and Vere 1995). In South Africa, southern Australia and New Zealand, *N. trichotoma* probably has potential to spread further, especially in degraded pasturelands.

Campbell (1998) has written that *N. trichotoma* 'has many ecological and biological features that explain its success as a weed. The ability to produce enormous numbers of seeds that are widely distributed by wind and establish readily facilitates invasion of land unprotected by vigorous pastures. Despite relatively slow seedling growth it invades because animals graze more palatable plants. Once established individual tussocks live for long periods and withstand grazing, drought, burning, infertile soils, unfavourable aspects and competition' (p. 80). Because the weed is not only unpalatable but also difficult to digest by both sheep and cattle, it can reduce the carrying capacity of invaded land by as much as 97% (Campbell 1998). In addition, *N. trichotoma* is expensive to control by herbicides or by pasture management. Currently, in Australia at least, the one chemical that had been effective (though expensive) has been taken off the market. A pathogenic fungus is known to occur on *N. trichotoma* in Argentina (Briese and Evans 1998). Biological control of a weedy grass species has yet to be achieved anywhere but because of its weediness and the intractability of *N. trichotoma* to existing control methods, it has been decided by southern Australian authorities as worthy of further attention. Studies on the effects of this fungus on growth and seed production of the grass are to start soon in Argentina, in the hope that it will be specific to *N. trichotoma* and not affect related Stipoid grasses native to Australia (or indeed of New Zealand or South Africa either). If biological control proves impractical, only afforestation of invaded grasslands remains as a feasible control method, at least for the large areas of relatively low fertility invaded by *N. trichotoma*.

CONCLUSIONS

The temperate grasslands of South America, South Africa, Australia and New Zealand are similar in many respects - the same genera and Tribes occur in them, similar functional groups are represented, the dominants are all tussock grasses, they have been burnt regularly by both aboriginal and European peoples, they have been grazed by native and domesticated animals and they are prone to invasion by weedy plants from other regions. They also differ among themselves but these differences are minor relative to the large differences between all temperate grasslands (both southern and northern) and their tropical equivalents in which summer-growing C_4 grasses predominate and fires may be annual (Shaw this volume). Many areas of temperate grasslands have been drastically modified for animal production and only fragments of the original grasslands remain. Such fragments are usually inadequate for nature conservation and are themselves being invaded by introduced annual grasses and forbs, thereby reducing their conservation value still further. Attempts at revegetation with native species have not yet been sufficiently successful to reverse the trend towards further ecosystem degradation. Southern temperate grasslands, like their northern equivalents, are urgently in need of greater research attention if they are to survive as one of the world's major ecosystems.

REFERENCES

Allen, H.H. (1935). Additions to the alien flora of New Zealand. *Transactions of the Royal Society of New Zealand* **65**, 2.

Briese, D.T. and Evans, H. C. (1998). Biological control of serrated tussock (*Nassella trichotoma*): Is it worth pursuing? *Plant Protection Quarterly* **13**, 94-97.

Campbell, M.H. (1998). Biological and ecological impact of serrated tussock (*Nassella trichotoma* (Nees) Arech.) on pastures in Australia. *Plant Protection Quarterly* **13**, 80-86.

Campbell, M.H. and Vere, D.T. (1995). *Nassella trichotoma* (Nees) Arech. In *The Biology of Australian Weeds*, Vol. 1 (eds R.H. Groves, R.C.H. Shepherd and R.G. Richardson), pp. 189-202. (R.G. & F.J. Richardson: Melbourne)

Connor, H.E. (1960). Nassella tussock in Argentina. *New Zealand Journal of Agriculture* **100**, 18-21.

Cross, D.O. (1937). Yass River tussock. *Agricultural Gazette of New South Wales* **48**, 546-548.

Gillison, A.N. (1993). Overview of the grasslands of Oceania. In: *Ecosystems of the World*, Vol. 8B, *Natural Grasslands. Eastern Hemisphere and Resume* (ed.R.T. Coupland), pp. 303-313. (Elsevier: Amsterdam)

Groves, R.H. (1965). Growth of *Themeda australis* tussock grassland at St Albans, Victoria. *Australian Journal of Botany* **13**, 291-302.

Groves, R.H. (1990). Native grassland species in revegetation programs. In: Proceedings of the Royal Australian Institute of Parks and Recreation, Victorian Region, *Management of Amenity and Sports Turf,* March 28-30, 1990, Ringwood, pp. 4- 12.

Hartley, W. (1950). The global distribution of Tribes of the Gramineae in relation to historical and environmental factors. *Australian Journal of Agricultural Research* **1**, 355-373.

Killick, D.J.B. (1963). An account of the plant ecology of the Cathedral Peak area of the Natal Drakensberg. *Memoirs of the Botanical Survey of South Africa* No. 34.

Moore, R.M. (1970). South-eastern temperate woodlands and grasslands. In: *Australia Grasslands* (ed. R.M. Moore) pp. 169-190. (Australian National University Press: Canberra)

Pyne, S.J. (1991). Burning Bush. A Fire History of Australia. 520 pp. (but see, especially, pp. 127-129). (Henry Holt and Company: New York.)

Soriano, A. (1992). Rio de la Plata grasslands. In: *Ecosystems of the World,* Vol. 8A. *Natural Grasslands. Introduction and Western Hemisphere* (ed. R.T. Coupland), pp. 367-407. (Elsevier: Amsterdam)

Tainton, N.M. and Walker, B.H. (1993). Grasslands of southern Africa. In *Ecosystems of the World,* Vol. 8B. *Natural Grasslands. Eastern Hemisphere and Resume* (ed. R.T. Coupland), pp. 265- 290. (Elsevier: Amsterdam)

Wardle, P. (1991). Vegetation of New Zealand. 672 pp. (Cambridge University Press: Cambridge.)

Wells, M.J. (1974). *Nassella trichotoma* (Nees) Hack. in South Africa. *Proceedings 1st South African Weeds Conference, Pretoria,* pp. 125-127.

Grasses: Systematics and Evolution. (2000). Eds S.W.L. Jacobs and J. Everett. (CSIRO: Melbourne)

FUTURE OF TEMPERATE NATURAL GRASSLANDS IN THE NORTHERN HEMISPHERE

Arthur W. Bailey

University of Alberta, Edmonton, Alberta T6G 2P5, CANADA.
  E-mail: awbailey@gpu.srv.ualberta.ca

Abstract

In the northern hemisphere, temperate natural grassland flora and fauna has evolved under the influence of cold, snow, heat, drought, fire, herbivory, glaciation, and are now adapting to man. There are wetland, prairie and steppe grassland ecosystems that have a 35 million year evolutionary history including climate change and glaciation within the past 1 million years. There are extensive residual tracts of natural grasslands in parts of North America and Asia. Three usage and perception models for native grassland management are presented. The first model is one of exploitation: the single use concept of intensive agriculture that destroys these grasslands due to the accepted perception that grasslands are wastelands which have no place in modern society. Intensive agriculture used the high quality, nutrient-rich soils that evolved under the natural grasslands to grow alien crops. The second model is one of preservation to 'protect' the remnant natural grasslands by legislating them into parks. Few preservationist models have been developed that recognize and comprehend the complex interactions of climate, fire, and herbivory required to maintain the high biodiversity and healthy productivity of temperate natural grasslands. The third model is one of conservation where there can be sustainable use of these natural grasslands through a conservative, multiple use approach to their management. This model requires the continuing development of an extensive, scientific knowledge base, and its practical implementation. One key to grassland management appears to be the application of a conservative grazing management strategy for domesticated or wild herbivory. There is a need for scientists and managers to recognize and implement effective fire management scenarios for these grasslands. Single purpose, industrial and agricultural use threaten the future wellbeing of temperate natural grasslands more than climate change, overgrazing, or the absence of grazing or fire. Preservationists and conservation-minded ranchers, land managers and scientists can work together to understand and implement improved management regimes for these ecosystems; there is an opportunity remaining to diminish the rate of exploitation of the residual temperate natural grassland ecosystems.

Key words: Temperate, grasslands, biogeography, grazing.

INTRODUCTION

The temperate thermal zone has both a summer and a winter, thus differentiating it from the summer-less climate of high latitudes and the winter-less climate of lower latitudes (Bailey 1996). In the northern hemisphere, temperate natural grasslands occur between the boreal forest to the north and the tropics to the south, in North America, Europe and Asia. Winter climates of these grasslands range from many months below zero°C in northern regions to a few weeks of freezing temperatures in southern parts of the temperate thermal zone. These grasslands are dominated in northern areas by grass species that use a C_3 photosynthetic pathway (Dix 1964). In southern regions there is a higher proportion of grass species using C$_4$ photosynthetic pathways. The grassland flora and fauna have evolved under the influence of cold, snow, heat, drought, fire, herbivory, glaciation, and are now adapting to man.

The north temperate natural grasslands are circumpolar, found in North America, Europe and Asia. The basic types of temperate natural grasslands are deserts, steppes, prairie and wetland. Bailey (1996) described the temperate deserts as having low rainfall and strong temperature contrasts between summer and winter. In North America, they are particularly common between the Pacific and the Rocky Mountains, usually in the rainshadow area of mountain ranges. In Asia they are in the south central area of the continent. The temperate steppes have about twice as much annual precipitation as the deserts, and have a semi-arid continental climate where evaporation usually exceeds precipitation, despite maximum rainfall occurring in summer. Winters are cold and dry, summers are warm to hot. Drought periods are common, particularly in mid to late summer. The steppes of both Asia and North America are primarily in the middle of each continent. The temperate prairies are associated with continental, mid-latitude subhumid climates with precipitation being about the same as evapo-transpiration. They generally occur north of the steppes in Russia and Eastern Europe and both north and east of the mid-continent North American steppes. Wetlands have a water table near or above the mineral soil for most of the thawed season, supporting a hydrophilic vegetation, and pools of open water (Zoltai and Pollett 1983).

EASTERN EUROPE AND ASIA

Most of the prairie (forb-grass) and steppes in eastern Europe and the Volga-Siberian-Kazakhstan region is in crop agriculture (Hart *et al.* 1996). Remnants are characterized by *Stipa pennata* subsp. *pennata* L., *S. tirsa* Stev. and other *Stipa* species with *Poa*, *Agropyron* and *Bromus* species on more mesic sites. The forest-steppe zone described by (Walter 1973) represents a transition from deciduous forest to prairie. It is a mosaic of deciduous forest stands (*Quercus* and *Populus* species) that occupy well drained habitats on valley slopes, while the meadow steppe (prairie) occupies the less well drained, flat sites of heavy, deep, chernozemic soils. In previous centuries, fire and grazing by herds of wild herbivores favored the growth of steppe over forest.

Steppes

South of the prairie (meadow-steppe) is the feather-grass or tuft-grass steppes extending from western Russia to the Amur basin (Hart *et al.* 1996; Walter 1973). Remnants of these grasslands in European Russia consisting of various *Stipa* species growing on chernozemic soils that are not as thick, nor as high in organic matter as in the prairies to the north. Dominant grasses included *Stipa lessingiana* Trin. & Rupr., *S. capillata* L. and *Festuca rupicola* Heuff. In eastern Siberia, dominants are *F. rupicola*, *Koeleria* sp., *Poa attenuata* subsp. *botyroides* (Trin. Ex. Griseb.) Tzvel. and *Cleistogenes squarosa* (Trin.) Keng.

Sagebrush-grass steppes dominated by *Artemisia* species are distributed from the northern coast of the Black Sea to the western foothills of Altai (Hart *et al.* 1996). These are often found on chestnut soils. The bunchgrasses *Stipa capillata* and *Festuca rupi-*

cola grow amongst the shrubs. In Kazakhstan there are large areas of semi-desert comparable to the sagebrush- grass zone of North America (Walter 1973). Steppes cover vast areas in almost every Asian mountain system; *Stipa* and *Festuca* dominate and include *Carex humilis* (Hart *et al.* 1996).

The temperate natural grasslands of China occupy about 25% of the country and extend from the northeast to the Tibetan plateau (Hart *et al.* 1996). About 80% is steppe with the remainder being meadow (prairie). The major grasses of meadow are *Stipa baicalensis* Roshev. and *Leymus chinensis* (Trin.) Tzvel.; in steppe *Stipa grandis* P. Smirn., *S. kyylovii* Roshev., and *S. brevifolia* R.A. Phil. About 55% of Outer Mongolia is steppe. *Leymus chinensis* and *Stipa grandis* dominate the two main communities of the eastern steppe region on sandy-clay chestnut or dark chestnut soils; annual precipitation ranges from 180 - 500 mm.

NORTH AMERICA

Great Plains

Wetlands

Throughout the Great Plains, meadows, marshes, ponds (sloughs) and lakes occur and provide the habitat required by millions of waterfowl. The meadows hold water for only a few weeks in spring whereas the marshes hold water all summer and may or may not be dry in autumn. Development of the vegetation in the different kinds of wetlands is dependent upon water regime and salinity of the water (Walker and Coupland 1970). The marshes often have *Scirpus validus* Vahl, *Typha latifolia* L. and *Phragmites australis* (Cav.) Trin. ex Steud. Open water in fresh water situations (Zoltai and Pollett 1983) and meadows are often dominated by *Carex atherodes* Spreng. and various willows such as *Salix bebbiana* Sarg.

Fescue Prairie

The fescue prairie occurs along the northern and northwest fringe of the North American Great Plains (Hart *et al.* 1996; Coupland 1961). To the west is coniferous forest and to the north is mixed wood boreal forest, while to the south is mixed grass steppe. This prairie has primarily plants possessing the C$_3$ photosynthetic pathway (cool season plants). *Festuca campestris* Rydb. dominates the grassland along the Rocky Mountain foothills in Alberta and Montana in association with *Danthonia parryi* Scribn. and *Festuca idahoensis* Elmer. On the plains of central Alberta and Saskatchewan is a fescue grassland dominated by *Festuca hallii* Piper and *Hesperostipa curtiseta* (A.S. Hitchc.) Barkworth. Associated with these grasslands are groves of *Populus tremuloides* Michx. (Coupland 1961). Common to both areas are rich, black chernozemic soils that are highly prized for crop agriculture. These prairie grasslands have a more favorable precipitation regime than the steppes, having a P:E (precipitation:evaporation) ratio of about 1.0. Tree and shrub encroachment has been common following settlement and the cessation of natural prairie fires. Prescribed fires favour grasslands over shrubs and trees (Wright and Bailey 1982). About 150 species of higher plants are found in the fescue prairie with about 50 absent from the plains region (Moss 1955). The author has unpublished evidence of the presence of 24-28 higher plant species / 5 m^2 in probable foothill glacial refugia compared to 10-12 plant species / 5 m^2 on normal soils in northern plains *Festuca hallii* stands and 5-6 plant species / 5 m^2 on sands of the adjacent *Pinus banksiana- Festuca halli* stands of the southern boreal forest. Repeated continental glaciation has likely reduced plant species diversity by about 50% in the northern Great Plains fescue prairie as compared to the related fescue prairie in the foothills of the Rocky Mountains. The presence of nearby unglaciated refugia likely provided for the continued enrichment of the foothills fescue grassland, but less so for the more distant and centrally located *Festuca hallii* prairie. The growing season is from April to September but about 85% of the growth is completed by mid-June (Willms *et al*. 1988).

About 50% of the foothills prairie dominated by *Festuca campestris* remains intact, the remainder is in crop agriculture or within the rural subdivisions that are rapidly spreading along the foothills of the Rocky Mountains in both Canada and the United States. On the plains where the fescue prairie is dominated by *Festuca hallii*, about 85-90% of these rich, black chernozemic soil areas are now in crop agriculture, mostly wheat, barley, canola and the alien forages of *Bromus inermis* Leyss. and *Medicago sativa* L.

Tallgrass Prairie

The tallgrass prairie occupies the eastern portion of the Great Plains and is the southern extension of the mesic prairie (Hart *et al*. 1996). It is bounded by the fescue prairie to the north, the more arid mixed grass steppe to the west, and the *Quercus* forest to the east. The tallgrass prairie primarily has plants that possess the C$_4$ photosynthetic pathway (warm season plants). The grasses include *Andropogon gerardii* Vitman and *Schizachyrium scoparium* (Michx.) Nash, *Panicum virgatum* L., *Sorghastrum nutans* (L.) Nash and *Sporobolus heterolepis* A.Gray. *Andropogon gerardii* is the most widespread and abundant. Soils are mostly of the Chernozemic order and are highly prized by crop agriculture. About 90% of the original tallgrass prairie is in annual crop agriculture, mostly maize, wheat and related crops. Only soils too shallow or too steep are spared from cultivation. The growing season is from April to October with most rapid growth in June and July (Weaver and Tomanek 1951).

Fire has a tremendous impact in maintaining the tallgrass prairie. All fires reduce the invasion of trees and shrubs. Spring fires suppress cool season plants favoring warm season plants (Anderson *et al*. 1970), while summer fires may shift species composition to favor cool season plants (Ewing and Engle 1988).

Mixedgrass Steppes

These steppes are usually described as the mixed-grass prairie. In the north, the cool season grasses, *Hesperostipa comata* (Trin.& Rupr.) Barkworth, *Agropyron smithii* Rydb., *Koeleria macrantha* (Ledeb.) J.A. Schultes f. , and *Poa* L. species, with *Nassella viridula* (Trin.) Barkworth in moister areas (Hart *et al*. 1996) and *Hesperostipa curtiseta* on deeper soils of more northern areas adjacent to the fescue prairie. In the southern mixed-grass steppe, the warm season mid-grasses *Schizachyrium scoparium* and *Bouteloua curtipendula* (Michx.) Torr. replace the northern cool season grasses. The mixed-grass steppe has a drier climate than the tall-

grass prairie. Most of the deeper, richer soils are now in crop agriculture leaving the driest, stoniest, roughest topography and saline sites as rangelands used by herbivores.

Shortgrass Steppe

This association is usually described as the shortgrass prairie but it is more appropriately described as steppe. It lies south of the northern mixed-grass steppe and west of the southern mixed-grass steppe in the rainshadow of the Rocky Mountains. The warm season shortgrasses *Bouteloua gracilis* (Willd. ex H.B.K.) Lag. ex Griffiths and *Buchloe dactyloides* (Nutt.) Engelm. are the principle dominant grasses (Hart *et al*. 1996). Cool season grasses are a minor component. The short grass steppe has a drier climate than the mixed-grass steppe. Many of the deeper soil areas are cultivated for wheat production.

Intermountain Region

East of the western mountain ranges that parallel the Pacific Ocean lies a vast area that has less precipitation because of the rainshadow effect of the western mountains. Much of it is occupied by temperate natural shrub-steppe.

Sagebrush-Grass Steppe

This steppe is widely distributed from southern British Columbia to Nevada and eastwards to Wyoming. The principal sagebrush species is *Artemisia tridentata* Nutt. and the main understory cool season bunchgrass is *Agropyron spicatum* (Pursh) Scribn. & Smith (Hart *et al*. 1996; Wright and Bailey 1982). There are other *Artemisia* species and also a number of cool season grasses. Annual precipitation ranges from 200-500 mm. Generally, as the density of shrubs increase, the productivity and density of grasses and forbs decreases. *Artemisia tridentata* is a non-sprouter and certain types of fires can kill most plants. Unlike the grasses on the Great Plains, *Agropyron spicatum* did not evolve under periodic heavy grazing (Hart *et al*. 1996). It is extremely sensitive to defoliation during spring. Overgrazing or frequent wildfire have allowed the alien annual grass *Bromus tectorum* L. to dominate millions of hectares (Wright and Klemmedson 1965).

Prairie

Only a small part of this region is occupied by prairie. The Palouse prairie is a comparatively small area in eastern Washington State of rich silt (loess) soils originally dominated by *Agropyron spicatum* and *Festuca idahoensis*. At least 90% of this natural grassland is now in crop agriculture, mostly growing wheat.

A fringe of *Festuca campestris-Agropyron spicatum* prairie acts as an ecotone at the upper margin of the Sagebrush-Grass steppe and the adjacent coniferous forest in south central British Columbia and adjacent Washington State. This prairie is closely related to the *Festuca campestris* prairie of the eastern Rocky Mountain foothills in Alberta and Montana. These prairies occur on black chernozemic soils, are very productive, have high fuel loads and are fire climax grasslands. A continuing invasion of scrubby conifers is occurring in this unique prairie during a century of fire suppression motivated by forestry and agricultural interests.

ORIGINS

Temperate natural grasslands of North America originated from early forested floras described by Dix (1964) as the Arctotertiary Flora of northern origin, which was composed of C_3 plants, and the Neotropical Tertiary Flora of southern origin, which was composed of C_4 plants. Drought assumed a major role in determining which species would survive the drier conditions during the Tertiary Period. As the Rocky Mountains arose 35 million years BP, a rainshadow was created on the plains to the east. Over 20 million years, the flora and fauna adapted to form grasslands; by 15 million years BP diverse herds of adapted herbivores grazed the plains (Nikiforuk 1994). It is assumed that the western grasslands, shrub-steppes, and forests of the intermountain and cordilleran region originated in a similar manner.

Mountain and continental glaciers of fairly recent geological time interrupted the long period of grassland domination in the northern half of North America, Asia and Eastern Europe. Once the ice began to recede, tundra was established (Pielou 1991) and many areas were flooded by melt water. Later, the grassland organisms that had survived in mountain refugia and to the south of the glaciers dispersed across the plains and interior valleys of western North America, and probably in a similar manner in Asia and eastern Europe. It remains unknown what proportion of the flora and fauna was lost due to glaciation.

The native temperate grasslands of North America have evolved and been influenced primarily by climate, herbivory and fire (Wright and Bailey 1982). In the Great Plains, extreme winter cold and moderate summer temperatures are the norm. Cycles of drought and non-drought have been very significant to the evolution of an adapted flora and fauna. Herbivory has always been on the plains (Nikiforuk 1994) and the vegetation is adapted to it. In more recent geological time periods, bison and grasshoppers have been the principle herbivores. West of the Rocky Mountains, there were frequently lower populations of herbivores and the grasslands are not as well adapted to grazing (Tisdale 1982). Thus, there is evidence that the temperate native grasslands of North America have had a long period of evolutionary history with only a few temporary periods of cold temperatures interrupting the northern sector with glaciation, tundra and forest vegetation (Dix 1964; Pielou 1991; Nikiforuk 1994). These grasslands have evolved without the aid or expense of cultural inputs such as energy and chemical fertilizers.

RECENT HISTORY

The Blackfoot confederacy were the undisputed rulers of the northern half of the Great Plains for centuries (Cruise and Griffiths 1996). They prevented white man's penetration of the temperate natural grasslands for 200 years. In the winter of 1792-93, the surveyor Peter Fidler traveled and lived with the Peigan Indians on the Alberta grasslands (Fidler 1792-93). He was impressed with the aboriginal survival skills and their extensive use of fire to manage the grasslands. Similarly, the Palliser expedition of 1857-1860 also observed the extensive use of fire by aboriginal man in the temperate grasslands (Spry 1968). James Hector, a member of the Palliser expedition described the high quality of forage provided by the native *Festuca hallii* grasslands

for wintering horses in central Alberta. In contrast, John Macoun, a professor from Ontario, and a member of a Geological Survey of Canada expedition, was anxious to plant 'improved' European hay and pasture species, thus replacing what he considered the inferior native species of the plains grasslands (Macoun 1882).

PERCEPTION AND MANAGEMENT MODELS

Model One: Exploitation (Modern Intensive Agriculture = an Old Roman Philosophy)

The current views and philosophies of modern agriculture date back over 2000 years. Since Roman times, European agricultural man has successfully destroyed the deciduous forests replacing them with cultivated crops in humid, maritime temperate climates where drought is rare. Plant and animal genetic material was successfully manipulated, and the soil was fertilized to raise the productivity and economic value of the agricultural products required by an 'advanced civilization'. This agricultural model is now being developed further by modern biotechnology methods. This intensive agriculture philosophy and mindset would not tolerate the Plains Indian tribes nor the natural grasslands, bison, grizzly bears, wolves and prairie fire that dominated the North American Great Plains. All of this 'nature in the raw' was 'uncivilized' and had to be subjugated or eliminated just as the Romans did in western Europe 2000 years ago (Reed 1954; Rackham 1990; Bailey and Bailey 1994).

North American temperate native grasslands were usually mismanaged by European colonizers. These rangelands have suffered from policy maker and settler assumptions and cultural biases (Bailey and Bailey 1994). Such biases include the assumption that the culture and concepts brought from the humid north temperate climates of Europe, eastern Canada or the eastern United States were superior and applicable to the continental climate of the plains and intermountain regions of the West. The West was assumed to be 'uncivilized'. They also assumed that natural grasslands were wastelands and this land must be cultivated and intensively managed for it to be valuable. A most potent assumption was that natural grasslands and native herbivores were inferior to introduced 'improved' cultivars or breeds. Macoun (1882) looked forward to the day when the Canadian prairie 'wasteland' would be civilized by European settlement and crop farming.

The Palliser Expedition of the 1850's warned the Canadian government regarding the high frequency of droughts on the Great Plains, and recommended against intensive agriculture in the region (Spry 1968). Federal policy makers essentially ignored their recommendations. The land grant and land valuation policies used in western Canada ensured that most natural grasslands would be plowed (Martin 1973). The land management policies applied to the settlement of the temperate natural prairie and steppes of the Great Plains and the Intermountain Regions of the West were designed by people who had little knowledge of these ecosystems. The European agricultural model was applied to plains grassland ecosystems. Negative repercussions resulting from many of these ill-advised decisions helped to create the environmental deteriorations of the 1930's and continue to affect the remnant temperate natural grasslands to this day.

The remnant temperate natural grasslands of western North America are important now to the range livestock industry, as habitat for all wildlife and as space for the grassland itself, for watersheds and for recreational opportunities. Traditionally, over the last 40 years their replacement by *Agropyron cristatum* (L.) Gaertn. and *Elymus junceus* Fisch. and other European grasses was considered an improvement (Dormaar *et al.* 1995). However, research has demonstrated that these monocultures reduce soil quality due to a shift in carbon partitioning from below ground to above ground, increased nutrient export, reduced energy flow and decreased organic matter input. Dormaar *et al.* (1995) recognized the value of these planted alien monocultures to the livestock industry but recommended that there be limits placed on acreages since soil sustainability is more important than short-term forage production.

The percentage of natural grasslands left uncultivated is very low. Today, about 30-60% of temperate native grasslands remain uncultivated in the drier ecosystems while as little as 1-15% remain in most higher moisture native grassland areas. However, Harrison *et al.* (1997) have found that on the most productive soils derived from grasslands in Saskatchewan, as little as 0.01% remains in the natural grassland state.

Model Two: Preservation

The preservation movement is currently very popular in the western world and its proponents refer to themselves as 'environmentalists'. The parks services and private preservation agencies have the objective of containing ('preserving') a large portion of the remnant temperate natural grasslands, including both prairies and steppes, in North America. This goal of preserving natural ecosystems is a noble concept but it often lacks resource management and ecological expertise. Hummel (1995) argued that '…We have a once-only opportunity to ensure that significant parts of our country remain in a wild, natural state, changing only at the hands of nature, and serving as benchmarks for measuring the changes we are making to so much of the rest of our lands and waters…'. Preservationists tend to ignore the current exploitation of many 'conservation areas' that are no longer in a wild, natural state. In Canada at Jasper and Banff National Parks as well as in Yellowstone National Park in the U.S.A., concerned range managers frequently demonstrate wild ungulate overgrazing of the natural grasslands. Overgrazing has predisposed these temperate natural grasslands and shrublands to reduced biomass productivity, reduced biodiversity, reduced fuel load, reduced fire frequency, increased alien weed infestations, increased domination by European grasses, and increased soil erosion. This is the case where herd animals such as bison, wapiti and big horn sheep overgraze during the growing season. In spite of predation in the mountain parks by large wolf and cougar populations as well as by black and grizzly bears, the populations of these wild ungulates is rarely within the carrying capacity of the natural grassland rangeland ecosystems. Vegetation and soil deterioration will not stabilize until management limits wild ungulates to the carrying capacity of the ecosystem. Such management changes a preservation model into an exploitation model.

It is apparent that preservation without management is no longer a choice because man has interfered with too many natural eco-

logical processes. Simulation of some ecological processes is required even in parks. The temperate natural grasslands evolved under specific ecological and environmental conditions that were always dynamic, and never static. These grasslands were not only adapted to cycles of drought and non-drought for about 35 million years, they also came and went after various glaciations, occupation by tundra and forest during cold cycles (Pielou 1991) and also endured low and high population cycles of mammalian, insect and rodent herbivory, as well as prairie fire. It was not all lightning-caused wildfire either, aboriginal peoples manipulated fire for their own purposes and the vegetation adapted to it over thousands of years. The existing national parks rarely have sufficient safeguards to prevent temperate natural grasslands from continuous overgrazing in the growing season by wild ungulate herbivores. Ecosystem preservation without adequate implementation of rangeland resource management policies is not responsible conservation.

Model Three: Conservation and Stewardship

Between the two extremes of exploitation by agronomists and cropland farmers (Model 1) on the one hand and preservationists (Model 2) on the other, lies a more moderate, less vocal group of conservationists, that is described as Model 3. Conservationists recognize the natural worth of the temperate natural grasslands, and also recognize the practical necessity of these ecosystems to produce some kind of goods and services. This recognition is both ecologically based and practically oriented. Herbivory and fire are necessary parts of the natural grassland ecosystem and humans must be wise and responsible stewards of the land. On the other hand, the land must be managed in a manner that is within the ecological amplitude of the ecosystem. The concern remains for these grassland ecosystems to be wisely used without either destroying them by cultivation or exploiting them by overgrazing. Conservationists consider themselves to be responsible rangeland managers and they represent both the private and the public sector.

Throughout the 100 to 150 year occupation of temperate natural grasslands by European man in Canada, there have been a few who placed great value on these grasslands. One of the first was Dr. James Hector of the Palliser Expedition who observed in 1858 that the lush native *Festuca hallii* prairie grasslands of central Alberta were of exceptional forage quality for providing sustenance to the expedition's horses throughout the five to six month winter period (Spry 1968).

Today, there are increasing numbers of conservationists who have become impatient and frustrated with both the disregard of agronomists and crop agriculture towards the merits of natural grasslands, and preservationists not conversant with the ecological requirements of temperate natural grasslands. Management of a temperate natural grassland requires adequate knowledge of ecosystem process and function. This includes a willingness to actively manage these grasslands. It is unfortunate that the proponents of Models 1 and 2 generally have not been willing to invest the time to understand the ecological requirements of temperate natural grassland. The flora and fauna have been manipulated by millions of years of evolution, rather than by scientists in laboratories or small plots. Ironically, periodic over-population of bison on the plains in previous centuries probably contributed towards the biological resilence and biodiversity of today's temperate natural grasslands.

THE FUTURE OF TEMPERATE NATURAL GRASSLANDS

At a recent conference in Edmonton Dr. Dennis Avery of the Center for Global Food Issues at Hudson Institute in Minneapolis, Minnesota U.S.A. argued that without high-yield, modern agriculture, there is no environment, no wildlife, there is nothing to save. In contrast, the urban studies professor Frank Popper of Rutgers University advocated that the Great Plains be allowed to revert to their natural 'pre-white' condition by destocking the crop farmers and replanting short grass and restocking buffalo (Harrison 1990). Popper believes that the Great Plains will become almost totally depopulated as a result of '…the largest, longest-running agricultural and environmental miscalculation in American history…'. Avery appears to have no concept of natural grasslands or habitat. Popper does not acknowledge the substantial destruction of a high proportion of Great Plains temperate natural grassland, and the demise of the bison, wolves, grizzly bear and the nomadic tribal people along with their use of fire to manage the resource. These may be considered extreme views of the future of temperate natural grasslands.

In Asia where traditional pastoral people, such as in Mongolia, continue to use these grasslands as has occurred for centuries, the future looks bright for them and the grasslands. A key reason is the hostile climate and the short growing season that produce failure in intensive agricultural crop experiments. In Europe, only a few remnant stands will remain largely known and hopefully untouched.

In North America the greatest hope for the maintenance of broad expanses of temperate natural grasslands lies with the traditional ranchers who need to carry on managing these rangelands as has happened for centuries. Ranchers, however, have problems with low cash flow in relation to artificially high land value pressures coming from wealthy urban areas that enables the rich to purchase these lands for investment or recreational purposes. There will likely be great efforts by naturalist groups to lock up more examples of temperate natural grasslands in parks where they are often unwisely managed. The failure of these groups to understand the necessity of periodic grazing and burning on these grasslands and the need to maintain herbivory within the carrying capacity for the range will continue to produce expensive and unnecessary failures. There is a serious threat to some grasslands posed by alien weeds including forb species of the genus *Centaurea* L., and *Euphorbia esula* L., as well as the annual grasses *Bromus tectorum* L. and *Elymus caput-medusae* L. These alien weeds tend to exclude natural grassland species. Biological control is widely touted, but to date there are more failures than successes.

Only in recent decades has serious research been supported enabling the development of a greater understanding of the sustainability, productivity, forage quality and biodiversity of temperate natural grasslands. There needs to be an expansion of research support available to demonstrate the economic and ecological value of these ancient, and sustainable, natural grasslands. The

heritage value of these grasslands for future generations needs to be impressed upon both the general population and the leadership of each country.

In most first world countries there is little legislation or government policy in place to prevent private landowners, corporations and government agencies from destroying temperate natural grasslands. Similarly, little or no legislation or policy is in place to prevent government agencies, such as highways departments, and private utility corporations, from expropriating land occupied by natural grasslands for the expansion of new highway, pipeline or utility corridors, or for dam construction. Nor is there adequate legislation and policy governing the protection of rare natural grasslands from urban and industrial expansion. It is necessary for conservation and preservation lobbyists to motivate the initiation of government legislation and policy favourable to the maintenance of existing temperate natural grasslands.

Future generations will look back with disbelief at the extremely low level of recognition given by this generation, and our European ancestors, to the remnant temperate natural grasslands. Concerned citizenry need to arm themselves with knowledge to prevent the exploitation of the last remnants of temperate natural grasslands.

Private preservationist groups continue to buy remnant natural grasslands and the urban public generally favours more parks in these grasslands. Normally, accompanying such acquisitions is the absence of an effective management plan to simulate the environmental forces required to maintain these ecosystems. Although temperate natural grasslands are at the top of the list for many preservation activists, these same activists have little knowledge or appreciation of the role current grazing management and prescribed burning on private and public natural grasslands. More discussions about methods of actively managing these private grasslands are necessary, if the ecosystems are to survive.

In future, more attention needs to be placed upon overgrazing of temperate natural grasslands, especially by uncontrolled herbivore populations in national parks. Overgrazing can precipitate the permanent loss of topsoil through the forces of erosion. It also increases the proportion of shallow-rooted species, that are less productive but more resistant to grazing pressure, and there is an accompanying decline in soil organic matter (Dormaar and Willms 1990). Overgrazing causes reductions in litter, reduced water infiltration, higher soil temperatures (Naeth *et al.* 1991), and thus artificial drought. Overgrazing during the growing season has more serious effects on grasslands than do similar grazing pressures during the dormant season.

Prescribed and uncontrolled natural fire needs to be reintroduced where feasible because these grasslands evolved under a burning regime. Burning promotes high biodiversity, a concept recognized as being important in ecosystems (West 1993) by both preservationists and conservationists alike. The effects of 100 years of no burning is unknown. However, it is known that burning does kill many over-wintering insects and diseases. Although preservationists wish to promote high species biodiversity, some of their actions have the opposite effect. The locking up of pristine native grasslands may not lead to their successful conservation. These grasslands are not as fragile as some preservationists think, and moderate grazing use and burning may conserve their diversity more effectively than either non-use or extreme overgrazing.

Preservationists generally tend to lobby big government to legislate protection as represented by the Endangered Species Act in the United States and similar legislation soon to be enacted in Canada. The effectiveness of this type of legislation has been questioned by many. As West (1997) indicated 'The Endangered Species Act is another of the top-down, command and control approaches to conservation that came largely from Washington D.C. Such legislation and the bureaucratic attitudes they generate could be foisted off on public land states in the West, whether the locals liked it or not.' This is much harder to do where there are vast tracts of private land, as on the Great Plains.

In 'conservation areas', policies and management decisions that have caused a deterioration of temperate native grasslands need to be revised to accommodate the requirements of entire ecosystems, as opposed to the over emphasis of a few large herbivore or predator species, of tourism business interests, or of cultural biases of the country's human population. There is an urgent need to re-educate national parks policy makers, managers and support groups in basic range management principles in order for them to understand the urgency of effective conservation of temperate native grasslands.

The developed world is experiencing an incredibly rapid rate of change in technology that affects all facets of human life. Many people are reacting in anger and frustration as they find themselves victims of changing times. Recently, there has been great upheaval for users of temperate natural grasslands. As a result of the declining population of commercial rangeland users, such as ranchers and residents of rural areas, more and more influence is being felt by government decision-makers from urban-based lobby groups whose personal livelihoods are not directly affected by the changes for which they lobby. Nevertheless, new information and new societal priorities will precipitate changes in both legislation and policies regarding the use of temperate natural grasslands.

The Search for Truth

Scientists generally confine their search for new knowledge and for 'the truth' to a relatively narrowly defined discipline. Normally, this is the discipline in which they have received their postgraduate education. How many scientists regularly conduct an intensive personal scrutiny into the assumptions and biases within their discipline, as well as within their own mind, and moral values? Few scientists are aware of their personal biases. Most scientists feel that because they have been trained in a discipline, and have used the scientific method in their research endeavours, that they are honorable and consistently tolerant of other points of view. Is this attitude sometimes in error? Should each scientist face the reality that each of us comes with our own unique upbringing, social and cultural orientation, perceptions, assumptions and biases? Specialists in one discipline frequently misunderstand another specialist's point of view because each has failed to identify and communicate adequately. For example, most agronomists assume that their new grain cultivar or forage cultivar will consistently produce more grain or forage, and have no negative effects on the soil and other parts of the ecosystem.

Thus, this justifies their position that it was correct for crop agriculture to destroy most of the temperate natural grasslands. However, Dormaar et al. (1995) have demonstrated that some of the most widely used temperate grass monocultures reduce soil quality. Similarly, most preservationists assume that temperate natural grasslands can be locked up for decades in the absence of appropriate herbivory and fire management regimes without losing biodiversity. Leach and Givinish (1996) have demonstrated otherwise. Preservationists seem to see nothing wrong with encouraging the overpopulation of their favorite mammalian herbivore, to excessively graze in national parks to the extent of the elimination of dominant natural grasses, the invasion of alien weeds and subsequent high levels of erosion. In North America, agronomists, preservationists and most conservationists alike seem to be petrified regarding the natural role of fire in the temperate natural grassland ecosystems. They are taking no leadership role in encouraging the development of suitable legislation, policy and the municipal infrastructure that is necessary to enable the safe and effective reintroduction of prescribed fire into the natural ecosystems, whether these be on private ranchland, rural acreages, private or public parks.

CONCLUSIONS

In North America, there were decisions made by policy makers based on eastern American and Canadian self-interest, which resulted in the demise of a much larger proportion of the temperate natural grasslands than was economically and ecologically warranted. The settlers who farmed the Great Plains and intermountain valleys did need to plant some crops to earn a livelihood and to provide winter fodder for livestock. However, the attitudes expressed so bluntly by Macoun (1882) were based upon assumptions that have caused difficulties for every European colonial power.

The tide is turning from agrarian disregard for temperate natural grasslands (Model 1) towards an acceptance of these ecosystems as being economically viable as well as of heritage value. Many preservationists (Model 2) are eager to lock up more native grasslands in parks where grassland management is often inadequate. There is also a more moderate group (Model 3) who are concerned about both conservation and wise utilization of this valuable resource. Greater interest is being displayed towards the utilization of native plant species to re-establish more temperate native grasslands. However, as the human population continues to increase, policy changes should be made to require the planning and management of all temperate native grassland ecosystems regardless of ownership. It needs to be acknowledged that fire and herbivory are the primary ecological tools available for the sustainable management of all temperate native grassland ecosystems. It has been this way since time began. The most sensitive and difficult policy issues for managers on both private and public natural grasslands is related to the coordination of the diverse interest groups demanding access to, and influence on, the management of these resources.

ACKNOWLEDGEMENTS

Professor R.D.B. Whalley, Department of Botany, University of New England, Armidale, N.S.W. is acknowledged for his organization of the Grassland of the World session, and for his efforts in obtaining financial assistance for the author. The Cooperative Research Centre for Weed Management Systems is acknowledged for partial funding of the author's travel to Australia. The editors of the symposium are acknowledged for their patience and understanding in dealing with the controversy associated with this subject matter.

REFERENCES

Anderson, K.L.., Smith, E.F., and Owensby, C.E. (1970). Burning bluestem range. *Journal of Range Management* **28**, 81-92.

Bailey, R.G. (1996). Ecosystem geography. (Springer-Verlag: New York.)

Bailey, A.W., and Bailey, P.G. (1994). The traditions of our ancestors influence rangeland management. *Rangelands* **16**, 29-32.

Coupland, R.T. (1961). A reconsideration of grassland classification in the Northern Great Plains of North America. *Journal of Ecology* **49**, 135-167.

Cruise, D., and Griffiths, A. (1996).The great adventure: how the Mounted Police captured the west. (Penguin: Toronto.)

Dix, R.L. (1964). A history of biotic and climatic changes within the North American grassland. In 'Grazing in terrestrial and marine environments'. (Ed. D.J. Crisp) pp. 71-89 (Blackwell Scientific Publications: Oxford.)

Dormaar, J.F., Naeth, M.A., Willms, W.D., and Chanasyk, D.S. (1995). Effect of native prairie, crested wheatgrass (*Agropyron cristatum* (L.) Gaertn.) and Russian wildrye (*Elymus junceus* Fisch.) on soil chemical properties. *Journal of Range Management* **48**, 258-263.

Dormaar, J.F., and Willms, W.D. (1990). Sustainable production from the rough fescue prairie. *Journal of Soil and Water Conservation* **45**, 137-140.

Ewing, A.L., and Engle, D.M. (1988). Effects of late summer fire on tallgrass prairie microclimate and community composition. *American Midland Naturalist* **120**, 212-223.

Fidler, P. (1991). Journal of a journey over land from Buckingham House to the Rocky Mountains in 1792 & 93. In 'A southern Alberta bicentennial: a look at Peter Fidler's journal, 1792-93'. (Ed. B. Haig) (Historic Trails West Ltd: Lethbridge.)

Harrison, E. (1990). Return Great Plains to nature, prof urges. Chicago Sun-Times, July 1, 1990.

Harrison, T., Reimer, G., and Lynn, N. (1997). An assessment of prairie remnants in highly cultivated ecoregions of Saskatchewan, p.76. In 'Abstracts, 50th Annual Meeting, Society For Range Management, Rapid City, South Dakota, February 16-21, 1997'.

Hart, R.H., Willms, W.D., and George, M.R. (1996). Cool-season grasses in rangelands. In 'Cool-season forage grasses'. (Eds L.E. Moser, D.R. Buxton, and M.D. Casler) pp. 357-381. Agronomy Monograph 34. (American Society of Agronomy, Crop Science Society of America, and Soil Science Society of America. Madison.)

Hummel, M. (ed.) (1995). Protecting Canada's endangered spaces. (Key Porter Books: Toronto.)

Leach, M.K., and Givnish, T.J. (1996). Ecological determinants of species loss in remnant prairies. *Science* **273**, 1555-1558.

Macoun, John. (1882). Manitoba and the Great North-west: field for investment; the home of the emigrant, being a full and complete history of the country. (World Pub. Co.)

Martin, Chester. (1973). 'Dominion Lands' policy. (Carlton Library No. 69., McClelland and Stewart: Toronto.)

Moss, E.H. (1955). The vegetation of Alberta. *Botanical Review* **21**, 493-567.

Naeth, M.A., Chanasyk, D.S., Rothwell, R.L., and Bailey, A.W. (1991). Grazing impacts on soil water in mixed prairie and fescue grassland ecosystems of Alberta. *Canadian Journal of Soil Science* **71**, 313-325.

Nikiforuk, A. (ed.) (1994). The land before us: a geological history of Alberta. (Tyrell Museum of Paleontology. Red Deer College Press: Red Deer.)

Pielou, E.C. (1991). After the ice age: the return of life to glaciated North America. (University of Chicago Press: Chicago.)

Rackham, O. (1990). Trees and woodland in the British landscape. (Dent: London.)

Reed, J.L. (1954). Forests of France. (Faber: London.)

Spry, I. M. (ed.) (1968). The papers of the Palliser expedition 1857-1860. (The Champlain Society: Toronto.)

Tisdale, E.W. (1982). Grasslands of western North America: the Pacific Northwest bunchgrass. In 'Grassland ecology and classification, Symposium Proceedings'. (Eds A.C. Nicholson, A. McLean, and T.E. Baker) pp. 223-245 (British Columbia Ministry of Forests Publication No.R28-82060.)

Walker, B.H., and Coupland, R.T. (1970). Herbaceous wetland vegetation in the aspen grove and grassland regions of Saskatchewan. *Canadian Journal of Botany* **48**, 1861-1878.

Walter, H. (1973). Vegetation of the earth in relation to climate and the ecophysiological conditions. (Springer-Verlag: New York.)

Weaver, J.E., and Tomanek, G.W. (1951). Ecological studies in a midwestern range: The vegetation and effects of cattle on its composition and distribution. Nebraska University Conservation & Survey Division Bulletin No. 31.

West, N.E. (1997). Book review of F.B. Samson and F.L. Knopf. (ed.) Prairie conservation: preserving North America's most endangered ecosystems. Island Press. *Journal of Range Management* **50**, 110.

West, N.E. (1993). Biodiversity of rangelands. *Journal of Range Management* **46**, 2-13.

Willms, W.D., Smoliak, S., and Schaalje, G.B. (1988). Forage production and utilization in various topographic zones of the fescue grasslands. *Canadian Journal of Animal Science* **68**, 211-223.

Wood, V.A. (1951). Alberta's public land policy, past and present. *Journal of Farm Economics* **33**, 735-749.

Wright, H. A., and Bailey, A.W. (1982). Fire ecology of the United States and southern Canada. (Wiley: New York.)

Wright, H.A., and Klemmedson, J.O. (1965). Effects of fire on bunchgrasses of sagebrush-grass region in southern Idaho. *Ecology* **46**, 680-688.

Zoltai, S.C., and Pollett, F.C. (1983). Wetlands in Canada: their classification, distribution, and use. In 'Mires: swamp, bog, fen and moor'. (Ed. A.J.P. Gore) (Elsevier: Amsterdam.)

ENDEMIC GRASS GENERA OF MAHARASHTRA STATE, PENINSULAR INDIA

C.B. Salunkhe

Department of Botany, Krishna Mahavidyalaya, Shivnagar, Rethare (BK.) 415 108 (M.S.) India.

Abstract

The ubiquitous grasses, which form one of the largest families of flowering plants, comprise about 700 genera and nearly 10,000 species. The family Poaceae is represented in India by about 262 genera and 1110 species of which 225 species and 21 varieties are endemic to Peninsular India. At generic level, as many as 14 genera are endemic to Peninsular India of which 11 are monotypic. The monotypic genera are *Chandrasekharania, Danthonidium, Hubbardia, Indopoa, Limnopoa, Manisuris, Pogonachne, Pseudodichanthium, Silentvalleya, Trilobachne* and *Triplopogon*. The other three genera are *Bhidea* with three species, *Glyphochloa* with eight species and three varieties and *Lophopogon* with two species. Most of the genera are palaeoendemics of phytogeographical importance. The monotypic genus *Hubbardia* may have become extinct in the recent past. Most of the endemic genera of Peninsular India are retracting and confined to small geographical areas (ecological niches) without any sign of active evolution, indicating their last phase of natural death and extinction. However, some of them, viz. *Bhidea* and *Glyphochloa*, have become active epibiotics under changing climatic and edaphic conditions in Peninsular India. Taxonomically difficult genera like *Arthraxon, Dimeria, Isachne* and *Ischaemum* have a number of closely related species with intermediate forms between relatives that is indicative of their diverse genetic stock and evolution at work. The present paper describes morphology, taxonomy, interrelationships and evolution of some of the endemic grass genera of Maharashtra State, Peninsular India.

Key words: Grasses, India, Andropogoneae, biogeography.

INTRODUCTION

The family Poaceae, represented by about 700 (793 according to Watson and Dallwitz 1992) genera and nearly 10,000 species, is distributed throughout the world encompassing over 24% of the earth's vegetation cover. In addition to its great economic importance and role in human life and civilization, it is known for its ubiquitous distribution, wide ecological amplitude, complex reproductive structures, great adaptability and little understood evolutionary relationships. Of the 700 genera 390 are monotypic or ditypic (Watson and Dallwitz 1992). The taxonomic position, interrelationships and evolution of a number of such isolated genera are little known and understood. Comparative molecular studies may provide more definitive answers for many of these monotypic genera. About 262 genera and 1110 species of grasses grow in India (Karthikeyan *et al.* 1989) of which 30% (360 species) are endemic to India (Jain 1986). There are 17 genera endemic to India of which three grow in the Himalayan Region, while the remaining 14 grow in Peninsular India. Of the 17 endemic genera 14 are monotypic, one is ditypic, one tritypic and another has eight species and three varieties [Appendix]. About 225 species (Nayar 1996) and 21 varieties (Ahmedullah and Nayar 1987) are endemic to Peninsular India.

About 50% (93 species) of the endemic species of Peninsular India belong to the tribe Andropogoneae representing a high degree of endemism of the tribe in the region. The majority of the endemic genera of Peninsular India are poorly known in their taxonomy, anatomy, embryology, relationships, phylogenetic position and evolution. The endemic genera of Peninsular India are: *Bhidea* (3), *Chandrasekharania* (1), *Danthonidium* (1), *Glyphochloa* (8), *Hubbardia* (1), *Indopoa* (1), *Limnopoa* (1), *Lophopogon* (2), *Manisuris* (1), *Pogonachne* (1), *Pseudodichanthium* (1), *Silentvalleya* (1), *Trilobachne* (1) and *Triplopogon* (1). Many of the endemic grass genera of Peninsular India have a restricted distribution, some of them are known only from their type locality and a few of them have not been collected for a long time and probably have become extinct before we had the chance to understand their taxonomy, relationships and evolution. The present paper deals with the nine endemic grass genera that grow in the State of Maharashtra, their distribution, taxonomy and ecology towards a better understanding of their systematics and evolution. Genera like *Pogonachne, Pseudodichanthium, Trilobachne*, and *Triplopogon* are exclusively known from Maharashtra.

I) BHIDEA Stapf ex Bor

B. burnsiana Bor Fig. 1.

Annuals. Culms 15-35 cm high, tufted, erect, upper nodes villous, lower glabrous. Leaves: sheaths compressed; ligule membranous, truncate; blades lanceolate, 2-10 cm long, 0.1-0.3 cm wide, acute. Racemes 2, 2-5 cm long, shortly exserted from the spathes; joints linear-turbinate, 2.5-3 mm long, densely villous on one side. Spikelets 2-nate. Sessile spikelets lanceolate, 7.5-8.5 mm long (including an arista), awned; callus 0.8-1 mm long, villous. Lower glume coriaceous, lanceolate, 6-7 mm long, obscurely 3-nerved, keels broadly winged; wings hyaline, unequal, acuminate. Upper glume subcoriaceous, oblong-elliptic, 7-8 mm long (including 4-5 mm long arista), obscurely 3-nerved; apex 3-lobed, deeply cleft, 1-keeled, keel running into a 4-5 mm long arista from a sinus. Lower lemma hyaline, ovate-lanceolate, 3.2-4 mm long, 3-nerved, 2-keeled; keels broadly winged, acute; epaleate. Upper lemma subcoriaceous, oblong-elliptic, 3-4 mm long, nerves obscure; apex notched, awned in the cleft; awn geniculate, 15-22 mm long, scaberulous. Palea hyaline, ovate, 1-1.5 mm long, obtuse. Lodicules 2. Stamens 3; anthers 1-2 mm long. Ovary linear, 0.5 mm long; styles 1 mm long; stigmas 1-1.5 mm long. Grain oblong-elliptic, 1.5-2 mm long. Pedicelled spikelets lanceolate, oblique, empty, 6-8 mm long; pedicels linear, 2-3 mm long, densely villous on one side. Lower glume chartaceous, oblique-lanceolate, 6-7 mm long, obscurely 3-5-nerved, broadly winged on one side, acute. Upper glume chartaceous, lanceolate, 7-8 mm long, obscurely 3-nerved, keeled, acuminate. Lower lemma hyaline, ovate-lanceolate, 3-4 mm long, 1-nerved, obtuse. Data on fruit morphology and cytology is wanting.

Ecology, Geography and Regional Distribution

Bhidea is represented by three species in Peninsular India. They grow on lateritic plateaus ranging between 50-1200 m altitude. They mainly grow in crevices of lateritic rocks and in the sparsely accumulated soil layer on lateritic slabs especially in coastal plains of Konkan, Goa , Karnataka and Kerala. They grow in extremely poor soils of lateritic plateaus rich in iron.

II) DANTHONIDIUM C.E. Hubb.

D. gammiei (Bhide) C.E. Hubb. Fig. 2.

Danthonia gammiei Bhide

Annuals. Culms tufted, 10-45 cm high; nodes glabrous. Leaves: sheaths glabrous, striate; ligule a fimbriate membrane; blades linear, 1.5-7 cm long, 0.1-0.26 cm wide. Inflorescence spiciform, ovate-oblong, 1-5 cm long; peduncles puberulous; rachis hairy. Spikelets oblong-lanceolate, 10-20 mm long; pedicels short. Lower glume chartaceous, oblong-lanceolate, 10-20 mm long, 5-nerved, aristate. Upper glume chartaceous, oblong-lanceolate, 8-12 mm long, 3-nerved, aristate. Lemmas coriaceous, oblong, 2.5-5 mm long, 7-9-nerved, convolute, base tapering, bearded with long hairs, 3-awned at apex; lateral awns shorter, 4-7 mm long, median awn geniculate, 10-25 mm long with tuft of hairs on either side, column twisted, bristle scaberulous. Palea hyaline, oblong-lanceolate, 2.4-4 mm long, 2-keeled; keels ciliate above; apex notched, 2-toothed. Rachilla spathulate, 1-1.8 mm long, ciliate along the margins. Lodicules 2. Stamens 3; anthers 1-2 mm long. Ovary linear, 1-1.2 mm long; styles 1 mm long; stigmas 1-1.2 mm long. Grain elliptic-obovate, 2.6-3 mm long. Data on fruit morphology and cytology is wanting.

Ecology, Geography and Regional Distribution:

Bhidea, Danthonidium and *Glyphochloa* grow together; mainly in crevices of lateritic rocks or on thin soils accumulated on lateritic plateaus of coastal plains of Konkan, Goa, Karnataka and Kerala. They grow in the rainy season and, as rain recedes, the plants flower and complete their life cycle. *Danthonidium gammiei* is sparsely distributed.

III) GLYPHOCHLOA W.D. Clayton

G. forficulata (Fischer) W.D. Clayton Fig. 3.

Manisuris forficulata Fischer

Annuals. Culms tufted, 5-30 cm high, slender, erect or ascending, glabrous or sparsely villous; nodes glabrous. Leaves: sheaths compressed, glabrous or sparsely villous; ligule membranous, truncate; blades linear-lanceolate, 3-15 cm long, 0.2-0.5 cm wide, glabrous or densely covered with tubercle-based hairs. Racemes spicate, solitary, 2-6 cm long; peduncles villous; joints clavate, 1.5-2.5 mm long. Sessile spikelets ovate-oblong, 4-6 mm long (including wings); callus glabrous. Lower glume coriaceous in lower half, ovate-oblong, 4-6 mm long, 5-7-nerved with or without tubercles in lower half, hooked on the margins, broadly winged on both sides above the middle, 2-cleft, 2-awned; awn 3.5-6.5 mm long. Upper glume chartaceous, oblong-elliptic, 2.5-3.5 mm long, 3-nerved, acute. Lower lemma empty, hyaline, oblong-lanceolate, 2.5-3 mm long, obscurely nerved, obtuse. Palea hyaline, oblong, 2-2.8 mm long, obtuse. Upper lemma hyaline, elliptic-ovate, 2-2.5 mm long, obscurely nerved, subacute. Palea hyaline, elliptic-lanceolate, 1.5-2 mm long, subacute. Lodicules 2. Stamens 3; anthers 0.5-0.8 mm long. Ovary oblong, 0.25-0.6 mm long; styles 0.6-1 mm long; stigmas 0.6-1 mm long. Grain elliptic, 1.6-2.8 mm long. Pedicelled spikelets oblong-lanceolate, 4-6 mm long (including wings), awned; pedicels fused with the joint. Lower glume chartaceous, oblong-

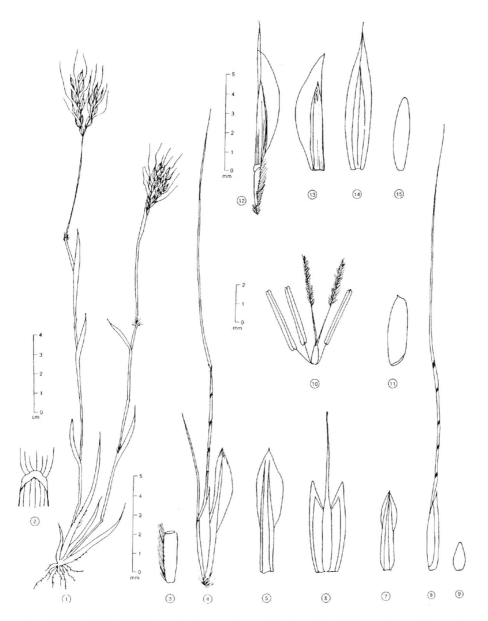

Fig. 1. *Bhidea burnsiana* 1) Habit 2) Ligule 3) Joint. 4-11: Sessile spikelet 4) Sessile spikelet 5) Lower glume 6) Upper glume 7) Lower lemma 8) Upper lemma 9) Palea 10) Stamens and pistil 11) Caryopsis. 12-15: Pedicelled spikelet 12) Pedicelled spikelet 13) Lower glume 14) Upper glume 15) Lower lemma.

lanceolate, 4-6 mm long (excluding awn), 5-7-nerved, margins winged on one side, awned; awn 5-8 mm long. Upper glume chartaceous, boat-shaped, 3.5-5 mm long, 3-nerved, keeled; keel winged; awned. Lower lemma empty, hyaline, oblong-lanceolate, 2-3 mm long, obscurely 2-nerved, obtuse. Palea hyaline, ovate-lanceolate, 1.4-1.6 mm long, subacute. Upper lemma hyaline, oblong-lanceolate, 2-2.5 mm long, obscurely nerved, obtuse. Palea oblong, 1.8-2 mm long, subacute. Data on anatomy and cytology is wanting.

Ecology, Geography, Regional Distribution

Glyphochloa is represented by eight species and three varieties. Of the eight species *G. forficulata* has a wide ecological amplitude, growing up to 1200 m altitude. All the species grow on lateritic rocks on extremly thin and poor soil. Various species have succulent leaves, an adaptation for aridity during dry spells in the rainy season. All the eight species grow in Ratnagiri and Sindhudurg districts of Maharashtra. Various species show great variations with respect to size, shape and ornamentation of the lower glume.

IV) *INDOPOA* Bor

I. paupercula (Stapf) Bor Fig. 4.

Tripogon pauperculus Stapf in Hook.

Annuals. Culms densely tufted, 5-12.5 cm high, erect; nodes glabrous. Leaves: sheaths compressed, striate; ligule thinly membranous, obtuse; blades involute, setaceous, 1.5-6 cm long, 0.1-0.2 cm wide, acute. Racemes spiciform, 3-8 cm long; rachis

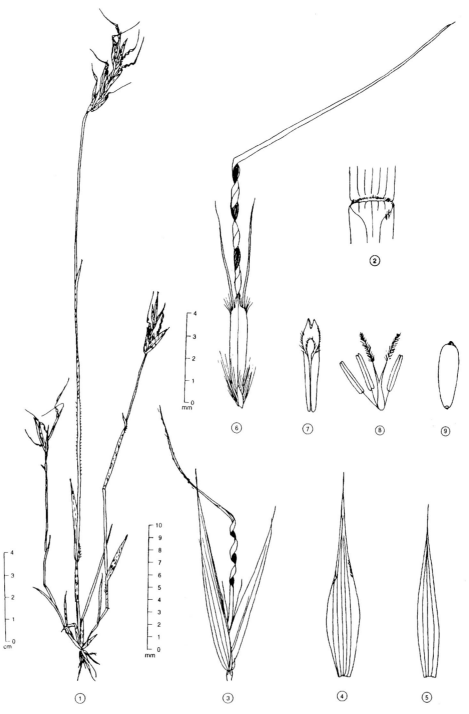

Fig. 2. *Danthonidium gammiei* 1) Habit 2) Ligule 3) Spikelet 4) Lower glume 5) Upper glume 6) Lemma 7) Palea with rachilla 8) Stamens and pistil 9) Caryopsis.

triquetrous. Spikelets ovate-linear, 3-10, distant, 7-11 mm long (excluding awns), 2-5-flowered. Lower glume thinly membranous, lanceolate, 2.5-4 mm long, 1-nerved, keeled, asymmetric at the tip, acute. Upper glume thinly membranous, lanceolate, 6-7 mm long, 1-nerved, keeled, mucronate. Lemmas thinly membranous, linear-oblong, 5-6.2 mm long, bearded at base; apex 2-lobed, deeply cleft, 3-awned; median awn in the cleft geniculate, twisted below the knee, 12-13.2 mm long, minutely scaberulus; lateral awns shorter. Palea hyaline, oblanceolate, 4.2-4.6 mm

long. obscurely 2-nerved, 2-keeled; keels ciliolate, obtuse. Lodicules 2. Stamens 3, anthers 0.3-0.4 mm long, subglobose. Ovary oblong, 0.4-0.5 mm long; styles 0.4-0.5 mm long; stigmas white, 0.3-0.4 mm long. Grain linear, 2-3.5 mm long.

Ecology, Geography and Regional Distribution

I. paupercula grows with *Bhidea, Danthonidium* and *Glyphochloa*. It extends from the coastal line of the Arabian Sea to the Western Ghats. It grows on lateritic rocks with or without accumulated

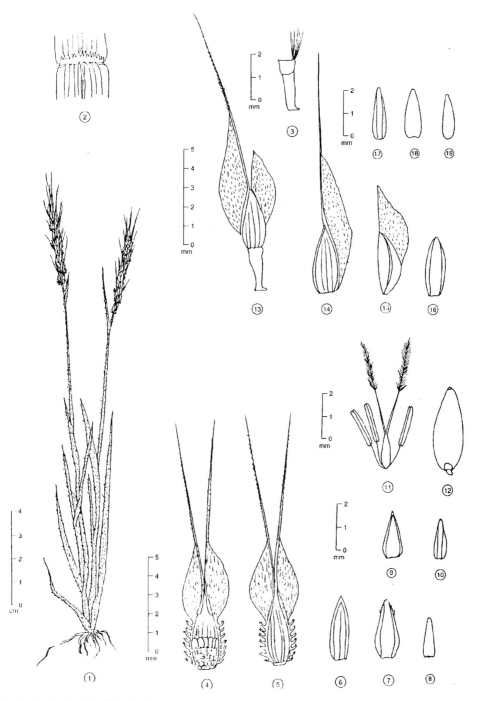

Fig. 3. *Glyphochloa forficulata* 1) Habit 2) Ligule 3) Joint. 4-12: Sessile spikelet 4) Sessile spikelet 5) Lower glume 6) Upper glume 7) Lower lemma 8) Palea 9) Upper lemma 10) Palea 11) Stamens and pistil 12) Caryopsis. 13-19: Pedicelled spikelet 13) Pedicelled spikelet 14) Lower glume 15) Upper glume 16) Lower lemma 17) Palea 18) Upper lemma 19) Palea.

soil. It is also adapted to environmental extremes during dry spells in the rainy season. It is a very common grass on rocky substrates throughout the Western Ghats, coastal plains of Maharashtra, Goa and Karnataka.

V) LOPHOPOGON Hackel

Lophopogon tridentatus (Roxb.) Hack. in Engl. and Prantl Fig. 5.

Andropogon tridentatus Roxb.

Annuals or perennials. Culms tufted, 10-25cm high, very slender; nodes glabrous. Leaves: sheaths terete; ligule a ciliate membrane; blades linear, flat or involute, 2-10 cm long, 0.15-0.25 cm wide, glabrous or pilose, acuminate. Inflorescence spicate. Racemes 2, closely appressed to form solitary ovate head, 1-1.5 cm long (excluding awns), erect; peduncles very slender; joints of the rachis very short, slender, glabrous. Spikelets densely imbricate. Sessile spikelets obovate-oblong, 4.5-5 mm long, unawned; callus short with ferruginous hairs. Lower glume chartaceous, cuneate-oblong, 3.6-4.2 mm long, 5-nerved,

373

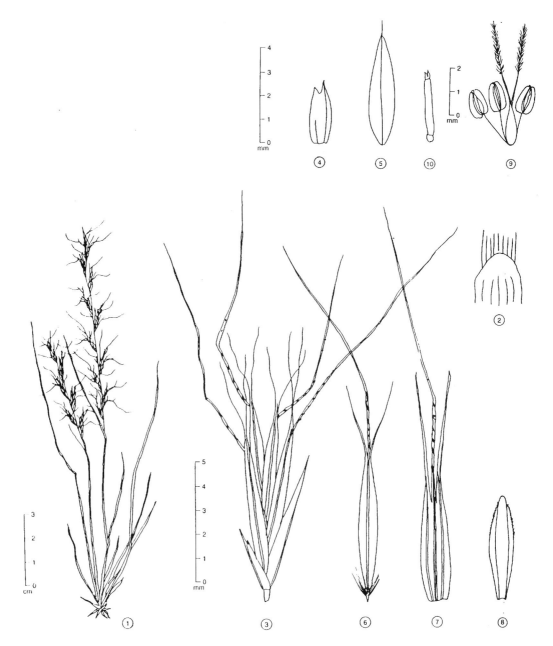

Fig. 4. *Indopoa paupercula* 1) Habit 2) Ligule 3) Spikelet 4) Lower glume 5) Upper glume 6) Lemma 7) Palea 8) Stamens and pistil 9) Caryopsis.

glabrous, truncate with 2 long lateral teeth and 1 or 2 shorter in between. Upper glume membranous, lanceolate, 4.5-5 mm long (excluding an awn), 3-nerved, 2-keeled; keels hirsute with ferruginous hairs above the middle, awned at apex; awn bristle-like, 2.5-3 mm long. Lower lemma hyaline, linear-lanceolate, 3.6-4 mm long, nerveless, muticous, epaleate. Lodicules absent. Stamens 2; anthers 2.5-3 mm long. Ovary elliptic-oblong, 0.5-0.6 mm long; styles 2-2.2 mm long; stigmas 2.8-3 mm long. Grain obovate-elliptic, 2.5-3 mm long, purple. Upper lemma absent. Pedicelled spikelets linear-lanceolate, 5-5.5 mm long, awned; pedicels short, 0.5-0.6 mm long. Lower glume chartaceous, elliptic-lanceolate, 3.8-4 mm long, 5-nerved, with tufts of ferruginous hairs about the middle; margins hairy near the base; apex 3-toothed. Upper glume membranous, lanceolate, 4.5-5.2 mm long, 5-nerved, 2-keeled; keels hirsute with ferruginous hairs

above the middle; apex 2-fid with bristle like 4.5-5 mm long awn. Lower lemma hyaline, elliptic-lanceolate, 3.4-3.6 mm long, nerveless, truncate, epaleate. Lodicules absent. Stamens 2; anthers 2.8-3 mm long. Upper lemma hyaline, linear-lanceolate, 4-4.2 mm long, 1-nerved; apex 2-lobed with 14-16 mm long geniculate awn from the sinus, Epaleate. Data on fruit morphology and anatomy is wanting.

Ecology, Geography and Regional Distribution

Lophopogon has two species. *L. tridentatus* is an annual as well as perennial grass and widely distributed throughout Peninsular India. It grows on disturbed poor soils of hillocks in the plains and the Eastern to Western Ghats. It seems that it does not grow on lateritic soils of western India. It is tufted and grows in great abundance wherever it grows.

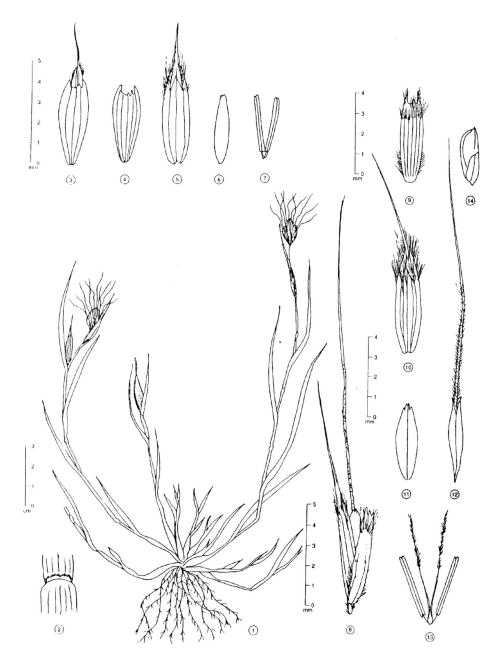

Fig. 5. *Lophopogon tridentatus* 1) Habit 2) Ligule 3-8: Sessile spikelet 3) Sessile spikelet 4) Lower glume 5) Upper glume 6) Lower lemma 7) Stamens and pistil 8) Caryopsis. 9-14: Pedicelled spikelet 9) Pedicelled spikelet 10) Lower glume 11) Upper glume 12) Lower lemma 13) Stamens 14) Upper lemma.

VI) POGONACHNE BOR

P. racemosa Bor Fig. 6.

Annuals or perennials. Culms tufted, 50-120 cm high, rooting at lower nodes; nodes glabrous. Leaves: sheaths terete, glabrous or sparsely hairy, margins sparsely ciliate; ligule membranous, 2.5-3 mm long, truncate; blades linear - lanceolate, 5-35 cm long, 0.5-2 cm wide, glabrous or with sparse tubercle-based hairs, acuminate. Racemes solitary, 4-15 cm long, more or less exserted from the spathes; spathes boat-shaped, keeled, sparsely hairy; joints linear, ciliate on one margin. Spikelets solitary, elliptic-lanceolate or ovate-lanceolate, 7-11 mm long, laterally compressed, awned;

pedicels compressed, linear, ciliate on one margin; callus bearded. Lower glume coriaceous, ovate-lanceolate, 7-11 mm long, many-nerved, 2-keeled at apex; sparsely ciliate towards the tip, acute. Upper glume subcoriaceous, ovate-oblong, 7-11 mm long, 5-nerved, with a tuft of hairs above the middle on the back; margins hyaline, keeled; keels ciliate at apex, subacute. Lower lemma empty, hyaline, ovate-oblong, 5-7 mm long, 3-nerved, muticous. Palea hyaline, linear-lanceolate, 3.5-5 mm long, acute. Upper lemma hyaline, oblong, 3.5-4.5 mm long, 1-nerved, cleft at apex into two subacute lobes, awned from the sinus; awn 2.5-4.3 cm long, geniculate, scaberulous. Palea hyaline, linear-oblong, 3.5-4 mm long, 2-nerved, obtuse. Lodicules 2. Stamens 3; anthers 2-3

375

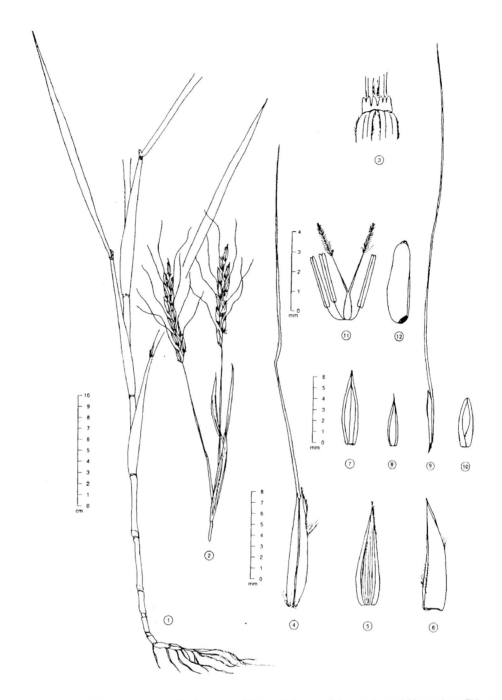

Fig. 6. *Pogonachne racemosa* 1) Basal part of plant 2) Inflorescence 3) Ligule 4) Spikelet 5) Lower glume 6) Upper glume 7) Lower lemma 8) Palea 9) Upper lemma 10) Palea 11) Stamens and pistil 12) Caryopsis.

mm long. Ovary ovate-oblong, 0.8-1.2 mm long; styles 1.5-2 mm long; stigmas 1-1.5 mm long. Grain oblong-elliptic, 4-5 mm long. Data on fruit morphology, anatomy and cytology is wanting.

Ecology, Geography, and Regional Distribution
P. racemosa is a tall, annual or perennial grass restricted to higher ranges of the Western Ghats between 700-1200 m altitude. It grows on rocky substrates in streams and other such moist places in the Western Ghats of Maharashtra. It produces prop roots as an adaptation to the sloping habitat and continuous flow of stream water. It also grows on hanging rocks of the Ghat area.

VII) *PSEUDODICHANTHIUM* BOR

P. serrafalcoides (Cooke et Stapf) Bor Fig. 7.

Andropogon serrafalcoides Cooke et Stapf

Dichanthium serrafalcoides (Cooke et Stapf) Blatt. et McCann

Annuals. Culms 25-60 cm high, weak, straggling, slender. Leaves: sheaths terete, ligule membranous, short, truncate; blades linear, 3.5-15 cm long, 0.3-0.7 cm wide, sparsely hairy; margins scaberulous, acuminate. Racemes solitary, 1.5-2.5 cm long (excluding awns); peduncles capillary; joints linear, ciliate on one

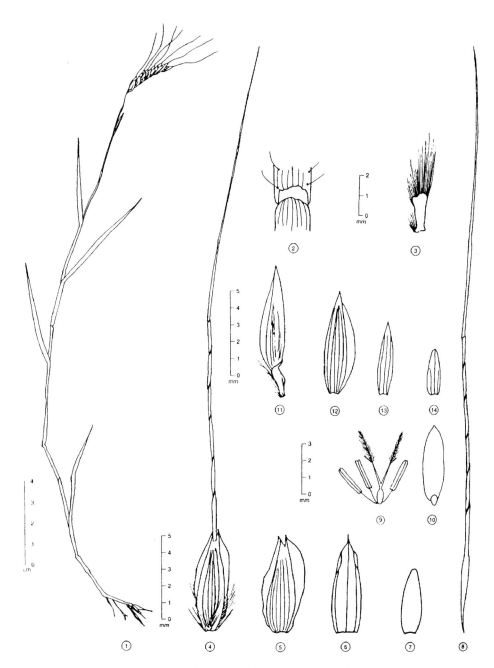

Fig. 7. *Pseudodichanthium serrafalcoides* 1) Habit 2) Ligule 3) Joint. 4-10: Sessile spikelet 4) Sessile spikelet 5) Lower glume 6) Upper glume 7) Lower lemma 8) Upper lemma 9) Stamens and pistil 10) Caryopsis. 11-14: Pedicelled spikelet 11) Pedicelled spikelet 12) Lower glume 13) Upper glume 14) Lower lemma.

margin. Sessile spikelets elliptic-oblong, 4.5-7 mm long, awned; callus bearded. Lower glume chartaceous, elliptic-oblong, 4.5-7 mm long, obscurely 7-9-nerved, 2-keeled; keels winged, wings unequal; apex bifid. Upper glume thinly membranous, lanceolate, 5-6 mm long, 3-nerved, 2-keeled, acute. Lower lemma empty, hyaline, 3-4 mm long, nerveless, obtuse, epaleate. Upper lemma reduced to a hyaline base of awn, epaleate; awn 2.6-4 cm long, geniculate. Lodicules 2. Stamens 3; anthers 1.2-1.5 mm long. Ovary oblong, 0.5 mm long; styles 1 - 1.2 mm long; stigmas 1.5-2.5 mm long. Grain elliptic, 2.5-3 mm long. Pedicelled spikelets elliptic-oblong, 7-11 mm long, unawned; pedicels lower shorter, upper longer, ciliate on one margin. Lower glume chartaceous, oblong-elliptic, 7–11 mm long, 7-9-nerved, 2-keeled; keels winged, wings unequal, acute. Upper glume thinly membranous, oblong, 4.5-7 mm long, 3-nerved, acute. Lower lemma empty, hyaline, elliptic, 3.5-4 mm long, 1-nerved, obtuse, epaleate. Upper lemma absent. Data on fruit morphology, anatomy and cytology is wanting.

Ecology, Geography and Regional Distribution

P. serrafalcoides is a wiry annual growing between 100 and 1200 m altitude. It grows on steep slopes of the Western Ghats in

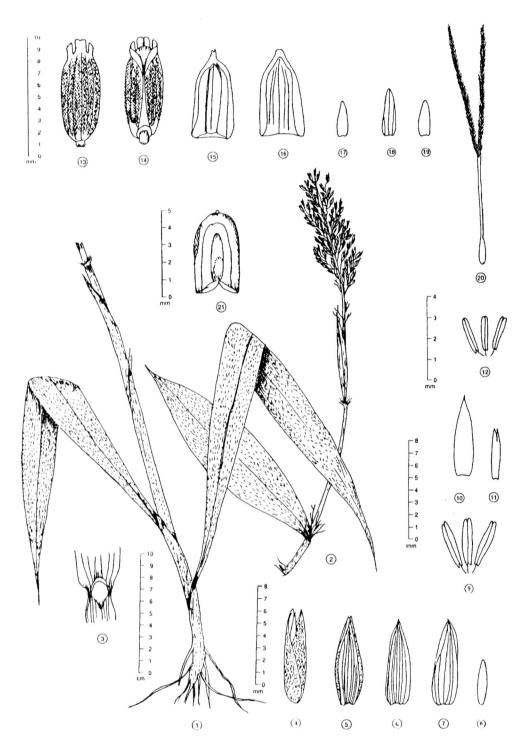

Fig. 8. *Trilobachne cookei* 1) Basal part of plant 2) Inflorescence 3) Ligule. 4-12: Male spikelet 4) Male spikelet 5) Lower glume 6) Upper glume 7) Lower lemma 8) Palea 9) Stamens 10) Upper lemma 11) Palea 12) Stamens. 13-21: Female spikelet 13) Female spikelet 14) Lower glume 15) Upper glume 16) Lower lemma 17) Palea 18) Upper lemma 19) Palea 20) Pistil 21) Caryopsis.

moist, cool places. *Tripogon lisboae* is a common associate. *Arthraxon jubatus, Ischaemum diplopogon, I. raizadae, Pogonachne, Pseudodichanthium,* and *Tripogon lisboae* are typical grasses of slopes of the Western Ghats which remain moist, shaded and cool.

P. serrafalcoides has subsequently been collected in Arabia by T. A. Cope (pers. comm.).

VIII) *TRILOBACHNE* SCHENCK EX HENR.

T. cookei (Stapf) Schenck ex Henr. Fig. 8.

Polytoca cookei Stapf in Hook.

Annuals. Culms 50-200 cm high, erect; nodes bearded. Leaves: sheaths subcompressed, hispid with bulbous-based hairs; ligule short, a rim of hairs; blades linear-lanceolate, 15-45 cm long,

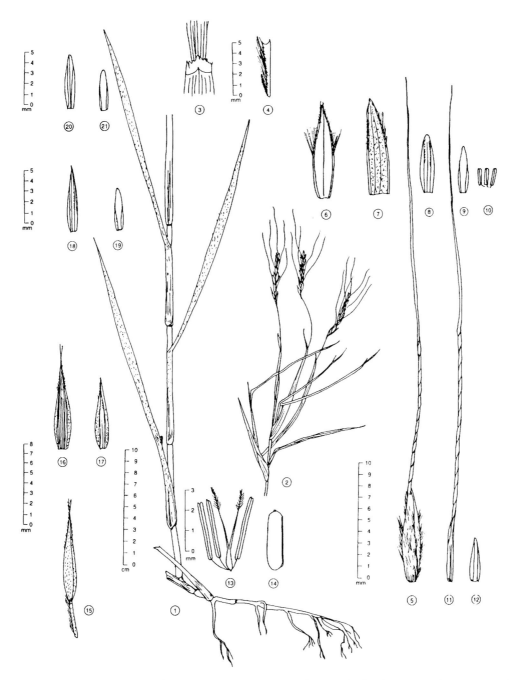

Fig. 9. *Triplopogon ramosissimus* 1) Basal part of plant 2) Inflorescence 3) Ligule 4) Joint. 5-14: Sessile spikelet 5) Sessile spikelet 6) Lower glume 7) Upper glume 8) Lower lemma 9) Palea 10) Stamens 11) Upper lemma 12) Palea 13) Stamens and pistil 14) Caryopsis. 15-21: Pedicelled spikelet 15) Pedicelled spikelet 16) Lower glume 17) Upper glume 18) Lower lemma 19) Palea 20) Upper lemma 21) Palea.

3.5-6 cm wide, hispid on both surfaces with bulbous-based hairs, margins serrulate, acuminate. Inflorescence fascicled, crowded in the axil of upper leaves, 10-20 cm long. Male spikelets binate, one shortly pedicelled and other with long pedicels; pedicels slender, sparsely hairy. Spikelets lanceolate, 8-12 mm long, pubescent, acuminate. Lower glume membranous, lanceolate, 8-12 mm long, many-nerved, pubescent, 2-keeled; keels scabrid; margins inflexed; apex acuminate, 2-toothed. Upper glume chartaceous, lanceolate, 7.5-11.5 mm long, 11-13-nerved, pubescent, 2-keeled, margins winged, acute. Lower lemma hyaline, lanceolate, 7-9.5 mm long, 5-nerved, 2-keeled, acute. Palea hyaline, linear-oblong, 4-5 mm long, obscurely 1-nerved, obtuse.

Lodicules 2. Stamens 3; anthers 4-6 mm long. Upper lemma hyaline, linear-lanceolate, 7-8 mm long, 3-nerved, 2-keeled; keels winged, acute. Palea hyaline, linear-oblong, 4-5 mm long, obscurely 2-nerved, apex notched, 2-lobed. Lodicules 2. Stamens 3; anthers 3.5-4.5 mm long. Female spikelets oblong, 8-12 mm long. Lower glume crustaceous, oblong, 8-12 mm long, 11-12-nerved, pubescent, keeled, embracing the upper glume; apex 3-lobed, lateral lobes shorter, middle lobe larger, emarginate. Upper glume chartaceous, oblong, 7.5-11.5 mm long, 11-13-nerved, puberulous, keeled, grooved on the back; apex 2-toothed. Lower lemma chartaceous, oblong, 7.5-11.5 mm long, 9-11-nerved, glabrous, truncate at apex. Palea hyaline, linear-oblong,

379

2.5-3.5 mm long, nerveless, obtuse. Upper lemma hyaline, oblong-lanceolate, 3.5-4.5 mm long, obscurely nerved, acute. Palea hyaline, linear-oblong, 2.5-3.5 mm long, nerveless, obtuse. Lodicules absent. Ovary elliptic,1-1.5 mm long; styles flattened; stigmas 2. Grain elliptic, 4.5-5 mm long, longitudinally grooved. Data on anatomy is wanting.

Ecology, Geography and Regional Distribution
T. cookei is a tall, annual herb growing in the Western Ghats between 600 and 1200 m altitude. It is known from Maharashtra and the Northern part of Karnataka. It grows along forest borders and the populations are being reduced every year. It has 20 chromosomes with a highly asymmetrical karyotype with smaller to largest and telocentric to metacentric chromosomes. This extensively-evolved genome is dwindling and deteriorating, leading to natural extinction. It represents a blind end in the evolutionary line of the tribe Maydeae.

IX) TRIPLOPOGON Bor

T. ramosissimus (Hack.) Bor Fig. 9.

Ischaemum ramosissimum Hack. in DC.

Ischaemum spathiflorum Hook. f.

Sehima spathiflorum (Hook. f.) Blatt. et McCann

Annuals or perennials. Culms tufted, 60-300 cm high, stilt-rooted, erect; nodes glabrous. Leaves: sheaths subcompressed; ligule membranous, truncate, 2-2.5 mm long; blades linear-lanceolate, 10-45 cm long, 0.7-2 cm wide, sparsely hairy, acuminate. Racemes 4-5 cm long (excluding awns), more or less exserted from the spathes; joints linear, compressed, ciliate on one margin. Sessile spikelets lanceolate, 6-7.5 mm long, awned; callus bearded. Lower glume coriaceous, oblong-lanceolate, 6-7.5 mm long, obscurely nerved, silky hairy on the back and with a tuft of hairs on either side above the middle, with a slit-like groove on the back; apex 2-toothed. Upper glume coriaceous to chartaceous, oblong-lanceolate, 6-7.5 mm long, obscurely nerved, silky hairy on the back and with a tuft of hairs above the middle, keeled; keels ciliate; margins sparsely ciliate; apex emarginate. Lower lemma hyaline, oblong-lanceolate, 5-5.5 mm long, 3-nerved, margins sparsely ciliate, obtuse. Palea hyaline, oblong-lanceolate, 4-4.5 mm long, obtuse. Lodicules 2. Stamens 3; anthers 1-1.5 mm long. Upper lemma hyaline, linear, 5-5.5 mm long, 3-nerved, cleft at apex into 2 acute lobes, awned from the sinus; awn 3-4.5 cm long, geniculate. Palea hyaline, lanceolate, 2.5-3.5 mm long, obtuse. Lodicules 2. Stamens 3; anthers 2.5-3.5 mm long. Ovary linear-oblong, 1 mm long; styles 2 mm long; stigmas 1-1.5 mm long. Grain elliptic, 2.8-3.2 mm long. Pedicelled spikelets lanceolate, 9-12 mm long, unawned; pedicels linear, ciliate on one margin. Lower glume chartaceous, lanceolate, 9-12 mm long, many-nerved, sparsely silky, 2-keeled; keels scaberulus; apex acuminate, notched and with 2 short unequal awns. Upper glume chartaceous, lanceolate, 6-7.5 mm long,

obscurely 3-5-nerved, sparsely silky, margins ciliolate, acute. Lower lemma hyaline, oblong, 5.5-7 mm long, 3-nerved, margins ciliolate, acute. Palea hyaline, oblong, 3.5-4.5 mm long, obtuse. Lodicules 2. Stamens 3; anthers 2.5-3 mm long. Upper lemma hyaline, oblong, 5.5-6 mm long, 3-nerved, acute. Palea hyaline, oblong-linear, 4-5 mm long, obtuse. Data on cytology is wanting.

Ecology, Geography and Regional Distribution
T. ramosissimus is a tall, annual or perennial grass growing to a height of 3 m. It grows on the slopes of the Western Ghats on basaltic rocks. It is a gregarious grass; locally abundant at places. It grows at altitude between 100 to 1200 m, in tufts forming pure stands in the Kolaba district. It produces prop roots in response to its sloping habitat and water flow in the region during the rainy season.

CONCLUSIONS

Genera like *Bhidea, Danthonidium, Glyphochloa* and *Indopoa* grow in more or less similar edaphic and ecological conditions. The soils on which species of these genera grow are mainly lateritic in origin, rich in iron and poor in nutrients. All of them are monsoon annuals and complete their life cycle during the short span of the rainy season of about three months. Another group of genera, *Pogonachne, Trilobachne* and *Triplopogon,* grow in shady and exposed slopes of the Western Ghats. The soil is varied. They are adapted to a more or less similar ecological niche in the Western Ghats. *Lophopogon* is the only genus that grows in the interior parts of the country on poor soils and on rocky substrates in hilly regions.

ACKNOWLEDGEMENTS

I am thankful to Dr. J. F. Veldkamp, Rijksherbarium, Leiden, The Netherlands; Dr. T. A. Cope, Royal Botanical Gardens, Kew, England; Dr. V. N. Naik, Dr. B. A. Marathwada University, Aurangabad and Dr. S. R. Yadav, Shivaji University, Kolhapur, India for their valuable suggestions in preparation of manuscript of the paper and encouragement.

REFERENCES

Ahmedullah, M., and Nayar, M.P. (1987). Endemic plants of the Indian Region Vol. I Peninsular India. (Botanical Survey of India: Calcutta.)

Jain, S.K. (1986). The grass flora of India - A synoptic account of uses and phytogeography. *Bulletin of Botanical Survey of India* **28**, 229-240.

Karthikeyan, S., Jain, S. K., Nayar, M.P., and Sanjappa, M.(1989). Florae Indicae Enumeratio: Monocotyledonae. (Botanical Survey of India: Calcutta.)

Nayar, M.P. (1996). Hot spots of Endemic plants of India, Nepal & Bhutan. (Tropical Botanical Garden and Research Institute: Thiruvananthapuram.)

Watson, L., and Dallwitz, M.J. (1992). The Grass Genera of the World. (CAB International: Wallingford, U.K.)

Appendix

Name of the Genus and Species	Taxonomic Position	Distribution in India
Bhidea B. borii B. burnsiana B. fischeri	Panicoideae; Andropogonodae; Andropogoneae, Andropogoninae	Maharashtra, Karnataka, Goa, Kerala.
Danthonidium D. gammiei	Arundinoideae; Danthonieae	Maharashtra, Karnataka, Goa, Kerala.

Glyphochloa
G. acuminata var. acuminata
  var. stocksii
  var. woodrowii
G. divergens
G. forficulata
G. goaensis
G. mysorensis
G. ratnagirica
G. santapaui
G. talbotii
	Panicoideae; Andropogonodae;
	Andropogoneae; Rottboelliineae
		Maharashtra, Karnataka, Goa, Kerala.

Indopoa I. paupercula	Chloridoideae	Maharashtra, Karnataka, Goa.
Lophopogon L. duthiei L. tridentatus	Panicoideae; Andropogonodae; Andropogoneae; Andropogoninae	Peninsular India.
Pogonachne P. racemosa	Panicoideae; Andropogonodae; Andropogoneae; Andropogoninae	Maharashtra.
Pseudodichanthium P. serrafalcoides	Panicoideae; Andropogonodae; Andropogoneae; Andropogoninae	Maharashtra.
Trilobachne T. cookei	Panicoideae; Andropogonodae; Maydeae	Maharashtra, Karnataka.
Triplopogon T. ramosissimus	Panicoideae; Andropogonodae; Andropogoneae; Andropogoninae	Maharashtra.

INDEX

Entries in **bold** indicate that a major treatment on that subject begins on that page. No attempt has been made to indicate synonymy, or to rationalise the application of names between contributions.